P9-BJD-844

NUCLEAR PHYSICS

To Ruth

This book is in the

ADDISON-WESLEY SERIES IN
NUCLEAR SCIENCE AND ENGINEERING

Herbert Goldstein
Consulting Editor

NUCLEAR PHYSICS

by

IRVING KAPLAN

Department of Nuclear Engineering
Massachusetts Institute of Technology

SECOND EDITION

ADDISON-WESLEY PUBLISHING COMPANY, INC.

READING, MASSACHUSETTS · PALO ALTO · LONDON

Printed in the United States of America

Library of Congress Catalog Card No. 62–9402
Second Edition
Second Printing—March 1964

PREFACE TO THE FIRST EDITION

I have tried to write an elementary, yet coherent, account of nuclear physics suitable as an introduction to this field. The book is elementary in the sense that the reader is assumed to have had no previous exposure to atomic and nuclear physics. It is supposed that he has had a two-year college course in physics covering approximately the material contained in Sears' books: *Mechanics, Heat, and Sound; Electricity and Magnetism;* and *Optics,* in the Addison-Wesley Physics Series. It is also assumed that the reader is familiar with the differential and integral calculus; a one-year course in general chemistry would be helpful, if not essential. This book should therefore be useful in an advanced undergraduate course in nuclear physics, to engineers interested in the large-scale applications of nuclear physics grouped together under the name "nuclear engineering," or to anyone else with the indicated preparation who might be interested in nuclear physics.

The concepts and techniques of nuclear physics are not elementary in the sense that they are easily mastered. They have evolved through a great deal of experimental and theoretical research and cannot be expressed or explained in simple nontechnical terms. I have tried, therefore, to show how our present knowledge of atoms and nuclei has been developed, and how some of this knowledge is applied. In treating the experiments that yield information about atoms and nuclei and the ways in which these experiments are interpreted in terms of theory, I have tried to avoid both the devil of experimental complication and the deep blue sea of mathematical detail. The emphasis is on the principles underlying the experiments and on the experimental results rather than on the apparatus, on the physical ideas rather than on the details of the calculations. Thus, derivations have been included when they involve straightforward applications of physical principles in terms of mathematics not beyond the integral calculus. When it is necessary to use the results of more advanced methods, such as quantum mechanics, the details of the derivations are omitted; but the physical ideas contained in the results are discussed in some detail and related to the experimental data.

The plan of the book is based on the advice of the King of Hearts: "Begin at the beginning," the King said gravely, "and go on till you come to the end: then stop." The need for a reasonable limit on the size of the book makes it necessary, however, to define the beginning and end arbitrarily and to leave some things out on the way. The first part of the book, consisting of the first seven chapters, is devoted to the background of nuclear physics and begins with the chemical foundations of atomic theory. The second part, Chapters 8 through 17, treats the physics of the nucleus in a way that seems to me to be logical, yet consistent with the elementary nature of the book. The third part, Chapters 18 through 22, deals with special topics and applications and includes neutron physics and nuclear fission, which do not fit conveniently into the scheme of the second part and at the same time lead into the most spectacular application of nuclear physics—"nuclear energy." The subjects of charged-particle accelerators and isotope separation, although not really nuclear physics, are closely related to important branches of this field; they are included for this reason and because they form, along with nuclear energy, the main part of nuclear engineering. I have not discussed such subjects as cosmic rays, mesons, and nuclear moments because I think that they need a more advanced treatment than the material covered.

I have included, for several reasons, a large number of literature references. For the skeptical or curious reader, the references will supply the experimental and theoretical details omitted from the text, as well as proofs of results that are stated without proof. A more important reason for including these references stems from the fact that atomic and nuclear physics are recently developed fields, and the original literature is more readily available than that of classical physics. The student can read for himself the original papers and books of the pioneers in this field—Rutherford, Bohr, Millikan, Moseley, Aston, Chadwick, Fermi, and others. The articles in which the basic discoveries were first published give, more than any textbook can, a feeling for the imagination and beauty inherent in modern physics. The current literature adds a feeling for the gradual, and sometimes painful, accumulation of experimental and theoretical information, and for the problems which have yet to be solved.

This book has been developed from the notes for a course of lectures given at the Brookhaven National Laboratory. I am indebted to colleagues who attended the lectures or used the notes for many valuable comments and suggestions, and to many workers in the field of nuclear physics for permission to cite and use their published results. Above all, I find it hard to express adequately my indebtedness and gratitude to Jean Harless for her unfailing patience, good humor, and skill in preparing the manuscript.

IRVING KAPLAN

East Patchogue, New York
June, 1954

PREFACE TO THE SECOND EDITION

I have tried to bring the book up-to-date and to improve the presentation of the material. The rapid accumulation of information during the past seven years has made it necessary to change many of the numbers and some of the ideas, while the use of the book as a text, by friends and colleagues as well as by myself, has resulted in many suggestions for improvement and clarification.

For the convenience of those who used the first edition, the major changes are listed. Section 2–8, on the detection and measurement of radiation, has been expanded. In the treatment of special relativity, in Chapter 6, derivations of the Lorentz transformation equations and of the formula for the variation of mass with velocity have been included. Section 7–9 has been added to Chapter 7, in which the Schroedinger equation is applied to the problem of a particle in a one-dimensional box; it is hoped that this addition will help the student develop a little more insight into some of the elementary ideas of wave mechanics. Two new sections have been added to Chapter 14: Section 14–8 deals with the neutrino and its detection; Section 14–9 with symmetry laws and the nonconservation of parity in beta-decay. The last two sections of Chapter 15, dealing with gamma-decay, nuclear energy levels and isomerism, have been completely rewritten. A short section, 16–7, on the limitations of the compound nucleus theory has been added to Chapter 16. In Chapter 17, the treatment of nuclear models has been extended; in particular, the shell and collective models are discussed in somewhat more detail than in the first edition, and the optical model for nuclear reactions is introduced. A short section, 17–4, on the nuclear radius has been added. Chapter 19, on nuclear fission, has been modernized to include material which was classified at the time the first edition was written but is now part of the open literature. Corresponding changes have been made in Chapter 20, with the result that some of the elementary properties of nuclear reactors can be introduced in a more satisfactory way; a short section on controlled thermonuclear reactions has also been added. To compensate in part for the new material, Chapter 22 on isotope separation has been dropped; this subject seems to attract little interest in courses in nuclear physics, and is now treated thoroughly in other books. Finally, I have tried to improve the problems by adding to them and by replacing some of the "substitution" problems by "thought" problems.

I am indebted to many friends, colleagues, and students who have pointed out errors and suggested changes, and I regret that I cannot list them individually because of the lack of space. I have, however, exercised the author's privilege of disregarding some advice and criticism, and have only myself to blame for the shortcomings of the revised book.

I am indebted to Rachel Sprinsky for her invaluable assistance in the preparation of the manuscript.

Irving Kaplan

Cambridge, Mass.
January, 1962

CONTENTS

PART I. THE BACKGROUND OF NUCLEAR PHYSICS

Part I

The Background of Nuclear Physics

CHAPTER 1

THE CHEMICAL FOUNDATIONS OF ATOMIC THEORY

The foundations of modern atomic theory were laid in the late 18th and 19th centuries in the attempt to understand the chemical properties of matter. Two of the fundamental quantities of atomic and nuclear physics—atomic weight and atomic number—had their origins in the correlation of the results of chemical experiments and in the systemization of the properties of the chemical elements. The laws of chemical combination were unified at the beginning of the 19th century by Dalton's atomic theory, which introduced the concept of atomic weight. The development of methods for determining the atomic weights of the elements and the investigation of chemical reactions were major contributions of 19th century chemistry. The systemization of the atomic weights and properties of the elements then led to the formulation of the periodic system and to the concepts of atomic number and atomic structure.

1–1 The laws of chemical combination. The experimental information which gave rise to the atomic theory of matter can be summarized in a few basic laws. The first of these was deduced from investigations such as those of Lavoisier, who showed that when tin is made to react with air in a closed vessel, the weight of the vessel and its contents before and after heating is the same. This constancy of the weight, which has been found to be true for all chemical reactions, is expressed in the *law of conservation of mass*, which states that the mass of a system is not affected by any chemical change within the system.

It was found that when various metals are oxidized in excess air, one part by weight of oxygen always combines with 1.52 parts by weight of magnesium, 2.50 parts of calcium, 1.12 parts of aluminum, 3.71 parts of tin, 3.97 parts of copper, and so on. If, as turns out to be more convenient, the weight of oxygen is taken to be 8, the following combining weights are obtained.

Oxygen	*Magnesium*	*Calcium*	*Aluminum*	*Copper*	*Tin*
8	12.16	20.04	8.96	31.76	29.68

The same combining weights are obtained when the oxides are prepared by methods other than simple oxidation. These experimental results, and many others like them, are expressed by the *law of definite proportions*, which states that a particular chemical compound always contains the same elements united together in the same proportions by weight.

It was also found that two elements can combine to form more than one compound. For example, at least five distinct oxides of nitrogen are known in which the relative proportions by weight of nitrogen and oxygen are listed in the table.

Nitrogen	Oxygen
14	$8 = 1 \times 8$
14	$16 = 2 \times 8$
14	$24 = 3 \times 8$
14	$32 = 4 \times 8$
14	$40 = 5 \times 8$

The different weights of oxygen which can combine with the same weight of nitrogen are integral multiples of 8. Results of this kind are summarized by the *law of multiple proportions*, which states that if two elements combine to form more than one compound, the different weights of one which combine with the same weight of the other are in the ratio of small whole numbers.

The further study of the quantitative relations between the elements led to a fourth law of chemical combination. The *law of reciprocal proportions* states that the weights of two (or more) substances which react separately with identical weights of a third are also the weights which react with each other, or simple multiples of them. In other words, if each of two substances A and B combines with a substance C, then A and B can combine with each other only in those proportions in which they combine with C, or in some simple multiple of those proportions. This law is illustrated by the reactions which take place between oxygen and sulfur, oxygen and zinc, and sulfur and zinc. Thus, 8 parts by weight of oxygen combine with 8.015 parts of sulfur to form an oxide of sulfur, and with 32.69 parts of zinc to form an oxide of zinc; in the formation of zinc sulfide, $65.38(= 2 \times 32.69)$ parts of zinc combine with $32.06(= 4 \times 8.015)$ of sulfur.

Another generalization deduced from the analysis of chemical compounds is that of the combining weight or chemical equivalent of an element. This concept was touched upon in the discussion of the oxides of various metals in connection with the law of definite proportions. Consider a number of chemical compounds such as those tabulated on the next page. If oxygen $= 8$ is taken as the standard, the amount of each of the other elements which combines with this standard amount of oxygen can be calculated. Thus, 28.53 parts by weight of oxygen combine with 71.47 parts of calcium, and

$$28.53:8 = 71.47:x, \qquad \text{or} \qquad x = 20.04 \text{ for calcium,}$$

Compound	Percent	Percent
1. Calcium oxide	Calcium 71.47	Oxygen 28.53
2. Water	Hydrogen 11.19	Oxygen 88.81
3. Hydrogen chloride	Hydrogen 2.76	Chlorine 97.23
4. Magnesium chloride	Magnesium 25.53	Chlorine 74.47
5. Silver chloride	Silver 75.26	Chlorine 24.71
6. Silver iodide	Silver 45.94	Iodine 54.06

where x is the amount of the element, in this case calcium, which combines with 8 parts of oxygen. From the experimental data for water,

$$88.81:8 = 11.19:x, \qquad \text{or} \qquad x = 1.008 \text{ for hydrogen.}$$

In hydrogen chloride, if the weight of hydrogen is taken as 1.008, that of chlorine is 35.45; in magnesium chloride, if the weight of chlorine is 35.45, that of magnesium is 12.16. When this procedure is continued, it is found that a number can be assigned to each element which represents the number of parts by weight of the given element that can combine with 8 parts by weight of oxygen or 1.008 parts by weight of hydrogen.

Oxygen	*Calcium*	*Hydrogen*	*Chlorine*	*Magnesium*	*Silver*	*Iodine*
8	20.04	1.008	35.45	12.16	107.88	126.9

The value obtained for magnesium by the above indirect process is the same as that obtained earlier from data on the direct oxidation of magnesium. The numbers obtained for the different elements are called the *combining weights* or *equivalent weights* of the elements.

The four laws of chemical combination, conservation of mass, definite proportions, multiple proportions, and reciprocal proportions, together with the concept of combining weights, summarized the basic experimental facts of chemical combination as known about 1800, and led to the atomic hypothesis proposed early in the 19th century by Dalton.

1–2 Dalton's atomic hypothesis. Dalton's atomic hypothesis was proposed (1803) to account for the facts expressed by the laws of chemical combination and was based on the following postulates:

1. The chemical elements consist of discrete particles of matter, atoms, which cannot be subdivided by any known chemical process and which preserve their individuality in chemical changes.

2. All atoms of the same element are identical in all respects, particularly in weight or mass; different elements have atoms differing in weight. Each element is characterized by the weight of its atom, and the combining weights of the elements represent the combining weights of their respective atoms.

3. Chemical compounds are formed by the union of atoms of different elements in simple numerical proportions, e.g., 1:1, 1:2, 2:1, 2:3.

It is easy to show that the laws of chemical combination can be deduced from these postulates. Since atoms undergo no change during a chemical process, they preserve their masses, and the mass of a compound is the sum of the masses of its elements; the result is the law of conservation of mass. Since all atoms of the same element are identical in weight, and a compound is formed by the union of atoms of different elements in a simple numerical proportion, the proportions by weight in which two elements are combined in a given compound are always the same, giving the law of definite proportions.

Consider next the case in which two elements A and B can form two different compounds. Suppose that the first compound contains m atoms of A and n atoms of B, and that the second compound contains p atoms of A and q atoms of B. If a is the weight of an atom of A, and b is the weight of an atom of B, then the first compound contains ma parts of A and nb parts of B, and one part of A combines with nb/ma parts of B. Similarly, in the second compound, one part of A combines with qb/pa parts of B. Hence, the weights of B combined with a fixed weight of A are in the proportion

$$\frac{n}{m} : \frac{q}{p}, \qquad \text{or} \qquad np:mq.$$

According to the third postulate, n, m, p, and q are all small integers; the products np and mq are therefore also small integers and the weights of B combining with a fixed weight of A are in the ratio of two small integers. This treatment can be extended to any number of compounds formed by two elements, giving the law of multiple proportions. Similar arguments lead to the law of reciprocal proportions. Compounds of the elements A and B are formed according to the scheme m atoms of A and n atoms of B. Compounds of the elements A and C contain p atoms of A and q atoms of C, and compounds of the elements B and C contain x atoms of B and y atoms of C. If a, b, and c are the weights of the atoms A, B, and C, respectively, then one part of A combines with nb/ma parts of B, and qc/pa parts of C. It follows that

$$\frac{mq}{np} : \frac{b}{c}.$$

In the compound formed by B and C, the proportion by weight of B and C is

$$xb:yc, \qquad \text{or} \qquad \frac{y}{x} : \frac{b}{c}.$$

Hence,

$$\frac{y}{x} = \frac{mq}{np}.$$

Since, by the third postulate, all of the quantities in the last equation are small integers, then y, x are the same as q, n or small integral multiples of them. This is the law of reciprocal proportions.

The laws of chemical combination have thus been deduced from Dalton's postulates, and the latter form the basis of an atomic theory of matter. Dalton's theory was incomplete, however, because it provided no way of determining even the relative weights of the atoms of the different elements. This difficulty arose because Dalton had no way of finding out how many atoms of each element combine to form a compound. If W_1 and W_2 are the weights of two elements which combine to form a compound, then

$$\frac{W_1}{W_2} = \frac{n_1 A_1}{n_2 A_2},$$

where A_1 and A_2 are the atomic weights and n_1 and n_2 are the whole numbers of atoms of each element which enter into combination. When the ratio $n_1:n_2$ is known, the value of the ratio $W_1:W_2$ can fix only the *ratio* of the atomic weights. To apply his theory, Dalton was forced to make arbitrary assumptions; he assumed, for example, that if only one compound of two elements is known, it contains one atom of each element. Water was regarded as a compound of one atom of hydrogen and one atom of oxygen, the existence of hydrogen peroxide being unknown at the time. Dalton's assumption was simple, but wrong, and led to many difficulties in the application of his atomic theory to the rapidly growing field of chemistry.

1-3 Avogadro's hypothesis and the molecule. One of the difficulties met by the atomic hypothesis arose as a result of studies of the combining properties of gases. Gay-Lussac showed (1805–1808) that when chemical reactions occur between gases, there is always a simple relation between the volumes of the interacting gases, and also of the products, if these are gases. When the reacting gases and the products are under the same conditions of temperature and pressure

1 volume of hydrogen + 1 volume of chlorine
$\longrightarrow$ 2 volumes of hydrogen chloride;
2 volumes of hydrogen + 1 volume of oxygen
$\longrightarrow$ 2 volumes of steam;
3 volumes of hydrogen + 1 volume of nitrogen
$\longrightarrow$ 2 volumes of ammonia.

It follows that if elements in a gaseous state combine in simple proportions by volume, and if the elements also combine in simple proportions by atoms, then the numbers of atoms in equal volumes of the reacting gases must be simply related. Dalton assumed that equal volumes of different gases under the same physical conditions contain an equal number, say n, of atoms. Under this assumption, when 1 volume (n atoms) of hydrogen reacts with 1 volume (n atoms) of chlorine, then 2 volumes ($2n$ "compound atoms") of hydrogen chloride are formed. Every atom of hydrogen and chlorine must then be split in half to form two compound atoms of hydrogen chloride, a result which contradicts that postulate of Dalton's hypothesis according to which atoms cannot be subdivided by any chemical process.

Avogadro (1811) showed that the difficulty could be resolved if a distinction is made between elementary atoms and the small particles of a gas. He assumed that the latter are aggregates of a definite number of atoms, and called the aggregates *molecules* in order to distinguish them from the elementary atoms. He then postulated that equal volumes of all gases under the same physical conditions contain the same number of *molecules*. This hypothesis made it possible to interpret Gay-Lussac's law of combining volumes in terms of the atomic hypothesis. Assume that each molecule of hydrogen or chlorine consists of two elementary atoms, and suppose that 1 volume of hydrogen or chlorine contains n molecules. These molecules react to form $2n$ molecules of hydrogen chloride, each molecule containing one atom of hydrogen and one atom of chlorine. Although the atoms cannot be split so that one atom of hydrogen or chlorine enters into the composition of two molecules of hydrogen chloride, one molecule of hydrogen and one molecule of chlorine can be divided between two molecules of hydrogen chloride. In this way, Avogadro's hypothesis allowed Gay-Lussac's law to be reconciled with Dalton's atomic hypothesis. The detailed analysis of many reactions between gases has shown that the molecules of the gaseous elements hydrogen, oxygen, chlorine, and nitrogen contain two atoms; the inert gases helium, neon, argon, krypton, and xenon have one atom per molecule.

1–4 Molecular and atomic weights of gaseous elements. The relative molecular weights of gaseous substances can be determined from measurements of relative densities. The relative density of a gas is defined as the ratio of the weight of a given volume of the gas to that of an equal volume of a standard gas measured at the same temperature and pressure. In view of Avogadro's hypothesis, it is possible to write for any gaseous substance

$$\frac{\text{Molecular weight of gas}}{\text{Molecular weight of standard gas}} = \frac{\text{Weight of any volume of gas}}{\text{Weight of equal volume of standard}} = \frac{\text{Density of gas}}{\text{Density of standard gas}}.$$

TABLE 1–1

MOLECULAR WEIGHTS OF GASEOUS ELEMENTS

Element	Density, g/l at N.T.P.	Molecular weight		Molecular weight M (oxygen) = 32*
		M (hydrogen) = 2	M (oxygen) = 32	
Hydrogen	0.08988	2.000	2.013	2.016
Oxygen	1.42904	31.816	32.000	32.000
Nitrogen	1.25055	27.828	28.000	28.016
Fluorine	1.696	37.738	37.977	38.00
Chlorine	3.214	71.52	71.97	70.914
Helium	0.17847	3.971	3.996	4.003
Neon	0.90035	20.034	20.161	20.183
Argon	1.7837	39.690	39.942	39.944
Krypton	3.708	82.510	83.03	83.7
Xenon	5.851	130.19	131.02	131.3

* Corrected for deviations from ideal gas laws.

The choice of the standard gas is arbitrary and may be, for example, hydrogen, oxygen, or air. If hydrogen is chosen as the standard, and if its atomic weight is arbitrarily taken to be unity, the relative molecular weight of a gas is

$$\text{Molecular weight} = 2 \times \text{density of gas relative to that of hydrogen},$$

since a molecule of hydrogen contains two atoms. When the relative molecular weight of a gaseous element has been determined in this way, its relative atomic weight can be determined when the number of atoms per molecule is known.

The results for some molecular weight determinations based on this method are shown in Table 1–1. The densities of the gases are given in grams per liter measured under standard conditions (0°C and 760 mm mercury pressure, abbreviated as N.T.P.). The third column gives the molecular weight relative to that of hydrogen taken as 2. Oxygen, with its molecular weight arbitrarily taken as 32, may also be used as a standard, and the molecular weights relative to this standard are given in the fourth column. High-precision work on the combining volumes of gases has shown that there are slight deviations from Gay-Lussac's law which have been traced to deviations from the ideal gas laws. When corrections are made for these deviations, the results of the fifth column are obtained for the molecular weights relative to that of oxygen taken as 32. Of the gaseous elements listed, hydrogen, oxygen, nitrogen, fluorine, and chlorine are diatomic and their relative atomic weights are half of the molecular weights; the relative atomic weights of the inert gases are the same as their molecular weights.

1–5 The standard atomic weight. It has been emphasized that the values of the molecular and atomic weights determined from density measurements are relative, not absolute, values. For practical purposes, it is necessary to fix a standard for atomic weights, and this standard is taken as oxygen with an atomic weight of 16 and a molecular weight of 32. The choice of oxygen was based on the criterion of convenience. Oxygen is a useful standard for chemists because nearly all the elements form stable compounds with oxygen, and it was thought that the determination of the atomic weight of an element should be connected with the standard as closely as possible. Hydrogen was sometimes used as a standard, but it forms very few compounds with metals which are suitable for atomic weight determinations; consequently hydrogen was rejected in favor of oxygen.

The choice of the value of the atomic weight of oxygen was based on the desire for convenient values for the other atomic and molecular weights. Experimental evidence shows that 1.008 parts by weight, say grams, of hydrogen combine with 8 gm of oxygen. It has also been found that two atoms of hydrogen combine with one atom of oxygen to form one molecule of water. Hence, 1.008 gm of hydrogen contain twice as many atoms as 8 gm of oxygen. A possible choice for the atomic weight of oxygen is its combining, or equivalent, weight, 8 units. If this choice were made, then 1.008 (of the same units) would be the weight of two atoms of hydrogen, and the atomic weight of hydrogen would be 0.504. Now, hydrogen is the lightest of all the elements. If the standard is chosen so that the atomic weight of hydrogen is equal to or greater than unity, there will be no element with an atomic weight smaller than unity. The choice of 16 for oxygen makes the atomic weight of hydrogen equal to 1.008, and also leads to relative weights of heavier atoms and molecules which are not inconveniently large. The atomic weight of oxygen is then equal to twice the equivalent or combining weight. In general, the atomic weight of an element is a small integral multiple of the equivalent weight, and this integer is called the *valence* of the element:

$$\frac{\text{Atomic weight}}{\text{Equivalent weight}} = \text{Valence.}$$

The equivalent weight and the valence are useful because their values can be determined by chemical methods and lead to the determimation of atomic weights.

TABLE 1–2

THE ATOMIC WEIGHT OF CARBON AS FOUND FROM THE RELATIVE DENSITIES OF CARBON COMPOUNDS

Compound	Relative density, Hydrogen = 1	Molecular weight	Weight of carbon in one molecular weight of compound
Methane	8	16	12
Ethane	15	30	24 = (2 × 12)
Alcohol	23	46	24 = (2 × 12)
Ether	37	74	48 = (4 × 12)
Benzene	39	78	72 = (6 × 12)
Carbon monoxide	14	28	12
Carbon dioxide	12	44	12

1–6 Atomic weights of nongaseous elements. The method of relative densities can be used to find the approximate atomic weight of a nonvolatile element which has gaseous compounds. The molecular weights of as many volatile compounds of the element as possible are found from the relative densities. The weights of the particular element in the molecular weights of the various compounds are found by chemical analysis. These weights must be integral multiples of the atomic weight and if the number of compounds used is large enough, at least one of the weights of the element present in the molecular weights of its compounds will probably be the atomic weight itself. The smallest weight of the element present in one molecular weight of all its known volatile compounds is assumed to be the atomic weight of the element relative to oxygen = 16. In the case of carbon, the results of Table 1–2 are obtained, and the atomic weight of carbon is taken to be 12. The values given in the table are approximate; more refined measurements and corrections for deviations from the ideal gas laws must be made in order to get accurate values of the atomic weight. In general, accurate values of the molecular and atomic weights are found from careful chemical analysis of the compounds, and the relative density measurements are used only to decide between various possible molecular weights.

Approximate values of the atomic weights of solid elements which have few, or no, volatile compounds can be found with the aid of the rule of Dulong and Petit. According to this rule, the product of the atomic weight and the specific heat is roughly the same for many solid elements. When the specific heat is expressed in calories per gram, the numerical value of the product is usually between 6 and 7, with an average of about 6.4. If the specific heat of a solid element is measured, an approximate

TABLE 1–3

ATOMIC WEIGHTS OF COMMON ELEMENTS RELATIVE TO OXYGEN = 16

Element	Atomic weight	Element	Atomic weight
Aluminum	26.98	Lead	207.21
Barium	137.36	Mercury	200.61
Boron	10.82	Nitrogen	14.008
Calcium	40.08	Oxygen	16.000
Carbon	12.011	Potassium	39.100
Chlorine	35.457	Silicon	28.09
Copper	63.54	Silver	107.880
Fluorine	19.00	Sulfur	32.066
Gold	197.0	Uranium	238.07
Hydrogen	1.0080	Zinc	65.38
Iron	55.85		

value of the atomic weight is found. This value can then be used to determine which of the different values represented by the combining proportions by weight is the best choice for the atomic weight. For example, the specific heat of copper is 0.095 cal/gm. By the rule of Dulong and Petit,

$$\frac{6.4}{0.095} = 67.3$$

is the approximate atomic weight of copper. Chemical analyses of compounds of copper with oxygen show that 63.6 and 127.2 parts of this element combine with 16 parts of oxygen. Since 67.3 is much closer to 63.6 than 127.2, there can be little doubt that 63.6 is the atomic weight of copper. More refined chemical analysis gives the value 63.54.

Various other methods have been developed for determining accurate values of molecular and atomic weights, and a list of the presently accepted values of the atomic weights of some of the elements is given in Table 1–3.

1–7 Weights and sizes of atoms and molecules. When the standard for atomic and molecular weights has been chosen, it is possible to calculate the volume occupied at standard temperature and pressure by a gram-molecular weight of a gas, that is, by the number of grams of gas numerically equal to the molecular weight; this volume is called the *gram-molecular volume* or *molar volume*. Since the atomic weight of hydrogen is taken to be 1.008, its molecular weight is 2.016, and a gram-molecular weight of hydrogen is 2.016 gm. The density of hydrogen (Table 1–1) is very nearly 0.09 gm/liter at N.T.P., so that the gram-molecular volume is

$$\frac{2.016}{0.09} = 22.4 \text{ liters.}$$

Experiments with a large number of gases have shown that the same volume, 22.4 liters, is occupied by one gram-molecular weight, or mole, of *any gas* at N.T.P. According to Avogadro's hypothesis, this volume contains the same number of gas molecules regardless of the chemical nature of the gas. It follows, therefore, that the gram-molecular weights of all gases at N.T.P. contain the same number of molecules; this number is called the *Avogadro number* or *Avogadro's constant.* Its value has been determined by several independent methods, with results which agree very well; one method will be described in Chapter 2. The accepted value at present is 6.0249×10^{23}. This number is also the number of atoms in a gram-atomic weight of a gaseous element, and hence of any element. The weight in grams of a single atom or molecule can be found by dividing the gram-atomic or gram-molecular weight by the Avogadro number. The weight of the lightest atom, hydrogen, is 1.67×10^{-24} gm, while that of the heaviest naturally occurring element, uranium, is 3.95×10^{-22} gm.

An estimate of the sizes of molecules and atoms can be made by considering water, which has a molecular weight of 18 and a density of one gram per cubic centimeter. One gram-molecular weight of water occupies 18 cm^3 and contains 6.0×10^{23} molecules. If the water molecules were cubic in shape so that they packed together without any free space, the volume of a single molecule would be 3×10^{-23} cm^3. If the molecules are assumed to be spherical, their volume is about one-third smaller, or 2×10^{-23} cm^3. The radius of a water molecule is then given by the relation

$$\tfrac{4}{3}\pi r^3 = 2 \times 10^{-23}\ \text{cm}^3,$$

or

$$r = 1.7 \times 10^{-8}\ \text{cm}.$$

Values of the radii of gas molecules can also be calculated with the aid of the kinetic theory of gases from measured values of the viscosity, the diffusion constant, and the thermal conductivity. The radii of hydrogen and oxygen molecules obtained in this way are 1.37×10^{-8} cm and 1.81×10^{-8} cm, respectively, and are quite close to the value of the radius of the water molecule given by the above rough calculation. Similar values, in the neighborhood of 1×10^{-8} cm to 3×10^{-8} cm, are obtained for the molecular radii of most of the common gaseous elements and compounds. Since the number of atoms per molecule of these substances is small, 10^{-8} cm may be taken as the order of magnitude of atomic radii.

An estimate of the atomic radius of a solid element can be made from the atomic volume, that is, the volume of a gram-atom of the element, at the melting point. In the solid phase at the melting point, the atoms are closely packed and the distance between them is not much greater than the atomic dimensions. Solid lead, at its melting point, has a density of 11 gm/cm^3 and its atomic volume is 18.9 cm^3. The volume per atom is

3.15×10^{-23} cm^3, and if the atom is assumed cubical (no free space), the atomic dimension would be about 3.2×10^{-8} cm. If the lead atom is assumed to be spherical and the space between atoms is taken into account, the radius is about 1.6×10^{-8} cm. With better methods for determining the atomic radius, the following values have been obtained for some of the solid elements.

Carbon:	0.77×10^{-8} cm	Cesium:	2.6×10^{-8} cm
Aluminum:	1.45×10^{-8} cm	Tin:	1.4×10^{-8} cm
Sodium:	1.9×10^{-8} cm	Bismuth:	1.5×10^{-8} cm

These values are similar to those obtained for the atomic radii of the gaseous elements. They indicate that atoms have dimensions of the order of 10^{-8} cm regardless of the atomic weight, since elements of low, intermediate, and high atomic weight are included in the list. Cesium is the atom with the largest radius.

1–8 The periodic system of the elements. As information about the chemical and physical properties of the elements was accumulated, it became apparent that there are similarities in the properties of certain groups of elements, and attempts were made to classify the elements according to such similarities. The most successful attempt was that of Mendeléeff, who proposed his *periodic law* in 1869. This law may be summarized in the following statements made by Mendeléeff.

1. The elements, if arranged according to their atomic weights, show a periodicity of properties.
2. Elements which are similar as regards their chemical properties have atomic weights which are either approximately the same (e.g., platinum, iridium, osmium) or which increase regularly (e.g., potassium, rubidium, cesium).
3. The arrangement of the elements or of groups of elements in the order of their atomic weights corresponds with their valences.
4. The elements which are the most widely distributed in nature have small atomic weights, and all the elements of small atomic weights are characterized by sharply defined properties.
5. The *magnitude* of the atomic weight determines the character of an element.
6. The discovery of many yet unknown elements may be expected, for instance, elements analogous to aluminum and silicon.
7. The atomic weight of an element may sometimes be corrected with the aid of knowledge of the weights of adjacent elements.
8. Certain characteristic properties of the elements can be predicted from their atomic weights.

Mendeléeff's general conclusions have been verified and extended. Among the properties which show the kind of periodic variation noted by Mendeléeff are atomic volume (atomic weight/density), thermal expansion, thermal and electrical conductivity, magnetic susceptibility, melting point, refractive index, boiling point, crystalline form, compressibility, heats of formation of oxides and chlorides, hardness, volatility, ionic mobility, atomic heats at low temperature, valence, and others.

Mendeléeff constructed a table in which the elements were arranged horizontally in the order of their atomic weights and vertically according to their resemblances in properties. The early tables were imperfect, sometimes because of errors in the atomic weights, and because of ignorance of the inert gases. In spite of these imperfections, Mendeléeff was able to predict the discovery of certain elements, as well as the properties of these elements. Modern versions of the periodic table, such as that shown in Table 1–4, have rectified these errors. The table consists of a number of horizontal rows called *periods*, containing 2, 8, 8, 18, 18, 32, and 18 (incomplete) elements, respectively. In each period there is a definite and characteristic gradation of properties from one element to the next. The vertical columns, labeled by the Roman numerals I to VIII and by the letter O are called groups. Groups I to VII are each divided into subgroups; the subgroup on the left side of the column is usually called the "a" subgroup, that on the right side the "b" subgroup. In each subgroup, the elements have similar properties, although there is a steady but gradual variation with increasing atomic weight. The alkali metals constitute the subgroup 1a, the alkaline earth metals 2a, oxygen-like elements 6a, halogens 7a, and inert gases group O. Within a group, similarities of a minor character exist between the members of the a subgroup and those of the b subgroup.

For four pairs of elements, the order assigned in the periodic table departs from that of the atomic weights, argon and potassium, cobalt and nickel, tellurium and iodine, thorium and protoactinium. In each case the order, as determined by the atomic weights, is reversed so that these elements may fall into the places which correspond to their properties. This was done for the first three pairs by Mendeléeff himself, who believed that the accepted values of the atomic weights were inaccurate. Later research has shown that the chemical atomic weights were not seriously in error, and that the real property underlying the periodic classification is the *atomic number*, to which the atomic weight is approximately proportional.

The basis of Mendeléeff's classification of the elements, namely, the requirement that the elements should be arranged in the order of increasing atomic weight, offered no explanation of the different features of the periodic table. So far as was known in Mendeléeff's time, atoms of dif-

Table 1–4. Periodic Table of the Elements

Atomic weights are based on the most recent values adopted by the International Union of Chemistry. (For artificially produced elements, the approximate atomic weight of the most stable isotope is given in brackets.)

Group→		I	II	III	IV	V	VI	VII	VIII			O
Period	Series											
1	1	1 H 1.0080										2 He 4.003
2	2	3 Li 6.940	4 Be 9.013	5 B 10.82	6 C 12.011	7 N 14.008	8 O 16.0000	9 F 19.00				10 Ne 20.183
3	3	11 Na 22.991	12 Mg 24.32	13 Al 26.98	14 Si 28.09	15 P 30.975	16 S 32.066	17 Cl 35.457				18 A 39.944
4	4	19 K 39.100	20 Ca 40.08	21 Sc 44.96	22 Ti 47.90	23 V 50.95	24 Cr 52.01	25 Mn 54.94	26 Fe 55.85	27 Co 58.94	28 Ni 58.71	
	5	29 Cu 63.54	30 Zn 65.38	31 Ga 69.72	32 Ge 72.60	33 As 74.91	34 Se 78.96	35 Br 79.916				36 Kr 83.80
5	6	37 Rb 85.48	38 Sr 87.63	39 Y 88.92	40 Zr 91.22	41 Nb 92.91	42 Mo 95.95	43 Tc [99]	44 Ru 101.1	45 Rh 102.91	46 Pd 106.4	
	7	47 Ag 107.880	48 Cd 112.41	49 In 114.82	50 Sn 118.70	51 Sb 121.76	52 Te 127.61	53 I 126.91				54 Xe 131.30
6	8	55 Cs 132.91	56 Ba 137.36	57–71 Lanthanide series*	72 Hf 178.50	73 Ta 180.95	74 W 183.86	75 Re 186.22	76 Os 190.2	77 Ir 192.2	78 Pt 195.09	
	9	79 Au 197.0	80 Hg 200.61	81 Tl 204.39	82 Pb 207.21	83 Bi 209.00	84 Po 210	85 At [210]				86 Rn 222
7	10	87 Fr [223]	88 Ra 226.05	89– Actinide series**								

*Lanthanide series:	57 La 138.92	58 Ce 140.13	59 Pr 140.92	60 Nd 144.27	61 Pm [147]	62 Sm 150.35	63 Eu 152.0	64 Gd 157.26	65 Tb 158.93	66 Dy 162.51	67 Ho 164.94	68 Er 167.27	69 Tm 168.94	70 Yb 173.04	71 Lu 174.99
**Actinide series:	89 Ac 227	90 Th 232.05	91 Pa 231	92 U 238.07	93 Np [237]	94 Pu [242]	95 Am [243]	96 Cm [245]	97 Bk [249]	98 Cf [249]	99 E [253]	100 Fm [255]	101 Md [256]	102 No	103

ferent elements had nothing in common and were unrelated. Since the properties of an element are the properties of its atoms, it was difficult to see why, for example, lithium, sodium, and potassium atoms should have similar properties; it was evident that the atomic weights alone could not explain the features of the table. The periodicity of the properties of the elements stimulated speculation about the structures of the atoms of different elements. The gradual changes of properties from group to group suggested that some unit of atomic structure is added in successive elements until a certain portion of the structure is completed, a condition which occurs, perhaps, in the inert gases. Beginning with the element following an inert gas, a new portion of the structure is started, and so on. Speculations like these introduced a new idea, that of the structure of atoms. But before this idea could have any real physical meaning, experimental information bearing on the possible structure of atoms was needed. The methods and techniques of classical chemistry could not supply this information, but new discoveries and techniques in the field of physics opened the way to the understanding of atomic structure and the atomic number.

References

GENERAL

J. W. Mellor, *Modern Inorganic Chemistry*, revised and edited by G. D. Parkes. London: Longmans, Green and Co., 1951, Chapters 3, 4, 5, 6, 8.

F. T. Bonner and M. Phillips, *Principles of Physical Science*. Reading, Mass., Addison-Wesley, 1957, Chapters 5, 6, 7, 8, 9.

G. Holton and D. H. D. Roller, *Foundations of Modern Physical Science*. Reading, Mass., Addison-Wesley, 1958, Chapters 21, 22, 23, 24, 25.

T. G. Cowling, *Molecules in Motion*. London: Hutchinson and Co., 1950; New York: Harper Torchbooks, 1960.

R. D. Present, *Kinetic Theory of Gases*. New York: McGraw-Hill, 1958.

D. C. Peaslee, *Elements of Atomic Physics*. New York: Prentice-Hall, 1955, Chapters 1, 2.

PROBLEMS

1. Show how the five oxides of nitrogen listed below illustrate the law of multiple proportions.

Compound	Percentage composition	
	Nitrogen	Oxygen
Nitrous oxide	63.63	36.37
Nitric oxide	46.67	53.33
Nitrous anhydride	36.84	63.16
Nitrogen tetroxide	30.44	69.56
Nitric anhydride	25.93	74.04

2. The molecular weights of the five oxides listed in the above problem are 44.02, 30.01, 76.02, 46.01, and 108.02, respectively; calculate the atomic weight of nitrogen from these data and the data of Problem 1.

3. A certain element X forms several volatile compounds with chlorine, and the lowest weight of the element is found in a compound which contains 22.54% of X and 77.46% of chlorine. One liter of this compound at N.T.P. weighs 6.14 gm. If the atomic weight of chlorine is 35.457, find (a) the equivalent weight of X, (b) the approximate molecular weight of the compound, (c) the approximate atomic weight of X, (d) the valence of X, (e) the exact atomic weight of X, (f) the exact molecular weight of the compound. Which element is X? What is the compound?

4. Chemical analysis shows that platinic chloride contains 48.8 parts of platinum combined with 35.46 parts of chlorine. The specific heat of platinum at room temperature is 0.0324; what is the atomic weight of platinum?

5. A metal forms a chloride which contains 65.571% of chlorine. The specific heat of the metal is 0.112. Find (a) the approximate atomic weight of the metal, (b) the equivalent weight, (c) the valence, (d) the exact atomic weight. What is the compound?

6. The *Handbook of Chemistry and Physics*, published by the Chemical Rubber Company, Cleveland, Ohio, contains tables of various properties of the elements. From information contained in the *Handbook*, calculate the atomic volumes of the solid elements, and plot the values obtained as a function of the position of the element in the periodic table. Plot the melting points of the solid elements as a function of position in the periodic table. What periodic properties are observed in the two figures?

7. In 1947 it was announced that one of the long-lived radioactive products of the fission of U^{235} was a new chemical element different from any other elements then known. The new element formed a volatile oxide whose vapor density was 9.69 times that of oxygen gas at the same temperature and pressure; the oxide contained 36.2% of oxygen by weight. What is the molecular weight of the oxide? What is the combining weight of the new element? Where does the new element fit into the periodic system? What element would it be expected to resemble in chemical properties? What is the approximate atomic weight of the new element?

CHAPTER 2

ATOMS, ELECTRONS, AND RADIATIONS

2–1 Faraday's laws of electrolysis and the electron. It was seen in the first chapter that the combining properties of the elements can be interpreted in terms of an atomic theory of matter. From the work of Faraday on the electrolysis of aqueous solutions of chemical compounds, it appeared that electricity is also atomic in nature. Faraday discovered two laws of fundamental importance.

1. The chemical action of a current of electricity is directly proportional to the absolute quantity of electricity which passes through the solution.
2. The weights of the substances deposited by the passage of the same quantity of electricity are proportional to their chemical equivalents.

Careful measurements have been made of the amount of electricity needed to liberate one equivalent weight of a chemical substance, say 107.88 gm of silver, 1.008 gm of hydrogen, or 35.46 gm of chlorine. This amount is 96,490 coul or one *faraday*, and is denoted by F.

Faraday inferred from his experimental results that one and the same amount of electricity is associated in the process of electrolysis with one atom of each of these substances. He thought of this charge as carried by the atom, or in some cases by a group of atoms, and he called the atom, or group of atoms, with its charge, an *ion*. The current in an electrolytic process results from the movement of the ions through the solution, negative ions moving toward the anode and positive ions toward the cathode. Faraday's results implied that there is an elementary unit, or *atom*, of electricity. This idea, however, did not seem to fit in with other electrical phenomena such as metallic conduction, and both Faraday and Maxwell hesitated to accept the idea of the atomic nature of electricity. In 1874 Stoney ventured the hypothesis that there is a "natural unit of electricity," namely, that quantity of electricity which must pass through a solution in order to liberate, at one of the electrodes, one atom of hydrogen or one atom of any univalent substance. He suggested the name *electron* for this quantity, and tried to calculate the magnitude of the electronic charge. Stoney's calculation was based on the fact that one faraday of electric charge liberates the number of atoms in one gram-atom of a univalent element; this number is just Avogadro's number N_0, so that $F = N_0e$, where e represents the electronic charge. From the known value of F and the rough value of N_0 available at the time, Stoney obtained: $e = 0.3 \times 10^{-10}$ electrostatic unit of charge (esu). Although this is only $\frac{1}{16}$ of the presently accepted value, it served to indicate the order of magnitude of the electronic charge.

Faraday's experiments alone yield no certain information about the quantity of electricity represented by one electron, but they do give precise information about the ratio of the ionic charge to the mass of the atom with which it is associated in a given solution. For hydrogen, this ratio is

$$\frac{e_{\mathrm{H}}}{m_{\mathrm{H}}} = \frac{96490 \text{ coul}}{1.008 \text{ gm}} = \frac{96490 \times 3 \times 10^9}{1.008} \frac{\text{esu}}{\text{gm}}$$
$$= 2.87 \times 10^{14} \text{ esu/gm}.$$

2–2 The conduction of electricity in gases. Under normal conditions a gas is a poor conductor of electricity, but if ionization can be produced in the gas, it becomes a conductor; x-rays, discovered by Roentgen in 1895, provided a convenient means of producing ionization. Following this discovery, the mechanism of gaseous conduction and the nature of gaseous ions were investigated on an increased scale by many workers. Theoretical and experimental studies were made of the mobility of gaseous ions and the coefficient of diffusion of these ions. As a result of these studies, the value of the quantity of electric charge associated with one cubic centimeter of a gas could be calculated. This quantity can be written as ne_g, where n is the number of molecules of gas per cubic centimeter under some standard conditions of temperature and pressure, and e_g is the average charge on a gaseous ion. The value of ne_g was found to be 1.2×10^{10} esu for air at 15°C and 760 mm of mercury. The volume occupied by one gram-molecular weight of gas under these conditions is

$$2.24 \times 10^4 \left(\frac{288}{273}\right) \text{cm}^3 = 2.36 \times 10^4 \text{ cm}^3.$$

The total amount of electric charge associated with one gram-molecular weight of gas is then

$$1.2 \times 10^{10} \frac{\text{esu}}{\text{cm}^3} \times 2.36 \times 10^4 \text{ cm}^3 = 2.83 \times 10^{14} \text{ esu}.$$

Now, the amount of electricity carried by the univalent ions in one gram-molecular weight of a salt in solution is one faraday:

$$F = 96{,}490 \text{ coul} = 2.90 \times 10^{14} \text{ esu},$$

which is very close to the value found for the ionized gas. If the molecules of the gas are assumed to be singly ionized, they seem to carry, on the average, the same charge as that carried by univalent ions in solution. This result also points to the existence of a unit electrical charge.

2–3 Cathode rays. The experimental proof of the independent existence of particles of negative electricity followed from the study of the conduction of electricity in rarefied gases. When a gas is enclosed in a

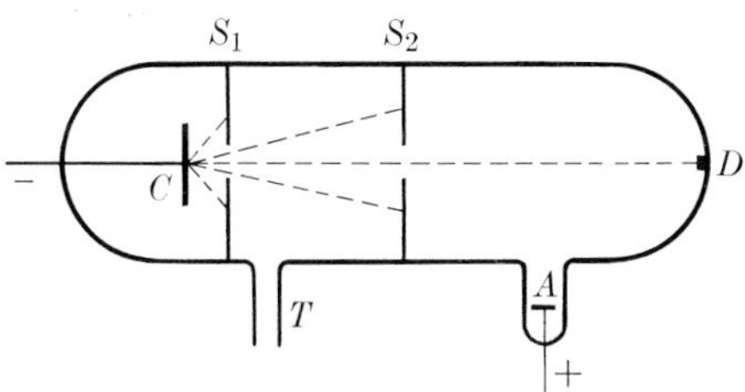

FIG. 2–1. Diagram of a tube for the production of cathode rays.

glass tube with two electrodes to which a sufficiently high potential is applied, electricity flows through the gas, which becomes strongly illuminated. When the pressure of the gas is reduced below 0.1 mm of mercury the illumination gradually disappears. At very low pressures, of the order of 10^{-4} or 10^{-5} mm of mercury, a fluorescent glow appears on the wall of the glass tube opposite the cathode. The presence of the glow was correctly attributed by early workers to rays from the cathode, and these rays were called *cathode rays*. A schematic diagram of a tube for the production of cathode rays is shown in Fig. 2–1; C is the cathode, A is the anode, and T is a side tube through which the gas can be exhausted.

The cathode rays were found to have several interesting properties.

1. The rays travel in straight lines. When screens S_1 and S_2, pierced with holes, are put into the tube, the glow is confined to a spot D on the end of the tube; or, if a body is placed in the path of the rays, it casts a shadow on the fluorescent part of the glass. From the geometrical relations it can be inferred that the rays travel in straight lines.

2. The rays can penetrate small thicknesses of matter. When a "window" of thin aluminum or gold foil is built into the end of the tube struck by the rays, the passage of the rays through the foil is shown by the presence of luminous blue streamers in the air on the far side of the foil.

3. The rays carry a negative electric charge. When the rays are caught in an insulated chamber connected to an electroscope, the electroscope becomes negatively charged.

4. The rays are deflected by an electrostatic field. When two parallel plate electrodes with their planes parallel to the path of the rays are introduced into the cathode-ray tube, the rays are deflected toward the positively charged plate.

5. The rays are deflected by a magnetic field. When an ordinary bar magnet is brought up to the cathode-ray tube, the fluorescent spot opposite the cathode moves in one direction or another depending on which magnetic pole is presented to the rays.

6. The cathode rays carry considerable amounts of kinetic energy. When a metal obstacle is placed in the path of the rays, it can quickly become incandescent.

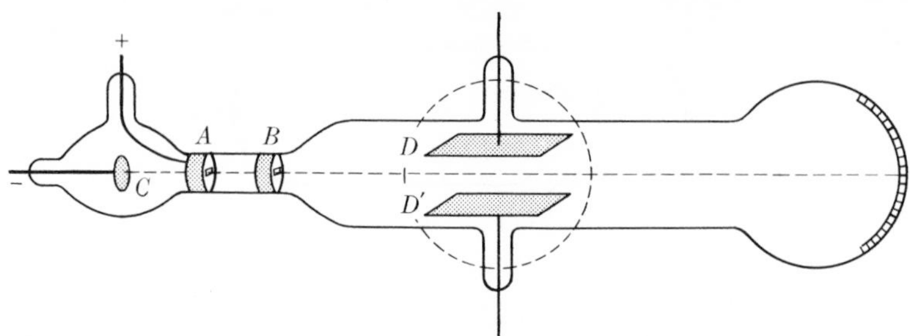

FIG. 2–2. The apparatus of J. J. Thomson for measuring the ratio e/m for cathode rays.

The only reasonable interpretation of these experimental facts is that the cathode rays consist of negatively charged particles.

Thomson, in his fundamental experiments (1897), confirmed the view that the cathode rays consist of negatively charged particles and determined the ratio charge/mass = e/m for the particles. For these experiments, the tube of Fig. 2–1 was modified so that the rays could be made to pass through an electrostatic field. In the apparatus shown schematically in Fig. 2–2, a potential difference of a few thousand volts is maintained between the anode A and the cathode C. A narrow beam of rays from the cathode passes through a slit in the anode and through a second slit in the metal plug B. The end of the tube is coated on the inside with a fluorescent material and the point of impact of the cathode-ray stream appears as a bright spot. Inside the tube are two metal plates D and D' between which a vertical electric field can be set up. By means of an external electromagnet, a magnetic field can be established, perpendicular to the plane of the diagram, within the region indicated by the dotted circle. If the electric field is upward (D' positively charged), and if there is no magnetic field, the cathode-ray stream is deflected downward. The electric and magnetic deflections can be made to cancel each other by the proper adjustment of the electric and magnetic fields.

Suppose that the cathode rays consist of negatively charged particles of charge e and mass m, and let E be the magnitude of the electrostatic field, and H that of the magnetic field. If the electric and magnetic fields are so adjusted that their effects cancel, and the particle stream passes through the tube without being deflected, then the electric force on the particles must be equal in magnitude and opposite in direction to the magnetic force. The electric force is given by eE, and the magnetic force by Hev, where v is the velocity of the particles.[(1)*] Then,

$$eE = Hev, \qquad \text{or} \qquad v = \frac{E}{H}, \tag{2–1}$$

* Superscript numbers in parentheses refer to the references at the end of the chapter.

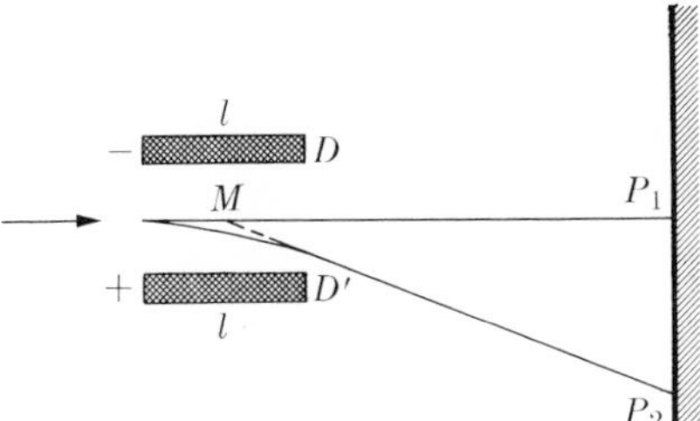

FIG. 2–3. The deflection of cathode-ray particles in an electrostatic field.

and this experiment can be used to measure the velocity of the cathode-ray particles. If E, e, and H are expressed in electromagnetic units, the velocity has the units centimeters per second.

Next, the magnetic field is removed and the deflection of the particles caused by the electrostatic field is measured as shown in Fig. 2–3. In this diagram, the distance $\overline{P_1P_2}$ is the deflection caused by the electrostatic field as observed on the fluorescent screen; D and D' are the parallel plates as before, and l is the length of the plates. As the particles traverse the field, they are subjected to a vertical acceleration in a direction parallel to that of the field during the time interval l/v, so that the velocity produced in the vertical direction is $(eE/m)\,(l/v)$. The horizontal velocity of the particles is not changed by the field. On leaving the field, the velocity of the particles has the components v in the horizontal direction and $(eE/m)\,(l/v)$ in the vertical direction. Hence, if P_2 is the deflected position of the ray, $\overline{P_1P_2}/\overline{MP_1}$ is the ratio of the vertical and horizontal velocities:

$$\frac{\overline{P_1P_2}}{\overline{MP_1}} = \frac{eE}{m}\,\frac{l}{v^2},$$

or

$$\frac{e}{m} = \frac{\overline{P_1P_2}}{\overline{MP_1}}\,\frac{v^2}{El}. \tag{2–2}$$

In Eq. (2–2), the quantities l, $\overline{MP_1}$ and E are known from the experimental setup; the deflection $\overline{P_1P_2}$ is measured, and v is known from the experiment in which the effects of the electric and magnetic fields cancelled each other.

Thomson's experiments showed that the cathode-ray particles move with very high speeds—of the order of a tenth of the velocity of light. For e/m, he found the value 1.7×10^7 emu/gm or, since 1 emu of charge $= 3 \times 10^{10}$ esu, $e/m = 5.1 \times 10^{17}$ esu/gm. This value was independent of the cathode materials which he used (aluminum, iron, and platinum), and of the gas (air, hydrogen, and carbon dioxide) in the discharge tube.

The value of e/m for the cathode-ray particles is about 1800 times as large as that found for hydrogen ions in electrolysis. If the cathode-ray particles were assumed to have the same mass as the smallest atom, hydrogen, they would have to carry a charge 1800 times as large as that on a hydrogen ion. Furthermore, since the value of e/m is independent of the materials in the cathode tube, and materials with different ionic masses were used, it would be necessary to imagine some mechanism which would make the charge on the particles directly proportional to the mass. If, as an alternative, the charge on the particle is assumed to be the same as that on the hydrogen ion, then the mass of the particle is about 1/1800 of that of the hydrogen ion. The latter assumption seemed more reasonable to Thomson, who therefore considered that the cathode-ray particles are a new kind of *negative corpuscle.* Later work proved that Thomson's conclusion was correct.

2–4 The electron: the determination of its charge. Thomson's experiments on cathode rays demonstrated the independent existence of very small negatively charged particles, but the *assumption* that the charge on one of these particles is the same as the charge on a univalent ion in solution and on a gaseous ion in a conducting gas presented a basic problem. Although the assumption seemed logical, a proof of the equality of the charges was needed. The direct way to solve this problem would be to measure the charge on the particles. This direct measurement could not be made because of the great speed of the cathode-ray particles, but other sources of similar negatively charged particles were available. Toward the end of the 19th century it was found that when ultraviolet light falls on certain metals, negatively charged particles with small velocities are emitted (the photoelectric effect). Thomson determined the value of e/m for these particles and found it to be practically the same as that for the cathode-ray particles. The same result was obtained for the negatively charged particles emitted by an incandescent filament (the thermionic effect). In view of the constancy of e/m for the negatively charged particles produced in different ways, it was concluded that all of these particles are identical. There still remained, however, the problem of proving experimentally that the charge on the different particles (monovalent electrolytic and gaseous ions, photoelectric particles, thermionic particles) is the same.

Several measurements of the charge were made around 1900, and their results may be summarized briefly as follows. For the average charge on an electrolytic ion, the value of 2×10^{-10} esu was obtained, based on an improved but still approximate value of the Avogadro number; later work (1905–1908) gave a value of 4.25×10^{-10} esu. The average charge on gaseous ions was found to lie between 3×10^{-10} esu and 5×10^{-10} esu, and similar values were found for the photoelectric particles. Experimental evidence of this kind pointed to the identity of the various negative electrical charges, and this view was gradually adopted. The term *electron* was applied to the negative particles themselves, rather than to the magnitude of the charge, and this usage has become general. The work on cathode rays and on the photoelectric and thermionic effects also indicated that electrons are a constituent part of all atoms, and the electron was regarded as a fundamental physical particle. In all of the experimental work discussed so far, either values of e/m or of the *average* charge of the various particles were measured. Although it was difficult to argue against the conclusion that electrons exist and have a certain elementary unit of charge, the correctness of this conclusion had not been proved experimentally beyond any doubt. The final proof of the atomic or particulate nature of electricity and the first really precise determination of the value of the electronic charge was supplied by the work of Millikan (1909–1917).

Millikan's experiments will be discussed in some detail because of their fundamental place in modern physics. The earlier experiments on the determination of the electronic charge were based on the fact that electrically charged atoms (ions) act as nuclei for the condensation of water vapor. Water vapor was allowed to condense on ions, forming a cloud, and the total charge carried by the cloud was measured with an electrometer. The number of water drops in the cloud was calculated by weighing the water condensed from the cloud and dividing by the average weight of a single drop. This average weight could be found by measuring the rate of fall of the drops through the air, because the weight of a drop and its rate of fall were known to be related through Stokes' law of fall.(2) The total charge on the cloud divided by the number of drops gave the average value of the charge per drop. This method involves several assumptions: (1) that the number of ions is the same as the number of drops, (2) that Stokes' law, which had never been tested experimentally, is valid, and (3) that the drops are all alike and fall at a uniform rate not affected by evaporation. These assumptions were all questionable, and their validity had to be tested in careful experiments.

The above method, used by Townsend (1897) and Thomson (1896), was improved by Wilson (1903), who formed water clouds between two parallel plates which could be connected to the terminals of a battery. The rate of fall of the top surface of the cloud was first determined under the influence of gravity. The plates were then charged so that the droplets were forced downward by both gravity and the electric field, and the new rate of fall was measured. When the two rates were compared, a value of the electronic charge could be obtained. In Wilson's method, it was not necessary to assume that the number of drops is equal to the number of ions. Only the rate of fall of the top of the cloud was observed and, since the more heavily charged drops were driven down more rapidly by the field than the less heavily charged ones, the actual experiments were made on the least heavily charged drops. Wilson obtained the value 3.1×10^{-10} esu, but the other assumptions remained, and there was considerable uncertainty in this result.

Millikan improved Wilson's method by using oil drops, thus avoiding errors caused by evaporation. He also noticed that a single drop could be held stationary between the charged plates by adjusting the voltage on them so that the weight of the drop was just balanced by the electrostatic field between the plates. From the relationship between the forces, the value of the charge of the electron could be obtained. In addition, while working with balanced drops, Millikan noticed that occasionally a drop would start moving up or down in the electric field. Such a drop had evidently picked up an ion, either positive or negative, and it was possible to determine the change in the charge carried by a drop without regard to the original charge on the drop.

The essential parts of Millikan's apparatus are shown in Fig. 2–4; P_1 and P_2 are two accurately parallel horizontal plates, 22 cm in diameter and 1.5 cm apart, housed in a large metal box to avoid air currents. Oil is sprayed in fine droplets from an atomizer above the upper plate and a few of the droplets fall through a small hole in the plate. A beam of light is directed between the plates, and a telescope is set up with its axis transverse to the light beam. The oil drops, illuminated by the light beam, look like tiny bright stars when viewed through the telescope, and move slowly under the combined influence of their weight, the buoyant force of the air, and the viscous force opposing their motion. The oil droplets in the spray from the atomizer are electrically charged, probably because of frictional effects; the charge is usually negative. Additional ions, both positive and negative, can be produced in the space between the plates by ionizing the air by means of x-rays. If the upper plate is positively charged to a potential of about 1000 volts, and the lower plate negatively charged, there is a uniform electric field between the plates, and there is an electric force on the droplet which affects its motion.

When a spherical body falls freely in a viscous medium, it accelerates until a terminal velocity is reached such that the net downward force on the body equals the viscous force. The latter is proportional to the terminal velocity of the particle, and

$$mg = kv_g, \tag{2–3}$$

where m is the apparent mass of the body, g the acceleration due to gravity, k a proportionality constant, and v_g the terminal velocity of free fall. If,

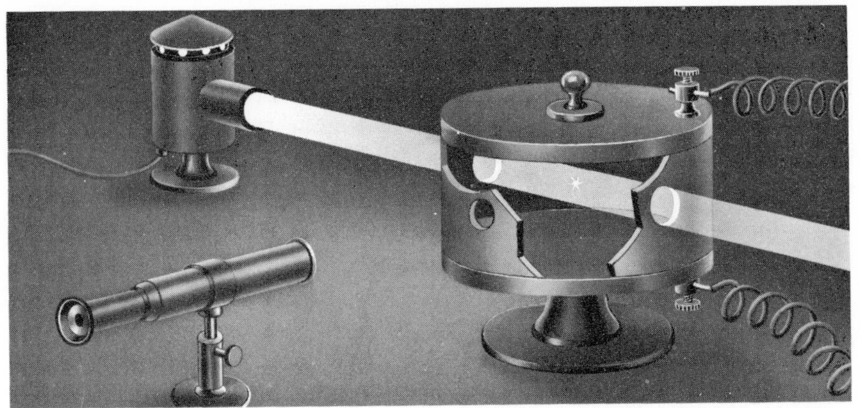

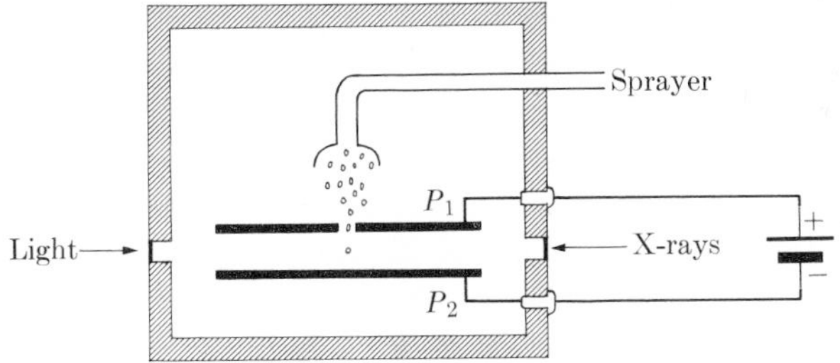

Fig. 2-4. Millikan's oil drop experiment.

in addition, the body carries a charge q and is subjected to the action of an electric field of intensity E, a new terminal velocity v_E is reached, given by

$$qE - mg = kv_E, \tag{2-4}$$

when the field acts counter to gravity. The ratio of the velocities is

$$\frac{v_g}{v_E} = \frac{mg}{qE - mg}. \tag{2-5}$$

The charge q is then

$$q = \frac{mg}{E} \frac{(v_g + v_E)}{v_g}. \tag{2-6}$$

It is evident that q can be evaluated if m and E are known and if the ratio $(v_g + v_E)/v_g$ is observed experimentally. The main difficulty lies in the determination of m, the mass of the drop, which is too small to be measured directly.

It is now possible to settle a fundamental question, namely, that of the atomic nature of electricity. If there is a real unit charge e, then the charge q on an oil drop must be some integral multiple of e. Suppose that a measurement is made, such as that leading to Eq. (2–6), and that the charge on the drop then changes from q to q' because another ion is caught. Then q' is given by

$$q' = \frac{mg}{E} \frac{(v_g + v'_E)}{v_g}. \tag{2–7}$$

Experimentally, ions supplied by the ionization of the air by x-rays can be caught by an oil drop while under observation, and the behavior of a single drop carrying different charges can therefore be observed in the same electric field E. The change in charge is

$$\Delta q = q' - q = \frac{mg}{E} \frac{(v'_E - v_E)}{v_g}. \tag{2–8}$$

Millikan found, in many such experiments, that Δq was never smaller than a certain minimum value, and that all values of Δq were exact integral multiples of this value. By taking the least common divisor of all the values of Δq for a large number of changes of charge on the same drop, Millikan obtained the *smallest value of the change in charge,* and he identified this value with Stoney's electron. Millikan's results provide direct proof that the elementary electric charge, the electron, is not an average charge, but rather that all electric charges on ions have either exactly the same value or else small, exact integral multiples of this value. If this value is denoted by e, then $q = ne$, where n is a small integer. Millikan also showed, by studying drops carrying positive ions, that the elementary unit of positive charge has the same magnitude as that of the negative electronic charge, and that all static charges both on conductors and insulators are built up of units of the electronic charge.

The precise evaluation of the electronic charge depends, as has been noted, on the mass of the oil droplet. The mass can be obtained when the value of the proportionality factor k in Eq. (2–3) is known, and k is given by Stokes' law of fall. From the equations of hydrodynamics, Stokes derived a theoretical expression for the rate of fall of a spherical body in a viscous medium,

$$\tfrac{4}{3}\pi a^3 g(\rho - \rho') = 6\pi\eta a v_g, \tag{2–9}$$

or

$$v_g = \frac{2}{9} \frac{ga^2}{\eta} (\rho - \rho'), \tag{2–9a}$$

where a is the radius of the body, ρ the density of the body, ρ' the density of the medium, η the viscosity of the medium, and v_g the terminal or equilibrium rate of fall of the body. Equation (2–9) represents the balance between the gravitational and viscous forces on the body and can be applied to the fall of an oil drop in air. The value of v_g is measured experimentally, and a is given by

$$a = \left[\frac{9}{2}\frac{\eta v_g}{g(\rho - \rho')}\right]^{1/2}, \tag{2–10}$$

so that

$$m = \tfrac{4}{3}\pi a^3(\rho - \rho') = \frac{4\pi}{3}\left(\frac{9}{2}\frac{\eta v_g}{g}\right)^{3/2}\left(\frac{1}{\rho - \rho'}\right)^{1/2}. \tag{2–11}$$

When the expression for m is inserted into Eq. (2–6) for the charge on an oil drop, the result is

$$q = e_n = \frac{4\pi}{3}\left(\frac{9\eta}{2}\right)^{3/2}\left[\frac{1}{g(\rho - \rho')}\right]^{1/2}\frac{(v_g + v_E)}{E}v_g^{1/2}. \tag{2–12}$$

In the last equation, e_n has been written for q because the charge on the oil drop has been proven to be an integral multiple of the electronic charge. If many observations of the quantity $(v_g + v_E)$ are made for changes of charge on the drop, the greatest common divisor, denoted by $(v_g + v_E)_0$, should correspond to the value of the electronic charge, denoted temporarily by e_1, so that

$$e_1 = \frac{4\pi}{3}\left(\frac{9\eta}{2}\right)^{3/2}\left[\frac{1}{g(\rho - \rho')}\right]^{1/2}\frac{(v_g + v_E)_0}{E}v_g^{1/2}. \tag{2–13}$$

Millikan found that the value of e_1 was constant for large drops, but increased as the drops grew smaller. Careful studies then showed that Stokes' law in the form Eq. (2–9a) does not hold for the free fall of very small drops. It is necessary in this case to replace Eq. (2–9a) by

$$v_g = \frac{2}{9}\frac{ga^2}{\eta}(\rho - \rho')\left(1 + \frac{b}{pa}\right),$$

where p is the pressure of the air, a is again the radius of the drop, and b is an experimentally determined constant. The electronic charge is then given by the expression:

$$e = \frac{e_1}{[1 + (b/pa)]^{3/2}},$$

or

$$e = \frac{4\pi}{3}\left(\frac{9\eta}{2}\right)^{3/2}\left[\frac{1}{g(\rho - \rho')}\right]^{1/2}\frac{(v_g + v_E)_0}{E[1 + (b/pa)]^{3/2}}v_g^{1/2}. \tag{2–14}$$

Millikan obtained the value $e = 4.774 \times 10^{-10}$ esu, which was accepted for some years. Later determinations of e by means of indirect but highly precise methods gave results which disagreed with Millikan's value. The discrepancy was caused by an error in the value of the viscosity of air used by Millikan. The use of more recent data for this parameter leads to the value $e = 4.8036 \pm 0.0048 \times 10^{-10}$ esu, which agrees well with the values of e obtained by other methods.

The constants e, F, and N_0, together with other quantities which will be met later, are called the *fundamental constants of physics* because of their great importance. The measurement of these constants is one of the basic problems of physics, and many highly precise methods have been developed for measuring them. The constants are often related, for example, $F = N_0 e$; and statistical methods have been developed for the analysis of the experimental results. This analysis yields *best values* of the constants which are consistent with the entire body of present knowledge;[(3)] a partial list of these values is given in Appendix II. The following best values have been obtained for constants which have been discussed so far:

$$\begin{aligned} e &= (4.80286 \pm 0.00009) \times 10^{-10}\,\text{esu}, \\ e/m &= (1.75890 \pm 0.00002) \times 10^{7}\,\text{emu/gm} \\ &= (5.27305 \pm 0.00007) \times 10^{17}\,\text{esu/gm}, \\ m &= (9.1083 \pm 0.0003) \times 10^{-28}\,\text{gm}, \\ F &= (2.89366 \pm 0.00003) \times 10^{14}\,\text{esu/gm-equivalent weight} \\ &= (96521.9 \pm 1.1)\,\text{coul/gm-equivalent weight}. \end{aligned}$$

The Avogadro number is

$$N_0 = (6.02486 \pm 0.00016) \times 10^{23}/\text{gm-mole}.$$

The mass of an individual hydrogen ion, which is usually referred to as a *proton*, is

$$M_{\text{H}} = (1.67239 \pm 0.00004) \times 10^{-24}\,\text{gm},$$

and the ratio of the mass of the proton to that of the electron is

$$M_{\text{H}}/m = 1836.12 \pm 0.02.$$

The values quoted for e/m and m are correct provided that the speeds of the particles are very small compared with the speed of light. It will be seen in Chapter 6 that the mass of the electron increases with its speed; the value given above for the mass of the electron is called the *rest mass* and is often denoted by m_0.

2-5 Positive rays. The fact that rays consisting of negatively charged particles are produced in a gas discharge tube suggested that rays of positively charged particles are also formed. Rays of this kind were discovered by Goldstein (1886), who observed that when the cathode of a discharge tube was pierced by narrow holes, streamers of light appeared behind it. He assumed that the luminosity was caused by rays which traveled in the opposite direction to the cathode rays and passed through the holes in the cathode. These rays, which were called *canal rays*, were deflected by electric and magnetic fields, and from the directions of the deflections it was concluded that the rays consist of positively charged particles. This result gave rise to the more generally used term *positive rays*.

As in the case of the cathode rays, it was possible to measure the velocity and charge-to-mass ratio of the positive-ray particles by means of the deflections in electric and magnetic fields. The results were quite different from those for the cathode-ray particles. Both the velocities and the e/m values were found to be much smaller than those for electrons. The value of e/m was found to depend on the atomic weight of the gas in the discharge tube, and to decrease with increase in atomic weight. The largest value of e/m was obtained for hydrogen and was very close to the value obtained in electrolysis. These results suggested that the positive rays are streams of positive ions produced by the ionization of atoms or molecules in the strong electric field of the discharge tube. The positive ion is formed by the removal of one or more electrons from the neutral atom, and the mass of the ion is practically the same as that of the atom. Hence, any measurement of the mass of positive-ray particles should yield direct information about the masses of atoms of elements or molecules of compounds. Furthermore, this information should refer to the atoms or molecules individually rather than give average values for a large number of particles. It is for this reason that the accurate measurement of the masses of positive-ray particles is of great importance in modern physics. In fact, measurements of this kind resulted in the proof of the existence of isotopes and in the determination of isotopic weights.

The subject of positive rays will be treated in greater detail in Chapter 9 in connection with the determination of isotopic masses and abundances.

2-6 X-rays. In 1895, Roentgen discovered x-rays, which are produced when a beam of cathode rays strikes a solid target. He found that the operation of a cathode-ray tube produced fluorescence in a screen covered with barium platinocyanide and placed at some distance from the tube. The effect was traced to radiation coming from the walls of the cathode-ray tube. In studying the properties of this new type of radiation, Roentgen noticed that if materials opaque to light were placed between the tube and the screen the intensity of the fluorescence decreased but did not disappear, showing that x-rays can penetrate substances which are opaque to ordinary light. It was also found that the x-radiation can blacken a photographic plate and produce ionization in any gas through which it passes; the last property is used to measure the intensity of the radiation. The x-rays were found to travel in straight lines from the source and could not be deflected by electric or magnetic fields, from which it was concluded that they are not charged particles. Later work showed that x-rays can be reflected, refracted, and diffracted, and there is convincing evidence that these rays are an electromagnetic radiation like light but of much shorter wavelength.

X-rays have turned out to be a valuable tool in atomic research and a great deal of information about atomic structure has been obtained from studies of the scattering and absorption of x-rays by atoms, as will be shown in Chapter 4.

2–7 Radioactivity: alpha-, beta-, and gamma-rays. In 1896, Becquerel discovered that crystals of a uranium salt emitted rays which were similar to x-rays in that they were highly penetrating, could affect a photographic plate, and induced electrical conductivity in gases. Becquerel's discovery was followed by the identification by the Curies (1898) of two other *radioactive* elements, polonium and radium. The activity of radium as measured by the intensity of the emitted rays was found to be more than a million times that of uranium.

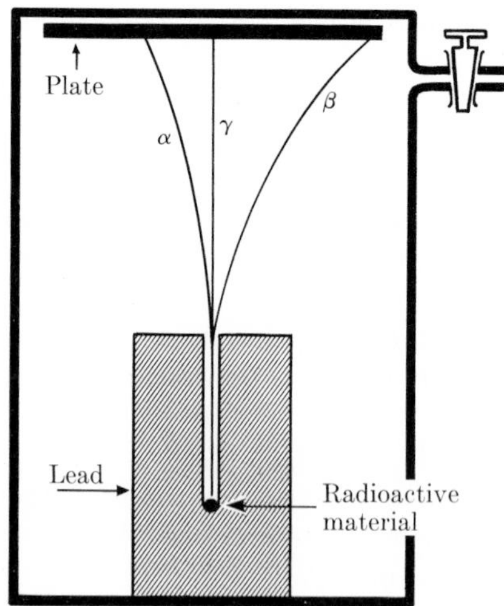

FIG. 2–5. The three radiations from radioactive materials and their paths in a magnetic field perpendicular to the plane of the diagram (schematic).

It can be shown, by means of experiments in magnetic fields, that there are three kinds of radiations from naturally occurring radioactive substances. In these experiments a collimated beam of rays is supplied by a small piece of radioactive material at the bottom of a long groove in a lead block, as shown in Fig. 2–5. A photographic plate is placed some distance above the lead block, and the air is pumped out of the chamber. A weak magnetic field is applied at right angles to the plane of the diagram, and directed away from the reader. When the plate is developed, two distinct spots are found, one in the direct line of the groove in the lead block, and one deflected to the right. The undeviated spot is caused by neutral rays called γ-rays; the spot to the right is caused by rays of negatively charged particles, called β-rays. If the weak magnetic field is replaced by a strong one in the same direction, the undeviated spot is again found together with a spot deflected to the left; the latter is caused by positively charged particles called α-rays. In the strong field, the path of the β-rays is bent so much that it doesn't reach the plate. The need for a strong field indicated that the mass of the positively charged particles is much greater than that of the negatively charged particles. The different electrical properties of the radiations are shown schematically in Fig. 2–5.

It has been shown that all three radiations are not emitted simultaneously by all radioactive substances. Some elements emit α-rays, others emit β-rays, while γ-rays sometimes accompany one and sometimes the other. The activity of a given material is not affected by any simple physical or chemical process, such as change in temperature or chemical combination with nonradioactive substances. These results, together with Mme. Curie's demonstration that the activity of any uranium salt is directly proportional to the quantity of uranium in the salt, showed that radioactivity is an atomic phenomenon.

The β-rays, being negatively charged particles, could be deflected in electric and magnetic fields, and their velocities and e/m values were measured. The results, although somewhat more complicated than in the case of cathode rays, showed that the β-rays consist of swiftly moving electrons and differ from cathode rays only in velocity. The β-rays from radium were found to have velocities up to within a few percent of that of light, while cathode rays usually have velocities considerably less than half that of light. The e/m values of the β-particles were of the same order of magnitude as those of the cathode-ray particles, but decreased with increasing velocity of the particles. Even before the Millikan oil-drop experiment, it was considered that electric charge is atomic in nature and should not vary with velocity, and it was concluded that the mass of a β-particle increases with the speed with which the particle is moving. The value of e/m for the slower β-particles was found to be almost identical with that obtained for cathode-ray particles. The β-rays are much more penetrating than α-rays, being able to pass through one millimeter of aluminum, whereas the α-rays are completely stopped by 0.006 cm of aluminum or by a sheet of ordinary writing paper. The ionization density produced in a gas by β-rays is much less intense, however, than that produced by α-rays.

Alpha-rays, because of their positive electrical charge, can be deflected by electric and magnetic fields, and their velocities and e/m values determined in the usual way. The velocities range from about 1.4×10^9 to 2.2×10^9 cm/sec, or less than $\frac{1}{10}$ of the velocity of light. The value of e/m is close to 4820 emu/gm $= 1.45 \times 10^{14}$ esu/gm, or just one-half of the value, 2.87×10^{14} esu/gm, for the singly charged hydrogen ion. It is evident that if the α-particle had the same charge as the univalent hydrogen ion, its mass would be twice that of a hydrogen ion or atom. If the α-particle carried a charge twice that of the hydrogen ion, its mass would be four times as great and would correspond closely to that of the helium atom, which has an atomic weight of 4 units. The exact nature of the α-particle could not be determined until either its charge or mass could be measured separately. Rutherford showed, in an ingenious experiment, that when a radioactive substance emits α-rays, helium is produced. A small quantity of radon, a gaseous radioactive element, was sealed in a small glass tube with extremely thin walls, so thin that the α-rays emitted could pass through them into an outer glass tube, while the radon was kept in. The outer tube was provided with electrodes, and the spectrum produced on passing a discharge was observed at regular intervals. After a few days it was found that the spectral lines of helium appeared; control experiments showed that ordinary helium gas could not penetrate the thin-walled tube, proving that the helium was produced from the α-rays. Thus, the α-particles must be helium ions and, since their mass is 4 times that of a hydrogen ion, the charge on an α-particle must be twice that on the univalent hydrogen ion. The α-particle is, therefore, a doubly ionized helium atom.

The charge on an α-particle was also determined separately by counting the number of α-particles emitted in a known time interval from a known weight of radioactive material, and at the same time measuring the total positive charge emitted. It will be shown in the next section that methods have been devised for counting individual α-particles. Two independent experiments, in which different counting methods were used, gave 9.3×10^{-10} esu and 9.58×10^{-10} esu, respectively, for the α-particle charge, and these values are very close to twice the electronic charge.

The γ-rays, which are not deflected by electric or magnetic fields, have been shown to consist of electromagnetic waves, and are therefore similar in nature to light and x-rays. The γ-rays are roughly 10 to 100 times as penetrating as the β-rays, but produce less ionization than do the β-rays.

The ionization produced by α-, β-, and γ-rays is roughly in the order 10,000, 100, and 1, a useful rule of thumb.

The detailed chemical study of the radioactive substances has shown that each such substance is an element, and that its radioactivity is caused by a spontaneous disintegration of radioactive atoms into other atoms. This process occurs according to definite laws. The transformation of one atom into another atom with the emission of either an α-particle or a β-particle, led to the idea that the atoms of the chemical elements are complex structures built of smaller particles. The study of atomic structure became an important branch of physics with the main object of discovering the laws according to which atoms are built up. The discovery of natural radioactivity, together with the discovery of x-rays and the proof of the independent existence of the electron, provided new and powerful methods for attacking the problem of atomic structure.

2-8 The detection and measurement of radiation. The study of radioactivity and the successful use of radiations as research tools or for other purposes depend on the quantitative detection and measurement of radiation. The quantities most often needed are the number of particles (e.g., electrons, protons, α-particles) arriving at a detector per unit time and their energies. When a charged particle passes through matter, it causes excitation and ionization of the molecules of the material. This ionization is the basis of nearly all of the instruments used for the detection of such particles and the measurement of their energies. Similar instruments can be used for uncharged radiations (x-rays or γ-rays) because these impart energy to charged particles, which then cause ionization. The different types of instrument differ in the material within which the ionization is produced and in the way in which it is observed or measured. Many instruments are based on the production of ionization in a gas. It is necessary, in this case, to separate and collect the positive and negative ions because, if these remained close together, they would recombine and no electrical information as to their presence could be obtained. The separation and collection of the ions requires the presence of an electrostatic field, and different instruments result depending on whether the field is small, large, or intermediate in magnitude. The ionization may also be produced in a liquid or in a crystal. When it is produced in a gas supersaturated with vapor or in a photographic emulsion, the tracks of the particles can be made visible. When particles strike certain liquid or solid materials called *phosphors*, which have the property of luminescence, part of the energy used up in molecular excitation and ionization is re-emitted as visible or ultraviolet light. Sometimes this light can be observed by eye, or it may be detected by means of more sensitive devices.

We shall consider some of the commonly used detection methods because of their great importance in nuclear physics, although a detailed treatment is beyond the scope of this book. Such treatments are given in some of the references at the end of the chapter.

The scintillation method. It was found, about 1900, that radiations, especially α-particles, can produce luminescence in zinc sulphide, barium platino-cyanide, and diamond. This luminescence, when produced by α-particles, is not uniform, but consists of a large number of individual flashes, which can be seen under a magnifying glass. Careful experiments have shown that each α-particle produces one scintillation, so that the number of α-particles which fall on a detecting screen per unit time is given directly by the number of scintillations counted per unit time. A screen can be prepared by dusting small crystals of zinc sulfide, containing a very small amount of copper impurity, on a slip of glass. The counting can be done with a microscope with a magnification of about 30 times, and good precision can be obtained, but with difficulty. This method was especially useful for counting α-particles in the presence of other radiations, because the zinc sulfide screen is comparatively insensitive to β- and γ-rays.

The visual scintillation counter has the disadvantages that the flashes are quite weak and can only be seen in a darkened room, and that visual counting is limited physiologically to about 60 scintillations/min and is very tedious. Nevertheless, the method was used in many of the basic experiments of nuclear physics between 1908 and about 1935 when it was replaced by electrical methods.

Since 1947, the use of scintillation counters has been revived and greatly extended because of the discovery of new phosphors which are also sensitive to β- and γ-rays, and the development of highly efficient photomultiplier tubes. The phosphors include: inorganic salts, primarily the alkali halides, containing small amounts of impurities as activators for luminescence, for example, sodium or potassium iodide activated with thallium; crystalline organic materials such as naphthalene, anthracene and stilbene; and solutions of organic compounds such as terphenyl dissolved in xylene. The photomultiplier tube, which replaces the microscope and observer, converts the scintillations from the phosphor into amplified electrical pulses which can be counted or otherwise analysed with suitable electronic equipment. The modern scintillation counter can detect and record many millions of flashes per second and can be used with intense radiations.

A schematic diagram of a photomultiplier tube is shown in Fig. 2-6. Light from the phosphor strikes the cathode, which is usually made of antimony and cesium, and ejects electrons (photoelectric effect). The tube has several electrodes called *dynodes* to which progresssively higher

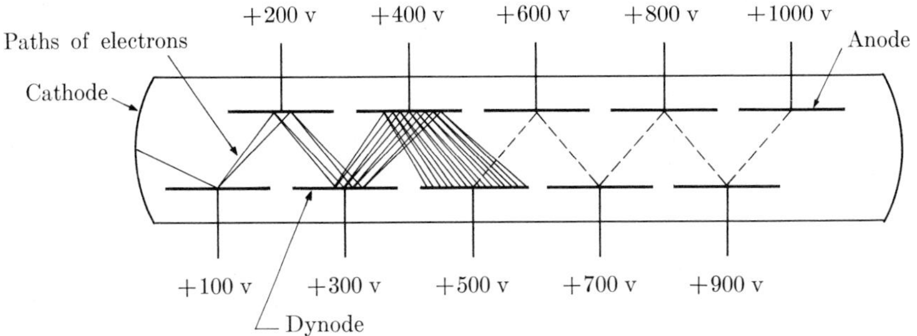

FIG. 2-6. Photomultiplier tube (schematic).

potentials are applied. The photoelectrons are accelerated in the electrostatic field between the cathode and the first dynode, which is at a positive potential relative to that of the cathode, and strike the dynode. The accelerated electrons impart enough energy to electrons in the dynode to eject some of them. There may be as many as ten secondary electrons for each electron which strikes the dynode. This process is repeated, and the electron current is amplified as the electrons are accelerated from dynode to dynode. The output current, or pulse, at the anode may be more than a million times as great as the current originally emitted from the cathode. Each particle incident on the phosphor produces a pulse, and the pulses are fed to an electronic system where they are counted. Electronic systems have also been developed which measure the energy of the incident particles; the resulting instrument is called a *scintillation spectrometer*.

A schematic diagram of a scintillation detector is shown in Fig. 2-7.

The electroscope. It has been mentioned that many of the instruments used for the measurement of radiation depend for their operation on the

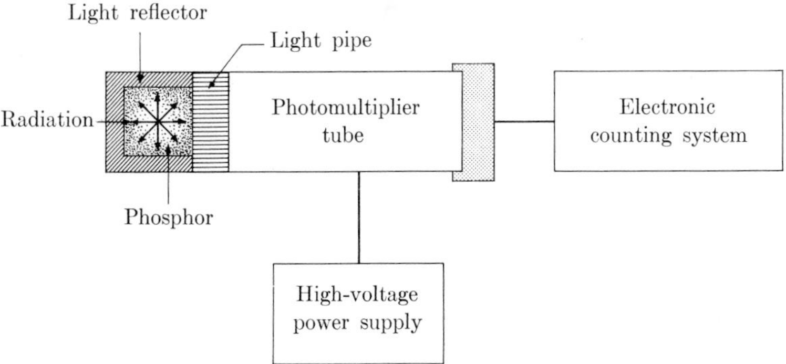

FIG. 2-7. Schematic diagram of a scintillation detector.

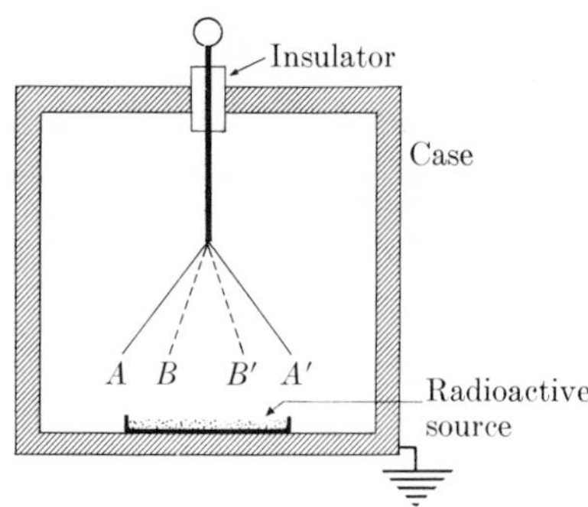

FIG. 2–8. Diagram of an α-ray electroscope.

ionization of a gas. A given sample of radioactive substance emits a definite number of α-, β-, or γ-rays per unit time. The number of ions which these rays can produce in air or in another gas at a fixed temperature and pressure is directly proportional to the number of particles. Hence the extent of the ionization is a direct measure of the amount of radioactive substance. One of the first and simplest instruments used in work on radioactivity is the electroscope, an example of which is shown in Fig. 2–8. This instrument is a condenser, one of whose elements is the outside case; the other element ends in a pair of gold leaves. When the two elements are connected across a battery, they become oppositely charged and the gold leaves (A and A') separate because of the electrostatic repulsion of their like charges. If the battery is removed and the gas in the condenser is ionized by a radioactive substance, gaseous ions migrate to the oppositely charged elements of the condenser. The net charge on the elements and the potential difference between them decrease, and the leaves collapse to the position B, B'. The rate of collapse of the gold leaves is a measure of the rate at which the gas in the chamber is ionized and hence of the intensity of the ionizing radiation.

If the radioactive substance emits α-, β-, and γ-rays, they all contribute to the ionization but, since the α-rays are by far the strongest ionizing agents, the instrument is really an α-ray electroscope. A β-ray electroscope is made by placing an aluminum sheet about 0.01 cm thick over the active material to absorb all of the α-rays, and the ionization observed is then caused by the β- and γ-rays. Since the γ-rays are much weaker ionizing agents than the β-rays, the collapse of the leaves is caused almost entirely by the β-rays. To measure γ-ray activity alone, the radioactive sample must be surrounded by enough material (2 or 3 mm of lead) to absorb completely the α- and β-rays. Sometimes the entire electroscope is built of, or surrounded by, lead and the sample is placed outside the instrument. Electroscopes have been developed which are much more

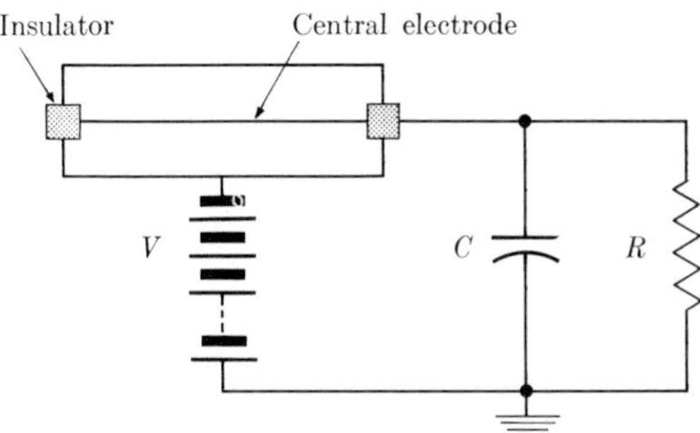

Fig. 2–9. Schematic diagram for pulse operation of a gas-filled chamber.

sensitive than the one just described. For example, the Lauritsen electroscope has as its moving element a very fine metal-coated quartz fiber. The movement of the fiber is observed through a microscope with a scale in the eyepiece.

Electrical instruments: ionization chamber, proportional counter, and Geiger-Mueller counter. Each of these three detectors is based on the production of ionization in a gas and the separation and collection of the ions by means of an electrostatic field. The differences in the three systems can be explained with the aid of Fig. 2–9, which shows a cylindrical conducting chamber containing a central conducting electrode located on the axis of the chamber and insulated from it. The chamber is filled with a gas at a pressure of one atmosphere or less. A voltage V is applied between the wall and the central electrode through the resistance R shunted by the capacitor C; the central electrode is at a positive potential relative to that of the chamber wall.

We suppose that some ionization occurs in the gas because of the passage of a charged particle. Each ion pair consists of a positive ion and an electron and we want to know, for a given initial ionization, how many ion pairs are collected, or how many electrons reach the central electrode as the applied voltage is varied. To make the discussion less abstract, we assume first that 10 ion pairs are formed, corresponding to curve a of Fig. 2–10 where curves of total ion collection are plotted as functions of applied voltage. For convenience, the logarithm to the base 10 of the number of ion pairs n has been used as the ordinate. The discussion is based on that of Wilkinson (see general reference at end of chapter). If there is no voltage across the electrodes, the ions will recombine, and no charge will appear on the capacitor. As the voltage is increased, say to a few volts, there is competition between the loss of ion pairs by recombination and the removal of ions by collection on the electrodes. Some electrons, less than 10, will reach the central electrode. At a voltage V_1 of,

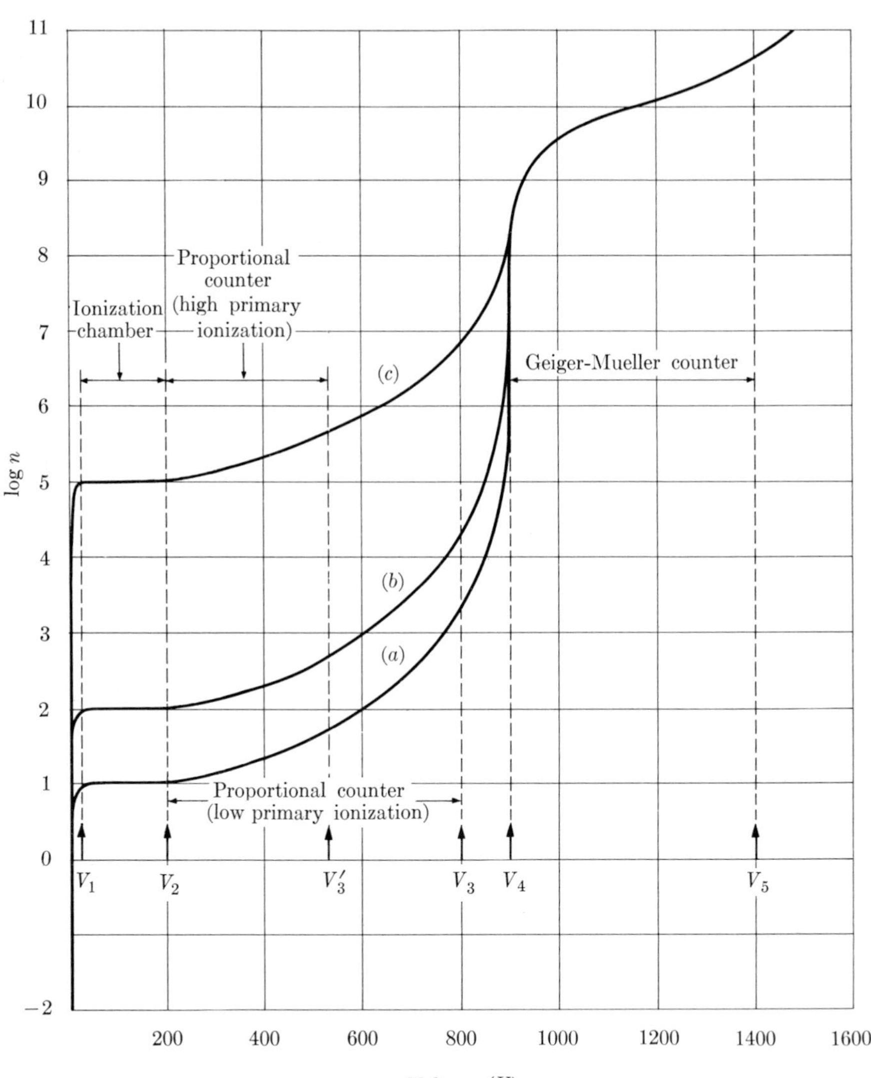

FIG. 2–10. Curves of total ion collection versus applied voltage to illustrate ionization, proportional, and Geiger-Mueller regions of operation of electrical radiation detection instruments.

perhaps, 10 volts the loss of ions by recombination becomes negligible, and 10 electrons reach the central electrode; log n reaches the value unity. As V is increased, n stays constant at 10 until a voltage V_2 is reached which may be some tens or hundreds of volts depending on the conditions of the experiment. The region between V_1 and V_2, in which the number of ion pairs collected is independent of the applied voltage and the curve is horizontal, is called the *ionization chamber region.*

When the voltage is increased above V_2, n increases above 10 because of a phenomenon called *gas multiplication* or *gas amplification.* The electrons released in the primary ionization acquire enough energy to produce additional ionization when they collide with gas molecules, and n increases roughly exponentially with V. Each initial electron produces a small "avalanche" of electrons with most of these secondary electrons liberated close to the central electrode. If 100 ion pairs are formed initially, curve b results; it is parallel to curve a in the ionization chamber region and is separated from it by one unit on the log n scale. The behavior of the two curves above V_2 is interesting. For some range of voltages, up to V_3, each electron acts independently and gives its own avalanche, not being affected by the presence of the other electrons. The initial 100 electrons always yield 10 times more final electrons than do the 10 initial electrons, and curves a and b continue parallel with a difference between them of one unit on the log n scale. Between V_2 and V_3, the number of ion pairs collected is then proportional to the initial ionization. This is the region of *proportional counter operation.*

Above V_3, the gas multiplication effect continues to increase very rapidly, and, as more electrons produce avalanches, the latter begin to interact with one another; the positive-ion space charge of one avalanche inhibits the development of the next avalanche. The discharge with 100 initial electrons is affected before the one with 10 initial electrons and increases less rapidly than the latter; curves a and b approach each other and eventually meet at V_4. The region between V_3 and V_4 is the region of *limited proportionality.* Above V_4 the charge collected becomes independent of the ionization initiating it, and the curves a and b become identical. The gas multiplication increases the total number of ions to a value that is limited by the characteristics of the chamber and the external circuit. The region above V_4 is the region of *Geiger-Mueller counter operation.* It ends at a voltage V_5 where the discharge tends to propagate itself indefinitely; V_5 marks the end of the useful voltage scale, the region above being that of the *continuous discharge.*

If the initial ionization is large, say 10^5 ion pairs, curve c is obtained. It is similar to curves a and b, and parallel to them between V_2 and V_3'; the proportional counter region ends sooner, at V_3' instead of at V_3. Thus, the extent of that region depends on the initial ionization.

As a result of the behavior of the ions of the gas in the electrostatic field, three detection instruments have been developed.

1. The *ionization chamber*, which operates at voltages in the range V_1 to V_2, is characterized by complete collection, without gas amplification, of all the electrons initially liberated by the passage of the particle. Subject to certain conditions, it will give a pulse proportional to the number of these electrons. Figure 2–9 is a schematic diagram of a chamber which could be used as an ionization chamber between 10 volts and 200 volts, approximately. The numbers in Fig. 2–9 would be appropriate for a counter with an outer cylinder 1 cm in radius, a wire central electrode 0.01 cm in radius, and filled with gas to a pressure of a few centimeters of Hg (e.g., argon at about 6 cm of Hg plus a little alcohol). The counter length would be about 10 cm.

The electrodes of an ionization chamber may also be parallel plates. A schematic diagram of an instrument of this type is shown in Fig. 2-11.

2. The *proportional counter*, which operates in the voltage region V_2 to V_3, is characterized by a gas multiplication independent of the number of initial electrons. Hence, although gas multiplication is utilized, the pulse is always proportional to the initial ionization. The use of this counter permits both the counting and the energy determination of particles which do not produce enough ions to yield a detectable pulse in the region V_1 to V_2. The proportional counter, therefore, offers advantages for pulse-type measurements of beta radiation, an application for which ionization chambers are not sensitive enough.

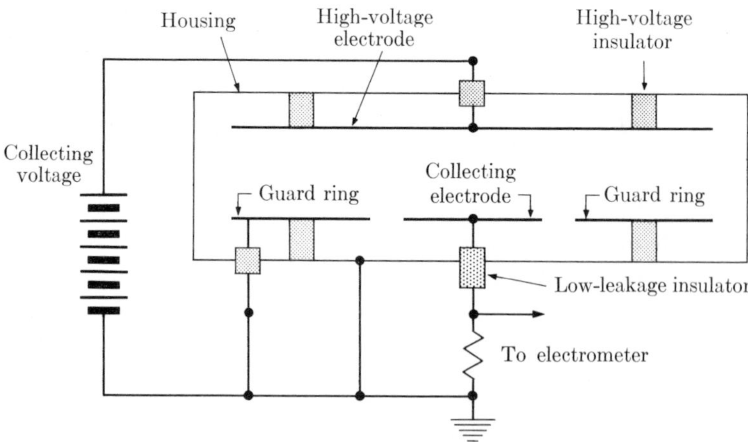

FIG. 2–11. Schematic diagram of a parallel plate ionization chamber (Reprinted by permission from W. J. Price, *Nuclear Radiation Detection.* McGraw-Hill, 1958.)

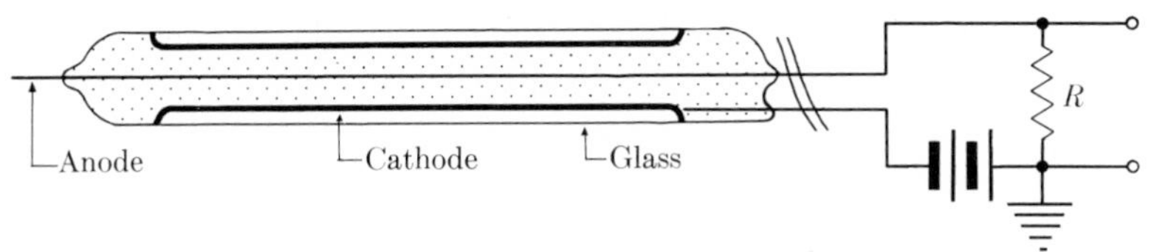

FIG. 2–12. Diagram of a Geiger-Mueller counter tube.

3. *The Geiger-Mueller counter,* also known as *Geiger,* or G-M, *counter,* which operates in the voltage region V_4 to V_5, is characterized by the spread of the discharge throughout the entire length of the counter, resulting in a pulse size independent of the initial ionization. It is especially useful for the counting of lightly ionizing particles such as β-particles or γ-rays. The G-M counter usually consists of a fine wire (e.g., tungsten) mounted along the axis of a tube which contains a gas at a pressure of about 2 to 10 cm Hg. The tube may be a metal such as copper, or a metal cylinder may be supported inside a glass tube as in Fig. 2–12; a mixture of 90% argon and 10% ethyl alcohol is suitable. A potential difference, which may be between 800 and 2000 volts, is applied to make the tube negative with respect to the wire.

Geiger-Mueller counters have been designed for the measurement of α-, β-, and γ-rays, as well as x-rays. In the form described, they are used mainly for β- and γ-rays, in part because of the difficulty of making tubes with windows thin enough to be penetrated by α-rays. The principle of the Geiger-Mueller counter was first used by Rutherford and Geiger in 1908 to count α-particles, with the object of determining their charge. The instrument that they used, the Geiger point counter, is shown schematically in Fig. 2–13, and instruments like it are still used to detect α-particles. It has a smooth fine point P at the end of a fine wire supported inside a tube T by an insulating plug. The tube contains dry air at atmospheric pressure, and the particles enter through a thin foil window W.

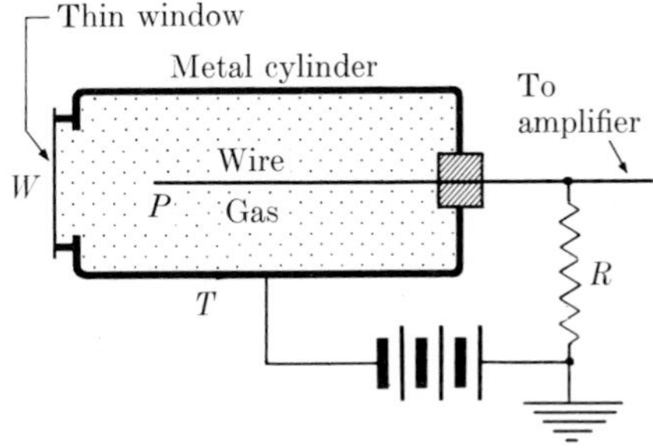

FIG. 2–13. Diagram of a Geiger point counter.

In other respects, the operation of this counter tube is similar to that of the Geiger-Mueller counter. By comparing the activity of a radioactive substance as measured with a Geiger point counter and with a zinc sulfide screen, Rutherford and Geiger showed that each α-particle that strikes the screen causes one scintillation, a fact which increases the usefulness of the screen for quantitative measurements.

In the electrical instruments discussed, a particle causes a small amount of ionization. The ions, either multiplied in number or not, are collected and produce a voltage pulse which may be as small as 10 microvolts. An electronic pulse amplifier accepts these small voltages and amplifies them to a level usually in the range of 5 to 50 volts. The amplified voltage pulses must then be counted in some way so that their rate can be measured. For very slow rates (less than about one per second), the pulses can be made to operate a relay-type recorder. For large rates, mechanical devices either fail completely or miss many counts. The response of an ionization chamber can be made sufficiently rapid so that α-particles arriving at intervals of 10^{-4} sec or less can be detected and the counting must then be made very efficient. A scaling unit, or *scaler*, is an electric device which selects precisely every mth pulse and passes it on to the recorder. The scaling ratio is usually a power of 2 or 10 and very high counting rates can be achieved with the aid of a scaler. The number of pulses indicated on the counter per unit time multiplied by the appropriate factor of the scaler gives the number of α-particles entering, per unit time, the space between the electrodes in the ionization chamber. Other electronic devices are also available for measuring the magnitude of the pulse, which gives the energy of the incident particle. Thus, it is often desired not only to record the occurrence of a pulse but, also, to sort pulses according to their size (by means of an electronic discriminator circuit) or to sort them according to the time intervals during which they arrive (by means of an electronic timing circuit). The detector then forms part of a circuit with appropriate electronic instrumentation. Scintillation detectors, of course, also require electronic instrumentation.

The cloud chamber. One of the most important instruments for basic research on radiations is the cloud chamber. This instrument, first used in 1912, is based on the discovery by C. T. R. Wilson (1897) that ions act as nuclei for the condensation of supersaturated water vapor. There are two types of cloud chamber differing in the method used for obtaining supersaturation. In the expansion chamber,[(4)] the gas, saturated with vapor, is made to expand by the quick motion of a piston. The expansion is adiabatic and lowers the temperature. The cooling is more than sufficient to overcome the effect of the increase in volume, and the air becomes supersaturated with water vapor. If an ionizing ray enters the chamber, the ions formed act as nuclei for the condensation of the vapor, and the

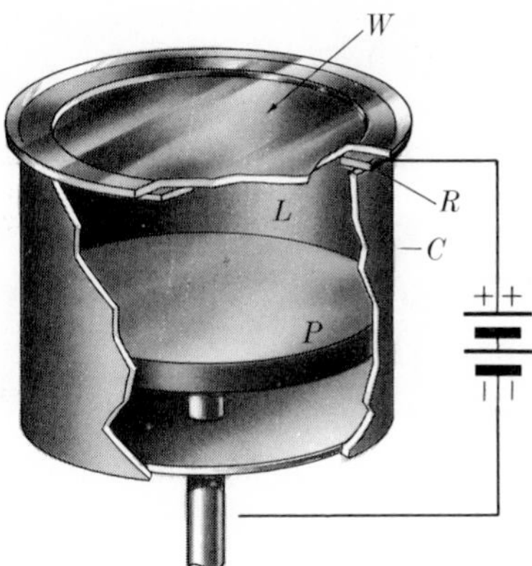

FIG. 2–14. A simple expansion cloud chamber.

path of the ray appears as a thin track of fog. In most expansion chambers, the gas-vapor mixture is air with water, or argon with ethyl alcohol, at atmospheric pressure.

A simplified diagram to indicate the principle of the expansion chamber is shown in Fig. 2–14. A cylinder of glass C is closed at one end by a glass window W and at the other end by a metal piston P to form the chamber. A small amount of water in the chamber keeps the air saturated. When the piston is pulled down, the air becomes supersaturated as described above and, in the presence of ionizing radiation, fog tracks are formed. The tracks can be illuminated by light L from the side and viewed or photographed through the window. The ions can then be removed by means of an electric field between the piston and the metal ring R. The piston is returned to its original position and the chamber is ready for another burst of radiation.

The Wilson cloud chamber makes it possible to study the behavior of individual atoms, to photograph the actual paths of ionizing radiations, and to analyze at leisure the complicated interactions which may take place between charged particles and individual atoms. Many modifications of the original cloud chamber have been made. For example, in order to obtain large numbers of photographs, automatic arrangements have been made so that the expansion can be repeated rapidly and photographs taken continuously on motion picture film. If two stereoscopic pictures are taken simultaneously, the path of the particle in space can be reconstructed.

When a cloud chamber is placed between the pole pieces of an electromagnet, it is possible to distinguish between positively and negatively charged particles. From the curvature of the path of a particle in the magnetic field, the sign of the charge and the magnitude of the momentum of the particle can be determined.

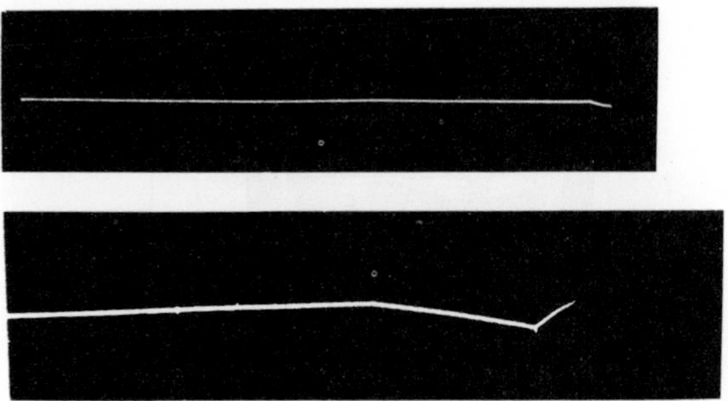

FIG. 2–15. Tracks of α-particles in air. The top photograph shows a complete track; the bottom photograph is an enlargement of the end portion of a track showing two deflections caused by collisions with atoms of the air. [Reprinted by permission from Rutherford, Chadwick, and Ellis, *Radiations from Radioactive Substances*. Cambridge University Press (Macmillan Co.) 1930.]

Different particles produce different types of tracks. Thus, heavy, slow particles such as α-particles produce broad, densely packed, straight line tracks with an occasional sharp, small angle bend, especially near the end of the track. Slow electrons produce narrow, beaded, tortuous tracks, while fast particles, both light and heavy, produce narrow, beaded straight tracks. Tracks of α-particles in air are shown in Fig. 2–15. The top photograph is that of the complete track of an α-particle from radon; the bottom photograph is an enlargement of the end portion of a track showing two deflections caused by collisions with atoms of the air.

The diffusion cloud chamber [(5)] operates by the diffusion of a condensable vapor from a warm region, where it is not saturated, to a cold region, where it becomes supersaturated. The vapor is usually methyl or ethyl alcohol. Air is satisfactory for the gas at pressures from 20 cm Hg to 4 atmospheres, and hydrogen from 10–20 atmospheres. A schematic diagram of a diffusion chamber is shown in Fig. 2–16. The vapor is introduced at the top, the warm end, and diffuses downward continuously through a region in which a steady vertical temperature gradient is maintained by cooling the bottom of the chamber. The diffusion chamber is continuously sensitive, unlike the expansion chamber, which is sensitive only for a short time interval after the expansion. This property, together with its relative mechanical simplicity, gives the diffusion chamber advantages over the expansion chamber. The latter, however, is still used in a large portion of serious cloud-chamber work because of the greater experience with it and because of certain other desirable properties which it has.

Photographic emulsions. An ionizing particle traveling through the emulsion of a photographic plate leaves a track containing a number of developable silver bromide grains. Special photographic emulsions, called nuclear

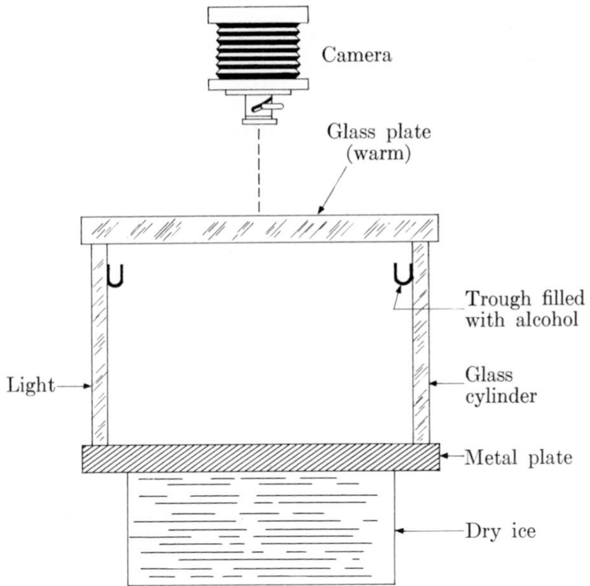

FIG. 2–16. Schematic diagram of a diffusion cloud chamber.

emulsions, have been developed. They are distinguished from optical emulsions in their high silver bromide content, which may be as much as four times as great as in photographic plates, in the grain size, and in the thickness of the emulsion. Like the cloud chamber, the photographic plate, upon development, records the path of the particle, and a variety of information can be obtained from the study of the tracks. Counting of the individual paths gives a measure of the number of particles entering the plate, and study of the detailed structure of the tracks yields information about the mass, charge, and energy of the particles. The photographic emulsion offers advantages over the cloud chamber in that the emulsion is solid so that the tracks are short, and its sensitivity is permanent rather than restricted to infrequently repeated short intervals. On the other hand, the cloud-chamber expansions can be controlled by means of a counter and can be made to record specific events. An emulsion is limited in the chemical elements which it can contain.

Finally, other instruments such as the bubble counter and the Cerenkov detector have been developed for use with particles of very high energy (they are discussed in the book by Price listed in the references).

References

GENERAL

J. D. STRANATHAN, *The "Particles" of Modern Physics.* Philadelphia: Blakiston, 1944.

J. A. CROWTHER, *Ions, Electrons, and Ionizing Radiations,* 8th ed. London: Edward Arnold, 1949.

R. A. MILLIKAN, *Electrons (+ and −), Protons, Photons, Mesotrons, and Cosmic Rays.* Chicago: University of Chicago Press, 1947.

J. B. HOAG and S. A. KORFF, *Electron and Nuclear Physics,* 3rd ed. New York: Van Nostrand, 1948.

A. P. FRENCH, *Principles of Modern Physics.* New York: Wiley, 1958.

E. R. COHEN, K. M. CROWE, and J. W. M. DUMOND, *Fundamental Constants of Physics.* New York: Interscience, 1957.

RADIATION DETECTION

H. H. STAUB, "Detection Methods," *Experimental Nuclear Physics,* Vol. I, Part I, E. Segrè, ed. New York: Wiley, 1953.

D. H. WILKINSON, *Ionization Chambers and Counters.* Cambridge: University Press, 1950.

J. B. BIRKS, *Scintillation Counters.* London: Pergamon Press, 1953.

S. A. KORFF, *Electron and Nuclear Counters,* 2nd ed. New York: Van Nostrand, 1955.

J. SHARPE, *Nuclear Radiation Detectors.* London: Methuen; New York: Wiley, 1955.

W. J. PRICE, *Nuclear Radiation Detection.* New York: McGraw-Hill, 1958.

PARTICULAR

1. F. W. CONSTANT, *Theoretical Physics: Thermodynamics, Electromagnetism, Waves and Particles.* Reading, Mass.: Addison-Wesley, 1958, Chapter 9.
2. F. W. SEARS, *Mechanics, Wave Motion, and Heat.* Reading, Mass.: Addison-Wesley, 1958, Chapter 15.
3. E. R. COHEN, K. M. CROWE, and J. S. M. DUMOND, *op. cit.* gen. ref., p. 267; also E. R. COHEN, J. W. M. DUMOND, J. W. LAYTON, and J. S. ROLLETT, "Analysis of Variance of the 1952 Data on the Atomic Constants and a New Adjustment," 1955. *Revs. Mod. Phys.* **27,** 363 (1955).
4. J. G. WILSON, *The Principles of Cloud-Chamber Technique.* Cambridge: University Press, 1951; also N. N. DAS GUPTA and S. K. GHOSH, "A Report on the Wilson Cloud Chamber and its Applications in Physics," *Revs. Mod. Phys.,* **18,** 225 (1946).
5. M. SNOWDEN, "The Diffusion Cloud Chamber" *Progress in Nuclear Physics,* Vol. 3, O. R. Frisch, ed. London: Pergamon Press; New York: Academic Press, 1953.

Problems

1. Take 1 faraday as 96,500 coul and the valences of chromium, copper, indium, iron, mercury, nickel, and silver as 6, 2, 3, 3, 2, 2, and 1, respectively. Show first that 1 amp-hr corresponds to 0.03731 chemical equivalents. Then calculate (a) the number of grams of each element deposited by 1 amp-hr of charge, (b) the number of coulombs of charge needed to deposit 1 mg of each element, (c) the charge in esu associated with 1 gm of each element.

2. In an attempt to measure the value of the charge-to-mass ratio of cathode rays, a Thomson apparatus with the following dimensions was used:

distance from condenser plates to screen = 32.0 cm,
length of plates = 8.0 cm,
distance between plates = 2.75 cm.

When the potential across the plates was 1100 volts and the magnetic field strength was 12.50 gauss, the cathode rays passed through the apparatus without being deflected. When the magnetic field was turned off, the rays were displaced 19.5 cm, on striking the screen, from their undeflected path. What was the speed of the rays? What value was obtained for e/m?

3. With the apparatus of Problem 2, the potential across the plates was changed to 1350 volts. What magnetic field was needed to keep the beam from being deflected? What displacement of the beam was observed when the magnetic field was turned off?

4. When cathode rays with a speed v move in a direction perpendicular to a uniform magnetic field of strength H, their path is a circle of radius R given by the relationship $Hev = mv^2/R$. Show that the value of e/m can be obtained from the equation $e/m = E/H^2R$, where $E/H = v$. In the experiment of Problem 2, suppose that the electric field is turned off instead of the magnetic field. What is the radius of the path of the cathode rays?

5. Another measurement of e/m for cathode rays was made by means of two experiments. In the first, the cathode rays passed between condenser plates 1 cm apart, with a potential of 730 volts across them. The effect of the electric field was just balanced by a magnetic field of 28 gauss. In the second experiment, the rays were bent into a path of 12.00 cm radius by a magnetic field of 12.50 gauss. What was the speed of the rays? What was the value obtained for e/m?

6. In experiments with cathode rays, the source often emits rays with a wide range of speeds, but only rays with a particular speed are desired. How could the Thomson apparatus be modified to serve as a *velocity filter*, i.e., to permit only the passage of rays with a certain desired speed? In the apparatus of Problem 2, suppose that the potential across the plates is fixed and that the magnetic field strength can be varied; how would you obtain a monoenergetic beam of rays with a speed of 2.0×10^9 cm/sec?

7. A uniform, radially directed, electric field of magnitude E is produced in the space between parallel plates which are closely spaced, concentric, cylindrical sections of mean radius R. Show that the kinetic energy of a particle of

charge q, mass M, and speed v, moving in the space between the plates is $\frac{1}{2}qER$. Why does such a system provide an *energy filter* for charged particles? If the mean radius is 10 cm, what field strength is needed to permit a beam of cathode rays with a speed of 1.0×10^9 cm/sec to pass through the filter?

8. In one of his experiments, Millikan observed the motion of an oil drop in a constant electric field. As the drop captured or lost ions, the time needed for it to traverse a given distance, 0.5222 cm, changed. Eight successive time intervals noted were, in seconds, 12.45, 21.5, 34.7, 85.0, 34.7, 16.0, 34.7, and 21.85. Explain, with the aid of Eq. (2–8), how these data show that the electric charge on an ion is an exact integral multiple of an elementary charge.

9. An oil drop of radius 2.76×10^{-4} cm and density 0.9199 gm/cm^3 falls in air at 23°C and 76 cm Hg. Under these conditions, the viscosity of air is 1.832 $\times 10^{-4}$ cgs and its density is 0.0012 gm/cm^3. Calculate the rate of fall, first from the uncorrected Stokes formula, then from the corrected formula; in the latter, the constant b is 0.000625 (cm) (cm Hg).

10. An oil-drop measurement of the charge on the electron was done under the following conditions:

distance between the condenser plates = 1.60 cm,
potential difference across the plates = 5085 volts,
air temperature = 23°C,
pressure = 76 cm Hg,
density of oil = 0.9199 gm/cm^3,
radius of oil drop = 2.76×10^{-4} cm.

The speed of free fall of the drop was 0.08571 cm/sec. The greatest common divisor $(v_g + v_E)_0$ measured for the rise of the drop through a certain fixed distance was 0.005480 cm/sec. Calculate (a) the uncorrected value e_1 of the electron charge, (b) the corrected value e.

CHAPTER 3

THE NUCLEAR ATOM

3–1 The Thomson atom. The discovery of radioactivity, together with Thomson's proof of the independent existence of the electron, provided a starting point for theories of atomic structure. The fact that atoms of a radioactive element are transformed into atoms of another element by emitting positively or negatively charged particles led to the view that atoms are made up of positive and negative charges. If this view is correct, the total negative charge in an atom must be an integral multiple of the electronic charge and, since the atom is electrically neutral under normal conditions, the positive and negative charges must be numerically equal. The emission of electrons by atoms under widely different conditions was convincing evidence that electrons exist as such inside atoms. The first modern theories of atomic structure were, therefore, based on the hypothesis that atoms are made up of electrons and positive charges. No particular assumptions could be made about the nature of the positive charges because the properties of the positive particles from radioactive substances and from gas discharge tubes did not have the uniformity shown by the properties of the negative particles.

Two important questions then arose: (1) how many electrons are there in an atom, and (2) how are the electrons and the positive charges arranged in an atom? Information about the first question was obtained experimentally by studying the way in which x-rays interact with atoms, and this problem will be treated in some detail in the next chapter. It will suffice, for the present, to state that early experiments of this kind indicated that the number of electrons per atom is of the order of the atomic weight. It was known that the mass of an electron is about one two-thousandth of the mass of a hydrogen atom, which has an atomic weight very close to unity. Hence, the total mass of the electrons in an atom is only a very small part of the mass of the atom, and it was logical to assume that practically the entire mass of an atom is associated with the positive charge.

In the absence of information about the way in which the positive and negative charges are distributed in an atom, Thomson proposed a simple model. He assumed that an atom consisted of a sphere of positive electricity of uniform density, throughout which was distributed an equal and opposite charge in the form of electrons. It was remarked that the atom, under this assumption, was like a plum pudding, with the negative electricity dispersed like currants in a dough of positive electricity. The diameter of the sphere was supposed to be of the order of 10^{-8} cm, the magnitude found for the size of an atom. With this model, Thomson was able to calculate theoretically how atoms should behave under certain conditions, and the theoretical predictions could be compared with the results of experiments. It became clear when this comparison was made that Thomson's theory was inadequate; but the failure of the Thomson model in the particular case of the scattering of α-particles proved to be most profitable because it led to the concept of the nuclear atom. This concept is fundamental to atomic and nuclear physics, and the scattering of α-particles will therefore be discussed in some detail.

When a parallel beam of rays from a radioactive substance or from a discharge tube passes through matter, some of the rays are deflected, or scattered, from their original direction. The scattering process is a result of the interaction between the rays of the beam and the atoms of the material, and a careful study of the process can yield information about the rays, the atoms, or both. The scattering of α-particles was first demonstrated by Rutherford, who found that when a beam of α-particles passed through a narrow slit and fell on a photographic plate, the image of the slit had sharply defined edges if the experiment was performed in an evacuated vessel. When the apparatus contained air, the image of the slit on the photographic plate was broadened, showing that some of the rays had been deflected from their original path by the molecules in the air. Alpha-particles are also scattered by a very thin film of matter such as gold or silver foil. When a beam of particles passes through a small circular hole and falls on a zinc sulfide screen, scintillations are seen over a well-defined circular area equal to the cross section of the beam. If a very thin foil is placed in the path of the beam, the area over which the scintillations occur becomes larger, and its boundary is much less definite than in the absence of the foil, showing again that some of the particles have been deflected from their original direction.

The scattering of charged particles such as α-particles can be described qualitatively in terms of the electrostatic forces between the particles and the charges which make up atoms. Since atoms contain both positive and negative charges, an α-particle is subjected to both repulsive and attractive electrostatic forces in passing through matter. The magnitude and direction of these forces depend on how near the particle happens to approach to the centers of the atoms past which or through which it moves. When a particular atomic model is postulated, the extent of the scattering of the α-particles can be calculated quantitatively and compared with experiment. In the case of the Thomson atom, it was shown that the average deflection caused by a single atom should be very small. The mean deflection of a particle in passing through a thin foil of thickness t should be, according to Thomson's theory,[1]

$$\phi_m = \theta(\pi n a^2 t)^{1/2}, \tag{3–1}$$

where θ is the average deflection caused by a single atom, n is the number of atoms per cubic centimeter, a is the radius of the atom in centimeters, and t is the thickness of the foil in centimeters. If the scatterer is a gold foil 4×10^{-5} cm thick, and if a is assumed to be 10^{-8} cm, then ϕ_m turns out to be about 30θ. The scattering of α-particles by a thin foil, according to the Thomson theory, is the result of a relatively large number of small deflections caused by the action of a large number of atoms of the scattering material on a single α-particle. This process is called *compound*, or *multiple*, scattering.

In an experiment, it is convenient to count the number of α-particles scattered through a certain angle. Rutherford showed that the number of α-particles N_ϕ scattered through an angle equal to, or greater than, ϕ should be given by

$$N_\phi = N_0 e^{-(\phi/\phi_m)^2}, \tag{3–2}$$

where N_0 is the number of particles corresponding to $\phi = 0$, and ϕ_m is the average deflection after passing through the foil. It was found experimentally by Geiger[2] that when a gold foil 4×10^{-5} cm thick was used, the most probable angle of deflection of a beam of α-particles was about 1°. If this value is used for ϕ_m in Eq. (3–2), it can be seen that the probability that an α-particle is scattered through a large angle becomes vanishingly small. For example, the number of α-particles which should be scattered through an angle of 10° or more is

$$N_{10°} = N_0 e^{-100} \approx 10^{-43} N_0.$$

Geiger found that the scattering agreed with that predicted by Eq. (3–2) for very small angles, that is, for very small values of ϕ, but the number of particles scattered through large angles was much greater[3] than that predicted by the Thomson theory. In fact, one out of about every 8000 α-particles was scattered through an angle greater than 90°, which means that a significant number of α-particles in a beam incident on a foil had their directions changed to such an extent that they emerged again on the side of incidence. The experimental scattering of α-particles at large angles could not possibly be reconciled with the theoretical predictions based on multiple scattering by a Thomson atom, and it was necessary to look for a better model for the atom.

3–2 Rutherford's theory of the scattering of alpha-particles. Rutherford[4] (1911) proposed a new theory of the scattering of α-particles by matter; this theory was based on a new atomic model and was successful in describing the experimental results. Rutherford suggested that the deflection of an α-particle through a large angle could be caused by a single encounter with an atom rather than by multiple scattering. Photographs of the tracks of α-particles in a cloud chamber showed that α-particles often traveled in a straight line for a considerable distance and then were deflected suddenly through a large angle. Rutherford's suggestion was in agreement with this kind of experimental evidence. For large angle scattering to be possible, it was necessary to suppose that there is an intense electric field near an atom. Rutherford proposed a simple model of the atom which could provide such a field. He assumed that the positive charge of the atom, instead of being distributed uniformly throughout a region of the size of the atom, is concentrated in a minute center or nucleus, and that the negative charge is distributed over a sphere of radius comparable with the atomic radius. On this model, an α-particle can penetrate very close to the nucleus before the repulsive force on it becomes large enough to turn it back, but the repulsive force can then be very large, and can result in a large deflection. At the same time, when the α-particle is near the nucleus, it is relatively far from the negative charges, which are spread over a much larger volume, so that the attractive forces exerted on the α-particle by the electrons can be neglected. For purposes of calculation, Rutherford assumed that the nuclear and α-particle charges act as point charges and that the scattering is caused by the repulsive electrostatic force between the nucleus and the α-particle. If the magnitude of the α-particle charge is $2e$ and that of the nucleus is Ze, where Z is an integer, and if r is the distance between the two charges, the magnitude of this force is

$$F = \frac{2Ze^2}{r^2}. \tag{3–3}$$

In his first calculation, Rutherford treated the case of an atom sufficiently heavy so that the nucleus could be considered to remain at rest during the scattering process. With the above assumptions, the calculation of the orbit of the α-particle is reduced to a familiar problem of classical mechanics, that of the motion of a highly energetic particle under a repulsive inverse square law of force.(5) The orbit is one branch of a hyperbola with the nucleus of the atom as the external focus,* as shown in Fig. 3–1, and the formula for the deflection can be derived from geometrical and physical relationships.

The first step in the derivation of the Rutherford scattering formula is to write down some useful geometrical relationships. In Fig. 3–1, the

* This statement is proved in Appendix IV, where a second derivation of the Rutherford scattering formula is given which contains certain points of special interest.

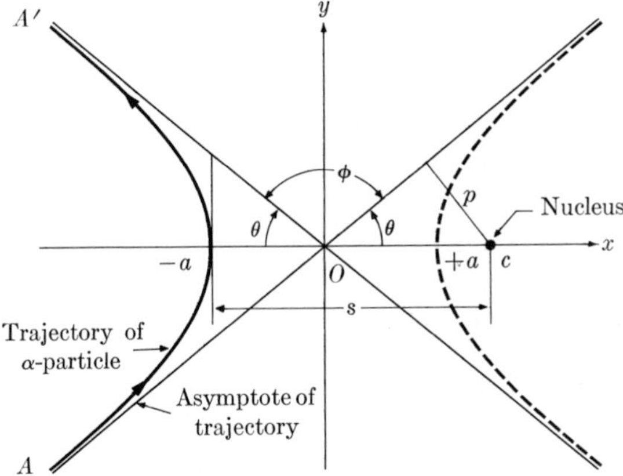

FIG. 3–1. The scattering of an α-particle by the Rutherford nuclear atom.

origin of the coordinate system is taken at the center of the hyperbola, one of whose branches is the path of the α-particle. The initial path of a particle, when the latter is far from the nucleus, is along AO, one of the asymptotes of the hyperbola; the perpendicular distance from the nucleus, situated at the external focus $x = c$, to the line OA is called the *impact parameter* and is denoted by p. The α-particle, on nearing the nucleus, is deflected through the angle ϕ and approaches the asymptote OA'. The vertex of the orbit is at $x = -a$, while that of the second branch of the hyperbola is at $x = a$. The equation of the hyperbola is, from analytic geometry,

$$\frac{x^2}{a^2} - \frac{y^2}{c^2 - a^2} = 1, \tag{3–4}$$

where a is the major semiaxis of the hyperbola. It can also be seen from the figure that $p^2 = c^2 - a^2$, so that p is the minor semiaxis. The eccentricity ϵ is defined as the ratio $\epsilon = c/a$, and the angle between the x-axis and the initial direction of the α-particle is denoted by θ. The angle of deviation ϕ is then equal to $\pi - 2\theta$ radians. Now, let s denote the distance of the nucleus from the vertex of the orbit of the α-particle; the magnitude of s is

$$s = c + a = c\left(1 + \frac{a}{c}\right) = c(1 + \cos\theta).$$

Also, $c = p/\sin\theta$, so that

$$s = \frac{p(1 + \cos\theta)}{\sin\theta} = p\cot\frac{\theta}{2}\cdot \tag{3–5}$$

The next step is to find a relationship between the impact parameter p and the scattering angle ϕ, which can be done by applying the laws of conservation of energy and angular momentum. According to the former, the sum of the kinetic energy and the potential energy is constant. When the α-particle is at a great distance from the nucleus, its potential energy, which is inversely proportional to this distance, is practically zero. If the velocity of the α-particle is V at this large separation, which is assumed to be infinite, its total energy is equal to its initial kinetic energy, $MV^2/2$, where M is the mass. This energy must also be equal to the total energy of the α-particle when it is just at the vertex of the hyperbola. If the velocity at this point is V_0, then

$$\tfrac{1}{2}MV^2 = \tfrac{1}{2}MV_0^2 + \frac{2Ze^2}{s}. \tag{3–6}$$

The second term on the right represents the potential energy of the α-particle, at the vertex of its orbit, in the electric field of the nucleus. If the last equation is divided through by $\frac{1}{2}MV^2$ and if a new quantity b is introduced, $b = 4Ze^2/MV^2$, the result is

$$\frac{V_0^2}{V^2} = 1 - \frac{b}{s}. \tag{3–7a}$$

When the expression for s from Eq. (3–5) is inserted into the last equation,

$$\frac{V_0^2}{V^2} = 1 - \frac{b}{p}\,\frac{\sin\theta}{(1+\cos\theta)}. \tag{3–7b}$$

It follows from the law of conservation of angular momentum that

$$MVp = MV_0s, \tag{3–8}$$

or

$$\frac{V_0}{V} = \frac{p}{s} = \frac{\sin\theta}{1+\cos\theta},$$

and

$$\frac{V_0^2}{V^2} = \frac{\sin^2\theta}{(1+\cos\theta)^2} = \frac{1-\cos\theta}{1+\cos\theta}. \tag{3–9}$$

When this value for $(V_0/V)^2$ is put into Eq. (3–7), it is found that

$$p = \frac{b\tan\theta}{2}, \tag{3–10}$$

and, since $\phi = \pi - 2\theta$,

$$p = \frac{b}{2}\cot\frac{\phi}{2}. \tag{3–11}$$

Equation (3–11) is the desired relation between the impact parameter and the scattering angle.

It is now possible to calculate the fraction of the α-particles scattered through a given angle ϕ. Suppose that the beam of α-particles is incident perpendicularly on a thin foil of material of thickness t, containing n atoms per unit volume. It is assumed that the foil is so thin that the particles pass through without any significant change in velocity and that, with the exception of a few particles which are scattered through a large angle, the beam passes perpendicularly through the foil. Then the chance that a particle passes within a distance p of a nucleus is

$$q = \pi p^2 nt. \tag{3–12}$$

A particle which moves so as to pass within a distance p of the nucleus is scattered through an angle greater than ϕ, where ϕ is given by Eq. (3–11). Hence the fraction of the total number of α-particles deflected through an angle greater than ϕ is obtained by inserting for p, from Eq. (3–11), and

$$q = \tfrac{1}{4}\pi ntb^2 \cot^2 \frac{\phi}{2}\cdot \tag{3–13}$$

Similarly, the probability of deflection through an angle between ϕ and $\phi + d\phi$ is equal to the probability of striking between the radii p and $p + dp$, and is given by

$$dq = 2\pi pnt\ dp.$$

Then,

$$\begin{aligned} dq &= \tfrac{1}{4}\pi ntb^2 \cot \frac{\phi}{2} \operatorname{cosec}^2 \frac{\phi}{2}\, d\phi \\ &= \tfrac{1}{4}\pi ntb^2 \cos \frac{\phi}{2} \operatorname{cosec}^3 \frac{\phi}{2}\, d\phi \\ &= \tfrac{1}{8}\pi ntb^2 \sin\phi \operatorname{cosec}^4 \frac{\phi}{2}\, d\phi. \end{aligned} \tag{3–14}$$

In the experiments made to test the theory, the scattering was determined by counting the number of α-particles incident perpendicularly on a constant area of a zinc sulfide screen placed at a distance R from the foil. The fraction of the scattered α-particles falling on an element of area of the screen at a distance R is given by

$$\frac{dq}{2\pi R^2 \sin\phi\, d\phi} = \frac{ntb^2 \operatorname{cosec}^4(\phi/2)}{16R^2}\cdot \tag{3–15}$$

If now, Q is the total number of α-particles incident on the foil, and if Y is the number of α-particles scattered to unit area of the zinc sulfide screen placed at a distance R from the foil and at an angle ϕ with the original direction of the particles, then

$$Y = \frac{Qntb^2 \operatorname{cosec}^4(\phi/2)}{16R^2}. \tag{3-16}$$

According to Rutherford's theory, the number of α-particles falling on a unit area of the zinc sulfide screen at a distance R from the point of scattering should be proportional to

1. $\operatorname{cosec}^4(\phi/2)$, where ϕ is the scattering angle,
2. t, the thickness of the scattering material,
3. $1/(MV^2)^2$, or to the reciprocal of the square of the initial energy of the α-particle,
4. $(Ze)^2$, the square of the nuclear positive charge.

3–3 The experimental test of the Rutherford scattering theory. Rutherford's nuclear theory of the scattering of α-particles was tested point by point in 1913 by Geiger and Marsden.[(6)] The dependence of the scattering on the four quantities listed at the end of the last section will be considered in order.

1. *The dependence of the scattering on the angle of deflection.* The effect of varying the angle of deflection ϕ was studied in the apparatus shown schematically in Fig. 3–2. In the diagram, R represents a radioactive substance which is the source of the α-particles, F is a very thin foil of scattering material, and S is a zinc sulfide screen rigidly attached to a microscope M. The source and foil were held fixed, while the screen and microscope could be rotated in an airtight joint, varying the angle of deflection. The entire apparatus was enclosed in a metal box which could be evacuated. The number of α-particles reaching unit area of the screen in a chosen time interval was obtained by counting the scintillations. In the experiment, the angle ϕ was varied while all of the other variables in Eq. (3–16) were held constant. The number of scintillations counted, N, is proportional to Y, or to $\operatorname{cosec}^4(\phi/2)$; hence, the ratio $N/\operatorname{cosec}^4(\phi/2)$ should be constant for a given foil under the conditions of the experiment.

The results of two sets of experiments, one with a silver scattering foil, the other with a gold foil, are given in Table 3–1. The first column gives the values of the angle ϕ between the direction of the incident beam of α-particles and the direction in which the scattered particles were counted; the second column gives the corresponding values of $\operatorname{cosec}^4(\phi/2)$. Colums III and V give the observed numbers N of scintillations for silver and gold respectively; columns IV and VI show the value of the ratio $N/\operatorname{cosec}^4(\phi/2)$. The variation in the value of the ratio is very small

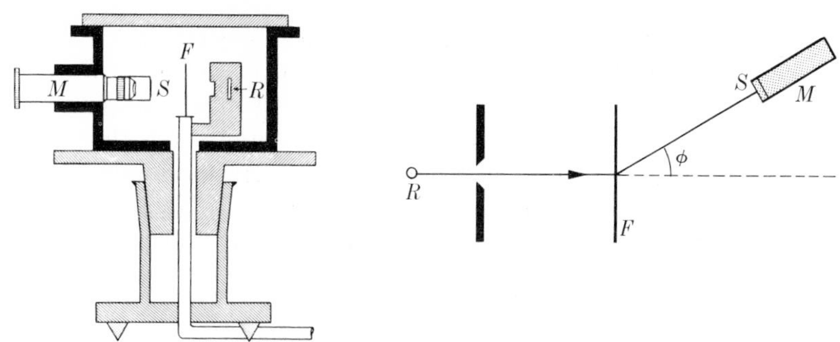

FIG. 3–2. Apparatus of Geiger and Marsden for testing the angular dependence of α-particle scattering.[6]

TABLE 3–1*

THE DEPENDENCE OF THE SCATTERING OF ALPHA-PARTICLES ON THE ANGLE OF DEFLECTION

I	II	III Silver	IV Silver	V Gold	VI Gold
Angle of deflection ϕ	$\text{cosec}^4(\phi/2)$	Number of scintillations N	$\frac{N}{\text{cosec}^4(\phi/2)}$	Number of scintillations N	$\frac{N}{\text{cosec}^4(\phi/2)}$
150°	1.15	22.2	19.3	33.1	28.8
135	1.38	27.4	19.8	43.0	31.2
120	1.79	33.0	18.4	51.9	29.0
105	2.53	47.3	18.7	69.5	27.5
75	7.25	136	18.8	211	29.1
60	16.0	320	20.0	477	29.8
45	46.6	989	21.2	1435	30.8
37.5	93.7	1760	18.8	3300	35.3
30	223	5260	23.6	7800	35.0
22.5	690	20,300	29.4	27,300	39.6
15	3445	105,400	30.6	132,000	38.4
30	223	5.3	0.024	3.1	0.014
22.5	690	16.6	0.024	8.4	0.012
15	3445	93.0	0.027	48.2	0.014
10	17,330	508	0.029	200	0.012
7.5	54,650	1710	0.031	607	0.011
5	276,300	....	.....	3320	0.012

* From Geiger and Marsden.[6]

compared with that of cosec^4 $(\phi/2)$, for angles between $\phi = 15°$ and $\phi = 150°$. For smaller angles, it was found desirable to reduce the number of scintillations counted; the value of the ratio was practically constant between $\phi = 5°$ and $\phi = 30°$. The results for the smaller angles can be compared with those of the larger angles by noting that in the case of the gold foil, the number of scintillations was reduced by about 2500. When the results are fitted to those for the larger angles, it is clear that the value of the ratio changes little over the entire range of values of ϕ, while the value of cosec^4 $(\phi/2)$ varies by a factor of 250,000. The deviations of the ratio from constancy were thought to be within the experimental error and it was concluded that the theory predicts the correct dependence of the scattering on the angle of deviation.

2. *The dependence of the scattering on the thickness of the scattering material.* The dependence of the scattering on the thickness of the scattering material was tested by fixing the angle of deflection and using foils of different thicknesses and also of different materials. The results of several experiments are shown in Fig. 3–3, in which the number of particles per minute scattered through an angle of 25° is plotted as ordinate and the thickness t of the scattering foil is plotted as abscissa. The thickness of the foil is expressed in terms of the equivalent length of path in air, that is, the thickness of air which produces the same loss in energy of the α-particles traversing it as that produced by the material being studied. The equivalent path length in air often serves as a useful standard for comparison

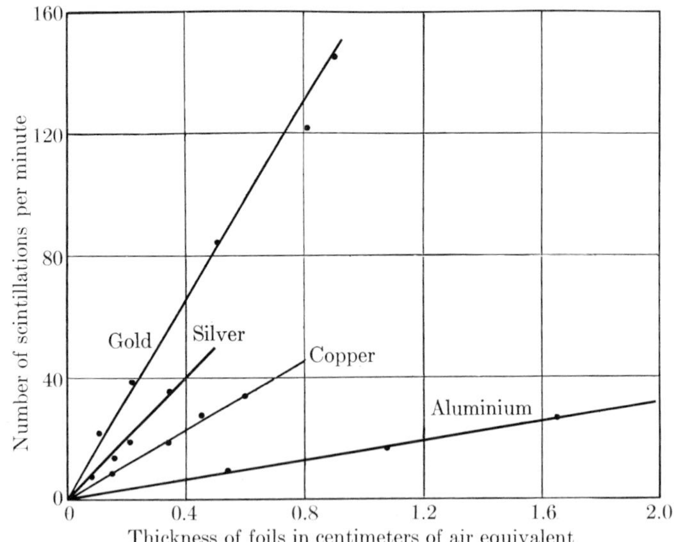

FIG. 3–3. The variation of the scattering of α-particles with the thickness of the scattering material.[(6)]

in work with α-particles. The straight lines show that for any one element, the number of particles scattered through a given angle per unit time is directly proportional to the thickness of the foil, in agreement with the theory.

3. *The dependence of the scattering on the energy or the velocity of the α-particles.* According to the Rutherford scattering formula, the number of α-particles scattered through an angle ϕ should be inversely proportional to the square of the energy of the particles, or to the fourth power of the velocity. In a series of experiments, Geiger and Marsden varied the velocity of the α-particles from a given source by placing screens of mica between the source and the scattering foil; the thicker the screen, the slower were the particles which reached the scattering foil. The velocities of the particles were determined by finding how far they traveled in air. This distance, called the *range* of the α-particles, could be determined in several ways, which will be discussed in Chapter 13. The range R was known to be related to the velocity by the empirical formula,

$$R = aV^3, \tag{3–17}$$

where a is a known constant. When the velocity of the α-particles which passed through a given thickness of mica screen had been obtained in this way, the scattering through a known angle was measured by counting the number N of scintillations. The product NV^4 should be constant when V is varied. The results of a typical experiment are shown in Table 3–2. The fourth column gives the number N of scintillations per minute under fixed conditions, when α-particles of the ranges given in the second column were used. The relative values of $1/V^4$ are given in the

TABLE 3–2*

THE VARIATION OF ALPHA-PARTICLE SCATTERING WITH VELOCITY

Number of sheets of mica	Range of α-particles (cm)	Relative values of $1/V^4$	Number N of scintillations per minute	NV^4
0	5.5	1.0	24.7	25
1	4.76	1.21	29.0	24
2	4.05	1.50	33.4	22
3	3.32	1.91	44	23
4	2.51	2.84	81	28
5	1.84	4.32	101	23
6	1.04	9.22	255	28

* From Geiger and Marsden.(6)

third column. The product NV^4 given in the last column is very nearly constant over the range of values of V studied, showing that the scattering varies inversely as the fourth power of the velocity, as predicted by Rutherford's theory.

4. *The dependence of the scattering on the nuclear charge.* The scattering angle, the thickness of the scatterer, and the velocity of the incident α-particles are quantities which could be measured directly and their effect on the scattering determined. The nuclear charge, unlike the other parameters, could not be measured directly, and a direct comparison between theory and experiment could not be made in this case. It is evident from Eq. (3–16), however, that the value of Z can be found by counting the number of α-particles in the beam incident on the scattering foil and the number in the scattered beam under fixed geometric conditions. Some information about Z could also be obtained from experiments on the scattering by different foil materials. From these two types of experiments it was found that for elements heavier than aluminum, the positive charge Ze on the nucleus was approximately $\frac{1}{2}Ae$, that is, $Z \approx A/2$, where A is the atomic weight and e is the electronic charge. These experiments were not accurate enough to provide a reliable determination of the nuclear charge Z. It was not until 1920 that Chadwick,[(7)] using improved scattering techniques, succeeded in measuring the nuclear charge with good precision. For platinum, silver, and copper foils, he obtained

$$\begin{aligned} \text{copper:}\quad Z &= 29.3 \pm 0.5, \\ \text{silver:}\quad Z &= 46.3 \pm 0.7, \\ \text{platinum:}\quad Z &= 77.4 \pm 1. \end{aligned}$$

These results are not precise enough to determine unique, integral values of Z, but, as will be seen in the next chapter, they agree well with the values 29, 47, and 78 for the three elements, obtained by an entirely independent method. Thus, all four tests of the Rutherford scattering theory were met successfully and constitute the earliest if not the greatest single piece of experimental evidence for the nuclear model of the atom.

3–4 Some characteristics of the atomic nucleus. The remarkably good agreement between the predictions of Rutherford's theory and the experimental results was interpreted as establishing the correctness of the concept of the nuclear atom. Since 1913 the atom has therefore been considered to consist of a minute, positively charged nucleus around which is distributed, in some way, an equal and opposite negative charge in the form of electrons.

So far, the atomic nucleus itself is a vague concept. It has been described as "minute" or "very small" and has been treated mathematically as a point; at the same time, it is supposed to contain practically all of the mass of the atom. It is clear that quantitative information is now needed about the size of the nucleus. The first information of this kind was obtained from the experiments on the scattering of α-particles; it came from the consideration of the distance of closest approach of an α-particle to a nucleus, and of the range of validity of the Coulomb force law.

For any hyperbolic orbit, the distance of closest approach is s, the distance from the vertex of the hyperbola to the nucleus as given by Eq. (3–5). The smallest value that s can have is that for a head-on collision, when the α-particle is deflected through an angle of 180°. In such a collision, the velocity of the α-particle at the turning point is just zero. It follows from Eq. (3–7a), that for V_0 to vanish, s must be equal to b; furthermore, since V_0 can never be negative, this is the smallest value which s may have. Hence, the quantity b, defined by

$$b = \frac{4Ze^2}{MV^2}, \tag{3–18}$$

gives the closest distance which an α-particle of velocity V can approach to a nucleus of charge Z. The magnitude of this distance can be estimated by calculating a typical value of b. Consider the case of a copper nucleus bombarded by α-particles from radon. Copper has an atomic weight of 63.5; if the results of Geiger and Marsden for Z as given by scattering experiments are used, then Z for copper is approximately half of 63.5, or 32. An α-particle has a mass four times that of the hydrogen atom, or $4 \times 1.67 \times 10^{-24}$ gm; the velocity of an α-particle from radon is close to 1.6×10^9 cm/sec. With $e = 4.8 \times 10^{-10}$ esu, the result for b is

$$b = \frac{4(32)(4.8)^2 10^{-20}}{(1.67)4(10^{-24})(1.6)^2(10^{18})} \approx 1.7 \times 10^{-12} \text{ cm.}$$

The above calculation depends on the assumption that the Coulomb force law between the α-particle and the nucleus is still valid at such small distances from the nucleus. The validity of this assumption was borne out by the agreement between the Rutherford scattering theory and the results of the experiments of Geiger and Marsden. By using faster α-particles, Rutherford and others extended the experiments to see how close to the nucleus the $1/r^2$ force law holds. The results showed that for silver the Coulomb law held down to 2×10^{-12} cm, for copper down to 1.2×10^{-12} cm, and for gold down to 3.2×10^{-12} cm. It might be expected that if an α-particle approaches more closely to a nucleus, the inverse square law would eventually break down. If this happens, the forces between the α-particle and the nucleus should begin to change very rapidly with the distance, and the scattering of α-particles should depart widely from the predictions of the theory. If the nucleus is *defined* as the region of deviation from the Coulomb force law, then for the elements mentioned, the radii of the respective nuclei are smaller than the distances listed. Thus, the nuclei of these elements are about 10^{-12} cm in radius, and are indeed very small compared with an atom, with its radius of 10^{-8} cm. It will be seen in later chapters that other methods of estimating values of nuclear radii give results in good agreement with that just obtained.

It is seen from Eq. (3–18) that for α-particles of a given energy the distance of closest approach is proportional to the nuclear charge Z. On the basis of the finding that Z was approximately proportional to the atomic weight, it was expected that α-particles might come closer than 10^{-12} cm to light nuclei, and that it might be possible to find departures from the Coulomb force law. Theoretical and experimental studies showed that such departures from the inverse square law do indeed exist.(8) In the case of aluminum, the inverse square law was found to break down at about 6 to 8×10^{-13} cm, with similar results for other light elements. The deviations from the inverse square law scattering showed that very close to the nucleus, the repulsion was smaller than that calculated from the Coulomb force alone. These results provided the first evidence of the existence of a nonelectrical, specifically nuclear force.

The study of the scattering of α-particles has continued to be an important source of information concerning the atomic nucleus.(9,10,11) By 1935, data had been collected on the elastic scattering of α-particles from most of the light elements through aluminum. In each case departures from Coulomb scattering were observed. With the aid of newer theoretical methods, these data could be used to make quantitative estimates of nuclear radii. It was shown(9) that the data could be interpreted in a consistent way if the radius of the nucleus is assumed to be approximately proportional to the cube root of the atomic weight, that is, if

$$r = r_0 A^{1/3},$$

where A is the atomic weight, and $r_0 = 1.4$ to 1.5×10^{-13} cm. This property will be discussed further in Chapter 13.

References

GENERAL

RUTHERFORD, CHADWICK and ELLIS, *Radiations from Radioactive Substances.* New York: Macmillan, 1930, Chapters 8, 9.

R. D. EVANS, *The Atomic Nucleus.* New York: McGraw-Hill, 1955, Appendix B.

PARTICULAR

1. J. J. THOMSON, "On the Scattering of Rapidly Moving Electrified Particles," *Proc. Cambridge Phil. Soc.*, **15,** 465 (1910).
2. H. GEIGER, "The Scattering of the α-Particles by Matter," *Proc. Roy. Soc.* (*London*), **83A,** 492 (1910).
3. H. GEIGER and E. MARSDEN, "On the Diffuse Reflection of the α-Particles," *Proc. Roy. Soc.* (London), **82A,** 495 (1909).
4. E. RUTHERFORD, "The Scattering of α- and β-Particles by Matter and the Structure of the Atom," *Phil. Mag.*, **21,** 669 (1911).
5. K. R. SYMON, *Mechanics.* Reading, Mass.: Addison-Wesley, 1960, Chapter 3.
6. H. GEIGER and E. MARSDEN, "The Laws of Deflection of α-Particles Through Large Angles," *Phil. Mag.*, **25,** 604 (1913).
7. J. CHADWICK, "Charge on the Atomic Nucleus and the Law of Force; Validity of the Inverse Square Law for the Pt Atom," *Phil Mag.*, **40,** 734 (1920).
8. RUTHERFORD, CHADWICK, and ELLIS, *op. cit.* gen. ref., Chapter 9.
9. E. POLLARD, "Nuclear Potential Barriers: Experiment and Theory," *Phys. Rev.*, **47,** 611 (1935).
10. N. F. MOTT and H. S. W. MASSEY, *The Theory of Atomic Collisions*, 2nd ed. Oxford, England: Clarendon Press, 1950.
11. R. M. EISBERG and C. E. PORTER, "Scattering of Alpha Particles," *Revs. Mod. Phys.*, **33,** 190 (1961).

Problems

1. A beam of α-particles (kinetic energy 5.30 Mev) from polonium, of intensity 10,000 particles/sec, is incident normally on a gold foil of density 19.3 gm/cm^3 and 1×10^{-5} cm thick. An α-particle counter with an aperture 1 cm^2 in area is placed at a distance of 10 cm from the foil in such a way that a line from the center of the counter to the center of the area at which the beam strikes the foil makes an angle of ϕ degrees with the direction of the beam. Calculate the number of counts per hour for $\phi = 5, 10, 15, 30, 45$, and 60°.
2. Repeat the calculation of Problem 1, but with 8.00-Mev α-particles.
3. Repeat the calculation of Problem 1 with a silver foil instead of a gold foil. The density of silver is 10.5 gm/cm^3.
4. In the setup of Problem 1 a copper foil 1×10^{-5} cm thick and of density 8.90 gm/cm^3 is used instead of the gold foil. When $\phi = 10°$, the counting rate is 820 counts/hr. Calculate the atomic number of copper from these data.

5. What is the distance of closest approach of 5.30-Mev α-particles to nuclei of the following elements: gold, silver, copper, lead, and uranium? How close would 7.00-Mev α-particles come to the same nuclei?

6. Show that the fraction of α-particles scattered through an angle between 90° and 180° is given by $\frac{1}{4}\, nt\pi b^2$. What fraction of the α-particles of Problem 1 are scattered through an angle greater than 90°?

7. What is the general expression for the distance of closest approach of a positively charged particle to a nucleus? What is the distance of closest approach of a 2-Mev proton to a gold nucleus? How does this distance compare with those for a deuteron and an α-particle of the same energy?

8. A beam of protons of 5-Mev kinetic energy traverses a gold foil; one particle in 5×10^6 is scattered so as to hit a surface 0.5 cm^2 in area at a distance 10 cm from the foil and in a direction making an angle of 60° with the initial direction of the beam. What is the thickness of the foil?

CHAPTER 4

X-RAYS AND ATOMIC STRUCTURE

The work of Geiger and Marsden on the scattering of α-particles confirmed Rutherford's concept of the nuclear atom, and showed that the number of elementary positive charges on the atomic nucleus is approximately equal to half of the atomic weight. This number must be the same as the number of electrons in the atom, because the atom is electrically neutral. Independent evidence for the number of electrons in the atom was supplied by the pioneer work of Barkla on the absorption and scattering of x-rays by matter. The relationship between the nuclear charge and the position of an element in the periodic table was determined by Moseley's work on the characteristic x-ray spectra of the elements. The study of atomic properties by means of x-rays, with the work of Barkla and Moseley as examples, has been one of the most fruitful sources of information about atoms, and will be discussed in the present chapter. In order to do so, it is necessary to review some of the properties of x-rays in somewhat greater detail than was done in the second chapter.

4–1 Some properties of x-rays. X-rays are produced when swiftly moving electrons strike a solid target. According to classical electrodynamics, a moving charged particle emits electromagnetic radiation when it is accelerated; the sudden stopping of an electron gives rise to a pulse of radiation which takes the form of x-rays. In practice, x-rays are sometimes produced in a low pressure, gas-filled cathode-ray tube in which a metal *anticathode* is situated opposite the cathode. The anticathode serves as a target for the electrons emitted from the cathode and as the source of x-rays. In the more frequently used Coolidge tube (Fig. 4–1), the cathode is a wire which is heated to such a temperature that it emits thermoelectrons. The tube is evacuated until there is no appreciable amount of gas remaining, so that all of the current through the tube is carried by the thermoelectrons. The anticathode is usually a metal of high atomic weight such as tungsten, because the energy carried by the x-rays from heavy metals has been found to be greater than the energy of the x-rays from light metals.

One of the most important properties of x-rays is, of course, their strong penetrating power. The extent to which a beam of x-rays will penetrate into a substance depends on the nature of the x-rays as well as on that of

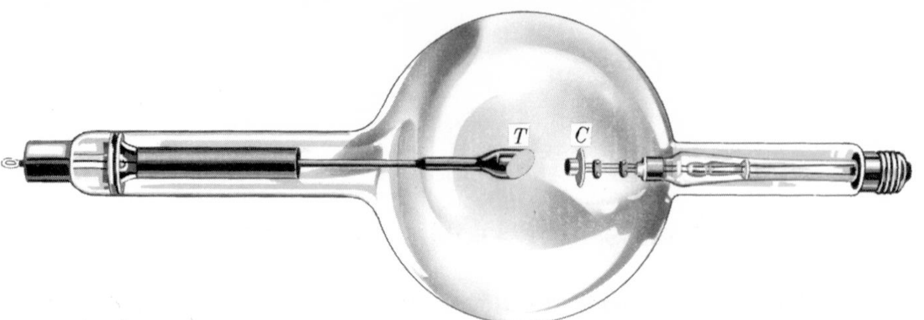

FIG. 4–1. Coolidge x-ray tube.

the substance, and to some extent x-rays can be characterized by their penetrating power in a given material. A rough differentiation may be made into *soft* rays, which have only small penetrating power and are easily absorbed, and *hard* rays, which have relatively high penetrating power. A distinction may also be made, on the basis of their absorption, between *homogeneous* and *heterogeneous* x-rays. The absorption of a *homogeneous* radiation can be described by the formula

$$I = I_0 e^{-\mu d}, \tag{4–1}$$

where I_0 represents the observed intensity of the beam incident normally on the absorbing material, I is the intensity after passing through a thickness d, and μ is a constant, called the *absorption coefficient*, which depends on the absorbing material and on the radiation but is independent of the initial intensity of the x-rays. The absorption equation (4–1) is valid only for homogeneous x-rays, and really defines what is meant by homogeneous x-rays. It does not hold, for example, for the radiation from an ordinary x-ray tube. In that case, no constant value of μ is obtained. With increasing thickness of the absorber the value of μ continually decreases, and finally approaches a limiting value. The variation of μ with thickness of the absorber indicates that the x-rays from the tube are heterogeneous and only become approximately homogeneous when the softer radiations are filtered out of the beam.

The process of absorption of x-rays is complicated and involves several phenomena closely connected with the properties of atoms. When a beam of *primary* x-rays coming from the anticathode of an x-ray tube falls on a plate of some chosen element, part of the radiation goes on through the plate, while the rest is transformed into heat, or into radiation of another sort. The rays going out from the plate consist in part of primary x-rays, the transmitted beam, and in part of rays excited by the primary beam and called collectively the *secondary radiation.* Figure 4–2 is a schematic representation of what happens. The secondary radia-

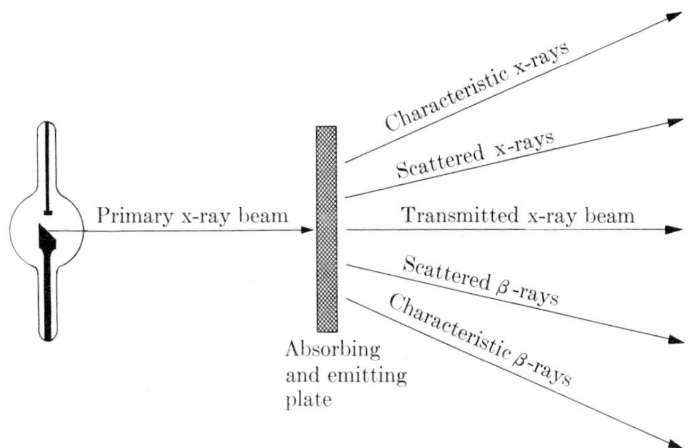

FIG. 4-2. Interaction between x-rays and matter.

tion contains four distinct, important types of radiation: (a) scattered x-rays, (b) characteristic x-rays, (c) scattered β-rays, (d) characteristic β-rays. The β-rays are not of interest for the present and will be discussed in a later chapter. The x-rays of the secondary radiation have turned out to be a fruitful source of information about atoms.

The scattered x-rays have very nearly the same absorption coefficient in a given material as the primary rays; they seem to be primary rays which have merely had their direction changed by the material through which they pass, and their character is independent of the nature of the absorbing and scattering material. Since the primary rays which come from an x-ray tube are heterogeneous, the scattered radiation is also heterogeneous provided the absorbing plate is not too thick.

The characteristic x-rays are so called because they are typical of the material of the plate and, in addition, are homogeneous. They always have the same properties for a given chemical element, independently of the hardness of the primary rays; the necessary condition that the radiation shall appear at all is, however, that the primary radiation shall have a certain minimum hardness. Barkla and his co-workers, during the early years of the 20th century, made detailed measurements of the absorption coefficients of the characteristic x-radiation from various elements. They found that the elements have, in general, two types of characteristic radiation which differ greatly in their absorption coefficients. These are called the K- and L-radiations; the K-radiation is the harder and may have an absorption coefficient several hundred times smaller than that of the L-radiation. The characteristic K- and L-radiations of the different elements can be compared in terms of their absorption coefficients in aluminum. The absorption coefficient for each of the radiations decreases with increasing atomic weight; in other words, the characteristic radiation of the elements increases in hardness with increasing atomic weight.

With the above information about x-rays as background, we can now consider some of the experiments with x-rays which added to the knowledge of atomic structure.

4–2 The scattering of x-rays by atoms and the number of electrons per atom. The scattering of the primary x-rays by a material plate was treated by Thomson on the basis of the classical theory of electrodynamics. According to this theory, radiation incident on a charged particle which can move freely exerts a force on the particle; the latter is accelerated and emits electromagnetic radiation. When a beam of x-rays strikes the atoms in the plate, the latter are accelerated. The electrons, because of their small mass, are accelerated much more strongly than are the atomic nuclei, and the radiation produced by the latter may be neglected. The electrons seem to abstract energy from the primary beam and re-radiate, or scatter it, as a secondary beam. The process takes place in such a way that, for elements of relatively low atomic weight and x-rays of moderate hardness, the scattered and incident radiations differ only in direction. The magnitude of the energy scattered by a free electron was calculated by Thomson,[(1)] and can be written as

$$I = \frac{8\pi}{3}\left(\frac{e^2}{mc^2}\right)^2 I_0. \tag{4–2}$$

Here, I_0 is the intensity of the primary radiation and has the units of energy per square centimeter, but I denotes the entire scattered energy. The quotient I/I_0 must have the dimensions of an area; it is often denoted by ϕ_0, and is called the *classical scattering coefficient for a free electron.* Of the incident radiation that falls on unit area of a surface drawn perpendicular to the beam, a fraction ϕ_0 is scattered, and the electron may be said to scatter as much radiation as falls on an area equal to ϕ_0. For this reason, ϕ_0 is also called the *classical cross section for scattering by a free electron.* If the values $e = 4.80 \times 10^{-10}$ esu, $m = 0.911 \times 10^{-27}$ gm, and $c = 3.0 \times 10^{10}$ cm/sec are inserted, then

$$\phi_0 = \frac{8\pi}{3}\left(\frac{e^2}{mc^2}\right)^2 = 6.65 \times 10^{-25}\ \text{cm}^2. \tag{4–3}$$

If ϕ_0 is regarded as the *effective cross section* of the electron, then the *effective radius* of the electron is given by

$$a = \left(\frac{8}{3}\right)^{1/2} \frac{e^2}{mc^2} = 4.60 \times 10^{-13}\ \text{cm}. \tag{4–4}$$

The *effective radius* of the electron is therefore of the same order of magnitude as the radius of the lighter nuclei. The quantity $e^2/mc^2 = 2.82 \times 10^{-13}$ cm is usually referred to as the *classical radius* of the electron.

When a beam of x-rays passes through a thin sheet of material, some of the energy is removed from the beam, and the intensity of the beam is reduced. If there are n electrons per atom, and N atoms per unit volume, and if it is assumed that all of the electrons scatter independently, then the fractional diminution $-\Delta I/I$ of the incident beam in going a distance Δx is given by

$$-\frac{\Delta I}{I} = nN\phi_0\,\Delta x. \tag{4–5}$$

Since ϕ_0 is independent of x, both sides of this equation can be integrated directly giving, for the intensity I of the beam after traversing a thickness x of the material,

$$I = I_0 e^{-nN\phi_0 x}. \tag{4–6}$$

It follows from the last equation that

$$n = -\frac{1}{N\phi_0 x}\ln\frac{I}{I_0}\cdot \tag{4–7}$$

In Eq. (4–7), N and ϕ_0 are known, and x is the chosen thickness of the scatterer; I_0 and I can be measured experimentally, for example, with ionization chambers. Hence, n, the number of electrons per atom, can be found from measurements of the scattering of x-rays.

Experiments of this type were performed by Barkla and his collaborators;[(2,3)] actually, other processes besides scattering were present in these experiments, but the necessary corrections could be made. Barkla found that within certain limits of hardness of the x-rays, the scattering could be represented by

$$nN\phi_0 = 0.2\rho, \tag{4–8}$$

independently of the hardness and of the scattering material. In Eq. (4–8), ρ is the density in grams per cubic centimeter.

Hence, with the known value of ϕ_0,

$$nN = 3.0 \times 10^{23}\rho.$$

Now, the number of atoms per cubic centimeter of an element is

$$N = \frac{N_0\rho}{A},$$

where N_0 is Avogadro's number $= 6.02 \times 10^{23}$ atoms per gram atomic weight, and A is the gram atomic weight of the element. Then,

$$nN = \frac{nN_0\rho}{A} = 3.0 \times 10^{23}\rho,$$

and

$$\frac{n}{A} = \frac{1}{2}, \tag{4-9}$$

that is, the number of electrons per atom is very close to half the atomic weight.

This result must be considered to be approximate for several reasons. First, it has been assumed that the laws of classical electromagnetics hold, an assumption which is valid only within certain limits of hardness of the x-rays and for scattering materials of low atomic weight. Second, the relationship (4–8) is approximate. For these reasons, the result that the number of electrons is half the atomic weight was a useful qualitative, or at best semiquantitative, result. It was, however, in agreement with the equally semiquantitative results for the charge on the nucleus obtained from α-scattering measurements. The fact that two completely independent methods gave the same result was a strong argument for its validity. The strongest evidence came from Moseley's work on x-ray spectra, but before that work could be done, some additional information was needed on the optical properties of x-rays.

4–3 The diffraction of x-rays and Bragg's law. Early work on x-rays indicated that they are electromagnetic radiations with wavelengths several thousand times smaller than those of visible light. Hence, the methods of ordinary spectroscopy could not be used to measure the wavelength or frequencies of x-rays. In 1912, von Laue and his co-workers discovered that crystals act as gratings for diffracting x-rays. Diffraction occurs because ordinary x-rays have wavelengths between 10^{-8} and 10^{-9} cm, while the average distance between atoms in a solid is between 10^{-7} and 10^{-8} cm. Furthermore, in a crystal there must be some atomic or molecular unit arranged in a regular repeating order which results in the observed crystal symmetry. The properties of x-rays and crystals result in conditions analogous to those which occur when visible light traverses an optical grating—regularly spaced discontinuities separated by distances several times the wavelength of the incident radiation. It is possible, therefore, to use the diffraction of x-rays by a crystal to make quantitative measurements of the wavelengths of x-rays. The way in which this could be done was shown by Bragg.

Following Bragg,[4,5] suppose that a train of monochromatic x-ray waves strikes a crystal which consists of a regular arrangement of atoms

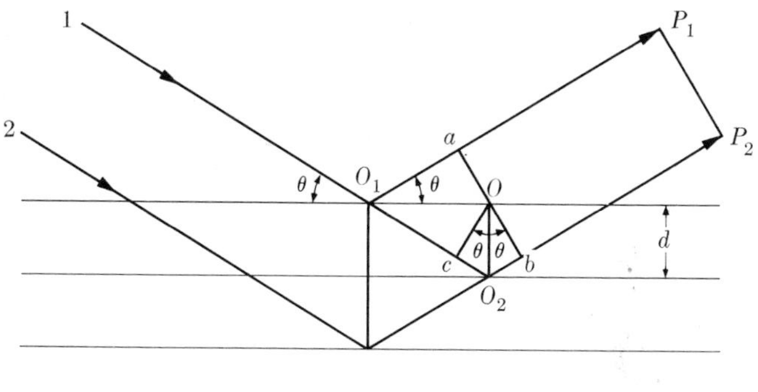

FIG. 4–3. The reflection of x-rays by crystal planes, illustrating the derivation of the Bragg equation.

or molecules. The crystal acts as a series of parallel reflecting planes, as in Fig. 4–3. If the incident or glancing angle θ has the appropriate relationship to the distance d between the reflecting planes and to the wavelength λ of the incident waves, the reflections from the various planes reinforce each other and the resulting reflection is exceptionally strong. The reflections are then said to be in phase. But if the angle θ does not satisfy the condition for the different reflections to be in phase, the latter interfere with one another and the resulting beam is weak. As the angle of incidence is changed, a series of reflections is seen which show alternate maxima and minima of intensity, and the diffraction caused by the reflection of x-rays from crystal planes becomes apparent.

The condition for obtaining reflection maxima for x-rays, the Bragg equation, can be derived by referring to Fig. 4–3. Consider a ray which meets two successive crystal planes at O_1 and O_2 respectively, and let a line drawn from O_2 perpendicular to the planes cut the first plane at O. Draw O_1P_1 and O_2P_2 representing rays reflected from the two planes, and draw aOb perpendicular to O_1P_1 and O_2P_2 to represent a wave front of the reflected beam. There will be reinforcement if the path O_1O_2b taken by waves scattered at O_2 is longer than the path O_1a for waves scattered at O_1 by an integral number of wavelengths. Let Oc be perpendicular to O_1O_2; then $O_1a = O_1c$ and the difference in path is $cO_2b = 2d \sin \theta$. The condition that there be a reinforced reflected beam is then

$$n\lambda = 2d \sin \theta, \tag{4-10}$$

where n is an integer. If the difference in path length is equal to one wavelength, then there will be a reflected beam at the position θ_1 which satisfies the condition

$$\lambda = 2d \sin \theta_1. \tag{4-11}$$

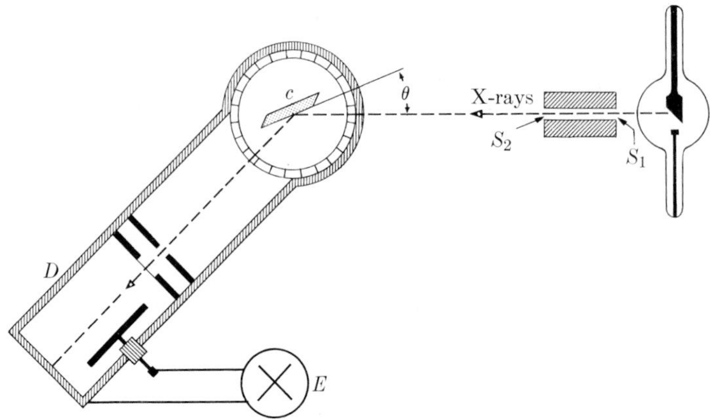

FIG. 4–4. Diagram of a Bragg x-ray spectrometer.

This beam gives the reflection or spectrum of the *first order*. If the difference in path length is 2, 3, ..., i wavelengths, then reflections of the second, third, ..., ith orders will be found at the positions $\theta_1, \theta_2, \ldots, \theta_i$.

These considerations formed the basis for the design, by Bragg,[(5,6)] of an x-ray spectrometer. A schematic diagram of this instrument is shown in Fig. 4–4. The x-rays from a tube are collimated into a narrow beam by the slits S_1 and S_2 cut in lead plates, and then impinge on the crystal c, which acts as a diffraction grating. The angular position of c is read by means of a vernier. After diffraction, the x-rays enter an ionization chamber D, filled with methyl iodide, which absorbs x-rays strongly. The electrometer E records the intensity of ionization in the chamber.

The validity of Bragg's law, Eq. (4–10), was shown by an experiment in which the glancing angle θ was varied; the angle between the ionization chamber and the primary beam was kept equal to 2θ in order to receive the beam reflected from the crystal, and the ionization was measured as a function of θ. The results of such an experiment are shown in Fig. 4–5. It is seen that instead of varying uniformly with the glancing angle, the ionization passes through peaks at certain sharply defined angles; the three peaks A_1, B_1, and C_1 represent x-ray spectrum lines. Secondary peaks, A_2, B_2, and C_2, are seen at angles whose sines are twice those of the angles corresponding to the first order reflections. According to Eq. (4–10), A_1, B_1, and C_1 are lines of the first order spectrum, while A_2, B_2, and C_2 belong to the second order spectrum. Not only are the angles at which the lines of the second order spectrum are seen just what they should be according to Eq. (4–10), but their relative intensities also are in the same ratio as those of the corresponding lines in the first order.

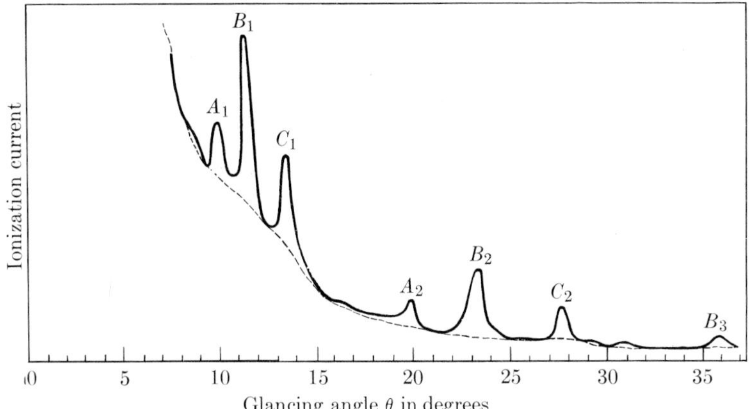

FIG. 4–5. The first x-ray spectrum (Bragg, [(6)]).

The spectral lines are superimposed on a *continuous* spectrum represented by the partially dotted line in the figure.

The x-ray spectra obtained in this way are characteristic of the target from which the x-rays are emitted, as is shown by the fact that if the anticathode in the x-ray tube is changed, an entirely different line spectrum is obtained. On the other hand, if the crystal is changed but the target is kept the same, the same lines appear with about the same relative intensity, but the angles at which they appear are changed, showing that the grating spacing d between the layers of atoms is different for different crystals.

When the grating spacing d is known accurately, the wavelengths of the spectrum lines can be measured with good precision. The grating spacing of the sodium chloride crystal can be calculated, because x-ray studies of this crystal show that the sodium and chlorine atoms occupy alternately the corners of the elementary cube of the lattice. Then, on the average, to every atom there corresponds a volume d^3 with mass ρd^3, where ρ is the density, equal to 2.165 gm/cm^3. Also, one gram-mole of NaCl weighs 58.454 gm and contains N_0 molecules or $2N_0$ atoms, where N_0 is the Avogadro number. The average mass per atom is then $58.454/2N_0$ gm; and

$$\rho d^3 = \frac{58.454}{2N_0}, \qquad \text{or} \qquad d^3 = \frac{58.454}{2(2.165)(6.025)10^{23}},$$

and $d = 2.820 \times 10^{-8}$ cm. The wavelengths of x-ray lines can now be measured accurately with ruled gratings and, with the known wavelengths, the lattice spacing of crystals can be determined experimentally. The best value obtained in this way for NaCl is 2.8197×10^{-8} cm and is in good agreement with the calculated value. Crystals other than NaCl can also be used and, in fact, calcite ($CaCO_3$) is now used as the *standard* crystal; it has a lattice spacing of 3.03567×10^{-8} cm.

4–4 Characteristic x-ray spectra. Moseley's law. The results discussed in the last section showed that the diffraction method could be applied to the analysis of the x-ray spectra of the elements, and the characteristic radiations could be described quantitatively by the wavelengths of the spectral lines. It was soon found that these radiations are more complex than was supposed from Barkla's absorption measurements, but they could still be separated into a number of sharply defined lines.

The first systematic study of the x-ray spectra of the elements was made by Moseley[(7)] in 1913–1914. He used an x-ray spectrometer with a crystal of potassium ferrocyanide, and detected the reflected x-rays with a photographic plate rather than with an ionization chamber. Thirty-eight different elements were used as targets in his x-ray tube. Moseley's photographs showed that the spectral lines emitted by these elements belonged to two different series which were identified with the K- and L-types of characteristic radiation previously observed by Barkla. The K-radiation was found to consist of two lines denoted by K_α and K_β; this

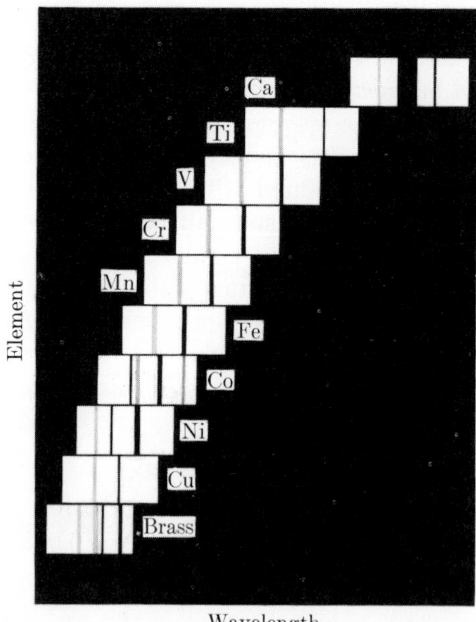

FIG. 4–6. Positions of the K_α and K_β x-ray spectrum lines of a sequence of elements as obtained by Moseley.

result can be seen from Fig. 4–6, which also indicates the relative positions in which these lines appeared in the photographic plates. The lines are those of the first sequence of elements studied by Moseley, the elements from calcium to copper, and it is clear that the wavelengths decrease in a regular way as the atomic weight increases. The gap between the calcium and titanium lines represents the positions of the lines of scandium, which occurs between those two elements in the periodic system and on which no measurements were made.

Moseley's quantitative analysis of some of his experimental data on the K_α-lines is shown in Table 4–1. The wavelengths of the K_α-lines of a sequence of elements are listed in the second column of Table 4–1. The third column of the table gives the values of a quantity Q_K defined by the relationship

$$Q_K = (\nu/\tfrac{3}{4}\nu_0)^{1/2}, \tag{4–12}$$

where ν is the frequency of the spectrum line, equal to c/λ, and c is the

TABLE 4–1

MOSELEY'S MEASUREMENTS OF WAVELENGTHS OF THE K_α LINES

Element	Wavelength, $\times 10^8$ cm	$Q_K = (\nu/\frac{3}{4}\nu_0)^{1/2}$	Atomic weight	Atomic number, Z
Aluminum	8.364	12.05	26.97	13
Silicon	7.142	13.04	28.06	14
Chlorine	4.750	16.00	35.457	17
Potassium	3.759	17.98	39.096	19
Calcium	3.368	19.00	40.08	20
Titanium	2.758	20.99	47.90	22
Vanadium	2.519	21.96	50.95	23
Chromium	2.301	22.98	52.01	24
Manganese	2.111	23.99	54.93	25
Iron	1.946	24.99	55.85	26
Cobalt	1.798	26.00	58.94	27
Nickel	1.662	27.04	58.69	28
Copper	1.549	28.01	63.54	29
Zinc	1.445	29.01	65.38	30
Yttrium	0.838	38.1	88.92	39
Zirconium	0.794	39.1	91.22	40
Niobium	0.750	40.2	92.91	41
Molybdenum	0.721	41.2	95.95	42
Ruthenium	0.638	43.6	101.7	44
Palladium	0.584	45.6	106.7	46
Silver	0.560	46.6	107.88	47

velocity of light; ν_0 is a constant, which represents a certain fundamental frequency, important in the study of line spectra. If the frequency is replaced by the wave number $\bar{\nu}$ (the reciprocal of the wavelength, $\bar{\nu} = 1/\lambda$), ν_0 is replaced by another constant R, called the *Rydberg constant*, for which Moseley used the value 109,720 cm^{-1}. The quantity Q_K then becomes

$$Q_K = (\bar{\nu}/\tfrac{3}{4}R)^{1/2}. \tag{4–13}$$

The Rydberg constant will be discussed at greater length in Chapter 7 and the basis for Moseley's choice of the constant R will be shown. For the present purposes it is enough to note that Q_K is proportional to the square root of the frequency of the K_α line.

Moseley saw that Q_K increases by a constant amount in passing from one element to the next when the elements are in their order in the periodic system. Except in the case of nickel and cobalt, this order is the same as the order of the atomic weights of the elements listed in Table 4–1. Although, with this exception, the atomic weights increase, they do so in a much less regular way than Q_K. Moseley concluded that there is in the atom a fundamental quantity which increases by regular steps from one element to the next. He considered that this quantity could only be the positive charge on the nucleus. It will be remembered that the conclusion had been drawn from α-particle and x-ray scattering that the number of unit charges on the nucleus of an atom is approximately half the atomic weight. Moseley then noted that atomic weights increase by approximately two units, on the average, in going from element to element and this suggested to him that the number of charges increases from atom to atom by a single electronic unit. Moseley concluded that the experiments led to the view that the number of unit charges on the nucleus is the same as the number of the place occupied by the element in the periodic system, and that both of these numbers could be represented by a quantity which he called the *atomic number*.

Moseley extended his experiments as far as gold. In order to do so, he worked with the lines of the L series because the wavelengths of the K series of the heavier elements became too small for precise analysis. His results for the L_α-lines of a series of elements are shown in Table 4–2. In this case, Moseley used the quantity Q_L, defined by

$$Q_L = (\bar{\nu}/\tfrac{5}{36}R)^{1/2}, \tag{4–14}$$

instead of Q_K, for reasons which will be explained in Chapter 7. The same conclusions can be drawn from the variation of Q_L as were drawn in the case of Q_K, so that experimental results were available for 38 elements between aluminum and gold. On the basis of these results, Moseley assigned a value of the *atomic number* to each element from aluminum to

TABLE 4-2

MOSELEY'S MEASUREMENTS OF WAVELENGTHS OF THE L_α LINES

Element	Wavelength, $\times 10^8$ cm	$Q_L = (\nu/\frac{5}{36}\nu_0)^{1/2}$	Atomic weight	Atomic number, Z
Zirconium	6.091	32.8	91.22	40
Niobium	5.749	33.8	92.91	41
Molybdenum	5.423	34.8	95.95	42
Ruthenium	4.861	36.7	101.7	44
Rhodium	4.622	37.7	102.91	45
Palladium	4.385	38.7	106.7	46
Silver	4.170	39.6	107.88	47
Tin	3.619	42.6	118.70	50
Antimony	3.458	43.6	121.76	51
Lanthanum	2.676	49.5	138.92	57
Cerium	2.567	50.6	140.13	58
Praeseodymium	2.471	51.5	140.92	59
Neodymium	2.382	52.5	144.27	60
Samarium	2.208	54.5	150.43	62
Europium	2.130	55.5	152.0	63
Gadolinium	2.057	56.5	156.9	64
Holmium	1.914	58.6	164.94	66
Erbium	1.790	60.6	167.2	68
Tantalum	1.525	65.6	180.88	73
Tungsten	1.486	66.5	183.92	74
Osmium	1.397	68.5	190.2	76
Iridium	1.354	69.6	193.1	77
Platinum	1.316	70.6	195.23	78
Gold	1.287	71.4	197.2	79

gold. Aluminum, the first element in the periodic system studied by Moseley, was assigned the atomic number 13, because 12 elements were known to precede it in the system.

The order of the atomic numbers turned out to be the same as that of the atomic weights except where the latter were known to disagree with the order of the chemical properties of the elements. For example, cobalt has a greater atomic weight than nickel, but its chemical properties are such that it should precede nickel in the periodic system. The relative wavelengths and values of Q_K (Table 4-1) show that cobalt should indeed precede nickel in the periodic system in spite of the discrepancy in atomic weights. When elements were skipped in Moseley's work, corresponding gaps in the values of Q_K or Q_L were found. Known elements were found to correspond with all the numbers between 13 and 79 except for four numbers, 43, 61, 72 and 75, which corresponded to still undiscovered elements. These missing elements have since been discovered and identified by their x-ray spectra. In addition, the atomic numbers of copper, silver, and platinum turned out to be 29, 47, and 78, respectively. As mentioned in Section 3–3, Chadwick, a few years later, obtained directly from α-scattering measurements the values 29.3 ± 0.5, 46.3 ± 0.7, and 77.4 ± 1, respectively.

It can be seen from Tables 4–1 and 4–2 that

$$Q_K = (\nu/\tfrac{3}{4}\nu_0)^{1/2} = Z - 1, \tag{4–15a}$$

$$Q_L = (\nu/\tfrac{5}{36}\nu_0)^{1/2} = Z - 7.4. \tag{4–15b}$$

Thus, if the square root of the frequency, or the wave number, for either the K_α or L_α line is plotted against the atomic number, a straight line should result. The quantity Q_K, which is proportional to the square root of the frequency of the K_α line, is plotted against the atomic number in Fig. 4–7. The points, represented by the dots in the figure, do indeed fall on a straight line. When Q_K is plotted against the atomic weight, the points (represented by crosses) show much greater scatter. Similar results are obtained for the L_α lines in Fig. 4–8. The figures show in a striking way that the

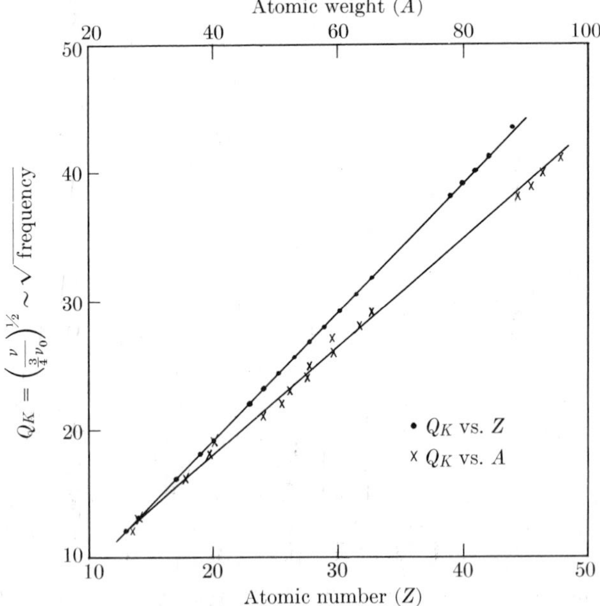

FIG. 4–7. Moseley's relationship between the atomic number and the square root of the frequency of the K_α line.

correlation between the frequency of the characteristic x-ray lines and the atomic number is much more satisfactory than that between the frequency and the atomic weight.

The frequencies may also be expressed by the empirical formulas

$$K_\alpha: \quad \nu = \left(\frac{1}{1^2} - \frac{1}{2^2}\right)\nu_0(Z - 1)^2, \tag{4–16a}$$

$$L_\alpha: \quad \nu = \left(\frac{1}{2^2} - \frac{1}{3^2}\right)\nu_0(Z - 7.4)^2; \tag{4–16b}$$

in terms of the wave numbers,

$$K_\alpha: \quad \bar{\nu} = \left(\frac{1}{1^2} - \frac{1}{2^2}\right) R(Z - 1)^2, \tag{4–17a}$$

$$L_\alpha: \quad \bar{\nu} = \left(\frac{1}{2^2} - \frac{1}{3^2}\right) R(Z - 7.4)^2. \tag{4–17b}$$

These relationships will be useful later for comparison with theory.

Moseley's work has been extended to other elements, and the methods of x-ray spectroscopy have been improved. Additional types of char-

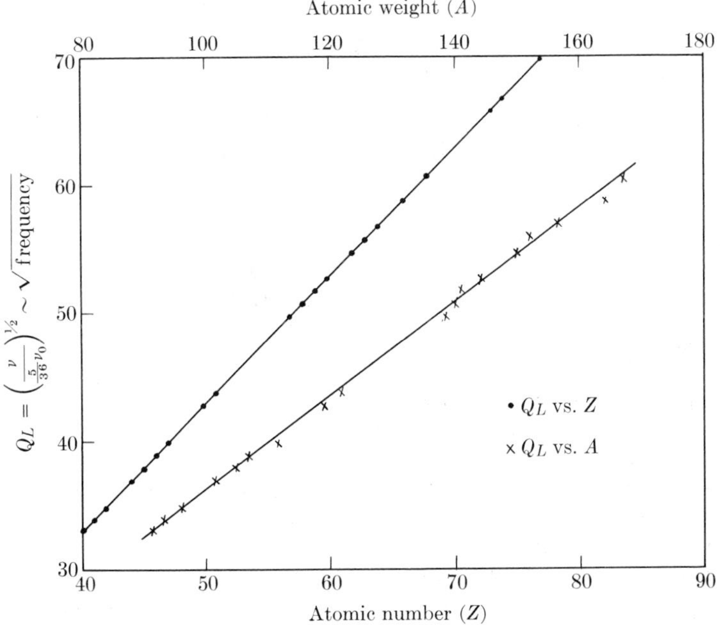

FIG. 4–8. Moseley's relationship between the atomic number and the square root of the frequency of the L_α line.

acteristic radiation have been found, such as the M and N series of lines, which are even softer than the L-radiations. The conclusions reached by Moseley have been confirmed, and atomic numbers have been assigned to all of the chemical elements. The atomic number of an element is the same as the number of unit charges on the nucleus, and is equal to the number of electrons in the atom; it is also the same as the ordinal number of the element in the periodic system. The identity of the latter number with the nuclear charge number and the number of electrons had been suggested by van den Broek[(8)] but could not be considered to be proven until the work of Moseley. The fact that the position of an element in the periodic system is more directly related to the atomic number than to the atomic weight shows that the chemical properties are directly related to the nuclear charge, and are not determined by the magnitude of the atomic weight, as was thought by Mendeléeff.

References

GENERAL

A. H. Compton and S. K. Allison, *X-Rays in Theory and Experiment,* 2nd ed. New York: Van Nostrand, 1935.

M. Siegbahn, *The Spectroscopy of X-Rays.* Oxford: Oxford University Press, 1925.

M. Siegbahn, *Spektroscopie der Rontgenstrahlen,* 2nd ed. Berlin: Julius Springer, 1931.

G. L. Clark, *Applied X-Rays.* 3rd ed. New York: McGraw-Hill, 1942.

F. K. Richtmyer, E. H. Kennard, and T. Lauritsen, *Introduction to Modern Physics,* 5th ed. New York: McGraw-Hill, 1955, Chapter 8.

PARTICULAR

1. A. H. Compton and S. K. Allison, *op. cit.* gen ref., Chapter III.
2. C. G. Barkla and C. A. Sadler, "The Absorption of Roentgen Rays," *Phil. Mag.* **17,** 739 (1909).
3. C. G. Barkla, "Note on the Energy of Scattered X-Radiation," *Phil. Mag.,* **21**, 648 (1911).
4. W. L. Bragg, "The Diffraction of Short Electromagnetic Waves by a Crystal," *Proc. Cambridge Phil. Soc.,* **17**, 43 (1912).
5. W. L. Bragg, *The Crystalline State: Vol. I, General Survey.* London: G. Bell, 1949, Chapter 3.
6. W. H. Bragg and W. L. Bragg, "The Reflection of X-Rays by Crystals," *Proc. Roy. Soc.* (London), **A88,** 428 (1913); **A89,** 246 (1914).
7. H. G. J. Moseley, "The High Frequency Spectra of the Elements," *Phil. Mag.,* **26,** 1024 (1913); **27,** 703 (1914).
8. A. van den Broek, "Die Radioelemente, das periodische System und die Konstitution der Atome," *Physik. Z.,* **14,** 32 (1913).

Problems

1. The intensity of a beam of x-rays is reduced to 90% of its initial value in passing through a slab of beryllium 0.30 cm thick and of density 1.84 gm/cm^3. How many electrons are there per atom of beryllium (a) as calculated directly from the attenuation, (b) as calculated from Barkla's relationship, Eq. (4–8)? (c) What is the value of the absorption coefficient?

2. A beam of x-rays is incident on a sodium chloride crystal (lattice spacing $= 2.820 \times 10^{-8}$ cm; 1×10^{-8} cm $=$ 1 angstrom). The first order Bragg reflection is observed at a *grazing* angle of 8° 35′. What is the wavelength of the x-ray? At what angles would the second and third order Bragg reflections occur?

3. The grating spacings of calcite, quartz, and mica are 3.036 A, 4.255 A, and 9.963 A, respectively. At what grazing angle would the first reflection of a beam of 1-A x-rays be observed in rock salt, calcite, quartz, and mica, respectively?

4. In his experiments, Moseley used a crystal of potassium ferrocyanide having a grating spacing of 8.408 A. Find the positions of the first reflections of the K_α x-rays of chromium, manganese, iron, cobalt, nickel, copper, and zinc, respectively.

5, Calculate the frequencies and the wave numbers of the K_α lines listed in Table 4–1.

6. Technecium is one of the new elements discovered among the products of the fission of uranium. The wavelength of the K_α x-ray line from this element has been determined by measuring the glancing angle as 7.0° for first-order Bragg reflection from a cleavage face of NaCl. What is the wave length of the x-ray? What is the atomic number of technicium?

7. The K_α line for copper has a wavelength of 1.5412 A as measured with a ruled grating. The first-order Bragg reflection from a cleavage face of NaCl is found in an experiment at an angle of 15° 53′. What is the lattice spacing of NaCl? Find Avogadro's number, given that the density of NaCl is 2.164 gm/cm^3 and the molecular weight is 58.454.

8. The wavelength of the L_α x-ray lines of Ag and Pt are 4.1538 A and 1.3216 A, respectively. An unknown substance emits L_α x-rays with a wavelength of 0.966 A. With these data, and with the known atomic numbers 47 and 78 for Ag and Pt, respectively, determine the atomic number of the unknown substance.

CHAPTER 5

THE QUANTUM THEORY OF RADIATION

5–1 The failure of classical physics to describe atomic phenomena. In the first four chapters, the fundamental experiments bearing on the problem of the nature of atoms were discussed. These experiments were interpreted in terms of certain concepts which form the basis of present ideas of atomic structure. In 1913, an atom was supposed to consist of a positively charged nucleus about 10^{-12} cm in radius, surrounded by electrons distributed over a volume about 10^{-8} cm in radius. The charge on the nucleus was known to be Ze, where Z is a positive integer and e is the magnitude of the electronic charge; the number of electrons was known to be equal to Z because the atom is electrically neutral under normal conditions. Moseley's work showed that the number Z is also equal to the atomic number, which represents the position of the element in the periodic system. In this system, the elements are ordered according to their properties, so that the latter are directly related to the nuclear charge and the number of electrons in the atom. To describe in detail how these properties depend on the nuclear charge, a more detailed model of the atom was needed, and it would be expected that such a model would include a description of the way in which electrons are arranged in the atom.

Attempts were made to construct a theoretical atomic model on the basis of classical physics, but they all failed, and it was realized eventually that classical physics (Newtonian mechanics, Maxwellian electromagnetics, and thermodynamics) could not explain or describe atomic phenomena. This failure may be illustrated by means of two elementary examples. The first example is that of an atom consisting of stationary positive and negative charges. Consider an atom consisting of a nucleus with two positive elementary charges and with two electrons somewhere outside the nucleus. If the electrons are each at a distance a from the nucleus, and $2a$ from each other, the repulsive electrostatic force between the electrons is $e^2/4a^2$; each electron is attracted to the nucleus by a force equal to $2e^2/a^2$, eight times as great as the repulsive force. Hence, the electrons would fall into the nucleus and there would be no mechanical stability. This example is a special case of a theorem which states that an electric charge cannot be in equilibrium, at rest, under the action of electric forces alone; and the example can be extended to any atom.

Suppose next that the electrons revolve in some way about the nucleus, and consider for simplicity the case of a hydrogen atom which has one electron and a nucleus with one positive charge. Assume that the electron revolves about the nucleus in a circular orbit of radius a, and that the velocity of the electron is such that the attractive force between the electron and the nucleus provides just the centripetal force required. The system, nucleus and electron, should then be mechanically stable. The electron, however, is subject to a constant acceleration toward the nucleus, and, according to electromagnetic theory, the electron should radiate energy. The energy of the system should then decrease; the electron should gradually spiral in toward the nucleus, emitting radiation of constantly increasing frequency, and should eventually fall into the nucleus. These predictions of classical physics are in strong disagreement with the experimental facts. It had long been known that when the atoms of an element are excited, in the case of hydrogen, for example, by passing an electric discharge through the gas, and if the resulting light (radiation) emitted by the element is dispersed by a prism, a line spectrum characteristic of the element is observed. Hydrogen always gives a set of lines with the same wavelengths, helium gives another set, sodium still another, and so on, in contrast to the prediction of the emission of radiation of increasing frequency. Actually, there is no atomic catastrophe, and atoms seem to last for a long time. Theory has again failed to account for the stability of atoms. Moreover, in spite of many attempts, classical physics could not account for the characteristic optical spectra of the elements, nor could it account for x-ray spectra.

After many attempts like those cited, it became apparent that the description of atomic phenomena required a new kind of physical theory, one which could describe atomic phenomena in terms of new, nonclassical concepts. Bohr's quantum theory of atomic structure was the first successful attempt to fill the need, and although it eventually was found to be unsatisfactory, it prepared the way for the modern, more successful theory of quantum mechanics. The Bohr theory was based on the quantum theory of heat radiation introduced by Planck in 1901 and applied to light by Einstein in 1905. Planck's quantum theory lies at the foundation of present ideas about atoms and nuclei, and it would be futile to try to continue into atomic and nuclear physics without first having some grasp of the basic ideas of this theory. The quantum theory actually arose from the failure of classical physics to explain some of the experimental facts of thermal radiation. In particular, classical physics could not explain the dependence of the intensity of the radiant energy emitted by a blackbody on the wavelength of the radiation. To explain this phenomenon, it was necessary to develop a theory for the emission of radiation which was based on a concept entirely opposed to the ideas of classical physics. The new concept, that of *quanta* or discrete corpuscles of energy, must be applied to all problems involving the emission and absorption of electromagnetic radiation; but these are the basic problems of atomic and nuclear physics, and this is the reason for the importance of quantum theory in these fields.

The concept of quanta of energy is so different from classical physical ideas that only a careful study of the problem of thermal radiation can show the need for a quantum theory, and make clear what is meant by quanta of energy. The problem of thermal radiation will therefore be considered in some detail.

5–2 The emission and absorption of thermal radiation. It is well known that a hot body emits radiation in the form of heat. Thermal radiation consists of electromagnetic waves and differs from visible light and x-rays in having longer wavelengths. Thus, radiations with wavelengths between 7000×10^{-8} cm and 4000×10^{-8} cm are usually considered to be light, because the human eye can see these wavelengths. The length 10^{-8} cm is called the angstrom unit, and is abbreviated as A. Waves longer than 7000 A and shorter than 0.01 cm are the *infrared* or heat waves; wavelengths between 4000 A and about 50 A constitute the ultraviolet radiation, and those shorter than 50 A but longer than about 0.01 A are usually classified as x-rays. Gamma-rays have still shorter wavelengths.

At any temperature, the emitted heat energy is distributed over a continuous spectrum of wavelengths, and this spectral distribution changes with temperature. At low temperatures, the rate of radiation is small and the energy is chiefly of relatively long wavelength (infrared radiation). At temperatures between 500 and 550°C, bodies begin to radiate visible light, which means that the distribution of energy among the different wavelengths has shifted so that a large enough portion of the radiant energy has wavelengths within the visible spectrum. As the temperature rises, the fraction of visible radiation increases until at 3000°C, approximately the temperature of an incandescent lamp filament, the radiation contains enough of the shorter wavelengths so that the body appears "white hot."

The consideration of measurements by Tyndall on the radiation from hot platinum wires led Stefan (1879) to suggest an empirical rule that can be written in the form

$$W = e\sigma T^4, \tag{5–1}$$

where W is the rate of emission of radiant energy per unit area and is expressed in ergs per square centimeter per second, and T is the absolute temperature in °K; W is called the *total emissive power*, or *total emittance*. The quantity e is called the *emissivity* of the surface and has a value between zero and unity, depending on the nature of the surface; σ is a constant called the *Stefan-Boltzmann constant.* In 1884, Boltzmann derived Eq. (5–1) from thermodynamics, and the equation is therefore known as the Stefan-Boltzmann law. If a body of emissivity e, at the temperature T_1, is surrounded by walls at the temperature T_2, smaller than T_1, the net rate of loss of energy by the body is given by

$$W_{\text{net}} = e\sigma(T_1^4 - T_2^4). \tag{5–2}$$

Equation (5–2) is familiar because of its application to practical problems of heat transfer. In the late 19th and early 20th centuries, physicists were more concerned with the way in which the emissive power of a body varies with temperature *and wavelength.* Their interest was concentrated on "explaining" or "understanding" the variation in terms of the fundamental concepts and theories of physics. By "explaining" or "understanding" is meant fitting the experimental information into the scheme of existing physical theories or, if this cannot be done, modifying the theories so that they include the experimental data. Although this interest may not appear to be a matter of great practical importance, the difficulties to which it led gave birth to the greatest revolution in physical thought during the 20th century.

Physicists chose as the subject of their studies the thermal radiation from a *blackbody.* To understand what is meant by the term *blackbody* we first define the *absorptivity* of a body as the fraction of the radiant energy, incident on the surface of the body, which is absorbed. A blackbody is, by definition, one with absorptivity equal to unity, i.e., a body which absorbs all of the radiant energy falling upon it. A very simple relation has been shown to exist between the absorptivity of a body and its total emissive power. This relation, called Kirchhoff's law, states that the ratio of the emissive power to the absorptivity is the same for all bodies at the same temperature, and is equal to the emissive power of a blackbody at this temperature; this law holds for each wavelength, and has been confirmed by experiment. It follows from Kirchhoff's law that no body can emit radiant energy at a greater rate than a blackbody, for the maximum value of the absorptivity, namely that for a blackbody, is unity, and any smaller value of the absorptivity necessarily implies a smaller value of the emissive power. Hence, the blackbody, which is the most efficient absorber of radiant energy, is also the most efficient emitter. The emissivity of a blackbody, the quantity e in Eq. (5–1), is equal to unity, and the total emissive power of a blackbody depends only on the temperature and not on the nature of the body. The spectral distribution of the energy radiated by a blackbody has, therefore, a special interest.

A blackbody does not really exist in nature, but some substances such as lampblack, flat black lacquer, rough steel plate, or asbestos board reflect only a few percent of the incident radiation and approximate a blackbody. It has also been shown that the radiation coming out of the small opening of an almost completely closed, uniformly heated, hollow enclosure, or cavity, is a good substitute for the radiation of a blackbody. As a source of blackbody radiation, an electric furnace may be used, consisting of a long tube, preferably with blackened walls, heated by an electric current flowing in a wire wound around the tube. The temperature of the central part of the interior is measured with a thermometer of some kind. A small hole is made through the wall and the radiation coming through the hole is observed. The spectral distribution of the radiation can be analyzed by means of an optical spectrometer with the photographic plate replaced by some energy-measuring device. The radiations from the blackbody are refracted by a salt prism which may be rock salt, quartz, or fluorite. Salt is used instead of glass because glass absorbs too much infrared radiation. The energy-measuring device may be a linear-type thermopile connected to a galvanometer with high voltage sensitivity. The radiation falling on one junction of the thermopile causes the temperature at that junction to rise above that of the second junction, and the resulting electric current causes a deflection of the galvanometer needle. The deflection caused by the absorption of radiation in a given range of wavelength depends on the emissive power of the radiator in that range. Hence, the emissive power can be measured as a function of wavelength. Another useful energy-measuring device is the line bolometer, which consists of a blackened, very thin metal strip with electrical connections. This strip is the receiver of the radiations, and is connected as one arm of a balanced Wheatstone bridge. Radiation falling on the strip raises its temperature, causing a change in electrical resistance, as measured by a sensitive bridge galvanometer. The bolometer can detect very small changes in temperature and, therefore, very small differences in radiant energy, so that it is a sensitive and useful instrument for measuring the spectral distribution of thermal radiation.

When the emissive power of a heated body is measured as a function of wavelength, at a fixed temperature, a curve like that in Fig. 5–1 is obtained. The abscissa is the wavelength in microns (1 micron $= 10^{-4}$ cm). The ordinates are relative values of the *monochromatic emissive power*, or *spectral emittance*. This quantity, denoted by W_λ, is the emissive power per unit range of wavelength, or the radiant energy emitted per unit area per unit time in the range of wavelengths between λ and $\lambda + d\lambda$. The monochromatic emissive power is related to the total emissive power by the equation

$$W = \int_0^\infty W_\lambda \, d\lambda. \tag{5–3}$$

It is seen from the figure that for very short and also for very long wave-

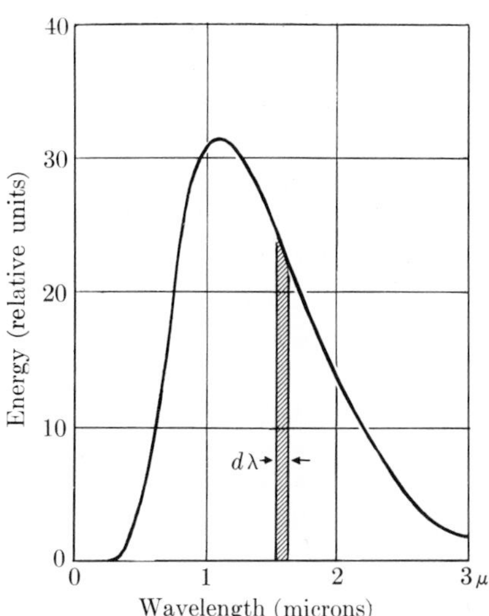

FIG. 5-1. The distribution of energy in the spectrum of a tungsten lamp. (By permission from F. K. Richtmyer and E. H. Kennard, *Introduction to Modern Physics*, 4th ed. New York: McGraw-Hill, 1947.)

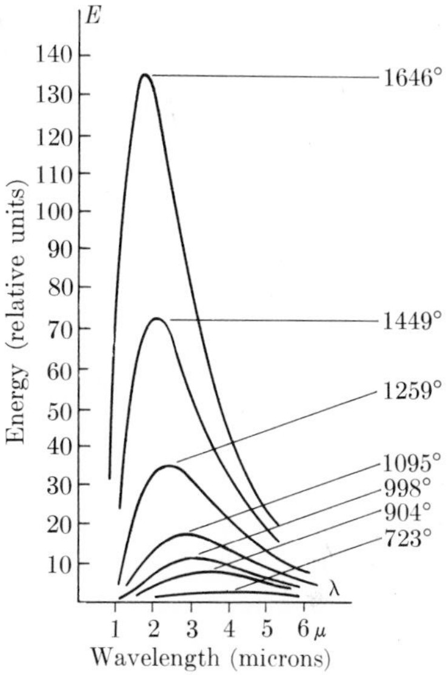

FIG. 5-2. The distribution of energy in the spectrum of a blackbody at different temperatures.

lengths the monochromatic emissive power is very small. In the intermediate wavelength range, it has a maximum value at a definite wavelength λ_m. Measurements made at different temperatures by Lummer and Pringsheim (1899) yielded a set of curves like that of Fig. 5-2. The curves are similar in shape, but as the temperature increases, the height of the maximum increases and the position of the maximum is shifted in the direction of smaller wavelengths or greater frequencies. The total radiant energy for a given temperature is represented by the area between the curve and the horizontal axis. This area increases as the fourth power of the absolute temperature, according to the Stefan-Boltzmann law. The experimental results contained in the curves of Fig. 5-2 are typical of the phenomena which must be explained by theory.

5-3 The classical theory of thermal radiation. Wien (1893) showed that it was possible to predict some features of the spectral distribution of thermal radiation from the laws of classical physics. His results will be discussed in some detail in order to understand the difficulties which arose and which could only be resolved by the quantum theory.

In the theoretical treatment of the problem of radiation it is convenient to use a quantity u_λ, called the monochromatic energy density, rather than the monochromatic emissive power W_λ. The energy density represents, not the energy falling on one square centimeter per second, but the energy contained in a cubic centimeter of volume. The energy density and the emissive power are related geometrically in a simple way. If radiation is propagated in a definite direction, the amount of it which will strike unit cross section per second is the amount contained in a prism whose base is 1 cm^2 in area and whose slant height along the direction of propagation is the velocity of the radiation c. This amount is the volume of the prism ($c \cos \theta$, if θ is the angle between the direction of propagation and the normal to the surface) multiplied by the energy of the radiation per unit volume. Hence, if the energy density is known, the amount of light falling on 1 cm^2/sec can be found by integrating over all directions of propagation. The amount of radiation of a given wavelength falling on 1 cm^2 of a body per second at a given temperature is equal to the amount which 1 cm^2 of surface of the body would radiate in one second at the same temperature (Kirchhoff's law). Since this amount is just the monochromatic emissive power, the latter is proportional to the energy density; the proportionality factor contains only geometrical factors and the velocity of light, neither of which depends on wavelength or temperature. We shall not need this relationship and shall therefore not show it explicitly; it will suffice to remember that the monochromatic energy density and the monochromatic emissive power can be used interchangeably so long as absolute values are not required. The energy density can also be written as u_ν rather than u_λ to show that it can be expressed in terms of the frequency rather than of the wavelength. By definition, $u_\nu d\nu$ is the radiant energy per unit volume in the frequency range from ν to $\nu + d\nu$; $u_\lambda d\lambda$ is the radiant energy per unit volume in the range of wavelength from λ to $\lambda + d\lambda$. Frequency and wavelength are related by the expression

$$\nu\lambda = c,$$

where c is the velocity of light. Now, $u_\nu d\nu$ must be equal in magnitude to $u_\lambda d\lambda$, since each frequency corresponds to one wavelength, and

$$u_\nu = u_\lambda \left| \frac{d\lambda}{d\nu} \right|, \tag{5–4a}$$

where the bars indicate the absolute magnitude of the quantity. The absolute magnitude must be used because the energy density is positive. From $\nu\lambda = c$, $|d\lambda/d\nu| = \lambda^2/c$, and

$$u_\nu = \frac{\lambda^2}{c} u_\lambda. \tag{5–4b}$$

According to Maxwell's electromagnetic theory of light, radiation in an enclosure exerts a pressure on the walls of the enclosure which is proportional to the energy density of the radiation. Wien, therefore, treated the radiation as a thermodynamic engine to which the first and second laws of thermodynamics could be applied. He considered radiation of a single wavelength and analyzed the problem of what would happen to this radiation in an adiabatic expansion of the enclosure that it occupied. He found that the wavelength and temperature before and after the adiabatic change would be connected by the relationship

$$\frac{T}{T_0} = \frac{\lambda_0}{\lambda}, \tag{5–5}$$

where T_0, λ_0 are the values of the temperature and wavelength of the radiation before the adiabatic change, and T, λ are the values after the change. Equation (5–5) may be written in the more general form

$$\lambda T = \text{constant}. \tag{5–6}$$

This equation expresses the fact that if radiation of a particular wavelength whose intensity corresponds to a definite temperature is changed adiabatically to another wavelength, then the absolute temperature changes in the inverse ratio. Wien also obtained another important result, namely, that the monochromatic energy density and temperature before and after the adiabatic change are connected by the relationship

$$\frac{u_\lambda}{u_{\lambda_0}} = \frac{T^5}{T_0^5}, \tag{5–7}$$

where the subscript zero again refers to conditions before the change. Equation (5–7) may also be written in the more general form

$$\frac{u_\lambda}{T^5} = \text{constant}, \tag{5–8a}$$

or, with the aid of Eq. (5–6),

$$u_\lambda \lambda^5 = \text{constant}. \tag{5–8b}$$

In order that Eqs. (5–6) and (5–8) hold simultaneously, $u_\lambda \lambda^5$ or u_λ/T^5 must be a function of the product λT:

$$u_\lambda \lambda^5 = Cf(\lambda T), \tag{5–9}$$

or,

$$u_\lambda = \frac{C}{\lambda^5} f(\lambda T), \tag{5–10}$$

where C is a constant and the function $f(\lambda T)$ is, as yet, undetermined.

TABLE 5–1*

EXPERIMENTAL VERIFICATION OF WIEN'S DISPLACEMENT LAW

Observed temperature, T: deg. K	Wavelength for maximum energy density, $\lambda_m \times 10^4$ cm	Maximum relative energy density, $u_{\lambda m}$	$\lambda_m T$, cm-deg	$u_{\lambda m} T^{-5}$, $\times 10^{17}$
621.2	4.53	2.026	0.2814	2190
723	4.08	4.28	0.2950	2166
908.5	3.28	13.66	0.2980	2208
998.5	2.96	21.50	0.2956	2166
1094.5	2.71	34.0	0.2966	2164
1259.0	2.35	68.8	0.2959	2176
1460.4	2.04	145.00	0.2979	2184
1646	1.78	270.6	0.2928	2246

* From Preston, *Theory of Heat.*

Equations (5–6) and (5–7), or (5–10), constitute *Wien's displacement law.* The reason for this name is that if the change from one temperature to a higher one is supposed to take place by means of an adiabatic compression, then, by Wien's law, the new distribution curve (cf. Fig. 5–2) is obtained from the first one by displacing each abscissa toward the origin in the ratio of distances, $\lambda/\lambda_0 = T_0/T$, and increasing the ordinate in the ratio T^5/T_0^5. Thus, to a maximum $(W_{\lambda m})_0$ or $(u_{\lambda m})_0$ in the first curve corresponds the maximum ordinate $u_{\lambda m}$ of the second; and if λ_m is the abscissa of a maximum ordinate,

$$\lambda_m T = \text{constant} = a, \qquad u_{\lambda m}/T^5 = \text{constant} = b. \tag{5–11}$$

These predictions can be compared with the experimental data of Lummer and Pringsheim, shown in Fig. 5–2. The results of the comparison are listed in Table 5–1; the fourth and fifth columns of the table give the experimental value of the products $\lambda_m T$ and $u_{\lambda m} T^{-5}$, respectively. It is evident that there is no significant variation in either product over a wide range of values of the different quantities, and Wien's displacement law holds. The average value of the product obtained by Lummer and Pringsheim was 0.2940 cm-deg. More recent and more precise experiments give for the best value of this quantity, the Wien displacement constant, the value 0.2884 ± 0.0006 cm-deg. The relation (5–10) has also been verified by the experiments of Lummer and Pringsheim. Since, from Eq. (5–6), $\lambda = \text{constant}/T$, Eq. (5–10) can be written as

$$u_\lambda/T^5 = C'f(\lambda T), \tag{5–12}$$

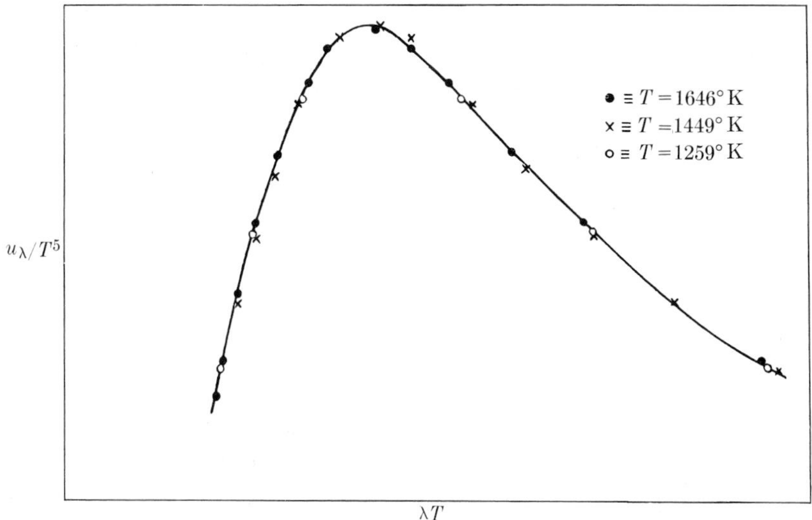

FIG. 5-3. The experimental verification of the displacement law for blackbody radiation. (By permission from F. K. Richtmyer and E. H. Kennard, *Introduction to Modern Physics*, 4th ed. New York: McGraw-Hill, 1947.)

where C' is a new constant. If the experimental values of the quantity u_λ/T^5 are plotted against the values of the product λT, a single curve should result in accordance with the functional relationship (5-12); in Fig. 5-3 is shown the composite curve obtained by combining the data taken at three different temperatures. The theoretical prediction is seen to be fulfilled, and a single curve serves to represent the spectral distribution of blackbody radiation at all temperatures. The problem of the spectral distribution of blackbody radiation has therefore been reduced to the determination of the single function $f(\lambda T)$ in Eq. (5-12)—the function that fixes the shape of the spectral distribution curve.

5-4 The failure of the classical theory of thermal radiation. The derivation of Wien's law was based on thermodynamics and did not involve the nature of the radiator or the mechanism of emission of thermal radiation. Consequently, Wien's law gives the general dependence of the monochromatic energy density on wavelength and temperature. The detailed, explicit form of the dependence is affected by the choice of a model for the radiator. This choice is arbitrary, and any model can be postulated which appears reasonable, leads to a distribution law of the general form of Eq. (5-10) or Eq. (5-12), and agrees with experiment.

The first attempt to determine the function $f(\lambda T)$ was by Wien (1896), who assumed that the radiation was produced by emitters or oscillators of molecular size, that the frequency of the radiation was proportional to the kinetic energy of the oscillator, and that the intensity in any particular wavelength range was proportional to the number of oscillators with the requisite energy. With these assumptions, Wien derived the following expression for the distribution:

$$u_\lambda = \frac{c_1}{\lambda^5} e^{-c_2/\lambda T}, \tag{5–13}$$

where c_1 and c_2 are constants. This distribution has the required general form and was found to fit the experimental curves quite well at short wavelengths, say from 1 to 3×10^{-4} cm, but at longer wavelengths it predicts values of the energy density which are too small. The distribution formula also fails at high temperatures because u_λ (and W_λ) should increase beyond all limits when the temperature increases to an unlimited extent, which is not the case in Eq. (5–13). The exponential function, however, is so satisfactory at short wavelengths that any successful radiation law must reduce approximately to such a function in this region of the spectrum.

Another attempt to obtain a distribution law resulted in the Rayleigh-Jeans formula. This formula was derived in a much more general way than Wien's expression, and was shown to be a necessary consequence of classical dynamics and statistics. It was, therefore, just about the best that classical physics could do. A brief outline of the ideas involved follows. Consider a hollow cavity with perfectly reflecting walls and, for convenience, assume that the cavity is rectangular. In such a cavity there can be standing electromagnetic waves; these waves, in fact, constitute the thermal radiation. If the average energy carried by each wave is known, and if the number of standing waves with frequency between ν and $\nu + d\nu$ can be calculated, the energy density of the radiation can be obtained. The number of waves has been calculated and is equal to

$$N(\nu)\, d\nu = \frac{8\pi\nu^2\, d\nu}{c^3}. \tag{5–14}$$

The average energy carried by each wave can also be calculated. Each standing wave of radiation may be considered to be caused by an electric dipole acting as a linear harmonic oscillator with a frequency ν. The thermal radiation consists, then, of the electromagnetic waves emitted by a large number of such oscillators. The number of these oscillators with frequency between ν and $\nu + d\nu$ is given by Eq. (5–14) and each oscillator has an energy ϵ which may take on any value between 0 and ∞. There is, however, a certain average value $\bar{\epsilon}$ of the energy of the oscillator that can be obtained from classical statistical mechanics. According to that theory, when the oscillators are at equilibrium, the value ϵ for any energy of the oscillator occurs with the relative probability $e^{-\epsilon/kT}$, where k is Boltzmann's constant. The average energy $\bar{\epsilon}$ is obtained by averaging over all values of ϵ, with this weight factor. Set $\beta = 1/kT$, for convenience; then

$$\bar{\epsilon} = \frac{\int_0^\infty \epsilon e^{-\beta\epsilon}\, d\epsilon}{\int_0^\infty e^{-\beta\epsilon} d\epsilon} = -\frac{d}{d\beta} \log \int_0^\infty e^{-\beta\epsilon}\, d\epsilon$$

$$= -\frac{d}{d\beta} \log \frac{1}{\beta} = \frac{1}{\beta} = kT.$$

The energy density of the radiation is, then,

$$u_\nu\, d\nu = N(\nu)\bar{\epsilon}\, d\nu = \frac{8\pi\nu^2 kT}{c^3}\, d\nu. \tag{5–15}$$

Equation (5–15) is the Rayleigh-Jeans formula, and may also be written in the form

$$u_\lambda = \frac{8\pi kT}{\lambda^4}. \tag{5–16}$$

The Rayleigh-Jeans formula agrees well with the experimental intensity distribution for long wavelengths; in this region the intensity of the radiation increases with the square of the frequency, as in Eq. (5–15). For large values of the frequency, i.e., small values of λ, however, the formula fails. According to Eq. (5–16), the energy radiated by a blackbody in a given range of wavelength increases rapidly as λ decreases, and approaches infinity as the wavelength becomes very small. The experimentally observed radiation curve is in complete disagreement with this conclusion, since for very small wavelengths the energy density (or monochromatic emissive power) actually becomes vanishingly small. Furthermore, the energy carried by all wavelengths would be

$$\int_0^\infty u_\lambda\, d\lambda = \int_0^\infty \frac{8\pi kT}{\lambda^4}\, d\lambda,$$

and the integral is infinite for any value of T other than $T = 0$, which would mean that the total energy radiated per unit time per unit area is infinite at all finite temperatures. This conclusion is false, because the total energy radiated at any temperature is actually finite. Hence, the Rayleigh-Jeans formula also fails to account for the observed dependence of radiation on temperature.

5–5 Planck's quantum theory of thermal radiation. The problem of the spectral distribution of thermal radiation was solved by Planck in 1901 by means of a revolutionary hypothesis. Planck postulated that a linear harmonic oscillator like that discussed in the last section does not have an energy that can take on any value from zero to infinity, but can only take on values equal to 0 or ϵ_0 or $2\epsilon_0$ or $3\epsilon_0$, . . . , or $n\epsilon_0$, where ϵ_0 is a discrete, finite amount, or *quantum*, of energy and n is an integer. The average energy of an oscillator is obtained in an analogous way to that used in deriving the Rayleigh-Jeans law except that sums are used instead of integrals. The average value is now

$$\bar{\epsilon} = \frac{\sum_{n=0}^{\infty} n\epsilon_0 e^{-\beta n\epsilon_0}}{\sum_{n=0}^{\infty} e^{-\beta n\epsilon_0}} = -\frac{d}{d\beta}\log\sum_{n=0}^{\infty} e^{-\beta n\epsilon_0} = -\frac{d}{d\beta}\log\frac{1}{1-e^{-\beta\epsilon_0}}$$

$$= \frac{\epsilon_0 e^{-\beta\epsilon_0}}{1-e^{-\beta\epsilon_0}} = \frac{\epsilon_0}{e^{\beta\epsilon_0}-1} = \frac{\epsilon_0}{e^{\epsilon_0/kT}-1}.$$

The energy density is then, with the aid of Eq. (5–14),

$$u_\nu\,d\nu = N(\nu)\bar{\epsilon}\,d\nu = \frac{8\pi\nu^2}{c^3}\,\frac{\epsilon_0}{e^{\epsilon_0/kT}-1}\,d\nu, \tag{5–17}$$

or

$$u_\nu = \frac{8\pi\nu^2}{c^3}\,\frac{\epsilon_0}{e^{\epsilon_0/kT}-1}. \tag{5–18}$$

This formula must have the same general form as Wien's law, Eq. (5–10), and the temperature must appear in the combination λT, or T/ν, or ν/T. Hence, ϵ_0 must be proportional to ν, or

$$\epsilon_0 = h\nu, \tag{5–19}$$

where h is a new universal constant, called *Planck's constant.* Planck's distribution law for thermal radiation is therefore

$$u_\nu = \frac{8\pi h\nu^3}{c^3}\,\frac{1}{(e^{h\nu/kT}-1)}, \tag{5–20}$$

or, in terms of wavelength,

$$u_\lambda = \frac{8\pi hc}{\lambda^5}\,\frac{1}{(e^{hc/k\lambda T}-1)}. \tag{5–21}$$

It is interesting to compare Planck's radiation law with the Wien distribution Eq. (5–13) and the Rayleigh-Jeans formula Eq. (5–15). At low frequencies, $h\nu/kT \ll 1$, and the exponential function in the denominator of Eq. (5–20) can be expanded. Then

$$u_\nu = \frac{8\pi h\nu^3}{c^3} \frac{1}{[1 + (h\nu/kT) + \cdots] - 1} = \frac{8\pi\nu^2}{c^3} kT + \cdots,$$

and at low frequencies (long wavelengths) the Planck formula reduces to the Rayleigh-Jeans formula (5–15), known to be valid at long wavelengths. At frequencies such that $h\nu/kT \gg 1$, the 1 in the denominator of Eq. (5–21) can be neglected in comparison with the exponential, and

$$u_\lambda = \frac{8\pi hc}{\lambda^5} e^{-hc/k\lambda T},$$

which is Wien's formula (5–13), known to be valid at short wavelengths. Thus, Planck's radiation formula reduces to forms that agree with experiment at the extremes of the wavelength scale. In the intermediate wavelength range, the Planck distribution gives a maximum value for the energy density. The position of the maximum can be found from the condition $du_\lambda/d\lambda = 0$, with u_λ given by Eq. (5–21). Omitting the details of the calculations, we get the result

$$\lambda_m T = \frac{ch}{k} \frac{1}{4.965}. \tag{5–22}$$

In Eq. (5–22), c is the velocity of light, k is the Boltzmann constant, and h is Planck's constant. Hence, the product $\lambda_m T$ is a constant, in agreement with experiment.

It can also be shown that Planck's radiation law leads to the Stefan-Boltzmann law. The total energy density is given by

$$u = \int_0^\infty u_\lambda \, d\lambda = 8\pi ch \int_0^\infty \frac{1}{(e^{ch/k\lambda T} - 1)} \frac{d\lambda}{\lambda^5}.$$

This integral can be evaluated, and the resulting expression for u is

$$u = aT^4, \tag{5–23}$$

the familiar fourth power law, with

$$a = \frac{8}{15} \frac{\pi^5 k^4}{c^3 h^3}. \tag{5–24}$$

Equations (5–22) and (5–24) are two relations containing the three constants, c, h, and k. In 1901, the value of c was known quite well; also, the experimental value of the product $\lambda_m T$ was known to be 0.294 from the experiments of Lummer and Pringsheim (Table 5–1). The value of the constant a was known from experimental studies of the Stefan-Boltzmann law, $a = 7.061 \times 10^{-15}$ erg/cm^3(°K)^4. With the known values of a, $\lambda_m T$, and c, Planck obtained

$$h = 6.55 \times 10^{-27} \text{ erg-sec},$$

and

$$k = 1.346 \times 10^{-16} \text{ erg/deg}.$$

This was the first calculation of Planck's constant, and the best, up to that time, of Boltzmann's constant. More recent and more precise experimental determinations of h and k by a variety of methods give

$$h = (6.62517 \pm 0.00023) \times 10^{-27} \text{ erg-sec},$$

$$k = (1.38042 \pm 0.00007) \times 10^{-16} \text{ erg/deg}.$$

When the above values are inserted into Eq. (5–20) or Eq. (5–21), the shape of the spectral distribution curve is completely determined. Planck's distribution law can then be compared directly with experiment; this is done in Fig. 5–4 for the case of blackbody radiation at 1600°K. The points are the experimental results for the energy distribution at different wavelengths, and the solid line is the distribution according to the Planck radiation law; the agreement is excellent over the entire wavelength range

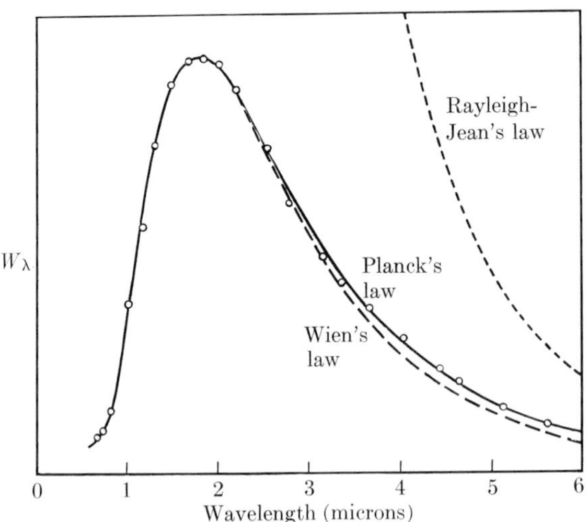

FIG. 5–4. A comparison of the Planck, Wien, and Rayleigh-Jeans radiation laws with experiment at 1600°K. The ordinate represents W_λ on an arbitrary scale, and the circles represent experimental points. (By permission from F. K. Richtmyer and E. H. Kennard, *Introduction to Modern Physics*, 4th ed. New York: McGraw-Hill, 1947.)

of these experiments. For comparison, the theoretical predictions of the Wien distribution (5–13) and the Rayleigh-Jeans formula (5–16) are also included. These results are typical of those found over a wide range of temperatures, and there is no doubt that the Planck radiation law is the only one which satisfactorily accounts for the experiments.

The magnitude of the energy associated with a quantum of radiation can be obtained as a function of wavelength. Equation (5–19) can also be written

$$\epsilon_0 = \frac{hc}{\lambda}. \tag{5–25}$$

Then ϵ_0 is given, in ergs, by

$$\begin{aligned} \epsilon_0 \text{ (ergs)} &= \frac{6.62 \times 10^{-27} \times 3.00 \times 10^{10}}{\lambda} \\ &= \frac{1.99 \times 10^{-16}}{\lambda}, \end{aligned}$$

where λ is expressed in centimeters. For a gamma-ray with a wavelength of 10^{-10} cm, the quantum is 1.99×10^{-6} erg. For a visible light quantum with a wavelength of 5000 A $= 5 \times 10^{-5}$ cm, the energy is 4.0×10^{-12} erg. For a short radio wave, say $\lambda = 10$ cm, the quantum is 1.99×10^{-17} erg, and so on.

The revolutionary nature of Planck's theory is contained in the postulate that the energy of an oscillator can vary only in discrete jumps. If this is so, then the emission and absorption of radiation must be discontinuous processes. Emission can take place only when an oscillator makes a discontinuous transition from a state in which it has one particular energy to another state in which it has another energy different from the first by an amount which is an integral multiple of $h\nu$, where ν is the frequency of the emitted radiation. We are thus led to the idea that an oscillator, or any other physical system capable of emitting electromagnetic radiation, has, in general, a discrete set of possible energy values or levels; energies intermediate between these allowed values never occur. The idea of discrete energy levels will be seen to be a fundamental one in atomic and nuclear physics, and is a consequence of Planck's postulate of quanta of energy. This idea is not an outgrowth of classical physics, but represents a radical, empirical modification of classical ideas, needed in order to bring theory and experiment into agreement.

5–6 Quantum theory and the photoelectric effect. The concept of discrete energy states and quanta of energy was so different from the classical ideas to which physicists were accustomed that Planck's theory met with strong opposition. The theory was successful, however, not only in explaining the phenomena of thermal radiation, but also in the solution of other problems for which classical physics failed. One of these problems was that of the photoelectric effect.

Hertz (1887) discovered that a metallic surface can emit electricity when light of very short wavelength falls on it, and in 1898 Thomson showed that the e/m value of the emitted particles was the same as that for cathode rays. It is now known, of course, that the particles are electrons. The photoelectric effect was studied very carefully and the most important experimental results may be summarized as follows:

1. If light of a given frequency can liberate electrons from a surface, the electron current is proportional to the intensity of the light.

2. For a given metallic surface, there is a smallest value of the frequency for which the incident light can liberate electrons; light of smaller frequency cannot eject electrons no matter how long it falls on the surface or how great is its intensity.

3. Light of frequency greater than this critical value causes the immediate emission of electrons; the time interval between the incidence of the light on the metallic surface and the appearance of electrons is not more than 3×10^{-9} sec.

4. The maximum kinetic energy of the emitted electrons is a linear function of the frequency of the light which causes their emission, and is independent of the intensity of the incident light.

These experimental facts could not be explained on the basis of the classical electromagnetic wave theory of light. There was no way in which a train of light waves spread out over a large number of atoms could, in a very short time interval, concentrate enough energy on one electron to knock it out of the metal. Furthermore, the wave theory of light was unable to account for the fact that the maximum energy of the ejected electrons increases linearly with the frequency of the light but is independent of the intensity. In 1905, Einstein proposed a mechanism for the photoelectric effect, based on Planck's idea of quanta of energy, that could account for the experiments. He assumed that the energy emitted by any radiator not only kept together in quanta as it traveled through space, but that a given source could emit and absorb radiant energy only in units which are all exactly equal to $h\nu$. According to Einstein, light itself consists of quanta or corpuscles of energy $h\nu$ which move through space with the velocity of light. This hypothesis is completely at variance with the wave theory of light but, as will be seen, accounts successfully for the photoelectric effect. With this hypothesis, Einstein deduced his photoelectric equation

$$\tfrac{1}{2}mv^2 = h\nu - A. \tag{5–26}$$

This equation states that the maximum kinetic energy of the emitted electron is equal to the energy $h\nu$ of the incident light quantum minus a quantity A which represents the amount of work needed by the electron to get free of the surface.

It is evident that Eq. (5–26) agrees qualitatively with the experimental facts. If the frequency is so small that $h\nu \leq A$, no electrons can be emitted, so that the frequency has a threshold value, for a given metal, below which no photoelectric emission can occur. The energy of the electrons varies linearly with the frequency and is independent of the intensity. If the frequency is above the threshold value, a light quantum striking the metal and colliding with one of its electrons can give up all of its energy to the electron and knock it out of the metal. But before the electron emerges, it loses a part A of this energy needed to remove it from the metal. The number of electrons ejected is proportional to the number of incident light quanta and therefore to the intensity of the light falling on the metal.

Although the Einstein photoelectric equation accounted for the experiments qualitatively, it was not verified quantitatively until 1916 when Millikan performed a very beautiful experiment. He studied the effect of light of a range of frequencies on sodium, potassium, and lithium. The energy of the ejected electrons was measured by applying to the metallic surface a positive potential just strong enough to prevent any of the electrons from carrying a charge to an electrometer. If this potential is equal to V volts, then the kinetic energy is given by

$$\tfrac{1}{2}mv^2 = Ve,$$

where e is the electronic charge. Combining this relation with Eq. (5–25), we get

$$Ve = h\nu - A,$$

or

$$V = (h/e)\,\nu - (A/e)\,\cdot \tag{5–27}$$

If the potential V is measured as a function of the frequency of the incident light, the result should be a straight line with a slope whose magnitude is h/e. The experimental results gave, in all cases, a good straight line, as predicted by Einstein's theory. Since the value of e had already been determined by Millikan, the value of h could be determined; it was found to be in good agreement with the value obtained by Planck from the measurements on blackbody radiation. Einstein's photoelectric law has also been found to be valid for the electrons ejected by x-rays and γ-rays, and there is no doubt of its general validity.

A quantum of light or, more generally, of radiation is also called a *photon;* the photon may be described as a "bundle" or "particle" of radiation. Thus, a photon of radiation of frequency ν carries an amount $h\nu$ of energy.

The quantum theory was also applied successfully by Einstein (1907) to the problem of the specific heat of solids and polyatomic gases, where again there was experimental information which could not be explained by classical theory. These applications of the quantum theory by Einstein helped establish the theory, aided its acceptance, and laid the groundwork for the application of quantum theory to the problem of atomic structure.

References

GENERAL

M. Planck, *Theory of Heat.* London: Macmillan Co., 1929, Parts III, IV.

T. Preston, *The Theory of Heat,* 4th ed., J. R. Cotter, ed. London: Macmillan Co., 1929, Chapter 6.

J. K. Roberts, *Heat and Thermodynamics,* 4th ed., revised by A. R. Miller. Glasgow: Blackie; New York: Interscience, 1951, Chapters 20, 21.

A. Sommerfeld, *Thermodynamics and Statistical Mechanics.* New York: Academic Press, 1956, Sections 20, 35.

F. K. Richtmyer, E. H. Kennard, and T. Lauritsen, *Introduction to Modern Physics,* 5th ed. New York: McGraw-Hill, 1955, Chapter 4.

G. P. Harnwell and J. J. Livingood, *Experimental Atomic Physics.* New York: McGraw-Hill, 1933, Chapter 2.

R. A. Millikan, *Electrons, (+ and —), Protons, Photons, Mesotrons, and Cosmic Rays.* Chicago: University of Chicago Press, 1947.

Problems

1. Calculate the total emissive power of a blackbody at the following temperatures: 0°C, 100°C, 300°C, 500°C, 1000°C, 1500°C, 2000°C, and 3000°C. The Stefan-Boltzmann constant σ has the value 0.5669×10^{-4} erg-cm^{-2}-sec^{-1} (°K)$^{-4}$. Give the results in terms of ergs and calories.

2. Suppose that the blackbody of Problem 1 is surrounded by walls kept at a temperature of 0°C. Calculate the net rate of energy loss by the body to the walls at each of the temperatures listed in Problem 1.

3. Derive, from Planck's radiation law, the expression (5–22) for the wavelength at which the radiation density has its maximum value. The best value of the Wien displacement law constant is now thought to be 0.2884 cm-°K. Calculate the value of h. Use $k = 1.3804 \times 10^{-16}$ erg/°K and $c = 2.99793 \times 10^{10}$ cm/sec.

4. Derive an expression for the total energy density in a blackbody enclosure by integrating Planck's distribution law over all possible values of the frequency. Note that

$$\int_0^\infty \frac{x^3\,dx}{e^x - 1} = \frac{\pi^4}{15}.$$

The best value for the total energy density is given by $a = 7.563 \times 10^{-15}$ erg/cm^3. Calculate the value of h; use the values of k and c given in Problem 3.

5. Calculate the frequency and the amount of energy associated with a quantum of electromagnetic radiation at each of the following wavelengths: 1 km, 1 m, 1 cm, 1 mm, 10^{-4} cm, 5000 A, 1000 A, 1 A, 10^{-3} A, 10^{-13} cm, 10^{-15} cm. Give the energy in ergs and in electron volts.

6. The photoelectric work functions for samples of some metals are Ag, 4.5 ev; Ba, 2.5 ev; Li, 2.3 ev; Pt, 4.1 ev; Ni, 5.0 ev. (a) What is the wavelength at the threshold in each case? (b) What is the maximum velocity of the ejected photoelectrons in each case, when the samples are illuminated with ultraviolet light of wavelength 2000 A? (c) What is the stopping potential, in volts, for the photoelectrons of maximum velocity?

7. The first precise determination of h was made by Millikan, who measured the energy with which photoelectrons are ejected as a function of the frequency of the incident light. He found that the potential V needed to keep the electrons from sodium from reaching an electrometer increased with frequency by 4.124×10^{-15} volt/cycle/sec. What is the resulting value of h? Note that 300 volts = 1 esu of potential, and use $e = 4.803 \times 10^{-10}$ esu.

8. Show that the velocity of the fastest photoelectrons is related to the stopping potential by the equation

$$v_{\text{max}} = 5.93 \times 10^7 \times \sqrt{V},$$

where v_{max} is in centimeters per second and V is in volts.

9. Photoelectrons emitted from a photocell by light with a wavelength of 2500 A can be stopped by applying a potential of 2 volts to the collector. (a) What is the work function (in ev) of the surface? (b) What is the maximum wavelength that the light can have and eject any photoelectrons? (c) If an electrometer is connected across the photocell, to what potential will it be charged when ultraviolet light of wavelength 1000 A is used?

10. The emission of x-rays as a result of the incidence of high-speed electrons on a metal surface, as in an x-ray tube, may be regarded as an *inverse photoelectric effect*. The maximum frequency of the emitted radiation is given by the relation $h\nu_m = eV$, where V is the anode voltage. Show that λ_m (A) = $12395/V$ (volts). What are the wavelength, frequency, and energy (in ev) of the x-rays from a Coolidge tube operated at 50,000 volts? At 10^5 volts? At 5×10^5 volts? At 10^6 volts?

CHAPTER 6

THE SPECIAL THEORY OF RELATIVITY

6–1 The role of the special theory of relativity in atomic and nuclear physics. The second great theory of 20th century physics that is indispensable to the development of atomic and nuclear physics is the special theory of relativity proposed by Einstein in 1905. The first applications of the relativity theory which will be met in atomic and nuclear physics depend on two closely related ideas. One idea is that of the variation of the mass of a particle with its velocity; the second is that of the proportionality between mass and energy. The latter relationship, expressed by the equation $E = mc^2$, has achieved a certain degree of notoriety in recent years and has even, on occasion, occupied positions of prominence in the public press.

The mass-energy equation is used often in nuclear physics, and its use should be accompanied by some understanding of its origin and meaning. A real understanding of the relationship can come only from a careful study of the relativity theory; but some familiarity with the ideas involved in the mass-energy equation can be obtained from a short discussion of the background of experiment and theory which led to relativity. Such a discussion should dispel some of the aura of magic that surrounds the relativity theory and even its simplest applications, and should allow the student to use mass and energy as interchangeable quantities without feeling uncomfortable.

The theory of relativity may be introduced in either of two ways. One way is to examine very carefully what is meant by such concepts as space, time, and simultaneity and, as a consequence of this examination, to modify the commonly accepted meanings of these ideas. This logical process, together with certain postulates, leads to the special theory of relativity. The theory can then be applied to physical problems to see how it meets the test of comparison with experiment. The second method is to consider certain problems which led to serious dilemmas in physical thought at the end of the 19th century, and to see how these dilemmas could be resolved only by the revolutionary ideas contained in the relativity theory. The second method is the one that will be adopted here because it is more closely related to experiment and is less abstract than the first method. Two problems will be considered; these are (1) the ether and the problem of absolute velocity, (2) the problem of the invariant form of physical theories.

6–2 The ether and the problem of absolute velocity. The interpretation of the phenomena of the reflection, refraction, and diffraction of light gave rise to the wave theory of light. The propagation of light waves seemed to require a medium, just as water or sound waves cannot be separated from their respective media. It was known from astronomical data that light can pass through space practically devoid of matter. For these reasons, physicists felt it necessary to postulate the existence of a medium in which light waves could be propagated even in the absence of matter. This medium, called the *ether*, was supposed to be a weightless substance permeating the entire universe. Because of the success of the wave theory of light in accounting for optical phenomena, the ether gradually came to be accepted as a physical reality. The assumption that there is an ether raised problems in connection with the motion of material bodies. The ether was assumed to be stationary, and a body such as the earth was supposed to move through it without producing any disturbance. These assumptions were not contradicted by astronomical evidence, and it was inferred that it should be possible to determine the *absolute* velocity of a body, i.e., its velocity relative to the stationary ether. Since the ether was supposed to be the medium for the propagation of light, it was thought that the measurement of absolute velocity should depend on the effect of motion on some phenomena involving light. Thus, if light and the earth both move with definite velocities relative to the stationary ether, it should be possible to devise an experiment, involving the propagation of light, from which the absolute velocity of the earth could be deduced. Many such experiments were proposed; the theoretical basis of the most famous one can easily be derived.

Suppose that the earth travels through the stationary ether with a velocity v, and that light waves travel through the ether with a velocity c. Consider a source, fixed on the earth, from which light is emitted. If the direction of propagation of the light is the same as that of the earth's motion, the velocity of the light relative to the earth will be $c - v$. If the light travels in the direction exactly opposite to that of the earth's motion, its velocity relative to the earth is $c + v$. The time taken for the light to travel a certain distance L in the direction of the earth's motion is $L/(c - v)$, while the same distance in the opposite direction will require the time $L/(c + v)$. Therefore, if a ray of light were to travel a distance L in the direction of the earth's motion, and then be reflected back and travel the same distance in the opposite direction, the time required would be

$$t_{||} = \frac{L}{c + v} + \frac{L}{c - v} = \frac{2cL}{c^2 - v^2}.$$

Consider next a ray of light propagated in a direction perpendicular to that of the earth's motion (Fig. 6–1). While this ray travels a distance L, equal to AB, the earth has moved from A to A', and the point B has moved to B'. The actual path of the light ray is AB'. If the time required is t, then $AB' = ct$. In the same time interval, A has moved to A' with the velocity v, so that $AA' = vt$. In the right-angled triangle $AA'B'$, we have $c^2t^2 = L^2 + v^2t^2$, or $t^2(c^2 - v^2) = L^2$, or $t = L/\sqrt{c^2 - v^2}$. The same length of time will be needed for the return trip, i.e., to A''. Hence, the time required to travel a distance L and back in a direction perpendicular to that of the earth's motion is

$$t_\perp = \frac{2L}{\sqrt{c^2 - v^2}}.$$

The ratio of the times required is

$$\frac{t_{||}}{t_\perp} = \frac{2cL}{c^2 - v^2} \cdot \frac{\sqrt{c^2 - v^2}}{2L} = \frac{1}{\sqrt{1 - (v^2/c^2)}}. \tag{6–1}$$

It should, therefore, take longer for light to traverse the reflected path in the direction parallel to that of the earth's motion than to traverse the same distance in the direction perpendicular to that of the earth's motion.

From an experimental viewpoint, the difference between $t_{||}$ and $t_\perp$ is important. This difference is

$$t_{||} - t_\perp = \frac{2L}{c}\left[\frac{1}{1 - (v^2/c^2)} - \frac{1}{\sqrt{1 - (v^2/c^2)}}\right]. \tag{6–2}$$

If we neglect terms of order higher than v^2/c^2, we may write

$$\frac{1}{1 - (v^2/c^2)} = 1 + \frac{v^2}{c^2} + \cdots,$$

$$\frac{1}{\sqrt{1 - (v^2/c^2)}} = 1 + \frac{1}{2}\frac{v^2}{c^2} + \cdots.$$

Hence,

$$\begin{aligned} t_{||} - t_\perp &= \frac{2L}{c}\left(1 + \frac{v^2}{c^2} - 1 - \frac{1}{2}\frac{v^2}{c^2} + \cdots\right) \\ &= \frac{L}{c}\cdot\frac{v^2}{c^2}. \end{aligned} \tag{6–3}$$

The difference $t_{||} - t_\perp$ and the ratio $t_{||}/t_\perp$ both depend on the quantity $(v/c)^2$. The orbital velocity of the earth is 30 km/sec, so that $v/c = 10^{-4}$ and $(v/c)^2 = 10^{-8}$, and the effect should be difficult to detect.

6–3 The Michelson-Morley experiment. The theoretical relations derived in the last section were tested experimentally by Michelson and Morley (1887) who used an ingenious instrument, the interferometer, shown schematically in Fig. 6–2. In the Michelson-Morley experiment, light from a source Q is divided into two perpendicular beams by a half-silvered mirror P. One beam passes through P to a mirror S_1 and is reflected back to P; the other beam is reflected by P to the mirror S_2 and is then reflected back to P. The two reflected beams pass through P to a telescope or photographic plate F. If the distances PS_1 and PS_2 are equal, and if one arm of the interferometer is parallel to the direction of the earth's motion, the two light beams should arrive at F at slightly different times, causing an interference pattern (interference fringes). The position of the pattern should depend, according to Eq. (6–3), on the absolute velocity of the earth and should be displaced from the position corresponding to $v = 0$ by an amount proportional to $(L/c)(v^2/c^2)$. If the interferometer is rotated through an angle of 90°, the arm which was originally in the direction of the earth's motion is now perpendicular thereto, and the interference pattern should be displaced to the opposite side of the position corresponding to $v = 0$. If the position of the pattern is observed during the rotation of the interferometer, the fringes should be shifted by an amount corresponding to the time interval $2(L/c)(v^2/c^2)$. The frequency of the light used in the experiment is ν, and its period (inverse frequency) is $T = 1/\nu$. Then the ratio of the displacement to the period is given by $(2L/cT)(v^2/c^2)$. Since the wavelength $\lambda = cT$, the ratio of displacement to wavelength is given by $(2L/\lambda)(v^2/c^2)$.

In the experiment of Michelson and Morley, the path length L was 1.1×10^3 cm, as a result of several reflections of the light beam. The wavelength of the light used was about 5.9×10^{-5} cm, and $v^2/c^2 = 10^{-8}$.

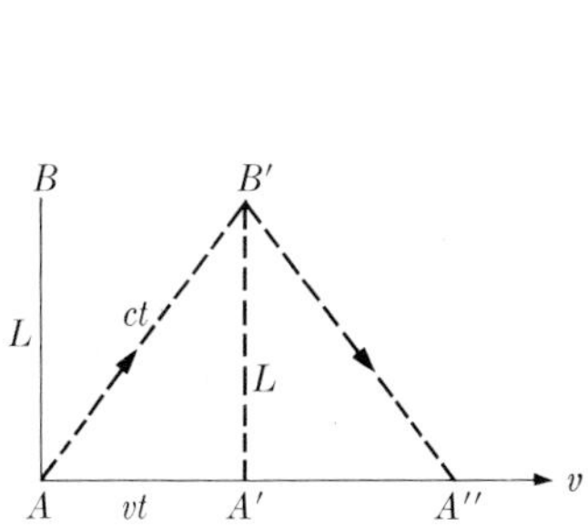

FIG. 6–1. Basis of an experiment on the velocity of the earth relative to the ether.

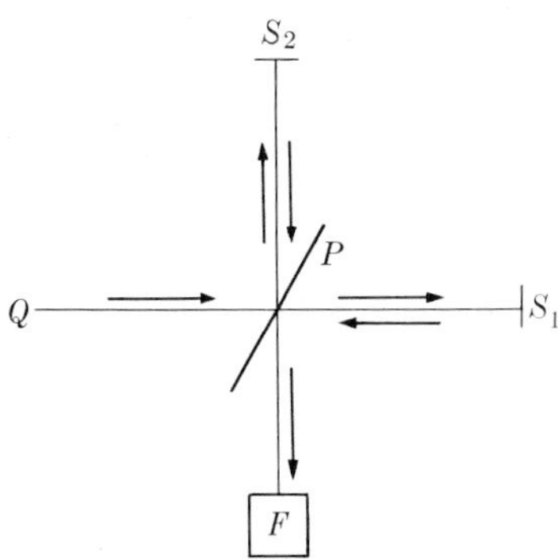

FIG. 6–2. The Michelson-Morley experiment: schematic diagram.

Hence, $(2L/\lambda)(v^2/c^2) = [2(1.1)10^3 \times 10^{-8}]/[5.9 \times 10^{-5}] = 0.37$. Thus, when the interferometer was rotated through 90°, the interference fringes should have been displaced by more than $\frac{1}{3}$ of a wavelength. In the experiment, a displacement of one-hundredth of this amount could have been observed, but no significant displacement was observed in the original experiment, or in many later repetitions. That is, there was no significant difference between the rates at which the light traveled in the two directions at right angles to each other, and the Michelson-Morley experiment showed that the ratio of the times of travel in Eq. (6–1) is unity. The simplest interpretation of this result seemed to be that v, the velocity of the earth with respect to the ether, is zero, a conclusion which led to serious difficulties.(1)

The theoretical value of the ratio of displacement to wavelength was derived by a simple and straightforward application of some of the fundamental ideas of classical physics, and this fact made it very difficult to explain the negative result of the Michelson-Morley experiment. The simplest explanation which could be proposed was that the earth carries the ether with it, in which case the velocity of the earth relative to the ether should be zero. This explanation was unattractive because every moving body would presumably carry its own ether with it, and the uniqueness of "the stationary ether" as the medium for the propagation of light would disappear.

There were also experimental objections to the idea that the earth carries the ether along with it. Suppose that in space there is an observer who is looking at a light source on the earth, and that light waves from this source are transmitted through the ether and eventually reach the observer. If the earth, together with the ether, were moving toward the observer, the velocity of the light waves from the source would appear to be greater than if the earth were moving away from the observer. This result follows from the general behavior of waves in a moving medium; the velocity of the waves relative to a stationary observer should be equal to the sum of the velocity of the waves in the medium and the velocity of the medium. Thus, the velocity of light as measured by the assumed observer would depend on the velocity of its source. This conclusion is contradicted by all observations on the velocity of light, especially by astronomical studies of double star systems and stellar aberration. The observed velocity of light is always the same regardless of whether or not the emitting source moves, or how it moves. The assumption that the earth carries the ether along with it led, therefore, to a serious contradiction with experiment, and had to be rejected.

Fitzgerald (1893) introduced a hypothesis to account for the negative result of the Michelson-Morley experiment while preserving the idea of the stationary ether. He assumed that when a body travels in a direction parallel to that of the earth's motion it contracts in length, becoming shorter by the factor $\sqrt{1 - (v^2/c^2)}$; he also assumed that no such contraction occurs in a direction perpendicular to that of the earth's motion. According to this hypothesis, the arm of an interferometer that is parallel to the direction of the earth's motion contracts in length from L to $L\sqrt{1 - (v^2/c^2)}$, but the length of the perpendicular arm remains equal to L. The light does not have to travel as far in the parallel direction as in the perpendicular direction, and the difference is such as to make the theoretical value of the shift of the interference pattern equal to zero. The contraction in the length of one arm of the interferometer would not be observed, because a measuring instrument placed along the arm would shrink in the same way.

The Fitzgerald contraction seems absurd at first sight because it is not the result of any forces acting on a body but depends only on the fact that the body is in motion. Fitzgerald's hypothesis was purely *ad hoc*, and its main object was to preserve the concept of a stationary ether. Lorentz, who was at the same time trying to develop a consistent electromagnetic theory of the electron, found the hypothesis useful in his work and incorporated it into his theories, so that the hypothesis is now known as the Lorentz-Fitzgerald contraction hypothesis. It had previously been shown that under certain assumptions, the mass of an electron could be expressed by the formula

$$m_0 = 2e^2/3r_0c^2, \tag{6-4}$$

where r_0 is the radius of the electron when it is at rest, and e is its charge. Lorentz assumed that a moving electron contracts in the direction of its motion by the Fitzgerald factor $\sqrt{1 - (v^2/c^2)}$, so that the radius becomes $r_0\sqrt{1 - (v^2/c^2)}$ when the electron moves with a velocity v. Then, if the mass of a moving electron is m, and if the mass when the electron is at rest is m_0, it follows that

$$m = \frac{m_0}{\sqrt{1 - (v^2/c^2)}}\cdot \tag{6-5}$$

According to Eq. (6-5), the mass of an electron should increase with its velocity, especially when the latter reaches a significant fraction of the velocity of light. Early experiments on the value of e/m for β-rays indicated that the value of e/m decreases with increasing velocity of the rays. Since there was no reason to suppose that the charge varies with velocity, it was concluded that the mass of the electron does indeed increase with its velocity. Thus, although the Lorentz-Fitzgerald contraction hypothesis seemed physically unreasonable, it led to theoretical results consistent with experiment. But the hypothesis lacked a sound theoretical basis, and classical physics was once again confronted by a problem (the negative result of the Michelson-Morley experiment) that it could not really solve.

The Michelson-Morley experiment has been repeated under different conditions and with modifications; other tests of the possible motion of the earth relative to the ether have also been made. No significant motion has been detected.(1) In 1958, in an extremely sensitive test which involved the use of *molecular clocks*,(2) it was found that the maximum velocity of the earth with respect to the ether is less than 1/1000 of the earth's orbital velocity, or less than 0.03 km/sec. These results all show that the hypothesis of a stationary ether through which the earth moves is not verified by experiment.

6–4 The problem of the invariant form of physical theories. The concept of the invariant form of physical theories is a fundamental one in science. To see what this means, let us consider what is meant by a physical theory. A physical theory consists, essentially, of equations or formulas which relate measurable quantities such as distance and time to invented quantities such as energy, momentum, potential, field strength, and so on. Numbers can be assigned to the invented quantities, and when they are inserted into the formulas of the theory, and when the appropriate mathematical operations are performed, the calculated values of the measurable quantities should agree with the values measured in suitable experiments. The measured values can usually be expressed in terms of positions in some coordinate system such as x, y, z, and of time intervals t. It is then highly desirable that the formulas which make up a theory be the same for different observers. For example, suppose that there is an observer who is at rest with respect to some system of reference S (Fig. 6–3). He measures position in terms of the coordinate system (x, y, z) and measures time intervals t on a clock. Suppose that there is another observer who is at rest with respect to another system of reference S' which moves with uniform velocity v along the x-axis, the direction of which coincides with the direction of the x'-axis. At time

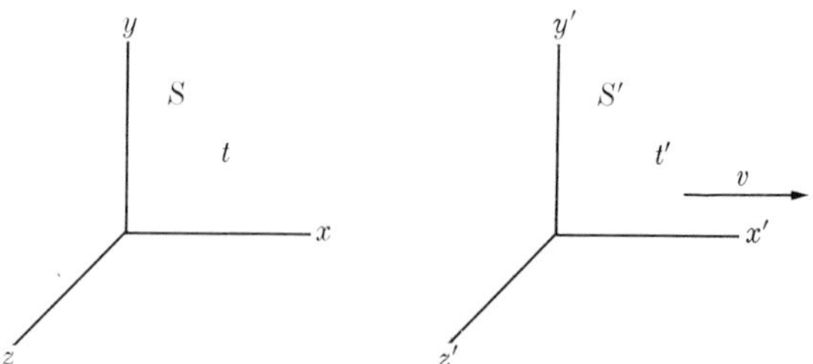

FIG. 6–3. Two Cartesian reference systems in uniform relative translatory motion.

$t = 0$, the two reference frames are assumed to coincide. The second observer uses the spatial coordinates x', y', z' and the time coordinate t'. Now let both observers S and S' perform an experiment which involves the motion of a body. Each observer will make measurements appropriate to the experiment, and will describe the results by means of formulas. In the formulas of the observer S, the quantities x, y, z, and t will appear; in those of S', the quantities x', y', z', and t' will appear. It was found long ago that the laws of classical mechanics as expressed by the two observers S and S' have the same form provided that the coordinates used by the two observers are related by means of the transformation equations:

$$x' = x - vt, \qquad y' = y, \qquad z' = z, \qquad t' = t. \tag{6–6}$$

Equations (6–6) connect the variables of the two systems S and S' in such a way that if an experiment involving the motion of a body has been measured by an observer moving with the system S' and has been described in terms of the quantities x', y', z', t', the substitutions (6–6) give a correct description of the same experiment in terms of the variables x, y, z, t, used by an observer moving with the system S. In more formal language, the laws of classical mechanics are said to be invariant with respect to the transformation (6–6).

When the transformation (6–6) is applied to the theory of electromagnetics, the formulas of the theory are not invariant; the formulas of an observer in the moving system S' are not the same as those of an observer in the S system. Lorentz, who developed a theory of electrons based on classical electromagnetics, solved the problem of what transformation equations would lead to the same formulas for his electron theory for two observers, one of whom is moving with a uniform velocity v with respect to the other. The system of transformation equations that he derived is known as the *Lorentz transformation*, and is given by the following relationships:

$$x' = \frac{x - vt}{\sqrt{1 - (v^2/c^2)}}, \qquad z' = z,$$
$$y' = y, \qquad t' = \frac{t - (vx/c^2)}{\sqrt{1 - (v^2/c^2)}}. \tag{6–7}$$

These equations are remarkable because they indicate that space and time are not independent entities but are related; time as measured by an observer in the S' system is different from that measured by an observer in the S system, and depends on the velocity with which the S' observer is moving with respect to the S observer as well as on the space coordinate

x. Classical physicists found it impossible to accept this relationship between the space and time variables because it was contrary to the fundamental classical idea of the independence of space and time. The relation between x' and x, however, does lead to the Lorentz-Fitzgerald contraction, as can be shown quite easily.

Suppose that a body, when at rest in S, has a length L_0 in the direction of the x-axis. Let it be set into motion relative to S at such a speed that it is at rest in S'; that is, it moves with a velocity v relative to S, and this velocity v is the same velocity with which S' moves with respect to S. The length will then be L_0 as measured in S', since the length must have the same value in any reference system in which the body is at rest. But what value does the observer S measure for the length of the body which is now moving with respect to him? He measures a value L given by

$$L = x_2 - x_1,$$

where x_2 and x_1 are the coordinates of the ends of the body in the S system. The S' observer measures the length as

$$L_0 = x_2' - x_1'.$$

If we substitute for x_2' and x_1' the values calculated from the first of Eqs. (6–7), we get

$$L_0 = \frac{x_2 - vt}{\sqrt{1 - (v^2/c^2)}} - \frac{x_1 - vt}{\sqrt{1 - (v^2/c^2)}}$$

$$= \frac{x_2 - x_1}{\sqrt{1 - (v^2/c^2)}} = \frac{L}{\sqrt{1 - (v^2/c^2)}},$$

or

$$L = L_0\sqrt{1 - (v^2/c^2)}.$$

According to this relation, a rigid body, when it is in uniform motion relative to a stationary observer, appears contracted in the direction of its relative motion by the factor $\sqrt{1 - (v^2/c^2)}$, while its dimensions perpendicular to the direction of motion are unaffected. When this relation is applied to the interferometer arms in the Michelson-Morley experiment, it gives just the contraction needed to account for the negative result of that experiment. Thus, the requirement that the equations of electrodynamics be invariant under the transformation equations (6–7) leads to the Lorentz-Fitzgerald contraction. The difficulty with accepting these results was caused by the lack of a consistent physical basis for them, and it became necessary to revise some of the fundamental ideas of physics.

6–5 The special theory of relativity; the variation of mass with velocity. The problems raised by the Michelson-Morley experiment and the requirement of invariance were solved in 1905 by Einstein, who proposed the special, or restricted, theory of relativity. Einstein interpreted the negative results of the Michelson-Morley experiment to mean that it is indeed impossible to detect any absolute velocity through the ether. The ether was as close to an "absolutely stationary" reference frame as classical physics could come. Hence, according to Einstein, only relative velocities can be measured, and it is impossible to ascribe any absolute meaning to different velocities. It follows that the general laws of physics must be independent of the velocity of the particular system of coordinates used in their statement because, if this were not so, it would be possible to ascribe some absolute meaning to different velocities. But this is just another way of stating the requirement that the equations of a physical theory be invariant with respect to coordinate systems moving with different velocities. In formulating the relativity theory, Einstein adopted this requirement of invariance as one of the two basic postulates of the theory.

In his first statement of the theory, Einstein was restricted by mathematical difficulties to the consideration of reference systems moving at a constant velocity relative to each other. He therefore took as his first postulate:

POSTULATE 1. The laws of physical phenomena are the same when stated in terms of either of two reference systems moving at a constant velocity relative to each other.

For his second postulate, Einstein took:

POSTULATE 2. The velocity of light in free space is the same for all observers, and is independent of the relative velocity of the source of light and the observer.

The second postulate may be regarded as a result of the Michelson-Morley experiment and other optical experiments and of astronomical observations, i.e., it is believed to represent an experimental fact.

The term "special theory of relativity" refers to the restriction in the first postulate to reference systems moving at a constant velocity relative to each other. It was not until 1916 in the general theory of relativity that Einstein was able to show that physical laws can be expressed in a form that is valid for any choice of space-time coordinates. The general theory is not needed to describe most atomic and nuclear phenomena and therefore will not be discussed.

From his two postulates, Einstein deduced the Lorentz transformation equations as well as other kinematical relationships. The Michelson-

Morley experiment could be interpreted directly in terms of these relationships and there was no need for the Fitzgerald contraction hypothesis or for the ether. The interrelationship between space and time coordinates also follows from the postulates and represents a more advanced treatment of these ideas than does the earlier classical theory. The interdependence between the space and time coordinates means that time must be treated in the same way as the space coordinates. This result gave rise to the ideas of a four-dimensional space-time continuum as well as to popular misconceptions about the "fourth dimension." It must be pointed out, however, that when v is much smaller than c, the Lorentz transformation equations (6–7) reduce to the classical equations (6–6), so that for ordinary mechanical phenomena, the classical or Newtonian theory is adequate. The relativity theory does not "overthrow" the classical theory but rather extends and modifies it. The need for the relativity theory in atomic and nuclear physics comes in part from the fact that fundamental particles such as electrons can travel with velocities approaching that of light, with the result that the kinematic and dynamical consequences of the Lorentz equations are important.

The Lorentz transformation equations can be derived quite easily from Einstein's postulates if we consider measurements, by two observers, of the distance traveled by a light ray. Suppose that the system S' is moving at uniform speed v in the x-direction relative to the system S, and that at time $t = t' = 0$ a light ray starts from the (then coincident) origins of coordinates, $x = x' = y = y' = z = z' = 0$. At some later instant, the distance traveled by the light ray is measured by observers in the two systems. In the S system the result is

$$x^2 + y^2 + z^2 = c^2t^2; \tag{6–8a}$$

according to Einstein's postulates, the result in the S' system must be

$$x'^2 + y'^2 + z'^2 = c^2t'^2, \tag{6–8b}$$

with c, the speed of light, the same in the two systems. It is easily verified that Eqs. (6–6) do not lead from Eq. (6–8b) to Eq. (6–8a), and another transformation is needed. As a more general transformation we try the equations

$$x' = \gamma(x - vt), \qquad y' = y, \qquad z' = z, \qquad t' = ax + bt, \tag{6–9}$$

where γ, a, b are constants to be determined. The first relation is chosen because it is similar to the first of the classical equations (6–6) and should reduce to that equation when v is small; it is also desirable that the transformation equations be linear. The second and third equations are chosen because there is no reason to suppose that the relative motion in the x-direction should affect the y- and z-coordinates. The fourth relation is chosen because it is linear in x and t.

If we now insert Eqs. (6–9) into Eq. (6–8b) we get

$$(\gamma^2 - a^2c^2)\, x^2 + y^2 + z^2 = (c^2b^2 - \gamma^2v^2)\, t^2 + 2xt\,(abc^2 + v\gamma^2). \tag{6–10}$$

Since this equation is to be identical with Eq. (6–8a), we must have

$$\gamma^2 - a^2c^2 = 1, \qquad c^2b^2 - \gamma^2v^2 = c^2, \qquad abc^2 + v\gamma^2 = 0. \tag{6–11}$$

On setting $A = ac$, and $\beta = v/c$, we get

$$\gamma^2 - A^2 = 1, \tag{6–12a}$$

$$b^2 - \beta^2\gamma^2 = 1, \tag{6–12b}$$

$$Ab + \beta\gamma^2 = 0 \tag{6–12c}$$

Elimination of A between Eq. (6–12a) and Eq. (6–12c) gives

$$\gamma^2 - \frac{\beta^2\gamma^4}{b^2} = 1 = \frac{\gamma^2}{b^2}\,(b^2 - \beta^2\gamma^2), \tag{6–13}$$

and comparison with Eq. (6–12b) yields

$$\gamma^2 = b^2. \tag{6–14}$$

If we now substitute this result in Eq. (6–12b), we get

$$\gamma^2 = b^2 = \frac{1}{1 - \beta^2} = \frac{1}{1 - (v^2/c^2)}\,; \tag{6–15}$$

then

$$A^2 = a^2c^2 = \gamma^2 - 1 = \frac{\beta^2}{1 - \beta^2}\cdot \tag{6–16}$$

In finding γ, b, and A, we choose the signs of the square roots so that they are consistent with Eq. (6–12c), and lead to the necessary results $x = x'$ and $t = t'$ when $v = 0$. This procedure yields

$$\gamma = b = \frac{1}{\sqrt{1 - (v^2/c^2)}}, \tag{6–17a}$$

$$a = -\frac{v/c^2}{\sqrt{1 - (v^2/c^2)}}\cdot \tag{6–17b}$$

When these relations are inserted into Eqs. (6–9), we get the Lorentz transformation equations (6–7). The latter can be solved for x and t, with the result

$$x = \frac{x' + vt'}{\sqrt{1 - (v^2/c^2)}}, \qquad t = \frac{t' + (v/c^2)x'}{\sqrt{1 - (v^2/c^2)}}, \tag{6–18}$$

$$y = y', \qquad z = z'.$$

Equations (6–18) are exactly similar in form to Eqs. (6–7) except that v, the velocity of S' relative to S, is replaced by $-v$, the velocity of S relative to S'.

The transformation equations for the components of the velocity of a moving point can be found by taking the derivatives of the Lorentz transformation equations with respect to t and t'. If we set

$$u'_x = \frac{dx'}{dt'}, \qquad u_x = \frac{dx}{dt},$$

with similar definitions for u'_y, u'_z, u_y, and u_z, we have,

$$u'_x = \frac{dx'}{dt}\,\frac{dt}{dt'} = \frac{u_x - v}{1 - (u_x v/c^2)}, \tag{6–19a}$$

$$u'_y = \frac{dy}{dt}\,\frac{dt}{dt'} = \frac{\sqrt{1 - (v^2/c^2)}}{1 - (u_x v/c^2)}\, u_y, \tag{6–19b}$$

$$u'_z = \frac{\sqrt{1 - (v^2/c^2)}}{1 - (u_x v/c^2)}\, u_z, \tag{6–19c}$$

$$\frac{dt'}{dt} = \frac{1 - (v u_x/c^2)}{\sqrt{1 - (v^2/c^2)}}. \tag{6–19d}$$

The inverse transformation equations are again obtained by replacing v by $-v$:

$$u_x = \frac{u'_x + v}{1 + (u'_x v/c^2)}; \quad u_y = \frac{\sqrt{1 - (v^2/c^2)}}{1 + (u'_x v/c^2)}\, u'_y; \quad u_z = \frac{\sqrt{1 - (v^2/c^2)}}{1 + (u'_x v/c^2)}\, u'_z. \tag{6–20}$$

The transformation equations for the components of the velocity lead to an important addition theorem for velocities. In Newtonian kinematics, relative velocities are found by simple addition or subtraction; in relativistic kinematics, the composition of velocities is more complicated. Because of the denominator in Eqs. (6–20), the sum of two velocities which are separately less than the velocity of light c can never exceed c. Thus, in the first of Eqs. (6–20), if u'_x and v both approach c, u_x also approaches c; but for values of u'_x and v small compared with c, the denominator approaches unity and u_x approaches $u'_x + v$, the classical expression for the addition of velocities. Although the relativistic equations seem, at first to be contrary to our physical intuition, they yield correct results when applied to such problems as the Doppler effect of a light wave and the aberration of starlight.

It is not surprising, in view of the interrelationships between space and time coordinates as expressed in the Lorentz transformation equations, that relativistic mechanics differs from classical, Newtonian mechanics in certain fundamental ways. One of the most important results of the relativity theory is the deduction that the mass of a body varies with its velocity. The relationship between mass and velocity can be derived by considering a simple conceptual experiment of a type first discussed by Tolman. The derivation which follows is due to Born,[(3)] and is highly instructive because it shows how some of the basic ideas of the relativity theory are used.

Imagine an observer A located on the y-axis of a reference system S, and another observer B on the y'-axis of a second system S', with S' moving at a constant velocity v, relative to S, along the x-axis. Suppose that A throws a ball with velocity U along the y-axis, and that B throws a ball with velocity $-U$ along the y'-axis. The ball thrown by A has,

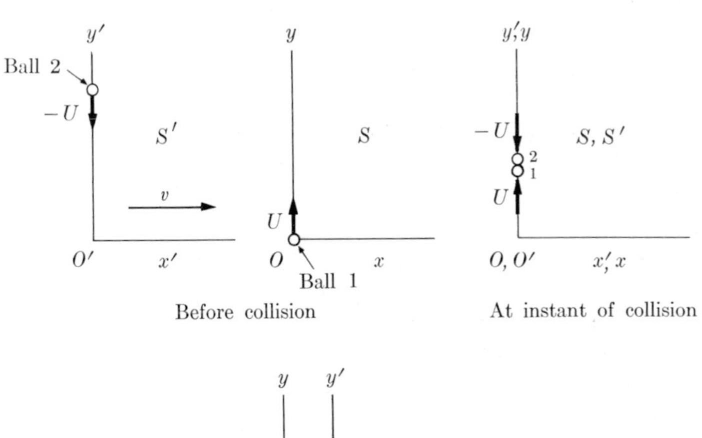

FIG. 6–4. Elastic collision of two identical balls, considered relativistically.

when at rest in the S system, the same mass as the ball thrown by B when it is at rest in the S' system. Suppose, further, that the balls are thrown so that they collide, with the line of centers at the instant of impact in the direction of the y-axis, as shown in Fig. 6–4. Let the subscript 1 denote the ball thrown by A and the subscript 2 the ball thrown by B; let u_1 and u_2 be the velocities of the balls before the collision, as measured by A, and let u_1' and u_2' be the corresponding velocities as measured by B. We use the subscript l to denote the longitudinal component of a velocity, that is, the component parallel to the x or x' direction, and the subscript t for the transverse component, parallel to the y- or y'-direction. The components of the velocity of the first ball, before the collision, as seen by the observer in the S system are then

$$u_{1l} = 0; \qquad u_{1t} = U. \tag{6–21a}$$

The components of the velocity of the second ball, before the collision, as seen by the observer in the S' system, are

$$u_{2l}' = 0; \qquad u_{2t}' = -U. \tag{6–21b}$$

The components of the velocity of the second ball relative to the S system can be obtained by using the first and second of the velocity-transformation equations (6–20), and taking $u_y' = u_{2t}' = -U$ and $u_x' = u_2' = 0$. The result is

$$u_{2l} = v; \qquad u_{2t} = -U\sqrt{1 - (v^2/c^2)}. \tag{6–21c}$$

We calculate next the total momentum of the two balls relative to the S observer. It will *not* be assumed, however, that the two balls appear to that observer to have the same mass; it will, in fact, be seen that their masses must appear different. We, therefore, denote the masses of the two balls, before the collision, and relative to the S system, by m_1 and m_2, respectively. We shall assume that, as in classical mechanics, the momentum is defined as the product of mass and velocity, and that the momentum vector has the same direction as the velocity vector. Then the total momentum before the collision, as calculated by A (the S observer), has the components

$$p_l = m_1 u_{1l} + m_2 u_{2l} = m_2 v, \tag{6–22a}$$

$$p_t = m_1 u_{1t} + m_2 u_{2t} = m_1 U - m_2 U\sqrt{1 - (v^2/c^2)}. \tag{6–22b}$$

We now consider the effect of the collision. If the balls are assumed to be perfectly smooth and to undergo an elastic collision, they cannot exert tangential forces on each other, the x-components of their velocities are not changed by the collision, and the collision must take place in a *symmetrical* way. The observer A must see his ball undergo the same change in motion that the observer B sees his ball undergo. The first ball must take on a transverse velocity $-\overline{U}$ opposite in direction to its velocity before the collision, while the second ball must take on a velocity $\overline{U}$, equal in magnitude but opposite in direction to that of the first ball. We shall not assume that the mass and velocity of either ball must be the same after the collision as before but shall see what the theory tells us. If we label quantities after the collision with bars, we have

$$\overline{u}_{1l} = 0, \qquad \overline{u}_{1t} = -\overline{U} \tag{6–23a}$$

$$\overline{u}'_{2l} = 0, \qquad \overline{u}'_{2t} = \overline{U}. \tag{6–23b}$$

The components of the velocity of the second ball as calculated by A are, from Eqs. (6–20),

$$\overline{u}_{2l} = v, \qquad \overline{u}_{2t} = \overline{U}\sqrt{1 - (v^2/c^2)}. \tag{6–23c}$$

The components of the total momentum after the collision as calculated by A are then,

$$\overline{p}_l = \overline{m}_1\overline{u}_{1l} + \overline{m}_2\overline{u}_{2l} = \overline{m}_2 v, \tag{6–24a}$$

$$\overline{p}_t = \overline{m}_1\overline{u}_{1t} + \overline{m}_2\overline{u}_{2t} = -\overline{m}_1\overline{U} + \overline{m}_2\overline{U}\sqrt{1 - (v^2/c^2)}. \tag{6–24b}$$

We now require that the *total momentum of the two balls be conserved* in the collision, in analogy with classical mechanics; this requirement is fundamental in relativistic mechanics. The longitudinal and transverse components of the momentum must be conserved separately, and comparison of Eqs. (6–22) and 6–24) yields

$$m_2 v = \overline{m}_2 v, \tag{6–25a}$$

$$m_1 U - m_2 U\sqrt{1 - (v^2/c^2)} = -\overline{m}_1\overline{U} + \overline{m}_2\overline{U}\sqrt{1 - (v^2/c^2)}. \tag{6–25b}$$

If the mass were constant, $m_1 = m_2 = \overline{m}_1 = \overline{m}_2$, the first equation would be identically correct; but the second would lead to a contradiction because it would require that

$$(U + \overline{U})\,[1 - \sqrt{1 - (v^2/c^2)}] = 0,$$

which is impossible because, by hypothesis, U and v are different from zero and positive, and $\overline{U}$ must be positive because of the symmetry of the collision.

If conservation of momentum is required, we must give up the idea that the mass of either ball is constant and replace it by the assumption that the mass is variable and depends on its velocity (since there is no other parameter in the problem on which it can depend). More precisely, we assume that the mass of a body as measured in a chosen reference system depends on the magnitude of the body's velocity relative to that system.

The magnitude of a velocity u is given by

$$u = \sqrt{u_l^2 + u_t^2}.$$

The speeds of the two balls relative to the S system are then, before the collision,

$$u_1 = U, \tag{6–26a}$$

$$u_2 = \sqrt{v^2 + U^2[1 - (v^2/c^2)]}; \tag{6–26b}$$

after the collision,

$$\overline{u}_1 = \overline{U}, \tag{6–26c}$$

$$\overline{u}_2 = \sqrt{v^2 + \overline{U}^2[1 - (v^2/c^2)]}. \tag{6–26d}$$

Now Eq. (6–25a) requires that $m_2 = \overline{m}_2$; if the mass varies only with the velocity, then m_2 and $\overline{m}_2$ can be equal only if the corresponding velocities u_2 and $\overline{u}_2$ are equal. This condition requires that

$$v^2 + U^2\left(1 - \frac{v^2}{c^2}\right) = v^2 + \overline{U}^2\left(1 - \frac{v^2}{c^2}\right),$$

from which it follows that $U = \overline{U}$. Equations (6–26a) and (6–26c) then show that $u_1 = \overline{u}_1$, that is, the speed of the first ball relative to observer A is not changed by the collision; hence, $m_1 = \overline{m}_1$, and the mass of the first ball relative to A is also not changed by the collision. Equation (6–25) can, therefore, be written,

$$m_1U - m_2U\sqrt{1 - (v^2/c^2)} = -m_1U + m_2U\sqrt{1 - (v^2/c^2)},$$

or

$$m_1 - m_2\sqrt{1 - (v^2/c^2)} = 0.$$

It follows that

$$m_2 = \frac{m_1}{\sqrt{1 - (v^2/c^2)}}. \tag{6–27a}$$

According to Eq. (6–27a), the two balls, which were assumed to have the same mass when at rest in their respective reference systems, appear to have different masses when one reference system moves with constant velocity v relative to the other. When the system S' moves with constant velocity v relative to S, the mass m_2 appears larger than m_1 to an observer stationary in S by the factor $1/\sqrt{1 - (v^2/c^2)}$. If we now imagine that the velocity U with which the balls are thrown is made smaller, then, according to Eqs. (6–26), $u_1 = 0$ and $u_2 = v$ in the limit $U = 0$. The velocity U is made to approach zero only to simplify the derivation and is not a necessary assumption (see Problem 8). The mass m_1, which corresponds to zero velocity is called the *rest mass* and is denoted by m_0, while m_2 is the mass which corresponds to velocity v, and is denoted by m. We have, finally,

$$m = \frac{m_0}{\sqrt{1 - (v^2/c^2)}}. \tag{6–27b}$$

The derivation has shown that when a body moves with constant velocity v with respect to a stationary observer, it has a mass which depends on the velocity and is larger than the rest mass m_0 characteristic of the body. Equation (6–27b) shows that no material body can have a velocity equal to, or greater than the velocity of light.

Equation (6–27b) had been obtained earlier for the mass of an electron; according to the relativity theory, it holds for any moving body. The variation in mass becomes important when the speed of the body begins to approach that of light, as is evident from the following values.

v/c	m/m_0	v/c	m/m_0
0.00	1.000	0.90	2.294
0.01	1.000	0.95	3.203
0.10	1.005	0.98	5.025
0.50	1.155	0.99	7.089
0.75	1.538	0.998	15.819
0.80	1.667	0.999	22.366

For bodies moving at speeds up to about one percent that of light, the variation in mass is negligible, but in the case of more swiftly moving particles the correction is appreciable. For this reason, β-rays from radioactive substances have been used to test the relativistic formula for the mass.

One of the methods used to test the formula depends on the measurement of the value of the specific charge, e/m, as a function of the speed of the beta particles. According to the relativity theory, the electric charge is a quantity which is independent of velocity and has the same value for all observers. Consequently, the value of e/m should depend on the velocity only because of the dependence of the mass on the velocity, and

$$\frac{e}{m} = \frac{e}{m_0}\sqrt{1 - \frac{v^2}{c^2}}, \tag{6–28a}$$

or

$$\frac{e}{m_0} = \frac{e/m}{\sqrt{1 - (v^2/c^2)}}, \tag{6–28b}$$

where e/m_0 is the value corresponding to the rest mass of the electron. If e/m and v are measured, the measured value of e/m divided by $\sqrt{1 - (v^2/c^2)}$ should be a constant, which is e/m_0, and the mass ratio m/m_0 is also obtained.

Several experiments of this type have been made,[(4)] the most famous of which is probably that of Bucherer in 1909. In this experiment, a small radioactive source S of β-rays was placed at the center between two very closely spaced metal discs forming an electrical condenser C. The condenser was placed inside a cylindrical box which could be evacuated, and a photographic film P was wrapped around the inner surface of the box. The box was placed in a homogeneous magnetic field H parallel to the plane of the discs, as shown schematically in Fig. 6–5(a). In the region of the electric field between the discs the electric and magnetic deflecting forces must balance each other exactly for a particle to escape. The condition for escape is $eE = evH$, or $v = E/H$, and the velocity of an emerging β-particle can be determined. The speed of an emerging β-particle is a function of the azimuthal angle θ at which it is emitted, and only particles with a single speed v emerge at a given angle θ, as indicated in Fig. 6–5(b). After emergence from between the plates, the β-particle is deflected in the magnetic field alone, and the result of many particles over a wide range of

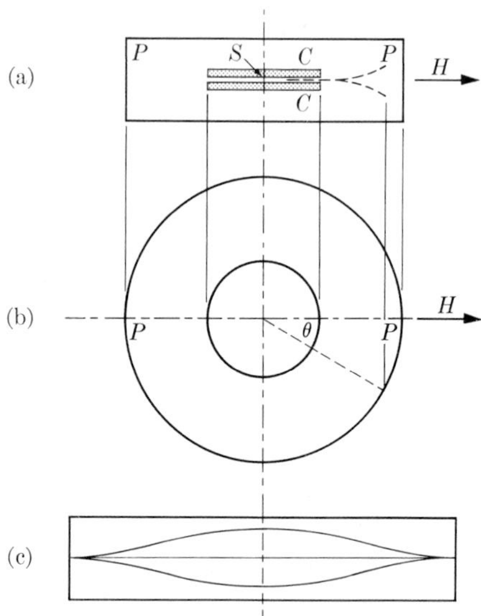

FIG. 6–5. Schematic view of Bucherer's apparatus for measuring e/m for electrons with different speeds. (a) elevation view (b) plan view (c) traces on photographic film. (Reprinted by permission from E. R. Cohen, K. M. Crowe, and J. W. M. Dumond, *Fundamental Constants of Physics*. New York: Interscience, 1957.)

speeds and emerging at different angles θ is a continuous curve on the photographic film. A typical trace corresponding to half the circumference of the box is shown by one of the curves in Fig. 6-5(c); the second curve on the film is formed when the electric and magnetic fields are reversed simultaneously. The value of e/m corresponding to a velocity v is obtained from the relation

$$Hev = \frac{mv^2}{r}, \qquad \text{or} \qquad \frac{e}{m} = \frac{v}{rH};$$

if we insert for v its value $v = E/H$, then

$$\frac{e}{m} = \frac{E}{rH^2}.$$

The field strengths E and H are known and r, the radius of the circular path of a β-particle of speed v, can be obtained from an analysis of the photographic trace and geometrical relationships among the various parts of the apparatus.

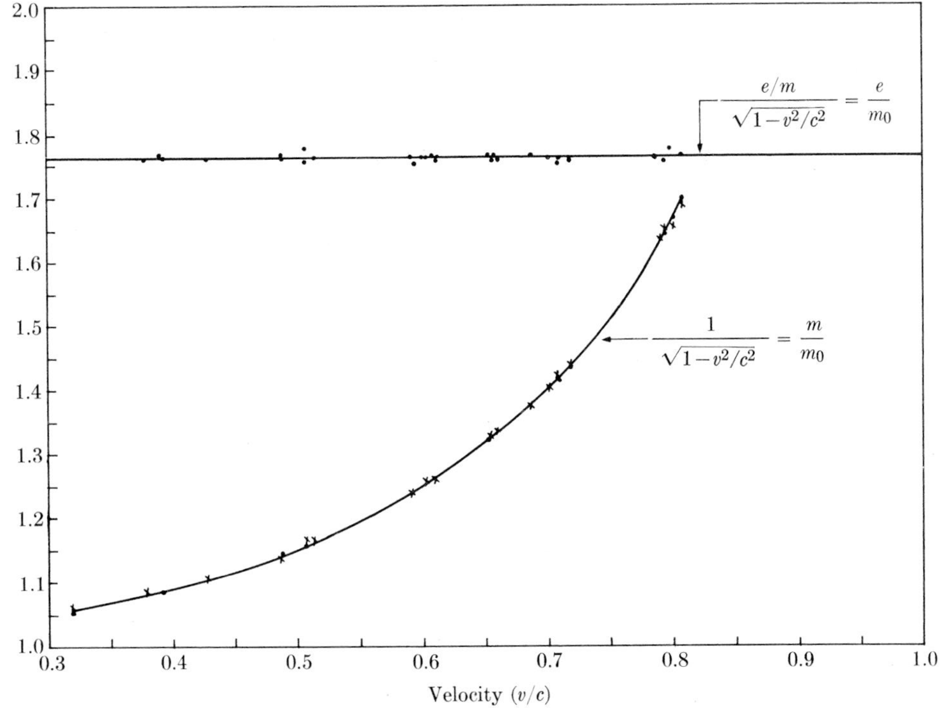

FIG. 6-6. The specific charge and mass of the electron as functions of velocity. Verification of the relativistic mass formula; e/m_0 is in units of 10^7 emu/gm.

Later experiments have used improved methods[5,6] and some of the results obtained by Bucherer and others are shown in Fig. 6–6. The solid curves represent the theoretical expressions for m and e/m_0 as functions of velocity, and the points represent the experimental results. There is no doubt that the theoretical predictions are confirmed for swiftly moving β-particles. The mass formula has also been confirmed for other particles.

6–6 The relationship between mass and energy. The variation of mass with velocity leads to modifications of our ideas about energy. In classical physics, force is defined as the time rate of change of momentum, and

$$F = \frac{d}{dt}(mv) = m\frac{dv}{dt}, \tag{6–29}$$

since mass is constant. This definition of force is kept in relativistic mechanics, but now mass is also a variable and

$$F = \frac{d}{dt}(mv) = m\frac{dv}{dt} + v\frac{dm}{dt}. \tag{6–30}$$

The kinetic energy T of a particle is defined in terms of force and distance, as in classical mechanics, and can be found in the usual way by calculating the work done in setting the particle into motion:

$$dT = F\,dx. \tag{6–31}$$

On insertion of Eq. (6–30) for F, there follows

$$\begin{aligned} dT &= m\frac{dv}{dt}\,dx + v\frac{dm}{dt}\,dx \\ &= mv\,dv + v^2\,dm. \end{aligned} \tag{6–32}$$

Differentiating both sides of Eq. (6–27b) we get

$$\begin{aligned} dm &= \frac{m_0}{c^2}\,\frac{v\,dv}{[1 - (v^2/c^2)]^{3/2}} \\ &= \frac{mv\,dv}{c^2 - v^2}. \end{aligned} \tag{6–33}$$

Substitution for $mv\,dv$ from Eq. (6–33) in Eq. (6–32) yields

$$dT = c^2\,dm. \tag{6–34}$$

Thus, a change in kinetic energy dT is directly proportional to a change in mass dm, and the proportionality factor is the square of the velocity of light. The integral form of this result can also be obtained easily. If we substitute for dm from Eq. (6–33) in Eq. (6–32), we get

$$dT = \frac{m_0 v}{\sqrt{1 - (v^2/c^2)}}\, dv + \frac{m_0 v^3/c^2}{[1 - (v^2/c^2)]^{3/2}}\, dv$$

$$= \frac{m_0 v\, dv}{[1 - (v^2/c^2)]^{3/2}}.$$

Integration of both sides of the last equation gives for the kinetic energy

$$T = m_0 c^2 \left[\frac{1}{\sqrt{1 - (v^2/c^2)}} - 1\right]$$

$$= mc^2 - m_0 c^2 = c^2(m - m_0). \tag{6–35}$$

According to this result, the kinetic energy of a body can be expressed in terms of the increase in the mass of the body over the rest mass. This relationship may be interpreted as meaning that the rest mass m_0 is associated with an amount of energy m_0c^2, which may be called the *rest energy* of the body. The *total energy* E of the body is then the sum of the kinetic energy and the rest energy, or

$$E = T + m_0 c^2$$

$$= mc^2 = \frac{m_0 c^2}{\sqrt{1 - (v^2/c^2)}}. \tag{6–36}$$

Hence, associated with a mass m, there is an amount of energy mc^2; conversely, to an energy E, there corresponds a mass given by

$$m = \frac{E}{c^2}. \tag{6–37}$$

Another important consequence of the proportionality between mass and energy has to do with the momentum associated with the transfer of energy. If a quantity of energy is transferred with a velocity v, then we can write for the magnitude of the associated momentum p

$$p = mv = \frac{Ev}{c^2}. \tag{6–38}$$

This relationship can be applied, for example, to the energy and momentum carried by a light quantum (photon). The energy carried by a photon is $h\nu$ and its velocity is c. The momentum is, by Eq. (6–38),

$$p = \frac{E}{c^2}\, c = \frac{E}{c} = \frac{h\nu}{c}. \tag{6–39}$$

The rest mass of a photon must be zero; otherwise the mass of a photon traveling with the velocity of light would be infinite, because the denominator of Eq. (6–27) becomes zero for $v = c$. Although the rest mass of a photon is zero, the photon has a *mass associated with its kinetic energy*, according to Eq. (6–37), and this mass is $h\nu/c^2$; corresponding to this mass is the momentum $h\nu/c$.

The interconversion of mass and energy can illustrated by several numerical examples which will be useful in later calculations. If the mass is in grams and the velocity is in centimeters per second, the energy is in ergs, and

$$\begin{aligned} E(\text{ergs}) &= m(\text{gm}) \times (2.998 \times 10^{10})^2 \text{ cm}^2/\text{sec}^2 \\ &= m(\text{gm}) \times 8.99 \times 10^{20} \text{ cm}^2/\text{sec}^2. \end{aligned} \tag{6–40}$$

One gram of mass is associated with nearly 10^{21} ergs of energy, or about 8.5×10^{10} Btu, since 1 Btu $= 1.055 \times 10^{10}$ ergs. This is a very large amount of energy indeed, but the practical conversion of mass to energy is a difficult problem. In an ordinary chemical process, mass is always converted into energy. For example, if 100 gm of a chemical substance take part in a reaction, with the result that 10^6 calories of heat are liberated, the mass-energy balance is (1 cal $= 4.184 \times 10^7$ ergs)

$$\begin{aligned} \Delta m &= \frac{\Delta E}{c^2} = \frac{4.184 \times 10^7 \times 10^6}{8.99 \times 10^{20}} \\ &= 4.65 \times 10^{-8} \text{ gm}. \end{aligned}$$

The loss of mass equivalent to the energy involved in the reaction is 4.65×10^{-10} gm/gm of substance, a decrease which cannot be detected by even the most sensitive chemical balances. This is the reason why the mass-energy effect has not yet been observed in chemical reactions.

It will be seen in later chapters that the mass-energy conversion can be detected in nuclear phenomena because the amounts of energy involved per atom are very much greater than those involved in chemical reactions. It is useful to obtain now some of the conversion factors which will be needed in later work. In atomic and nuclear calculations, the erg is an inconvenient unit of energy because the amounts of energy involved in single events usually are only small fractions of an erg, say 10^{-12} to 10^{-6} erg. It is therefore the practice to express energies in units of *electron volts*, abbreviated as ev. The electron volt is the energy acquired by any charged particle carrying a unit electronic charge when it falls through a potential of one volt; it is equivalent to 1.602×10^{-12} erg. For convenience, two other units are also used; one, equal to 10^3 electron volts, is represented by kev and the other, equal to 10^6 electron volts, is abbreviated by Mev. One kev is therefore equal to 1.602×10^{-9} erg, and one Mev is equal to 1.602×10^{-6} erg. The mass-energy relation becomes

$$E\text{ (ev)} = m(\text{gm}) \times \frac{8.99 \times 10^{20}\text{ cm}^2/\text{sec}^2}{1.602 \times 10^{-12}\text{ erg/ev}}$$

$$= m(\text{gm}) \times 5.61 \times 10^{32}, \tag{6–41a}$$

$$E\text{ (kev)} = m(\text{gm}) \times 5.61 \times 10^{29}, \tag{6–41b}$$

$$E\text{ (Mev)} = m(\text{gm}) \times 5.61 \times 10^{26}. \tag{6–41c}$$

The rest mass of an electron is 9.108×10^{-28} gm; by Eq. (6–41c), the rest energy is 0.511 Mev. For a hydrogen atom, the rest mass is 1.674×10^{-24} gm and the corresponding energy is 939 Mev. It is often convenient to use the energy corresponding to one atomic mass unit. This mass is 1.66×10^{-24} gm, or 931.141 Mev.

$$E\text{ (Mev)} = m(\text{atomic mass units}) \times 931.141. \tag{6–42}$$

The values of the kinetic energy which will be used most often range from electron volts to millions of electron volts. For example, an α-particle may have a velocity of 2×10^9 cm/sec, or about one-fifteenth of the velocity of light. The relativistic mass correction for this velocity is sufficiently small so that it may be safely neglected. The rest mass m_α of an α-particle is very closely four times that of a hydrogen atom. Hence, the kinetic energy of this α-particle is

$$\tfrac{1}{2}m_\alpha v^2 = \tfrac{1}{2} \times 4 \times 1.674 \times 10^{-24} \times (2 \times 10^9)^2\text{ erg}$$

$$= 13.39 \times 10^{-6}\text{ erg}$$

$$= \frac{13.39 \times 10^{-6}}{1.602 \times 10^{-6}}\text{ Mev} = 8.36\text{ Mev}.$$

6–7 The Compton effect. Some of the ideas treated in Chapter 5 and in the present chapter can be illustrated by discussing an effect discovered by A. H. Compton in 1923 during the course of his studies of the scattering of x-rays by matter. The theoretical treatment of the Compton effect involves Einstein's quantum theory of light as well as some of the ideas of the relativity theory. The effect is also of practical importance in the absorption of x-rays by matter and will be treated in further detail from that viewpoint in Chapter 15. Finally, some of the ideas used by Compton in his explanation of the effect play an important part in the modern quantum theory of atomic structure and will be needed in Chapter 7. For all of these reasons, it is convenient to discuss the Compton effect now.

In the treatment of the scattering of x-rays by electrons in Chapter 4, it was assumed that the wavelength of the scattered radiation was the same as that of the incident radiation. This kind of scattering is called "Thomson" scattering; its effect is to change the direction of the incident radiation. With improved techniques, Compton (1923) was able to show that when a beam of monochromatic x-rays was scattered by a light element such as carbon, the scattered radiation consisted of two components, one of the same wavelength as that of the incident beam, the second of slightly longer wavelength. The wavelengths of the two lines were measured with a Bragg x-ray spectrometer. The difference in wavelength $\Delta\lambda$ between the two scattered radiations was found to vary with the scattering angle and to increase rapidly at large angles of scattering. When the angle between the incident and scattered radiation was 90°, the difference in wavelength was found to be 0.0236×10^{-8} cm, independent of the wavelength of the primary beam and of the nature of the scattering material.

To account for the presence of the shifted component and for the amount of the shift, Compton assumed that the scattering process could be treated as an elastic collision between a photon and a free electron, and that in this collision energy and momentum are conserved. According to Einstein's quantum theory of the photoelectric effect, the primary x-rays, being electromagnetic radiations like light, are propagated as quanta with energy $h\nu$. Along with their energy, they carry momentum $h\nu/c$ in accordance with the ideas developed in the last section. The scattered quantum or photon moves in a different direction from that of the primary photon and carries a different momentum. In order that momentum be conserved, the electron which scatters the photon must recoil with a momentum equal to the vector difference between that of the incident and that of the scattered photon, as in Fig. 6–7. The energy of this recoiling electron is taken from that of the primary photon, leaving a scattered photon which has less energy—and hence a lower frequency, or longer wavelength—than that of the primary photon.

The requirement that energy be conserved gives

$$h\nu_0 = h\nu + m_0c^2\left[\frac{1}{\sqrt{1-(v^2/c^2)}} - 1\right], \tag{6–43}$$

where ν_0 is the frequency of the incident x-ray, ν that of the ray scattered by the electron, the recoil velocity of the electron is v, and the relativistic kinetic energy of the electron is used because the velocity of the electron may be great enough for relativistic effects to be significant. The require-

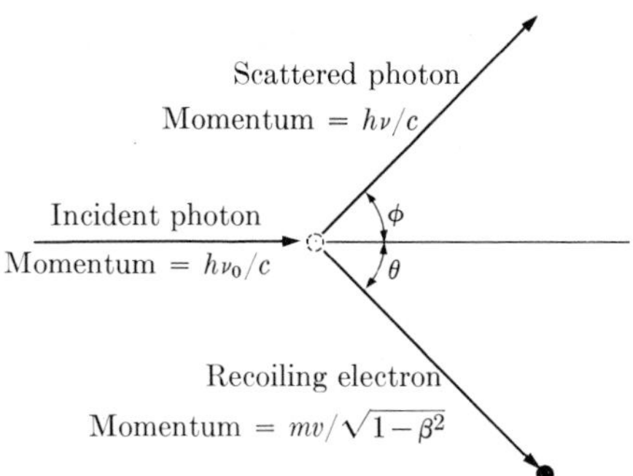

FIG. 6-7. The Compton effect.

ment that momentum be conserved gives two equations, one for the x-component of the momentum, and one for the y-component

$$x\text{-component:}\quad \frac{h\nu_0}{c} = \frac{h\nu}{c}\cos\phi + \frac{m_0 v}{\sqrt{1-(v^2/c^2)}}\cos\theta, \tag{6-44}$$

$$y\text{-component:}\quad 0 = \frac{h\nu}{c}\sin\phi - \frac{m_0 v}{\sqrt{1-(v^2/c^2)}}\sin\theta. \tag{6-45}$$

To solve these equations, it is convenient to set

$$\beta = v/c, \tag{6-46}$$

and rewrite the last three equations. This gives

$$h\nu_0 = h\nu + m_0c^2\left[\frac{1}{\sqrt{1-\beta^2}} - 1\right], \tag{6-47}$$

$$\frac{h\nu_0}{c} = \frac{h\nu}{c}\cos\phi + \frac{m_0\beta c}{\sqrt{1-\beta^2}}\cos\theta, \tag{6-48}$$

$$0 = \frac{h\nu}{c}\sin\phi - \frac{m_0\beta c}{\sqrt{1-\beta^2}}\sin\theta. \tag{6-49}$$

For a particular angle of scattering ϕ, the last three equations contain three unknowns, ν, β, and θ, and expressions for these quantities can be found by a straightforward solution of the equations. For comparison with experiment, however, we are more interested in the shift in wavelength. We therefore introduce the wavelengths

$$\lambda_0 = \frac{c}{\nu_0} \quad \text{and} \quad \lambda = \frac{c}{\nu}$$

of the incident and scattered radiations, respectively. Then Eqs. (6–48) and (6–49) become:

$$\frac{h}{\lambda_0} - \frac{h}{\lambda}\cos\phi = \frac{m_0\beta c}{\sqrt{1-\beta^2}}\cos\theta,$$

$$\frac{h}{\lambda}\sin\phi = \frac{m_0\beta c}{\sqrt{1-\beta^2}}\sin\theta.$$

Squaring these two equations and adding, we have

$$\frac{h^2}{\lambda_0^2} + \frac{h^2}{\lambda^2} - \frac{2h^2\cos\phi}{\lambda_0\lambda} = \frac{m_0^2\beta^2c^2}{1-\beta^2} = \frac{m_0^2c^2}{1-\beta^2} - m_0^2c^2. \qquad (6\text{–}50)$$

Similarly, Eq. (6–47) can be written

$$\frac{h}{\lambda_0} - \frac{h}{\lambda} + m_0c = \frac{m_0c}{\sqrt{1-\beta^2}}\cdot$$

On squaring, we get

$$\frac{h^2}{\lambda_0^2} + \frac{h^2}{\lambda^2} - \frac{2h^2}{\lambda_0\lambda} + 2m_0ch\left(\frac{1}{\lambda_0} - \frac{1}{\lambda}\right) + m_0^2c^2 = \frac{m_0^2c^2}{1-\beta^2}\cdot \qquad (6\text{–}51)$$

Subtracting Eq. (6–50) from Eq. (6–51), we get

$$\frac{2h^2}{\lambda_0\lambda}(\cos\phi - 1) + 2m_0ch\left(\frac{1}{\lambda_0} - \frac{1}{\lambda}\right) = 0.$$

Therefore,

$$\Delta\lambda = \lambda - \lambda_0 = \frac{h}{m_0c}(1 - \cos\phi). \qquad (6\text{–}52)$$

If the values $h = 6.624 \times 10^{-27}$, $m = 0.9107 \times 10^{-27}$, and $c = 2.998 \times 10^{10}$ are inserted, then

$$\begin{aligned}\Delta\lambda = \lambda - \lambda_0 &= 0.0242 \times 10^{-8}(1 - \cos\phi)\text{ cm}\\ &= 0.0242(1 - \cos\phi)\text{ A}. \qquad (6\text{–}53)\end{aligned}$$

Equation (6–53) states than when an incident x-ray of wavelength λ_0 is scattered through an angle ϕ by a free electron, the wavelength λ of the scattered x-ray should be greater than that of the incident x-ray by the amount $0.0242(1 - \cos\phi)$ A. For a given value of the scattering angle ϕ, the shift in wavelength is independent of the wavelength of the incident radiation. For $\phi = 90°$, $\Delta\lambda = 0.0242$ A, which agrees very well with the observed value of 0.0236 A. The predicted dependence of $\Delta\lambda$ on the angle ϕ was also verified by experiment.

The kinetic energy T of the recoil electron can be calculated without much difficulty, and it is found that

$$T = h\nu_0 \frac{(1 - \cos\phi)\alpha}{1 + \alpha(1 - \cos\phi)}, \tag{6–54}$$

where $\alpha = h\nu_0/m_0c^2$.

The recoil electrons were sought and found, and their observed energies agreed with the values predicted by the theory.

The theory just developed explains the observed shift in wavelength, but does not account for the presence of the unshifted line. It will be recalled that it was assumed that the electrons are free. Actually, electrons are bound to atoms more or less tightly, and a certain amount of energy is needed to shake an electron loose from an atom. If the amount of energy given to the electron by the photon is much larger than the work needed to detach it from the atom, the electron acts like a free electron, and Eq. (6–53) is valid. If the collision is such that the electron is not detached from the atom, but remains bound, the rest mass m_0 of the electron in Eq. (6–53) must be replaced by the mass of the atom, which is several thousand times greater. The calculated value of $\Delta\lambda$ then becomes much too small to be detected. A photon which collides with a bound electron, therefore, does not have its wavelength changed, and this accounts for the presence of the unshifted spectral line.

It will be noted that in the Compton formula, Eq. (6–52), the only trace of the relativity theory that appears is the zero subscript denoting the rest mass of the electron. It would seem as though the same effect would have been obtained if the problem had been set up without taking into account the relativistic effects, and this is actually the case so far as the shift in wavelength is concerned. The more refined details of the Compton effect, however, demand a rigorous relativistic treatment. Hence, even the more elementary problems are usually set up relativistically.

References

GENERAL

A. Einstein, *Relativity: The Special and General Theory*, 15th ed. London: Methuen, 1954; New York: Crown Publishers, 1961.

A. Einstein and L. Infeld, *The Evolution of Physics*. New York: Simon and Schuster, 1938, Part III.

A. Einstein, *The Meaning of Relativity*, 5th ed. Princeton: Princeton University Press, 1956.

C. V. Durell, *Readable Relativity*. London: G. Bell and Sons, 1926; New York: Harper and Brothers (Harper Torch Books/The Science Library), 1960.

A. EDDINGTON, *Space, Time and Gravitation.* Cambridge: University Press, 1920; New York: Harper and Brothers (Harper Torch Books/The Science Library), 1959.

H. DINGLE, *The Special Theory of Relativity.* London: Methuen and Co., 1940; New York: Wiley.

W. H. MCCREA, *Relativity Physics,* 4th ed. London: Methuen and Co., 1954; New York: Wiley.

P. G. BERGMANN, *Introduction to the Theory of Relativity.* New York: Prentice-Hall, 1942.

C. MOLLER, *Theory of Relativity.* Oxford University Press, 1952.

W. PAULI, *Theory of Relativity.* London: Pergamon Press, 1958.

R. C. TOLMAN, *Relativity, Thermodynamics, and Cosmology.* Oxford University Press, 1934.

The Principle of Relativity, A Collection of Memoirs on The Special and General Theory of Relativity, by H. A. Lorentz, A. Einstein, H. Minkowski, and H. Weyl. London: Methuen and Co., 1923; New York: Dover Publishing Co.

W. K. H. PANOFSKY and M. PHILLIPS, *Classical Electricity and Magnetism.* Reading, Mass.: Addison-Wesley, 1955, Chapters 14, 15, 16.

A. H. COMPTON and S. K. ALLISON, *X-Rays in Theory and Experiment,* 2nd ed. New York: Van Nostrand, 1935, Chapter 3.

PARTICULAR

1. W. K. H. PANOFSKY and M. PHILLIPS, *op. cit.* gen. ref., Chapter 14. "The experimental basis for the theory of special relativity.")
2. CEDARHOLM, BLAND, HAVENS, and TOWNES, "New Experimental Test of Special Relativity," *Phys. Rev. Letters,* **1**, 342 (1958).
3. M. BORN, *Die Relativitätstheorie Einsteins und ihre Physikalische Grundlagen.* Dritte Auflage. Berlin: Springer, 1922, Chapter 7, Section 7.
4. P. S. FARAGO and L. JANOSSY, "Review of the Experimental Evidence for the Law of Variation of the Electron Mass with Velocity", *Nuovo cimento,* Vol. V, No. 6, 1411 (1957).
5. COHEN, CROWE, and DUMOND, *Fundamental Constants of Physics.* New York: Interscience, 1957, pp. 130–142.
6. ROGERS, MCREYNOLDS, and ROGERS, "A Determination of the Masses and Velocities of Three Radium B β-particles; The Relativistic Mass of the Electron," *Phys. Rev.* **57,** 379 (1940).

PROBLEMS

1. Suppose that a clock at rest in the reference system S emits signals at time intervals $\Delta t = t_2 - t_1$. At what intervals will the signals appear to an observer at rest in the system S' moving along the x-axis with uniform velocity v with respect to the S system? Suppose next that there is a source of homogeneous x-rays at rest in the S' system, and that the wavelength of the x-rays as measured by the S' and S observers is λ' and λ, respectively. What is the relationship between λ' and λ?

2. Derive an expression for the quantity $u'^2 = u_x'^2 + u_y'^2 + u_z'^2$ in terms of u_x, u_y, and u_x. Show that if $v < c$, and $u = c$, then u' is also equal to c.

3. Suppose that two electrons, ejected by a heated filament stationary in S, move off with equal speeds of magnitude $0.9c$, one moving parallel to the $+x$-direction, the second parallel to the $-x$-direction. What is their speed relative to each other in the S system? Suppose that the S' system moves at a speed $v = 0.9c$ in the $-x$-direction, so that it keeps up with the second electron. What is the velocity of the second electron relative to the first, as measured in the S'system?

4. Calculate the ratio of the mass of a particle to its rest mass when the particle moves with speeds that are the following fractions of the speed of light: 0.1, 0.2, 0.5, 0.7, 0.9, 0.95, 0.99, 0.995, 0.999, 0.9999.

5. Calculate the kinetic energy, in ergs and in Mev, of the particle of Problem 5 at each speed; assume first that the particle is an electron, then that it is a proton.

6. Show, by expanding the expression $(1 - v^2/c^2)^{-1/2}$ in powers of v/c, that the kinetic energy can be written

$$T = \frac{1}{2} m_0 v^2 + \frac{3}{8} m_0 \frac{v^4}{c^2} + \cdots.$$

Find the values of v/c for which the second term of the expression is 1%, 10%, and 50%, respectively, of the first term.

7. Charged particles can be accelerated to high energies by appropriate machines (Chapter 21). A constant frequency cyclotron can accelerate protons to kinetic energies of about 20 Mev; synchrocyclotrons have been built which accelerate protons to kinetic energies of about 400 Mev, and the Cosmotron at the Brookhaven National Laboratory has produced 2.5 Bev protons. What is the mass of the protons relative to the rest mass in each of the three cases cited? What is the velocity relative to that of light? What is the total energy?

8. In connection with the derivation of the formula for the variation of mass with velocity, show that Eq. (6–27a),

$$m_2 = \frac{m_1}{\sqrt{1 - (v^2/c^2)}},$$

is valid generally, for $U \neq 0$.

9. Derive the following useful relationships concerning the relativistic properties of moving particles (the symbols are defined in Section 6–6).

(a) $(pc)^2 = (mc^2)^2 - (m_0c^2)^2$ (b) $p = \frac{1}{c}\sqrt{T^2 + 2m_0cT}$

(c) $m_0 = \frac{1}{c^2}\sqrt{E^2 - p^2c^2}$ (d) $T = \sqrt{(m_0c^2)^2 + (pc)^2} - m_0c^2$

10. Consider a body with rest mass M in a given frame of reference; the body is composed of two parts with rest masses M_1 and M_2. Suppose that the body breaks up spontaneously into its two parts with velocities v_1 and v_2, respectively.

(a) Write an equation for the conservation of *total* energy. (b) Show that the total energies of the two parts are given by

$$E_1 = c^2 \frac{M^2 + M_1^2 - M_2^2}{2M}; \qquad E_2 = c^2 \frac{M^2 - M_1^2 + M_2^2}{2M}.$$

11. A particle of rest mass m_1 and velocity v collides with a particle of mass m_2 at rest, after which the two particles coalesce. Show that the mass M and velocity V of the composite particle are given by

$$M^2 = m_1^2 + m_2^2 + \frac{2m_1m_2}{\sqrt{1-(v^2/c^2)}}; \qquad V = \frac{m_1 v}{m_1 + m_2\sqrt{1-(v^2/c^2)}}.$$

12. Show that the following relations hold in a Compton scattering process. (a) The maximum energy transfer to the electron is given by

$$T_{\max} = \frac{h\nu_0}{1 + (1/2\alpha)}$$

and

$$h\nu_0 = \frac{1}{2} T_{\max}\left(1 + \sqrt{1 + \frac{2m_0c^2}{T_{\max}}}\right)$$

(b) For very large energy of the incident photon, $\alpha \gg 1$, the energy of the backscattered photon approaches $\frac{1}{2} m_0c^2$.

(c) The angles θ and ϕ, which the directions of the recoil electron and scattered photon make with respect to that of the incident neutron, obey the relation

$$\cot\theta = (1+\alpha)\frac{1-\cos\phi}{\sin\phi} = (1+\alpha)\tan\frac{\phi}{2}.$$

(d) Show also that the *Compton wavelength*, h/m_0c, is the wavelength of radiation having a quantum energy equivalent to the rest energy of an electron.

13. What are the greatest wavelength and energy a photon can have and be able to transfer half its energy to the recoil electron during a Compton collision? What are the direction and magnitude of the velocity of the recoil electron in this case? How important is the relativistic variation of mass with velocity in this problem?

14. A beam of homogeneous x-rays with a wavelength of 0.0900 A is incident on a carbon scatterer. The scattered rays are observed at an angle of 54° with the direction of the incident beam. Find (a) the wavelength of the scattered rays, (b) the energies of the incident and scattered photons, (c) the momenta of the incident and scattered photons, (d) the energy, momentum, and velocity of the recoil electron, (e) the angle at which the electron recoils.

CHAPTER 7

ATOMIC SPECTRA AND ATOMIC STRUCTURE

7–1 Atomic spectra. The first problem that must be solved by a theory of atomic structure is that of the line spectra of the elements. This problem was discussed briefly in the fifth chapter; it will now be treated in greater detail. A *line,* or *emission, spectrum* is produced when light from a gas through which an electric discharge is passing, or from a flame into which a volatile salt has been put, is dispersed by a prism or grating spectrometer. Instead of a continuous band of colors, as in a rainbow or as in the spectrum from an incandescent solid, only a few colors appear, in the form of bright, isolated, parallel lines. Each line is an image of the spectrometer slit, deviated through an angle which depends on the frequency of the light forming the image. The wavelengths of the lines are characteristic of the element emitting the light, and each element has its own particular line spectrum, which is a property of the atoms of that element. For this reason, line spectra are also called *atomic spectra.* The lighter atoms such as hydrogen and helium yield fairly simple spectra with a relatively small number of lines, but for some of the heavier atoms the spectra may consist of hundreds of lines. The lines may be in the ultraviolet or infrared as well as in the visible range of wavelengths.

The spectrum of atomic hydrogen is especially interesting for historical and theoretical reasons. In the visible and near ultraviolet regions, it consists of a series of lines whose relative positions are shown in Fig. 7–1. There is an apparent regularity in the spectrum, the lines coming closer together as the wavelength decreases until the limit of the series is reached at 3646 A. In 1885, Balmer showed that the wavelengths of the nine lines then known in the spectrum of hydrogen could be expressed by the very simple formula

$$\lambda = B\,\frac{n^2}{n^2 - 4}, \tag{7–1}$$

where B is a constant equal to 3646 A, and n is a variable integer which takes on the values 3, 4, 5, . . . , for the H_α, H_β, H_γ, . . . lines, respectively. Rydberg, a few years later, found that the lines of the visible and near ultraviolet hydrogen spectrum, now called the Balmer series, can be described in a more useful way in terms of the wave number, or reciprocal of the wavelength. The wave number is the number of waves in one centimeter of light path in vacuum and is expressed in reciprocal centimeters, cm^{-1}. For example, the wavelength of the H_α line is 6562.8×10^{-8}

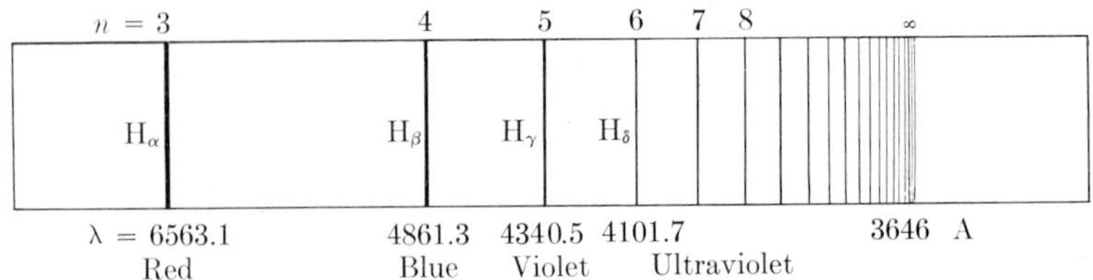

FIG. 7–1. Diagram of the lines of the Balmer series of atomic hydrogen.

cm, and the wave number $\bar{\nu} = 1/\lambda = 15237.4\ \text{cm}^{-1}$. According to Rydberg, the wave numbers of the Balmer lines are given by

$$\bar{\nu} = R_{\text{H}}\left(\frac{1}{2^2} - \frac{1}{n^2}\right), \tag{7–2}$$

where R_{H} is a constant with the dimension cm^{-1}, called the *Rydberg constant for hydrogen,* and n takes on the values 3, 4, 5, ... The value of the Rydberg constant for hydrogen has been determined empirically from spectroscopic measurements and is

$$R_{\text{H}} = 109{,}677.576 \pm 0.012\ \text{cm}^{-1}; \tag{7–3}$$

with this value of R_{H}, the wave numbers and thus the wavelengths of the lines of the Balmer series are given with great accuracy, as can be seen

TABLE 7–1

COMPARISON OF OBSERVED AND CALCULATED WAVELENGTHS OF THE LINES OF THE BALMER SERIES IN THE HYDROGEN SPECTRUM

Line	n	Wavelength measured in air, λ: A	Measured wavelength corrected to vacuum, λ: A	Wavelength calculated from the Rydberg formula
H_α	3	6562.85	6564.66	6564.70
H_β	4	4861.33	4862.68	4862.74
H_γ	5	4340.47	4341.69	4341.73
H_δ	6	4101.74	4102.89	4102.93
H_ϵ	7	3970.07	3971.19	3971.23
H_ζ	8	3889.06	3890.16	3890.19
H_η	9	3835.40	3836.48	3836.51
H_θ	10	3797.91	3798.98	3799.01
H_ι	11	3770.63	3771.85	3771.74
H_κ	12	3750.25	3751.31	3751.26
H_λ	13	3734.37	3735.43	3735.46
H_μ	14	3721.95	3723.00	3723.03
H_ν	15	3711.98	3713.03	3713.06

from Table 7-1. The wavelength measured in air has been corrected to vacuum by means of the relation $\lambda_{\text{vac}} = \mu\lambda_{\text{air}}$, where μ, the index of refraction of air, is a known, tabulated function of wavelength. The wavelength in vacuum is then compared with the value calculated from the Rydberg formula.

The results illustrate the accuracy with which spectroscopic measurements can be made as well as the remarkable agreement between the values of the wavelengths measured experimentally and those given by the simple Rydberg formula with the empirically determined value of the Rydberg constant for hydrogen. They also show that a theory of atomic structure, in order to be acceptable, must predict the values of the wavelengths and the Rydberg constant with great accuracy.

In addition to the Balmer series, the hydrogen spectrum contains four other series, one of which is in the ultraviolet region, and three in the infrared. Thus, for hydrogen, the following spectral series, named after their discoverers, are now known.

$$\bar{\nu} = R_{\text{H}}\left(\frac{1}{1^2} - \frac{1}{n^2}\right), \quad n = 2, 3, 4, \ldots \quad \text{(Lyman series, ultraviolet)}$$

$$\bar{\nu} = R_{\text{H}}\left(\frac{1}{2^2} - \frac{1}{n^2}\right), \quad n = 3, 4, 5, \ldots \quad \text{(Balmer series, visible and near ultraviolet)}$$

$$\bar{\nu} = R_{\text{H}}\left(\frac{1}{3^2} - \frac{1}{n^2}\right), \quad n = 4, 5, 6, \ldots \quad \text{(Paschen series, infrared)}$$

$$\bar{\nu} = R_{\text{H}}\left(\frac{1}{4^2} - \frac{1}{n^2}\right), \quad n = 5, 6, 7, \ldots \quad \text{(Brackett series, infrared)}$$

$$\bar{\nu} = R_{\text{H}}\left(\frac{1}{5^2} - \frac{1}{n^2}\right), \quad n = 6, 7, 8, \ldots \quad \text{(Pfund series, infrared)}$$

Rydberg noted that for each line in the Balmer series (the only series known at the time) the wave number could be expressed as the difference between two terms, one fixed and one varying. He looked for similar regularities in the spectra of elements other than hydrogen, and found that although the spectra of the heavier elements are more complex than that of hydrogen, a series of spectral lines can always be represented by the general formula

$$\bar{\nu} = \frac{R}{(m + a)^2} - \frac{R}{(n + b)^2}. \tag{7-4}$$

In this equation, R is a constant for the element, a and b are characteristic constants for a particular series, m is an integer which is fixed for a given series, and n is a varying integer whose values give the different lines of the series. The line spectrum of any element consists of several series of lines, and each line can be expressed as the difference between two terms.

The emission spectrum of an alkali metal is an example of a spectrum of this type; the following four series are found,

principal series: $$\bar{\nu} = \frac{R}{(1+s)^2} - \frac{R}{(n+p)^2}, \quad n = 2, 3, 4, \ldots,$$

sharp series: $$\bar{\nu} = \frac{R}{(2+p)^2} - \frac{R}{(n+s)^2}, \quad n = 2, 3, 4, \ldots,$$

diffuse series: $$\bar{\nu} = \frac{R}{(2+p)^2} - \frac{R}{(n+d)^2}, \quad n = 3, 4, 5, \ldots,$$

Bergmann series: $$\bar{\nu} = \frac{R}{(3+d)^2} - \frac{R}{(n+f)^2}, \quad n = 4, 5, 6, \ldots.$$

In these series, R is the Rydberg constant for the element, and s, p, d, f, are certain constants which are characteristic of the alkali atoms. The principal series is so called because it contains the brightest and most persistent lines; the lines of the sharp series appear relatively sharp in a spectrograph picture, while the lines of the diffuse series are relatively broad. In each case, as n becomes very large the lines of the series converge to a limit given by the constant term.

The lines of the spectra of other elements also can be represented as differences between terms, and the formula for the wave number can be written generally as

$$\bar{\nu} = T_1 - T_2. \tag{7–5}$$

For hydrogen the terms are given by

$$T_n = \frac{R_{\mathrm{H}}}{n^2}, \qquad n = 1, 2, 3, \ldots. \tag{7–6}$$

For other elements, the terms usually have a more complicated form, and the first and second members of the formula are obtained from different series. Ritz (1908) extended Rydberg's formulation of wave numbers by means of his combination principle, which states that, with certain limitations, the difference between any two terms gives the wave number of a spectral line of the atom. For example, the difference between T_3 and T_8 for hydrogen gives the fifth line of the Paschen series. As a result of the discovery of this combination principle, one of the chief problems of spectroscopy became the representation of the lines of a spectrum as differences between terms, with as few terms as possible. The clarification of the meaning of the terms then became a major theoretical problem.

In addition to emission spectra, the elements also have characteristic absorption spectra. The latter are obtained when light with a continuous spectrum (as that from a filament lamp) is passed through an absorbing layer of the element and then analyzed with a spectrograph. In a photograph of an absorption spectrum, light lines (absorption lines) appear on a dark continuous background. The absorption lines coincide exactly in wavelength with corresponding emission lines and, in principle, every emission line can also occur as an absorption line.

The field of atomic spectra is one of the great branches of physics, and the spectra of the elements have been analyzed in great detail. Only some of the simplest features of atomic spectra have been presented here; e.g., the *multiplet structure* of spectral lines has not been mentioned. In many elements, each line in the spectrum is single, but more commonly the lines form doublets, triplets, or groups of even more components. In such cases, a separate Rydberg formula must be written for each component line; in the spectra of the alkali metals, for example, doublets occur, so that six formulas instead of three are required for a complete representation of the series in the visible region. There are also other effects, such as the changes in the spectrum of an element when the atoms radiating the light are in a magnetic or electric field. When a light source is brought into a magnetic field, each emitted spectral line is split into a number of components (the Zeeman effect). The extent of the splitting depends on the strength of the field and on the nature of the spectral lines, whether they are singlets or multiplets. A splitting of the spectral lines is also observed if the radiating atoms are in an electric field. But in spite of the vast amount of information which has been collected about atomic spectra, the simple way in which spectral series can be expressed points to the existence within atoms of a relatively simple and universal method by which the characteristic atomic radiations are emitted.

7-2 The Bohr theory of atomic spectra and atomic structure. In a series of epoch-making papers written between 1913 and 1915, Bohr developed a theory of the constitution of atoms which accounted for many of the properties of atomic spectra and laid the foundations for later research on theoretical and experimental atomic physics. Bohr applied the quantum theory of radiation as developed by Planck and Einstein to the Rutherford nuclear atom. His theory is based on the following postulates.

1. An atomic system possesses a number of states in which no emission of radiation takes place, even if the particles are in motion relative to each other, although such an emission is to be expected according to ordinary electrodynamics. These states are called the *stationary* states of the system.

2. Any emission or absorption of radiation will correspond to a transition between two stationary states. The radiation emitted or absorbed in

a transition is homogeneous, and its frequency ν is determined by the relation

$$h\nu = W_1 - W_2, \tag{7–7}$$

where h is Planck's constant and W_1 and W_2 are the energies of the system in the two stationary states.

3. The dynamical equilibrium of the system in the stationary states is governed by the ordinary laws of mechanics, but these laws do not hold for the transition from one state to another.

4. The different possible stationary states of a system consisting of an electron rotating about a positive nucleus are those for which the orbits are circles determined by the relation

$$p = n\ (h/2\pi)\ , \tag{7–8}$$

where p is the angular momentum of the electron, h is Planck's constant, and n is a positive integer, usually called the *quantum number*.

These postulates are a combination of some ideas taken over from classical physics together with others in direct contradiction to classical physics. Bohr solved the problem of the stability of a system of moving electric charges simply by postulating that the cause of the instability, the emission of radiation, did not exist so long as the electron remained in one of its allowed, or *stationary* orbits. The emission of radiation was associated with a jump of the system from one stationary state (energy level) to another, in accordance with the ideas of Planck's quantum theory. Also, the laws of classical mechanics were presumed to be valid for an atomic system in a stationary state, but not during a transition from one such state to another. Finally, the fourth postulate, which separates the allowed orbits of the electron from the forbidden ones by means of the quantum condition, Eq. (7–8), has been shown to be in harmony with Planck's condition for the energy of an oscillator. The angular momentum is said to be *quantized* in these orbits. Thus, Bohr's theory is a hybrid, containing some classical ideas and some quantal ideas. The justification for the postulates was the astonishing success of the theory in accounting for many of the experimental facts about atomic spectra. But, as will be seen later, the hybrid nature of the theory eventually led to serious difficulties.

Starting with Bohr's postulates, we can calculate the values of the energy that correspond to the stationary states of an atom consisting of a nucleus and a single electron. Consider an atom with nuclear charge Ze, where Z is the atomic number and e is the magnitude of the electronic charge in electrostatic units. If $Z = 1$, the atom is the neutral hydrogen atom; for $Z = 2$, the system is the singly ionized helium atom He^+; for $Z = 3$, the system is the doubly ionized lithium atom Li^{++}.

In accordance with the fourth postulate, the electron revolves in a circle about the nucleus and can have the angular momentum p, given by

$$p = mvr = n\frac{h}{2\pi}, \tag{7–9}$$

where m is the mass of the electron, v is the linear velocity of the electron, r is the radius of the orbit, and n is a positive integer. By the third postulate, the motion of the electron in an allowed orbit must satisfy the classical mechanical laws. Hence, the centripetal force on the electron is provided by the attractive electrostatic force on the electron:

$$\frac{mv^2}{r} = \frac{Ze^2}{r^2}. \tag{7–10}$$

Equations (7–9) and (7–10) can be solved for the radius r and the velocity v, and it is found that

$$r = \frac{n^2h^2}{4\pi^2me^2Z}, \tag{7–11a}$$

and

$$v = \frac{2\pi Ze^2}{nh}. \tag{7–11b}$$

The radii of the allowed orbits are proportional to n^2, and increase in the proportions 1, 4, 9, 16, . . . , from orbit to orbit. For hydrogen, $Z = 1$, and if we insert the known values of m (rest mass), e, and h, we get, for the smallest orbit in hydrogen ($n = 1$),

$$r = 5.29 \times 10^{-9} \text{ cm},$$

a value that is in good agreement with estimates of atomic radii obtained by other methods. The velocity of the electron in the smallest orbit is about 2.2×10^8 cm/sec, and decreases as the radius of the orbit increases, so that m may be taken as the rest mass of the electron.

The energy of the electron is partly kinetic and partly potential. If the state of zero energy of the electron is arbitrarily chosen as that state in which the electron is at rest at a very great distance ("infinitely far") from the nucleus, then the potential energy of the electron is, by the usual electrostatic formula,

$$U = -\frac{Ze^2}{r}. \tag{7–12}$$

The kinetic energy of the electron is

$$T = \frac{1}{2}mv^2 = \frac{1}{2}\frac{Ze^2}{r}, \tag{7–13}$$

by Eq. (7–10), and $T = -\frac{1}{2}U$, a relation typical of motion under an inverse-square law of force. The total energy is $W = T + U$ or, for the stationary state defined by the quantum number n,

$$W_n = -\frac{1}{2}\frac{Ze^2}{r} = -\frac{2\pi^2 me^4 Z^2}{n^2 h^2}; \tag{7–14}$$

this is the energy of the atom when the electron is in its nth stationary or quantum state. The quantum number n, it will be remembered, may take on any integral value; $n = 1, 2, 3, \ldots$.

From Eq. (7–14) it is seen that the larger the value of n, the smaller numerically, but the larger in algebraic value, is the energy of the system. The lowest value of W_n is that corresponding to the first orbit. This is known as the *normal* or *ground* state of the atom, since it should be the most stable state and the one ordinarily occupied by the electron.

According to Bohr's second postulate, a hydrogen atom radiates energy when the electron jumps from one stationary state to another state of lower energy. The difference in the two energies is emitted as a single quantum of radiant energy whose frequency is given by the condition

$$\nu = \frac{W_{n_1} - W_{n_2}}{h}. \tag{7–15}$$

If we insert for W_{n_1} and W_{n_2} the values given by Eq. (7–14) for the states with quantum numbers n_1 and n_2, we get for the frequency of the line emitted when the electron jumps from state n_1 to state n_2

$$\nu = \frac{2\pi^2 me^4}{h^3} Z^2 \left(\frac{1}{n_2^2} - \frac{1}{n_1^2}\right). \tag{7–16}$$

In terms of wave numbers this becomes

$$\bar{\nu} = R_\infty Z^2 \left(\frac{1}{n_2^2} - \frac{1}{n_1^2}\right), \tag{7–17}$$

where we have defined the constant R_∞ as

$$R_\infty = \frac{2\pi^2 me^4}{ch^3}. \tag{7–18}$$

In the case of hydrogen, $Z = 1$ and Eq. (7–17) becomes

$$\bar{\nu} = R_\infty \left(\frac{1}{n_2^2} - \frac{1}{n_1^2}\right). \tag{7–19}$$

Since $n_2 < n_1$, the wave number is positive. If we set $n_2 = 2$,

$$\bar{\nu} = R_\infty\left(\frac{1}{2^2} - \frac{1}{n_1^2}\right), \qquad n_1 = 3, 4, 5, \ldots. \tag{7–20}$$

This equation has exactly the same form as the empirical equation (7–2), which was found to represent the lines of the Balmer series of the hydrogen spectrum. Bohr's theory also gives the value of the constant coefficient R_∞ in terms of fundamental constants. Hence, R_∞ can be calculated theoretically, and a crucial test of Bohr's theory is the agreement or disagreement between the theoretical value of the coefficient and the experimental value of the Rydberg constant.

Before this comparison can be made, it is necessary to correct for an approximation which has been made in the treatment so far. It has been tacitly assumed in the derivation of the formula for the wave number that the nucleus does not move, that it is infinitely heavy in comparison with the electron. The subscript ∞ has been used to denote the value of the coefficient under this assumption. Actually, for hydrogen, the nucleus of the atom is known to be about 1840 times as heavy as the electron rather than infinitely heavy, and it does have a slight motion. The precision of spectroscopic measurements is so great that in testing the theory, the motion of the nucleus cannot be neglected even though the nucleus is so much heavier than the electron. The effect of the motion of the nucleus is to replace the mass m of the electron by the *reduced mass*, $mM/(M + m)$, where M is the mass of the nucleus.

The formula for the reduced mass is one which is important in many problems involving the motion of two particles, and it can be derived quite easily. If neither of the particles can be assumed to be at rest, both particles will move about the center of mass of the two particles. In the case of a one-electron atom, if the electron revolves about the nucleus, its path in space will be a circle about the center of mass of the combined system. At the same time, the nucleus revolves about the center of mass in a smaller circle. Both particles must have the same angular velocity ω, since the center of mass must always lie on the line between the two particles.

Let r be the distance of the electron from the nucleus, and $r - x$ its distance from the center of mass; x is then the distance of the nucleus from the center of mass. Taking moments about the center of mass, we have

$$Mx = m(r - x),$$

or

$$x = \frac{mr}{M + m}; \qquad r - x = \frac{Mr}{M + m}.$$

The total angular momentum of the system about the center of mass is

$$m(r - x)^2\omega + Mx^2\omega = \left[\frac{mM^2}{(m + M)^2} + \frac{Mm^2}{(m + M)^2}\right]r^2\omega = \left(\frac{mM}{m + M}\right)r^2\omega.$$

The system acts as though a mass $m' = mM/(m + M)$ is revolving in a circle of radius r, with an angular velocity ω about the center of mass. The mass m of the electron must therefore be replaced by m', that is, m must be multiplied by the factor $M/(m + M)$, and the reduced mass of the electron is $mM/(m + M)$. Since M varies from atom to atom, the factor $1/[1 + (m/M)]$ varies slightly, and this accounts for the small variation in the Rydberg constant for different elements.

The constant R_∞ in Eqs. (7–17) through (7–20) should be replaced therefore by the new constant

$$\begin{aligned} R_{\mathrm{H}} &= \frac{2\pi^2 me^4}{ch^3}\,\frac{1}{[1 + (m/M_{\mathrm{H}})]} \\ &= R_\infty\,\frac{1}{[1 + (m/M_{\mathrm{H}})]}, \end{aligned} \tag{7–21}$$

where M_{H} is the mass of the nucleus. If we insert into Eq. (7–21) the presently accepted values of the constants e, m (rest mass), c, h, and m/M_{H}, we get R_{H} (theor) $= 109681$ as compared with the experimental value of 109677.58. The two values agree to within 3 parts in 100,000, which is considerably smaller than the uncertainty in the theoretical value. This agreement represents a remarkable feat of theoretical physics and is one of the great triumphs of the Bohr theory.

When the Bohr theory was proposed, only the Balmer and Paschen series for hydrogen were known. The theory [Eq. (7–19)] suggested that for values of n_2 different from 2 and 3, additional series should exist. The search for these series yielded the Lyman series (1916), the Brackett series (1922), and the Pfund series (1924), and in each series the wave numbers of the lines were found to be those predicted by the theory.

Some of the circular orbits of the electron in the hydrogen atom are shown on a relative scale in Fig. 7–2 together with the transitions that give rise to typical lines of the different series in the hydrogen spectrum.

The spectrum of singly ionized helium, He^+, was also successfully explained by the Bohr theory. The experimental value of the Rydberg constant for helium is $R_{\mathrm{He}} = 109722.269$, and is in good agreement with the theoretical value. A very important result is obtained when the experimental values of the Rydberg constants for hydrogen and ionized helium are compared. For helium

$$R_{\mathrm{He}} = \frac{R_\infty}{[1 + (m/M_{\mathrm{He}})]} = \frac{R_\infty}{[1 + (m/4M_{\mathrm{H}})]}, \tag{7–22}$$

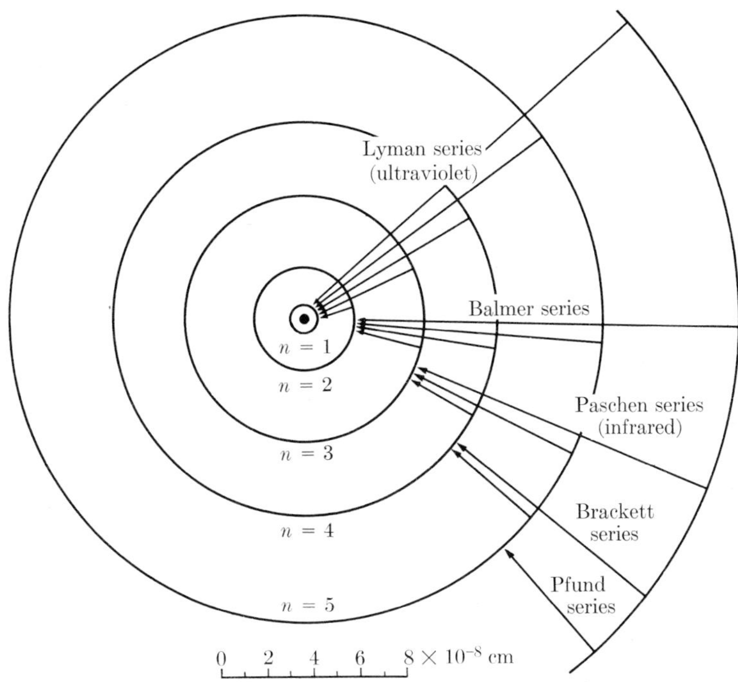

FIG. 7-2. Scheme of transitions between stable Bohr orbits in the hydrogen atom (approximately to scale).

since the mass of the helium nucleus is very nearly equal to 4 times that of the hydrogen nucleus. From Eqs. (7-21) and (7-22) it can be shown that

$$\frac{M_{\mathrm{H}}}{m} = \frac{R_{\mathrm{H}} - \frac{1}{4}R_{\mathrm{He}}}{R_{\mathrm{He}} - R_{\mathrm{H}}} = 1841. \tag{7-23}$$

This result is very close to that obtained in Section 2-4 by other methods, and represents another triumph for the Bohr theory.

7-3 The stationary states of an atom. The Bohr theory, in addition to accounting for the spectra of one-electron atoms, provides a physical interpretation of the spectral terms. In the frequency condition, Eq. (7-7) or Eq. (7-15), on substituting the wave number $\bar{\nu} = \nu/c$ for the frequency, we get

$$\bar{\nu} = \frac{W_1}{hc} - \frac{W_2}{hc}\cdot \tag{7-24}$$

Comparison with the Rydberg series formula, Eq. (7-5), shows that a term is equal to an energy state of a particular electron orbit divided by hc. The experimental determination of the spectral terms of an element, therefore, gives the values of the energy for the stationary states of the atom. It will be remembered that the energy scale was chosen in such a way that the electron has the energy $W = 0$ when it is completely removed from the nucleus. Hence, the work that must be done in order to remove the electron from its nth orbit to infinity is $-W_n$. This amount of work is called the *separation energy* and is a positive quantity. Apart from the factor hc, the terms are equal to the separation energies of the electron in the given states. For the lowest state ($n = 1$) of the atom, the ground state, the separation energy is called the *ionization energy*, or the *ionization potential*, which is, accordingly, equal to the largest term value of the atom. The states which correspond to quantum numbers greater than unity are called *excited states* of the atom because their energies are greater than those of the ground state. A transition from the ground state to an excited state can occur only when an atom receives an amount of energy equal to the excitation energy for the excited state, for example, by electron bombardment in a gas discharge tube.

The connection between term values and energies was demonstrated experimentally (1914) by the work of Franck and Hertz[(2)] on collisions between slow electrons and mercury atoms. They observed that electrons could not transfer energy to the mercury atoms unless the former had a kinetic energy of at least 4.9 ev. This should mean, according to the theory, that the mercury atom has an excited stationary state 4.9 ev above the ground state. If there is such a state, the energy of the excited atom should be emitted as a light quantum whose frequency or wavelength is determined by the Bohr frequency condition, Eq. (7–7). Now, an energy difference of 1 ev is equivalent to a certain wave number obtained by dividing 1 ev $= 1.602 \times 10^{-12}$ erg by hc; the result is 8066.1 cm^{-1}. An excitation energy of 4.9 volts should correspond therefore to a wave number $\bar{\nu} = 4.9 \times 8066.1 = 39523$ cm^{-1}, or a wavelength of $\lambda = 1/\bar{\nu} = 2530$ A, which is a line in the ultraviolet region. There is a line of wavelength 2535 A in the spectrum of mercury and the experimental result can therefore be interpreted in accordance with the Bohr theory. Thus, by electron impact, an excitation energy of 4.9 ev was transferred to the mercury atom, and when an electron jumped back to the ground state this excitation energy was emitted as a photon with the theoretically predicted wavelength. Many similar results have since been obtained with other lines and gases, and Bohr's relationship between the excited states and the term values of the spectral lines has been completely verified. In this way, the validity of Bohr's assumption about stationary states has been verified by direct experiments, and the concept of the stationary states of an atom has been strongly confirmed.

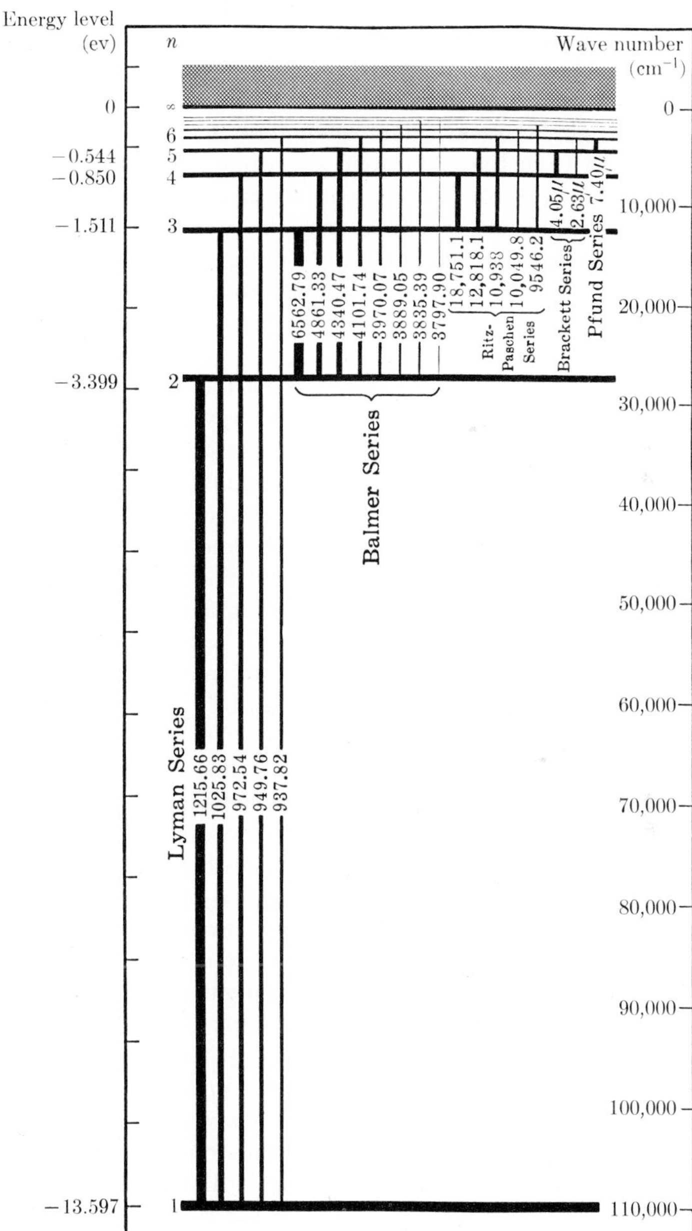

FIG. 7-3. Energy level diagram of the hydrogen atom.

The stationary states, the terms, and the spectral lines of hydrogen can be represented graphically in a very useful way by means of an energy level diagram, as shown in Fig. 7-3. In this figure, the ordinates give the energy, and the energy levels or terms R/n^2 are drawn as horizontal lines. At the right, the wave number or term scale is given in units of cm^{-1}, increasing from top to bottom; the value zero corresponds to the complete separation of the nucleus and electron ($n = \infty$). At the left is a scale in electron volts beginning with the ground state as zero. The electron-volt scale can be used directly to obtain the ionization potential of a given level. Thus, to remove the electron from the ground state ($n = 1$) to infinity ($n = \infty$), i.e., to ionize the atom in its ground state, 13.597 ev are needed.

A spectral line results from the transition of the atom from one energy level to another and is represented in the energy diagram by a vertical line joining the two levels. The length of the line connecting the two levels is directly proportional to the wave number of the spectral line (right-hand scale). A rough indication of the intensity of the spectral lines may also be given by means of their thickness. Toward the top of the diagram, the separation of the horizontal lines, which represent the energy levels or terms, decreases and converges to the value zero as $n \to \infty$. A continuous spectrum joins the term series here, indicated by cross-hatching. The theoretical explanation for the continuous spectrum will be given later. Spectra of other elements can be represented by similar, but more complicated, diagrams.

The Bohr model can account for the lines of absorption spectra as well as for those of emission spectra if it is assumed that an orbital electron can absorb a light quantum only if the energy so absorbed raises the electron into another allowed orbit. The absorption of light is then the exact inverse of emission and every absorption line should correspond exactly to an emission line, in agreement with experiment. Every emission line, however, is not found to correspond to an absorption line. For example, the absorption spectrum of hydrogen gas under ordinary conditions shows only the lines of the Lyman series. The Bohr theory can account for this experimental result in the following way. Under normal conditions, the atoms in hydrogen gas are in the most stable state (ground state), with $n = 1$. The energy differences between that level and all other levels correspond to lines in the Lyman series; photons with less energy, as all those corresponding to lines in the Balmer series, cannot be absorbed because they cannot make the electron go even from the innermost ($n = 1$) orbit to the second orbit. According to the theory, therefore, only the lines of the Lyman series should appear in the absorption spectrum of normal hydrogen. Only if the gas is initially excited, and there is an appreciable number of electrons in orbits with $n = 2$, should absorption lines corresponding to the Balmer series appear. Balmer lines are found in the absorption spectra of some stars in which the temperature of the stellar atmospheres is so high that an appreciable fraction of the hydrogen atoms is in the first excited state ($n = 2$).

7–4 Extension of the Bohr theory: elliptic orbits. The Bohr theory in its simplest form predicted with great accuracy the positions of the spectral lines of the neutral hydrogen atom and the singly ionized helium atom. Refined spectroscopic analysis showed, however, that these lines are not simple but have what is called *fine structure*, i.e., they consist of a number of component lines lying close together. In terms of energy levels, the existence of fine structure means that instead of a simple energy level corresponding to a given value of the quantum number n, there are actually a number of energy levels lying very close to one another. Sommerfeld succeeded in part in accounting for these levels and for the resulting fine structure of the spectral lines of hydrogen and He^+ by postulating the existence of elliptic orbits as well as circular orbits, and by taking into account the relativistic variation of the electron mass with velocity.

When it is assumed that an electron can travel in an elliptic orbit, there are two coordinates which vary. In polar coordinates these are r, the distance of the electron from the nucleus, which is at the focus of the ellipse, and ϕ, the azimuthal angle. The two coordinates represent two degrees of freedom to be quantized. The quantum condition that has previously been discussed is not enough to fix unambiguously both axes of the ellipse, and two quantum conditions are needed to fix the orbit. Sommerfeld and Wilson introduced a new and more general postulate than the original one of Bohr, namely, for the stationary states the so-called action integral, extended over one period of the motion, must be an integral multiple of h. This condition is expressed mathematically by the relation

$$\int p_i \, dq_i = n_i h, \tag{7–25}$$

where p_i is the momentum associated with the coordinate q_i. The integral has the units of *action*, i.e., erg-seconds, as does Planck's constant. When $q_i = \phi$, the angle of revolution, p_i is the orbital angular momentum of the system, denoted by p_ϕ to distinguish it from the radial momentum p_r. According to classical mechanics, the angular momentum of any isolated system is a constant. Then,

$$\int p_\phi \, d\phi = p_\phi \int_0^{2\pi} d\phi = 2\pi p_\phi,$$

and

$$p_\phi = n_\phi \frac{h}{2\pi}. \tag{7–26}$$

This condition is the same as that used by Bohr for his circular coordinates. The quantity n_ϕ, called the *azimuthal quantum number*, has been used instead of n for reasons which will soon become apparent.

The radial momentum p_r is not constant and the integral in the second quantum condition,

$$\int p_r \, dr = n_r h, \tag{7–27}$$

can be calculated only by more complicated methods. The result obtained is

$$\int p_r \, dr = 2\pi p_\phi \left(\frac{1}{\sqrt{1 - \epsilon^2}} - 1 \right) = n_r h. \tag{7–28}$$

In this equation, ϵ is the eccentricity of the ellipse with semimajor and semiminor axes a and b, respectively. The quantity n_r is called the *radial quantum number*. From Eqs. (7–28) and (7–26), it follows that

$$1 - \epsilon^2 = \frac{n_\phi^2}{(n_\phi + n_r)^2} = \frac{b^2}{a^2}. \tag{7–29}$$

The eccentricity of the allowed ellipses is thus determined by the ratio of the quantum numbers n_ϕ and n_r. The quantity $1 - \epsilon^2$ can take on values only between 0 and 1 because of the geometrical properties of the ellipse. For $1 - \epsilon^2 = 1$, $\epsilon = 0$, $b = a$, and the ellipse becomes a circle. For $1 - \epsilon^2 = 0$, $n_\phi = 0$, $b = 0$, and the ellipse becomes a straight line. Physically, the last result means that the electron would oscillate on a straight line through the nucleus. This case was excluded in the Bohr theory as having no physical meaning, so that n_ϕ was not allowed to be equal to zero. Hence, n_ϕ may have the values 1, 2, 3, . . . , and n_r may have the values 0, 1, 2, 3, ... The sum $n_r + n_\phi$ may have the values 1, 2, 3, . . . ; it is therefore set equal to another integer n,

$$n = n_\phi + n_r; \qquad n = 1, 2, 3, \ldots, \tag{7–30}$$

and n is called the *principal quantum number*. For any value of n, n_ϕ can take on the values 1, 2, 3, . . . n, and n_r is given by $n - n_\phi$.

The discussion so far has clarified the relations between the quantum numbers needed for the elliptical orbits and has defined the eccentricity of the allowed orbits, but has not told anything about the size of the ellipses, that is, a and b, or about the energies in the orbits. The latter quantities are determined, as in the case of the circular orbits, by combining the quantum conditions with the expression for the centripetal force provided by the attractive electrostatic force on the electron. The

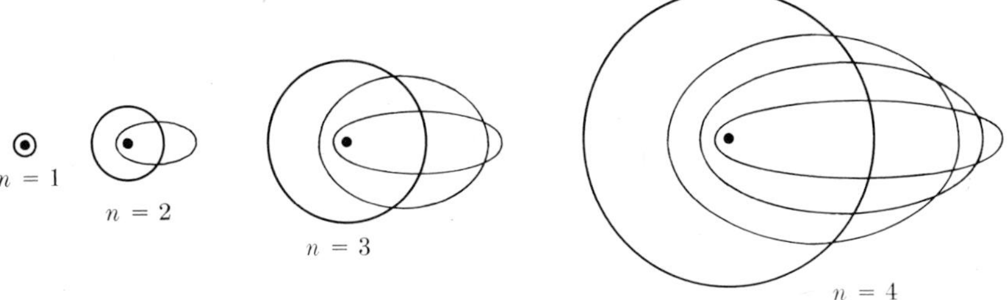

FIG. 7-4. Relative positions and dimensions of the elliptical orbits corresponding to the first four values of the quantum number n. (Sommerfeld's modification for the hydrogen atom.)

analysis is more complicated than in the circular case and the details are omitted. The results are

$$a = \frac{n^2h^2}{4\pi^2m'e^2Z}, \tag{7-31a}$$

$$b = a\,\frac{n_\phi}{n}, \tag{7-31b}$$

$$W_n = -\frac{2\pi^2m'e^4Z^2}{n^2h^2}, \tag{7-32}$$

where m' is the reduced mass of the electron. The values of the semi-major axis a and the energy W_n are determined by the principal quantum number n, and are the same as the values for the circular orbit of radius a and quantum number n. The *shape* of the orbit is determined by the ratio n_ϕ/n, and the number of possible orbit shapes of the same energy is increased over the case in which only circular orbits are considered. Figure 7-4 shows the circular and elliptic orbits (drawn to scale) for hydrogen with various n values, for $n_\phi = 1$, 2, and 3. According to this model of the atom, a given spectral line may be produced by a number of different but equivalent transitions between which no distinction can so far be made. Although the system has two degrees of freedom, *one* quantum number suffices to determine the permitted energy values. Such a system is said to be *degenerate*.

Sommerfeld showed that in the case of a one-electron atom, the degeneracy could be removed. That is, the previously equivalent states could take on different energies when the special theory of relativity was applied to the motion of the electron. The velocity of an electron in an elliptic orbit varies (in a circle it is constant), being greater when the electron is near the nucleus, and smaller when it is relatively far away. In accordance with the special theory of relativity, the mass of the electron varies in the orbit. If this relativistic effect is taken into account, the energy

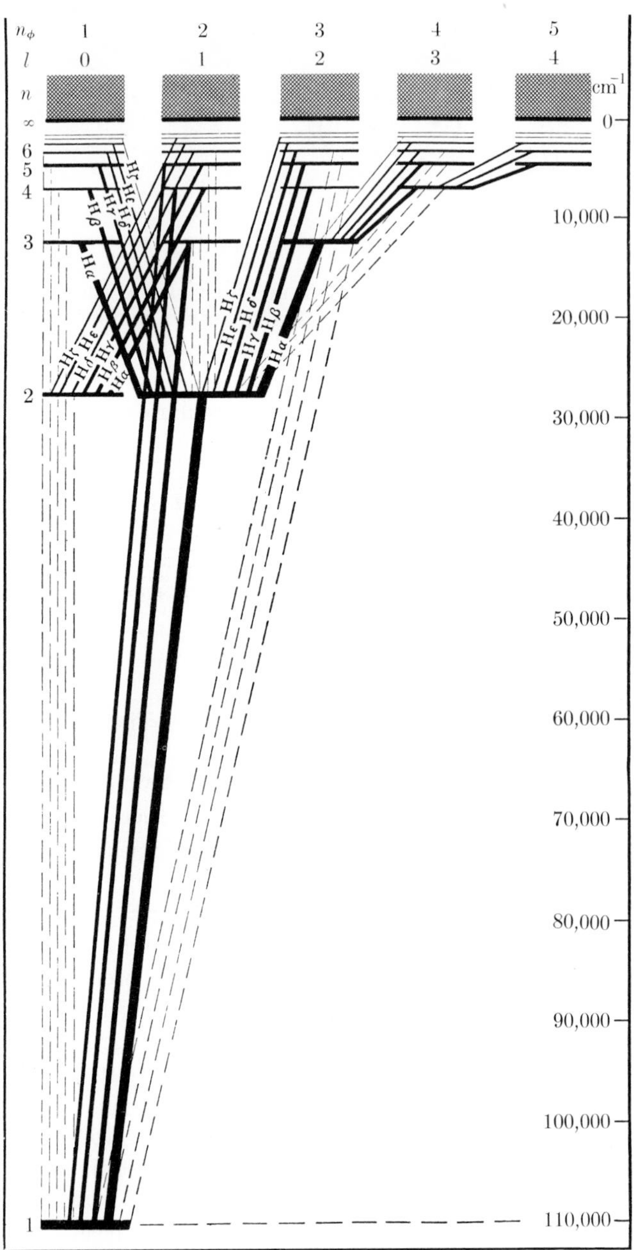

FIG. 7–5. Energy level diagram of the hydrogen atom, including fine structure.

of the electron is found to depend on both n and n_ϕ. The mathematical analysis is again quite complicated; the result is Sommerfeld's formula

$$W_{n,n_\phi} = -\frac{2\pi^2 m' e^4}{n^2 h^2} Z^2 \left[1 + \frac{\alpha^2 Z^2}{n}\left(\frac{1}{n_\phi} - \frac{3}{4n}\right)\right], \qquad (7\text{–}33)$$

where $\alpha = 2\pi e^2/hc = 1/137$ is the *fine structure constant.* The dependence on n_ϕ is small because of the smallness of the fine-structure constant, and the relativistic treatment of the elliptic orbits has the effect of replacing a single energy level by several levels lying close together. Hence, the frequency of the radiation emitted in the transition of an electron from a level given by n_2 to the level n_1 will be slightly different according to the various values of n_ϕ which are possible. The small variations in energy will result in small energy and frequency differences, with the result that in the hydrogen spectrum, a group of slightly separated lines appears, rather than a single line.

When the predictions of Sommerfeld's theory were compared with the experimental results on the resolution of the Balmer lines and of several He^+ lines, it was found that the theory predicted too many components for the fine structure. Agreement between theory and experiment was obtained by the introduction of a *selection rule* which limited the number of transitions. In the present case, the selection rule stated that only those transitions are possible for which the quantum number n_ϕ changes by $+ 1$ or $- 1$:

$$\Delta n_\phi = \pm 1. \qquad (7\text{–}34)$$

Rules like this one were found to be highly important in the theory of atomic spectra. They were not an intrinsic part of the Bohr theory but could be obtained by means of Bohr's *correspondence principle.*(3) It was known that for very large quantum numbers n, the behavior of the atom tends to that which would be expected on the basis of classical electrodynamics. For example (cf. Chapter 5), for radiation of very small frequency or very long wavelength, classical radiation theory was satisfactory. But classical theory made possible the calculation of allowed transitions, while the Bohr atomic theory did not. According to the correspondence principle, the selection rules calculated classically, and valid for very large quantum numbers, were also assumed to be correct for small quantum numbers. This method for deriving the selection rules worked, although it had the disadvantage of not being an integral part of the Bohr theory.

The fine structure of the lines of the hydrogen spectrum can be illustrated in an energy level diagram somewhat more complicated than that of Fig. 7–3. In Fig. 7–5, the levels with different values of n_ϕ are now drawn side by side at the same height, while states with equal values of

n_ϕ and different values of n are drawn above one another. For a given value of n, each level has n sublevels corresponding to the n values which n_ϕ can have. In place of the vertical lines of Fig. 7–3 to represent transitions from one value of n to another, there are oblique lines corresponding to changes in n_ϕ of ± 1. The wave number for a transition is still given by the *vertical* distance between two levels. To show the fine structure lines, the energy and wave number scales would have to be very large, because of the very small spacing between the sublevels.

In addition to circular and elliptic orbits, it is possible for electrons to have hyperbolic orbits; these are associated with the continuous spectrum. The energy of a one-electron atom has been defined in such a way that it has the value zero when the electron is at rest completely removed from the nucleus. If the free electron has kinetic energy, its total energy is positive and, by classical mechanics, its orbit is a hyperbola. According to the Bohr theory, only the circular and elliptic orbits are quantized, although an electron does not radiate in any orbit. Hence, all positive values of the energy are possible and, extending from the limit of the discrete energy levels, there is a continuous region of positive energy values. Radiation is emitted when an electron with positive energy jumps to a lower state of positive energy or to a quantized state with negative energy, and the series of discrete lines in the spectrum is followed by a continuous spectrum.

7–5 Failure of the Bohr theory: wave mechanics and the correct quantum numbers. Although the use of relativistic, elliptic orbits in the Bohr atom helped explain the fine structure of the spectra of hydrogen and singly ionized helium, serious difficulties were met in the further application of the Bohr theory. For two-electron atoms such as neutral helium and singly ionized lithium, there were serious discrepancies between theory and experiment. To try to account for the splitting of spectral lines in a magnetic field, it was necessary to introduce a third quantum number, the *magnetic quantum number*, m. As a result, three-dimensional, or spatial, quantization was necessary; the new quantum number was related to the orientation of the elliptic orbits in space. In the presence of a magnetic field, there is a component of the angular momentum in the direction of the field. This component, whose direction may be taken to be parallel to the z-axis of a three-dimensional coordinate system, was quantized according to the condition

$$p_z = m\frac{h}{2\pi}, \tag{7–35}$$

where m could have the values $-n_\phi, -n_\phi+1, \ldots 0, 1, \ldots, n_\phi$, or $2n_\phi + 1$ values in all. The mathematical treatment of the problem of the stationary states again led to too many energy levels and the selection rule,

$$\Delta m = 0, \pm 1, \tag{7-36}$$

had to be introduced. The success of the theory in dealing with the effect of a magnetic field on spectral lines was, however, still only partial.

Additional difficulties arose in connection with the multiplet structure of the spectral lines of many elements. Multiplets are different in character from the fine-structure lines in that the multiplets can be quite widely separated. The well-known sodium lines with wavelengths of 5895.93 A and 5889.96 A, respectively, are an example of a doublet, as are the other members of the series to which the D lines belong. In the magnesium spectrum, triplets are found, and the lines become more complex as the elements move to the right in the periodic table. In the absence of a magnetic field, multiplets could not be accounted for on the basis of the quantum numbers n and n_ϕ.

The difficulties discussed above (and others not mentioned) could not be resolved by further changes in the Bohr theory. It gradually became apparent that the difficulties of the theory were caused by an essential failure of the atomic model that was being used. The solution to the problem came in 1925 and the years following in the form of *wave mechanics* or *quantum mechanics*. Some of the basic ideas of the newer theory will be discussed later in this chapter. For the present it will suffice to state that wave mechanics provides a theory which does not depend on the classical picture of electrons revolving in planetary orbits around the nucleus. The new theory yields all of the correct ideas and results of the Bohr theory, such as stationary states with the correct energy values, the explanation of the emission and absorption of radiation in terms of transitions between stationary states, and the Rydberg-Ritz combination principle for spectral lines. Moreover, wave mechanics solves many of the problems which could not be handled by the Bohr theory. One of the most important results of the new theory has to do with the quantum numbers. It is found that in the case of the one-electron atom treated three-dimensionally, three quantum numbers appear in a direct way in the mathematical solution of the problem. These are the principal quantum number n, the orbital quantum number l, which takes the place of n_ϕ, and the magnetic quantum number m_l, which replaces m. The values which the new numbers can take on are

$$\begin{aligned} n &= 1, 2, 3, \ldots, \\ l &= 0, 1, 2, 3, \ldots, n-1, \\ m_l &= -l, -l+1, \ldots, 0, 1, \ldots, l-1, l. \end{aligned} \tag{7-37}$$

The necessary selection rules also turned out to be a direct result of the mathematical treatment and did not have to be introduced in an artificial way.

Wave mechanics has two disadvantages. First, it does not allow the structure of the atom to be depicted in a familiar way. The concepts from which the theory is built are more abstract than those of the Bohr theory and it is quite difficult to get an intuitive feeling for the problem of atomic structure. Second, the mathematical methods are considerably more advanced than those of the older theory. Hence, Bohr's mode of description, from which the essential features of atomic phenomena can readily be made out, is still often used. It is possible to go quite a long way with the Bohr model without coming into disagreement with facts, provided that the quantum numbers of the wave mechanical theory are used and that corresponding corrections are made in the conclusions resulting from the Bohr theory. In this sense, the Bohr theory may be considered to be a first approximation to wave mechanics, although both its conceptual basis and its mathematical methods have been replaced by those of the newer theory.

To account for multiplet spectra and for the so-called "anomalous Zeeman effect," wave mechanics was supplemented by the electron spin hypothesis of Uhlenbeck and Goudsmit (1926) according to which the electron was assumed to have a spin of magnitude $\frac{1}{2}h/2\pi$ and a magnetic moment equal to $(e/2mc)\,(h/2\pi)$. The spin angular momentum of the electron was represented by a quantum number that could have the values $+\frac{1}{2}$ or $-\frac{1}{2}$, and it was necessary to introduce the new quantum condition,

$$\text{spin angular momentum} = m_s\,(h/2\pi)\,, \tag{7–38}$$

where m_s can have the values $\frac{1}{2}$, $-\frac{1}{2}$. Thus, four quantum numbers are needed to account for atomic spectra. The electron spin was, at first, an arbitrary addition to wave mechanics, and not an intrinsic part of the theory, but this difficulty was removed in 1928 by Dirac. It had been known that the Schroedinger equation, the fundamental equation of wave mechanics, was not relativistically invariant. Dirac succeeded in obtaining an equation which satisfied the requirements of the special theory of relativity, and showed that all four quantum numbers followed from his theory without the introduction of any arbitrary assumptions. The result was a complete, consistent theory of atomic spectra and of many other atomic properties.

7–6 Atomic theory and the periodic table. One of the most important applications of modern atomic theory is to the interpretation of the structure of the periodic table in terms of the physical ideas that have been developed about atomic structure. As a result of the wave mechanical theory of the atom, it is possible to assign to any electron in an atom a value for each of the four quantum numbers n, l, m_l, and m_s, and an electron can be identified by the particular set of values that it has. The possible quantum numbers that an electron can have are limited by a fundamental rule, the *Pauli exclusion principle*, which states that no two electrons in an atom can have the same four quantum numbers. When this principle is applied to the electrons in an atom, two deductions can be made. First, there can be no more than $2n^2$ electrons with the same quantum number n. Second, there can be only $2(2l + 1)$ electrons with a given value of l. These two rules can be shown to account completely for the form of the periodic table.

Consider first the case $n = 1$. The only possible value of l is zero, and m_l must also be zero. The quantum number m_s may be either $+\frac{1}{2}$ or $-\frac{1}{2}$, so that there can be at most two electrons with $n = 1$. All the electrons which have the same total quantum number n are said to belong to a shell, the shells being denoted by K, L, M, N, . . . , for $n = 1, 2, 3, 4, \ldots$, respectively. The first or K-shell is full when it has two electrons. For the second or L-shell, $n = 2$ and the shell can contain up to 8 electrons ($2n^2 = 8$). It follows from the second rule that this shell can have no more than two electrons with $l = 0$, and no more than 6 electrons with $l = 1$. As a result of the second rule, *subgroups* can form within each shell, all of the electrons in a subgroup having the same values of n and l. The subgroups are designated by letters, as in Table 7–2.

An electron with $n = 3$ and $l = 2$ is called a 3d electron, and so on, the number giving the value of n, the letter indicating the value of l. All the electrons belonging to the same subgroup are called *equivalent electrons,* and it is usual to denote the number of such electrons in an atom by an index. Thus, two 1s electrons are written as $1s^2$, four 2p electrons as $2p^4$, seven 4d electrons as $4d^7$, and so on.

The structure of the periodic table can now be described in terms of the possible numbers of electrons in the subgroups. The basic idea underlying this description is that the number of electrons in an atom is equal to the atomic number. The chemical properties of an atom are completely determined by the number of outer electrons, and this number is in turn fixed by the nuclear charge. Any atomic electron structure is formed by

TABLE 7–2

Orbital angular momentum quantum number (l):	0	1	2	3	4	5	6
Electron designation:	s	p	d	f	g	h	i
Atomic state:	S	P	D	F	G	H	I

TABLE 7–3

ELECTRON SUBGROUPS FOR VALUES OF THE PRINCIPAL QUANTUM NUMBER UP TO $n = 4$

Shell	K	L	M	N
n	1	2	3	4
Possible electrons	1s	2s 2p	3s 3p 3d	4s 4p 4d 4f
Maximum number of electrons	2	2 6	2 6 10	2 6 10 14
Total, $2n^2$	2	8	18	32

the addition of a single electron to the electronic structure characteristic of the preceding atom in the periodic table. The quantum numbers chosen for the added electron are such as to place this electron into the most tightly bound state possible. This state is the one that, for the given number of electrons, leads to the smallest value for the total potential energy of the atom. The values of the potential energy can be calculated from the wave mechanical theory of the atom.

The subgroups of electrons which are possible are shown in Table 7–3 for values of n up to 4.

In the case of hydrogen, which has one electron, it is clear from Table 7–3 that this electron in the ground state occupies a 1s orbit; if the atom is excited, the electron can occupy one of the higher orbits (states).

Helium has two electrons, both of which can go into 1s orbits, making the potential energy a minimum. The configuration of the normal helium atom is then $1s^2$, and the K-shell is closed. The third electron needed for the formation of the lithium atom goes into a 2s orbit (to keep the potential energy a minimum); the electron configuration for this atom is $1s^22s$. Beryllium, with four electrons, has the configuration $1s^22s^2$; the K-shell is full, as is the s-electron subgroup of the L-shell. The process of adding electrons continues until, at neon, the entire L-shell is filled. The process can be continued until the entire periodic table is accounted for; the electron configurations of the elements are listed in Table 7–4.

The filling of the K- and L-shells at helium and neon, respectively, closes the first two periods of the periodic table (cf. Table 1–4). At argon the first two subgroups of the M-shell are filled; at krypton the first two subgroups of the N-shell are filled, and at xenon the first two subgroups of the O-shell are filled. Thus, the rare gases are associated with the completion of either a shell or a subgroup. After each rare gas a new subgroup begins, as required by the chemical properties of the elements. Not only are the chemical properties of the atoms accounted for, but the arrangement of subgroups explains the observed spectra of the elements.

For example, Table 7–4 shows that each alkali metal has one or more closed shells and subgroups and a single outer electron. The electron occupies the 2s, 3s, 4s orbits for Li, Na, K, respectively. All of the alkalis have the same chemical valence and similar chemical properties, and all have similar doublet spectra, as expected from the electron configuration. The alkali earths have two outer electrons, and so on through the columns of the periodic table. The last column in Table 7–4 gives the experimental values of the ionization potential or separation energy; these values offer clear evidence for the structure of the electron shells. The second electron in the helium atom has an ionization energy of 24.5 ev as compared with 13.6 ev for hydrogen, and the helium electrons must therefore be much more closely bound than the single electron in the hydrogen atom. The third electron which is built into the lithium atom has the small ionization potential of 5.4 ev; it must be at a considerably greater distance from the nucleus and is, consequently, loosely bound. With this electron, the construction of the second, or L-shell, begins. This shell is filled in the neon atom with an ionization potential of 21.5 ev, and the M-shell starts with sodium, with a potential of 5.1 ev. The trend of the experimental values of the ionization potential is exactly that predicted by the theory. It can now be said that modern atomic theory has solved the problem of the interpretation, in basic physical terms, of the chemical properties of the elements.

7–7 Atomic theory and characteristic x-ray spectra. The importance of Moseley's work on the characteristic x-ray spectra of the elements was discussed at some length in the fourth chapter. His results can be interpreted in terms of modern atomic theory and, in fact, the Bohr theory can be used to give a very neat picture of the production of the characteristic x-ray lines. It was seen in Section 4–4 that the wave numbers of the K_α and L_α lines of the x-ray spectrum are given by the formulas

$$K_\alpha\text{: } \bar{\nu} = R(Z - 1)^2\left(\frac{1}{1^2} - \frac{1}{2^2}\right), \tag{7–39}$$

$$L_\alpha\text{: } \bar{\nu} = R(Z - 7.4)^2\left(\frac{1}{2^2} - \frac{1}{3^2}\right). \tag{7–40}$$

Equation (7–39) suggests that a K_α line is the result of the jump of an electron from the L-shell ($n = 2$) to the K-shell ($n = 1$). Such a jump can occur only if one of the two electrons in the K-shell is missing. A vacancy in the K-shell can occur if one of the two electrons in that shell is struck by an energetic cathode ray and knocked either out of the atom or into an unfilled outer shell. Subsequently, an electron from the L-shell may drop into the vacancy in the K-shell with the emission of a quantum

Table 7–4

Electron Configurations and Ionization Potentials of Atoms

Z	Element	K	L		M			N				O			P			Q	Ionization potential, ev
		1s	2s	2p	3s	3p	3d	4s	4p	4d	4f	5s	5p	5d	6s	6p	6d	7s	
1	H	1																	13.595
2	He	2																	24.580
3	Li	2	1																5.390
4	Be	2	2																9.320
5	B	2	2	1															8.296
6	C	2	2	2															11.264
7	N	2	2	3															14.54
8	O	2	2	4															13.614
9	F	2	2	5															17.418
10	Ne	2	2	6															21.559
11	Na	2	2	6	1														5.138
12	Mg	2	2	6	2														7.644
13	Al	2	2	6	2	1													5.984
14	Si	2	2	6	2	2													8.149
15	P	2	2	6	2	3													10.55
16	S	2	2	6	2	4													10.357
17	Cl	2	2	6	2	5													13.01
18	Ar	2	2	6	2	6													15.755
19	K	2	2	6	2	6		1											4.339
20	Ca	2	2	6	2	6		2											6.111
21	Sc	2	2	6	2	6	1	2											6.56
22	Ti	2	2	6	2	6	2	2											6.83
23	V	2	2	6	2	6	3	2											6.74

24	Cr	2	2	6	2	6	5	1									6.764
25	Mn	2	2	6	2	6	5	2									7.432
26	Fe	2	2	6	2	6	6	2									7.90
27	Co	2	2	6	2	6	7	2									7.86
28	Ni	2	2	6	2	6	8	2									7.633
29	Cu	2	2	6	2	6	10	1									7.724
30	Zn	2	2	6	2	6	10	2									9.391
31	Ga	2	2	6	2	6	10	2	1								6.00
32	Ge	2	2	6	2	6	10	2	2								7.88
33	As	2	2	6	2	6	10	2	3								9.81
34	Se	2	2	6	2	6	10	2	4								9.75
35	Br	2	2	6	2	6	10	2	5								11.84
36	Kr	2	2	6	2	6	10	2	6								13.996
37	Rb	2	2	6	2	6	10	2	6				1				4.176
38	Sr	2	2	6	2	6	10	2	6			2					5.692
39	Y	2	2	6	2	6	10	2	6	1		2					6.377
40	Zr	2	2	6	2	6	10	2	6	2		2					6.835
41	Nb	2	2	6	2	6	10	2	6	4		1					6.881
42	Mo	2	2	6	2	6	10	2	6	5		1					7.131
43	Tc	2	2	6	2	6	10	2	6	(5)		(2)?					7.23
44	Ru	2	2	6	2	6	10	2	6	7		1					7.365
45	Rh	2	2	6	2	6	10	2	6	8		1					7.461
46	Pd	2	2	6	2	6	10	2	6	10							8.33
47	Ag	2	2	6	2	6	10	2	6	10		1					7.574
48	Cd	2	2	6	2	6	10	2	6	10		2					8.991
49	In	2	2	6	2	6	10	2	6	10		2	1				5.785
50	Sn	2	2	6	2	6	10	2	6	10		2	2				7.332

(Continued)

Table 7–4 (*Continued*)

Z	Element	*K*	*L*		*M*			*N*				*O*			*P*			*Q*	Ionization potential, ev
		1s	2s	2p	3s	3p	3d	4s	4p	4d	4f	5s	5p	5d	6s	6p	6d	7s	
51	Sb	2	2	6	2	6	10	2	6	10		2	3						8.639
52	Te	2	2	6	2	6	10	2	6	10		2	4						9.01
53	I	2	2	6	2	6	10	2	6	10		2	5						10.44
54	Xe	2	2	6	2	6	10	2	6	10		2	6						12.127
55	Cs	2	2	6	2	6	10	2	6	10		2	6		1				3.893
56	Ba	2	2	6	2	6	10	2	6	10		2	6		2				5.210
57	La	2	2	6	2	6	10	2	6	10		2	6	1	2				5.61
58	Ce	2	2	6	2	6	10	2	6	10	1	2	6	1	2				
59	Pr	2	2	6	2	6	10	2	6	10	3	2	6		2				
60	Nd	2	2	6	2	6	10	2	6	10	4	2	6		2				6.3
61	Pm	2	2	6	2	6	10	2	6	10	5	2	6		2				
62	Sm	2	2	6	2	6	10	2	6	10	6	2	6		2				5.6
63	Eu	2	2	6	2	6	10	2	6	10	7	2	6		2				5.67
64	Gd	2	2	6	2	6	10	2	6	10	7	2	6	1	2				6.16
65	Tb	2	2	6	2	6	10	2	6	10	8	2	6	1	2				
66	Dy	2	2	6	2	6	10	2	6	10	9	2	6	1	2	?			
67	Ho	2	2	6	2	6	10	2	6	10	10	2	6	1	2	?			
68	Er	2	2	6	2	6	10	2	6	10	11	2	6	1	2	?			
69	Tm	2	2	6	2	6	10	2	6	10	13	2	6		2				
70	Yb	2	2	6	2	6	10	2	6	10	14	2	6		2				6.22
71	Lu	2	2	6	2	6	10	2	6	10	14	2	6	1	2				6.15
72	Hf	2	2	6	2	6	10	2	6	10	14	2	6	2	2				5.5
73	Ta	2	2	6	2	6	10	2	6	10	14	2	6	3	2				7.7
74	W	2	2	6	2	6	10	2	6	10	14	2	6	4	2				7.98

75	Re	2	2	6	2	6	10	2	6	10	14	2	6	5		2				7.87
76	Os	2	2	6	2	6	10	2	6	10	14	2	6	6		2				8.7
77	Ir	2	2	6	2	6	10	2	6	10	14	2	6	7		2				9.2
78	Pt	2	2	6	2	6	10	2	6	10	14	2	6	9		1	?			9.0
79	Au	2	2	6	2	6	10	2	6	10	14	2	6	10		1				9.22
80	Hg	2	2	6	2	6	10	2	6	10	14	2	6	10		2				10.434
81	Tl	2	2	6	2	6	10	2	6	10	14	2	6	10		2	1			6.106
82	Pb	2	2	6	2	6	10	2	6	10	14	2	6	10		2	2			7.415
83	Bi	2	2	6	2	6	10	2	6	10	14	2	6	10		2	3			7.287
84	Po	2	2	6	2	6	10	2	6	10	14	2	6	10		2	4			8.43
85	At	2	2	6	2	6	10	2	6	10	14	2	6	10		2	5			
86	Rn	2	2	6	2	6	10	2	6	10	14	2	6	10		2	6			10.745
87	Fr	2	2	6	2	6	10	2	6	10	14	2	6	10		2	6		1	
88	Ra	2	2	6	2	6	10	2	6	10	14	2	6	10		2	6		2	5.277
89	Ac	2	2	6	2	6	10	2	6	10	14	2	6	10		2	6	1	2?	
90	Th	2	2	6	2	6	10	2	6	10	14	2	6	10		2	6	2	2?	
91	Pa	2	2	6	2	6	10	2	6	10	14	2	6	10	2	2	6	1	2?	
92	U	2	2	6	2	6	10	2	6	10	14	2	6	10	3	2	6	1	2	4
93	Np	2	2	6	2	6	10	2	6	10	14	2	6	10	4	2	6	1	2?	
94	Pu	2	2	6	2	6	19	2	6	10	14	2	6	10	5	2	6	1	2?	
95	Am	2	2	6	2	6	10	2	6	10	14	2	6	10	7	2	6		2	
96	Cm	2	2	6	2	6	10	2	6	10	14	2	6	10	7	2	6	1	2?	
97	Bk	2	2	6	2	6	10	2	6	10	14	2	6	10	8	2	6	1	2?	
98	Cf	2	2	6	2	6	10	2	6	10	14	2	6	10	9	2	6	1	2?	

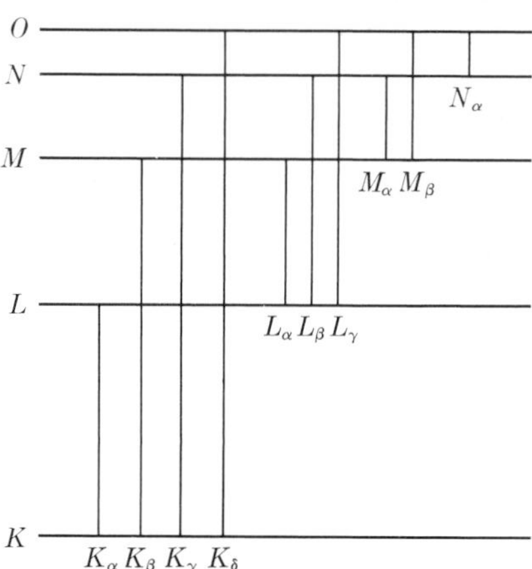

FIG. 7–6. Simple term scheme of x-ray levels showing transitions corresponding to x-ray lines.

of radiation whose frequency is that of the K_α line. The electron which goes from the L-shell to the K-shell moves in the electric field of the positive nuclear charge and the negative charge of the electron remaining in the K-shell. This electric field is equivalent to that of a positive charge of magnitude $(Z - 1)e$. The single K-shell electron tends to screen the nucleus and makes its effective nuclear charge about one unit less than the actual charge.

The explanation of the experimental results for the L_α line is, in principle, the same as that for the K_α line. This time the vacancy is in the L-shell and the radiation is emitted when an electron from the M-shell $(n = 3)$ drops into the vacancy. An M-shell electron is screened from the nucleus by the other 7 L-shell electrons and the K-electrons, and the nuclear charge is effectively reduced by about 7 units.

Just as for the optical spectra, the stationary energy levels and the spectral lines can be represented graphically in a term diagram, and a simple example of such a diagram is shown in Fig. 7–6.

7–8 The basic ideas of wave mechanics. Some of the basic ideas of wave mechanics are applied often in nuclear physics and a brief discussion of these ideas is in order. The fundamental idea of wave mechanics is that of the wave nature of matter. This concept arose as an extension of the treatment of light sometimes as waves and sometimes as particles. Light and x-rays have been treated as waves during the discussion of experiments involving reflection, refraction, and diffraction. The design and operation of optical and x-ray spectrometers as well as the interpretation of the results obtained with these instruments depend on properties typical of waves. On the other hand, in the photoelectric effect, light must be assumed to consist of particles (quanta, or photons). In the Compton effect, the interaction between an x-ray and an electron is treated as an elastic collision between two particles, and equations are written for the balance of energy and momentum, typical particle properties. The results of the calculation are expressed in terms of the shift in wavelength because this is the quantity that is measured experimentally. It is clear that some experiments involving radiation can only be interpreted on the hypothesis that radiation is wave-like in nature, while for other experiments radiation must be regarded as particle-like. This duality is shown in the expression for the momentum of a photon

$$p = \frac{h\nu}{c} = \frac{h}{\lambda}. \tag{7-41}$$

The momentum (a particle property) is related directly to the wavelength (a wave property).

In 1924, de Broglie[4] suggested that matter also has a dual (particle-like and wave-like) character. He assumed that the relation (7-41) holds for electrons and other particles as well as for radiation. Although this assumption was a bold and surprising one to make, it was amenable to experimental tests. Suppose that the particles are electrons which have been accelerated to a velocity v under the action of a potential difference V. If v is small compared with the velocity of light, then

$$\frac{1}{2} mv^2 = \frac{eV}{300}, \tag{7-42}$$

where m, v, and e are in cgs units and V is in volts. The wavelength associated with the electron is, when Eqs. (7-41) and (7-42) are combined,

$$\lambda = \frac{h}{\sqrt{meV/150}}. \tag{7-43}$$

If V is 100 volts, the value of λ should be 1.2×10^{-8} cm, of the order of magnitude of the distances between atomic planes in crystals. This fact suggested that the existence of electron waves might be shown by using crystals as diffraction gratings, as is done for x-rays, and experiments based on this idea were done in 1927 by Davisson and Germer.[5] On bombarding a single crystal of nickel with an electron beam, they found that the reflected beam showed maxima and minima which could be explained in terms of the diffraction of electron waves. The analysis of the experiments was similar to that used for x-ray diffraction by crystals, and the wavelength was determined from the Bragg equation. The quantitative agreement between the predictions of the de Broglie hypothesis and the experimental results was remarkably good.

Diffraction patterns were also obtained[(6)] when electrons passed through thin films of matter such as metal foil or mica, and the patterns were exactly analogous to those obtained in experiments on the diffraction of x-rays. In Fig. 7–7, the similarity between the effect of x-rays and that of electrons is so striking as to prove conclusively the wave properties of electrons. Particles of atomic size, such as helium atoms and helium molecules, have also been found to show diffraction effects, and there is no doubt about the correctness of de Broglie's hypothesis.

The dual wave and particle nature of matter can be expressed mathematically by means of a wave equation first derived by Schroedinger.[(7)] This equation can be derived quite easily for the case of a particle, or a beam of particles all with the same energy, moving in an electrostatic field of force. The usual three-dimensional wave equation in rectangular Cartesian coordinates is

$$\frac{\partial^2\Psi}{\partial x^2} + \frac{\partial^2\Psi}{\partial y^2} + \frac{\partial^2\Psi}{\partial z^2} = \frac{1}{u^2}\frac{\partial^2\Psi}{dt^2}, \tag{7–44}$$

where Ψ (x, y, z, t) is the amplitude of the wave associated with the particle and u is its velocity. If we desire a solution that will represent standing waves (such as those on a string fastened at both ends), we may write Ψ in the form

$$\Psi = \psi e^{2\pi i\nu t}, \tag{7–45}$$

where ψ is a function of x, y, and z, but not of the time t, and ν is the frequency. If the expression for Ψ is inserted into Eq. (7–44), we get

$$\frac{\partial^2\psi}{\partial x^2} + \frac{\partial^2\psi}{\partial y^2} + \frac{\partial^2\psi}{\partial z^2} = -\frac{4\pi^2\nu^2}{u^2}\psi = -\left(\frac{2\pi}{\lambda}\right)^2\psi, \tag{7–46}$$

where $\lambda = u/\nu$ is the wavelength. In view of the wave nature of particles, we may write for the energy W and the momentum p of a particle

$$W = h\nu, \tag{7–47a}$$

$$p = \frac{h}{\lambda}, \tag{7–47b}$$

and the wavelength is then

$$\lambda = \frac{h}{p}. \tag{7–47c}$$

We assume now that the speed v of the particles is much less than that of

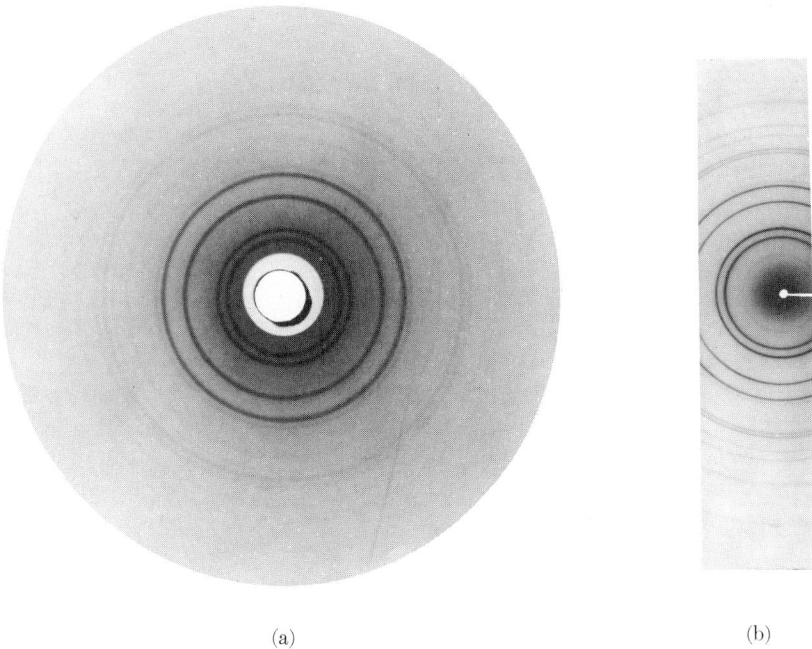

(a) (b)

FIG. 7-7. (a) Diffraction pattern obtained from a gold crystal with x-rays of wavelength 0.709 A. (Courtesy of Mr. O. F. Kammerer, Brookhaven National Laboratory.) (b) Diffraction pattern obtained from a gold crystal with electrons of momentum corresponding to a wavelength of 0.055 A. (Courtesy of Dr. S. Bauer, Cornell University.)

light. The kinetic energy of the particle is the difference between the total energy W and the potential energy U, and we can write

$$\tfrac{1}{2}mv^2 = W - U. \tag{7-48}$$

Then

$$p = mv = \sqrt{2m(W - U)}$$

and, from Eq. (7-47c),

$$\lambda = \frac{h}{\sqrt{2m(W - U)}}\cdot \tag{7-49}$$

On substituting this expression for λ in Eq. (7-46), we get

$$\frac{\partial^2\psi}{\partial x^2} + \frac{\partial^2\psi}{\partial y^2} + \frac{\partial^2\psi}{\partial z^2} + \frac{8\pi^2 m}{h^2}(W - U)\psi = 0. \tag{7-50}$$

Equation (7-50) is the Schroedinger equation, which gives the wave function as a function of the coordinates; the potential energy is also a function of the coordinates.

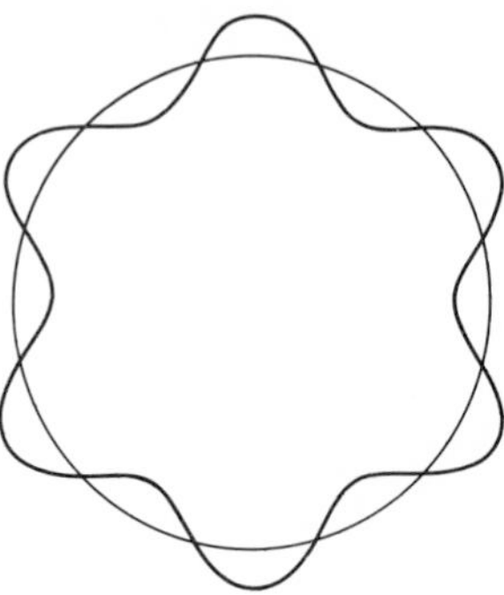

FIG. 7–8. Representation of a one-dimensional standing electron wave in a Bohr orbit.

The mathematical problem which must be solved consists of inserting the appropriate expression for the potential energy into the Schroedinger equation and solving for the wave function ψ. The one-electron atom is treated by setting $U = -Ze^2/r$, where $r = \sqrt{x^2 + y^2 + z^2}$; actually, the Schroedinger equation would be written in spherical polar coordinates in this case. It is found that physically significant solutions for ψ for bound states of the atom exist only when W has the values

$$W_n = -\frac{2\pi^2 me^4}{n^2h^2} Z^2, \tag{7–51}$$

where n is an integer. These values are the same as those found for the stationary states in the Bohr theory, but the result is obtained from the mathematical formulation of the experimentally proven idea of the wave nature of matter rather than from a hybrid set of classical-quantal assumptions.

Bohr's quantum conditions also follow from the idea of the wave nature of matter. As a simple example, consider the electron in the hydrogen atom as a standing wave extending in a circle around the nucleus. In order that the electron wave may just fill the circumference of a circle, as in Fig. 7–8, the circle must contain an integral number of wavelengths, or

$$2\pi r = n\lambda, \tag{7–52}$$

where r is the radius of the circle and n is an integer. By de Broglie's hypothesis, $\lambda = h/p = h/mv$, and Eq. (7–52) becomes

$$2\pi r = n\frac{h}{mv}, \qquad \text{or} \qquad mvr = n\frac{h}{2\pi}. \tag{7–53}$$

But mvr is the angular momentum of the electron regarded as a particle, and we see that the wave mechanical picture leads naturally to Bohr's postulate that the angular momentum is equal to an integral multiple of $h/2\pi$. This type of proof can be extended to the other quantum conditions, and the latter appear as a consequence of the condition for a standing wave rather than as arbitrary assumptions. The concept of the electron as a standing wave rather than as a particle revolving in an orbit also removes the difficulty that in the Bohr theory the electron travels in its orbit without radiating.

It can also be shown that in wave mechanics the emission of a photon of frequency ν is a logical consequence of an energy change of the atom; the mechanism of emission does not have to be introduced as a separate postulate, as in the Bohr theory. Thus, the difficulties inherent in Bohr's postulates are all satisfactorily removed by wave mechanics.

The concept of the dual particle and wave nature of matter leads to another important result. According to classical mechanics, a particle has a position and momentum which can, in principle, be determined with any desired accuracy. But a wave is extended throughout a region of space, and the location of an electron, considered as a wave, seems to present some difficulty. The question arises: If an electron must be described sometimes as a wave and sometimes as a particle, is it possible to know exactly where an electron is in space at some given instant? The answer to this question is given by *Heisenberg's uncertainty principle*,[(8)] which states that the position and momentum of a particle *cannot* be determined simultaneously with any arbitrary desired accuracy. These quantities can be determined only with accuracies limited by the relation

$$\Delta x \Delta p \sim h, \tag{7–54}$$

where Δx is the error in the determination of the position and Δp is the error in the momentum. The product of the two errors is at least of the order of magnitude of Planck's constant. A similar relation holds for the energy of a particle and the time,

$$\Delta t \Delta E \sim h. \tag{7–55}$$

The condition of Eq. (7–54) means that the more precisely the position of a particle is determined, the less precisely can the momentum be measured, and conversely. The last condition means that the more accurately the time at which an atomic event occurs is determined, the less accurately can the energy change of the atomic system be determined, and conversely.

The meaning of the uncertainty principle can be illustrated by an idealized experiment. Consider the following attempt to measure simultaneously the position and momentum of an electron with an imaginary microscope of very high resolving power. It has been shown in wave optics that the resolving power of a microscope is given approximately by

$$\Delta x \sim \frac{\lambda}{2 \sin A}, \tag{7–56}$$

where Δx is the distance between two points which can just be resolved by the microscope, λ is the wavelength of the light used, and A is the half-angle of the lens used, as in Fig. 7–9. Then Δx is the uncertainty in the determination of the position of the electron; to make Δx as small as possible the wavelength must be very small and either x-rays or γ-rays must be used. Now the position of an electron can be measured with a microscope only if at the very least one photon is deflected from its initial direction into the microscope. But when this photon is scattered it gives to the electron, because of the Compton effect, a momentum of the order of $h\nu/c$. The change, Δp, in the momentum of the electron cannot be determined exactly because the scattered photon can enter the microscope anywhere within the angle A. The uncertainty in the momentum is of the order

$$\Delta p \sim \frac{h\nu}{c} \sin A. \tag{7–57}$$

From Eqs. (7–56) and (7–57), it follows that

$$\Delta p \, \Delta x \sim h,$$

which is the Heisenberg relationship.

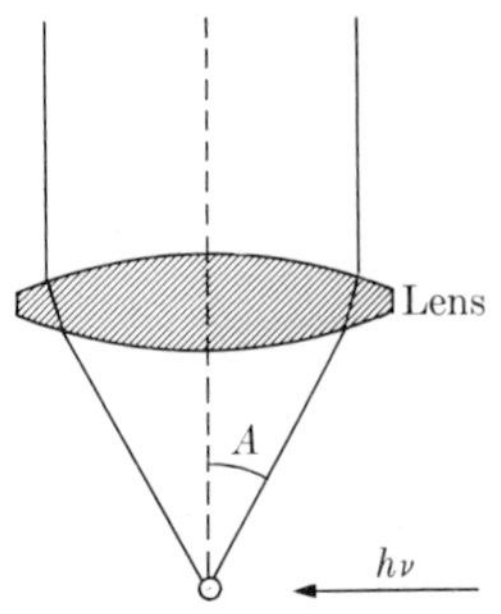

FIG. 7–9. Imaginary experiment in connection with the uncertainty principle.

This discussion shows that one of the causes of the uncertainty relationship is the fact that the process of measurement itself perturbs the particle whose properties are being measured. From an operational viewpoint, the exact location of an electron has therefore no real meaning, and there is no real point in thinking of an atom in the same way that we think about a macroscopic planetary system. This result gives a clue to the failure of Newtonian mechanics when it is applied to atomic physics.

The interpretation of the Schroedinger wave function ψ presented a problem until Born (1926) showed that logical consistency and agreement between theory and experiment could be obtained provided that ψ is regarded as a measure of the probability of finding a particle in a given region. Thus, $|\psi|^2 \, dx \, dy \, dz$, where the vertical lines denote the absolute magnitude of ψ, is the probability that the electron will be found in the volume element $dx \, dy \, dz$ about the point (x, y, z). Other physical results can be expressed in terms of *expectation values;* these are mathematical expectations, in the sense of probability theory, for the results of a single measurement, or for the average of a large number of experiments. This interpretation is in accord with the statistical nature of the results of experiments on atomic systems.

The success of wave mechanics is not limited to the examples that have been discussed here. The quantitative treatment of many problems in nuclear physics depends on the application of wave mechanics. Molecular physics, the physics of the solid and liquid states, statistical mechanics, the theory of chemical reaction rates, and many electric and magnetic phenomena can also be treated by means of the ideas and techniques of wave mechanics. Many examples of these applications will be found in the books cited in the general references at the end of this chapter.

7–9 The solution of the Schroedinger equation: some useful examples and results. 1. *Particle in a box.* The solution of the Schroedinger equation for significant physical problems usually involves mathematical methods beyond the scope of this book. There are, however, a few simple problems whose solutions show some of the basic differences between classical mechanics and quantum mechanics. One of these problems is that of a free particle in a "one-dimensional box." We consider a particle of mass m which can move freely over the range $0 < x < a$, but is reflected when it strikes either end of the range. These conditions can be represented in Schroedinger's equation by assuming that the potential energy U is zero for $0 < x < a$, and that the wave function ψ vanishes at $x = 0$ and at $x = a$. The Schroedinger equation for the particle is then

$$\frac{d^2\psi}{dx^2} + \frac{8\pi^2 mW}{h^2}\psi = 0. \tag{7–58}$$

If we set

$$k^2 = \frac{8\pi^2 mW}{h^2}, \tag{7–59}$$

the general solution is

$$\psi(x) = A \sin kx + B \cos kx. \tag{7–60}$$

The boundary condition $\psi = 0$ at $x = 0$ requires $B = 0$, and the condition at $x = a$ requires

$$\psi(a) = A \sin ka = 0. \tag{7–61}$$

We cannot take $A = 0$ because there would then be no solution. Hence, we must have $\sin ka = 0$, or

$$k = \frac{n\pi}{a}, \qquad n = 1, 2, 3, \ldots. \tag{7–62}$$

The only allowed solutions of the wave equation are the functions

$$\psi_n(x) = A \sin \frac{n\pi x}{a}, \tag{7-63}$$

with n related to k by Eq. (7–62). That relation and Eq. (7–59) require that the energy W can have only the values

$$W_n = \frac{n^2h^2}{8ma^2}. \tag{7-64}$$

An energy level W_n corresponds to a value of n, and to a wave function ψ_n given by Eq. (7–63); each value of W_n is called an *eigenvalue* or *proper value,* and the corresponding wave function ψ_n is called an *eigenfunction* or *proper function.* The interpretation of the wave function requires that

$$\int_0^a |\psi_n(x)|^2 \, dx = A^2 \int_0^a \sin^2 \frac{n\pi x}{a} \, dx = 1, \tag{7-65}$$

or that $A = \sqrt{2/a}$, and

$$\psi_n(x) = \sqrt{\frac{2}{a}} \sin \frac{n\pi x}{a}. \tag{7-66}$$

The wave functions are said to be normalized when they satisfy Eq. (7–65). The wave functions have another important property, that of *orthogonality:* for $n = m$, it is easily verified that

$$\int_0^a \psi_n(x)\psi_m(x) \, dx = 1, \tag{7-67}$$

while for $n \neq m$,

$$\int_0^a \psi_n(x)\psi_m(x) = 0.$$

Equation (7–64) is a quantization condition which introduces the quantum number n; it comes from the physical boundary conditions and gives rise to the quantized values W_n of the energy. The momentum is also quantized; for each energy value W_n, there is a momentum p_n given by

$$p_n^2 = 2mW_n,$$

or

$$p_n = \frac{nh}{2a} = \frac{\pi}{a} n \frac{h}{2\pi} = \frac{\pi}{a} n\hbar. \tag{7-68}$$

The first three wave functions are plotted in Fig. 7–10.

We can calculate the probability that the particle, in a given state of energy, will be located within any element dx of the one-dimensional box.

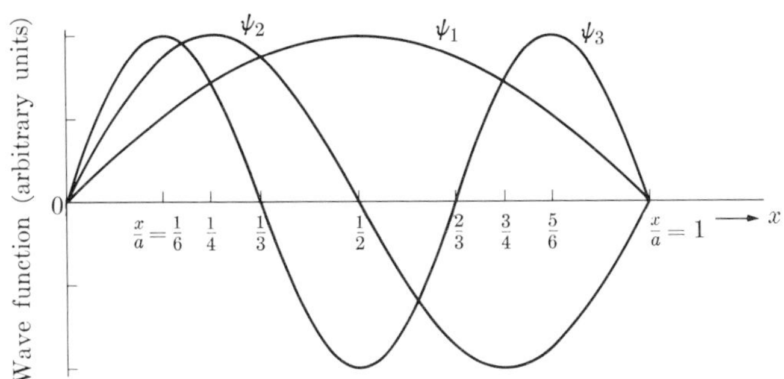

FIG. 7–10. The first three wave functions for a particle in a one-dimensional box.

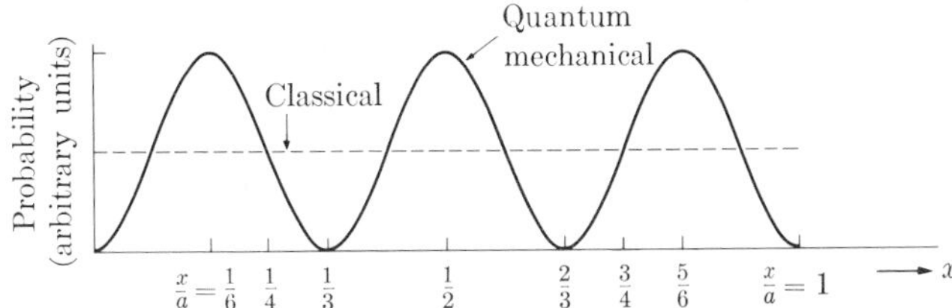

FIG. 7–11. The probability distribution for finding a particle of given energy at points in a one-dimensional box (case $n = 3$).

According to quantum mechanics, this probability is

$$|\psi(x)|^2\,dx = \frac{2}{a}\sin^2\frac{n\pi x}{a}\,dx, \tag{7–69}$$

which is a wave-like function; it is shown graphically in Fig. 7–11 for the state defined by $n = 3$, and $W_3 = 9h^2/8ma^2$. There are actually nodes, at $x/a = \frac{1}{3}$ and $x/a = \frac{2}{3}$ in this case, where the probability is zero. Thus, the quantum mechanical result is very different from the classical one for which the probability is the same for any element dx.

If the particle is confined within a three-dimensional box with edges of lengths a, b, c, parallel to the x-, y-, and z-axes, respectively, the normalized wave functions for the stationary states and the energy levels are found to be

$$\psi_{n_1,n_2,n_3}(x, y, z) = 2\sqrt{\frac{2}{abc}}\sin\frac{n_1\pi x}{a}\sin\frac{n_2\pi y}{b}\sin\frac{n_3\pi z}{c}, \tag{7–70}$$

$$W_{n_1,n_2,n_3} = \frac{h^2}{8m}\left(\frac{n_1^2}{a^2} + \frac{n_2^2}{b^2} + \frac{n_3^2}{c^2}\right), \tag{7–71}$$

where n_1, n_2, n_3 denote any set of three positive integers.

2. *The hydrogen atom: angular momentum; parity.* The hydrogen atom is treated in quantum mechanics by putting $U = -e^2/r$ into the Schroedinger equation (7–50). The solution of the equation is complicated; one of the results obtained has already been mentioned, namely, that the energy has the eigenvalues given by Eq. (7–51). Another very important result involves the angular momentum and its quantization. It is found that the angular momentum **L** (vector quantities are indicated by boldface letters), has the magnitude, L, given by

$$L = \sqrt{l(l+1)}\,\frac{h}{2\pi}, \tag{7–72}$$

where l is the orbital quantum number, which may take on the values given in the second of Eqs. (7–37). The angular momentum is a vector quantity and its direction must be specified, as well as its magnitude. The magnetic quantum number m_l, whose possible values are given in the third of Eqs. (7–37) defines a limited number of orientations which **L** may take up with respect to a specified axis. These orientations are shown, for $l = 3$, in Fig. 7–12; in this case, $L = \sqrt{12}\,h/2\pi$. The orbital angular momentum is said to be spatially quantized in terms of the quantum numbers m_l. The angular momentum vector **L** can never lie entirely along a chosen axis but only at a minimum inclination given by $\cos^{-1}(l/l+1)^{1/2}$, which approaches zero in the limit of large values of l. The *projection* of **L** along a specified axis, usually taken as the z-axis, has magnitudes which are given by

$$L_z = m_l\left(\frac{h}{2\pi}\right). \tag{7–73}$$

According to quantum mechanics, an electron also has a spin angular momentum S characterized by a spin quantum number which can take on the values $\pm\frac{1}{2}h/2\pi$. More precisely, the magnitude of the spin angular momentum is

$$S = \sqrt{s(s+1)}\left(\frac{h}{2\pi}\right), \qquad s = \frac{1}{2}; \tag{7–74}$$

the component of the spin angular momentum along a specified axis S_z is quantized according to the rule

$$S_z = m_s\,\frac{h}{2\pi}; \qquad m_s = \pm\frac{1}{2}. \tag{7–75}$$

The orbital and spin motions of the electron are combined, or *coupled,*

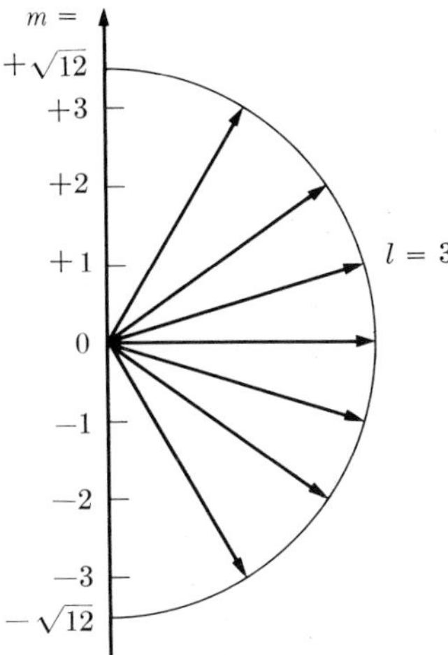

FIG. 7-12. Spatial quantization of the orbital angular momentum.

according to the rules of vector addition to give a resultant angular momentum **J** characterized by a quantum number j;

$$\mathbf{J} = \mathbf{L} + \mathbf{S}; \tag{7-76}$$

for a given value of l, j can take on only the values $l \pm \frac{1}{2}$.

When changes in atomic states are produced by the action of two or more electrons, the value of the total angular momentum of these electrons, denoted by **J**, depends on the coupling between the orbital and angular momenta of the individual electrons. The study of atomic spectra has shown that the predominant type of coupling is that in which all the orbital angular momentum vectors $\mathbf{L}_i$ of the electrons combine to form a resultant **L** and, independently, all the spin angular momentum vectors $\mathbf{S}_i$ combine to form a resultant **S**. The total angular momentum is then given by the relation

$$\mathbf{J} = \mathbf{L} + \mathbf{S} = \sum_i \mathbf{L}_i + \sum_i \mathbf{S}_i. \tag{7-77}$$

This type of coupling is called *L-S*, or Russell-Saunders coupling after its discoverers. The total angular momentum is quantized and there are appropriate rules for the possible values which may be taken on by the quantum numbers corresponding to **L**, **S**, and **J**.

We have seen that the quantum-mechanical study of the hydrogen-like atom has led to certain quantum numbers which apply to physical quantities, such as energy and angular momentum, well known in classical mechanics. There is another quantum number, the parity of a wave function ψ, which is related to the *symmetry* properties of ψ and has no relation to a classical physical quantity. The parity of a wave function refers to its behavior under a simultaneous reflection of the space coordinates, $x \to -x$, $y \to -y$, $z \to -z$. We may distinguish between two ideal cases

$$\psi(-x, -y, -z) = +\psi(x, y, z), \tag{7-78a}$$

$$\psi(-x, -y, -z) = -\psi(x, y, z); \tag{7-78b}$$

the wave function may be invariant under the reflection, or it may change sign. Equations (7–78) can be combined in the form

$$\psi(-x, -y, -z) = P\psi(x, y, z), \tag{7-79}$$

where $P = \pm 1$. We can treat P as a quantum number, and the property it defines is called the *parity* of the system. The parity is then quantized and has the quantum number P; when $P = 1$, the system is said to have *even* parity; when $P = -1$, the system is said to have *odd* parity. As for other quantum numbers, there are selection rules involving the change of parity in a transition.

In the case of a hydrogen-like atom, it is found that the parity is related to the orbital quantum number l,

$$P = (-1)^l. \tag{7-80}$$

Since the parity is defined in terms of the Schroedinger wave function ψ, which is a purely quantum-mechanical concept, it can have no classical analog.

References

GENERAL

F. K. Richtmyer, E. H. Kennard, and T. Lauritsen, *Introduction to Modern Physics,* 5th ed. New York: McGraw-Hill, 1955, Chapters 5, 6, 7.

A. P. French, *Principles of Modern Physics.* New York: Wiley, 1958, Chapters 5, 7, 8.

R. B. Leighton, *Principles of Modern Physics.* New York: McGraw-Hill, 1959, Chapters 2–8.

EXPERIMENTAL TECHNIQUES

G. P. Harnwell and J. J. Livingood, *Experimental Atomic Physics.* New York: McGraw-Hill, 1933, Chapters 5, 7, 8.

J. B. Hoag and S. A. Korff, *Electron and Nuclear Physics,* 3rd ed. New York: Van Nostrand, 1948, Chapters 3, 7.

G. R. Harrison, R. C. Lord, and J. R. Loofbourow, *Practical Spectroscopy.* New York: Prentice-Hall, 1948.

ATOMIC SPECTRA

L. PAULING and S. GOUDSMIT, *The Structure of Atomic Spectra.* New York: McGraw-Hill, 1930.

R. BACHER and S. GOUDSMIT, *Atomic Energy States.* New York: McGraw-Hill, 1932.

H. E. WHITE, *Introduction to Atomic Spectra.* New York: McGraw-Hill, 1934.

G. HERZBERG, *Atomic Spectra and Atomic Structure.* New York: Dover, 1946.

G. W. SERIES, *The Spectrum of Atomic Hydrogen.* Oxford: Clarendon Press, 1957.

ATOMIC THEORY AND ITS APPLICATIONS

N. BOHR, *Atomic Theory and the Description of Nature.* Cambridge: University Press, 1934

M. BORN, *Atomic Physics,* 6th ed. London and Glasgow: Blackie; New York: Hafner, 1957.

U. FANO and L. FANO, *Basic Physics of Atoms and Molecules.* New York: Wiley, 1959.

H. SEMAT, *Introduction to Atomic and Nuclear Physics,* New York: Rinehart, 1954.

D. C. PEASLEE, *Elements of Atomic Physics.* New York: Prentice-Hall, 1955.

G. P. HARNWELL and W. E. STEPHENS, *Atomic Physics.* New York: McGraw-Hill, 1955.

J. C. SLATER, *Quantum Theory of Matter.* New York: McGraw-Hill, 1951.

J. C. SLATER, *Quantum Theory of Atomic Structure.* New York: McGraw-Hill Vol. I, 1960; Vol. II, 1960.

W. KAUZMANN, *Quantum Chemistry.* New York: Academic Press, 1957.

WAVE MECHANICS (QUANTUM MECHANICS)

W. HEITLER, *Elementary Wave Mechanics,* 2nd ed. Oxford: Clarendon Press, 1956.

N. F. MOTT, *Elements of Wave Mechanics.* Cambridge: University Press, 1958.

R. H. DICKE and J. P. WITTKE, *Introduction to Quantum Mechanics.* Reading, Mass.: Addison-Wesley, 1960.

J. L. POWELL and B. CRASEMANN, *Quantum Mechanics.* Reading, Mass.: Addison-Wesley, 1961.

D. BOHM, *Quantum Theory.* New York: Prentice-Hall, 1951.

N. F. MOTT and I. N. SNEDDON, *Wave Mechanics and Its Applications.* Oxford: Clarendon Press, 1948.

L. I. SCHIFF, *Quantum Mechanics,* 2nd ed. New York: McGraw-Hill, 1955.

F. MANDL, *Quantum Mechanics,* 2nd ed. London: Butterworth; New York: Academic Press, 1958.

P. A. M. DIRAC, *The Principles of Quantum Mechanics,* 4th ed. Oxford: Clarendon Press, 1958.

L. D. LANDAU and E. M. LIFSHITZ, *Quantum Mechanics: Non-Relativistic Theory.* London: Pergamon Press; Reading, Mass.: Addison-Wesley, 1958.

PARTICULAR

1. N. Bohr, "On the Constitution of Atoms and Molecules," *Phil. Mag.* **26,** 1, 476, 857 (1913); "On the Quantum Theory of Radiation and the Structure of the Atom," *ibid.*, **30,** 394 (1915).

2. J. Franck and G. Hertz, *G. phys. Ges., Verhandlungen,* **16,** 457, 512 (1914); also A. P. French, *op. cit.* gen. ref., Chapter 12.

3. N. Bohr, *On the Application of the Quantum Theory to Atomic Structure. Part I. The Fundamental Postulates.* Cambridge: University Press, 1924; also *Z. Physik* **13,** 117 (1923).

4. L. de Broglie, in L. de Broglie and L. Brillouin, *Selected Papers on Wave Mechanics.* London: Blackie, 1929.

5. C. Davisson and L. H. Germer, "Diffraction of Electrons by a Crystal of Nickel," *Phys. Rev.* **30,** 705 (1927); *Proc. National Acad. Sci.,* **14,** 217, 619 (1928).

6. G. P. Thomson, "Experiment on the Diffraction of Cathode Rays," *Proc. Roy. Soc.* (London), **A117,** 600 (1928); **A119,** 663 (1928).

7. E. Schroedinger, *Collected Papers on Wave Mechanics.* London: Blackie, 1929.

8. W. Heisenberg, *The Physical Principles of the Quantum Theory.* Chicago: University of Chicago Press, 1929; New York: Dover.

Problems

1. Three lines in the Balmer series of hydrogen have the wavelengths 3682.82 A, 3669.43 A, and 3662.24 A, respectively, measured in air. What is the value of n for each line? Take R_{H} as 109,677.576, and the index of refraction of air as 1.0002837.

2. The hydrogen isotope of mass 2, deuterium, has a spectrum in which the Balmer lines are displaced slightly from the lines of ordinary hydrogen (mass 1). Find the value of the Rydberg constant for deuterium, then calculate the difference between the wavelengths of the first three Balmer lines of hydrogen and deuterium.

3. Show that the frequency of revolution of an electron in a Bohr orbit with quantum number n is

$$\nu_{\text{orb}} = \frac{4\pi^2 m' e^4 Z^2}{h^3 n^3}.$$

Show next that if an electron makes a transition from a state for which $n = n_1$ to a state for which $n = n_2$, with $n_1 - n_2 = 1$, then the frequency ν of the emitted radiation is intermediate between the frequencies of orbital revolution in the initial and final states. What is the relationship between the emitted and orbital frequencies when n_1 and n_2 are both very large?

4. Calculate the energy of each of the five lowest stationary states (energy levels) of the hydrogen atom, in wave-number units, ergs, electron volts, and calories per gram.

5. The best values of R_{H} and R_{He} are now considered to be $R_{\text{H}} = 109677.576$ cm^{-1} and $R_{\text{He}} = 109722.267$ cm^{-1}. Calculate accurately the value of the ratio

of the mass of the hydrogen atom to that of the electron. Use the result to calculate the value of e/m for the electron.

6. In an experiment of the Franck-Hertz type, hydrogen was bombarded by electrons with known kinetic energy. The temperature of the hydrogen was high enough so that the gas was dissociated into atomic hydrogen. It was observed that radiation was emitted when the electrons had the following discrete energies: 10.19 ev, 12.08 ev, 12.75 ev, 13.05 ev, 13.12 ev, 13.31 ev. What is the wavelength of the radiation corresponding to each electron energy? To which spectral series do the lines belong? What are the quantum numbers of the initial and final states in the transition represented by each spectral line?

7. From the constancy of angular momentum at $kh/2\pi$ and the constancy of total energy at $-2\pi^2 m' e^4 Z^2/n^2 h^2$, where m' is the reduced mass of the electron, find the semiaxes of the Sommerfeld elliptic orbits in terms of k and n. Prove that $b/a = k/n$. (See Born, *Atomic Physics*, Appendix XIV.)

8. The noble gases Ne, A, Kr, Xe, and Rn have electron configurations in which the s and p orbits of the outer shell are filled. Prepare a table showing the number of electrons in each shell and each orbit for these elements, and give the value of n and l associated with each shell and each orbit. Add to this table the electron configurations attributed to the alkalis, Na, K, Rb, Cs, and Fr, and to the halogens, F, Cl, Br, I, and At. How is the occurrence of 15 rare earth elements with similar chemical properties explained?

9. Calculate the wavelengths of the K_α x-ray lines of Al, K, Fe, Co, Ni, Cu, Mo, and Ag, and compare the values with those listed in Table 4–1. Use $R = 109737.31\ \text{cm}^{-1}$.

10. Show that the wavelength of an electron can be written in the form

$$\lambda = \frac{h}{m_0 c} \frac{\sqrt{1 - (v^2/c^2)}}{(v/c)}.$$

The quantity h/m_0c is called the *Compton wavelength of the electron*. What is its magnitude? Compute the wavelength of electrons for each of the speeds listed in Problem 5, Chapter 6. What is the Compton wavelength of a proton? What is the wavelength of protons at each of the above speeds?

11. Show that the wavelength of electrons moving at a velocity very small compared with that of light and with a kinetic energy of V electron volts can be written

$$\lambda = \frac{12.268}{V^{1/2}} \times 10^{-8}\ \text{cm}.$$

What is the wavelength of electrons with energies of 1, 10, 100, 1000 volts?

12. If V exceeds a few thousand electron volts, the effect of the relativistic increase of mass with velocity must be considered. Show that the wavelength is given by

$$\lambda = \frac{12.268(m/m_0)^{-1/2}}{V^{1/2}} \times 10^{-8}\ \text{cm},$$

where m/m_0 is the ratio of the moving mass of the electron to its rest mass. What is the wavelength of electrons accelerated through a potential of 10^4 volts? 10^5, 10^6, 10^7 volts?

13. What is the de Broglie wavelength of (a) a hydrogen molecule moving at a speed of 2.4×10^5 cm/sec? (b) a rifle bullet weighing 2.0 gm, and moving at a speed of 400 m/sec?

14. What relationship exists between the emission spectrum of singly ionized helium and that of atomic hydrogen? What are the wavelengths of the first four lines of the series which includes transitions to the state $n = 2$ of singly ionized helium? How does the spectrum of doubly ionized lithium compare with that of hydrogen? What is the radius of the first Bohr orbit of doubly ionized lithium?

15. It is known that a particle exists, called the *positron,* with the same mass as that of the electron and with a charge equal in magnitude but opposite in sign. A positron and an electron can form a system, called *positronium,* analogous to hydrogen, with a very short, but measurable, half-life. · (a) Show that the energy levels of positronium are analogous to those of hydrogen but that their magnitudes are smaller by the factor $\frac{1}{2}[1 + (m/M_{\mathrm{H}})]$. (b) Show that the radii of the Bohr orbits of positronium are larger than those of hydrogen by a factor which is the reciprocal of the factor of part (a). (c) What is the ground state energy of positronium in electron volts? What is the radius of the first Bohr orbit?

16. A particle called a *negative π-meson* has been shown to exist; its charge is the same as that of the electron and its mass is 273 times that of the electron. A proton and a negative π-meson can form short-lived stationary states analogous to those of the hydrogen atom. (a) How do the mesonic orbits compare with those of the corresponding electronic orbits? What is the radius of the mesonic orbit for $n = 1$? (b) What is the wavelength of the electromagnetic radiation emitted when a negative π-meson, initially at rest infinitely far from the proton, drops into the smallest orbit permitted by quantum theory?

Part II

The Nucleus

CHAPTER 8

THE CONSTITUTION OF THE NUCLEUS

In the last chapter, it was shown that the application of quantum theory to the nuclear model of the atom led to the development of a satisfactory theory for those properties of atoms that depend on the extranuclear electrons. The development of a theory of the nucleus is a more difficult problem. The density of matter in the nucleus offers a clue to the source of the difficulty. The work of Rutherford and his colleagues on the scattering of α-particles showed that the atomic nucleus has a radius of 10^{-12} to 10^{-13} cm, so that the volume of the nucleus is of the order of 10^{-36} cm^3 or less. Now, the mass of one of the lighter atoms is about 10^{-24} gm, and it is almost all concentrated in the nucleus, with the result that the density of the nucleus is at least 10^{12} gm/cm^3. A density of this magnitude is inconceivably large, and it is clear that in the atomic nucleus, matter is put together in a way which may not be amenable to ordinary experimental and theoretical methods of analysis. Consequently, the interpretation of the nuclear properties of atoms in terms of a theory of nuclear structure presents great problems.

There is now available a large amount of experimental information about nuclei, derived from work in several fields: (1) the precise measurement of the masses of atoms; (2) radioactivity, natural and artificial; (3) the artificial transmutation of nuclei by bombardment with particles from radioactive substances or with high-speed particles produced by laboratory methods; (4) optical spectroscopy in the visible and ultraviolet regions; (5) the direct measurement of certain nuclear properties, such as spin and magnetic moment. The main problems of nuclear physics are the collection and correlation of the experimental facts and their interpretation in terms of a theory or model of the nucleus. In this chapter, the problem of the constitution of the nucleus will be discussed, and some ideas will be developed which will be useful in the interpretation of experimental data in the following chapters. In addition, some new and unfamiliar properties of the nucleus will be considered.

8–1 The proton-electron hypothesis of the constitution of the nucleus. The fact that certain radioactive atoms emit α- and β-rays, both of which are corpuscular in nature, led to the idea that atoms are built up of elementary constituents. As early as 1816, on the basis of the small number of atomic weights then known, Prout suggested that all atomic weights are whole numbers, that they might be integral multiples of the atomic weight of hydrogen, and that all elements might be built up of hydrogen. Prout's hypothesis was discarded when it was found that the atomic weights of some elements are fractional, as for example, those of chlorine (35.46) and copper (63.54). Nevertheless, so many elements have atomic weights which are very close to whole numbers that there seemed to be some basis for Prout's hypothesis. The idea that all elements are built up from one basic substance received new support during the early years of the 20th century when the study of the radioactive elements led to the discovery of isotopes. It was found that there are atomic species which have different masses in spite of the fact that they belong to the same element and have the same atomic number and chemical properties; the different species belonging to the same element are called *isotopes*. For example, four radioactive isotopes of lead were discovered with atomic weights of 214, 212, 211, and 210, as well as three nonradioactive isotopes with atomic weights of 206, 207, and 208. The nuclei of these isotopes, varying in mass from 206 to 214, show a wide range of stability as measured by the extent of their radioactivity, although all have the same charge.

The proof of the existence of isotopes in the radioactive elements led to experiments to test whether some of the ordinary elements also consist of a mixture of isotopes. It was found that this is indeed the case. Most elements are mixtures of isotopes, and the atomic masses of the isotopes are very close to whole numbers. Chlorine, as found in nature, has two isotopes with atomic weights of 34.98 and 36.98, respectively; 75.4% of the chlorine atoms have the smaller mass, while 24.6% have the greater mass, and this distribution explains the atomic weight 35.46 of chlorine. Analogous results were obtained for copper. The different isotopes of an element have the same number and arrangement of extranuclear electrons, and consequently their spectra have the same general structure; they are distinguished from one another by their different atomic masses.

The fact that the atomic masses of the isotopes of an element are close to whole numbers led Aston to formulate his *whole number rule*. According to this rule, which is really a modified form of Prout's hypothesis, all atomic weights are very close to integers, and the fractional atomic weights determined by chemical methods are caused by the presence of two or more isotopes each of which has a nearly integral atomic weight. Much of the experimental work on isotopes involved the analysis of the positive rays from different substances; and in all the work of this kind the lightest positively charged particle that was ever found had the same mass as the hydrogen atom, and carried one positive charge equal in magnitude to the electronic charge, but of opposite sign. This particle is evidently the nucleus of a hydrogen atom and, as shown in Chapter 1, has a mass very close to *one* atomic mass unit. The combination of the whole number rule and the special properties of the hydrogen nucleus led to the assumption that atomic nuclei are built up of hydrogen nuclei, and the hydrogen nucleus was given the name *proton* to indicate its importance as a fundamental constituent of all atoms.

The whole number rule is actually an approximation, holding to an accuracy of about 1 part in 1000. The most precise experiments show that there are small but systematic departures from this rule over the whole range of elements. It will be shown in the next chapter that these variations are of great importance in adding to our knowledge of the structure of nuclei.

To account for the mass of a nucleus whose atomic weight is very close to the integer A, it was necessary to assume that the nucleus contained A protons. But if this were the case, the charge on the nucleus would be equal to A, nearly the same as the atomic weight and not equal to the atomic number Z, which is half, or less, of the atomic weight. To get around this difficulty, it was assumed that in addition to the protons, atomic nuclei contained $A - Z$ electrons; these would contribute a negligible amount to the mass of the nucleus, but would make the charge equal to $+Z$, as required. It was thus possible to consider the atom as consisting of a nucleus of A protons and $A - Z$ electrons surrounded by Z extranuclear electrons. The number A is called the *mass number* and is the integer closest to the atomic weight.

The proton-electron hypothesis of the nucleus seemed to be consistent with the emission of α- and β-particles by the atoms of radioactive elements. The interpretation of certain generalizations about radioactivity in terms of the nuclear atom showed that both the α- and β-particles were ejected from the nuclei of the atoms undergoing transformation; and the presence of electrons in the nucleus made it seem reasonable that under the appropriate conditions one of them might be ejected. It was also reasonable to assume that α-particles could be formed in the nucleus by the combination of four protons and two electrons. The α-particles could exist as such, or they might be formed at the instant of emission.

8–2 The angular momentum of the nucleus; failure of the proton-electron hypothesis. Although the hypothesis that nuclei are built up of protons and electrons had some satisfactory aspects, it eventually led to contradictions and had to be abandoned. One of the failures of the hypothesis was associated with a hitherto unknown property of the nucleus, the angular momentum. The discovery that the atomic nucleus has an angular momentum, or *spin*, with which is associated a magnetic moment, was the result of the detailed study of spectral lines. When individual components of multiplet lines were examined with spectral apparatus of the highest possible resolution, it was found that each of these components is split into a number of lines lying extremely close together; this further splitting is called *hyperfine structure*. The total splitting, in units of wave number, is only about 2 cm^{-1} or less. The hyperfine structure could not be accounted for in terms of the extranuclear electrons, and it was necessary to assume, as Pauli did in 1924, that it is related to properties of the atomic nucleus. The properties associated with the hyperfine structure are the mass and angular momentum of the nucleus.

It was shown in Chapter 7 that because of the simultaneous motion of nucleus and electron around the common center of gravity, the Rydberg constant of an element depends on the mass of the atomic nucleus. Hence, if an element has more than one isotope, each isotope has a slightly different value of the Rydberg constant, and corresponding spectral lines of different isotopes have slightly different wave numbers. This effect has been found experimentally, and the theoretical predictions have been confirmed. In many cases, however, the isotope effect is not enough to explain the hyperfine structure because the number of components is greater than the number of isotopes. Elements which have only one isotope, such as bismuth, also show hyperfine structure. In these cases, the hyperfine structures can be accounted for quantitatively if it is assumed, as for the extranuclear electrons, that the nucleus has an angular momentum.

A magnetic moment is associated with the angular momentum of the nucleus just as one is associated with the angular momentum of an electron. The two magnetic moments interact, and the interaction energy perturbs the total energy of the electrons; there is, therefore, a splitting of the atomic levels, which gives rise to the hyperfine structure of the lines of the atomic spectrum. The multiplicity and the relative spacings of the lines can be derived theoretically and depend on the magnitudes of the nuclear angular momentum and magnetic moment. The nuclear angular momentum can then be deduced from the experimentally determined multiplicity and relative spacings. Newer methods, which involve radiofrequency spectroscopy, microwave spectroscopy, or the deflection of molecular beams in magnetic fields, have been developed for measuring the nuclear magnetic moment, and a large body of experimental information has been built up concerning these nuclear properties.

The nuclear angular momentum has quantum mechanical properties analogous to those of the angular momentum of the electron. It is a vector, $\mathbf{I}$, of magnitude $\sqrt{I(I+1)}\, h/2\pi$, where I is the quantum number which defines the greatest possible component of $\mathbf{I}$ along a specified axis, according to the rule

$$I_Z = I\frac{h}{2\pi}. \tag{8–1}$$

The value of I has been found experimentally to depend on the mass number A of the nucleus; if A is even, I is an integer or zero; if A is odd, I has an odd half-integral value (half of an odd integer). In other words, if the mass number of the nucleus is even, I may have one of the values: 0, 1, 2, 3, ...; if A is odd, I may have one of the values $\frac{1}{2}$, $\frac{3}{2}$, $\frac{5}{2}$, ...

The rules just cited lead to one of the failures of the proton-electron hypothesis for the constitution of the nucleus. Nitrogen has an atomic number of 7 and a mass number of 14, and its nucleus would have 14 protons and 7 electrons under this hypothesis. The contribution of the protons to the angular momentum should be an integral multiple of $h/2\pi$, whether the angular momentum of a proton is an integral or odd half-integral multiple of $h/2\pi$. An electron has spin $\frac{1}{2}(h/2\pi)$ so that its total angular momentum is always an odd half-integral multiple of $h/2\pi$ (see Section 7–9). The contribution of 7 electrons is, therefore, an odd half-integral multiple of $h/2\pi$, and the total angular momentum of the nitrogen nucleus should be an odd half-integral multiple of $h/2\pi$. But the angular momentum of the nitrogen nucleus has been found experimentally to be $I = 1$, an integer, in contradiction to the value predicted by the hypothesis. Experiments show that the proton, like the electron, has an intrinsic spin of $\frac{1}{2}(h/2\pi)$, and it is possible with this additional information to predict the angular momenta of many other nuclei on the basis of the proton-electron hypothesis. Thus, the isotopes of cadmium ($Z = 48$), mercury ($Z = 80$), and lead ($Z = 82$) with odd mass numbers should each have an odd number of electrons and an even number of particles all together. The angular momenta of these nuclei should, therefore, be zero or integral; they have been found experimentally to be odd half-integers, and the hypothesis again predicts values which disagree with the experimental results.

The proton-electron hypothesis also fails to account for the order of magnitude of nuclear magnetic moments. Measurements of the magnetic moments of many nuclei have given values which are only about 1/1000 of the value of the magnetic moment of the electron. The magnitude of the latter is

$$\mu_B = \frac{eh}{4\pi mc}, \tag{8–2}$$

where m is the mass of the electron; this quantity is called the *Bohr magneton* and has the value 0.92×10^{-20} erg/gauss. All measured nuclear magnetic moments are of the order 10^{-23} erg/gauss, and their values can be expressed appropriately in terms of the quantity

$$\mu_N = \frac{eh}{4\pi M_H c} = 0.505 \times 10^{-23} \text{ erg/gauss}, \tag{8–3}$$

in which the electron mass has been replaced by the proton mass; the quantity μ_N is called the *nuclear magneton.* Measured values of the nuclear magnetic moment vary from zero to about 5 nuclear magnetons, and the proton has a magnetic moment of 2.7926 ± 0.0001 nuclear magnetons. If electrons were present in the nucleus, we should expect to find nuclear magnetic moments of the order of magnitude of the Bohr magneton, at least in those nuclei for which $A - Z$ is odd (and which would, on the proton-electron hypothesis, contain an odd number of electrons). The fact that nuclear magnetic moments are only of the order of magnitude of the nuclear magneton is, therefore, another strong argument against the existence of electrons inside the nucleus.

There is also a wave-mechanical argument against the existence of free electrons in the nucleus. According to the uncertainty principle (cf. Section 7–8),

$$\Delta x \Delta p \sim h, \tag{8–4}$$

where Δx is the uncertainty in the position of a particle and Δp is the uncertainty in its momentum. Suppose that the principle is applied to an electron in the nucleus. The uncertainty Δx in the position of an electron is roughly the same as the diameter of the nucleus, which is assumed here to be 2×10^{-12} cm. Then

$$\Delta p \sim \frac{h}{\Delta x} = \frac{6.6 \times 10^{-27}}{2 \times 10^{-12}} = 3.3 \times 10^{-15} \text{ erg-sec/cm.}$$

From the uncertainty in the momentum, it is possible to get a rough estimate of the energy of an electron in the nucleus. Relativistic formulas must be used because the electron moves very rapidly in the nucleus, as will be seen. From Eqs. (6–36) and (6–38), the total energy of a particle can be expressed in terms of the momentum,

$$E^2 = p^2c^2 + m_0^2c^4, \tag{8–5}$$

where m_0 is the rest mass of the electron, and c is the velocity of light. If it is now assumed that the momentum p of the electron is no larger than the values just found for the uncertainty Δp, then $p \simeq 3.3 \times 10^{-15}$, and $p^2c^2 = (3.3 \times 10^{-15})^2(3 \times 10^{10})^2 = 10^{-8}$ erg. This value is much greater than the term $m_0^2c^4 = (9 \times 10^{-28})^2(3 \times 10^{10})^3 \simeq 10^{-12}$, which can consequently be neglected. Then $E^2 \simeq 10^{-8}$ and

$$E \simeq 10^{-4} \text{ erg} = \frac{10^{-4}}{1.6 \times 10^{-12}} \simeq 6 \times 10^7 \text{ ev} = 60 \text{ Mev.}$$

According to this result, a free electron confined within a space as small as the nucleus would have to have a kinetic energy of the order of 60 Mev, and a velocity greater than $0.999c$. Experimentally, however, the electrons emitted by radioactive nuclei have never been found to have kinetic energies greater than about 4 Mev, or at least an order of magnitude smaller than that calculated from the uncertainty principle. Although the calculation is a rough one, similar results are obtained from more rigorous calculations, and in view of the large discrepancy it seems improbable that nuclei can contain free electrons.

The above argument does not apply to a proton in the nucleus because the proton mass (1.67×10^{-24} gm) is nearly 2000 times as great as the electron rest mass. For the proton in the nucleus, Δx is also about 2×10^{-12} cm, and $\Delta p \simeq 3.3 \times 10^{-15}$ erg-sec/cm. If it is assumed, as in the case of the electron, that the momentum is of the order of the uncertainty in the momentum, then from Eq. (8–5),

$$\begin{aligned} E^2 &\simeq 10^{-8} + (1.67 \times 10^{-24})^2(3 \times 10^{10})^4 \\ &= 10^{-8} + 2.3 \times 10^{-6}, \end{aligned}$$

and now the first term may be neglected in comparison with the second. Then

$$E \simeq 1.5 \times 10^{-3} \text{ erg} = \frac{1.5 \times 10^{-3}}{1.6 \times 10^{-12}} = 9.4 \times 10^8 \text{ ev} = 940 \text{ Mev}.$$

This value is only slightly greater than the rest energy of the proton, which is $1.008 \times 931 = 938$ Mev. Hence, the kinetic energy of a proton in the nucleus is of the order of a few Mev, and it should be possible for a free proton to be contained in the nucleus.

8–3 Nuclear transmutation and the discovery of the neutron. The failure of the proton-electron hypothesis of the nucleus was related to the properties of the free electron. It was proposed, therefore, that the electrons are bound to the positively charged particles and have no independent existence in the nucleus. One possibility, which had been suggested by Rutherford as early as 1920, was that an electron and a proton might be so closely combined as to form a neutral particle, and this hypothetical particle was given the name *neutron*. Now, all of the methods that have been discussed so far for detecting particles of nuclear size depend on effects of the particle's electric charge, as deflection in magnetic or electric fields and ionization. The presence of a neutron, which has no charge, would be very hard to detect, and many unsuccessful attempts were made to find neutrons. Finally, in 1932, as one of the results of research on the disintegration or transmutation of nuclei by α-particles, Chadwick demonstrated the existence of neutrons. This discovery opened up a vast field for further experimental work and led to the presently accepted idea of the constitution of the nucleus: that it is built of protons and neutrons.

It has been mentioned that the fact that the masses of the atoms of all elements, or their isotopes, are very close to whole numbers led to the hypothesis that the atomic nucleus is a composite structure made up of an integral number of elementary particles with masses close to unity. If so, it should be possible by suitable means to break down the nucleus and to change one element into another. Radioactivity is a naturally occurring process of this kind. Rutherford and his colleagues used energetic α-particles from radioactive substances as projectiles with which to bombard atomic nuclei in the attempt to produce artificial nuclear disintegrations. They found that when nitrogen was bombarded with α-particles, energetic protons were obtained. The protons were identified by magnetic deflection measurements, and their energies were considerably greater than those of the bombarding α-particles. They must have been shot out from the struck nucleus in such a way that part, at least, of their energy was derived from the internal energy of the nucleus. Cloud-chamber studies of the process showed that the α-particle was captured by the nitrogen nucleus, and then the proton was ejected. This explanation corresponds to the formation from nitrogen (atomic number 7, atomic weight 14) of an atom of atomic number 8 and atomic weight 17, i.e., the oxygen isotope of mass 17. The bombardment with α-particles was found by Rutherford to cause emission of protons from all the elements of atomic number up to 19 with the exception of hydrogen, helium, lithium, carbon, and beryllium. The most marked transmutation effects occurred with boron, nitrogen, and aluminum.

Closer investigation of the bombardment of boron and beryllium by α-particles gave some additional and unexpected results. Bothe and Becker (1930) discovered that these elements emitted a highly penetrating radiation when so bombarded. It was thought that this radiation might be a form of γ-ray of very high energy. Curie and Joliot (1932) found that when the radiation was allowed to fall on substances containing hydrogen, it caused the production of highly energetic protons. Chadwick (1931) was also able to show that the rays emitted from bombarded beryllium gave rise to rapidly moving atoms when allowed to fall on other substances, for example, He, Li, Be, C, O, and N. These results could not be explained under the assumption that the new radiation consisted of high-energy γ-rays. Chadwick finally (1932) proved that the energies of the protons ejected from hydrogenous materials, and of the other rapidly moving atoms, could only be explained on the view that the "rays" from bombarded beryllium actually consisted of particles with a mass close to that of the proton. These particles, unlike protons, produce no tracks in the cloud chamber and no ionization in the ionization chamber. These facts, together with the extremely high penetrating power of the particles, show that the charge of the latter must be zero. Since the new particle was found to be neutral and to have a mass close to unity, it was identified with Rutherford's neutron. Later measurements have shown that the mass of the neutron is 1.00898 amu, so that it is slightly heavier than the proton, with a mass of 1.00758 amu.

8-4 The proton-neutron hypothesis. The discovery of a particle, the neutron, with an atomic weight very close to unity and without electric charge, led to the assumption that every atomic nucleus consists of protons and neutrons. This hypothesis was used for the first time as the basis of a detailed theory of the nucleus by Heisenberg in 1932. Under the proton-neutron hypothesis, the total number of elementary particles in the nucleus, protons and neutrons together, is equal to the mass number A of the nucleus; the atomic weight is therefore very close to a whole number. The number of protons is given by the nuclear charge Z, and the number of neutrons is $A - Z$.

The new nuclear model avoids the failures of the proton-electron hypothesis. The empirical rule connecting mass number and nuclear angular momentum can be interpreted as showing that the neutron, as well as the proton, has a half-integral spin; the evidence is now convincing that the spin of the neutron is indeed $\frac{1}{2}h/2\pi$. If both proton and neutron have spin $\frac{1}{2}$ then, according to quantum theory, the resultant of the spins of A elementary particles, neutrons and protons, will be an integral or half-integral multiple of $h/2\pi$ according to whether A is even or odd. This conclusion is in accord with all the existing observations of nuclear angular momenta. The value of the magnetic moment of the neutron is close to -2 nuclear magnetons; it is opposite in sign to that of the proton, but not very different in magnitude. The values for both the proton and neutron are consistent with those measured for many different nuclei. Finally, since the mass of the neutron is very close to that of the proton, the argument showing that protons can be contained within the nucleus is also valid for neutrons.

The neutron-proton hypothesis is consistent with the phenomena of radioactivity. Since there are several reasons why electrons cannot be present in the nucleus, it must be concluded that in β-radioactivity, the electron is created in the act of emission. This event is regarded as the result of the change of a neutron within the nucleus into a proton, an electron, and a new particle called a *neutrino,* and both experimental and theoretical evidence offer strong support for this view. In β-radioactivity, then, the nucleus is transformed into a different one with one proton more and one neutron less, and an electron is emitted. An α-particle can be formed by the combination of two protons and two neutrons. It may exist as such in the nucleus, or it may be formed at the instant of emission; the latter possibility is now regarded as more likely.

It must be emphasized, however, that in considering the nuclei of different elements as being built of protons and neutrons, the neutron is not regarded as a composite system formed by a proton and an electron. The neutron is a fundamental particle in the same sense that the proton is. The two are sometimes called *nucleons* in order to indicate their function as the building blocks of nuclei.

One of the main problems of nuclear physics is that of understanding the nature of the forces holding the protons and neutrons together. Much of the research in nuclear physics is aimed at the clarification of the laws of interaction between nuclear particles. This problem will not be treated in a detailed, quantitative way in this book because of its complexity, but it will be discussed in a later chapter. The emphasis in the following chapters will be rather on the facts about nuclei, and on the transformation of one nuclear species into another by the rearrangement of its constituent nucleons. The information accumulated will be considered and interpreted in terms of the proton-neutron hypothesis. This aspect of nuclear physics is analogous to the application of atomic physics to the chemical properties of the elements, and is sometimes referred to among physicists as *nuclear chemistry*.

For the present, it will suffice to point out some of the qualitative properties of the forces between the protons and neutrons in the nucleus. Because of the positive charge of the proton, there must be repulsive, electrostatic forces between the protons tending to push the nucleus apart. It is apparent from the small size and great density of the nucleus that these forces must be very large in comparison with the forces between the nucleus and the extranuclear electrons. Hence, if stable complex nuclei are to exist, there must be attractive forces in the nucleus strong enough to overcome the repulsive forces. These attractive forces are the specifically *nuclear* forces between a proton and a neutron, between two neutrons, or between two protons. They seem to be more complex than the gravitational or electromagnetic forces of classical physics. The nuclear attractive forces must be very strong at distances of the order of the nuclear radius, i.e., they are *short-range* forces. Outside the nucleus, they decrease very rapidly, and the Coulomb repulsive forces responsible for the scattering of α-particles predominate. The magnitude of the nuclear forces is such that the work required to divide a nucleus into its constituent particles (the binding energy of the nucleus) is very much greater than the work needed to separate an extranuclear electron from an atom. Whereas the latter is of the order of electron volts, energy changes in the nucleus are of the order of millions of electron volts. It is the magnitude of the energies associated with nuclear transformations that is responsible for the large-scale applications of nuclear physics, and for the rise of the new field of nuclear engineering.

8–5 Magnetic and electric properties of the nucleus. It will be convenient to include in this chapter a few more remarks about the angular momentum of the nucleus, as well as a brief discussion of some magnetic and electric properties of the nucleus. These properties are important in the interpretation of many nuclear phenomena and in the theory of the nucleus, and they occupy a prominent place in nuclear physics. They are, however, among the less elementary aspects of the subject and will not be treated in detail in this book. Nevertheless, some familiarity with the ideas and terminology of these matters will be helpful to the reader.

Each proton and each neutron in the nucleus has an angular momentum which may be pictured as being caused by the particle's spinning motion about an axis through its center of mass. The magnitude of this spin angular momentum is $\frac{1}{2}h/2\pi$. The wave-mechanical properties of an angular momentum of this kind are such that its orientation in space can be described by only two states: the spin axis is either "parallel" or "antiparallel" to any given direction. The component of the spin along a given direction, say the z-axis, is either $\frac{1}{2}h/2\pi$ or $-\frac{1}{2}h/2\pi$. In addition, each nucleon may be pictured as having an angular momentum associated with orbital motion within the nucleus. According to quantum theory, the orbital angular momentum is a vector whose greatest possible component in any given direction is an integral multiple of $h/2\pi$. Each nucleon has a total angular momentum i about a given direction, with

$$i = l \pm s, \tag{8–6}$$

where l is the orbital angular momentum and s is the spin angular momentum. The spin of any single nucleon can add or subtract $\frac{1}{2}h/2\pi$ depending on its orientation with respect to the axis of reference, and i is therefore half-integral. For nuclei containing more than one particle, it is customary to write corresponding relationships between the momenta in capitals; the resultant total angular momentum of the nucleus is then

$$I = L \pm S, \tag{8–7}$$

where L is the total orbital angular momentum, and S is the total spin angular momentum. The total angular momentum is actually a vector, denoted by $\mathbf{I}$, and the scalar quantity I is defined as the maximum possible component of $\mathbf{I}$ in any given direction. The orbital angular momentum L is an integral multiple of $h/2\pi$; S is an even half-integral multiple of $h/2\pi$ if the number of nuclear particles is even, and an odd half-integral multiple if the number of particles is odd. Hence, I is an integral multiple of $h/2\pi$ when A is even, and an odd half-integral multiple when A is odd, in agreement with experimental results.

There are two possible sources of confusion which arise from careless usage. The term "spin" is often used for the total angular momentum of a nucleus rather than for the spin S alone. This incorrect usage was introduced before the problem of the internal structure of nuclei had attained its present importance and has been continued. In addition, the total orbital angular momentum of a nucleus is often denoted by l rather than by L. The meaning of the term "spin" and the symbol l can usually be inferred without difficulty from the context in which they appear.

The magnetic moment of a nucleus can be represented as

$$\boldsymbol{\mu}_I = \gamma_I \frac{h}{2\pi} \mathbf{I} = g_I \mu_N \mathbf{I}, \tag{8–8}$$

where γ_I and g_I are defined by Eqs. (8–3) and (8–8) and are called the *nuclear gyromagnetic ratio* and *nuclear g-factor,* respectively; μ_N is the nuclear magneton defined by Eq. (8–3), and has the numerical value 5.04929×10^{-24} erg/gauss. The quantity which measures the magnitude of $\boldsymbol{\mu}_I$, and is called the nuclear magnetic moment μ_I, is

$$\mu_I = \gamma_I \frac{h}{2\pi} I. \tag{8–9}$$

The details of the methods for measuring nuclear spins and magnetic moments will not be discussed in this book. The interested reader is referred to the works by Ramsey and Kopfermann listed at the end of the chapter, for a treatment of the methods and results. The spins and magnetic moments of many nuclei have been measured, particularly for the ground state of the nucleus, and certain patterns have been observed in the experimental results. Significant conclusions as to the structure of the nucleus may be obtained from data on nuclear spin and magnetic moment, just as knowledge of the arrangement of the extranuclear electrons was obtained from information about their angular momenta. It has been found, for example, that $I = 0$ for nuclei containing even numbers of protons and neutrons. It follows from Eq. (8–9) that a so-called even-even nucleus should have no magnetic moment, and this has been found experimentally to be the case. This generalization, and others, have proved very useful in the study of nuclear structure and other properties.

Another property which is highly important in connection with the *shape* of the nucleus is the *electric quadrupole moment.* This quantity, which cannot be discussed in a simple way, is a measure of the deviation of a nucleus from spherical symmetry. If a nucleus is imagined to be an ellipsoid of revolution whose diameter is $2b$ along the symmetry axis and $2a$ at right angles to this axis, and if the electric charge density is assumed uniform throughout the ellipsoid, the quadrupole moment Q is given by

$$Q = \frac{2}{5} Z(b^2 - a^2). \tag{8–10}$$

Its magnitude depends on the size of the nucleus, the extent of the deviation from spherical symmetry, and the magnitude of the charge; the sign may be positive or negative. Many nuclei have been found to have quadrupole moments; thus, deuterium, a nucleus with one proton and one neutron, has a Q-value of $+0.00274 \times 10^{-24}$ cm^2, while an isotope of lutecium containing 176 nucleons has a Q-value of 7×10^{-24} cm^2. The investigation of quadrupole moments along with spins and magnetic moments has led to important developments in the theory of nuclear structure, as will be shown in Chapter 17.

8–6 Additional properties of atomic nuclei. In addition to their electric and magnetic properties, nuclei have certain properties which are not obviously physical in nature. Although these properties will not be treated in detail, the reader should at least know of the existence and usefulness of the concepts involved. The properties which will be discussed very briefly are the *statistics* to which nuclei are subject, and the *parity*.

The concept of statistics in physics is related to the behavior of large numbers of particles. Thus, the distribution of energies or velocities among the molecules of a gas can be described by the classical Maxwell-Boltzmann statistics, as can many other macroscopic properties of gases. The properties of assemblies of photons, electrons, protons, neutrons, and atomic nuclei cannot, in general, be described on the basis of classical statistics, and two new forms of statistics have been devised, based on quantum mechanics rather than on classical mechanics. These are the Bose-Einstein statistics, and the Fermi-Dirac statistics. The question of which form of quantum statistics applies to a system of particles of a given kind is related to a particular property of the wave function which describes the system. This property has to do with the effect on the wave function of interchanging all of the coordinates of two identical particles, say of two protons in a nucleus. A nucleon is described by a function of its three space coordinates and the value of its spin, whether it is $\frac{1}{2}h/2\pi$ or $-\frac{1}{2}h/2\pi$. The Fermi-Dirac statistics apply to systems of particles for which the wave function of the system is *antisymmetrical*, i.e., it changes sign when *all* of the coordinates (three spatial and one spin) of two identical particles are interchanged. It follows from this property of the wave function that each completely specified quantum state can be occupied by only one particle; that is, the Pauli exclusion principle applies to particles obeying the Fermi-Dirac statistics. It has been deduced from experiments that electrons, protons, and neutrons obey the Fermi-Dirac statistics, as do all nuclei of odd mass number A. In the Bose-Einstein statistics, the wave function is symmetrical, i.e., it does not change sign when all the coordinates of any pair of identical particles are interchanged. Two or more particles may be in the same quantum state. All nuclei having even mass number A obey the Bose-Einstein statistics. There is a direct correlation between the total angular momentum of a nucleus and its statistics: Fermi-Dirac particles (odd A) have total nuclear angular momenta which are odd half-integral multiples of $h/2\pi$, while Bose-Einstein particles (even A) have momenta which are integral multiples of $h/2\pi$.

The last property that will be mentioned is the *parity*. To a good approximation, the wave function of a nucleus may be expressed as the product of a function of the space coordinates and a function depending only on the spin orientation. The motion of a nucleus is said to have *even parity* if the spatial part of its wave function is unchanged when the space coordinates (x,y,z) are replaced by $(-x, -y, -z)$. This transformation of coordinates is equivalent to a reflection of the nucleus' position about the origin of the x, y, z system of axes. When reflection changes the sign of the spatial part of the wave function, the motion of the nucleus is said to have *odd parity*. It has been shown that the parity of a nucleus in a given state is related to the value of the orbital angular momentum L; if L is even, the parity is even; if L is odd, the parity is odd. A system of particles will have even parity when the sum of the numerical values for L for all its particles is even, and odd parity when the sum is odd. Although the parity seems to be an abstract sort of property, the selection rules for many nuclear transitions involve conditions on the parity, as well as on the total angular momentum.

References

GENERAL

R. D. Evans, *The Atomic Nucleus*. New York: McGraw-Hill, 1955, Chapters 2, 4, 5.

E. U. Condon and H. Odishaw, eds., *Handbook of Physics*. New York: McGraw-Hill, 1958, Part 7, Chapters 1–6; Part 9, Chapter 3; by various authors.

N. F. Ramsey, "Nuclear Moments and Statistics," *Experimental Nuclear Physics*, E. Segrè, ed., New York: Wiley, 1953, Vol. I, Part III.

N. F. Ramsey, *Molecular Beams*. Oxford: Clarendon Press, 1956.

K. F. Smith, "Nuclear Moments and Spins," *Progress in Nuclear Physics*, O. R. Frisch, ed. London: Pergamon Press, Vol. 6, p. 52, 1957.

H. Kopfermann, *Nuclear Moments* (English version prepared from the second German edition by E. E. Schneider). New York: Academic Press, 1958.

C. H. Townes and A. L. Schawlow, *Microwave Spectroscopy*. New York: McGraw-Hill, 1955.

CHAPTER 9

ISOTOPES

9–1 Natural radioactivity and isotopes. The discovery of isotopes was one of the results of work on the radioactive elements. It was found that generally the product of radioactive decay is itself radioactive, and that each of the decay products behaves chemically in a different manner from its immediate parent and its daughter product. Radium, a metal with an atomic weight of 226, was shown to be the product of the α-disintegration of an element that was given the name *ionium.* Radium in turn emits an α-particle, forming the rare gas *radon,* which also is radioactive. The chemical properties of ionium turned out to be similar to those of thorium and, in fact, the two elements when mixed could not be separated chemically even with extremely sensitive methods. Moreover, ionium and thorium were found to be spectroscopically identical. The systematic study of the radioactive elements showed, however, that the atomic weight of ionium must be 230, while that of thorium was known to be 232. Other pairs of elements were also discovered which were chemically identical but differed in atomic weight, and Soddy, in 1913, suggested as a name for them the word *isotopes,* meaning the same place in the periodic table.

The way in which isotopes arise in the radioactive elements can be understood in terms of the effect of radioactive decay on the atomic number and atomic weight. Each time an α-particle is emitted, the charge of the nucleus of the radioactive atom decreases by two units, since the α-particle carries a positive charge of two units. At the same time, the atomic mass decreases by four units. Each time a β-particle is emitted, the nuclear charge increases by one unit because one negative charge is removed, but the mass is practically unchanged. Thus, the emission of an α-particle causes a decrease of two in the atomic number, i.e., a shift of two places to the left in the periodic table. The emission of a β-particle causes an increase of one in the atomic number, i.e., a shift of one place to the right in the periodic table. This law, first deduced by Soddy and Fajans in 1913, is known as the *displacement law of radioactivity.* Consider what happens as a result of the radioactive decay of the element called uranium I (UI) with an atomic number of 92 and an atomic weight (or mass number) of 238 units. This element emits an α-particle, and according to the displacement law, the product, which was called uranium X_1 (UX_1), has an atomic number of 90 and a mass number of 234. Uranium X_1 emits a β-particle giving the product uranium X_2 (UX_2) with an atomic number of 91 and a mass number of 234; UX_2, in turn, emits a β-particle giving the element uranium II (UII), with an atomic number of 92 and a mass number of 234. Hence, UI and UII have the same atomic number 92. It follows that these two elements must be entirely identical in all their chemical properties, and in fact it is impossible to separate these two elements from each other by chemical means; they are isotopes. From a chemical point of view, the mixture constitutes one element. Now, UII emits an α-particle giving the product ionium, with an atomic number of 90 and a mass number of 230. Naturally occurring thorium also has the atomic number 90, and has the mass number 232. Hence, thorium, ionium, and UX_1 are also isotopes. There are many other groups of isotopes among the radioactive elements, and they will be discussed in greater detail in the next chapter. The elements UX_1, UX_2, and UII, which are isotopes of thorium ($Z = 90$), protoactinium ($Z = 91$), and uranium ($Z = 92$), respectively, all have mass numbers of 234. They are chemically different elements with the same atomic weight or mass number, and are called *isobars*.

9–2 Positive-ray analysis and the existence of isotopes. After the existence of isotopes had been proven for radioactive elements, Thomson (1913), by deflection experiments on positive rays, proved that isotopes also occur among ordinary elements. Thomson's work was extended and improved by Aston, who developed a new type of positive-ray apparatus which he called a mass spectrograph. Aston showed that most of the elements are mixtures, having in some cases as many as nine or ten isotopes. He was also able to obtain fairly accurate measurements of the relative abundances of the isotopes of the different elements. The most important uses to which the mass spectrograph has been put by Aston and others is the identification of isotopes and the precise determination of their masses. The isotopic mass is one of the directly measurable quantities essential to an understanding of the atomic nucleus, and the importance of highly precise measurements of this quantity will become evident.

One form of apparatus used by J. J. Thomson(1) for the analysis of positive rays is shown schematically in Fig. 9–1. The discharge by which the positive rays are produced takes place in a large tube A. The anode is an aluminum rod D; the cathode K has a very small hole through which the positive ions pass. The cathode is kept cool during the discharge by means of the water jacket J. The gas to be studied is allowed to leak in through a fine glass capillary tube L, and after circulating through the apparatus is pumped off at F. The pressure in the tube is adjusted so that the discharge potential is 30,000 to 50,000 volts.

The positive ions emerge from the end of the cathode as a narrow beam of ions whose velocities depend on their charge and mass and on the dis-

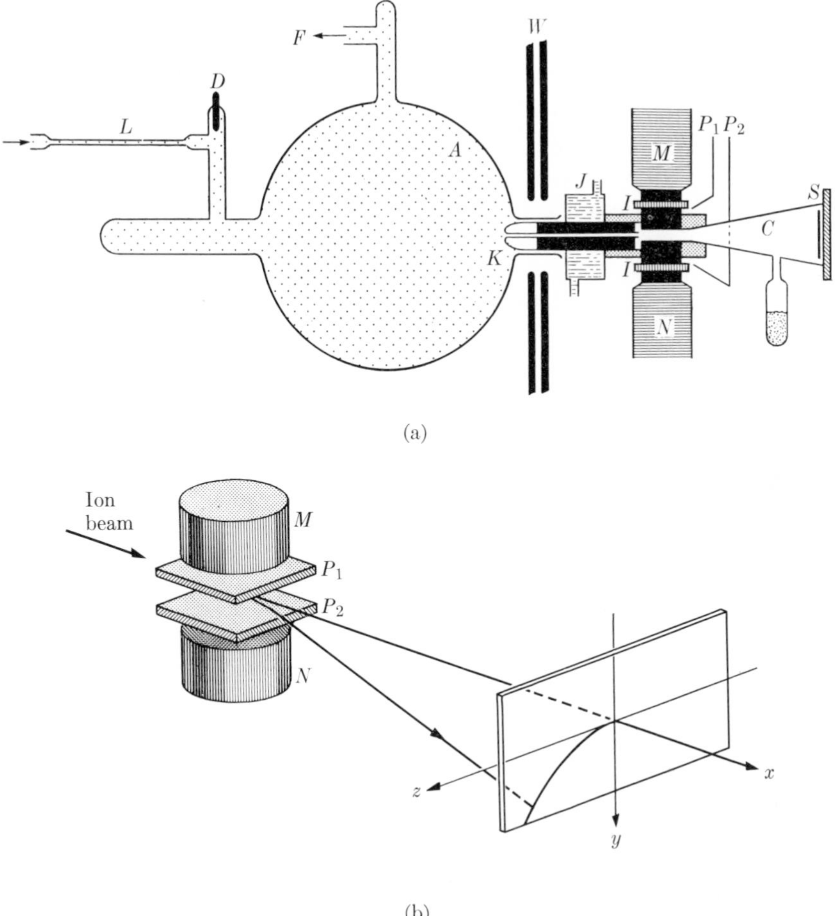

Fig. 9–1. Thomson's method for the analysis of positive rays. (a) Diagram of the apparatus. (b) Formation of Thomson's parabolas.

charge potential. The beam is analyzed by causing it to pass between the pieces of soft iron P_1 and P_2, which are placed between the poles M and N of an electromagnet; P_1 and P_2 are the pole pieces of the magnet but are electrically insulated from M and N by insulating strips I. Thus, P_1 and P_2 can be maintained at any desired potential difference, so that in the space between them there can be parallel magnetic and electric fields. A positive ion moving from left to right in this space undergoes a deflection in the plane of the paper because of the electric field, and perpendicular to the plane of the paper because of the magnetic field. After leaving the space between P_1 and P_2, the deflected particle moves in the field-free, evacuated "camera" C and falls on the fluorescent screen or photographic plate S. A soft iron shield W is placed between the magnet and the discharge tube to prevent the stray magnetic field from interfering with the discharge in the tube A.

Thomson found that the positive ions, after passing through the fields, formed a pattern of parabolas on a photographic plate. The equation of one of these parabolas can easily be derived, and it involves the charge-to-mass ratio of the ions in the beam. Suppose that an ion of mass M, charge q, and velocity v enters the space P_1P_2 along a path coincident with the x-axis of a system of rectangular coordinates with origin in the space P_1P_2. Let E and H be the electric and magnetic field strengths, respectively, let L be the length of the path through the fields, and assume that the latter are constant and terminate sharply. The electrostatic force on the ion is qE, so that the ion is subjected to an acceleration qE/M in the direction of the field. The time taken to pass through the field is L/v. From the usual formula for the distance traveled by a particle under a constant acceleration, the deflection y' from the x-axis is

$$y' = \frac{1}{2}\frac{qE}{M}\left(\frac{L}{v}\right)^2. \tag{9–1}$$

The magnetic field exerts a force Hqv on the particle, which consequently suffers a deflection z' in the z-direction, given by

$$z' = \frac{1}{2}\frac{Hqv}{M}\left(\frac{L}{v}\right)^2. \tag{9–2}$$

After passing through the fields, the particle moves in a straight line. If the distance of the photographic plate from the fields is large compared with L, the point where the particle strikes the plate will have coordinates y and z which are proportional to y' and z', respectively. The relation between z and y will therefore be the same as that between z' and y'. The latter relation is obtained by eliminating v from Eqs. (9–1) and (9–2),

$$z^2 = C\frac{q}{M}\frac{H^2}{E}y, \tag{9–3}$$

where C is a constant which depends on the dimensions of the apparatus. Equation (9–3) represents a parabola. For constant values of the fields E and H, ions with different values of the velocity v but with the same value of q/M should make a parabolic trace on the photographic plate, as Thomson found. Ions with different values of q/M produce different parabolas. If there are no fields, the beam of ions will strike the plate at a point in line with the hole through the cathode called the undeflected spot.

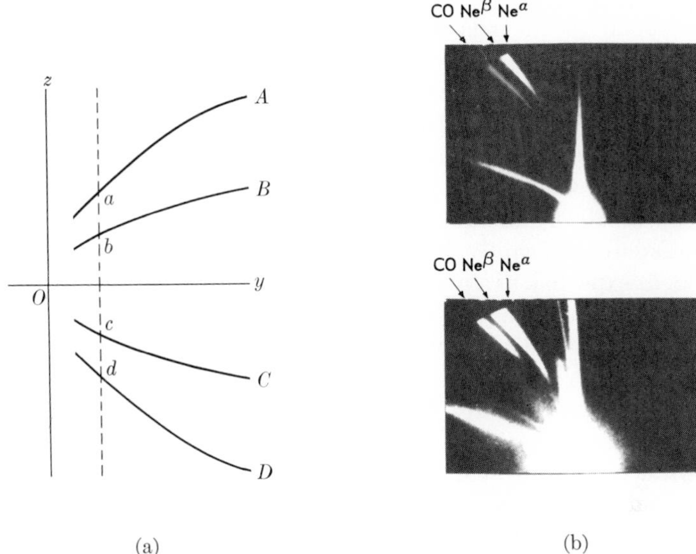

FIG. 9–2. Positive-ray parabolas obtained by the Thomson method. (a) Schematic diagram of the parabolas. (b) Actual parabolas obtained by Thomson (1913). Ne^α is Ne^{20}; Ne^β is Ne^{22}. The lower picture had a longer exposure than the upper, and brought out the weak Ne^{22} trace that is barely visible in the upper picture.

If the ions in a positive-ray beam have the same value of q/M but travel with different velocities, a parabola like that denoted by A in Fig. 9–2(a) is observed. If ions are present with the same value of q, but with a mass M' greater than M, a different parabola B will be seen. This parabola will be lower than A because the magnetic displacement is smaller for the heavier ions. If the mass M of the ions giving the parabola A is known accurately, then M' can be determined in the following way. The magnetic field is reversed after half the time interval of the exposure of the plate, and a pair of curves C and D is obtained which are the mirror images of A and B; the mid-line Oy is also obtained in this way. From Eq. (9–3), it follows that $M'/M = (ad/bc)^2$, and the latter ratio can be measured on the plate, giving the mass M' in terms of the standard atomic mass M. By a systematic analysis of the relative positions of the various parabolic traces appearing on a series of plates, Thomson was able to determine the origin of the traces. Parabolas corresponding to H^+, H_2^+, O^+, O_2^+, CO^+, and other ions with known masses were identified. These ions could then be used as standards from which the masses of ions producing other traces could be determined. If the positive rays consist of a mixture of different ions, the relative intensities of the individual parabolas correspond to the relative abundances of the various ions in the mixture.

In 1912, Thomson[2] analyzed the positive-ray parabolas formed when neon was used as the source of the ions [Fig. 9–2(b)]. With the apparatus that he used, the parabolas corresponding to masses differing by 10% could be clearly distinguished. The atomic weight of neon had been determined as 20.20, and neon was the lightest element whose atomic weight differed significantly from a whole number. An intense parabola was found, as expected, at a position corresponding to the atomic weight 20. A much less intense trace was also found corresponding to the atomic weight 22. The same two curves, with the same relative intensities, were obtained with neon samples of different purity and for different conditions of discharge and gas pressure. No element of atomic weight 22 was known; the trace could not be identified with any known molecular ion. The experimental facts suggested that neon could exist in two forms which could not be distinguished chemically, but with different atomic weights. The chemical atomic weight 20.20 would result if there were present nine times as many neon atoms with atomic weight 20 as with atomic weight 22, since

$$\frac{(9 \times 20) + (1 \times 22)}{10} = 20.2.$$

The suggestion that neon has two isotopes was so striking that Aston looked for further evidence that might bear on the problem. It was well known that a light gas diffuses through a porous partition more rapidly than a heavier gas, and Aston used this property of gases in an attempt to achieve a partial separation of the two constituents of neon. He passed a sample of neon through a pipe-clay tube, collected the portion of the gas which diffused through and allowed this to diffuse once more, and so on. After repeated diffusion and rediffusion, he obtained from 100 cm^3 of ordinary neon gas two extreme fractions of 2 to 3 cm^3 with atomic weights, calculated from their densities, of 20.15 and 20.28, respectively. The fraction with the smaller atomic weight was supposed to contain more neon of atomic weight 20 than ordinary neon; the fraction with the greater atomic weight was supposed to contain more neon of atomic weight 22 than ordinary neon. The changes in the proportions of the two constituents were large enough to produce appreciable changes in the relative intensity of the two traces (for the masses 20 and 22) in the positive-ray photograph. The changes in intensity corresponded approximately to the degree of enrichment or depletion of the mass 20 constituent. Although the proof was not complete, it was difficult to avoid the conclusion that there are two isotopes of neon, one with atomic weight 20, the other with atomic weight 22.

9-3 Isotopic masses and abundances: the mass spectrograph and mass spectrometer. The Thomson parabola method of analyzing positive rays was adequate for a general survey of masses and velocities, but it could not yield precise values of isotopic masses and abundances. The further quantitative study of the constitution of the elements required the determination of isotopic masses with a precision of at least one part in a thousand. To meet this need, Aston[(3)] (1919) designed the mass spectrograph. Aston's method of analysis was an improvement on that of Thomson in that greater dispersion was achieved (i.e., greater separation of ions of different masses) and all ions with a given value of q/M were brought to a focus instead of being spread out in a parabola. By these means, greater sensitivity and precision were attained.

A schematic diagram of Aston's first mass spectrograph is shown in Fig. 9-3. The positive rays from a discharge tube pass through two very narrow parallel slits S_1 and S_2 and enter the space between the metal plates P_1 and P_2. An electric field between these plates causes a deflection of the ions toward P_2, the amount of the deflection being greater the smaller the velocity of the ions. The narrow beam contains particles with a wide range of velocities, with the result that it is broadened as it passes through the field. A group of these particles is selected by means of the relatively wide diaphragm D. After passing through D, this diverging stream of ions enters a magnetic field, indicated by the circle at O, perpendicular to the plane of the paper. The magnetic field causes deflections, as shown in the diagram, the more slowly moving ions being deflected more than the faster ones. The paths of the slow-moving ions, therefore, intersect those of the faster-moving ions at some point F. If the instrument is properly designed, ions having the same value of q/M but slightly different energies can be brought to a single focus on a photographic plate. Other ions with the same range of energies but a different value of q/M are brought to a focus at a different point on the photographic plate. The

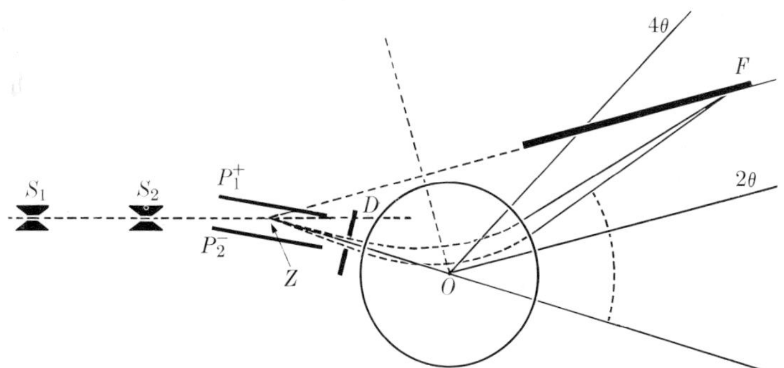

Fig. 9-3. Diagram of Aston's first mass spectrograph (1920).

focus for a particular value of q/M is actually a line, and the result of an analysis is a series of lines reminiscent of an optical line spectrum. Because of this similarity, the series of lines is called a *mass spectrum* and the apparatus a *mass spectrograph.* Some typical mass spectra are shown in Fig. 9–4.

The isotopic masses can be determined quantitatively in several different ways. In one method, for example, the positions of the lines caused by the masses in question are compared with the positions of the lines caused by standard substances whose masses are accurately known. With this new instrument, Aston was able to prove beyond a doubt that neon has two isotopes with masses very close to the integers 20 and 22, respectively.(4) He also showed that in the mass spectrum of chlorine there was no line corresponding to the chemical atomic weight (35.46) of chlorine. Instead, *two* lines were seen, corresponding very closely to the masses 35.0 and 37.0, the former being the more intense line. This result showed that chlorine has two isotopes of nearly integral atomic mass, and ordinary chlorine is a mixture of these two kinds of atoms in such proportions that the chemical atomic weight is 35.46.

Aston improved the design of the mass spectrograph, and in his second instrument(5) isotopic masses could be determined with an accuracy of 1 part in 10,000. With this instrument, Aston determined the masses of the isotopes of a large number of elements, as well as their abundances, and showed that the masses of atoms are very nearly, *but not quite,* integers, when the mass of oxygen is taken as 16. For example, the isotopes of chlorine were found to have the masses 34.983 and 36.980 rather than 35.0 and 37.0, respectively. Later instruments designed and built by Aston, Dempster, Bainbridge, Jordon, Mattauch, and others have yielded isotopic masses with accuracies approaching 1 part in 100,000.

Modern mass spectroscopic measurements are based on the *mass-doublet* technique in which the quantity actually determined is the difference in mass between two ions of the same mass number but having slightly different masses. The newer spectrometers yield high dispersion,

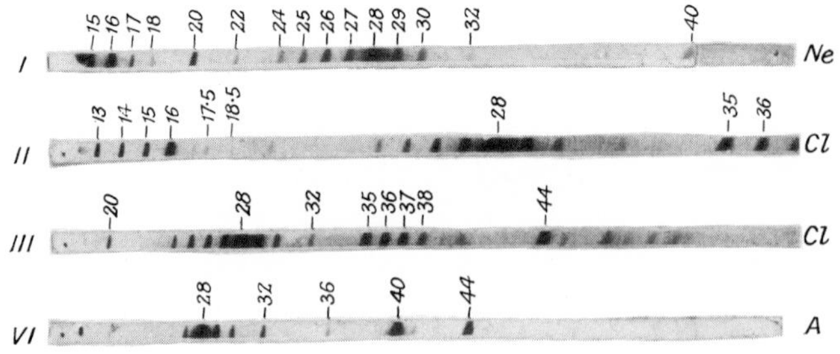

FIG. 9–4. Mass spectra obtained by Aston with his first spectrograph (1920).

that is, the distance between the two lines of a doublet are relatively large, and differences in mass can be measured with high precision. The mass of one of the members of the doublet must be known accurately. Hydrocarbon compounds (which give molecular ions) are used as sources of reference masses because of the relative ease with which fragments of almost any mass number can be obtained for comparison with other ions of the same mass number. The masses of C^{12}, the carbon isotope of mass number 12, and H^1, the hydrogen isotope of mass number 1, are used as secondary standards, with 0^{16} the primary standard.

The masses of C^{12} and H^1 relative to that of 0^{16} can be determined in a number of different ways. One method is to measure the following mass-doublet differences:

$$(0^{16})_2 - S^{32} = a,$$

$$(C^{12})_4 - S^{32}O^{16} = b,$$

$$C^{12}(H^1)_4 - O^{16} = c.$$

The three equations are solved simultaneously, with $O^{16} = 16$ atomic mass units (exactly), and the result is

$$S^{32} = 32 - a,$$

$$C^{12} = 12 + \frac{b - a}{4},$$

$$H^1 = 1 + \frac{a - b + c}{16}.$$

More complicated cycles involving a larger number of atoms can also be used, and highly precise results can be obtained; values of mass-doublet differences have been compiled,[6,7] and values of atomic masses determined by this method.[8–17]

It will be seen in Chapter 11 that information from nuclear reactions can also be used to determine atomic masses with precision comparable to that obtained with mass spectroscopic methods. Authoritative compilations of atomic masses usually combine results obtained with both methods.[18–21]

Other techniques have also been developed[22] which will only be mentioned here. The *chronotron* of Goudsmit[23–25] measures the time needed for ions to describe a number of revolutions in a uniform magnetic field. This time is proportional to the mass of the ion, with the result that the precision is practically constant for all masses, whereas the precision which can be obtained with the mass spectrograph decreases with increasing mass. Another device which depends on the angular

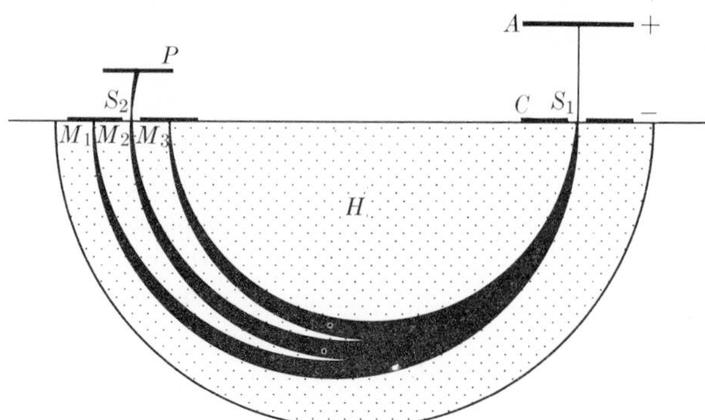

Fig. 9–5. Diagram of Dempster's mass spectrometer.

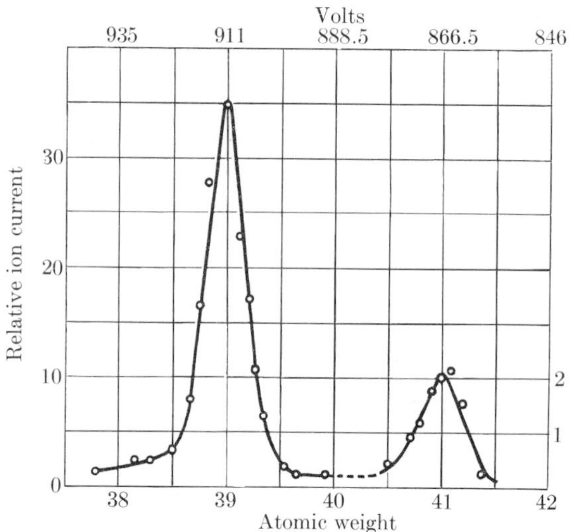

Fig. 9–6. Mass spectrum of potassium (Dempster, 1922).

motion of ions in a known magnetic induction is the *mass synchrometer*,[26] which has given good results. Microwave spectroscopy[27] has also been used successfully for the measurement of atomic masses.

In the systematic study of isotopes, it is essential to know not only the isotopic masses, but also the relative numbers of atoms of each isotope of an element. The mass spectrograph can be used for making abundance measurements, and indeed many of the isotopes now known were first discovered and their abundances measured by Aston by means of this instrument. When used for abundance measurements, however, a mass spectrograph is inconvenient because a photographic plate is used to record the different isotopic ions, and the procedure of determining abundances from the plate traces is both more tedious and less reliable than the direct measurements made with a somewhat simpler instrument, the *mass spectrometer*.

About the time of the development of the mass spectrograph by Aston, Dempster[(28)] built an instrument which was basically simpler and which was well suited for making abundance measurements, although it could not be used for making accurate mass measurements. It was called a mass spectrometer because the ion current was measured electrically rather than recorded on a photographic plate. A schematic diagram of one of Dempster's spectrometer models is shown in Fig. 9–5. Ions of the element to be analyzed are formed by heating a salt of the element, or by bombarding it with electrons. Upon emerging from the source A, the ions are accelerated through a potential difference V of about 1000 volts by an electric field maintained in the region between A and C. A slit S_1 in the plate C allows a narrow bundle of ions to pass into the region of the magnetic field H. In passing from A to C, the positive ions carrying a charge q acquire energy equal to qV. This energy may also be represented by $\frac{1}{2}Mv^2$, where M is the mass of the positive ion and v its velocity on emerging from the slit in C; consequently,

$$qV = \tfrac{1}{2}Mv^2. \tag{9–4}$$

If the magnetic field is perpendicular to the plane of the paper, the ions

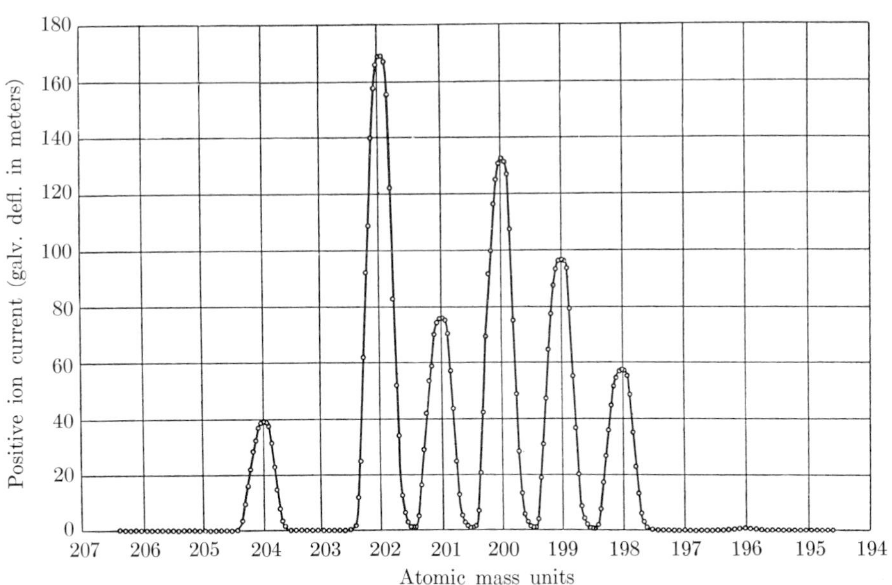

FIG. 9–7. Mass spectrum of mercury (Nier, 1937).

will be forced to move along a circular path defined by the relationship

$$Hqv = \frac{Mv^2}{R}, \tag{9–5}$$

where R is the radius of the circle. Since the radius of the circle must have a certain definite value in order that the ions enter the slit S_2 and be detected by an electrometer at P, it is clear that ions with only one particular value of q/M, say q/M', will be received for a given combination of accelerating potential and magnetic field. By varying the potential difference V, ions with different values of q/M are, in turn, made to pass through the second slit S_2 to the collector plate P. The current recorded by the electrometer is proportional to the number of positive ions reaching it per unit time and, since each accelerating potential corresponds to a definite mass of particle reaching the electrometer, the current can be plotted against the atomic weight. A typical curve obtained by Dempster

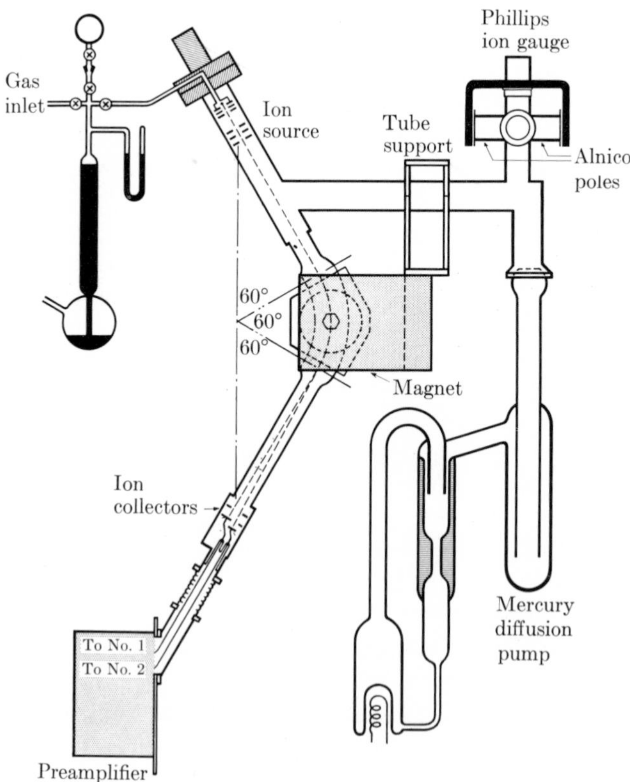

FIG. 9–8. Schematic diagram of a Nier mass spectrometer (Nier[30]).

for potassium with a slightly modified apparatus is shown in Fig. 9-6. There are two isotopes with atomic weights 39 and 41 units, the former being much more abundant; for the ratio of the abundances, Dempster obtained 18:1.

During recent years, owing largely to improved vacuum techniques and the development of new methods of electrical measurement, mass spectrometry has advanced rapidly. One of the most useful instruments was that of Nier,(29) which had extremely high resolving power and sensitivity and was especially designed for searching for rare isotopes and measuring relative abundances. It was a modification of the Dempster apparatus and gave results like those shown in Fig. 9-7. The figure shows a "mass spectrogram" of mercury, and the isotope abundances are closely proportional to the magnitude of the peaks of the positive ion current. A more recent form of spectrometer designed by Nier(30) is shown in Fig. 9-8. Ions are produced by the electron bombardment of the gas under investigation, and are accelerated by a potential drop of about 1000 volts. The beam of ions passes into a wedge-shaped magnetic field in which they suffer a deflection of 60°, rather than 180° as in the Dempster spectrometer. This deflection makes it possible to obtain high resolution with a simple magnet. In the diagram the ion beam is shown broken into two parts which fall on separate collectors and are measured by separate amplifiers. This instrument has been used to get highly precise isotopic abundances.

9-4 The stable isotopes of the elements and their percentage abundances. The mass spectra of the elements have been investigated in detail and the isotopic composition of the elements has been determined.(31,32,33) The results obtained for the isotopes of 83 elements are collected in Table 9-1. Before discussing these results, some remarks about terminology are in order. In recent years, the term *nuclide* has been widely accepted for a species of atom characterized by the constitution of its nucleus, i.e., by the numbers of protons and neutrons it contains. Thus, the atomic species listed in Table 9-1 may be referred to as the naturally occurring stable nuclides. Similarly, every radioactive species is a radioactive nuclide or radionuclide. An isotope is then one of a group of two or more nuclides having the same number of protons or, in other words, having the same atomic number. An element like beryllium or aluminum, of which only one species exists in nature, is said to form a single stable nuclide, rather than a single stable isotope, since the word isotope implies more than one species occupying the same place in the periodic system. A nuclide is usually indicated by the chemical symbol with a subscript at the lower left giving the atomic number, and a superscript at the upper right giving the mass number; in the symbol ${}_Z\mathrm{S}^A$, Z is the atomic number,

TABLE 9–1

THE STABLE NUCLIDES AND THEIR PERCENTAGE ABUNDANCE

Symbol	Atomic number, Z	Mass number, A	Relative abundance, %
H	1	1	99.9849–99.9861
D	1	2	0.0139–0.0151
He	2	3	$\sim 10^{-5}$–10^{-4}
		4	~ 100
Li	3	6	7.52
		7	92.48
Be	4	9	100
B	5	10	18.45–19.64
		11	80.36–81.55
C	6	12	98.892
		13	1.108
N	7	14	99.634
		15	0.366
O	8	16	99.759
		17	0.037
		18	0.204
F	9	19	100
Ne	10	20	90.92
		21	0.257
		22	8.82
Na	11	23	100
Mg	12	24	78.70
		25	10.13
		26	11.17
Al	13	27	100
Si	14	28	92.21
		29	4.70
		30	3.09
P	15	31	100
S	16	32	95.0
		33	0.760
		34	4.22
		36	0.014
Cl	17	35	75.529
		37	24.471
A	18	36	0.337
		38	0.063
		40	99.600
K	19	39	93.10
		40	0.012
		41	6.88
Ca	20	40	96.97
		42	0.64
		43	0.145
		44	2.06
		46	0.003
		48	0.185
Sc	21	45	100
Ti	22	46	7.93
		47	7.28
		48	73.94
		49	5.38
		50	5.34
V	23	50	0.24
		51	99.76

TABLE 9–1 (*Continued*)

Symbol	Atomic number, Z	Mass number, A	Relative abundance, %
Cr	24	50	4.31
		52	83.76
		53	9.55
		54	2.38
Mn	25	55	100
Fe	26	54	5.82
		56	91.66
		57	2.19
		58	0.33
Co	27	59	100
Ni	28	58	67.88
		60	26.23
		61	1.19
		62	3.66
		64	1.08
Cu	29	63	69.1
		65	30.9
Zn	30	64	48.89
		66	27.81
		67	4.11
		68	18.56
		70	0.62
Ga	31	69	60.4
		71	39.6
Ge	32	70	20.52
		72	27.43
		73	7.76
		74	36.54
		76	7.76
As	33	75	100
Se	34	74	0.87
		76	9.02
		77	7.58
		78	23.52
		80	49.82
		82	9.19
Br	35	79	50.54
		81	49.46
Kr	36	78	0.354
		80	2.27
		82	11.56
		83	11.55
		84	56.90
		86	17.37
Rb	37	85	72.15
		87	27.85
Sr	38	84	0.56
		86	9.86
		87	7.02
		88	82.56
Y	39	89	100
Zr	40	90	51.46
		91	11.23
		92	17.11
		94	17.40
		96	2.80
Nb	41	93	100
Mo	42	92	15.84
		94	9.04
		95	15.72
		96	16.53
		97	9.46
		98	23.78
		100	9.63
Ru	44	96	5.51
		98	1.87
		99	12.72
		100	12.62
		101	17.07
		102	31.61
		104	18.58

(*Continued*)

TABLE 9–1 (*Continued*)

Symbol	Atomic number, Z	Mass number, A	Relative abundance, %	Symbol	Atomic number, Z	Mass number, A	Relative abundance, %
Rh	45	103	100	Te	52	126	18.71
						128	31.79
Pd	46	102	0.96			130	34.49
		104	10.97				
		105	22.23	I	53	127	100
		106	27.33				
		108	26.71	Xe	54	124	0.096
		110	11.81			126	0.090
						128	1.919
Ag	47	107	51.35			129	26.44
		109	48.65			130	4.08
						131	21.18
Cd	48	106	1.215			132	26.89
		108	0.875			134	10.44
		110	12.39			136	8.87
		111	12.75				
		112	24.07				
		113	12.26	Cs	55	133	100
		114	28.86				
		116	7.58	Ba	56	130	0.101
						132	0.097
In	49	113	4.28			134	2.42
		115	95.72			135	6.59
						136	7.81
Sn	50	112	0.96			137	11.32
		114	0.66			138	71.66
		115	0.35				
		116	14.30	La	57	138	0.089
		117	7.61			139	99.911
		118	24.03				
		119	8.58	Ce	58	136	0.193
		120	32.85			138	0.250
		122	4.72			140	88.48
		124	5.94			142	11.07
Sb	51	121	57.25				
		123	42.75	Pr	59	141	100
Te	52	120	0.089	Nd	60	142	27.11
		122	2.46			143	12.17
		123	0.87			144	23.85
		124	4.61			145	8.30
		125	6.99			146	17.22

TABLE 9–1 (*Continued*)

Symbol	Atomic number, Z	Mass number, A	Relative abundance, %	Symbol	Atomic number, Z	Mass number, A	Relative abundance, %
Sm	62	144	3.09	Yb	70	172	21.82
		147	14.97			173	16.13
		148	11.24			174	31.84
		149	13.83			176	12.73
		150	7.44				
		152	26.72	Lu	71	175	97.40
		154	22.71			176	2.60
Eu	63	151	47.82	Hf	72	174	0.18
		153	52.18			176	5.20
						177	18.50
Gd	64	152	0.20			178	27.14
		154	2.15			179	13.75
		155	14.73			180	35.24
		156	20.47				
		157	15.68	Ta	73	181	100
		158	24.87				
		160	21.90	W	74	180	0.135
						182	26.41
Tb	65	159	100			183	14.40
						184	30.64
Dy	66	156	0.0524			186	28.41
		158	0.0902				
		160	2.294	Re	75	185	37.07
		161	18.88			187	62.93
		162	25.53				
		163	24.97	Os	76	184	0.018
		164	28.18			186	1.59
						187	1.64
Ho	67	165	100			188	13.3
						189	16.1
Er	68	162	0.136			190	26.4
		164	1.56			192	41.0
		166	33.41				
		167	22.94	Ir	77	191	37.3
		168	27.07			193	62.7
		170	14.88				
				Pt	78	190	0.0127
Tm	69	169	100			192	0.78
						194	32.9
Yb	70	168	0.135			195	33.8
		170	3.03			196	25.3
		171	14.31			198	7.21

(*Continued*)

TABLE 9-1 (*Concluded*)

Symbol	Atomic number, Z	Mass number, A	Relative abundance, %	Symbol	Atomic number, Z	Mass number, A	Relative abundance, %
Au	79	197	100	Pb	82	204	1.48
						206	23.6
Hg	80	196	0.146			207	22.6
		198	10.02			208	52.3
		199	16.84				
		200	23.13	Bi	83	209	100
		201	13.22				
		202	29.80	Th	90	232	100
		204	6.85				
				U	92	234	0.0056
Tl	81	203	29.50			235	0.7205
		205	70.50			238	99.2739

A is the mass number, and S is the chemical symbol. Thus, the symbol $_{50}Sn^{120}$ represents the nuclide with 50 protons and a mass number of 120; it is one of the ten isotopes of tin. The number of neutrons is, of course, equal to $A - Z$. Sometimes it is not necessary to show the number of protons explicitly, and the symbol for a nuclide is then shortened to S^A.

Not all of the nuclides listed in Table 9-1 are actually stable. Thorium and uranium are radioactive, but they occur in sufficient amounts and with sufficiently weak activity so that they can be handled in the same way as the stable elements. At least nine naturally occurring isotopes of "stable" elements show feeble radioactivity: K^{40}, Rb^{87}, In^{115}, La^{138}, Nd^{144}, Sm^{147}, Lu^{176}, Re^{187} and Pt^{190}. These nuclides are distinct from the families or chains of the heavy naturally occurring radionuclides and are much feebler in activity. It is therefore more convenient to include them with the stable elements than with the radioactive ones.

With the exceptions noted, Table 9-1 contains 284 stable nuclides divided among 83 elements. Twenty elements, about one-fourth in all, are single species; all the others consist of two or more isotopes. Hydrogen has two isotopes, the one with mass number 2 having a relative abundance of only about 0.015%. This rare isotope, however, has a mass about double that of the common isotope, so that the difference in mass is as great as the mass of the ordinary hydrogen atom itself. This relationship between the masses is an exceptional one and, as a result, the differences between the properties of the two isotopes are more marked than in any other pair of isotopes. The hydrogen isotope of mass 2 has therefore been given its own name, *deuterium*, with the occasionally used symbol D.

TABLE 9–2

SOME ISOTOPE STATISTICS

	Number of elements	Odd A	Even A	Total	Average number of isotopes
Odd Z	40	53	8	61	1.5
Even Z	43	57	166	223	5.2
Total	83	110	174	284	3.4

Carbon and nitrogen also have two isotopes, while oxygen has three, two of which are rare. Tin has the greatest number of isotopes, ten, while xenon has nine, cadmium and tellurium have eight each, and several elements have seven.

There are some striking regularities in Table 9–1. Nuclides of even Z are much more numerous than those of odd Z, and nuclides of even A are much more numerous than those of odd A. Nearly all nuclides with even A have even Z, the only common exceptions being ${}_1H^2$, ${}_3Li^6$, ${}_5B^{10}$, and ${}_7N^{14}$. The nuclides ${}_{19}K^{40}$ and ${}_{71}Lu^{176}$ have odd Z and even A but are weakly radioactive, while ${}_{23}V^{50}$ and ${}_{57}La^{138}$ are very rare. The numbers of nuclides with the various combinations of even and odd atomic and mass numbers are listed in Table 9–2. Of the 20 elements which have only a single nuclide, only beryllium has an even value of Z, while the other 19 have odd Z. Nineteen elements with odd Z have two isotopes apiece, and each of these nuclides has odd A. One element with odd Z, potassium, has three isotopes; two of these have odd mass numbers, and the only isotope with an even mass number is the weakly radioactive K^{40}. The four common elements which have odd values of Z and A, hydrogen lithium, boron, and nitrogen, have equal numbers of protons and neutrons. The elements which have more than two isotopes (apart from potassium) all have even values of Z.

An examination of the values of Z and A in Table 9–1 shows that in the stable nuclei, with the exception of H^1 and He^3, the number of neutrons is always greater than or equal to the number of protons. There is always at least one neutron for each proton. This property of the stable nuclides is shown in Fig. 9–9, in which the number of neutrons $A - Z$ is plotted against the number of protons. The number of neutrons which can be included in a stable nucleus with a given number of protons is limited. For example, tin with an atomic number of 50 has neutron numbers from 62 to 74, and the mass numbers of the tin isotopes lie between 112 and 124. Apparently the tin nucleus cannot contain less than 62 neutrons nor more than 74 neutrons and still remain stable. For the other elements (except

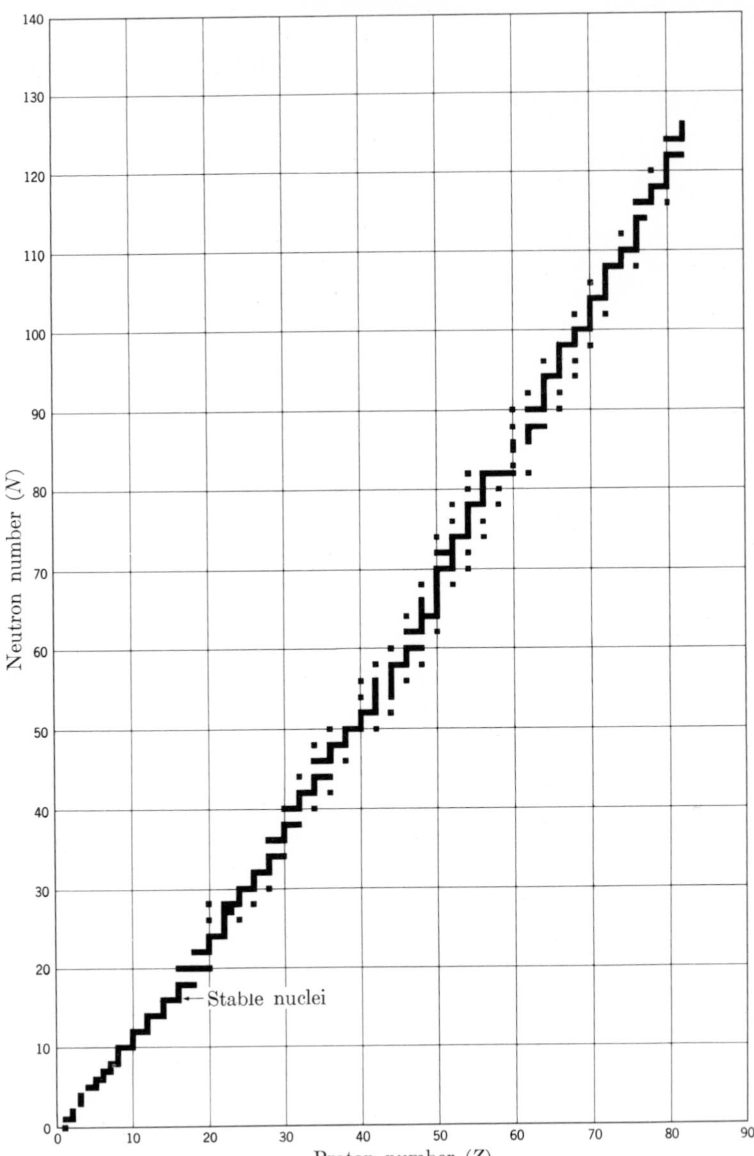

FIG. 9–9. Neutron-proton plot of the stable nuclides.

xenon) this range is smaller and the limits for the existence of stable nuclei are narrower.

The regularities that have just been noted are closely connected with the problem of nuclear stability and must eventually be accounted for in terms of the proton-neutron theory of the nucleus, in particular, in terms of the forces between nucleons. From this standpoint, the hydrogen isotope of mass two (deuterium) is especially important because its nucleus consists of one proton and one neutron. The properties of this nucleus yield information about the force between a proton and a neutron, and deuterium has a prominent place in theoretical nuclear physics.

In the preceding discussion it was assumed that the relative abundance of an isotope is constant in nature, that is, independent of the source of the sample that is measured. This assumption is true, in general. There are, however, some exceptions which, although interesting, do not affect any of the conclusions previously drawn from the consideration of abundances. The relative abundances of H^1 and H^2 depend somewhat on the source, and the spread in the values listed in Table 9–1 shows that the variation is small. The ratio of hydrogen to deuterium abundance has been determined as 6700 ± 50 for samples of tap water from London, Osaka, and various cities in the United States; this corresponds to a value of 0.0149% for the relative abundance of H^2 in tap water. The range 0.0139 to 0.0151% includes the variation over a wide range of substances such as water, snow, ice, organic compounds, and animal and mineral materials from many sources. There are wide variations in the ratio of He^3 and He^4 abundances; for example, the abundance of He^3 is approximately ten times as great in atmospheric helium as in well helium. Helium is formed in radioactive minerals because of alpha-decay, and such helium is all He^4. In nonradioactive ores, the He^3 content varies widely, and it has been suggested that He^3 can be formed as a result of the transformation of various nuclides in the air and on the ground by cosmic-ray bombardment. Cosmic rays are highly penetrating radiations which originate outside the earth and consist of protons, electrons, neutrons, photons, and other particles. The range of abundances of the boron isotopes shown in the table is equivalent to a variation in the B^{11}/B^{10} ratio from 4.27 to 4.42 or about 3%. Although this variation seems small, it affects the use of certain boron standards in nuclear physics; it is necessary, therefore, to cite the abundance ratios when measurements involving this standard are reported. The abundance of the carbon isotopes cited in Table 9–1 are those found in limestone and correspond to a C^{12}/C^{13} ratio of 89.2; in coal the ratio is 91.8. In general, the relative content of C^{12} seems to be somewhat greater in plant material than in limestone.

The relative abundances of the oxygen isotopes also vary, and the value of the O^{16}/O^{18} ratio in nature has a spread of about 4%. The values of the abundances quoted in the table are those for atmospheric oxygen and correspond to a value of the O^{16}/O^{18} ratio of 489.2 ± 0.7. This value is also correct for the oxygen from limestone, but for oxygen from water or iron ores the ratio may be 4% higher. The values used for the abundances of the oxygen isotopes affect the value of the factor for converting atomic weights from the physical scale to the chemical scale (see Section 9–5). The variation in the value of the conversion factor may be large enough to affect the precision of atomic weight determinations and, if the latter are to be made to more than five significant figures, the isotopic composition of the oxygen used as a reference must be specified.

Wide variations in the abundances of the lead isotopes are also found, and these are usually associated with the radioactive sources from which the different lead samples are derived. Even for common lead, it is not possible to give exact isotopic abundances without specifying the source of the material; the values in the table are for Great Bear Lake galena.

The abundance variations which have been discussed are the most important ones known and, for the remaining elements, either there are no significant variations, or else they do not seriously affect further work in nuclear physics.

9–5 Atomic masses: packing fractions and binding energies. Some atomic masses(18,19,36) are given in Table 9–3. The standard of mass used here is slightly different from that used for the chemical atomic weights. It is seen from Table 9–1 that oxygen actually has three isotopes, the most abundant of which has the mass number 16. The other two isotopes together constitute only about 0.2% of the oxygen atoms. In the determination of isotopic weights by the mass spectrograph it is the practice to take as the standard the value of 16.00000 for the weight of the common isotope of oxygen. The weights so obtained differ slightly from those based on the ordinary chemical atomic weight scale. In the latter case, the number 16.00000 is associated with ordinary atmospheric oxygen, which is a mixture of isotopes, whereas on the mass spectrographic, or physical, atomic weight scale, this is taken as the isotopic weight of the single, most abundant isotope. The relationship between the chemical and physical atomic weight scales may be determined in the following way. Atmospheric oxygen consists of 99.759% of the isotope of mass 16.00000, together with 0.037% of the isotope of mass 17.004529, and 0.204% of the isotope of mass 18.004840. The weighted mean of these values is 16.004462, and this is the atomic weight of atmospheric oxygen on the physical scale, as compared with the postulated value of 16.00000 on the chemical scale. Then

$$\frac{\text{Physical atomic weight}}{\text{Chemical atomic weight}} = \frac{16.004462}{16.000000} = 1.000279.$$

TABLE 9–3

ATOMIC MASSES, PACKING FRACTIONS, AND BINDING ENERGIES OF SOME OF THE STABLE NUCLIDES

Nuclide	Number of protons, Z	Number of neutrons, $A - Z$	Mass, amu	Packing fraction, $\times 10^4$	Binding energy total, Mev	Binding energy per nucleon, Mev
n^1	0	1	1.0089830 (±1.7)	89.8		
H^1	1	0	1.0081437 (±1.8)	81.4		
H^2	1	1	2.0147361 (±2.9)	73.7	2.225	1.113
He^4	2	2	4.0038727 (±2.1)	9.7	28.29	7.07
Li^7	3	4	7.018222 (±6)	26.0	39.24	5.61
Be^9	4	5	9.015041 (±5)	16.7	58.15	6.46
B^{11}	5	6	11.012795 (±5)	11.6	76.19	6.93
C^{12}	6	6	12.0038065 (±3.9)	3.2	92.14	7.68
C^{13}	6	7	13.0074754 (±4.1)	5.8	97.09	7.47
N^{14}	7	7	14.0075179 (±3.0)	5.4	104.63	7.47
O^{16}	8	8	16.0000000	0	127.58	7.97
O^{17}	8	9	17.0045293 (±3.9)	2.7	131.73	7.75
O^{18}	8	10	18.004840 (±9)	2.7	139.80	7.77
F^{19}	9	10	19.004447 (±7)	2.3	147.75	7.78
Ne^{20}	10	10	19.998765 (±10)	−0.6	160.62	8.03
Al^{27}	13	14	26.990080 (±14)	−3.7	224.92	8.33
Si^{28}	14	14	27.985777 (±16)	−5.1	236.51	8.45
P^{31}	15	16	30.983563 (±18)	−5.3	262.88	8.48
S^{32}	16	16	31.982190 (±20)	−5.6	271.74	8.49
Cl^{35}	17	18	34.979906 (±80)	−5.7	298.13	8.52
Cl^{37}	17	20	36.9776573 (±16)	−6.0	317.00	8.57
A^{40}	18	22	39.975088 (±4)	−6.2	343.71	8.59
Ca^{40}	20	20	39.975330 (±30)	−6.2	341.92	8.55
Fe^{56}	26	30	55.952722 (±6)	−8.4	492.11	8.79
Cu^{63}	29	34	62.949607 (±11)	−8.0	551.22	8.75
As^{75}	33	42	74.945510 (±100)	−7.3	652.23	8.70
Sr^{88}	38	50	87.933680 (±300)	−7.5	768.13	8.73
Mo^{98}	42	56	97.937240 (±350)	−6.3	845.73	8.63
Sn^{116}	50	66	115.938850 (±300)	−5.3	988.14	8.52
Sn^{120}	50	70	119.940330 (±140)	−5.0	1020.2	8.50
Xe^{130}	54	76	129.944810 (±30)	−4.3	1096.6	8.44
Xe^{136}	54	82	135.950420 (±25)	−3.7	1141.5	8.39
Nd^{150}	60	90	149.968490 (±70)	−2.1	1237.1	8.25
Hf^{176}	72	104	175.99650 (±800)	−0.2	1419.1	8.06
W^{184}	74	110	184.008300 (±600)	0.4	1473.5	8.01
Au^{197}	79	118	197.028000 (±1000)	1.4	1560.0	7.92
Pb^{206}	82	124	206.037900 (±500)	1.8	1623.7	7.88
Th^{232}	90	142	232.109800 (±500)	4.7	1768.0	7.62
U^{238}	92	146	238.124300 (±500)	5.2	1803.1	7.58

Hence, isotopic weights obtained by means of the mass spectrograph must be divided by 1.000279 in order to convert the results to the chemical atomic weight scale. The conversion factor is used only when it is necessary to compare mass spectrographic results with weights obtained by chemical methods; in nuclear work the atomic masses on the physical scale are used.

Estimates of the errors of the mass spectrographic measurements have been included in order to indicate the high precision with which atomic masses can now be determined. The error, given in parentheses, is expressed in units of 10^{-6} amu. This precision is essential in the study of nuclear reactions and transformations, as will be seen in later chapters. It is also important in the precise calculation of packing fractions and binding energies, as will now be shown.

It is seen from Table 9–3 that the isotopic masses are indeed very close to whole numbers. It seemed clear to the early workers in this field that the systematic study of the divergences of the masses of nuclides from whole numbers was an important problem. Aston expressed these divergences in the form of a quantity called the *packing fraction* defined by

$$\text{Packing fraction} = \frac{\text{Atomic mass} - \text{Mass number}}{\text{Mass number}}$$

$$= \frac{M_{Z,A} - A}{A}, \tag{9–6}$$

where $M_{Z,A}$ is the actual weight of a nuclide on the physical atomic weight scale, and A is the mass number. If the packing fraction is denoted by f, then

$$M_{Z,A} = A(1 + f). \tag{9–7}$$

Sample values of the packing fraction are listed in Table 9–3, and the variation of the packing fraction with mass number for a larger number of nuclides is shown graphically in Fig. 9–10. The packing fractions, with the exception of those for He^4, C^{12}, and O^{16}, fall on or near the solid curve. The values are high for elements of low mass number, apart from the nuclides mentioned. For O^{16}, the value is zero, by definition. As A increases, the packing fraction becomes negative, passes through a rather flat minimum and then rises gradually, becoming positive again at values of A of about 180. The packing fraction was very useful in the study of isotopic masses, but it does not have a precise physical meaning. The explanation for its usefulness will appear from the discussion of the binding energies of nuclei.

The atomic mass of a nuclide can be understood in terms of the masses of its constituent particles and a quantity called the *binding energy*. It

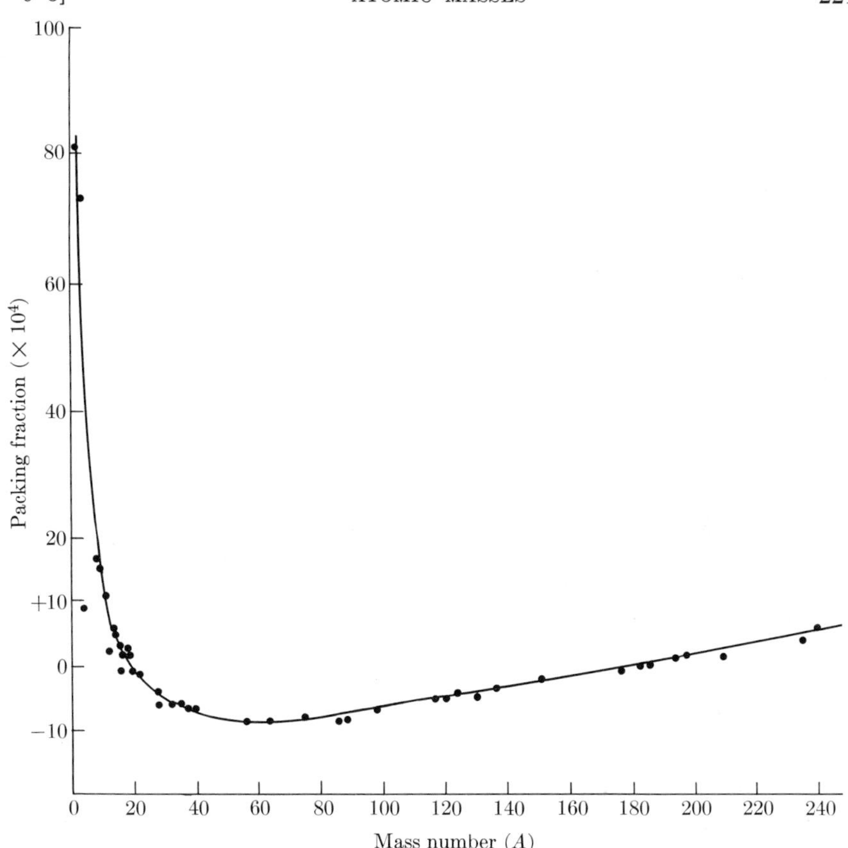

Fig. 9–10. Packing fraction as a function of mass number.

might at first be supposed that the mass of an atom should be the sum of the masses of its constituent particles. A survey of the atomic masses shows, however, that the atomic mass is less than the sum of its constituent particles in the free state. To account for this difference in mass, the principle of the equivalence of mass and energy, derived from the special theory of relativity, is used. If ΔM is the decrease in mass when a number of protons, neutrons, and electrons combine to form an atom, then the above principle states that an amount of energy equal to

$$\Delta E = c^2 \Delta M \tag{9–8}$$

is released in the process. The difference in mass, ΔM, is called the *mass defect;* it is the amount of mass which would be converted to energy if a particular atom were to be assembled from the requisite numbers of protons, neutrons, and electrons. The same amount of energy would be

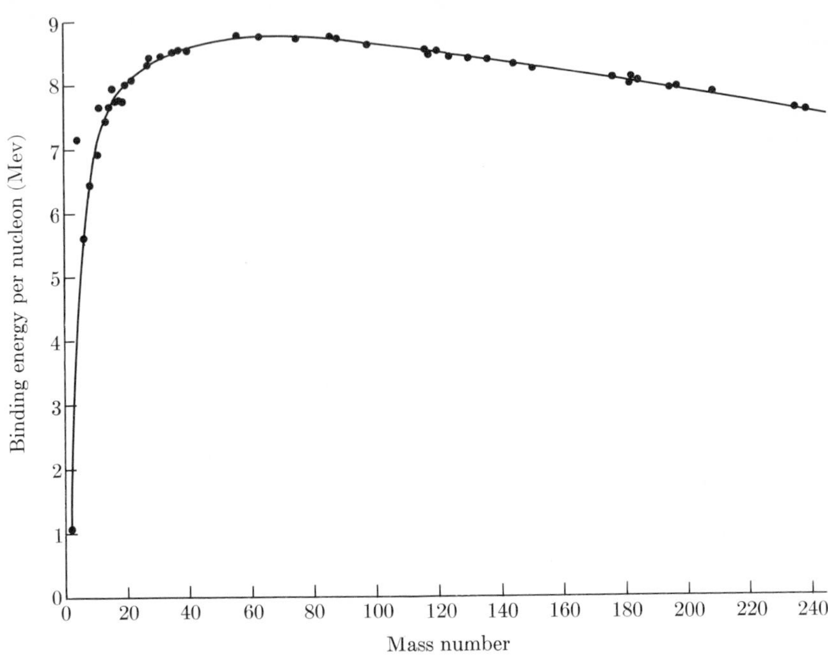

FIG. 9–11. Binding energy per nucleon as a function of mass number.

needed to break the atom into its constituent particles, and the energy equivalent of the mass defect is therefore a measure of the binding energy of the nucleus. The mass of the constituent particles is the sum of Z proton masses, Z electrons, and $A - Z$ neutrons. The proton and electron masses can be combined into the mass of Z hydrogen atoms because the minute change in mass which may accompany the formation of a hydrogen atom from a proton and electron is negligible. The mass defect can then be written

$$\Delta M = Zm_{\text{H}} + (A - Z)m_n - M_{Z,A}, \tag{9–9}$$

where m_{H}, the mass of the hydrogen atom is 1.0081437 mass units and m_n, the mass of the neutron, is 1.0089830 mass units. Then

$$\Delta M = 1.0081437Z + 1.0089830(A - Z) - M_{Z,A}. \tag{9–10}$$

Since one atomic mass unit is equivalent to 931.145 Mev, the binding energy of the nucleus is given by

$$\Delta E(\text{Mev}) = 931.145[1.0081437Z + 1.0089830(A - Z) - M_{Z,A}]. \tag{9–11}$$

The average binding energy per nucleon is obtained by dividing the total binding energy of the nucleus by the mass number A. Some values of the binding energy obtained by the method just outlined are listed in the last two columns of Table 9–3, and a graph of the binding energy per nucleon as a function of the mass number is shown in Fig. 9–11. With the exception of He^4, C^{12}, and O^{16}, the values of the binding energy per particle lie on or close to a single curve. The binding energies of some of the very light nuclides, such as H^2, are very small. The binding energies of He^4, C^{12}, and O^{16} are considerably greater than those of their neighbors, as is shown clearly by the values in the last column of Table 9–3. The binding energy per particle rises sharply, and reaches a maximum value of about 8.8 Mev in the neighborhood of $A = 50$. The maximum is quite flat, and the binding energy is still 8.4 Mev at about $A = 140$. For higher mass numbers, the value decreases to about 7.6 Mev at uranium.

The magnitude of the binding energy is enormous, as can be shown by converting from Mev per nucleon to more familiar units such as Btu per pound. One Mev is equivalent to 1.519×10^{-16} Btu. One pound of an atomic species contains $453.6/M$ gram atomic weights (M is the atomic weight), or

$$453.6 \times 6.023 \times 10^{23} \times A/M \text{ nucleons.}$$

The mass number A and the atomic weight M are practically the same, so that one pound contains 2.73×10^{26} nucleons. The unit "1 Mev/nucleon" is then practically the same as 4.15×10^{10} Btu/pound. In the neighborhood of the maximum of the binding energy curve, the binding energy is about 8.8 Mev/nucleon or 350 billion Btu/pound. This enormous value of the energy that would be needed to dissociate a nucleus into its constituent protons and neutrons is another indication of the magnitude of the nuclear forces.

The application of the principle of the equivalence of mass and energy and the introduction of the concept of binding energy have, thus far in our treatment, only a theoretical basis. In recent years, however, many nuclear transmutations have been accurately studied and careful measurements have been made of the changes of mass and energy in these reactions. The validity of the relativistic mass-energy relationship has been proven, as well as that of its application to the problems of nuclear physics. The binding energy of a nucleus is, therefore, a quantity with real physical meaning, and the masses and binding energies of nuclei yield useful information about the constitution and stability of nuclei. In addition, the analysis of nuclear transformations has provided an independent means of determining atomic masses which is at present even more powerful than the mass spectrographic method. These matters will be discussed in detail in later chapters.

The subject of atomic masses should not be left without a brief discussion of a proposal[(34)] to replace the mass standard, O^{16}, by C^{12}, i.e., to assign to C^{12} the atomic mass 12.0000000 units. The use of C^{12} as the standard would have several advantages in mass spectrometry. Carbon forms many more chemical compounds which can provide molecular ions for use in a spectrometer than does oxygen. Doubly, triply, and quadruply charged ions of C^{12} occur at integral mass numbers and can be paired in doublets with ions of mass number 6, 4, and 3, respectively, so that C^{12} would be a convenient standard for atoms of low mass number. No other element besides carbon forms molecular ions containing as many atoms of one kind, up to 10 or more; this property would permit many more doublet comparisons to be made directly with the reference nuclide than are now possible, and would yield masses with increased precision at intermediate and large values of A. Thus, carbon forms many compounds with hydrogen, providing easy reference lines for doublets with masses up to 120 units; for values of A between 120 and 240, doubly charged ions of heavier elements could be compared directly with singly charged ions of the type $(C^{12})_n$ or $(C^{12})_n\,(H^1)_m$.

A table of atomic masses has been prepared[(35)] based on C^{12} as the standard and the following results have been obtained.

Nuclide	O^{16} Standard	C^{12} Standard
n	1.0089861	1.0086654
H^1	1.0081456	1.0078252
H^2	2.0147425	2.0141022
C^{12}	12.0038150	12.0000000
O^{16}	16.0000000	15.9941949

The relation between the atomic mass units on the two scales is

$$1 \text{ amu } (C^{12}) = 1.00031792 \text{ amu } (O^{16}).$$

The conversion factor from mass to energy is

$$1 \text{ amu } (C^{12}) = 931.441 \text{ Mev},$$

as compared with 1 amu $(O^{16}) = 931.145$ Mev. The values of binding energies are unchanged; thus, the binding energy of the deuteron is 2.2247 Mev when either mass standard is used.

Throughout this book, we shall use O^{16} as the mass standard and the atomic masses given in Table 9–3, but the reader should be aware of the possibility that the standard may be changed within the next few years.

References

GENERAL

F. W. Aston, *Mass Spectra and Isotopes*, 2nd ed. London: Edward Arnold, 1942.

Sir J. J. Thomson, *Rays of Positive Electricity and Their Application to Chemical Analyses*, 1st ed. London: Longmans, Green and Co., 1913; 2nd ed., 1921.

M. G. Inghram, "Modern Mass Spectroscopy," *Advances in Electronics*, Vol. 1. New York: Academic Press, 1948.

J. D. Stranathan, *The "Particles" of Modern Physics*. Philadephia: Blakiston, 1944, Chapter 5.

K. T. Bainbridge, "Charged Particle Dynamics and Optics, Relative Isotope Abundances of the Elements, Atomic Masses," *Experimental Nuclear Physics*, E. Segrè, ed., Vol. I. New York: Wiley, 1953.

1959 Nuclear Data Tables. Nuclear Data Project, National Academy of Sciences—National Research Council. Washington, D.C.: United States Atomic Energy Commission. Table IV: Relative Isotopic Abundances; Table VI: Mass Differences and Ratios.

PARTICULAR

1. J. J. Thomson, Papers on positive rays and Isotopes, *Phil. Mag.*, **13,** 561 (1907); **16,** 657 (1908); **18,** 821 (1909); **20,** 752 (1910); **21,** 225 (1911); **24,** 209, 669 (1912).
2. F. W. Aston, *Mass Spectra and Isotopes, op. cit.* gen. ref., Chapter 4.
3. F. W. Aston, "A Positive Ray Spectrograph," *Phil. Mag.*, **38,** 707 (1919).
4. F. W. Aston, "The Constitution of Atmospheric Neon," *Phil. Mag.*, **39,** 449 (1920).
5. F. W. Aston, "A New Mass Spectrograph and the Whole Number Rule," *Proc. Roy. Soc. (London)*, **A115,** 487 (1927).
6. Duckworth, Hogg, and Pennington, "Mass Spectroscopic Atomic Mass Differences," *Revs. Mod. Phys.*, **26,** 463 (1954).
7. H. E. Duckworth, "Mass Spectroscopic Atomic Mass Differences II," *Revs. Mod. Phys.*, **29,** 767 (1957).
8. Quisenberry, Scolman, and Nier, "Atomic Masses of H^1, D^2, C^{12}, and S^{32}," *Phys. Rev.* **102,** 1071 (1956).
9. Scolman, Quisenberry, and Nier, "Atomic Masses of the Stable Isotopes, $10 \leq A \leq 30$," *Phys. Rev.* **102,** 1076 (1956).
10. M. E. Kettner, "Atomic Masses from C^{12} to Ne^{22}," *Phys. Rev.* **102,** 1065 (1956).
11. Quisenberry, Giese, and Benson, "Atomic Masses of H^1, C^{12}, and S^{32}," *Phys. Rev.* **107,** 1664 (1957).
12. C. F. Giese and J. L. Benson, "Atomic Masses from Phosphorus Through Manganese," *Phys. Rev.* **110,** 712 (1958).
13. Collins, Nier, and Johnson, "Atomic Masses in the Region about Mass 40 and from Titanium through Zinc," *Phys. Rev.*, **84,** 717 (1951); **86,** 408 (1952).
14. R. E. Halstead, "Atomic Masses from Palladium through Xenon," *Phys. Rev.*, **88,** 666 (1952).

15. H. E. Duckworth and R. S. Preston, "Some New Atomic Mass Measurements and Remarks on the Mass Evidence for Magic Numbers," *Phys. Rev.*, **82,** 468 (1951).

16. Duckworth, Kegley, Olson, and Stamford, "Some New Values of Atomic Masses, Principally in the Region of 82 Neutrons," *Phys. Rev.*, **83,** 1114 (1951).

17. Stamford, Duckworth, Hogg, and Geiger, "Masses of Pb^{208}, Th^{232}, U^{234}, and U^{238}," *Phys. Rev.*, **86,** 617 (1952).

18. J. Mattauch and F. Everling, "Masses of Atoms of $A < 40$," *Progress in Nuclear Physics*, O. R. Frisch, ed., Vol. 6, p. 233. London: Pergamon Press, 1957.

19. H. E. Duckworth, "Masses of Atoms of $A > 40$," *Progress in Nuclear Physics*, O. R. Frisch, ed., Vol. 6, p. 138. London: Pergamon Press, 1957.

20. A. H. Wapstra, "Atomic Masses of Nuclides," *Handbuch der Physik*, Vol. 38, Part 1, p. 1. Berlin: Springer Verlag, 1958.

21. *American Institute of Physics Handbook*, p. 8–6. New York: McGraw-Hill, 1957.

22. H. Hintenberger, ed., *Nuclear Masses and their Determination.* London: Pergamon Press (1957).

23. S. Goudsmit, "A Time-of-Flight Mass Spectrometer," *Phys. Rev.*, **74,** 622 (1948).

24. Hays, Richards, and Goudsmit, "Mass Measurements with a Magnetic Time-of-Flight Mass Spectrometer," *Phys. Rev.*, **84,** 824 (1951); **85,** 1065 (1952).

25. Richards, Hays, and Goudsmit, "Masses of Lead and Bismuth," *Phys. Rev.*, **85,** 630 (1952).

26. L. G. Smith, "Measurements of Light Masses with the Mass Synchrometer," *Phys. Rev.*, **111,** 1606 (1958).

27. S. Geschwind, "Determination of Atomic Masses by Microwave Methods," *Handbuch der Physik*, Vol. 38, Part 1, p. 38. Berlin: Springer Verlag, 1958.

28. A. J. Dempster, "A New Method of Positive-Ray Analysis," *Phys. Rev.*, **11,** 316 (1918); "Positive-Ray Analysis of Potassium, Calcium and Zinc," *Phys. Rev.*, **20,** 631 (1922).

29. A. O. Nier, "Mass Spectrograph Studies of the Isotopes of Various Elements," *Phys. Rev.*, **50,** 1041 (1936); **52,** 933 (1937).

30. A. O. Nier, "A Mass Spectrometer for Isotope and Gas Analysis," *Rev. Sci. Instr.*, **18,** 398 (1947).

31. K. T. Bainbridge and A. O. Nier, *Relative Isotopic Abundances of the Elements*, Preliminary Report No. 9, Nuclear Science Series, Division of Mathematical and Physical Sciences of the National Research Council, Washington, D.C. (1950).

32. Hollander, Perlman, and Seaborg, "Table of Isotopes," *Revs. Mod. Phys.*, **25,** 469 (1953).

33. K. Way, ed., *1959 Nuclear Data Tables.* National Academy of Sciences—National Research Council. Washington, D.C.: U.S. Government Printing Office, April 1959, Table V, pp. 66–88.

34. KOHMAN, MATTAUCH, and WAPSTRA, "New Reference Nuclide," *Science* **127,** 1431 (1958).

35. EVERLING, KONIG, MATTAUCH, and WAPSTRA, "Atomic Masses of Nuclides, $A \leq 70$," *Nuclear Physics,* **15,** 342 (1960); **18,** 529 (1960).

36. BHANOT, JOHNSON, and NIER, "Atomic Masses in the Heavy Mass Region," *Phys. Rev.,* **120,** 235 (1960).

PROBLEMS

1. In a mass spectrometer, a singly charged positive ion ($q = 1.602 \times 10^{-20}$ emu) is accelerated through a potential difference of 1000 volts. It then travels through a uniform magnetic field for which $H = 1000$ gauss, and is deflected into a circular path 18.2 cm in radius. What is (a) the speed of the ion? (b) the mass of the ion, in grams and atomic mass units? (c) the mass number of the ion?

2. Show that for a singly charged ion in a mass spectrometer, the following relation holds:

$$MV = 4.826 \times 10^{-5} H^2 R^2,$$

where R is expressed in centimeters, H in gauss, V in volts, and M is in atomic mass units relative to $0^{16} = 16.0000$. Use the formula to check the result of Problem 1.

3. Suppose that singly charged ions with masses close to 12 and 14 amu are accelerated through a potential difference of 1000 volts and then travel through a magnetic field of 900 gauss. Where should collector plates for the two different ions be located?

4. Plot the number of stable isotopes per element against the atomic number Z. At what values of Z are maxima observed?

5. Plot the number of stable nuclides against the neutron number $A - Z$. At what values of $A - Z$ are maxima observed?

6. From the values of the atomic masses listed in Table 9–3, calculate the binding energy of the *last* proton in C^{12}, N^{14}, F^{19}, Ne^{20}, and Si^{28}. Compare the results with the average binding energy per nucleon.

7. Suppose that C^{13}, N^{13}, N^{14}, and 0^{16} were formed by combining the appropriate atoms with C^{12}; how much energy would be liberated in each case? (For the mass of N^{13}, see Table 11–1.)

8. The following values of mass spectroscopic doublets were obtained, with $0^{16} = 16.000000$.

$$(H^1)_2 - H^2 = 0.0015483 \text{ amu}$$
$$(H^2)_3 - \tfrac{1}{2}\, C^{12} = 0.042298 \text{ amu}$$
$$C^{12}(H^1)_4 - 0^{16} = 0.036390 \text{ amu}$$

Calculate the atomic masses of H^1, H^2 and C^{12}.

9. Suppose that, in Problem 8, $C^{12} = 12.000000$ were taken as the mass standard. What would be the doublet values and the masses of H^1, H^2, and 0^{16}?

10. The following doublet values were obtained with $O^{16} = 16.000000$.

$$2(O^{16}) - S^{32} = 0.017762$$
$$4C^{12} - S^{32}O^{16} = 0.033027$$

What values are obtained for the masses of C^{12} and S^{32}?

11. Starting with Eq. (9–9), show that the average binding energy per nucleon for all but the lightest elements is close to 8 Mev.
[*Hint:* Note that Eq. (9–9) may be written

$$\frac{\Delta M}{A} = m_n - \frac{M_{Z,A}}{A} - \frac{Z}{A}(m_n - m_{\mathrm{H}}).]$$

12. The best value of the faraday is considered to be 96521.9 coul/gm-molecular weight, on the physical scale. What is the corresponding value on the chemical scale?

CHAPTER 10

NATURAL RADIOACTIVITY AND THE LAWS OF RADIOACTIVE TRANSFORMATION

Many of the ideas and techniques of atomic and nuclear physics are based on the properties of the radioactive elements and their radiations, and the study and use of radioactivity are essential to nuclear physics. It has been seen that the emission of α- and β-particles by certain atoms gave rise to the idea that atoms are built up of smaller units, and to the concept of atomic structure. The investigation of the scattering of α-particles by atoms led to the idea of the nuclear atom, which is fundamental to all of atomic theory. The analysis of the chemical relationships between the various radioactive elements resulted in the discovery of isotopes. The bombardment of atoms with swift α-particles from radioactive substances was found to cause the disintegration of atomic nuclei, and this led in turn to the discovery of the neutron and to the current theory of the composition of the nucleus. It will be shown in a later chapter that the transmuted atoms resulting from this kind of bombardment are often radioactive. This discovery of artificial, or induced, radioactivity by Joliot and Curie, in 1934, started a new line of research, and hundreds of radioactive nuclides have now been made by various methods. The investigation of the radiations from the natural and artificial radionuclides has shown that the nucleus has energy levels analogous to the atomic energy levels discussed in Chapter 7. *Nuclear spectroscopy*, which deals with the identification and classification of these levels, is an important source of information about the structure of the nucleus. Thus, radioactivity has been intimately connected with the development of nuclear physics, and it is impossible to conceive of nuclear physics as something separate from radioactivity.

The importance of radioactivity depends to a large extent on the ability to measure radioactive changes with high precision, and to describe them quantitatively by means of a straightforward theory. The laws of radioactive change were developed from information about the natural radioelements, but they are also valid for the artificial radionuclides. They can be applied, therefore, to any radioactive transformation, and are fundamental to a large part of the work to be discussed in the remaining chapters of this book.

10–1 The basis of the theory of radioactive disintegration. The first problem which will be considered is that of the quantitative description of radioactive growth and decay. A clue to the way in which one radio-

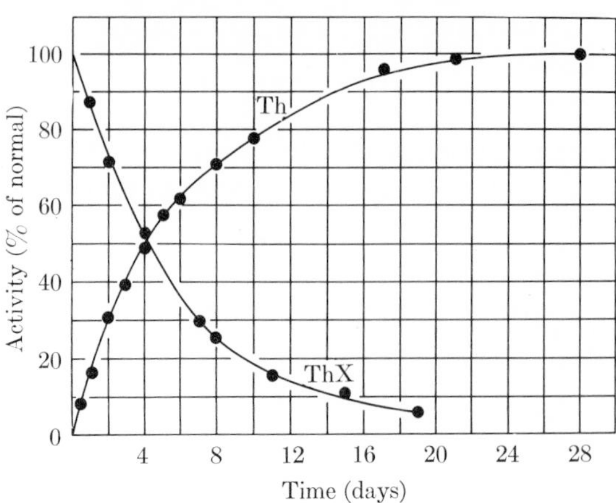

FIG. 10–1. The decay of thorium X activity and the recovery of thorium activity.

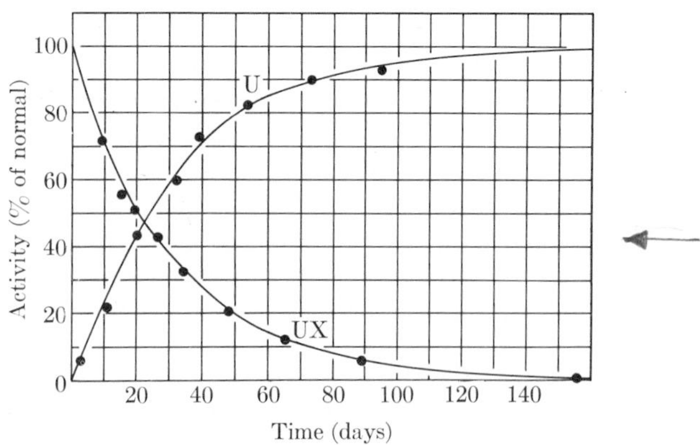

FIG. 10–2. The decay of uranium X activity and the recovery of uranium activity.

active substance is produced from another was provided by early experiments of Crookes, Becquerel, and Rutherford and Soddy.[1] Crookes (1900) found that if a uranium salt was precipitated from solution by the addition of ammonium carbonate, and then redissolved in excess of the reagent, a small residue was left. This residue, when removed from the solution, was found to be highly radioactive. The product obtained by evaporating the solution, which contained practically all the uranium, had very little activity. It appeared, therefore, that most of the observed activity of compounds of uranium was not caused by that element, but by another substance which could be separated from the uranium. The active substance was contained in the residue and was given the name uranium X (UX) to distinguish it from uranium. Becquerel then found that if the uranium X and uranium fractions were allowed to stand separately for some time, the activity of the UX decreased, while that of the uranium fraction increased. Rutherford and Soddy (1902)[2] obtained similar results with thorium salts; an active material which was called thorium X was separated, and the main body of the thorium was practically inactive. After a few days, it was noticed that the thorium X was losing its activity, while the thorium, which had been freed from thorium X, was recovering its activity.

Rutherford and Soddy studied quantitatively the rate of decay of the ThX activity and the rate of recovery of the thorium activity, and obtained the curves shown in Fig. 10–1. The experimental decay curve for the ThX was exponential in nature, i.e., the activity could be expressed as a function of time by the equation

$$A_x(t) = A_{x0}e^{-\lambda t}, \tag{10–1}$$

where A_{x0} is the initial activity of the ThX, $A_x(t)$ is the activity after a time t, and λ is a constant, called the *disintegration constant.* The recovery curve for the thorium was found to fit the formula

$$A(t) = A_0(1 - e^{-\lambda t}), \tag{10–2}$$

in which the constant λ has the same value as in Eq. (10–1); the decay and recovery curves are therefore symmetrical. The results for UX and U are shown in Fig. 10–2. The curves are similar to those for the thorium bodies except that the time scale is different. Thorium X loses half of its activity in about 3.5 days, while UX loses half of its activity in 24 days, so that the value of λ is greater for ThX than for UX.

These experimental observations enabled Rutherford and Soddy to formulate a theory of radioactive change. They suggested that the atoms of radioactive elements undergo spontaneous disintegration with the emission of α- or β-particles and the formation of atoms of a new element. Then the intensity of the radioactivity, which has been called the *activity*, is proportional to the number of atoms which disintegrate per unit time. The activity, A, measured by one of the methods discussed in Chapter 2, may then be replaced by the number of atoms N, and Eq. (10–1) may be written

$$N_x(t) = N_{x0}e^{-\lambda t}. \tag{10–3}$$

The notation may be simplified by dropping the subscript x, which was used only to distinguish between the X body and its parent substance. Then

$$N(t) = N_0e^{-\lambda t} \tag{10–4}$$

is the equation which represents the change with time of the number of atoms of a single decaying radioactive substance. Differentiation of both sides of Eq. (10–4) gives

$$\boxed{-\frac{dN}{dt} = \lambda N,} \tag{10–5}$$

where $N(t)$ has been abbreviated as N. According to Eq. (10–5), the decrease per unit time in the number of atoms of a radioactive element because of disintegration is proportional to the number of atoms which have not yet disintegrated. The proportionality factor is the disintegration constant, which is characteristic of a particular radioactive species.

Equation (10–5) is the fundamental equation of radioactive decay. With this equation, and with two assumptions, it was possible to account for the growth of activity in the thorium or uranium fractions from which the ThX or UX had been removed. The assumptions are (1) that there is a constant production of a new radioactive substance (say UX) by the radioactive element (uranium), and (2) that the new substance (UX) itself disintegrates according to the law of Eq. (10–5). Suppose that Q atoms of UX are produced per second by a given mass of uranium, and let N be the number of atoms of UX present at time t after the complete removal of the initial amount of UX. Then the net rate of increase of UX atoms in the uranium fraction is

$$\frac{dN}{dt} = Q - \lambda N. \tag{10–6}$$

The first term on the right side of Eq. (10–6) gives the rate of formation of UX atoms from U atoms; the second term gives the rate of disappearance of UX atoms because of their radioactive disintegration. To integrate Eq. (10–6), write it in the form

$$\frac{dN}{dt} + \lambda N = Q,$$

and multiply through by $e^{\lambda t}$. Then

$$e^{\lambda t}\frac{dN}{dt} + \lambda N e^{\lambda t} = Qe^{\lambda t}, \qquad \text{or} \qquad \frac{d}{dt}(Ne^{\lambda t}) = Qe^{\lambda t}.$$

The last equation can be integrated directly to give

$$Ne^{\lambda t} = \frac{Q}{\lambda}e^{\lambda t} + C, \qquad \text{or} \qquad N = \frac{Q}{\lambda} + Ce^{-\lambda t};$$

C is an integration constant determined by the condition that $N = 0$

when $t = 0$. This condition gives $C = -Q/\lambda$, and

$$N = \frac{Q}{\lambda}(1 - e^{-\lambda t}) = N_0(1 - e^{-\lambda t}), \tag{10–7}$$

with $N_0 = Q/\lambda$.

Equation (10–7) is the same as the recovery equation (10–2) so that the theory gives the correct result for the growth of activity in the uranium or thorium after the removal of the X body. Equation (10–7) also shows that the number of UX atoms in the mass of uranium approaches an equilibrium value for large values of t given by the ratio

$$\frac{Q}{\lambda} = \frac{\text{Number of atoms of UX produced from U per second}}{\text{Fractions of atoms of UX which decay per second}}.$$

The exponential law of decay was deduced by E. von Schweidler (1905) without any special hypothesis about the structure of the radioactive atoms or about the mechanism of disintegration. He assumed only that the disintegration of an atom of a radioactive element is subject to the laws of chance, and that the probability p for an atom to disintegrate in a time interval Δt is independent of the past history of the atom and is the same for all atoms of the same type. The probability of disintegration then depends only on the length of the time interval and, for sufficiently short intervals, is proportional to Δt. Then $p = \lambda \Delta t$, where λ is the disintegration constant characteristic of the particular radioactive substance. The probability that the given atom will not disintegrate during the short interval Δt is $1 - p = 1 - \lambda \Delta t$. If the atom has survived this interval, then the probability that it will not disintegrate in a second time interval Δt is again $1 - \lambda \Delta t$. The probability that the given atom will survive both the first and the second intervals is $(1 - \lambda \Delta t)^2$; for n such intervals, the probability of survival is $(1 - \lambda \Delta t)^n$. If the total time $n \Delta t$ is set equal to t, the probability of survival is $[1 - \lambda(t/n)]^n$. The probability that the atom will remain unchanged after time t is the limit of this quantity as Δt becomes vanishingly small, or as n becomes very large. Now, one of the definitions of the exponential functions is

$$e^{-x} = \lim_{n \to \infty} \left(1 - \frac{x}{n}\right)^n,$$

from which it follows that

$$\lim_{n \to \infty} \left(1 - \lambda \frac{t}{n}\right)^n = e^{-\lambda t}.$$

The statistical interpretation of this result is that if there are initially a large number N_0 of radioactive atoms, then the fraction remaining unchanged after a time t is $N/N_0 = e^{-\lambda t}$, where N is the number of unchanged atoms at time t.

The law of radioactive decay is thus a statistical law and is the result of a very large number of events subject to the laws of probability. The number of atoms which disintegrate in one second is, on the average, λN, but the number which break up in any second shows fluctuations around this value. The magnitude of these fluctuations can be calculated with the aid of the theory of probability, and the statistical considerations involved are important in the design and interpretation of experiments having to do with the measurement of radioactivity.[(3)]

The number of radioactive atoms N and the activity A have been used interchangeably so far on the grounds that the latter is proportional to the former. For a given radioactive substance, the two quantities are actually connected by the relationship

$$A = c\lambda N. \tag{10-8a}$$

The proportionality factor c, which is sometimes called the *detection coefficient*, depends on the nature and efficiency of the detection instrument and may vary considerably from one radioactive substance to another. For any one substance, the quantity in which we are usually interested is the ratio of the number of atoms at two values of the time. But, from Eq. (10–8a), it follows that

$$\frac{N(t_1)}{N(t_2)} = \frac{A(t_1)}{A(t_2)}, \tag{10-8b}$$

and the detection coefficient cancels out. Hence, the use of N and A as equivalent quantities usually leads to no confusion in the case of a single substance. When two different substances are considered, the measured activities are $A_1 = c_1\lambda_1 N_1$ and $A_2 = c_2\lambda_2 N_2$, respectively. If the number of atoms which disintegrate per unit time is the same for both substances, $\lambda_1 N_1 = \lambda_2 N_2$, but the measured activities, say in counts per minute, are not necessarily equal. They are equal only if the detection coefficients are equal. It will be assumed, with occasional exceptions, that the detection coefficients are all equal to unity, i.e., that each disintegration is detected. This condition usually cannot be achieved in practice, but it is adopted here to simplify the discussion. The activity will then be equal to the number of atoms disintegrating per unit time, $A = \lambda N$, unless otherwise noted.

10–2 The disintegration constant, the half-life, and the mean life. A radioactive nuclide may be characterized by the rate at which it disintegrates, and any one of three quantities, the disintegration constant, the half-life, or the mean life, may be used for this purpose. The disintegra-

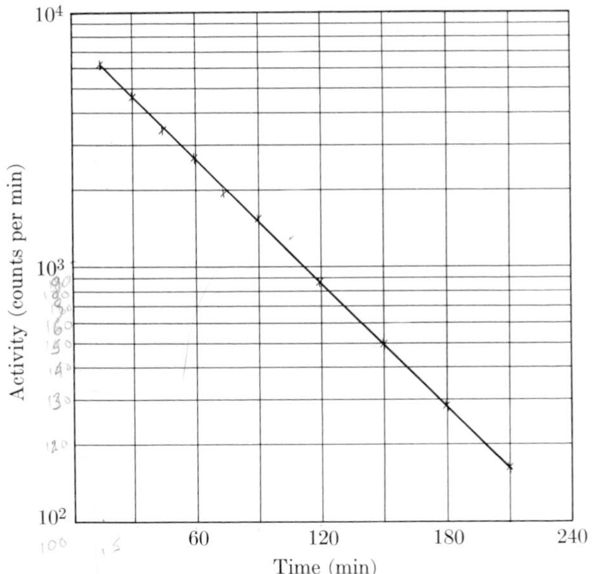

FIG. 10–3. Graphical method for determining the value of the disintegration constant.

tion constant λ can be determined experimentally, in many cases, with the help of Eq. (10–4), which may be written

$$\ln \frac{N(t)}{N_0} = -\lambda t, \tag{10–9}$$

where the symbol "ln" represents the natural logarithm. The latter can be transformed to the ordinary logarithm, denoted by "log," and Eq. (10–9) becomes

$$\log \frac{N(t)}{N_0} = -0.4343\lambda t, \tag{10–10}$$

since the logarithm to the base e is equal to 2.3026 times the logarithm to the base 10. The number of atoms $N(t)$ is proportional to the measured activity $A(t)$, so that $N(t)/N_0 = A(t)/A_0$, and Eq. (10–10) may be written

$$\log A(t) = \log A_0 - 0.4343\lambda t. \tag{10–11}$$

Hence, if the logarithm of the measured activity is plotted against the time, a straight line should result whose slope is equal to -0.4343λ. An example of this method of determining λ is shown in Fig. 10–3; for convenience, the plot is made on semilog paper. In the example shown, the slope is -0.00808, with the time expressed in minutes; λ is then 0.0186 min^{-1}, or $3.10 \times 10^{-4}\ \text{sec}^{-1}$.

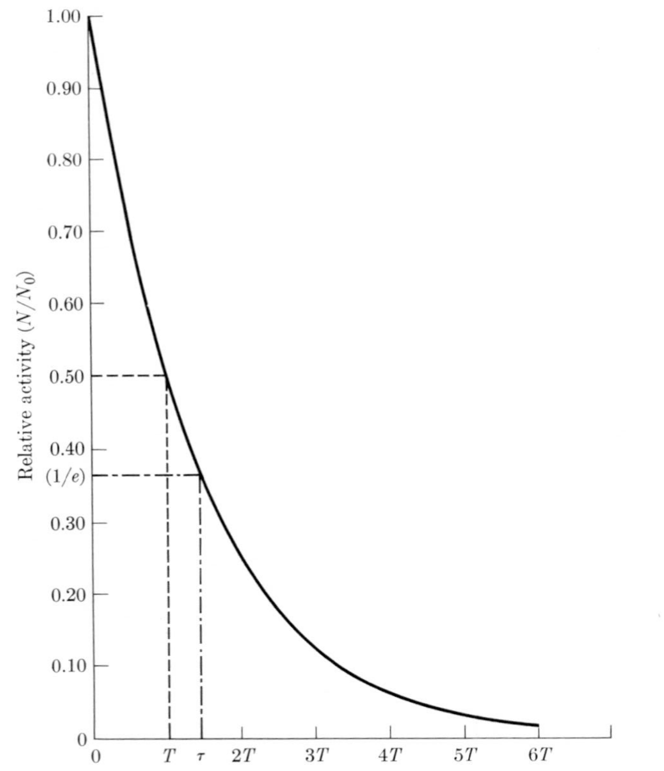

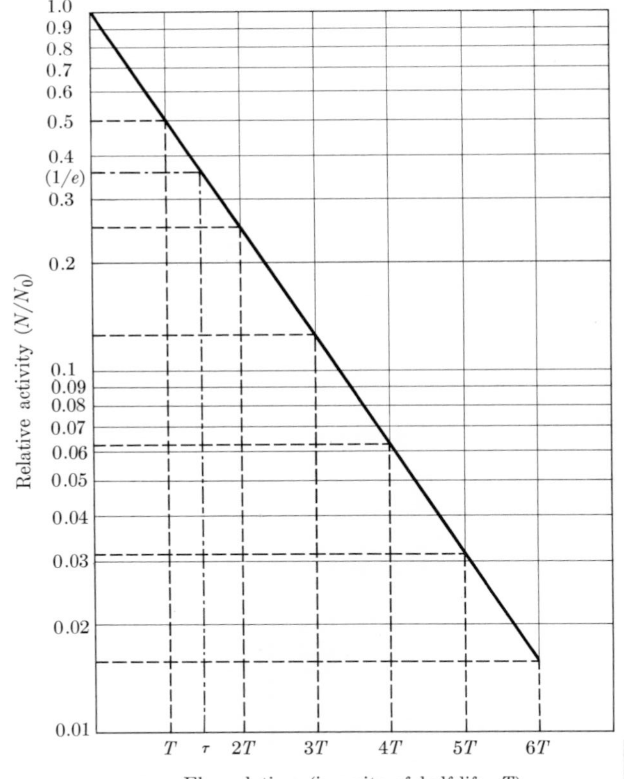

FIG. 10–4. Radioactive decay: the half-life. The curve on the left is plotted on linear graph paper, while the curve on the right is plotted on semilog paper.

Another quantity which is used to characterize a radionuclide is the *half-life*, T, the time needed for half of the radioactive atoms to disintegrate. After one half-life, $N(T)/N_0 = 0.5$, and it follows from Eq. (10–11) that $\log 0.5 = -0.4343\lambda T$ or $\log 2 = 0.4343\lambda T$. Since $\log 2 = 0.3010$,

$$\text{half-life} = T = 0.693/\lambda. \tag{10–12}$$

The relationship between the activity and the half-life is illustrated in Fig. 10–4. After n half-lives ($t = nT$), the fraction of the activity remaining is $(\frac{1}{2})^n$. This fraction never reaches zero, but it becomes very small; after seven half-lives the activity is 1/128, or less than one percent of the initial activity. After ten half-lives, the activity has fallen to 1/1024 or about 0.1% of the original amount, and is usually negligible in comparison with the initial value.

It is also possible to determine the *mean life*, or average life expectancy, of the atoms of a radioactive species. The mean life, usually denoted by τ, is given by the sum of the times of existence of all the atoms, divided by the initial number. Mathematically, it is found in the following way. The number of atoms which decay between t and $t + dt$ is

$$dN = \lambda N dt;$$

but the number of atoms still existing at time t is

$$N = N_0 e^{-\lambda t},$$

so that

$$dN = \lambda N_0 e^{-\lambda t}\, dt.$$

Since the decay process is a statistical one, any single atom may have a life from 0 to ∞. Hence, the mean life is given by

$$\tau = \frac{1}{N_0}\int_0^\infty N_0 \lambda t e^{-\lambda t}\, dt = \lambda \int_0^\infty t e^{-\lambda t}\, dt = \frac{1}{\lambda}, \tag{10–13}$$

and is simply the reciprocal of the disintegration constant. From Eqs. (10–12) and (10–13), it follows that the half-life and the mean life are proportional quantities:

$$T = 0.693\tau. \tag{10–14}$$

If the half-life of a single radioactive species has a value in the range from several seconds to several years, it can be determined experimentally by measuring the activity as a function of the time, as in the case of the disintegration constant. When the activity is plotted against the time on semilog paper, a straight line is obtained, and the half-life can be read off directly.

It often happens that two or more radioactive species are mixed together, in which case the observed activity is the sum of the separate activities. If the activities are independent, i.e., one component of the mixture does not give rise to another, the various activities can sometimes be distinguished, and the separate half-lives determined. When the total activity is plotted against the time on semilog paper, a curve like the solid one in Fig. 10–5 is obtained. The curve is concave upward because the shorter-lived components decay relatively rapidly, eventually leaving the long-lived components. After a sufficiently long time, only the longest-lived activity will remain, and the value of its half-life can be read from the late portion of the decay curve, which will be a straight line. If this straight-line portion is extrapolated back to $t = 0$, and if the values of the activity given by the line are subtracted from the total activity, the curve that remains will represent the decay of all the components of the mixture except the longest-lived. The example in Fig. 10–5 is a mixture of two activities, one with a half-life of 0.8 hour, the other with a half-life of 8 hours. The curve for the total activity is the sum of the two straight lines which represent the individual activities. Although, in principle, any complex decay curve can be analyzed into its components, practical difficulties may limit the usefulness of the method to three com-

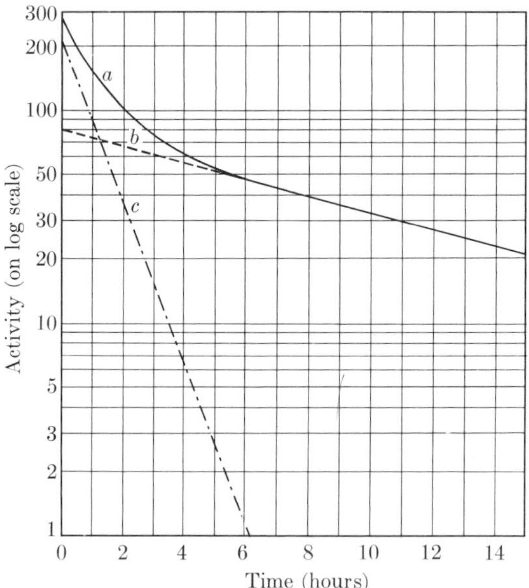

FIG. 10–5. Analysis of a composite decay curve. (a) Composite curve. (b) Longer-lived component ($T_{\frac{1}{2}} = 8.0$ hr). (c) Shorter-lived component ($T_{\frac{1}{2}} = 0.8$ hr). (Reprinted by permission from G. Friedlander and J. W. Kennedy, *Introduction to Radiochemistry*. New York: Wiley, 1949.)

ponents, and even a two-component mixture may be hard to resolve if the half-lives differ by less than a factor of about two.

If the half-life of a radionuclide is either very long or very short, methods different from those discussed so far must be used. When the half-life is very long, i.e., λ is very small, it may not be possible to detect a change in activity during the course of the measurement. The experimental activity, A, is equal to $c\lambda N$, where c gives the fraction of the disintegrating atoms detected by the measuring device. Then λ can be found from the relation

$$\lambda N = -\frac{dN}{dt} = \frac{A}{c},$$

provided that N, the number of atoms of the nuclide in the sample, is known, and that c is known as a result of an appropriate calibration. This method has been used successfully for long-lived α-emitters, and values of half-lives up to the order of 10^{10} years have been determined. For very short half-lives, other methods [4] must be used, the discussion of which is beyond the scope of this book; half-lives as short as 10^{-9} sec have been determined.

10–3 Successive radioactive transformations. It was found experimentally that the naturally occurring radioactive nuclides form three series. In each series, the parent nuclide decays into a daughter nuclide, which decays in turn, and so on, until finally a stable end product is reached. In the study of radioactive series, it is important to know the number of atoms of each member of the series as a function of time. The answer to a problem of this kind can be obtained by solving a system of differential equations. The procedure may be illustrated by treating the case of a radioactive nuclide, denoted by the subscript 1, which decays into another radioactive nuclide (subscript 2); the latter, in turn, decays into a stable end product (subscript 3). The numbers of atoms of the three kinds at any time t are denoted by N_1, N_2, and N_3, respectively, and the disintegration constants are λ_1, λ_2, and λ_3. The system is described by the three equations

$$\frac{dN_1}{dt} = -\lambda_1 N_1, \tag{10–15a}$$

$$\frac{dN_2}{dt} = \lambda_1 N_1 - \lambda_2 N_2, \tag{10–15b}$$

$$\frac{dN_3}{dt} = \lambda_2 N_2. \tag{10–15c}$$

These equations express the following facts: the parent nuclide decays according to the basic law Eq. (10–5); atoms of the second kind are formed at the rate $\lambda_1 N_1$ because of the decay of parent atoms, and disappear at the rate $\lambda_2 N_2$; atoms of the stable end product appear at the rate $\lambda_2 N_2$ as a result of the decay of atoms of the second kind.

It is instructive to solve this system of equations in detail because the procedure is one which is often used. The number of atoms N_1 can be written down immediately,

$$N_1(t) = N_1^0 e^{-\lambda_1 t}, \tag{10–16}$$

where N_1^0 is the number of atoms of the first kind present at the time $t = 0$. This expression for N_1 is inserted into Eq. (10–15b), and gives

$$\frac{dN_2}{dt} = \lambda_1 N_1^0 e^{-\lambda_1 t} - \lambda_2 N_2,$$

or

$$\frac{dN_2}{dt} + \lambda_2 N_2 = \lambda_1 N_1^0 e^{-\lambda_1 t}. \tag{10–17}$$

Multiply Eq. (10–17) through by $e^{\lambda_2 t}$; this gives

$$e^{\lambda_2 t} \frac{dN_2}{dt} + \lambda_2 N_2 e^{\lambda_2 t} = \lambda_1 N_1^0 e^{(\lambda_2 - \lambda_1)t},$$

or

$$\frac{d}{dt}(N_2 e^{\lambda_2 t}) = \lambda_1 N_1^0 e^{(\lambda_2 - \lambda_1)t}.$$

The last equation can be integrated directly to give

$$N_2 e^{\lambda_2 t} = \frac{\lambda_1}{\lambda_2 - \lambda_1} N_1^0 e^{(\lambda_2 - \lambda_1)t} + C,$$

where C is a constant of integration. Multiplying through by $e^{-\lambda_2 t}$ gives

$$N_2(t) = \frac{\lambda_1}{\lambda_2 - \lambda_1} N_1^0 e^{-\lambda_1 t} + C e^{-\lambda_2 t}. \tag{10–18}$$

The value of the integration constant is determined by noting that when $t = 0$, the number of atoms of the second kind has some constant value, or $N_2 = N_2^0$, with N_2^0 equal to a constant. Then

$$C = N_2^0 - \frac{\lambda_1}{\lambda_2 - \lambda_1} N_1^0.$$

Inserting this value into Eq. (10–18) and rearranging, we obtain the solution for N_2 as a function of time:

$$N_2 = \frac{\lambda_1}{\lambda_2 - \lambda_1} N_1^0 (e^{-\lambda_1 t} - e^{-\lambda_2 t}) + N_2^0 e^{-\lambda_2 t}. \tag{10–19}$$

The number of atoms of the third kind is found by inserting this expression for N_2 into Eq. (10–15c) and integrating, which gives

$$N_3 = \left(\frac{\lambda_1}{\lambda_2 - \lambda_1} N_1^0 - N_2^0\right) e^{-\lambda_2 t} - \frac{\lambda_2}{\lambda_2 - \lambda_1} N_1^0 e^{-\lambda_1 t} + D, \quad (10\text{–}20)$$

where D is an integration constant, determined by the condition $N_3 = N_3^0$ at $t = 0$. This condition gives

$$D = N_3^0 + N_2^0 + N_1^0.$$

When this expression for D is inserted into Eq. (10–20), the result is

$$N_3 = N_3^0 + N_2^0(1 - e^{-\lambda_2 t}) + N_1^0\left(1 + \frac{\lambda_1}{\lambda_2 - \lambda_1} e^{-\lambda_2 t} - \frac{\lambda_2}{\lambda_2 - \lambda_1} e^{-\lambda_1 t}\right). \quad (10\text{–}21)$$

Equations (10–16), (10–19), and (10–21) represent the solution of the problem.

One of the cases met most often in practice is that in which only radioactive atoms of the first kind are present initially. In this case, the constants N_2^0 and N_3^0 are both equal to zero, and the solutions for N_2 and N_3 reduce to

$$N_2 = \frac{\lambda_1}{\lambda_2 - \lambda_1} N_1^0(e^{-\lambda_1 t} - e^{-\lambda_2 t}), \quad (10\text{–}22)$$

$$N_3 = N_1^0\left(1 + \frac{\lambda_1}{\lambda_2 - \lambda_1} e^{-\lambda_2 t} - \frac{\lambda_2}{\lambda_2 - \lambda_1} e^{-\lambda_1 t}\right). \quad (10\text{–}23)$$

The curves of Fig. 10–6 show what happens in this case if it is assumed that the half-lives of the active species are $T_1 = 1$ hour and $T_2 = 5$ hours, respectively; the corresponding values of the disintegration constants are $\lambda_1 = 0.693\ \text{hr}^{-1}$, and $\lambda_2 = 0.1386\ \text{hr}^{-1}$, respectively. The ordinates of the curves represent the relative numbers of the substances 1, 2, and 3 as functions of the time when the initial number of atoms of the substance 1 is taken as $N_1^0 = 100$. The number of atoms, N_1, of substance 1, decreases exponentially according to Eq. (10–16) with a half-life of 1 hour; N_2 is initially zero, increases, and passes through a maximum after about three hours, and then decreases gradually. The number of atoms N_3 of the stable end product increases steadily with time, although slowly at first; when t becomes very large, N_3 approaches 100, since eventually all the atoms of the substance 1 will be converted to atoms of the stable end product.

The treatment just discussed can be extended to a chain of any number of radioactive products, and the solution of this problem is often useful.

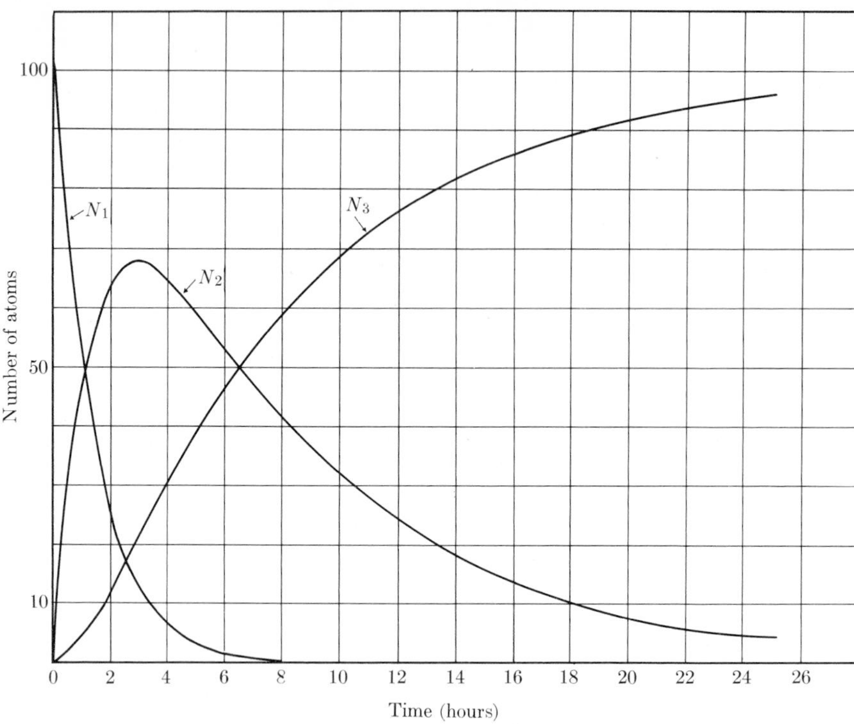

FIG. 10–6. A radioactive series with three members: only the parent ($T_{\frac{1}{2}}$ = 1 hr) is present initially; the daughter has a half-life of 5 hr, and the third member is stable.

The procedure is similar to that of the special case already considered except that the mathematics becomes more tedious and the expressions for the numbers of atoms become more complicated as the length of the chain increases. The differential equations of the system are

$$\left\{\begin{aligned} \frac{dN_1}{dt} &= -\lambda_1 N_1, \\ \frac{dN_2}{dt} &= \lambda_1 N_1 - \lambda_2 N_2, \\ \frac{dN_3}{dt} &= \lambda_2 N_2 - \lambda_3 N_3, \\ &\vdots \\ \frac{dN_n}{dt} &= \lambda_{n-1} N_{n-1} - \lambda_n N_n. \end{aligned}\right. \tag{10–24}$$

The solution of this system of equations under the assumption that at

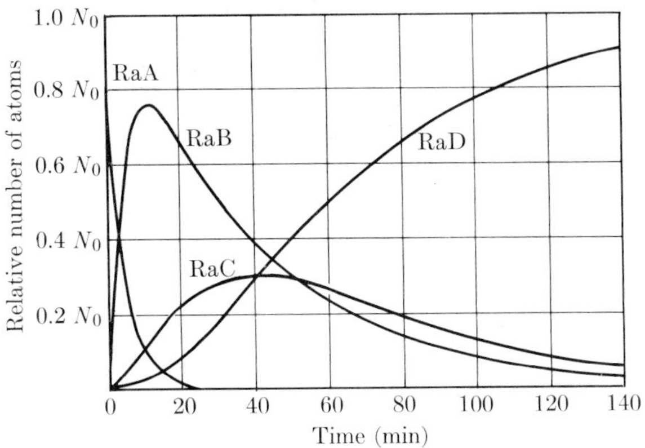

FIG. 10–7. The decay of Radium A. [Reprinted by permission from Rutherford, Chadwick, and Ellis, *Radiations from Radioactive Substances,* Cambridge University Press (Macmillan Co.), 1930.]

$t = 0$ only the parent substance is present was derived by Bateman.[(5)] The initial conditions are

$$t = 0; \qquad N_1 = N_1^0, \qquad N_2^0 = N_3^0 = \cdots = N_n^0 = 0. \tag{10–25}$$

The number of atoms of the nth member of the chain is given by

$$N_n(t) = C_1 e^{-\lambda_1 t} + C_2 e^{-\lambda_2 t} + C_3 e^{-\lambda_3 t} + \cdots + C_n e^{-\lambda_n t}, \tag{10–26}$$

with

$$\begin{aligned} C_1 &= \frac{\lambda_1 \lambda_2 \ldots \lambda_{n-1}}{(\lambda_2 - \lambda_1)(\lambda_3 - \lambda_1)\ldots(\lambda_n - \lambda_1)} N_1^0, \\ C_2 &= \frac{\lambda_1 \lambda_2 \ldots \lambda_{n-1}}{(\lambda_1 - \lambda_2)(\lambda_3 - \lambda_2)\ldots(\lambda_n - \lambda_2)} N_1^0, \\ &\vdots \\ C_n &= \frac{\lambda_1 \lambda_2 \ldots \lambda_{n-1}}{(\lambda_1 - \lambda_n)(\lambda_2 - \lambda_n)\ldots(\lambda_{n-1} - \lambda_n)} N_1^0. \end{aligned} \tag{10–27}$$

An example of the application of the Bateman equations, Eqs. (10–24) to (10–27), is shown in Fig. 10–7, taken from Rutherford, Chadwick, and Ellis' book *Radiations from Radioactive Substances.* The curves were obtained under the conditions which follow. A test body was exposed for a few seconds to radon (Em^{222}), and a certain number of atoms of the decay product of radon, RaA or Po^{218}, with a half-life of 3.05 min were deposited on the test body. The RaA decays into RaB (Pb^{214}) with a half-life of 26.8 min, which decays, in turn, into RaC (Bi^{214}) with a half-life of 19.7 min. Finally, the end product RaD or Pb^{210}, with a half-life of 22 years, is formed. The last half-life is sufficiently long so that the number of radium D atoms which disintegrate may safely be neglected. The number of RaA atoms decreases exponentially. The number of RaB atoms is initially zero, passes through a maximum about 10 min later, and then decreases with time. The number of RaC atoms passes through a maximum after about 35 min. The number of RaD atoms increases, reaching a maximum when the RaA and RaB have disappeared. Eventually the RaD would decay exponentially with a half-life of 22 years. The sum of all the atoms present at any time is N_0, the initial number of atoms of RaA.

Solutions can be obtained for other problems with different initial conditions, and the reader is referred to the book by Rutherford, Chadwick, and Ellis for additional examples of the application of the theory.

10–4 Radioactive equilibrium. The term *equilibrium* is usually used to express the condition that the derivative of a function with respect to the time is equal to zero. When this condition is applied to the members of a radioactive chain described by Eqs. (10–24), it means that the derivatives $dN_1/dt, dN_2/dt, \ldots, dN_n/dt$ are all equal to zero, or that the number of atoms of any member of the chain is not changing. The conditions for equilibrium are then

$$\begin{aligned}\frac{dN_1}{dt} &= -\lambda_1 N_1 = 0,\\ \lambda_1 N_1 &= \lambda_2 N_2,\\ \lambda_2 N_2 &= \lambda_3 N_3,\\ &\vdots\\ \lambda_{n-1} N_{n-1} &= \lambda_n N_n,\end{aligned} \tag{10–28}$$

These conditions cannot be satisfied rigorously if the parent substance is a radioactive substance because the first of Eqs. (10–28) implies that $\lambda_1 = 0$, which is a contradiction. It is possible to achieve a state very close to equilibrium, however, if the parent substance decays much more slowly than any of the other members of the chain; in other words, if the parent has a half-life very long compared with that of any of its decay products. This condition is satisfied by the naturally occurring radioactive chains. Uranium I has a half-life of 4.5×10^9 years and the fraction of UI atoms transformed during the life of an experimenter is indeed negligible. In such a case, the number of atoms N_1 can be taken to be constant, and the value of λ_1 is very much smaller than that of any of the other λ's in the chain. The first of Eqs. (10–28) is then a very good approximation, and the rest of the conditions are rigorously valid. This type of equilibrium is called *secular equilibrium*, and satisfies the condition

$$\lambda_1 N_1 = \lambda_2 N_2 = \lambda_3 N_3 = \cdots = \lambda_{n-1} N_{n-1} = \lambda_n N_n, \tag{10–29}$$

or, in terms of half-lives,

$$\frac{N_1}{T_1} = \frac{N_2}{T_2} = \frac{N_3}{T_3} = \cdots = \frac{N_{n-1}}{T_{n-1}} = \frac{N_n}{T_n}. \tag{10–30}$$

The relationships (10–29) and (10–30) may be applied whenever several short-lived products arise from successive decays beginning with a relatively long-lived parent. It is only necessary to be sure that the material has been undisturbed, that is, that no decay products have been removed or allowed to escape for a long enough time for secular equilibrium to be established. Secular equilibrium can also be attained when a radioactive substance is produced at a steady rate by some artificial method, such as a nuclear reaction in a cyclotron or chain-reacting pile. The term $\lambda_1 N_1$ in the second of Eqs. (10–24) is then constant, as in the case of a very long-lived parent, and the condition (10–29) is satisfied.

The relationships (10–30) can be used to find the half-life of a radionuclide whose half-life is very long. For example, uranium minerals in which secular equilibrium has been established have been shown to contain one atom of radium for every 2.8×10^6 atoms of uranium I. If uranium I is denoted by the subscript 1 and radium by the subscript 2, then at equilibrium $N_1/N_2 = 2.8 \times 10^6$. The half-life of radium is known from direct measurements to be 1620 years. Consequently, the half-life of uranium I is

$$T_1 = \frac{N_1}{N_2} T_2 = 2.8 \times 10^6 \times 1620 = 4.5 \times 10^9 \text{ years.}$$

We consider next an example of the approach to secular equilibrium. The case is that of a long-lived parent ($T \approx \infty$) and a short-lived daughter. It is assumed that the daughter has been separated from the parent, so that the latter is initially pure. The mathematical expressions for the number of atoms of parent and daughter may be obtained from Eqs. (10–16) and (10–22) if it is noted that $\lambda_1 \approx 0$, and $\lambda_1 \ll \lambda_2$. Then $e^{-\lambda_1 t} \approx 1$, and

$$N_1 \approx N_1^0, \tag{10–31}$$

$$N_2 \approx \frac{\lambda_1}{\lambda_2} N_1^0 (1 - e^{-\lambda_2 t}). \tag{10–32}$$

Equation (10–32) may be rewritten

$$\lambda_2 N_2 \approx \lambda_1 N_1^0 (1 - e^{-\lambda_2 t}). \tag{10–33}$$

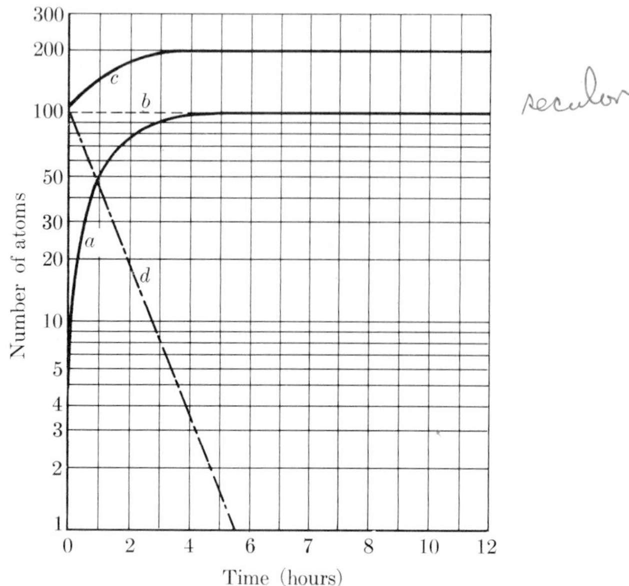

FIG. 10–8. Secular equilibrium. (a) Daughter activity growing in freshly purified parent fraction. (b) Activity of parent ($T_{\frac{1}{2}} = \infty$). (c) Total activity of an initially pure parent fraction. (d) Decay of freshly isolated daughter fraction ($T_{\frac{1}{2}} = 0.80$ hr). (Reprinted by permission from G. Friedlander and J. W. Kennedy, *Introduction to Radiochemistry*. New York: Wiley, 1949.)

The last equation gives the activity of the daughter as a function of the time, in terms of the (constant) activity of the parent. The total activity is given by

$$A_{\text{total}} = \lambda_1 N_1^0 + \lambda_2 N_2 \approx 2\lambda_1 N_1^0 - \lambda_1 N_1^0 e^{-\lambda_2 t}. \tag{10–34}$$

Equation (10–33) shows that as t increases the activity of the daughter increases, and after several half-lives, $\lambda_2 N_2$ approaches $\lambda_1 N_1^0$, satisfying the condition for secular equilibrium. These relationships are shown graphically in curves a and b of Fig. 10–8; curve c gives the total activity. For comparison, curve d shows how a freshly isolated daughter fraction would decay.

A somewhat different state of affairs, called *transient equilibrium*, results if the parent is longer-lived than the daughter ($\lambda_1 < \lambda_2$), but the half-life of the parent is not very long. In this case, the approximation $\lambda_1 = 0$ may not be made. If the parent and daughter are separated so that the parent can be assumed to be initially pure, the numbers of atoms are again given by Eqs. (10–16) and (10–22). After t becomes sufficiently large, $e^{-\lambda_2 t}$ becomes negligible compared with $e^{-\lambda_1 t}$, and the number of

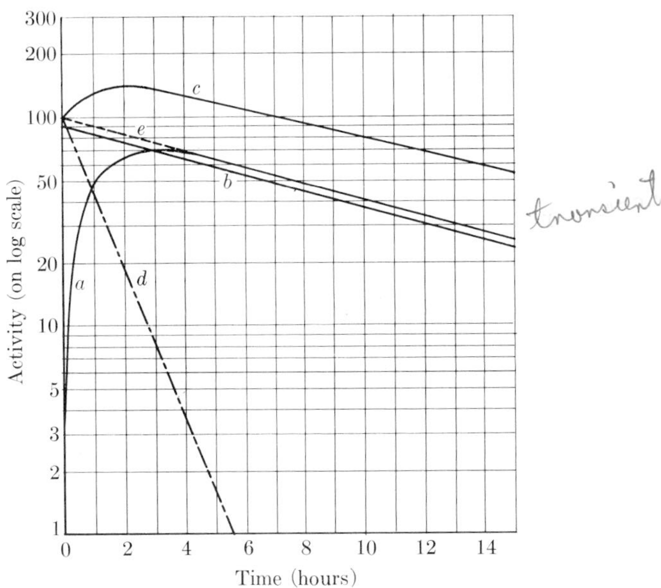

FIG. 10–9. Transient equilibrium. (a) Daughter activity growing in freshly purified parent fraction. (b) Activity of parent ($T_{\frac{1}{2}} = 8.0$ hr). (c) Total activity of an initially pure parent fraction. (d) Decay of freshly isolated daughter fraction ($T_{\frac{1}{2}} = 0.80$ hr). (e) Total daughter activity in parent-plus-daughter fractions. (Reprinted by permission from G. Friedlander and J. W. Kennedy, *Introduction to Radiochemistry*. New York: Wiley, 1949.)

atoms of the daughter becomes

$$N_2 \approx \frac{\lambda_1}{\lambda_2 - \lambda_1} N_1^0 e^{-\lambda_1 t}. \tag{10–35}$$

Thus, the daughter eventually decays with the same half-life as the parent. Since $N_1^0 e^{-\lambda_1 t} = N_1$, it follows from Eq. (10–35) that

$$\frac{N_1}{N_2} = \frac{\lambda_2 - \lambda_1}{\lambda_1}. \tag{10–36}$$

The ratio of the measured activities at equilibrium is

$$\frac{A_1}{A_2} = \frac{\lambda_1 N_1}{\lambda_2 N_2} = \frac{\lambda_2 - \lambda_1}{\lambda_2}, \tag{10–37}$$

and the daughter activity is greater than that of the parent by the factor $\lambda_2/(\lambda_2 - \lambda_1)$. The above results, which are characteristic of transient equilibrium between parent and daughter atoms, are shown graphically in Fig. 10–9.

When the parent has a shorter half-life than the daughter ($\lambda_1 > \lambda_2$), no state of equilibrium is attained. If the parent and daughter are separated initally, then as the parent decays, the number of daughter atoms will increase, pass through a maximum, and eventually decay with the half-life of the daughter.

10–5 The natural radioactive series. As a result of physical and chemical research on the naturally occurring radioactive elements, it was proved that each radioactive nuclide is a member of one of three long chains, or radioactive series, stretching through the last part of the periodic system. These series are named the uranium, actinium, and thorium series, respectively, after elements at, or near, the head of the series. In the uranium series, the mass number of each member can be expressed in the form (4n + 2), where n is an integer, and the uranium series is sometimes called the "4n + 2" series. In the actinium and thorium series, the mass numbers are given by the expressions 4n + 3 and 4n, respectively. There

TABLE 10–1

THE URANIUM SERIES

Radioactive species	Nuclide	Type of disintegration	Half-life	Disintegration constant, sec^{-1}	Particle energy, Mev
Uranium I (UI)	$_{92}U^{238}$	α	4.50×10^9 y	4.88×10^{-18}	4.20
Uranium X_1 (UX_1)	$_{90}Th^{234}$	β	24.1 d	3.33×10^{-7}	0.19
Uranium X_2 (UX_2)	$_{91}Pa^{234}$	β	1.18 m	9.77×10^{-3}	2.32
Uranium Z (UZ)	$_{91}Pa^{234}$	β	6.7 h	2.88×10^{-5}	1.13
Uranium II (UII)	$_{92}U^{234}$	α	2.50×10^5 y	8.80×10^{-14}	4.768
Ionium (Io)	$_{90}Th^{230}$	α	8.0×10^4 y	2.75×10^{-13}	4.68 m
Radium (Ra)	$_{88}Ra^{226}$	α	1620 y	1.36×10^{-11}	4.777 m
Ra Emanation (Rn)	$_{86}Em^{222}$	α	3.82 d	2.10×10^{-6}	5.486
Radium A (RaA)	$_{84}Po^{218}$	α, β	3.05 m	3.78×10^{-3}	α:5.998 β:?
Radium B (RaB)	$_{82}Pb^{214}$	β	26.8 m	4.31×10^{-4}	0.7
Astatine-218 (At^{218})	$_{85}At^{218}$	α	1.5–2.0 s	0.4	6.63
Radium C (RaC)	$_{83}Bi^{214}$	α, β	19.7 m	5.86×10^{-4}	α:5.51 m β:3.17
Radium C′ (RaC′)	$_{84}Po^{214}$	α	1.64×10^{-4} s	4.23×10^3	7.683
Radium C″ (RaC″)	$_{81}Tl^{210}$	β	1.32 m	8.75×10^{-4}	1.9
Radium D (RaD)	$_{82}Pb^{210}$	β	19.4 y	1.13×10^{-9}	0.017
Radium E (RaE)	$_{83}Bi^{210}$	β	5.0 d	1.60×10^{-6}	1.155
Radium F (RaF)	$_{84}Po^{210}$	α	138.3 d	5.80×10^{-8}	5.300
Thallium-206 (Tl^{206})	$_{81}Tl^{206}$	β	4.2 m	2.75×10^{-3}	1.51
Radium G (RaG)	$_{82}Pb^{206}$	Stable			

is no natural radioactive series of nuclides whose mass numbers are represented by $4n + 1$.

The members of the uranium series are listed in Table 10–1 together with the mode of disintegration, the half-life,[6] the disintegration constant, and the maximum energy of the emitted particles.[6] The changes in atomic number and mass number are shown in Fig. 10–10. Table 10–2 and Fig. 10–11 give the corresponding information about the actinium series, while Table 10–3 and Fig. 10–12 are for the thorium series. The first column of each table gives the old-fashioned, historical names of the radionuclides, while the second column gives the modern symbol. Both are included because both are found in the literature of physics, and it is helpful to have a code for translation readily available. The older name usually indicates the series to which a radioactive substance belongs and gives some idea of the relative position in the series; the modern symbol gives the atomic and mass numbers. The mode of decay, the value of the half-life (or disintegration constant), and the energy of the emitted particle characterize the radioactivity. The half-life may be in years, days, hours, minutes, or seconds, abbreviated as y, d, h, m, s, respectively.

Table 10–2

The Actinium Series

Radioactive species	Nuclide	Type of disintegration	Half-life	Disintegration constant, sec^{-1}	Particle energy, Mev
Actinouranium (AcU)	${}_{92}U^{235}$	α	7.10×10^8 y	3.09×10^{-17}	4.559 m
Uranium Y (UY)	${}_{90}Th^{231}$	β	25.6 h	7.51×10^{-6}	0.30
Protoactinium (Pa)	${}_{91}Pa^{231}$	α	3.43×10^4 y	6.40×10^{-13}	5.046 m
Actinium (Ac)	${}_{89}Ac^{227}$	α, β	21.6 y	1.02×10^{-9}	α:4.94 β:0.046
Radioactinium(RdAc)	${}_{90}Th^{227}$	α	18.17 d	4.41×10^{-7}	6.03 m
Actinium K (AcK)	${}_{87}Fr^{223}$	α, β	22 m	5.25×10^{-4}	β:1.2 α:5.34
Actinium X (AcX)	${}_{88}Ra^{223}$	α	11.68 d	6.87×10^{-7}	5.864
Astatine-219	${}_{85}At^{219}$	α, β	0.9 m	1.26×10^{-2}	α:6.27
Ac Emanation (An)	${}_{86}Em^{219}$	α	3.92 s	0.177	6.810 m
Bismuth-215	${}_{83}Bi^{215}$	α, β	8 m	1.44×10^{-3}	?
Actinium A (AcA)	${}_{84}Po^{215}$	α, β	1.83×10^{-3} s	3.79×10^2	α:7.37
Actinium B (AcB)	${}_{82}Pb^{211}$	β	36.1 m	3.20×10^{-4}	1.39
Astatine-215	${}_{85}At^{215}$	α	10^{-4} s	7×10^3	8.00
Actinium C (AcC)	${}_{83}Bi^{211}$	α, β	2.15 m	5.28×10^{-3}	α:6.617 m
Actinium C′ (AcC′)	${}_{84}Po^{211}$	α	0.52 s	1.33	7.442 m
Actinium C″ (AcC″)	${}_{81}Tl^{207}$	β	4.79 m	2.41×10^{-3}	1.44
Actinium D (AcD)	${}_{82}Pb^{207}$	Stable			

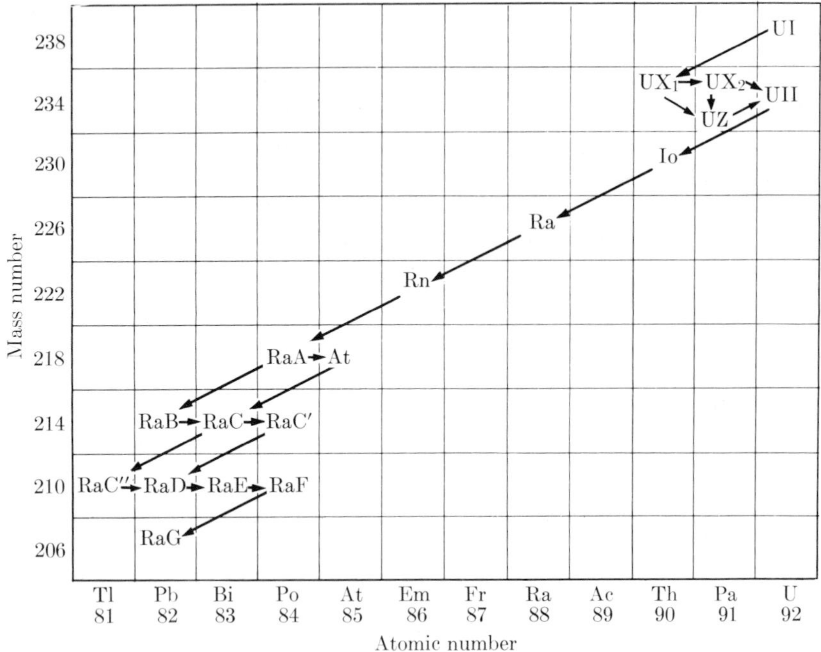

FIG. 10–10. The uranium (4n + 2) series.

TABLE 10–3

THE THORIUM SERIES

Radioactive species	Nuclide	Type of disintegration	Half-life	Disintegration constant, sec^{-1}	Particle energy, Mev
Thorium (Th)	$_{90}Th^{232}$	α	1.39×10^{10} y	1.58×10^{-18}	4.007
Mesothorium1(MsTh1)	$_{88}Ra^{228}$	β	6.7 y	3.28×10^{-9}	0.04
Mesothorium2(MsTh2)	$_{89}Ac^{228}$	β	6.13 h	3.14×10^{-5}	2.18
Radiothorium (RdTh)	$_{90}Th^{228}$	α	1.910 y	1.15×10^{-8}	5.423 m
Thorium X (ThX)	$_{88}Ra^{224}$	α	3.64 d	2.20×10^{-6}	5.681 m
Th Emanation (Tn)	$_{86}Em^{220}$	α	51.5 s	1.34×10^{-2}	6.280
Thorium A (ThA)	$_{84}Po^{216}$	α, β	0.16 s	4.33	6.774
Thorium B (ThB)	$_{82}Pb^{212}$	β	10.6 h	1.82×10^{-5}	0.58
Astatine-216 (At^{216})	$_{85}At^{216}$	α	3×10^{-4} s	2.3×10^{3}	7.79
Thorium C (ThC)	$_{83}Bi^{212}$	α, β	60.5 m	1.91×10^{-4}	α:6.086 m β:2.25
Thorium C′ (ThC′)	$_{84}Po^{212}$	α	3.0×10^{-7} s	2.31×10^{6}	8.780
Thorium C″ (ThC″)	$_{81}Tl^{208}$	β	3.10 m	3.73×10^{-3}	1.79
Thorium D (ThD)	$_{82}Pb^{208}$	Stable			

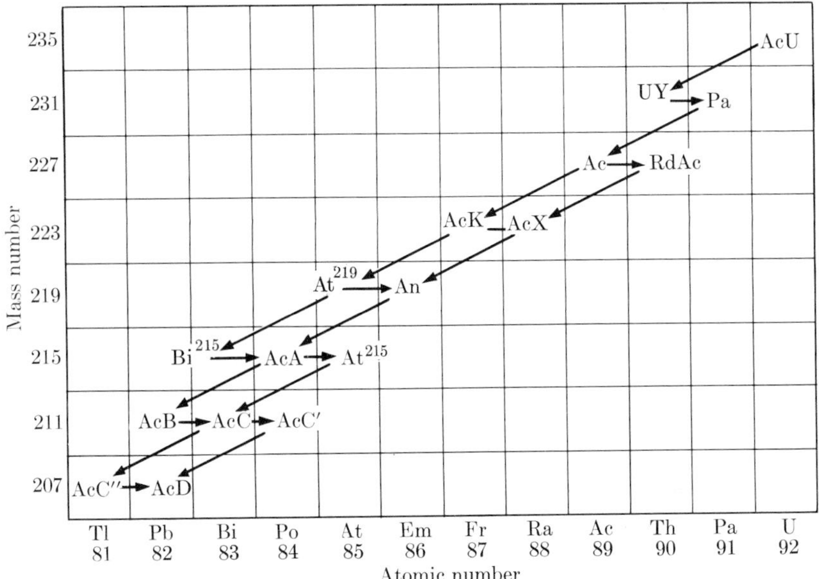

FIG. 10-11. The actinium (4n + 3) series.

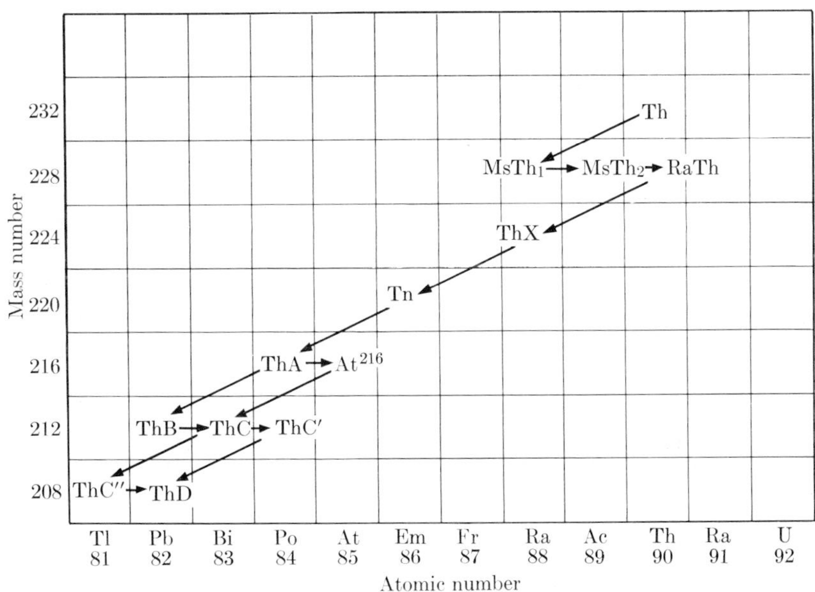

FIG. 10-12. The thorium (4n) series.

In the case of β-particles, the energy value listed is that of the most energetic particles. For some radionuclides, the α-particles are monoenergetic; for others, an α-particle may have one of several energy values, in which case the greatest value is listed and denoted by "m." Figures 10–10 through 10–12 illustrate Soddy's displacement law discussed in Section 9–1. It will be remembered that the emission of an α-particle decreases the charge of a nucleus by two units and the mass number by 4 units, while the emission of a β-particle increases the charge of the nucleus by one unit and leaves the mass number unchanged. In the figures, therefore, an α-disintegration is represented by an arrow sloping downward and to the left, and a β-transition by a horizontal arrow pointing to the right. All of the elements in any one column have the same atomic number and must occupy the same place in the periodic table. For example, RaA, RaC′, RaF, AcA, AcC′, ThA, and ThC′ all have the atomic number 84 and are isotopes of polonium. Similarly, RaB, RaD, RaG, AcB, AcD, ThB, and ThD all have the atomic number 82 and are isotopes of lead. Series schemes like those of Figs. 10–10 through 10–12 led Soddy (1913) to the discovery of the existence of isotopes.

In most of the disintegration processes that make up a radioactive series, each of the radionuclides breaks up in a definite way, giving one α- or β-particle and one atom of the product nuclide. In some cases, however, the atoms break up in two different ways, giving rise to two products with different properties. This kind of disintegration is called *branching decay* and is illustrated by the decay of Po^{218} (RaA), Bi^{214}(RaC), and Bi^{210} (RaE) in the uranium series; by Ac^{227}, Po^{215} (AcA), Bi^{211} (AcC), ${}_{87}Fr^{223}$ (AcK), and ${}_{85}At^{219}$ in the actinium series; and by Po^{216} (ThA) and Bi^{212} (ThC) in the thorium series. Each of these nuclides can decay either by α-emission or by β-emission. The probability of disintegration is the sum of the separate probabilities and $\lambda = \lambda_\alpha + \lambda_\beta$; the half-life is $T = 0.693/\lambda = 0.693/(\lambda_\alpha + \lambda_\beta)$. In most cases, one mode of decay is much more probable than the other. Radium A and AcA decay almost entirely by α-emission, with only a small fraction of one percent of the atoms disintegrating by β-emission. On the other hand, RaC, RaE, Ac, and AcC decay almost entirely by β-decay, with one percent or less of the atoms decaying by α-emission. The case of ThC (Bi^{212}) is especially interesting because 66.3% of the disintegrations are by β-emission, and 33.7% by α-emission, in contrast to the other cases mentioned. In each branched decay, the product atoms decay in turn to give the same nuclide. For example, ThC emits a β-particle to form ThC′ (Po^{212}) or an α-particle to form ThC″ (Tl^{208}); ThC′ then emits an α-particle to give stable Pb^{208} (ThD), while ThC″ emits a β-particle and also forms stable Pb^{208}.

10-6 Units of radioactivity. The intensity of radioactivity has so far been considered in terms of the number of atoms which disintegrate per unit time, or in terms of the number of emitted particles counted per unit time by a detector. As in other branches of physics, it is useful to have a standard quantitative unit with an appropriate name. In radioactivity, the standard unit is the *curie*, at present defined as that quantity of any radioactive material giving 3.70×10^{10} disintegrations/sec. The *millicurie*, which is one-thousandth of a curie, and the *microcurie*, equal to one-millionth of a curie, are also useful units; these would correspond to amounts of active materials giving 3.7×10^7 and 3.7×10^4 disintegrations/sec, respectively.

For various reasons, mainly historical, there has been some confusion about the use of the curie as the standard unit. Consequently, a new absolute unit of radioactive disintegration rate has been recommended, namely, the *rutherford* (rd), defined as that amount of a radioactive substance which gives 10^6 disintegrations/sec. The millirutherford (mrd) and microrutherford (μrd) correspond to 10^3 disintegrations/sec and 1 disintegration/sec, respectively.

To get an idea of some orders of magnitude, consider the problem of calculating the weight in grams of 1 curie and 1 rd of RaB (Pb^{214}), from its half-life of 26.8 min. The disintegration constant is

$$\lambda = 4.31 \times 10^{-4} \text{ sec}^{-1}.$$

If W is the unknown weight, then

$$\begin{aligned} -dn/dt &= \lambda N = 4.31 \times 10^{-4} \times W/214 \times 6.02 \times 10^{23} \\ &= 1.21W \times 10^{18} \text{ disintegrations/sec.} \end{aligned}$$

If $-dN/dt = 3.70 \times 10^{10}$ disintegrations/sec (1 curie), then

$$W = \frac{3.70 \times 10^{10}}{1.21 \times 10^{18}} = 3.1 \times 10^{-8} \text{ gm.}$$

For 1 rd,

$$W = \frac{10^6}{1.21 \times 10^{18}} = 8.3 \times 10^{-13} \text{ gm.}$$

Thus, for a substance with a short, but not very short, half-life very little material is needed to provide a curie of activity. When the half-life is a small fraction of a second (Po^{214}, Po^{215}, At^{215}, At^{216}, or Po^{212}), the amount of material which gives a curie of activity is almost unimaginably small. In the case of a nuclide with a very long half-life, such as U^{238} ($T = 4.50 \times 10^9$ y), $\lambda = 4.9 \times 10^{-18}$ sec^{-1}. Then

$$-\frac{dN}{dt} = \lambda N = 4.9 \times 10^{-18} \times \frac{W}{238} \times 6.02 \times 10^{23}$$

$$= \mathbf{1.24} \times 10^4 W \text{ disintegrations/sec.}$$

For 1 curie of activity,

$$W = \frac{3.70 \times 10^{10}}{1.14 \times 10^4} = 3.2 \times 10^6 \text{ gm},$$

and more than three metric tons are needed.

References

GENERAL

RUTHERFORD, CHADWICK, and ELLIS, *Radiations from Radioactive Substances.* New York: Macmillan, 1930. (Reprinted with corrections, 1951.)

G. HEVESY and F. A. PANETH, *A Manual of Radioactivity,* 2nd ed. Oxford University Press, 1938.

G. FRIEDLANDER and J. W. KENNEDY, *Nuclear and Radiochemistry.* New York: Wiley, 1955.

R. D. EVANS, *The Atomic Nucleus.* New York: McGraw-Hill, 1955, Chapters 15, 25, 26.

E. SEGRÈ, "Radioactive Decay," *Experimental Nuclear Physics,* E. Segrè, ed., Vol. III, Part IX. New York: Wiley, 1959.

J. L. PUTNAM, "Measurement of Disintegration Rate," *Beta- and Gamma-Ray Spectroscopy,* K. Siegbahn, ed. Amsterdam: North Holland Publishing Co.; New York: Interscience, 1955, Chapter 26.

G. J. HINE and G. L. BROWNELL, eds., *Radiation Dosimetry.* New York: Academic Press, 1956.

PARTICULAR

1. G. E. M. JAUNCEY, "Early Years of Radioactivity," *Am. J. Phys.* **14,** 226 (1946).
2. E. RUTHERFORD and F. SODDY, "The Cause and Nature of Radioactivity," *J. Chem. Soc.* **81,** 321, 837 (1902); *Phil. Mag.* **4,** 370, 569 (1902); *ibid* **5,** 576 (1903).
3. G. FRIEDLANDER and J. W. KENNEDY, *op. cit.* gen. ref., Chapter 9.
4. S. ROWLANDS, "Methods of Measuring Very Long and Very Short Half-Lives," *Nucleonics* **3,** No. 3, 2 (Sept. 1948).
5. H. BATEMAN, "The Solution of a System of Differential Equations Occurring in the Theory of Radio-active Transformations," *Proc. Cambridge Phil. Soc.* **16,** 423 (1910).
6. STROMINGER, HOLLANDER, and SEABORG, "Table of Isotopes," *Revs. Mod. Phys.* **30,** 585 (1958).
7. E. SEGRÈ, *op. cit.* gen. ref., pp. 8–16.

Problems

1. Find the values of the disintegration constant and half-life of a radioactive substance for which the following counting rates were obtained at different times.

Time, hr	Counting rate, counts/min	Time, hr	Counting rate, counts/min
0.0	....	6.0	1800
0.5	9535	7.0	1330
1.0	8190	8.0	980
1.5	7040	9.0	720
2.0	6050	10.0	530
3.0	4465	11.0	395
4.0	3295	12.0	290
5.0	2430		

What would have been the counting rate at $t = 0$?

2. The counting rates listed below were obtained when the activity of a certain radioactive sample was measured at different times. Plot the decay curve on semilog paper and determine the half-lives and initial activities of the component activities.

Time, hr	Counting rate, counts/min	Time, hr	Counting rate, counts/min
0.0	19100	4.0	3500
0.5	14500	5.0	2520
1.0	11410	6.0	1835
1.5	9080	8.0	985
2.0	7345	10.0	530
3.0	4985	12.0	290

3. The half-life of radon is 3.82 days. What fraction of a freshly separated sample of this nuclide will disintegrate in one day? In 2, 3, 4, 5, 10 days? If the sample contains, initially, one microgram of radon, how many atoms will disintegrate during the first day? During the fifth day? During the tenth day?

4. The half-life of UX_1 is 24.1 days. How long after a sample of UX_1 has been isolated will it take for 90% of it to change to UX_2? 95%? 99%? 63.2%?

5. What are the weights of (a) 1 curie and 1 rd of Ra? (b) 1 curie and 1 rd of Rn? (c) 1 microcurie and 1 rd of RaA? (d) 1 microcurie and 1 rd of RaC′?

6. A sample of ionium (Th^{230}) weighing 0.100 mg was found to undergo 4.32×10^6 disintegrations/min. What is the half-life of this nuclide? Ionium is formed by the α-decay of U^{234}. How many rutherfords of U^{234} would be needed to produce a 0.100-mg sample of Th^{230}?

7. A freshly separated sample of RaF (Po^{210}) contains 1.00×10^{-6} gm of that nuclide. How many disintegrations per second would the sample undergo immediately after separation? How many after 10, 30, 50, 70, 100, 300

days? How many curies of activity would the sample contain at these times? How many rutherfords?

8. Consider a radioactive series whose first two members have half-lives of 5 hours and 12 hours, respectively, while the third member is stable. Assume that there are initially 10^6 atoms of the first member, and none of the second and third members. Plot, as functions of time, the numbers of atoms of the three kinds. After how many hours will the number of atoms of the second member reach its maximum value?

9. A freshly purified sample of RaE contains 2.00×10^{-10} gm of that nuclide at time $t = 0$. If the sample is allowed to stand, what will be the greatest weight of RaF that it will ever contain? At what time will that amount be present? At that time, what will be the α-activity in disintegrations per second? The β-activity? How many rutherfords of RaF will be present? Plot the α- and β-activities as functions of time.

10. The nuclide Bi^{212} (ThC) decays both by α-emission and β-emission. (a) What are the partial disintegration constants for α-decay and β-decay? (b) What are the half-lives for α-decay alone and for β-decay alone? (c) At what rates will a freshly separated sample containing 10^{-7} gm of Bi^{212} emit α-particles? β-particles? (d) What will be the rates after three hours?

11. Natural samarium has been found to emit α-particles at a rate of 135 particles/gm/sec. The isotope Sm^{147} (abundance 15.0%) is responsible for the activity; what is its half-life in years?

12. The nuclide U^{235} has a half-life of 7.10×10^8 years; its daughter Th^{231} has a half-life of 24.6 hours. Suppose that the Th^{231} is separated chemically from a sample of U^{235}. Plot, as functions of time, the relative activities of the initially pure U^{235} fraction, the isolated Th^{231} fraction, the Th^{231} growing in the freshly purified U^{235} fraction, and of the U^{235} itself. What kind of equilibrium is reached, if any?

13. The nuclide Th^{234} (UX_1) is normally present in uranium compounds in secular equilibrium with its parent U^{238}, which has a half-life of 4.50×10^9 years. If the formation of UZ is neglected, the UX_1 decays with a half-life of 24.1 days to Pa^{234} (UX_2), which decays with a half-life of 1.18 min into U^{234} (UII); the latter has a half-life of 2.50×10^5 years. Suppose that at time $t = 0$, all the Th^{234} from 100 gm of uranium is removed. Plot the activities (in rutherfords) of Th^{234} and Pa^{234} as functions of time for (a) a 10-min period, (b) a 10-day period. Discuss the kinds of equilibrium involved, if any.

14. Natural uranium contains 0.72% by weight of U^{235} and 99.28% of U^{238}. In normal uranium ores, all radioactive nuclides produced from U^{238} and U^{235} are in secular equilibrium with these parent nuclides. Find the atomic concentrations, in parts per billion, of each daughter nuclide of uranium, with a half-life of one day or more. Neglect the less frequent type of decay when branching occurs.

15. The Pb^{206} (RaG) content of a uranium-containing mineral can be used to determine the age of the mineral. In the transition from U^{238} to Pb^{206}, the U^{238} loses eight α-particles,

$$U^{238} \rightarrow 8He + Pb^{206}.$$

Show that if the presence of U^{235} is neglected (and this has been shown to be valid), and if the decrease in the amount of U^{238} is neglected, then the age can be represented approximately by the formula

$$\text{Age} = (\text{Pb}^{206}/\text{U}) \times 7.5 \times 10^9 \text{ years},$$

where the quantity in parentheses is the weight in grams of Pb^{206} per gram of uranium. What is the age of a uranium-bearing rock which contains (a) 1.33×10^{-2} gm of Pb^{206} per gram of uranium? (b) 0.1 gm of Pb^{206} per gram of uranium? If the decrease in the amount of uranium because of disintegration is taken into account, the last formula may be replaced by the better approximation

$$\text{Age} = (\text{Pb}^{206}/\text{U} + 0.58\text{Pb}^{206}) \times 7.5 \times 10^9 \text{ years}.$$

How much difference does this correction make in the ages obtained in (a) and (b)?

16. Show that the disintegration of one gram of uranium per year, under the assumptions of the last problem, leads to 2.07×10^{-11} gm of helium, or 1.16×10^{-7} cm^3 of helium at N.T.P. Most of the helium produced in a uranium-bearing mineral is trapped and retained as gaseous helium. When a sample of the mineral is dissolved or fused, the helium can be collected and its volume measured. Show that the age of the mineral is given, approximately, by the formula

$$\text{Age} = (\text{He}/\text{U}) \times 8.6 \times 10^6 \text{ years},$$

where the quantity in brackets represents the quantity of helium in cubic centimeters at N.T.P. per gram of uranium. Show that the age may also be written

$$\text{Age} = 3.0(\text{He}/\text{Ra}) \text{ years}.$$

A given sample of a uranium-bearing mineral was found to contain 5.1×10^{-5} cm^3 of helium per gram of mineral, and 1.05×10^{-13} gm of radium per gram of mineral. What is the approximate age of the sample?

CHAPTER 11

ARTIFICIAL NUCLEAR DISINTEGRATION

11–1 Transmutation by alpha-particles: alpha-proton reactions. The fact that certain atoms undergo spontaneous disintegration led to speculation about the possibility of causing the disintegration of the ordinary inactive nuclides. It seemed possible that if atoms were bombarded with energetic particles, one of the latter might penetrate into a nucleus and cause a disruption. The radiations from the natural radionuclides could be used as projectiles, and α-particles seemed most likely to be effective because of their relatively great energy and momentum. Since most of the bombarding α-particles would probably be scattered, it was apparent that the probability of causing a nuclear disintegration would be small. The extent of the scattering could, however, be reduced and the probability of disintegration increased by using some of the lighter atoms as targets, thereby reducing the magnitude of the repulsive Coulomb forces between the target nuclei and the α-particles.

The first disintegration based on these ideas was made by Rutherford (1919),[(1)] who showed that the nuclei of nitrogen atoms emit swift protons when bombarded with α-particles from radium C. The apparatus that Rutherford used was simple but sensitive, and is shown schematically in Fig. 11–1. In one end E of a box B was cut an opening which was covered by a silver foil F. A zinc sulfide screen was placed at S, just outside the opening, and scintillations on the screen were observed by means of a microscope M. The source of the α-particles was radium C placed on a small disc D, whose distance from S could be varied. The silver foil F was thick enough to absorb the α-particles from the source. Different gases could be introduced into the box and removed through the side tubes T. When the box was filled with oxygen or carbon dioxide

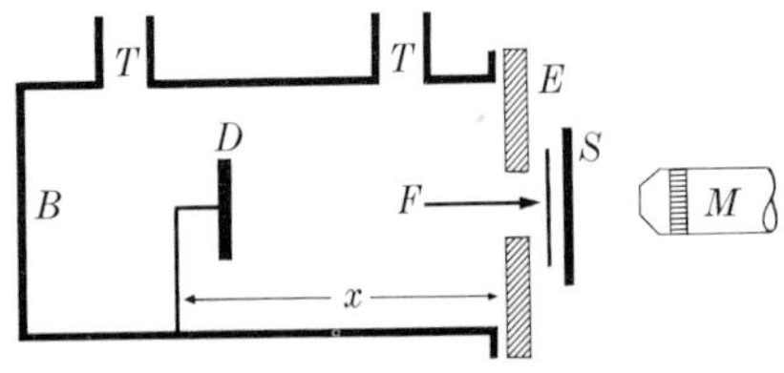

Fig. 11–1. Diagram of Rutherford's apparatus for the disintegration of nitrogen nuclei by α-particles. [Reprinted by permission from Rutherford, Chadwick, and Ellis, *Radiations from Radioactive Substances*, Cambridge University Press (Macmillan Co.), 1930.]

at atmospheric pressure, no scintillations were seen on the screen with the source 7 cm or more away. This thickness x of gas was great enough to absorb all of the α-particles from radium C even without the silver foil. When the gas in the box was nitrogen, however, scintillations were observed on the screen when the source of the α-particles was as much as 40 cm away. Since it was known that α-particles from radium C could not penetrate 40 cm of air, Rutherford concluded that the scintillations were caused by particles ejected from the nitrogen nucleus by the impact of an α-particle. Measurement of the magnetic deflection of the particles suggested that they were protons, and this surmise was confirmed by more precise work.

Rutherford ruled out, by means of careful experiments, the possibility that the protons came from hydrogen present in the nitrogen as an impurity, and concluded that artificial disintegration of nitrogen atoms had taken place. The disintegration was caused by the α-particles from the radium C, and one result was the emission of a highly energetic proton by the nitrogen nucleus. The experimental results also showed that the probability of disintegration was very small; one proton was produced for about one million α-particles passing through the gas. Rutherford and Chadwick[(2)] extended the work on nitrogen to other elements and found evidence of the disintegration of all of the light elements, from boron to potassium, with the exception of carbon and oxygen. They also found that in some cases the energy of the ejected protons was greater than that of the bombarding α-particles. This result provided additional evidence that the protons were emitted as the result of a disintegration process, the extra energy being acquired in the accompanying nuclear rearrangement.

Two hypotheses were suggested as to the nature of the nuclear process leading to the emission of the proton. They were: (a) The nucleus of the bombarded atom simply loses a proton as the result of a collision with a swift α-particle. (b) The α-particle is captured by the nucleus of the atom it hits, and the new, or compound, nucleus emits a proton. These two hypotheses for the disintegration process could be subjected to experimental test because in case (b) the α-particle should disappear, while in case (a) it should still exist after the collision. The choice between the two possibilities was settled in 1925 when Blackett[(3)] studied the tracks produced by α-particles passing through nitrogen in a cloud chamber. He showed, as can be seen in Fig. 11–2, that as a result of a disintegration, the only tracks which could be seen were those of the incident α-particle, a proton, and a recoil nucleus. The absence of a track corresponding to an α-particle after the collision proved that the α-particle disappeared completely. If the disintegration process had been the result simply of a disruption leading to the emission of a proton from the nitrogen nucleus, there should have been four tracks rather than the three actually

FIG. 11–2. A cloud chamber photograph showing the disintegration of a N^{14} nucleus by an α-particle with the formation of O^{17} and a proton. The long track of the proton and the short track of the recoiling oxygen nucleus can be seen. (Blackett.[(3)])

seen. It was concluded, therefore, that the α-particle entered the nucleus of the nitrogen atom with the formation of an unstable system which immediately expelled a proton.

The disintegration of a nitrogen atom by an α-particle may be represented by an equation analogous to those used for chemical reactions,

$$_7N^{14} + {}_2He^4 \rightarrow [_9F^{18}] \rightarrow {}_8O^{17} + {}_1H^1.$$

In this equation, the symbols on the left stand for the reacting nuclides. The symbol in brackets stands for the unstable nucleus formed as the result of the capture of the α-particle by the nitrogen nucleus; this kind of nucleus is often called a *compound nucleus.* The emitted proton and the final nucleus, the products of the reaction, are on the right side of the equation. The charge and mass numbers must be the same on the two sides of the equation, so that the nitrogen nucleus ($Z = 7$) must be transformed into the nucleus of an isotope of oxygen. In other words, an atom of nitrogen has been transformed, or transmuted, into an atom of oxygen. This transmutation may also be represented by the abbreviated notation $N^{14}(\alpha,p)O^{17}$. The transmutation of other nuclides by

α-particles from radioactive substances may be represented by equations similar to that for nitrogen.

$$_5B^{10} + {}_2He^4 \rightarrow [_7N^{14}] \rightarrow {}_6C^{13} + {}_1H^1$$

$$_{11}Na^{23} + {}_2He^4 \rightarrow [_{13}Al^{27}] \rightarrow {}_{12}Mg^{26} + {}_1H^1$$

$$_{13}Al^{27} + {}_2He^4 \rightarrow [_{15}P^{31}] \rightarrow {}_{14}Si^{30} + {}_1H^1$$

$$_{16}S^{32} + {}_2He^4 \rightarrow [_{18}A^{36}] \rightarrow {}_{17}Cl^{35} + {}_1H^1$$

$$_{19}K^{39} + {}_2He^4 \rightarrow [_{21}Sc^{43}] \rightarrow {}_{20}Ca^{42} + {}_1H^1$$

In each case, the charge of the nucleus is increased by one unit and the mass is increased by three units. The alpha-proton reaction may, therefore, be written in the form

$$_ZX^A + {}_2He^4 \rightarrow [_{Z+2}Cn^{A+4}] \rightarrow {}_{Z+1}Y^{A+3} + {}_1H^1, \qquad (11\text{–}1)$$

where X, Y, and Cn represent the target, product, and compound nuclei, respectively.

It is often convenient to refer to a nuclear reaction in terms of the incident and emitted particles, apart from the nuclei involved. Thus, the reactions mentioned so far are examples of (α,p) reactions.

11–2 The balance of mass and energy in nuclear reactions. A nuclear reaction such as that represented by Eq. (11–1) can be analyzed quantitatively in terms of the masses and energies of the nuclei and particles involved. The analysis of nuclear reactions is one of the main sources of information about nuclear properties and will therefore be discussed in some detail before more reactions are considered. The analysis is similar to that used for chemical reactions except that the relativistic relation between mass and energy must be taken into account. Consider a nuclear reaction represented by the equation

$$x + X \rightarrow Y + y, \qquad (11\text{–}2)$$

where X is the target nucleus, x the bombarding particle, Y the product nucleus, and y the product particle. In the only type of reaction considered so far, x is an α-particle, and y is a proton. In other reactions, to be discussed in later sections of this chapter, other bombarding and product particles will be met. It will be assumed that the target nucleus X is initially at rest so that it has no kinetic energy. Since the total energy of a particle or atom is the sum of the rest energy and the kinetic energy, the statement that the total energy is conserved in the nuclear reaction means that

$$(E_x + m_xc^2) + M_Xc^2 = (E_Y + M_Yc^2) + (E_y + m_yc^2). \qquad (11\text{–}3)$$

In Eq. (11–3), m_x, M_X, m_y, and M_Y represent the masses of the incident particle, target nucleus, product particle, and product nucleus, respectively; the E's represent kinetic energies. We now introduce the quantity Q, which represents the difference between the kinetic energy of the products of the reaction and that of the incident particle,

$$Q = E_Y + E_y - E_x. \tag{11–4}$$

The quantity on the right side of Eq. (11–4) can be expressed in terms of the masses because of the relationship (11–3),

$$E_Y + E_y - E_x = (M_X + m_x - M_Y - m_y)c^2. \tag{11–5}$$

Hence, from Eqs. (11–4) and (11–5),

$$Q = E_Y + E_y - E_x = (M_X + m_x - M_Y - m_y)c^2. \tag{11–6}$$

The quantity Q is called the *energy balance* of the reaction or, more commonly, the *Q-value;* it can be determined either from the energy difference or from the mass difference in Eq. (11–6).

If the value of Q is positive, the kinetic energy of the products is greater than that of the reactants; the reaction is then said to be *exothermic* or *exoergic.* The total mass of the reactants is greater than that of the products in this case. If the value of Q is negative, the reaction is *endothermic* or *endoergic.* It is apparent from Eq. (11–6) that the analysis of nuclear reactions involves information about nuclear masses and particle energies. Nuclear reactions can, therefore, be used to obtain information about the masses of nuclei, about particle energies, or about Q-values, depending on what information is available, and which quantities can be measured.

The term E_Y in Eq. (11–6) represents the recoil (kinetic) energy of the product nucleus. It is usually small and hard to measure, but it can

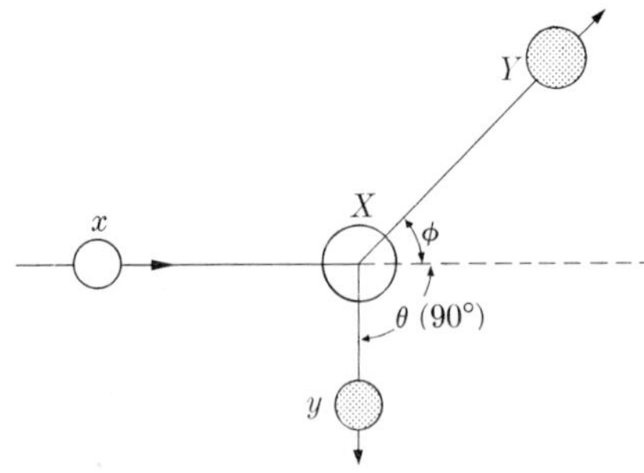

FIG. 11–3. Conservation of momentum in nuclear reactions. The outcoming particles are viewed at an angle of 90° with the direction of the incident particles.

be eliminated by taking into account the conservation of momentum. Consider, for simplicity, the special case in which the outcoming particles are observed at an angle of 90° with the direction of the beam of projectiles, taken as the x-axis, as in Fig. 11–3. Before the collision, the momentum vector is directed along the x-axis. Since the resultant momentum after the collision must also be directed along the x-axis, it follows that

$$m_x v_x = M_Y V_Y \cos\phi, \qquad m_y v_y = M_Y V_Y \sin\phi, \tag{11–7}$$

where v_x, v_y, and V_Y are the velocities of the incident particle, the ejected particle, and the recoil nucleus, respectively, and ϕ is the angle between the x-axis and the direction of recoil of the product nucleus. When the last two equations are squared and added, the result is

$$(m_x v_x)^2 + (m_y v_y)^2 = (M_Y V_Y)^2.$$

Setting $E_x = \frac{1}{2}m_x v_x^2$, $E_y = \frac{1}{2}m_y v_y^2$, $E_Y = \frac{1}{2}M_Y V_Y^2$, and solving for E_Y, we get

$$E_Y = \frac{m_x}{M_Y} E_x + \frac{m_y}{M_Y} E_y. \tag{11–8}$$

Insertion of this expression for E_Y into Eq. (11–6) gives

$$Q = E_y\left(1 + \frac{m_y}{M_Y}\right) - E_x\left(1 - \frac{m_x}{M_Y}\right). \tag{11–9}$$

The value of Q can be determined, therefore, if the energies of the incident and ejected particles are measured and if the mass number of the product nucleus is known.

In the general case of a reaction in which the outcoming particle is observed at an angle θ with the direction of the projectile beam, it can be shown that (see Problem 19 at the end of the chapter),

$$Q = E_y\left(1 + \frac{m_y}{M_Y}\right) - E_x\left(1 - \frac{m_x}{M_Y}\right) - \frac{2}{M_Y}\,(E_x E_y m_x m_y)^{1/2}\cos\theta. \tag{11–10}$$

The importance of the last term decreases as the mass of the target nucleus increases. If the masses are not known accurately, a good approximation to the value of Q can be obtained by using the mass numbers instead of the actual masses in Eqs. (11–9) and (11–10). Equation (11–10) reduces, of course, to Eq. (11–9) when $\theta = 90^0$.

In an endoergic reaction, the energy $-Q$ is needed to excite the compound nucleus sufficiently so that it will break up. This energy must be supplied in the form of kinetic energy of the incoming particle. But not all of that kinetic energy is available for excitation because some is used to impart momentum to the compound nucleus; this momentum is then distributed among the products of the reaction. Consequently, for $-Q$ to be available for excitation of the compound nucleus, we must supply some energy in addition to $-Q$. The amount of energy needed for an endoergic reaction is called the *threshold energy* and can be calculated easily.

If we let M_C and V_C denote the mass and velocity of the compound nucleus, conservation of momentum requires that

$$m_x v_x = M_C V_C, \qquad \text{or} \qquad V_C = \frac{m_x}{M_C} v_x.$$

The part of the kinetic energy of the incident particle needed for excitation of the compound nucleus is

$$-Q = \frac{1}{2} m_x v_x^2 - \frac{1}{2} M_C V_C^2 = \frac{1}{2} m_x v_x^2 \left(1 - \frac{m_x}{M_C}\right).$$

But, $M_C = M_X + m_x$, and

$$(-Q) = \frac{1}{2} m_x v_x^2 \left(\frac{M_X}{M_X + m_x}\right).$$

The threshold energy is then

$$E_{\text{th}} = \frac{1}{2} m_x v_x^2 = (-Q)\left(1 + \frac{m_x}{M_X}\right); \tag{11-11}$$

it can be determined experimentally and the result used to find the value of Q from Eq. (11–11). For a reaction induced by γ-rays, $m_x = 0$, and the threshold energy is just $-Q$.

The masses which appear in the equations of this section are nuclear masses. In actual calculations, however, they may be replaced by the masses of the neutral atoms. The electrons which must be added to the nuclei to form the neutral atoms cancel in the equation for a nuclear reaction because the number of electrons is the same on the two sides of the equation. As an example of the use of Eq. (11–6) consider the $N^{14}(\alpha,p)O^{17}$ reaction. The following values of the atomic masses are obtained from Table 11–1, Section 11–9.

$$M_X = M(N^{14}) = 14.007518 \text{ amu}$$

$$m_x = M(He^4) = 4.003873 \text{ amu}$$

$$M_Y = M(O^{17}) = 17.004529 \text{ amu}$$

$$m_y = m(H^1) = 1.008144 \text{ amu}$$

The value of Q as calculated from the masses is $Q = -0.001282$ amu $= -1.19$ Mev and the reaction is endoergic. A Q-value of -1.16 ± 0.04 Mev has been obtained experimentally from energy measurements and agrees very well with the value calculated from the masses; according to Eq. (11-11), the threshold energy is 1.49 Mev.

The equation for a nuclear reaction sometimes includes Q; for example,

$$X + x \rightarrow [Cn] \rightarrow Y + y + Q.$$

Although this is not necessary, it is a convenient way of showing the energy balance.

11-3 The neutron: alpha-neutron reactions. The discovery of the neutron has been discussed briefly in Section 8-3; it will now be treated in somewhat greater detail. The capture of an α-particle by a nucleus does not always result in the emission of a proton by the compound nucleus. When beryllium was bombarded by α-particles, one of the products of the reaction seemed to be a very penetrating kind of radiation, and no protons were observed. It was assumed that the radiation consisted of γ-ray photons produced by the reaction

$$_4Be^9 + {}_2He^4 \rightarrow [{}_6C^{13}] \rightarrow {}_6C^{13} + \gamma. \tag{11-12}$$

This assumption led, however, to difficulties. It was shown from absorption measurements that the energy of the "photons" should be about 7 Mev, but these radiations were able to knock protons out of hydrogenous materials and to impart to those protons energies of about 5 Mev. On the assumption that the protons were liberated by elastic collisions with γ-ray photons, calculations showed that each photon must have had an energy of about 50 Mev, a value much greater than that deduced from the absorption measurements. There was, therefore, a serious contradiction between the values of the photon energy given by the two methods. The energy that can be attained by the assumed γ-radiation can also be computed from known masses and energies. It is the sum of the Q-value for the reaction (11-12) and the kinetic energy of the incident α-particle minus the recoil energy of the product nucleus. The Q-value is given by the formula

$$Q = (M_X + m_x - M_Y)c^2,$$

since m_y, the mass of the photon, is zero. Inserting the values $M_X = 9.01504$, $m_x = 4.00387$, and $M_Y = 13.00748$, we get $Q = 10.5$ Mev. If the kinetic energy of the α-particle is assumed to be 5 Mev, the total energy available for the reaction is 15.5 Mev, part of which is retained as recoil energy of the product nucleus. The greatest possible energy of the assumed γ-radiation is, therefore, less than 15.5 Mev, which is again much smaller than 50 Mev.

The problem of the identity of the products of the bombardment of beryllium was solved by Chadwick,[4] as was mentioned in Chapter 8. In a series of experiments on the recoil of nuclei which were struck by the rays coming from the bombarded beryllium, he showed that if these rays were assumed to be γ-rays then the results of the experiments led to values for their energy which depended on the nature of the recoil nucleus. For example, protons ejected from paraffin had energies which led to a photon energy of 55 Mev, while recoiling nitrogen nuclei had energies which led to a photon energy of about 90 Mev. If the energy of the recoil nuclei was assumed to be produced by elastic collisions with photons, it turned out that the energy that had to be attributed to the photon increased with the mass of the recoil atom—a requirement that is contrary to the principles of conservation of energy and momentum in elastic collisions. Chadwick showed that all of the difficulties disappeared when the hypothesis was made that the radiation coming from beryllium bombarded with α-particles consists of particles (neutrons) with a mass very nearly equal to that of the proton, but having no charge. The nuclear reaction which produces these neutrons is

$$_4\mathrm{Be}^9 + {_2\mathrm{He}^4} \rightarrow [_6\mathrm{C}^{13}] \rightarrow {_6\mathrm{C}^{12}} + {_0\mathrm{n}^1}, \tag{11–13}$$

where $_0\mathrm{n}^1$ is the symbol for the neutron. The penetrating nature of the neutrons follows from the absence of a charge, and the energies of the recoil atoms in Chadwick's experiments could be completely accounted for on the basis of elastic collisions with an energetic particle of unit mass.

Chadwick also proved that the mass of the neutron is approximately equal to that of the proton, by means of the experiments on the collisions of neutrons with protons and nitrogen nuclei. Since the neutron is not a charged particle, its mass could not be determined from deflection measurements in electric and magnetic fields, and indirect methods had to be used. Chadwick's method was based on the fact that in a collision between a moving and a stationary particle, the velocity imparted to the latter is greatest in a head-on collision, one in which the struck particle moves in exactly the same direction as that in which the incident particle approached it. An expression for the maximum velocity can be derived from the equations of conservation of kinetic energy and momentum in a head-on collision. Suppose that a particle with mass m_1 and velocity v collides with a stationary particle of mass m_2, and let the velocity of the incident particle after the collision be v_1 and that of the struck particle be v_2. The equation of conservation of energy is

$$\tfrac{1}{2}m_1v^2 = \tfrac{1}{2}m_1v_1^2 + \tfrac{1}{2}m_2v_2^2, \tag{11–14}$$

and the equation of conservation of momentum is

$$m_1 v = m_1 v_1 + m_2 v_2. \tag{11–15}$$

The velocity v_1 can be eliminated from the last two equations, and a relationship is obtained between v_2 and v,

$$v_2 = \frac{2m_1}{m_1 + m_2}\, v. \tag{11–16}$$

If the incident particle is a neutron with known velocity v, its mass m_1 can be calculated from Eq. (11–16) if the mass m_2 and the maximum speed v_2 of the recoiling nucleus are known. Chadwick, however, did not have reliable information about the speed of the neutrons and had to use a less direct method. If a neutron with the same velocity v collides with a third particle of mass m_3, the maximum velocity imparted to the latter is given by

$$v_3 = \frac{2m_1}{m_1 + m_3}\, v, \tag{11–17}$$

where m_1 and v are the same as in Eq. (11–16). Upon dividing Eq. (11–16) by Eq. (11–17), we get

$$\frac{v_2}{v_3} = \frac{m_1 + m_3}{m_1 + m_2}. \tag{11–18}$$

If m_2, m_3, v_2, and v_3 are known, the mass m_1 of the neutron can be found. The maximum velocity imparted to hydrogen nuclei ejected from paraffin wax by the neutrons emitted by beryllium bombarded with polonium α-particles was found to be 3.3×10^9 cm/sec. The recoils of nitrogen nuclei after being struck by the same neutrons were observed in cloud-chamber experiments and their maximum velocity was found to be 4.7×10^8 cm/sec. Taking m_2 and m_3 to be the masses of the proton and nitrogen nuclei, equal to 1 amu and 14 amu respectively, and with the corresponding values $v_2 = 3.3 \times 10^9$ and $v_3 = 4.7 \times 10^8$, we get from Eq. (11–18)

$$\frac{m_1 + 14}{m_1 + 1} = \frac{v_2}{v_3} = \frac{3.3 \times 10^9}{4.7 \times 10^8},$$

and $m_1 = 1.15$. This result, although approximate because of errors in the determination of the maximum recoil velocities, showed that the mass of the neutron is roughly the same as that of the proton.

Chadwick obtained a somewhat better value of the neutron mass by considering the reaction

$$_5\mathrm{B}^{11} + {}_2\mathrm{He}^4 \rightarrow {}_7\mathrm{N}^{14} + {}_0\mathrm{n}^1.$$

The Q-value was found from the kinetic energies of the incident α-particles, of the nitrogen nuclei (measured in cloud-chamber experiments), and of the neutron (estimated from the maximum recoil energy of a proton after being struck by a neutron). From the Q-value and the known atomic masses of B^{11}, He^4, and N^{14}, the mass of the neutron can be obtained with the aid of Eq. (11–6). In this way, Chadwick found a value between 1.005 and 1.008 atomic mass units. The best methods now available for determining the neutron mass give 1.008983 amu.

The proof of the existence of the neutron and the development of methods for analyzing nuclear reactions in which one of the products is a neutron led to the discovery of other reactions similar to that represented by Eq. (11–13). Among these are

$$_3Li^7 + {}_2He^4 \rightarrow [_5B^{11}] \rightarrow {}_5B^{10} + {}_0n^1,$$

$$_7N^{14} + {}_2He^4 \rightarrow [_9F^{18}] \rightarrow {}_9F^{17} + {}_0n^1,$$

$$_{11}Na^{23} + {}_2He^4 \rightarrow [_{13}Al^{27}] \rightarrow {}_{13}Al^{26} + {}_0n^1,$$

$$_{13}Al^{27} + {}_2He^4 \rightarrow [_{15}P^{31}] \rightarrow {}_{15}P^{30} + {}_0n^1,$$

$$_{18}A^{40} + {}_2He^4 \rightarrow [_{20}Ca^{44}] \rightarrow {}_{20}Ca^{43} + {}_0n^1.$$

When these reactions are compared with those mentioned in Section 11–1, it is seen that some atoms, such as N^{14}, Na^{23}, and A^{27}, emit either a proton or a neutron when bombarded with α-particles. The compound nucleus, e.g., $[_9F^{18}]$, $[_{13}Al^{27}]$, $[_{15}P^{31}]$, may disintegrate by means of either process, giving a different product in the two cases. It will be seen in the next chapter that the products of these nuclear reactions are sometimes radioactive.

11–4 The acceleration of charged particles. The early artificial disintegrations were produced by bombarding materials with α-particles from natural radioactive substances. The neutrons from the bombardment of beryllium with α-particles were also used as projectiles and, as will be seen later, have produced many artificial disintegrations. The development of the field of nuclear transmutation was limited, however, because α-particles could be obtained only in beams of low intensity and with energies not greater than 7.68 Mev (RaC′), and transmutations by these particles were possible only with the lighter elements. Furthermore, it seemed possible that other particles, protons, deuterons, and even γ-rays, might yield interesting results when used as projectiles. Protons and deuterons with a single positive charge would experience smaller repulsive Coulomb forces than α-particles in the field of the nucleus, and it was thought that this effect might make disintegrations possible with low-energy protons. It became important, therefore, to develop laboratory methods of accelerating charged particles to high energies, and several successful methods were invented about 1930.

Among the earliest particle accelerators were the voltage multiplier of Cockcroft and Walton, the electrostatic generator of Van de Graaff, and the cyclotron of Lawrence and Livingston. By the suitable choice of a machine and its operating conditions, protons, deuterons, or α-particles with energies as high as several Mev could be obtained by 1936. The further development of these machines and the design of new accelerators such as the frequency-modulated cyclotron and the synchrotron made it possible to obtain particles with energies up to hundreds of Mev. The "Cosmotron" at the Brookhaven National Laboratory has produced protons with energies of 2.5 billion electron volts, and recently built machines have raised the energies to 25 Bev. Electrons can be accelerated to energies up to 300 Mev in a betatron, or in a linear accelerator. When these electrons strike a metal target, highly energetic x-rays are produced, which can also be used as projectiles. Finally, the discovery of nuclear fission and the development of chain-reacting piles, or nuclear reactors (cf. Chapter 20), have supplied highly intense sources of neutrons for nuclear bombardment.

With the available particles and accelerators, a large variety of nuclear reactions can be made to occur. The rest of this chapter will be limited to the commonest reactions which take place with particle energies up to about 20 Mev. These reactions will serve to show the most important features of artificial disintegration, and the ideas involved can then be extended to the more complex reactions which occur with particle energies of the order of 100 Mev. The particle accelerators used to produce projectiles for nuclear reactions will be discussed in Chapter 21. For the purposes of the present chapter, it is enough to realize that various bombarding particles of almost any desired energy can be obtained with the help of suitable devices, and that these projectiles can cause many different kinds of nuclear reactions.

11-5 Transmutation by protons. The first case of a nuclear disintegration brought about entirely by artificial means was one for which protons were used as the projectiles. Cockcroft and Walton[(5)] bombarded lithium with protons accelerated to energies of 0.1 to 0.7 Mev. Scintillations caused by particles ejected from the lithium were observed on a zinc sulfide screen placed a short distance away. The particles were proved to be α-particles by photographing their tracks in a cloud chamber,[(6)] and the lithium isotope that was disintegrated was proved to be Li^7. The cloud-chamber pictures showed that two α-particles leave the point of disintegration and proceed with equal energies in opposite directions. The reaction may be represented by the equation

$$_3\mathrm{Li}^7 + {_1\mathrm{H}^1} \rightarrow [_4\mathrm{Be}^8] \rightarrow {_2\mathrm{He}^4} + {_2\mathrm{He}^4}.$$

This reaction has a certain historical interest because it provided one of the earliest quantitative proofs[(7)] of the validity of the Einstein mass-energy relationship. It was a good reaction for this purpose because the energies of the products could be measured precisely, and the masses were known. The proof will be repeated here, with newer values of the various quantities. The values of the atomic masses are, from Table 11–1,

$$M_X = M(\mathrm{Li}^7) = 7.018222 \text{ amu},$$

$$m_x = m(\mathrm{H}^1) = 1.008144,$$

$$M_Y = m_y = m(\mathrm{He}^4) = 4.003873.$$

The value of Q obtained from the masses is $Q_m = 0.01860$ amu $=$ 17.34 Mev. The best experimental value of Q, obtained from the energies of the incident protons and the emergent α-particles, is 17.33 Mev. This agreement shows clearly that the theoretical expression for Q, based on the mass-energy relationship, agrees with experiment, and that there was a genuine release of energy from the lithium atom at the expense of its mass. Since 1932, many nuclear transformations have been studied in detail and the results invariably agree with the relationships deduced from the Einstein equation.

The disintegration that has just been discussed is an example of the general type

$$_ZX^A + {_1\mathrm{H}^1} \rightarrow [_{Z+1}Cn^{A+1}] \rightarrow {_{Z-1}Y^{A-3}} + {_2\mathrm{He}^4}. \qquad (11\text{–}19)$$

Other examples are

$$_3\mathrm{Li}^6 + {_1\mathrm{H}^1} \rightarrow [_4\mathrm{Be}^7] \rightarrow {_2\mathrm{He}^3} + {_2\mathrm{He}^4},$$

$$_4\mathrm{Be}^9 + {_1\mathrm{H}^1} \rightarrow [_5\mathrm{B}^{10}] \rightarrow {_3\mathrm{Li}^6} + {_2\mathrm{He}^4},$$

$$_9\mathrm{F}^{19} + {_1\mathrm{H}^1} \rightarrow [_{10}\mathrm{Ne}^{20}] \rightarrow {_8\mathrm{O}^{16}} + {_2\mathrm{He}^4},$$

$$_{13}\mathrm{Al}^{27} + {_1\mathrm{H}^1} \rightarrow [_{14}\mathrm{Si}^{28}] \rightarrow {_{12}\mathrm{Mg}^{24}} + {_2\mathrm{He}^4}.$$

An interesting reaction occurs when the target is B^{11},

$$_5\mathrm{B}^{11} + {_1\mathrm{H}^1} \rightarrow [_6\mathrm{C}^{12}] \rightarrow {_4\mathrm{Be}^8} + {_2\mathrm{He}^4};$$

Be^8 is highly unstable and breaks up into two more α-particles,

$$_4\mathrm{Be}^8 \rightarrow {_2\mathrm{He}^4} + {_2\mathrm{He}^4}.$$

The final result of the bombardment is three α-particles, and the reaction may be regarded as a case of multiple particle production.

Proton bombardment can also result in nuclear reactions in which one of the products is a neutron. The reaction is of the type

$$ {}_ZX^A + {}_1\text{H}^1 \rightarrow [{}_{Z+1}Cn^{A+1}] \rightarrow {}_{Z+1}Y^A + {}_0\text{n}^1. \qquad (11\text{–}20)$$

The effect of the transmutation is to increase the charge on the nucleus by one unit, moving it one place to the right in the periodic table; the mass number is not changed. Examples of this reaction are

$$ {}_5\text{B}^{11} + {}_1\text{H}^1 \rightarrow [{}_6\text{C}^{12}] \rightarrow {}_6\text{C}^{11} + {}_0\text{n}^1, $$
$$ {}_8\text{O}^{18} + {}_1\text{H}^1 \rightarrow [{}_9\text{F}^{19}] \rightarrow {}_9\text{F}^{18} + {}_0\text{n}^1, $$
$$ {}_{28}\text{Ni}^{58} + {}_1\text{H}^1 \rightarrow [{}_{29}\text{Cu}^{59}] \rightarrow {}_{29}\text{Cu}^{58} + {}_0\text{n}^1. $$

In the (p,n) reactions, the mass change is usually negative, so that the reactions are endoergic. The reaction

$$ {}_{29}\text{Cu}^{65} + {}_1\text{H}^1 \rightarrow [{}_{30}\text{Zn}^{66}] \rightarrow {}_{30}\text{Zn}^{65} + {}_0\text{n}^1 $$

is an example of a case for which the threshold energy has been measured.(8) The source of the protons was an electrostatic generator, and the relative number of neutrons produced was determined as a function of the energy of the bombarding protons. The experimental results in the neighborhood of the threshold energy were then analyzed, and the value of the threshold energy was found to be $E_{\text{th}} = 2.164 \pm 0.01$ Mev. The Q-value is obtained from Eq. (11–11), in which the masses M_X and m_x may be replaced by the mass numbers 65 and 1, respectively. Then

$$ Q = -E_{\text{th}} \times \tfrac{65}{66} = -2.164 \times \tfrac{65}{66} = -2.13 \text{ Mev}. $$

In some cases the bombarding proton is simply captured by a nucleus. The compound nucleus which is formed is again unstable, but becomes less unstable by emitting a γ-ray photon rather than a neutron or an α-particle. The reaction is of the type

$$ {}_ZX^A + {}_1\text{H}^1 \rightarrow [{}_{Z+1}Cn^{A+1}] \rightarrow {}_{Z+1}Y^{A+1} + \gamma, \qquad (11\text{–}21)$$

and some examples are

$$ {}_3\text{Li}^7 + {}_1\text{H}^1 \rightarrow [{}_4\text{Be}^8] \rightarrow {}_4\text{Be}^8 + \gamma, $$
$$ {}_6\text{C}^{12} + {}_1\text{H}^1 \rightarrow [{}_7\text{N}^{13}] \rightarrow {}_7\text{N}^{13} + \gamma, $$
$$ {}_9\text{F}^{19} + {}_1\text{H}^1 \rightarrow [{}_{10}\text{Ne}^{20}] \rightarrow {}_{10}\text{Ne}^{20} + \gamma, $$
$$ {}_{13}\text{Al}^{27} + {}_1\text{H}^1 \rightarrow [{}_{14}\text{Si}^{28}] \rightarrow {}_{14}\text{Si}^{28} + \gamma. $$

The γ-ray photons emitted in these reactions are often very energetic and can, in turn, be used to produce nuclear disintegrations. The bombardment of lithium yields photons with an energy of 17.2 Mev, far more energetic than the 2.6-Mev photons which are the most energetic available from the natural radioactive nuclides.

Another type of disintegration caused by protons is that in which deuterons are produced. Examples of this reaction are

$$_4\mathrm{Be}^9 + {}_1\mathrm{H}^1 \rightarrow [{}_5\mathrm{B}^{10}] \rightarrow {}_4\mathrm{Be}^8 + {}_1\mathrm{H}^2,$$

$$_3\mathrm{Li}^7 + {}_1\mathrm{H}^1 \rightarrow [{}_4\mathrm{Be}^8] \rightarrow {}_3\mathrm{Li}^6 + {}_1\mathrm{H}^2.$$

11–6 Transmutation by deuterons. A great many nuclear reactions have been observed with high energy deuterons as the bombarding particles. In most of these cases, the deuterons have been accelerated up to energies of several Mev in a cyclotron or in an electrostatic generator. One of the first deuteron-induced reactions studied was again that on a lithium atom,[(9)]

$$_3\mathrm{Li}^6 + {}_1\mathrm{H}^2 \rightarrow [{}_4\mathrm{Be}^8] \rightarrow {}_2\mathrm{He}^4 + {}_2\mathrm{He}^4.$$

Other examples of (d,α) reactions are

$$_8\mathrm{O}^{16} + {}_1\mathrm{H}^2 \rightarrow [{}_9\mathrm{F}^{18}] \rightarrow {}_7\mathrm{N}^{14} + {}_2\mathrm{He}^4,$$

$$_{13}\mathrm{Al}^{27} + {}_1\mathrm{H}^2 \rightarrow [{}_{14}\mathrm{Si}^{29}] \rightarrow {}_{12}\mathrm{Mg}^{25} + {}_2\mathrm{He}^4.$$

The mass change is usually positive so that the Q-values are positive, and the reactions are exoergic. The general reaction of this type is

$$_ZX^A + {}_1\mathrm{H}^2 \rightarrow [{}_{Z+1}Cn^{A+2}] \rightarrow {}_{Z-1}Y^{A-2} + {}_2\mathrm{He}^4. \qquad (11\text{–}22)$$

As in the case of proton bombardment, disintegrations produced by deuterons do not always yield α-particles. Deuteron-proton reactions are often found, such as

$$_6\mathrm{C}^{12} + {}_1\mathrm{H}^2 \rightarrow [{}_7\mathrm{N}^{14}] \rightarrow {}_6\mathrm{C}^{13} + {}_1\mathrm{H}^1,$$

$$_{11}\mathrm{Na}^{23} + {}_1\mathrm{H}^2 \rightarrow [{}_{12}\mathrm{Mg}^{25}] \rightarrow {}_{11}\mathrm{Na}^{24} + {}_1\mathrm{H}^1,$$

$$_{15}\mathrm{P}^{31} + {}_1\mathrm{H}^2 \rightarrow [{}_{16}\mathrm{S}^{33}] \rightarrow {}_{15}\mathrm{P}^{32} + {}_1\mathrm{H}^1.$$

The result of these transformations is to increase the mass of the nucleus by one unit, leaving the charge unchanged. The general (d,p) reaction is

$$_ZX^A + {}_1\mathrm{H}^2 \rightarrow [{}_{Z+1}Cn^{A+2}] \rightarrow {}_ZX^{A+1} + {}_1\mathrm{H}^1. \qquad (11\text{–}23)$$

The Q-values for these reactions are usually positive, so that the reactions are exoergic.

Neutrons are often produced as a result of deuteron bombardment, as in the following reactions:

$$_6C^{12} + {_1H^2} \rightarrow [_7N^{14}] \rightarrow {_7N^{13}} + {_0n^1},$$

$$_4Be^9 + {_1H^2} \rightarrow [_5B^{11}] \rightarrow {_5B^{10}} + {_0n^1},$$

$$_3Li^7 + {_1H^2} \rightarrow [_4Be^9] \rightarrow {_4Be^8} + {_0n^1}.$$

The general (d,n) reaction is represented by the equation

$$_ZX^A + {_1H^2} \rightarrow [_{Z+1}Cn^{A+2}] \rightarrow {_{Z+1}Y^{A+1}} + {_0n^1}. \tag{11-24}$$

One of the most interesting cases of deuteron bombardment is that in which the target contains deuterons. Deuterium targets have been made by freezing deuterium oxide (D_2O—"heavy water") onto a surface kept cold by liquid air. Both the (d,p) and the (d,n) reactions have been observed,

$$_1H^2 + {_1H^2} \rightarrow [_2He^4] \rightarrow {_1H^3} + {_1H^1},$$

$$_1H^2 + {_1H^2} \rightarrow [_2He^4] \rightarrow {_2He^3} + {_0n^1}.$$

The *excited* compound nucleus $[_2He^4]$ can disintegrate in two ways. In the first, a proton and a new isotope of hydrogen are formed; in the second, a neutron and an isotope of helium are formed. The new hydrogen isotope has a mass very nearly equal to 3 atomic mass units, and has been given the name *tritium*. It is unstable and has a half-life of about 12 years. The helium isotope of mass 3 is stable and is found in nature.

11-7 Transmutation by neutrons. Neutrons have proved to be especially effective in producing nuclear transformations. Since they have no electric charge, they are not subject to repulsive electrostatic forces in the neighborhood of a positively charged nucleus, and are therefore more likely to penetrate nuclei than are protons, deuterons, or α-particles. Not only are highly energetic neutrons capable of causing nuclear reactions, but slowly moving neutrons are also extremely effective. Because of these properties of the neutron many more disintegrations have been produced with neutrons than with any other particle. Before the development of the chain-reacting pile, or nuclear reactor, the main sources of neutrons were reactions such as $H^2(d,n)He^3$, $Be^9(d,n)B^{10}$ and $Be^9(\alpha,n)C^{12}$. The fast neutrons produced were slowed down by allowing them to pass through some hydrogen-containing substance such as water or paraffin. A neutron gives up a large fraction of its energy in a collision with a hydrogen nucleus and, after many collisions, the average energy of the neutrons is reduced to a few hundredths of an electron volt. These slow neutrons have been found to be especially useful from a practical standpoint, and their interactions with nuclei are particularly interesting.

The reaction between a neutron and a nucleus gives rise in most cases to an α-particle, a proton, a γ-ray photon, or to two neutrons. The (n,α) reaction[(10)] is represented by the general equation

$$_ZX^A + {}_0n^1 \rightarrow [_ZCn^{A+1}] \rightarrow {}_{Z-2}Y^{A-3} + {}_2He^4. \qquad (11\text{–}25)$$

Among the most interesting examples of this reaction are

$$_3Li^6 + {}_0n^1 \rightarrow [_3Li^7] \rightarrow {}_1H^3 + {}_2He^4,$$

$$_5B^{10} + {}_0n^1 \rightarrow [_5B^{11}] \rightarrow {}_3Li^7 + {}_2He^4,$$

$$_{13}Al^{27} + {}_0n^1 \rightarrow [_{13}Al^{28}] \rightarrow {}_{11}Na^{24} + {}_2He^4.$$

The first two reactions have a relatively high yield and are therefore often used to detect neutrons. In one method, an ionization chamber is lined with boron, usually in the form of a compound. The capture of a neutron by an atom of the B^{10} isotope causes the liberation of an α-particle, which is detected by the ionization it produces in the chamber.

In some cases, the compound nucleus formed by the capture of a neutron emits a proton. The (n,p) reaction is described by the equation

$$_ZX^A + {}_0n^1 \rightarrow [_ZCn^{A+1}] \rightarrow {}_{Z-1}Y^A + {}_1H^1. \qquad (11\text{–}26)$$

The effect of this reaction is to replace a proton in the nucleus by a neutron; the mass number is not changed, but the charge is decreased by one unit and the atom is moved one place to the left in the periodic table. Some examples of this reaction are

$$_7N^{14} + {}_0n^1 \rightarrow [_7N^{15}] \rightarrow {}_6C^{14} + {}_1H^1,$$

$$_{13}Al^{27} + {}_0n^1 \rightarrow [_{13}Al^{28}] \rightarrow {}_{12}Mg^{27} + {}_1H^1.$$

The first of these reactions can be induced by slow neutrons; with heavier nuclei, more energetic neutrons must be used, as in the case of Al^{27} and in the reaction

$$_{30}Zn^{64} + {}_0n^1 \rightarrow [_{30}Zn^{65}] \rightarrow {}_{29}Cu^{64} + {}_1H^1.$$

In another type of reaction, one neutron is captured by the nucleus and two neutrons are emitted. The (n,2n) reaction leaves the charge of the nucleus unchanged and decreases the mass number by one unit. The result is an isotope of the target nucleus with a mass number one unit smaller,

$$_ZX^A + {}_0n^1 \rightarrow [_ZCn^{A+1}] \rightarrow {}_ZX^{A-1} + {}_0n^1 + {}_0n^1. \qquad (11\text{–}27)$$

An example of this reaction is

$$_{13}\mathrm{Al}^{27} + {}_0\mathrm{n}^1 \rightarrow [{}_{13}\mathrm{Al}^{28}] \rightarrow {}_{13}\mathrm{Al}^{26} + {}_0\mathrm{n}^1 + {}_0\mathrm{n}^1.$$

The mass change in the (n,2n) reaction is always negative; the Q-value is negative and fast neutrons are needed to bring about this reaction.

The commonest process which results from neutron capture is *radiative capture,*[(11)] represented by

$$_ZX^A + {}_0\mathrm{n}^1 \rightarrow [{}_ZCn^{A+1}] \rightarrow {}_ZX^{A+1} + \gamma. \tag{11–28}$$

The compound nucleus emits one or more γ-ray photons, and the final nucleus is an isotope of the target nucleus with a mass number one unit greater. The (n,γ) process has been observed in nearly all of the elements. The Q-values are always positive, the excess energy being carried away by the γ-rays. The simplest (n,γ) reaction with slow neutrons occurs with hydrogen as the target nucleus,

$$_1\mathrm{H}^1 + {}_0\mathrm{n}^1 \rightarrow [{}_1\mathrm{H}^2] \rightarrow {}_1\mathrm{H}^2 + \gamma.$$

The product of the reaction is deuterium. When deuterium is bombarded with slow neutrons, tritium is formed,

$$_1\mathrm{H}^2 + {}_0\mathrm{n}^1 \rightarrow [{}_1\mathrm{H}^3] \rightarrow {}_1\mathrm{H}^3 + \gamma.$$

Other typical (n,γ) reactions are

$$_{13}\mathrm{Al}^{27} + {}_0\mathrm{n}^1 \rightarrow [{}_{13}\mathrm{Al}^{28}] \rightarrow {}_{13}\mathrm{Al}^{28} + \gamma,$$
$$_{49}\mathrm{In}^{115} + {}_0\mathrm{n}^1 \rightarrow [{}_{49}\mathrm{In}^{116}] \rightarrow {}_{49}\mathrm{In}^{116} + \gamma,$$
$$_{92}\mathrm{U}^{238} + {}_0\mathrm{n}^1 \rightarrow [{}_{92}\mathrm{U}^{239}] \rightarrow {}_{92}\mathrm{U}^{239} + \gamma.$$

The radiative capture of slow neutrons often results in product nuclei which are radioactive, and this reaction is one of the most important sources of artificial radioactive nuclides.

11–8 Transmutation by photons. Atomic nuclei can also be disintegrated by bombardment with high-energy photons, a process which is usually called *photodisintegration.* Since the photon has no mass, it can supply only its kinetic energy to a nuclear reaction. This energy must be at least as great as the binding energy of a nuclear particle before such a particle can be ejected from a nucleus. Photodisintegration reactions are, therefore, endoergic and usually have threshold energies of the order of 10 Mev. With only two exceptions, photodisintegration does not occur with γ-rays from natural radioactive substances. These exceptions are the deuteron, which has a binding energy of only 2.2 Mev, and the nuclide $_4\mathrm{Be}^9$, in which one neutron is loosely bound. In the case of the deuteron,[12] the reaction is

$$_1H^2 + \gamma \rightarrow [_1H^2] \rightarrow {}_1H^1 + {}_0n^1,$$

and is the reverse of the radiative capture of a neutron by a proton. In beryllium, the reaction is

$$_4Be^9 + \gamma \rightarrow [_4Be^9] \rightarrow {}_4Be^8 + {}_0n^1,$$

with a Q-value of -1.67 Mev.

It will be recalled from Section 11–5 that when Li is bombarded with protons, 17-Mev γ-rays are produced. These have been used successfully to cause the photodisintegration of other nuclides, as in the (γ,n) reaction

$$_{15}P^{31} + \gamma \rightarrow [_{15}P^{31}] \rightarrow {}_{15}P^{30} + {}_0n^1.$$

The (γ,p) reaction requires still higher energies, and has been observed with high-energy photons from a betatron.

11–9 Nuclear chemistry: nuclear masses. The transmutations which have been discussed are the commonest examples of a large variety of nuclear reactions brought about by charged particles, neutrons, and photons. Many other reactions have been discovered and analyzed. Some have not been considered here because they are relatively rare; others have been omitted because they take place only at very high energies, and their importance and interest lie outside the limits of elementary nuclear physics.

It has been seen that nuclides can be transmuted in a number of different ways. A given nuclide can be transformed to any one of several other nuclides, the nature of the newly formed substance depending on the particle used to produce the transmutation and upon the particle given off during the reaction. For example, the nuclide $_{13}Al^{27}$ has been made to undergo the following transformations.

(1) $_{13}Al^{27} + {}_2He^4 \rightarrow [_{15}P^{31}] \rightarrow {}_{15}P^{30} + {}_0n^1$

(2) $_{13}Al^{27} + {}_2He^4 \rightarrow [_{15}P^{31}] \rightarrow {}_{14}Si^{30} + {}_1H^1$

(3) $_{13}Al^{27} + {}_1H^2 \rightarrow [_{14}Si^{29}] \rightarrow {}_{12}Mg^{25} + {}_2He^4$

(4) $_{13}Al^{27} + {}_1H^2 \rightarrow [_{14}Si^{29}] \rightarrow {}_{13}Al^{28} + {}_1H^1$

(5) $_{13}Al^{27} + {}_1H^2 \rightarrow [_{14}Si^{29}] \rightarrow {}_{14}Si^{28} + {}_0n^1$

(6) $_{13}Al^{27} + {}_1H^2 \rightarrow [_{14}Si^{29}] \rightarrow {}_{11}Na^{24} + {}_1H^1 + {}_2He^4$

(7) $_{13}Al^{27} + {}_1H^1 \rightarrow [_{14}Si^{28}] \rightarrow {}_{12}Mg^{24} + {}_2He^4$

(8) ${}_{13}Al^{27} + {}_{1}H^{1} \rightarrow [{}_{14}Si^{28}] \rightarrow {}_{14}Si^{27} + {}_{0}n^{1}$

(9) ${}_{13}Al^{27} + {}_{1}H^{1} \rightarrow [{}_{14}Si^{28}] \rightarrow {}_{14}Si^{28} + \gamma$

(10) ${}_{13}Al^{27} + {}_{1}H^{1} \rightarrow [{}_{14}Si^{28}] \rightarrow {}_{11}Na^{24} + 3({}_{1}H^{1}) + {}_{0}n^{1}$

(11) ${}_{13}Al^{27} + {}_{0}n^{1} \rightarrow [{}_{13}Al^{28}] \rightarrow {}_{11}Na^{24} + {}_{2}He^{4}$

(12) ${}_{13}Al^{27} + {}_{0}n^{1} \rightarrow [{}_{13}Al^{28}] \rightarrow {}_{12}Mg^{27} + {}_{1}H^{1}$

(13) ${}_{13}Al^{27} + {}_{0}n^{1} \rightarrow [{}_{13}Al^{28}] \rightarrow {}_{13}Al^{28} + \gamma$

(14) ${}_{13}Al^{27} + \gamma \rightarrow [{}_{13}Al^{27}] \rightarrow {}_{13}Al^{26} + {}_{0}n^{1}$

(15) ${}_{13}Al^{27} + \gamma \rightarrow [{}_{13}Al^{27}] \rightarrow {}_{11}Na^{25} + {}_{1}H^{1} + {}_{1}H^{1}$

(16) ${}_{13}Al^{27} + \gamma \rightarrow [{}_{13}Al^{27}] \rightarrow {}_{11}Na^{24} + {}_{1}H^{1} + {}_{1}H^{1} + {}_{0}n^{1}$

The reactions numbered (10), (15), and (16) are included to give some idea of transmutations which are possible at higher energies. The protons in (10) and the γ-rays in (15) and (16) have energies of about 50 Mev. At still higher energies, still more particles are given off and the term *atom smashing* is really appropriate.

In the 16 reactions listed, 11 different nuclides are formed. One nuclide, ${}_{11}Na^{24}$, is formed from ${}_{13}Al^{27}$ in 4 different ways, in reactions (6), (10), (11), and (16). It can also be formed in 5 more reactions.

$$
\begin{aligned}
{}_{11}Na^{23} + {}_{1}H^{2} &\rightarrow [{}_{12}Mg^{25}] \rightarrow {}_{11}Na^{24} + {}_{1}H^{1} \\
{}_{11}Na^{23} + {}_{0}n^{1} &\rightarrow [{}_{11}Na^{24}] \rightarrow {}_{11}Na^{24} + \gamma \\
{}_{12}Mg^{24} + {}_{0}n^{1} &\rightarrow [{}_{12}Mg^{25}] \rightarrow {}_{11}Na^{24} + {}_{1}H^{1} \\
{}_{12}Mg^{25} + \gamma &\rightarrow [{}_{12}Mg^{25}] \rightarrow {}_{11}Na^{24} + {}_{1}H^{1} \\
{}_{12}Mg^{26} + {}_{1}H^{2} &\rightarrow [{}_{13}Al^{28}] \rightarrow {}_{11}Na^{24} + {}_{2}He^{4}
\end{aligned}
$$

Thus, the nuclide ${}_{11}Na^{24}$ can be formed in at least 9 different ways. These examples show the tremendous variety of nuclear reactions which can be induced with the nuclides, particles, accelerators, and other sources of projectiles now available. The study of these reactions is one of the great branches of nuclear physics and because of the problems involved in separating and identifying the products, this field is aptly called "nuclear chemistry."

One of the important applications of nuclear reactions is in the determination of nuclear, or atomic, masses, a problem which has a direct analog in chemistry. The determination of atomic masses by means of the mass spectrograph has been discussed in Chapter 9 and the results obtained by that method can now be combined with values obtained from nuclear reactions. According to Eq. (11–6), a Q-value can be expressed

TABLE 11–1*

THE MASSES OF LIGHT ATOMS UP TO CHLORINE-37

Nuclide	Atomic mass		Nuclide	Atomic mass		Nuclide	Atomic mass	
n^1	1.0089830	(±1.7)	C^{14}	14.0076845	(±3.1)	Mg^{23}	23.001409	(±15)
H^1	1.0081437	(±1.8)	N^{13}	13.0098617	(±4.8)	Mg^{24}	23.992638	(±13)
H^2	2.0147361	(±2.9)	N^{14}	14.0075179	(±3.0)	Mg^{25}	24.993747	(±14)
H^3	3.0169980	(±4.5)	N^{15}	15.0048627	(±4.5)	Mg^{26}	25.990796	(±15)
He^3	3.0169786	(±4.5)	N^{16}	16.01100	(±130)	Mg^{27}	26.992866	(±15)
He^4	4.0038727	(±2.1)	O^{14}	14.013052	(±42)	Al^{25}	24.99835	(±60)
He^5	5.013880	(±31)	O^{15}	15.007768	(±7)	Al^{27}	26.990080	(±14)
He^6	6.020818	(±29)	O^{16}	16.0000000	(±0)	Al^{28}	27.990768	(±15)
Li^5	5.01392	(±70)	O^{17}	17.0045293	(±3.9)	Si^{27}	26.995233	(±17)
Li^6	6.0170281	(±4.7)	O^{18}	18.004840	(±9)	Si^{28}	27.985777	(±16)
Li^7	7.018222	(±6)	O^{19}	19.00931	(±300)	Si^{29}	28.985661	(±16)
Li^8	8.025020	(±7)	F^{17}	17.0074989	(±4.1)	Si^{30}	29.983252	(±17)
Be^7	7.019149	(±6)	F^{18}	18.006635	(±10)	Si^{31}	30.985153	(±18)
Be^8	8.0078473	(±4.2)	F^{19}	19.004447	(±7)	P^{28}	28.00055	(±300)
Be^9	9.015041	(±5)	F^{20}	20.006341	(±10)	P^{30}	29.987885	(±36)
Be^{10}	10.016706	(±7)	Ne^{19}	19.007944	(±9)	P^{31}	30.983563	(±18)
B^9	9.016190	(±6)	Ne^{20}	19.998765	(±10)	P^{32}	31.984025	(±20)
B^{10}	10.016109	(±6)	Ne^{21}	21.000494	(±10)	S^{32}	31.982190	(±20)
B^{11}	11.012795	(±5)	Ne^{22}	21.998346	(±16)	S^{33}	32.981887	(±23)
B^{12}	12.018166	(±7)	Ne^{23}	23.001755	(±18)	S^{34}	33.97920	(±170)
C^{10}	10.02014	(±100)	Na^{22}	22.001396	(±17)	S^{35}	34.98014	(±80)
C^{11}	11.014924	(±6)	Na^{23}	22.997047	(±11)	Cl^{35}	34.97996	(±80)
C^{12}	12.0038065	(±3.9)	Na^{24}	23.998560	(±12)	Cl^{36}	35.97974	(±90)
C^{13}	13.0074754	(±4.1)	Na^{25}	24.99772	(±310)	Cl^{37}	36.997540	(±45)

as the difference between the masses of the reactants and products of a nuclear reaction. Many Q-values have been measured with high precision in recent years,[13,14] and it is now possible to obtain the masses of light nuclei directly in terms of the mass of O^{16} without the use of any mass-spectroscopic results.[15,16] The most recent compilations of atomic masses combine mass-spectroscopic and Q-value data. Nuclear reactions are, of course, an essential source of information concerning the masses of nuclides which do not occur naturally and can be obtained only in nuclear reactions. The atomic masses of nuclides[17] up to Cl^{37} are listed in Table 11–1. Both naturally occurring and artificially produced nuclides are included in the table. The uncertainty, given in parentheses, is expressed in units of 10^{-6} amu.

References

GENERAL

Rutherford, Chadwick, and Ellis, *Radiations from Radioactive Substances.* New York: Macmillan. 1930, Chapter 10 (reprinted with corrections 1951).

M. S. Livingston and H. A. Bethe, "Nuclear Physics, C. Nuclear Dynamics, Experimental," *Revs. Mod. Phys.* **9,** 245–390 (July 1937).

J. D. Stranathan, *The "Particles" of Modern Physics.* Philadelphia: Blakiston, 1944, Chapters 10, 11.

R. T. Beyer, ed., *Foundations of Nuclear Physics;* Facsimiles of Thirteen Fundamental Studies as they were originally reported in the Scientific Journals, with a Bibliography. New York: Dover, 1947.

R. D. Evans, *The Atomic Nucleus.* New York: McGraw-Hill, 1955, Chapters 12, 13.

PARTICULAR

1. E. Rutherford, "Collision of α-Particles with Light Atoms, Part IV." An Anomalous Effect in Nitrogen, *Phil. Mag.* **37,** 581 (1919). (In Beyer.)
2. E. Rutherford and J. Chadwick, "Artificial Disintegration of Light Elements," *Phil. Mag.* **42,** 809 (1921); **44,** 417 (1922); *Proc. Phys. Soc.* (London) **36,** 417 (1924).
3. P. M. S. Blackett, "Photography of Artificial Disintegration Collisions and Accuracy of Angle Determinations," *Proc. Roy. Soc.* (London) **A134,** 658 (1932).
4. J. Chadwick, "The Existence of a Neutron," *Proc. Roy. Soc.* (London) **A136,** 696 (1932). (In Beyer.)
5. J. D. Cockcroft and E. T. S. Walton, "Experiments with High Velocity Positive Ions." II. "The Disintegration of Elements by High Velocity Protons," *Proc. Roy. Soc.* (London) **A137,** 229 (1933).
6. P. I. Dee and E. T. S. Walton, "Transmutation of Lithium and Boron," *Proc. Roy. Soc.* (London) **A141,** 733 (1933).

7. K. T. Bainbridge, "The Equivalence of Mass and Energy," *Phys. Rev.* **44,** 123 (1933).

8. Shoupp, Jennings, and Jones, "Threshold for the Proton-Neutron Reaction in Copper," *Phys. Rev.* **73,** 421 (1948).

9. Lewis, Livingston, and Lawrence, "The Emission of α-Particles from Various Targets Bombarded by Deuterons of High Speed," *Phys. Rev.* **44,** 55 (1933).

10. N. Feather, "Collisions of Neutrons with Nitrogen Nuclei," *Proc. Roy. Soc.* (London) **A136,** 709 (1932).

11. Fermi, Amaldi, D'Agostino, Rasetti, and Segrè, "Artificial Radioactivity Produced by Neutron Bombardment," *Proc. Roy. Soc.* (London) **A146,** 483 (1934).

12. J. Chadwick and M. Goldhaber, "Nuclear Photoelectric Effect," *Nature* **134,** 237 (1934); *Proc. Roy. Soc.* (London) **A151,** 479 (1935).

13. W. W. Buechner, "The Determination of Nuclear Reaction Energies by Deflection Methods," *Progress in Nuclear Physics,* O. R. Frisch, ed., Vol. 5, p. 1, 1956.

14. D. M. Van Patter and W. Whaling, "Nuclear Disintegration Energies." I, *Revs. Mod. Phys.,* **26,** 402 (1954); II, *ibid.,* **29,** 757 (1957).

15. Li, Whaling, Fowler, and Lauritsen, "Masses of Light Nuclei from Nuclear Disintegration Energies," *Phys. Rev.* **83,** 512 (1951).

16. C. W. Li, "Nuclear Mass Determinations from Disintegration Energies: Oxygen to Sulphur," *Phys. Rev.* **88,** 1038 (1952).

17. J. Mattauch and F. Everling, "Masses of Atoms of $A < 40$," *Progress in Nuclear Physics,* O. R. Frisch, ed., Vol. 6, p. 233 (1957).

Problems

1. Complete the reactions listed below. Rewrite them in a way which shows the balance of atomic and mass numbers, and indicate the compound nucleus in each case: $H^1(n,\gamma)$; $H^2(n,\gamma)$; $Li^7(p,n)$; $Li^7(p,\alpha)$; $Be^9(p,d)$; $Be^9(d,p)$; $Be^9(p,\alpha)$; $B^{11}(d,\alpha)$; $C^{12}(d,n)$; $N^{14}(\alpha,p)$; $N^{15}(p,\alpha)$; $O^{16}(d,n)$.

2. Calculate the Q-values of each of the reactions of Problem 1. Which reactions are exoergic? Which are endoergic? Use the masses listed in Table 11–1.

3. In the four reactions which follow, the mass of the target nucleus Al is given as 26.99007 amu, and the Q-value of each reaction is given. Calculate the mass of the product nucleus.

(a) $Al^{27}(n,\gamma)$; $Q = 7.722$ Mev
(b) $Al^{27}(d,p)$; $Q = 5.497$ Mev
(c) $Al^{27}(p,\alpha)$; $Q = 1.594$ Mev
(d) $Al^{27}(d,\alpha)$; $Q = 6.693$ Mev

4. One of the reactions which occurs when boron is bombarded with 1.510-Mev deuterons is $B^{11}(d,\alpha)Be^9$. The α-particles coming off at an angle of 90° with the direction of the deuteron beam have an energy of 6.370 Mev. What is the Q-value of the reaction?

5. Under the conditions of Problem 4, the reaction $B^{11}(d,p)B^{12}$ was also found to occur, and to have a Q-value of 1.136 Mev. What is the energy of the protons observed at 90°?

6. The reactions $Be^9(p,n)B^9$, $C^{13}(p,n)N^{13}$, and $O^{18}(p,n)F^{18}$ were found to have the threshold energies 2.059 Mev, 3.236 Mev, and 2.590 Mev, respectively. What are the Q-values for the reactions?

7. Calculate, from the atomic masses, the threshold energies for the reactions $B^{11}(p,n)C^{11}$, $O^{18}(p,n)F^{18}$, and $Na^{23}(p,n)Mg^{23}$.

8. Write ten nuclear reactions which might be expected to take place with C^{12} as the target. Which of these reactions would be expected to be exoergic? Endoergic?

9. Write ten nuclear reactions which should give the product nucleus C^{12}. Which reactions should be exoergic? Endoergic?

10. A sample of silicon is bombarded with 1.8-Mev deuterons. Show that the following reactions can produce charged particles with energies greater than 4 Mev: $Si^{28}(d,p)Si^{29}$; $Si^{29}(d,p)Si^{30}$; $Si^{29}(d,\alpha)Al^{27}$; $Si^{30}(d,p)Si^{31}$; $Si^{30}(d,\alpha)Al^{28}$.

11. The reactions listed below have been found to have the indicated Q-values

(a) $O^{16}(d,\alpha)\,N^{14}$; $Q = 3.112$ Mev
(b) $N^{14}(d,p)\,N^{15}$; $Q = 8.615$ Mev
(c) $N^{15}(d,\alpha)\,C^{13}$; $Q = 7.681$ Mev
(d) $C^{13}(d,\alpha)\,B^{11}$; $Q = 5.160$ Mev
(e) $B^{11}(p,\alpha)\,Be^8$; $Q = 8.567$ Mev
(f) $Be^8(\alpha)\alpha$; $Q = 0.089$ Mev

Show that these data lead to a value of 4.003869 amu for the mass of He^4 if the mass of deuterium is taken as 2.014723 amu.

12. Suppose that, in separate experiments, C^{12} is bombarded with neutrons, protons, deuterons, and α-particles, each having kinetic energy of 10.00 Mev. (a) What would be the excitation energy of the compound nucleus in each case if it is assumed that the recoil energy of the compound nucleus can be neglected? (b) What is the recoil energy of the compound nucleus in each case when the recoil energy is not neglected?

13. With the aid of the data of Table 11–1, calculate the binding energy of the "last neutron" in each of the following sets of isotopes:

(a) C^{12}, C^{13}, C^{14}; (b) N^{14}, N^{15}, N^{16};
(c) O^{16}, O^{17}, O^{18}; (d) Ne^{20}, Ne^{21}, Ne^{22}.

14. With the aid of the data of Table 11–1, calculate the binding energy of the last proton in each of the following sets of isotopes:

(a) C^{11}, C^{12}, C^{13}; (b) N^{13}, N^{14}, N^{15};
(c) O^{14}, O^{15}, O^{16}, O^{17}; (d) Ne^{19}, Ne^{20}, Ne^{21}.

15. What hypotheses would you propose to account for those values of the binding energy in Problems 13 and 14 which are particularly small, i.e., smaller than 5 Mev?

16. Use the values of the mass differences: neutron-proton, and Be^7-Li^7 to determine the minimum proton energy needed to observe the reaction $Li^7(p,n)Be^7$.

17. Calculate the difference between the masses of Al^{28} and Si^{28} from the following data: $Al^{27}(d,p)Al^{28}$, Q = 5.48 Mev; $Si^{28}(d,p)$ Si^{29}, Q = 6.25 Mev; $Si^{29}(d,\alpha)Al^{27}$, Q = 5.99 Mev; $2H^1 - H^2 = 1.584$ millimass units; $2H^2 - He^4 = 25.600$ millimass units.

18. From the nuclear masses involved, calculate the range of energy of the neutrons produced when Be is bombarded with 5.30 Mev α-particles from Po^{210}.

19. Derive Eq. (11–10).

CHAPTER 12

ARTIFICIAL RADIOACTIVITY

12–1 The discovery of artificial radioactivity. The discovery by Curie and Joliot, in 1934, that the products of some induced nuclear transmutations are radioactive opened a new era in nuclear physics. The study of the radioactivity of the artificially made atomic nuclei has done much to clarify concepts of the constitution and stability of nuclei. Many of the nuclear reactions discussed in the last chapter yield radioactive nuclides and about 1200 radioactive species have been made up to 1958. These nuclides decay spontaneously according to the same laws that govern the disintegration of the naturally occurring radioactive elements, and the term *artificial radioactivity* refers to the way in which the new radionuclides are produced rather than to their decay.

Artificial radioactivity was discovered by Curie and Joliot while they were studying the effects of α-particles on the nuclei of light elements. When boron, magnesium, and aluminum were bombarded with α-particles from polonium, protons and neutrons were observed, as expected from the known (α,p) and (α,n) reactions. In addition to these particles, *positive electrons* or *positrons* were also observed.[1] The positive electron is a fundamental particle whose rest mass is the same as that of the electron, and whose charge has the same magnitude but opposite sign from that of the electron. It was discovered by Anderson[2] in 1932 in the course of a study of photographs of cosmic-ray tracks in a Wilson cloud chamber. In the presence of a magnetic field, some tracks were seen which could have been caused only by particles of electronic mass and charge, but from the direction of the curvature of these tracks it was evident that the particles producing them must have been positively charged. Anderson called these particles *positrons*. Throughout the rest of this book the word *electron* will be used to denote the negative electron and the word *positron* will be used for the positive electron. This usage is common and more convenient than the use of the terms "negative electron" and "positive electron." The word "negatron" which has been suggested for the negative electron has not been generally adopted.

The source of the positrons observed by Curie and Joliot was not clear at first and several hypotheses were suggested which proved to be incorrect. For example, it was suggested that the positrons might result directly from the disintegration by the α-particles, with the formation of a neutron and positron instead of the usual proton. Subsequent experiments by Curie and Joliot[3] showed, however, that the light-element targets continued to emit positrons after the source of the α-particles had been removed; the rate of emission of the positrons gradually decreased with time, finally approaching zero. When the rate was plotted against the time after the removal of the α-particle source, an exponential curve was obtained for each target similar to those obtained in the β-decay of natural radionuclides. The activity could be expressed accurately by the equation $A = A_0e^{-\lambda t}$, and characteristic half-lives were found, namely, 14 min, $2\frac{1}{2}$ min, and $3\frac{1}{4}$ min for boron, magnesium, and aluminum targets, respectively. The explanation offered by Curie and Joliot was that the product nucleus formed in the (α,n) reaction in each case was an unstable nuclide, which then disintegrated with the emission of a positron. The nuclear reactions when boron and aluminum were the targets were postulated to be

$$\text{(a)} \qquad {}_5\text{B}^{10} + {}_2\text{He}^4 \rightarrow [{}_7\text{N}^{14}] \rightarrow {}_7\text{N}^{13} + {}_0\text{n}^1,$$

$$ {}_7\text{N}^{13} \rightarrow {}_6\text{C}^{13} + {}_1\text{e}^0; \qquad T = 14 \text{ min},$$

$$\text{(b)} \qquad {}_{13}\text{Al}^{27} + {}_2\text{He}^4 \rightarrow [{}_{15}\text{P}^{31}] \rightarrow {}_{15}\text{P}^{30} + {}_0\text{n}^1,$$

$$ {}_{15}\text{P}^{30} \rightarrow {}_{14}\text{Si}^{30} + {}_1\text{e}^0; \qquad T = 3\tfrac{1}{4} \text{ min}.$$

The symbol ${}_1\text{e}^0$ or ${}_{+1}\text{e}^0$ is used to denote the positron, since its charge is the same as that of the proton, and its mass number is zero. The half-life of N^{13} was later found to be 10 min and that of P^{30} $2\frac{1}{2}$ min, rather than 14 min and $3\frac{1}{4}$ min, respectively, as first reported by Curie and Joliot. The nuclides N^{13} and P^{30} are not stable, as can be verified from Table 9–1, but the nuclides which remain after the emission of the positron are stable.

The explanation offered by Curie and Joliot was tested chemically(4) by separating the target element from the product element; radioactivity was found only in the latter. In the boron reaction, the target, boron nitride BN, was irradiated with α-particles for several minutes and then heated with caustic soda, liberating all the nitrogen as gaseous ammonia. The ammonia was found to have all the radioactivity. The half-life was the same as that found with other boron targets, and was not found with nitrogen targets. Hence, the boron must have been changed into a radioactive isotope of nitrogen. An aluminum target, after being irradiated with α-particles, was dissolved in hydrochloric acid, liberating any phosphorus present as the gas phosphine (PH_3), while the aluminum remained behind in the solution. The solution was then evaporated to dryness and the residue tested for positron activity. No activity was found. The gas, which contained the phosphorus, showed the characteristic positron activity, and the activity was therefore associated with an isotope of phosphorus. These tests provided the first definite chemical evidence for the artificial transformation of one element to another. The chemical identification of the product nucleus, together with the determination of the nature of the ejected particles by measurements of ionization in a cloud chamber and of the deflection in a magnetic field, proved the correctness of the equations proposed for the disintegrations.

After the discovery that the bombardment of light nuclides with α-particles can lead to radioactive products, it was found that nuclear reactions induced by protons, deuterons, neutrons, and photons can also result in radioactive products. As in the case of the natural radionuclides, an artificial radionuclide can be characterized by its half-life and by the radiation it emits. These properties have helped greatly in determining the chemical nature of the products of nuclear reactions. When the products are radioactive, they can be traced in chemical reactions by means of their characteristic half-lives or disintegration products. Otherwise they could not be traced because of the very small amounts in which they are present, and their chemical nature could only be inferred from indirect physical evidence, such as mass and energy balances. The special branch of chemistry that deals with these and similar problems is called *radiochemistry*, and has become an important part of nuclear science.

By radiochemical methods, the active products of nuclear reactions can be separated in forms suitable for the study of their properties. It is possible to determine the element with which a particular product is isotopic, and to assign a mass number. The latter assignment often depends on cross-bombardment, that is, the particular nuclide is obtained by several different nuclear reactions, from which the mass number of the given active nuclide can be inferred. When the nuclide has been obtained in suitable form, its decay rate can be determined, as well as the properties of its radiations. About 1200 artificially radioactive nuclides have been identified and their properties determined by combined chemical and physical techniques.

12–2 The artificial radionuclides. Electron and positron emission. Orbital electron capture. The constitution of the known nuclides, both natural and artificial, can be shown by means of a graph in which the number of neutrons is plotted against the number of protons, as in Fig. 12–1. The stable nuclides are indicated by black squares (as in Fig. 9–9). Alpha-emitters are indicated by open triangles and squares; the latter are β-stable, while the former may also undergo β-decay. Electron-emitters are indicated by open circles and positron emitters by crosses. A cross in an open circle indicates that the nuclide decays by β^--emssion, β^+-emission, and electron capture. The diagonal lines are lines of constant mass number, A, with $A = Z + N$; nuclides which lie on the same diagonal line are called *isobars* because they have the same mass number and, therefore, very nearly the same atomic weight. For the most part, the

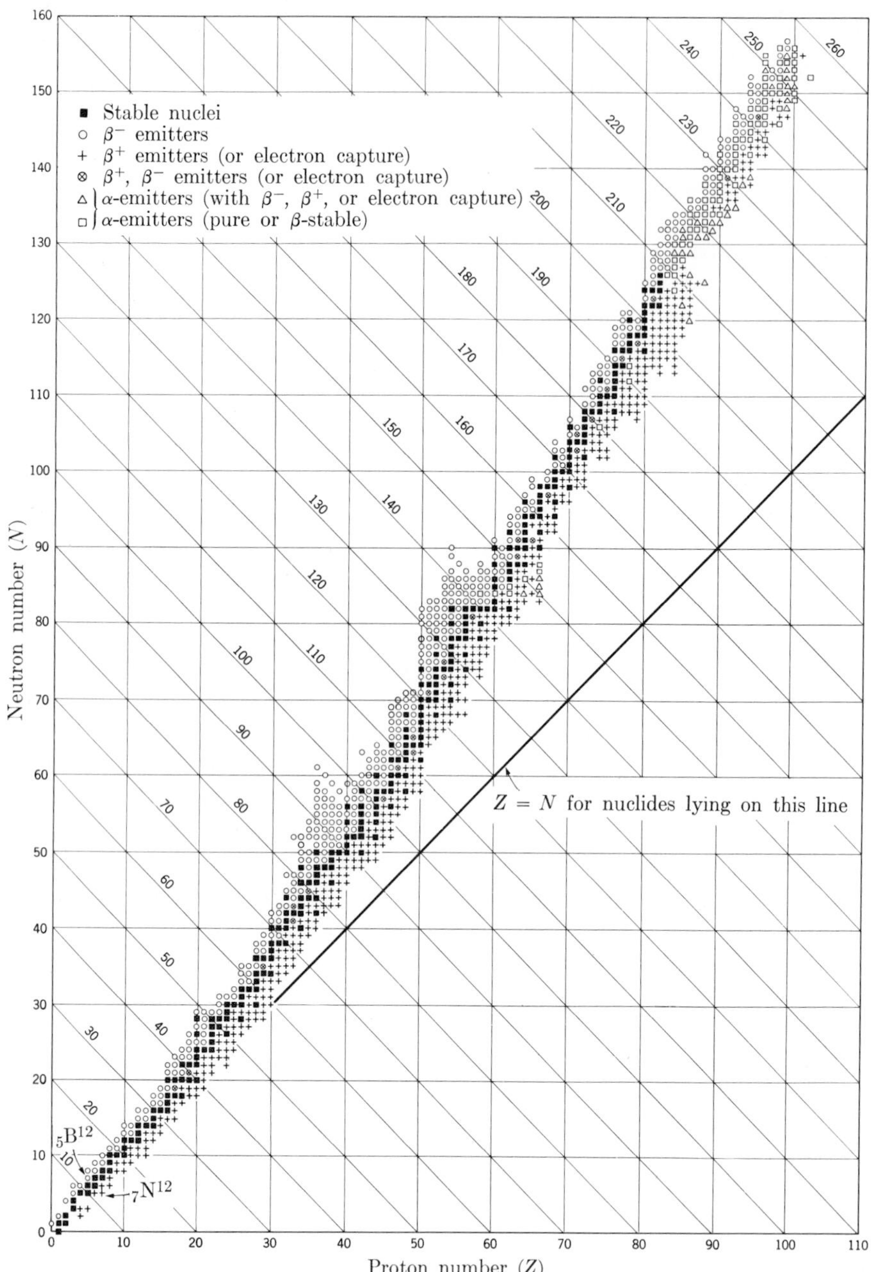

FIG. 12–1. Chart of the known nuclei. (Reprinted by permission from S. E. Liverhant, *Elementary Introduction to Nuclear Reactor Physics*. New York: Wiley, 1960)

artificially radioactive nuclides lie above or below the stable nuclides; they have either too many neutrons (not enough protons) or too few neutrons (too many protons) to be stable. A nucleus with too few protons tends to become stable by increasing the nuclear charge, while a nucleus with too many protons tends to become stable by decreasing the nuclear charge. One way by which the nuclear charge can be increased is by electron emission, as was seen during the treatment of natural radioactivity. The nuclear charge can be decreased by emission of a positron. Consequently, positron activity is usually seen in the isotopes of an element having mass numbers smaller than those of the stable isotopes, while electron activity usually occurs in isotopes with mass numbers greater than those of the stable isotopes.

The charge on the nucleus is also reduced if the nucleus captures an orbital electron. The extranuclear electrons, in the course of their motions, often approach close to the nucleus and, according to wave mechanics, may even penetrate it. The electrons which are most likely to do this are the K-electrons in atoms with a large value of Z. These electrons may be thought of as available for capture by the nucleus, especially if the latter is proton-rich. If an electron from the K-shell is captured, the process is called K-capture; less often an L-electron is captured. The vacancy in the K-shell or L-shell is usually filled by an electron from an outer shell with the emission of a K or L x-ray, respectively, characteristic of the product nucleus. In this process, no charged particle is emitted and the process can be observed only because of the x-ray emission. Orbital electron capture is sometimes accompanied by the emission of electrons from the extranuclear structure. These electrons, called *Auger electrons* after their discoverer, result from what might be described as an internal photoelectric effect. For example, the emission of a K x-ray may be replaced by the ejection of an L-electron with a kinetic energy equal to the difference between the K x-ray energy and the binding energy of the L-electron. The three types of disintegration are *isobaric transformations* because they involve changes in nuclear charge but not in mass number.

As in the case of natural radioactivity, some artificially radioactive nuclides emit γ-rays as well as electrons or positrons. About one-third of the active nuclides are simple in the sense that they give off only particles; the remainder also emit γ-rays.

The question of whether an artificial nuclide will decay by electron emission, positron emission, or orbital electron capture can be discussed in terms of the energy available for the disintegration. Consider first the case of electron emission, in which the reaction is

$$_ZX^A \rightarrow {}_{Z+1}Y^A + {}_{-1}e^0. \tag{12-1}$$

The energy balance for this reaction is given by the expression

$$Q/c^2 = M_n({}_ZX^A) - M_n({}_{Z+1}Y^A) - m, \tag{12–2}$$

where the M_n's represent the nuclear masses of the artificial nuclide and its decay product, and m is the electron mass. The last equation can be written in terms of the atomic masses if it is noted that

$$M_a({}_ZX^A) = M_n({}_ZX^A) + Zm,$$

where M_a is the atomic mass; the binding energy of the electrons is then included but leads to no difficulty because it always cancels. Then Eq. (12–2) becomes

$$\begin{aligned} Q/c^2 &= M_a({}_ZX^A) - Zm - M_a({}_{Z+1}Y^A) + (Z+1)m - m \\ &= M_a({}_ZX^A) - M_a({}_{Z+1}Y^A). \end{aligned} \tag{12–3}$$

The condition that electron emission be energetically possible is that the value of Q be positive, or

$$M_a({}_ZX^A) > M_a({}_{Z+1}Y^A), \tag{12–4}$$

i.e., the atomic mass of the artificial nuclide must be greater than that of its isobar with nuclear charge one unit greater.

In the case of positron emission, the reaction is

$${}_ZX^A \rightarrow {}_{Z-1}Y^A + {}_1e^0, \tag{12–5}$$

and the Q-value is given by

$$Q/c^2 = M_n({}_ZX^A) - M_n({}_{Z-1}Y^A) - m,$$

or,

$$\begin{aligned} Q/c^2 &= M_a({}_ZX^A) - Zm - M_a({}_{Z-1}Y^A) + (Z-1)m - m \\ &= M_a({}_ZX^A) - M_a({}_{Z-1}Y^A) - 2m. \end{aligned} \tag{12–6}$$

For positron emission to be energetically possible, the atomic mass of the artificial nuclide must be greater than the atomic mass of its isobar with nuclear charge one unit smaller by at least 2 electron masses, or 1.02 Mev,

$$M_a({}_ZX^A) > M_a({}_{Z-1}Y^A) + 2m. \tag{12–7}$$

The reaction in the case of orbital electron capture is

$${}_ZX^A + {}_{-1}e^0 \rightarrow {}_{Z-1}Y^A, \tag{12–8}$$

and the energy available is

$$Q/c^2 = M_a({}_ZX^A) - M_a({}_{Z-1}Y^A). \tag{12-9}$$

The condition that orbital electron capture be energetically possible is therefore

$$M_a({}_ZX^A) > M_a({}_{Z-1}Y^A), \tag{12-10}$$

i.e., the mass of the artificial nuclide must be greater than that of its isobar with nuclear charge one unit smaller.

Comparison of Eqs. (12–7) and (12–10) shows that the capture of an electron by a nucleus is an energetically more favorable process than the emission of a positron. But the capture of even a K-electron depends on the small probability that the electron be very close to, or "within," the nucleus. The result is that as soon as the energy available exceeds $2mc^2$ positron emission tends to happen more often than orbital electron capture. Hence, K-capture has been observed only when the condition (12–10) is satisfied, but it is not always observed when that condition is fulfilled.

There are many examples of the above rules. Thus, the nuclides C^{11}, N^{13}, O^{15}, F^{17}, Ne^{19}, Na^{21}, Mg^{23}, and Al^{25} all decay by positron emission. Each has a mass number smaller than those of the stable isotopes of the element, and is consequently proton-rich. In each case, the mass of the radioactive nuclide is greater than that of the decay product by more than 2 electron masses. For example, according to Table 11–1, C^{11} has an atomic mass of 11.014924 amu, and B^{11} has a mass of 11.012795 amu; the difference is 0.002129 amu, considerably greater than 2 electron masses, which are equal to 0.001098 amu. The nuclides C^{14}, N^{16}, F^{20}, Ne^{23}, Na^{24}, Mg^{27}, and Al^{28} decay by electron emission. Each has a mass number greater than those of the stable isotopes of the element and is therefore neutron-rich. It can be seen from Table 11–1 that, in each case, the mass of the active nuclide is greater than that of the product. For example, the mass of C^{14} is 14.0076845 amu, while that of N^{14} is 14.0075179 amu.

The isotopes of an element of intermediate atomic weight, such as iodine, show all three types of decay. Iodine has only one stable species, ${}_{53}I^{127}$, containing 53 protons and 74 neutrons. There are, however, 17 artificially radioactive isotopes, containing from 68 to 86 neutrons. Of these, I^{121} and I^{122}, with the smallest number of neutrons, decay by positron emission. The I^{123} nucleus captures an orbital electron, while I^{124} sometimes captures an orbital electron and sometimes emits a positron; I^{125} captures an orbital electron; I^{126} sometimes captures an orbital electron and sometimes emits a negative electron; I^{127} is stable; I^{128} decays by emission of an electron 95% of the time and either by orbital electron capture or positron emission 5% of the time. The eleven remaining neutron-rich iodine isotopes, from I^{129} to I^{139} inclusive, decay by the emission of an electron. The decay products of some of the iodine isotopes are also unstable and decay in turn; I^{133} decays into Xe^{133}, which is unstable and decays by electron emission into stable Cs^{133}.

There is some correlation between the radioactivity of an artificial nuclide and the transformation by means of which it is produced. Electron emission is common for activities produced by (n,γ), (n,p), (n,α), and (d,p) reactions, since these reactions decrease the charge-to-mass ratio. Examples are

$$
\begin{aligned}
&{}_{49}\mathrm{In}^{115}(\mathrm{n}, \gamma){}_{49}\mathrm{In}^{116}; && {}_{49}\mathrm{In}^{116} \rightarrow {}_{50}\mathrm{Sn}^{116} + {}_{-1}\mathrm{e}^0, && T = 13 \text{ sec},\\
&{}_{7}\mathrm{N}^{14}(\mathrm{n}, \mathrm{p}){}_{6}\mathrm{C}^{14}; && {}_{6}\mathrm{C}^{14} \rightarrow {}_{7}\mathrm{N}^{14} + {}_{-1}\mathrm{e}^0, && T = 5568 \text{ y},\\
&{}_{13}\mathrm{Al}^{27}(\mathrm{n}, \alpha){}_{11}\mathrm{Na}^{24}; && {}_{11}\mathrm{Na}^{24} \rightarrow {}_{12}\mathrm{Mg}^{24} + {}_{-1}\mathrm{e}^0, && T = 15.0 \text{ h},\\
&{}_{15}\mathrm{P}^{31}(\mathrm{d}, \mathrm{p}){}_{15}\mathrm{P}^{32}; && {}_{15}\mathrm{P}^{32} \rightarrow {}_{16}\mathrm{S}^{32} + {}_{-1}\mathrm{e}^0, && T = 14.3 \text{ d}.
\end{aligned}
$$

Positron emission is common for activities produced by (p,γ), (p,n), (α,n), (d,n), and (γ,n) reactions, which increase the charge-to-mass ratio of the nucleus. Examples are

$$
\begin{aligned}
&{}_{6}\mathrm{C}^{13}(\mathrm{p}, \gamma){}_{7}\mathrm{N}^{13}; && {}_{7}\mathrm{N}^{13} \rightarrow {}_{6}\mathrm{C}^{13} + {}_{1}\mathrm{e}^0, && T = 10 \text{ min},\\
&{}_{28}\mathrm{Ni}^{58}(\mathrm{p}, \mathrm{n}){}_{29}\mathrm{Cu}^{58}; && {}_{29}\mathrm{Cu}^{58} \rightarrow {}_{28}\mathrm{Ni}^{58} + {}_{1}\mathrm{e}^0, && T = 2.6 \text{ sec},\\
&{}_{7}\mathrm{N}^{14}(\alpha, \mathrm{n}){}_{9}\mathrm{F}^{17}; && {}_{9}\mathrm{F}^{17} \rightarrow {}_{8}\mathrm{O}^{17} + {}_{1}\mathrm{e}^0, && T = 70 \text{ sec},\\
&{}_{7}\mathrm{N}^{14}(\mathrm{d}, \mathrm{n}){}_{8}\mathrm{O}^{15}; && {}_{8}\mathrm{O}^{15} \rightarrow {}_{7}\mathrm{N}^{15} + {}_{1}\mathrm{e}^0, && T = 2.1 \text{ min},\\
&{}_{15}\mathrm{P}^{31}(\gamma, \mathrm{n}){}_{15}\mathrm{P}^{30}; && {}_{15}\mathrm{P}^{30} \rightarrow {}_{14}\mathrm{S}^{30} + {}_{1}\mathrm{e}^0, && T = 2.5 \text{ min}.
\end{aligned}
$$

The (α,p), (p,α), and (d,α) reactions usually lead to stable products. The charge and mass changes are such that the target and product nuclei are generally both within the band of stable nuclides, as in the examples

$$\mathrm{N}^{14}(\alpha,\mathrm{p})\mathrm{O}^{17}, \quad \mathrm{Al}^{27}(\alpha,\mathrm{p})\mathrm{Si}^{30}; \qquad \mathrm{F}^{19}(\mathrm{p},\alpha)\mathrm{O}^{16}, \quad \mathrm{Al}^{27}(\mathrm{p},\alpha)\mathrm{Mg}^{24};$$

and

$$\mathrm{O}^{16}(\mathrm{d},\alpha)\mathrm{N}^{14}, \quad \mathrm{Al}^{27}(\mathrm{d},\alpha)\mathrm{Mg}^{25}.$$

12–3 The transuranium elements. The most fruitful source of artificial radionuclides is the radiative capture of neutrons,[(5)] the (n,γ) reaction, which usually yields an electron-emitting product. In 1934, Fermi[(6)] suggested the possibility that the bombardment of uranium with neutrons might result in the production of elements with nuclear charge greater than 92. If U^{238} were to capture a neutron, the following reaction should occur:

$$_{92}U^{238} + {}_0n^1 \rightarrow [_{92}U^{239}] \rightarrow {}_{92}U^{239} + \gamma;$$

if the U^{239} were then to decay by electron emission, the result would be a nuclide with $Z = 93$, an isotope of a hitherto unknown element. Early experiments showed that when uranium was bombarded with neutrons, four different β-activities were detected, one of which was presumably associated with $_{92}U^{239}$. It was also considered possible that the new element with $Z = 93$ might decay by electron emission to form an isotope of another new element with $Z = 94$. Any elements which might have values of Z greater than 92 were called *transuranium elements* because they would lie beyond uranium in the periodic table. Between 1934 and 1939, many attempts were made to make and identify transuranium elements, but although some doubtless were produced, there was difficulty with their identification. The research on these elements led to the discovery of nuclear fission, and the work which followed this discovery resulted, among other things, in the systematic production and study of the transuranium elements.

In 1940, it was proved that the bombardment of uranium with slow neutrons does indeed produce a new isotope U^{239} which has a half-life of 23 min and is an electron emitter. It decays into an isotope of a new element with $Z = 93$ which, in turn, emits an electron and has a half-life of 2.3 days.(7) The element with $Z = 93$ was given the name *neptunium.* The new nuclide, Np^{239}, results from the decay

$$_{92}U^{239} \rightarrow {}_{93}Np^{239} + {}_{-1}e^0, \qquad T = 23 \text{ min}.$$

The decay of Np^{239} yields an isotope of a new element with $Z = 94$, called *plutonium*,(8,9)

$$_{93}Np^{239} \rightarrow {}_{94}Pu^{239} + {}_{-1}e^0, \qquad T = 2.3 \text{ d}.$$

This isotope of plutonium is manufactured in large amounts in nuclear reactors and can be used for the production of nuclear power or atomic bombs. It is an α-emitter with a half-life of 24,400 years. Soon after the discovery of $_{93}Np^{239}$, two other isotopes of neptunium were made. When uranium oxide is bombarded with fast deuterons from a cyclotron, the following (d,2n) reaction takes place:

$$_{92}U^{238} + {}_1H^2 \rightarrow {}_{93}Np^{238} + {}_0n^1 + {}_0n^1.$$

The product Np^{238} emits an electron to form Pu^{238}, and has a half-life of

2.0 days. When uranium is exposed to *fast* neutrons, an (n,2n) reaction occurs,

$$_{92}U^{238} + {}_0n^1 \rightarrow {}_{92}U^{237} + {}_0n^1 + {}_0n^1.$$

The isotope $_{92}U^{237}$ is an electron emitter with a half-life of 6.8 days, and decays into Np^{237}. The latter is an α-emitter with a half-life of 2.2×10^6 years, and is the longest-lived of the known transuranium nuclides. Isotopes of neptunium can also be obtained by bombarding uranium with deuterons with energies of about 100 Mev; reactions occur in which 4, 6, or even 9 neutrons are ejected. Alpha-particles can also be used. Some of the isotopes and the reactions by which they are made are

$$Np^{231}\text{: } U^{233}(d,4n); \quad U^{235}(d,6n); \quad U^{238}(d,9n),$$

$$Np^{236}\text{: } U^{235}(d,n); \quad U^{235}(\alpha,p2n); \quad U^{238}(d,4n),$$

$$Np^{238}\text{: } U^{238}(d,2n); \quad U^{238}(\alpha,p3n).$$

In all, 10 isotopes of neptunium are now known, with mass numbers from 231 to 240.

Isotopes of elements 95 (americium) and 96 (curium) can be made by bombarding Pu^{239} with neutrons. The reactions are

$$Pu^{239} + n \rightarrow Pu^{240} + \gamma,$$

$$Pu^{240} + n \rightarrow Pu^{241} + \gamma,$$

$$_{94}Pu^{241} \xrightarrow{\beta^-} {}_{95}Am^{241}.$$

At the same time, $_{96}Cm^{242}$ is formed,

$$Am^{241} + n \rightarrow Am^{242} + \gamma$$

$$_{95}Am^{242} \xrightarrow{\beta^-} {}_{96}Cm^{242}.$$

The nuclide Am^{241}, on bombardment with α-particles, gives an isotope of element 97 (berkelium),

$$_{95}Am^{241} + {}_2He^4 \rightarrow {}_{97}Bk^{243} + 2n.$$

These reactions were used in the discovery of berkelium,[(10)] and similar reactions were used in the first preparation and identification of an isotope of californium, $_{98}Cf^{244}$.[(11)] The new nuclides are identified, i.e., their Z- and A-values are determined, by separating them chemically and studying their radioactive properties together with those of their decay products. Thus, $_{97}Bk^{243}$ decays by orbital electron capture to $_{96}Cm^{243}$ which decays to Pu^{239} by α-emission; Bk^{243} also emits an α-particle (with

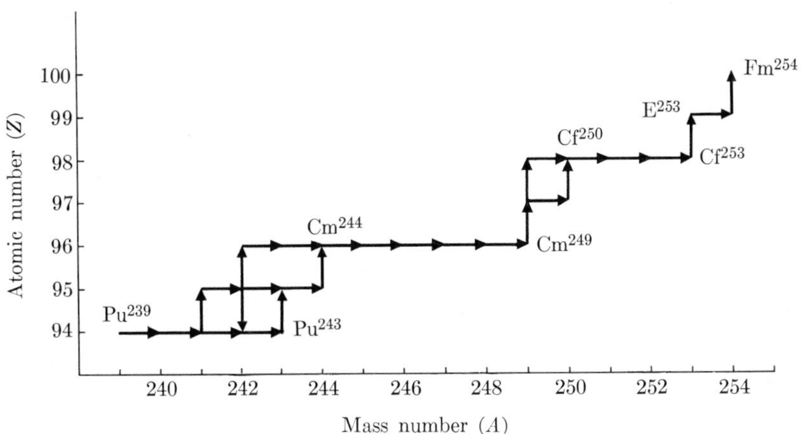

FIG. 12-2. Nuclear reaction sequences for production of heavy nuclides by intense slow neutron irradiation of Pu^{239}. (From Seaborg, *The Transuranium Elements*, gen. ref.).

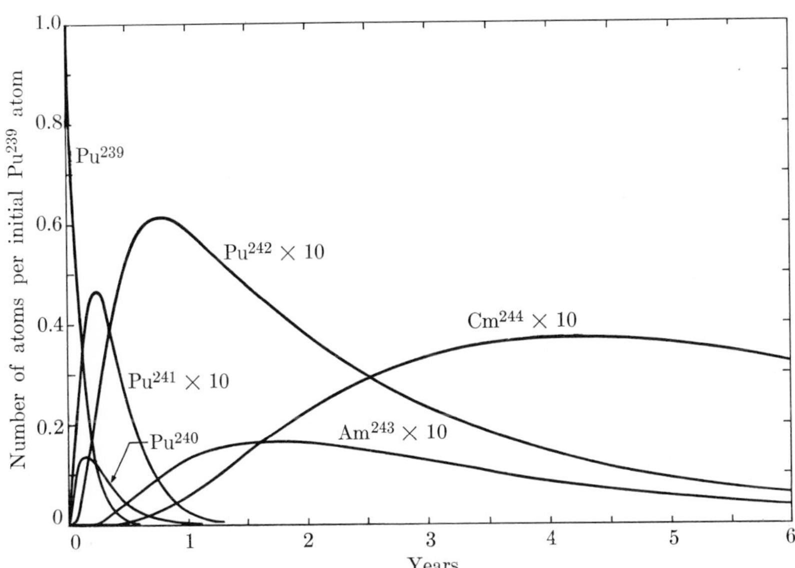

FIG. 12-3. Production of some heavy nuclides by irradiation of Pu^{239} at a flux of 3×10^{14} neutrons/cm^2/sec (from Seaborg, *The Transuranium Elements*, gen. ref.).

a branching ratio of less than one percent) to give $_{95}Am^{239}$ which captures an orbital electron to become Pu^{239}. Since the radioactive properties of Pu^{239} are known and it can be shown how the Pu^{239} is formed, the identity of $_{97}Bk^{243}$ can be established.

The bombardment of Pu^{239} with neutrons can be carried much further than has been indicated, and many transuranium nuclides which are β-stable or decay by negative β-emission can be built up through the absorption of successive neutrons. If the Pu^{239} is irradiated in a nuclear reactor with a high neutron flux, i.e., 3×10^{14} neutrons/cm^2/sec (see Section 18–4), it is possible to build up, in the course of several years, appreciable yields of nuclides which correspond to the absorption of as many as 15 neutrons. The sequences of neutron capture reactions and β-decays leading to nuclides as heavy as $_{100}Fm^{254}$ are shown in Fig. 12–2; the horizontal arrows indicate neutron capture, and the vertical arrows β-decay. Still more extensive reactions have been found to occur in a thermonuclear explosion and elements 99 and 100 were actually discovered in debris from an explosion in November 1952. Many isotopes of elements 95 to 100 have been identified as products of Pu^{239} irradiated either in a nuclear reactor or an explosion.[(12–20)]

The rate of buildup of higher nuclides can be analyzed mathematically by means of equations analogous to the Bateman equations of natural radioactivity.[(21)] Some results for the irradiation of Pu^{239} in a neutron flux of 3×10^{14} neutrons/cm^2/sec are shown in Fig. 12–3.

It is also possible to produce isotopes of transuranium elements by bombarding uranium with highly ionized heavy nuclei such as C^{12}, N^{14}, or O^{16}. It has been shown,[(22,23)] for example, that $C^{12}(6+)$ or $N^{14}(6+)$ accelerated to energies in the neighborhood of 100 Mev, can induce reactions in which the atomic number is increased by 6 or more units. Isotopes of californium, einsteinium and fermium have been produced by the bombardment of uranium with carbon, nitrogen and oxygen ions, respectively.[(24,25,26)] Among the reactions which occur are

$$_{92}U^{238} + {_6C^{12}} \rightarrow {_{98}Cf^{244}} + 6n,$$

$$_{92}U^{238} + {_7N^{14}} \rightarrow {_{99}E^{246}} + 6n,$$

$$_{92}U^{238} + {_8O^{16}} \rightarrow {_{100}Fm^{250}} + 4n.$$

An isotope of element 101 (mendelevium) has been prepared by bombarding a sample of $_{99}E^{253}$ with α-particles; the product was identified by means of its decay products as $_{101}Mv^{256}$.[(27)] An isotope of element 102 (as yet unnamed) has been made by bombarding a mixture of Cm^{244} and Cm^{246} with C^{12} atoms at energies of 60–100 Mev; analysis of the products indicates that 102^{254} was produced.[(28,29,30)] The production of a second isotope of element 102 by bombardment of a mixture of curium isotopes with C^{13} nuclei has also been reported.[31]

Work on the transuranium nuclides has so far resulted in the identification of 10 isotopes of neptunium ($Z = 93$), 15 isotopes of plutonium ($Z = 94$), 10 isotopes of americium ($Z = 95$), 13 isotopes of curium ($Z = 96$), 8 isotopes of berkelium ($Z = 97$), 11 isotopes of californium ($Z = 98$), 10 isotopes of einsteinium ($Z = 99$), 8 isotopes of fermium ($Z = 100$), one species of mendelevium ($Z = 101$) and one (or perhaps 2) species of element 102.

12–4 The artificial radionuclides: alpha-emitters. It was mentioned in the last section that some of the isotopes of the transuranium elements decay by emission of an α-particle. In 1939, only 24 α-emitters were known and these were members of the naturally occurring radioactive series. At the present time, about 150 α-decaying nuclear species are known.[32] About 50 of these are among the transuranium elements; others form part of a new radioactive series, the $4n + 1$, or neptunium, series. There are several so-called *collateral series*, and additional α-emitters are found among the neutron-deficient isotopes of bismuth ($Z = 83$), polonium ($Z = 84$), astatine ($Z = 85$), emanation ($Z = 86$), and francium ($Z = 87$).

The members of the neptunium series[33,34] and their properties are listed in Table 12–1, and the relationships between the members are

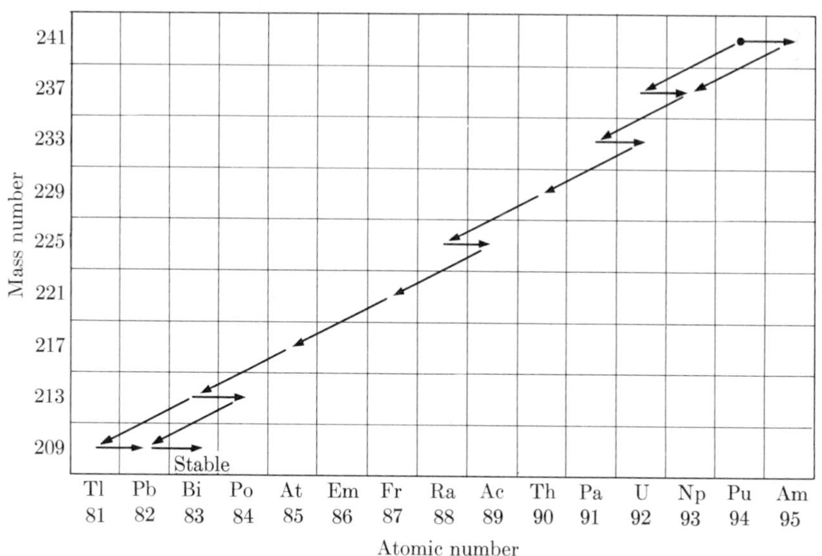

FIG. 12–4. The neptunium (4n + 1) series.

TABLE 12–1

THE NEPTUNIUM SERIES

Nuclide	Type of disintegration	Half-life	disintegration constant, sec^{-1}	particle energy, Mev
$_{94}Pu^{241}$	β, α	13.0 y	1.69×10^{-9}	α: 4.893 m β: 0.021
$_{95}Am^{241}$	α	458 y	4.80×10^{-11}	5.535 m
$_{92}U^{237}$	β	6.75 d	1.19×10^{-6}	0.248
$_{93}Np^{237}$	α	2.20×10^{6} y	9.99×10^{-15}	4.870
$_{91}Pa^{233}$	β	27.0 d	2.97×10^{-7}	0.568 m
$_{92}U^{233}$	α	1.62×10^{5} y	1.36×10^{-13}	4.816
$_{90}Th^{229}$	α	7340 y	3.0×10^{-12}	5.02 m
$_{88}Ra^{225}$	β	14.8 d	5.42×10^{-7}	0.32
$_{89}Ac^{225}$	α	10.0 d	8.02×10^{-7}	5.818 m
$_{87}Fr^{221}$	α	4.8 m	2.41×10^{-3}	6.33 m
$_{85}At^{217}$	α	0.018 s	3.85×10^{1}	7.05
$_{83}Bi^{213}$	β, α	47 m	2.46×10^{-4}	α: 5.86 β: 1.39
$_{84}Po^{213}$	α	4.2×10^{-6} s	1.65×10^{5}	8.35
$_{81}Th^{209}$	β	2.2 m	5.25×10^{-3}	1.99
$_{82}Pb^{209}$	β	3.30 h	5.84×10^{-5}	0.635
$_{83}Bi^{209}$	Stable			

shown graphically in Fig. 12–4. Like the naturally occurring radioactive series, the neptunium series shows branched disintegration near the end. The stable end product is the ordinary bismuth species of mass number 209, rather than an isotope of lead as in the uranium, thorium, and actinium series. The half-life of Np^{237}, the longest-lived member of the series, is 2.2×10^6 years. It is generally considered that the earth is about 5×10^9 years old. If it is assumed, as is probable, that neptunium was formed at the same time as the earth, then many half-lives have elapsed for this nuclide and the amounts still present would be so minute as to be beyond the possibility of detection. The absence of a naturally occurring $4n + 1$ series can therefore be understood; even if such a series did exist at one time, its members would long since have decayed to Bi^{209}.

In addition to the neptunium series, artificial α-emitters have been made which form chains collateral to the heavy radioactive series.[35] These chains include series collateral to each of the natural radioactive series, the uranium (4n + 2), thorium (4n), and actinium (4n + 3), as well as the artificial neptunium (4n + 1) series. As an example of such a series, consider the decay of $_{91}Pa^{227}$, formed by bombarding thorium with 80-Mev deuterons; the reaction is $Th^{232}(d,7n)Pa^{227}$. The nuclide Pa^{227} decays in the following way:

$$_{91}Pa^{227} \xrightarrow[T=38.3\text{ m}]{\alpha} {}_{89}Ac^{223} \xrightarrow[2.2\text{ m}]{\alpha} {}_{87}Fr^{219} \xrightarrow[0.025\text{ s}]{\alpha} {}_{85}At^{215} \xrightarrow[10^{-4}\text{ s}]{\alpha}$$

$$_{83}Bi^{211}(AcC) \xrightarrow[2.16\text{ m}]{\alpha} {}_{81}Tl^{207}(AcC'') \xrightarrow[4.79\text{ m}]{\beta} Pb^{207} \quad \text{(stable)}.$$

If this series is compared with the actinium (4n + 3) series in Fig. 10–11, it is seen that the new series runs parallel to part of the actinium series, and this is the reason for the name *collateral series.* There is a series starting with U^{227} which is also collateral to the actinium series; this chain decays by successive α-emissions as follows:

$$U^{227} \rightarrow Th^{223} \rightarrow Ra^{219} \rightarrow Em^{215} \rightarrow Po^{211} \rightarrow Pb^{207}.$$

The other collateral series are

$$Pa^{226} \rightarrow Ac^{222} \rightarrow Fr^{218} \rightarrow At^{214}(4n + 2 \text{ mass type}),$$

$$Pa^{228} \rightarrow Ac^{224} \rightarrow Fr^{220} \rightarrow At^{216}(4n \text{ mass type}),$$

$$U^{228} \rightarrow Th^{224} \rightarrow Ra^{220} \rightarrow Em^{216}(4n \text{ mass type}),$$

$$U^{229} \rightarrow Th^{223} \rightarrow Ra^{221} \rightarrow Em^{217}(4n + 1 \text{ mass type}).$$

12–5 Isotope tables and nuclide charts. The vast amount of information which has been collected about the different nuclear species, both stable and unstable, makes it necessary to have some convenient methods for recording nuclear data. Tables of isotopes are published periodically, such as those cited in the general references at the end of this chapter, and compilations of nuclear data are available like that prepared by the National Bureau of Standards. The most convenient method for quick reference is that of the *nuclide chart,* a portion of which is shown in Fig. 12–5. In this type of chart the atomic number Z is plotted against the number of neutrons, $A - Z$. Each nuclide occupies a square, and stable, naturally radioactive, and artificial nuclides are differentiated by the color or shading of the square. The symbol and mass number are shown in each case, as well as the abundance of the isotope if it is stable, the half-life, the type or types of decay, and the energy of the emitted particles

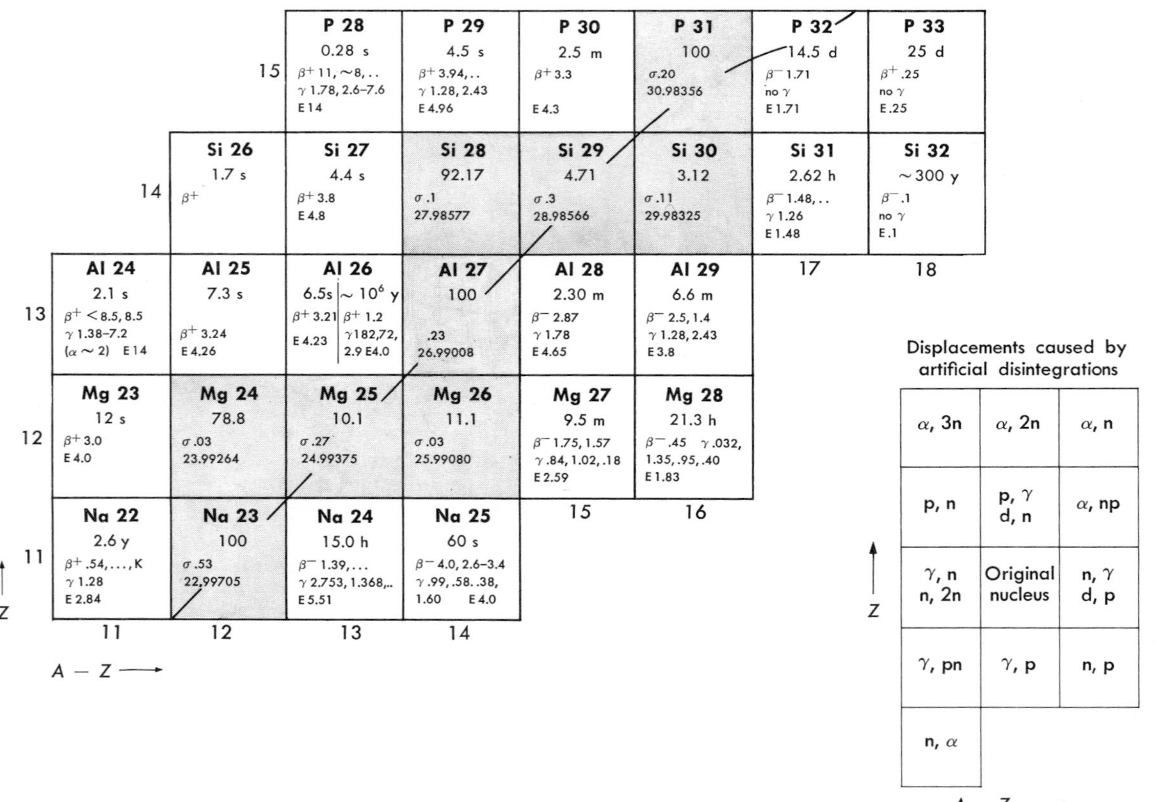

FIG. 12–5. Portion of the nuclide chart.

and γ-rays. In some charts, the atomic mass and other data are also included, such as the nuclear spin and magnetic moment. Sometimes all nuclides on the same horizontal line have the same atomic number, and nuclei with the same mass number (isobars) lie on a 45° diagonal line running from upper left to lower right. The example of Fig. 12–5 is of this type. A complete nuclide chart is inserted at the back of the book. In other charts, the isotopes lie along a diagonal line and isobars lie in the same vertical column.

The nuclide chart can be used to obtain quickly information about the products of artificial nuclear disintegrations. The displacements in charge and mass caused by various disintegrations are usually shown as in the lower right-hand corner of Fig. 12–5. It can be seen immediately that the product of an (α,n) reaction on Al^{27} is P^{30}, which is a positron emitter with a half-life of 2.5 min. The chart gives the energy of the swiftest positrons emitted, which is, in this case, 3.3 Mev; no γ-rays are emitted. Similarly, the effect of a (d,p) reaction on Al^{27} is seen immediately to be Al^{28}, which has a half-life of 2.3 min, and emits electrons with a maximum energy of 2.87 Mev, along with 1.78-Mev γ-rays.

The information shown in the squares of Fig. 12–5 has been limited to keep the figure clear. Even with the information shown, it should be apparent that the nuclide chart is very useful to anyone who needs easily available data about atomic nuclei.

The discussion of artificial radioactivity in this chapter is far from complete. Only the most common modes of decay have been discussed. One reason is that further discussion of the artificial nuclides and the information they give about nuclear stability and nuclear structure will require knowledge of the properties of the radiations from radioactive substances. These properties will be treated in the next three chapters.

References

GENERAL

M. S. Livingston and H. A. Bethe, "Nuclear Physics, C. Nuclear Dynamics, Experimental," *Revs. Mod. Phys.* **9,** 245–390 (July 1937).

Strominger, Hollander, and Seaborg, "Table of Isotopes," *Revs. Mod. Phys.* **30,** 585 (1958).

G. T. Seaborg, *The Transuranium Elements.* New Haven, Conn.: Yale University Press; Reading, Mass.: Addison-Wesley, 1958.

E. K. Hyde and G. T. Seaborg, "The Transuranium Elements," *Handbuch der Physik,* Vol. 43, p. 205, Berlin: Springer Verlag (1958).

PARTICULAR

1. I. Curie and F. Joliot, "Electrons Produced by Artificial Disintegration," *Compt. rend.* **196,** 1885 (1933); **198,** 254 (1934); *J. phys. et radium* **4,** 494 (1933).

2. C. D. Anderson, "The Positive Electron," *Science* **76,** 238 (1932); *Phys. Rev.* **43,** 491 (1933).

3. I. Curie and F. Joliot, "A New Type of Radioactivity," *Compt. rend.* **198,** 254 (1934); *Nature* **133,** 201 (1934); *J. phys. et radium* **5,** 153 (1934).

4. I. Curie and F. Joliot, "Chemical Separation of New Radioactive Elements," *Compt. rend.* **198,** 559 (1934).

5. Fermi, Amaldi, D'Agostino, Rasetti, and Segrè, "Artificial Radioactivity Produced by Neutron Bombardment," *Proc. Roy. Soc.* (London) **A146,** 483 (1934).

6. E. Fermi, "Possible Production of Elements of Atomic Number Higher than 92," *Nature* **133,** 898 (1934).

7. E. M. McMillan and P. H. Abelson, "Radioactive Element 93," *Phys. Rev.* **57,** 1185 (1940).

8. Seaborg, McMillan, Kennedy, and Wahl, "Radioactive Element 94 from Deuterons on Uranium," *Phys. Rev.* **69,** 366 (1946).

9. Kennedy, Seaborg, Segrè, and Wahl, "Properties of 94 (239)," *Phys. Rev.* **70,** 555 (1946).

10. Thompson, Ghiorso, and Seaborg, "The New Element Berkelium (Atomic Number 97)," *Phys. Rev.* **80,** 781 (1950).

11. Thompson, Street, Ghiorso, and Seaborg, "The New Element Californium (Atomic Number 98)," *Phys. Rev.* **80,** 790 (1950).

12. Thompson, Ghiorso, Harvey, and Choppin, "Transcurium Isotopes Produced in the Neutron Irradiation of Plutonium," *Phys. Rev.* **93,** 908 (1954).

13. Diamond, Magnusson, Mech, Stevens, Friedman, Studier, Fields, and Huizenga, "Identification of Californium Isotopes 249, 250, 251, and 252 from Pile-Irradiated Plutonium," *Phys. Rev.* **94,** 1083 (1954).

14. Choppin, Thompson, Ghiorso, and Harvey, "Nuclear Properties of some Isotopes of Californium, Elements 99 and 100," *Phys. Rev.* **94,** 1080 (1954).

15. Ghiorso, Thompson, Choppin, and Harvey, "New Isotopes of Americium, Berkelium, and Californium," *Phys. Rev.* **94,** 1081 (1954).

16. Studier, Fields, Diamond, Mech, Friedman, Sellers, Pyle, Stevens, Magnusson, and Huizenga, "Elements 99 and 100 from Pile-Irradiated Plutonium," *Phys. Rev.* **93,** 1428 (1954).

17. Ghiorso, Thompson, Higgins, Seaborg, Studier, Fields, Fried, Diamond, Mech, Pyle, Huizenga, Hirsch, Manning, Smith, and Spence, "New Elements 99 and 100," *Phys. Rev.* **99,** 1048 (1955).

18. Jones, Schuman, Butler, Cowper, Eastwood, and Jackson, "Isotopes of Einsteinium and Fermium Produced by Neutron Irradiation of Plutonium," *Phys. Rev.* **102,** 203 (1956).

19. Eastwood, Butler, Cabell, Jackson, Schuman, Rourke and Collins, "Isotopes of Berkelium and Californium Produced by Neutron Irradiation of Plutonium," *Phys. Rev.* **107,** 1635 (1957).

20. Fields, Studier, Diamond, Mech, Inghram, Pyle, Stevens, Fried, Manning, Ghiorso, Thompson, Higgins, and Seaborg, "Transplutonium Elements in Thermonuclear Test Debris," *Phys. Rev.* **102,** 180 (1956).

21. W. Rubinson, "The Equations of Radioactive Transformation in a Neutron Flux," *J. Chem. Phys.* **17,** 542 (1949).

22. Miller, Hamilton, Putnam, Haymond, and Rossi, "Acceleration of Stripped C^{12} and C^{13} Nuclei in the Cyclotron," *Phys. Rev.* **80,** 486 (1950).

23. Rossi, Jones, Hollander, and Hamilton, "The Acceleration of Nitrogen-14 (6+) Ions in a 60-inch Cyclotron," *Phys. Rev.* **93,** 256 (1954).

24. Ghiorso, Thompson, Street, and Seaborg, "Californium Isotopes from Bombardment of Uranium with Carbon Atoms," *Phys. Rev.* **81,** 154 (1951).

25. Ghiorso, Rossi, Harvey, and Thompson, "Reactions of U^{238} with Cyclotron-Produced Nitrogen Ions," *Phys. Rev.* **93,** 257 (1954).

26. Atterling, Forsling, Holm, Melander, and Åström, "Element 100 Produced by Means of Cyclotron-Accelerated Oxygen Ions," *Phys. Rev.* **95,** 585 (1954).

27. Ghiorso, Harvey, Choppin, Thompson, and Seaborg, "New Element Mendelevium, Atomic Number 101," *Phys. Rev.* **98,** 1518 (1955).

28. Ghiorso, Sikkeland, Walton, and Seaborg, "Element No. 102," *Phys. Rev. Letters.* **1,** 18 (1958).

29. A. Ghiorso and T. Sikkeland, "Heavy Ion Reactions with Heavy Elements," *Proceedings of the Second International Conference on the Peaceful Uses of Atomic Energy,* Geneva, 1958, P/2440, **14,** 158 (1958).

30. G. N. Flerov, "Heavy Ion Reactions," see Ref. 29, P/2299, **14,** 151 (1958).

31. Fields, Friedman, Milsted, Atterling, Forsling, Holm and Åström, "On the Production of Element 102," *Phys. Rev.* **107,** 1460 (1957); *Arkiv. f. Physik.* **15,** 225 (1959).

32. Strominger, Hollander, and Seaborg, "Table of Isotopes," *Revs. Mod. Phys.* **30,** 585 (1958).

33. Hagemann, Katzin, Studier, Ghiorso, and Seaborg, "The $(4n + 1)$ Radioactive Series: The Decay Product of U^{233}," *Phys. Rev.* **72,** 252 (1947).

34. English, Cranshaw, Demers, Harvey, Hincks, Jelley, and May, "The $(4n + 1)$ Radioactive Series," *Phys. Rev.* **72,** 253 (1947).

35. Meinke, Ghiorso, and Seaborg, "Artificial Chains Collateral to the Heavy Radioactive Families," *Phys. Rev.* **81,** 782 (1951); *Phys. Rev.* **85,** 429 (1952).

Problems

1. Deduce, from the data in the table of stable nuclides (Table 9–1), which of the following nuclides are electron emitters and which are positron emitters: Ga^{73}, Nb^{96}, Cs^{127}, Ir^{197}, Au^{198}, Br^{78}, V^{48}, Sc^{47}, Ag^{110}, Xe^{137}, Xe^{123}, Zn^{63}. Check your results by comparison with a nuclide chart.

2. Show, from the masses involved (Table 11–1), which of the following nuclides are stable, which are electron emitters, and which are positron emitters: Na^{23}, P^{32}, Si^{31}, Mg^{27}, Na^{22}, F^{18}, Be^{10}, He^{6}. Check your results by comparison with a nuclide chart.

3. Calculate the Q-value in Mev for each of the following nuclear changes: $He^6(\beta^-)Li^6$; $C^{14}(\beta^-)N^{14}$; $N^{13}(\beta^+)C^{13}$; $F^{18}(\beta^+)O^{18}$; $F^{20}(\beta^-)Ne^{20}$; $Na^{22}(\beta^+)Ne^{22}$; $Na^{24}(\beta^-)Mg^{24}$; $Al^{28}(\beta^-)Si^{28}$.

4. Derive a condition for the occurrence of α-decay analogous to those for electron emission and positron emission.

5. A sample of manganese is bombarded for 20 hr with deuterons in a cyclotron under conditions such that 5×10^8 atoms of Mn^{56} are formed per second as a result of the $Mn^{55}(d,p)Mn^{56}$ reaction. The Mn^{56} is an electron emitter with a half-life of 2.58 hr. Plot the number of Mn^{56} atoms present in the sample as a function of time, from the time the cyclotron is turned on until 40 hr later. What would be the number of atoms of Mn^{56} if secular equilibrium were attained? What fraction of this number is actually reached?

6. A sample of gold is exposed to a neutron beam with an intensity such that 10^{10} neutrons are absorbed per second because of the reaction $Au^{197}(n,\gamma)Au^{198}$. The nuclide Au^{198} decays by electron emission with a half-life of 2.70 days. How many atoms of Au^{198} will be present after 100 hr? After 10 days? How many atoms of Hg^{198} will be present at these times if it is assumed that the neutron beam does not affect this nuclide?

7. Under the conditions of Problem 6, how long would the irradiation have to be continued until the number of Au^{198} atoms reaches 95% of the value at secular equilibrium? Suppose that the sample is exposed for this length of time and then removed from the neutron beam. Plot the value of the number of atoms of Au^{198} during the period of the irradiation, and for 15 days thereafter.

8. A sample of iron is bombarded with deuterons in a cyclotron, and a radionuclide with a half-life of 46 days is obtained. Chemical analysis shows that the nuclide is an isotope of iron. When a cobalt target is bombarded with neutrons of moderate energy, the 46-day activity is observed again. Which isotope of iron is the nuclide?

9. When a zinc target is bombarded with α-particles, a new nuclide is formed as the result of an (α,n) reaction. This nuclide is a positron emitter with a half-life of 1.65 days. The same activity is found as a result of a (d,2n) reaction on gallium, and as a result of an (n,2n) reaction on germanium. The (d,p) reaction on the germanium isotope of mass number 70 gives a nuclide with a half-life of 11.4 days. Identify the unknown nuclide.

10. The following atomic masses are given.

Fluorine	Neon	Sodium	A
20.006341	19.998765	20.015236	20
21.006840	21.000494	21.004281	21

Which of these nuclides should decay by β^--emission? By β^+-emission? By orbital electron capture? Explain your answers and compare them with data in the "Table of Isotopes" of Strominger, Hollander, and Seaborg.

11. The following atomic masses are given.

Sulfur	Chlorine	Argon	A
34.98014	34.97996	34.98572	35
35.97844	35.97974	35.97892	36

Which of these nuclides should decay by β^--emission? By β^+-emission? By orbital electron capture? Explain your answers, and compare them with data in the "Table of Isotopes" of Strominger, Hollander, and Seaborg.

12. The following atomic masses are given: A^{40}: 39.975050; K^{40}: 39.976653; Ca^{40}: 39.975230; Sc^{40}: 39.990250. Which of these nuclides would you expect to be radioactive and how would you expect them to decay?

13. The nuclide Cu^{64} with a half-life of 12.8 hr decays by electron emission (39%), positron emission (19%) and orbital electron capture (42%). Calculate the partial disintegration constant and half-life for each mode of decay.

14. One gram of potassium emits 29 β^--particles/sec owing to the decay of K^{40}, which has an abundance ratio of 0.012 atom percent. Gamma-rays are also emitted, and the ratio of gammas to betas is 0.12. The γ-rays follow electron capture in K^{40}, and one photon is emitted for each orbital capture. What is the half-life of K^{40}?

15. Explain, with the aid of Q-values, why He^5, Li^5, and Be^8 are not found in nature.

16. Show that Li^6 must be stable by considering all the possible ways into which it might be imagined to split.

CHAPTER 13

ALPHA-DECAY

The properties of the radiations from radioactive substances have been studied diligently since the discovery of radioactivity. The first interest in these radiations was in connection with the series of transformations of uranium, thorium, and actinium. This interest has been extended to the information that the radiations give about the nucleus and the energy changes involved in its transformations. Accurate measurements of the energies of the radiations emitted by the natural radionuclides led to the idea of nuclear energy states analogous to atomic energy states, and the study of the radiations emitted by both the natural and artificial radionuclides has resulted in the accumulation of a large amount of information about nuclear levels. Theories of the emission of α-, β-, and γ-rays have been developed, and the combination of the experimental and theoretical knowledge of these processes forms one of the great branches of nuclear physics.

13–1 The velocity and energy of alpha-particles. The determination of the velocity and energy of α-particles will be discussed in some detail for several reasons. First, the accurate measurement of α-particle energies made it possible to determine energies which differ only by small amounts, and this led to the discovery that some radionuclides actually emit a *spectrum* of α-particles. Second, knowledge of the energies of the components of α-spectra makes it possible to assign certain nuclear energy levels with confidence. Third, the methods for determining the energies of α-particles are also used for protons and deuterons. These three charged particles are involved in many artificial disintegrations, and the accurate measurement of their energies yields accurate Q-values; from the Q-values, it is possible to determine nuclear masses and nuclear energy levels. Fourth, accurate values of α-particle energies are needed in the development and use of the theory of α-decay.

The method that gives the most precise results for the velocity and energy of α-particles depends on the measurement of the deflection of the paths of the particles in a magnetic field.[1,2,3] When a charged particle moves in a magnetic field, its orbit is a circle whose radius is determined by the relation

$$Hqv = \frac{Mv^2}{r}, \tag{13–1}$$

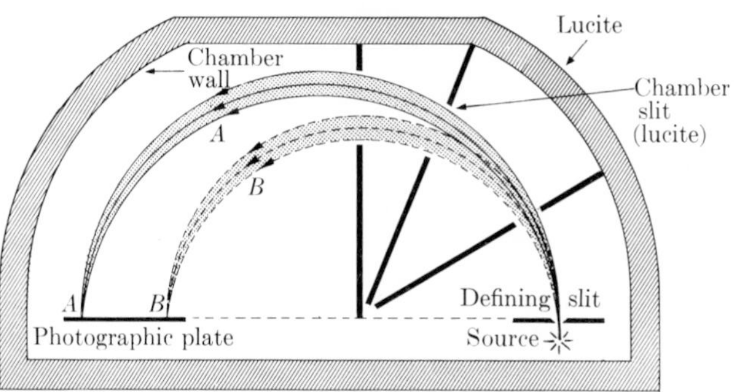

FIG. 13-1. Schematic diagram of the deflection chamber of a magnetic spectrograph for α-particles.

where H is the field strength, q and M are the charge and mass of the particle, respectively, and r is the radius of the orbit. Equation (13-1) may also be written

$$v = \frac{q}{M} Hr. \tag{13-2}$$

The velocity can be determined if the strength of the magnetic field is known and if the radius of the orbit is measured, since the value of the charge-to-mass ratio is well known.

A schematic diagram of an apparatus based on this principle is shown in Fig. 13-1. The α-particles from a radioactive source emerge in a narrow beam through a defining slit. A magnetic field of known strength, acting in a direction perpendicular to the plane of the diagram, bends the α-particles through an angle of 180°. The chamber slits help to reduce the scattering of α-particles from the top, bottom, and walls of the chamber. Particles with the same velocity have semicircular paths of the same radius; they may be detected with a photographic plate or with counters, and the radius of the path measured. In the figure, the paths of two groups of particles with different velocities are shown. In most instruments of this type, the α-particles can be bent into a semicircle of 40 or 50 cm maximum radius. The instrument is called a *magnetic spectrograph* and is designed on the principle of semicircular magnetic focusing.

The velocity v is obtained in centimeters per second when H is expressed in gauss, r in centimeters, and q/M in emu per gram. The charge q is, in emu,

$$q = \frac{2 \times 4.8029 \times 10^{-10}\ \text{esu}}{2.9979 \times 10^{10}\ \text{cm/sec}} = 3.2043 \times 10^{-20}\ \text{emu}.$$

The mass of the α-particle is equal to the atomic mass of the helium atom

TABLE 13–1

VELOCITIES AND ENERGIES OF ALPHA-PARTICLES OBTAINED BY THE MAGNETIC DEFLECTION METHOD[(2b)]

Radionuclide	Magnetic deflection, Hr: gauss-cm $\times 10^{-5}$	Velocity, cm/sec $\times 10^{-9}$	Energy, Mev
AcC(Bi^{211})	3.7067 $\pm$ 0.003	1.7846	6.620 $\pm$ 0.013
ThC(Bi^{212})	3.54232 $\pm$ 0.0008	1.7056	6.0466 $\pm$ 0.0027
RaF(Po^{210})	3.31649 $\pm$ 0.0008	1.5972	5.3007 $\pm$ 0.0026
ThC′(Po^{212})	4.26934 $\pm$ 0.0009	2.0514	8.7801 $\pm$ 0.004
Tn(Em^{220})	3.6108 $\pm$ 0.0003	1.7373	6.2823 $\pm$ 0.0013
Rn(Em^{222})	3.37401 $\pm$ 0.00020	1.6247	5.4861 $\pm$ 0.0007
ThX(Ra^{224})	3.4336 $\pm$ 0.0003	1.6533	5.6814 $\pm$ 0.0011
Ra(Ra^{226})	3.1490 $\pm$ 0.0016	1.5167	4.779 $\pm$ 0.005
RdTh(Th^{228})	3.3544 $\pm$ 0.0010	1.6154	5.4226 $\pm$ 0.003
Io(Th^{230})	3.118 $\pm$ 0.003	1.5018	4.685 $\pm$ 0.010

less two electron masses, or

$$\begin{aligned} M &= 4.003873 - 0.001098 = 4.00278 \text{ amu} \\ &= 4.00278 \times 1.6596 \times 10^{-24} = 6.6430 \times 10^{-24} \text{ gm.} \end{aligned}$$

Then

$$\frac{q}{M} = 4823.5 \text{ emu/gm,}$$

and

$$v \text{ (cm/sec)} = 4823.5\, Hr. \tag{13–3}$$

For α-particles from the natural radionuclides, the magnetic field strength used is generally of the order of 10,000 gauss, and the values of Hr are in the range 300,000 to 500,000 gauss-cm. The velocities, which can be conveniently measured by this method, vary from about 1.6×10^9 cm/sec to about 2.2×10^9 cm/sec. At these velocities, the relativistic mass correction is small and may often be neglected. In this case, the kinetic energy E may be taken equal to $\frac{1}{2}Mv^2$ and is given by

$$\begin{aligned} E = \tfrac{1}{2}Mv^2 &= \tfrac{1}{2}(6.643)10^{-24}v^2 \text{ erg} \\ &= 2.074 \times 10^{-18}v^2 \text{ Mev,} \end{aligned} \tag{13–4}$$

where v is given by Eq. (13–3). When the relativity correction must be taken into account, as in highly accurate work, Eqs. (13–3) and (13–4) are replaced by the more complicated formulas

$$v = Hr\frac{q}{M_0}\sqrt{1 - \frac{v^2}{c^2}}, \tag{13–5}$$

and

$$T = M_0c^2\left[\frac{1}{\sqrt{1 - (v^2/c^2)}} - 1\right], \tag{13–6}$$

where T represents the relativistic kinetic energy.

When the value of Hr is known, the velocity is obtained from Eq. (13–5); with the known value of v, the kinetic energy is then given by Eq. (13–6). Some typical results for natural α-emitters, from a critical survey,[2b] are listed in Table 13–1. The uncertainty in the measured value of the magnetic deflection is small enough so that the relativistic correction must be made. The uncertainty in the velocity is one unit in the fourth decimal place. If the relativistic correction were not made, the effect would be to reduce the energy by $\frac{1}{4}$ to $\frac{1}{2}$%, or 0.015 Mev in the case of ThX and 0.039 Mev in the case of ThC′.

Although the magnetic deflection method gives highly accurate results, it requires samples with a relatively large amount of activity, and cannot be used successfully with substances such as U^{238}, U^{235}, or Th^{232}. It is relatively expensive, and more convenient (if less accurate) methods of determining α-particle energies have also been developed.

13–2 The absorption of alpha-particles: range, ionization, and stopping power. The energies of charged particles, including α-particles, can be determined from measurements of their absorption by matter.[4,5] Before the methods and the results are discussed, some of the features of the absorption of α-particles will be treated.

Alpha-particles are easily absorbed; those emitted in radioactive disintegrations can generally be absorbed by a sheet of paper, by an aluminum foil 0.004 cm thick, or by several centimeters of air. If the particles emitted by a source in air are counted by counting the number of scintillations on a zinc sulfide screen, it is found that their number stays practically constant up to a certain distance R from the source, and then drops rapidly to zero. This distance R is called the *range* of the particles, and is related to the initial energy of the particles. If measured ranges are plotted against energies determined by magnetic deflection methods, the resulting range-energy curve can be used to find unknown energies from measured ranges. It is usually easier and cheaper to measure ranges than energies, and this method is used often.

Precise measurements of ranges in air can be made with an apparatus[5] like that shown in Fig. 13–2. The source is placed on a movable block

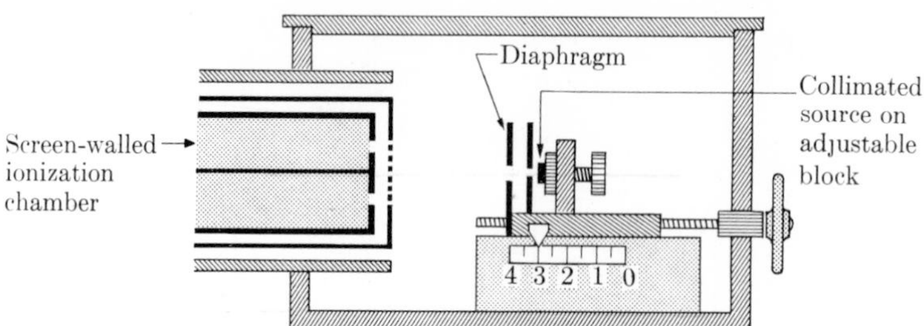

FIG. 13–2. An apparatus for precise measurements of the range of α-particles (Holloway and Livingston[5]).

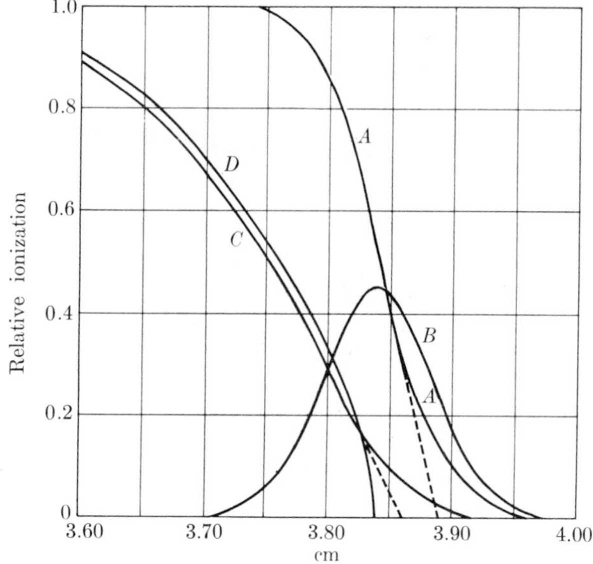

FIG. 13–3. Range curves for the α-particles from Po^{212} (Holloway and Livingston[5]). Curve *A:* Number-distance curve with the extrapolated range $R_e = 3.897$ cm. Curve *B:* Differential range curve, with the mean range $\overline{R} = 3.842$ cm. Curve *C:* End of a specific ionization curve with the ionization extrapolated range $R_i = 3.870$ cm. Curve *D:* End of a specific ionization curve for a single particle of mean range.

whose distance from a detector can be varied. A narrow beam of particles emerges through a collimating slit, passes through a known thickness of air, and reaches the detector. The latter is a thin, screen-walled ionization chamber 1 to 2 mm deep. Ions are formed in pulses in the chamber when individual α-particles pass through it. The voltage pulses induced on the chamber electrode are amplified electronically and counted. The counting rate is then determined as a function of the distance between source and

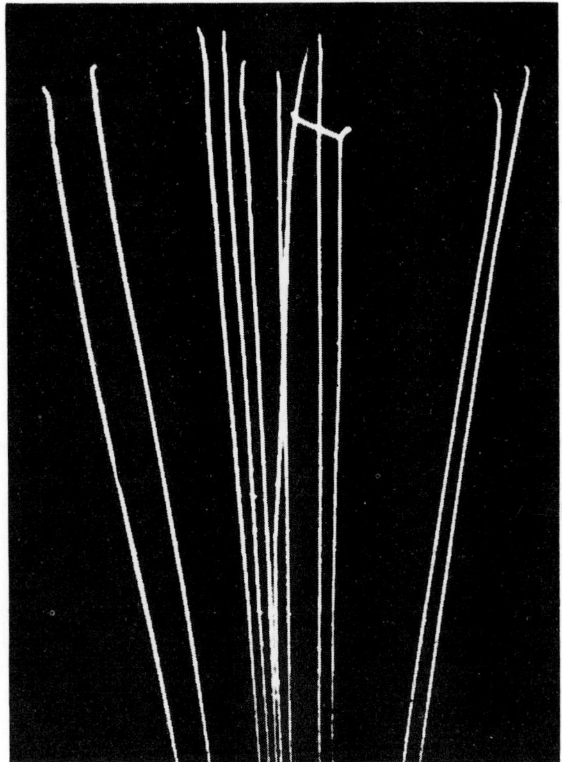

Fig. 13-4. A cloud chamber photograph showing the straggling of α-particles from Thorium C′. The last three centimeters of range are shown. [Reprinted by permission from Rutherford, Chadwick, and Ellis, *Radiations from Radioactive Substances*. New York: Macmillan, 1930.]

detector. The results of an experiment with α-particles from polonium are shown in Fig. 13-3. The ordinate is the relative number of particles, and the abscissa is the distance from the source. Curve A shows the fraction of particles in the beam detected at various distances from the source; only the results near the end of the path are shown. All of the particles which pass through the slit are counted until the detector is about 3.75 cm from the source; the fraction detected falls rapidly to 0.2 at 3.88 cm, and then decreases somewhat more slowly to zero. A quantity called the *extrapolated range*, R_e, is obtained by drawing the tangent to the curve at its inflection point and noting where the tangent crosses the distance axis. The dotted line in the figure is the tangent and gives the value 3.897 cm for the extrapolated range.

If the derivative of the number-distance curve (curve A) is computed at different distances from the source and then plotted against the distance, a *differential range curve* (curve B) is obtained. This curve gives the relative number of particles stopping at a given distance as a function of the distance from the source; the unit of the ordinate is so chosen that the area under the differential range curve is unity and all the particles are accounted for. The maximum ordinate of the differential range curve occurs at a value of the abscissa which is called the *mean range*, $\overline{R}$, defined so that half the particle track lengths exceed it, while half are shorter. In the case considered, the mean range is 3.842 cm. The results show that the track lengths of the particles in the beam are not all the same, but vary around an average value. This effect is called *straggling*, and is illustrated in Fig. 13–4, which is a cloud chamber photograph of α-particles from ThC′.

Alpha-particles lose a large fraction of their energy by causing ionization along their paths. The extent of the ionization caused by an α-particle depends on the number of molecules it hits along its path and on the way in which it hits them. Some particles hit more and others hit less than the average number of molecules in passing through a centimeter of air. Hence, the actual distance from the source at which their energy is completely used up is somewhat different for different particles, giving rise to straggling. Because of straggling, the actual range of an α-particle is not definite, and to avoid this indefiniteness either the extrapolated or the mean range is used. The values of the ranges are usually given for air at 15°C and 760 mm Hg.

The ionization caused by a beam of α-particles can be measured, and is related to the energy and range. An electron and the positive ion which results from its removal from an atom form an *ion pair*, and the intensity of the ionization caused by the particles is expressed by the *specific ionization*, defined as the number of ion pairs formed per millimeter of beam path. The apparatus of Fig. 13–2 can be used to measure the relative specific ionization produced by a beam of α-particles at different distances from the source. The amplifier can be designed so that the voltage height of its output pulse is very nearly proportional to the number of ion pairs formed in the chamber. Specific ionization-distance curves are shown in Fig. 13–5 for the particles from RaF (Po^{210}) and RaC′ (Po^{214}). The same type of curve is obtained for any gas and for any group of monoenergetic particles even though the magnitudes of the ionization and of the range vary considerably. As the distance of the α-particles from the source increases, the relative specific ionization increases, at first quite slowly and then more rapidly, reaches a maximum, and then drops sharply to zero. The effect of straggling is seen near the end of the range, and is shown clearly in curve C of Fig. 13–3. The *ionization extrapolated range* R_i is defined as the value of the abscissa at which the tangent to the curve at its inflection point crosses the horizontal axis; the value of R_i for Po^{210} was found to be 3.870 cm.

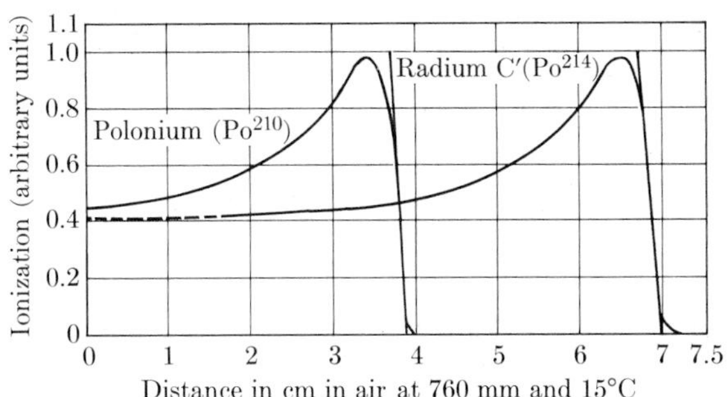

FIG. 13–5. Specific ionization of a beam of α-particles as a function of distance from the source. [Reprinted by permission from Rutherford, Chadwick, and Ellis, *Radiations from Radioactive Substances.* New York: Macmillan, 1930.]

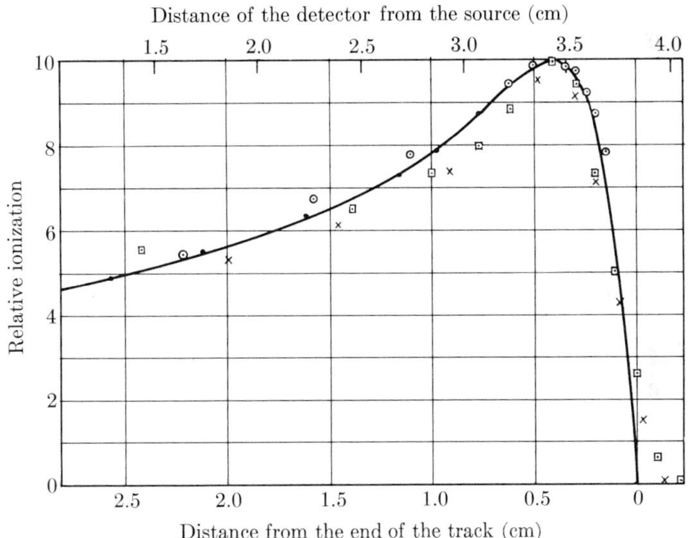

FIG. 13–6. Specific ionization of a single particle of mean range as a function of distance from the source and from the end of the track (Holloway and Livingston[5]).

It is possible to correct the relative ionization-distance curve of a beam of α-particles for the effect of straggling, and to obtain a specific ionization curve for a single particle of mean range; a curve of this kind is shown in Fig. 13–6. The specific ionization is usually measured from the end of the track, and the bottom abscissa scale gives the distance from the end of the track, while the top scale gives the distance of the detector from the source. A specific ionization curve for a single α-particle can be obtained directly by studying the photographic density of tracks in a cloud chamber;[6] the results for the α-particles from Po^{210} are plotted as in Fig. 13–6.

The total number of ions produced in a gas by the complete absorption of an α-particle of known energy can be estimated. An α-particle from RaC′ (Po^{214}) whose energy is 7.68 Mev and whose ionization extrapolated range is 6.95 cm produces a total of 2.2×10^5 ion pairs in air at 15°C and 760 mm of Hg before being stopped. It follows that the particle loses about 35 ev, on the average, for each ion pair formed. The ionization per millimeter of path for the average α-particle from RaC′ has been deduced from the ionization curve of Fig. 13–5. When the particle is just starting it produces about 2200 ion pairs/mm. The ionization increases very slowly at first as the α-particle loses energy, and is 2700 ion pairs/mm 3.0 cm from the end of the range. It then increases more and more rapidly and reaches a maximum of about 7000 ion pairs/mm when the particle is 4 or 5 mm from the end of its range. The specific ionization finally decreases very rapidly in the last few millimeters of its range. The shape of the ionization curve depends on the change in speed of an α-particle as it traverses its path. In producing ion pairs, the particle loses energy and its speed decreases. When it moves more slowly it spends more time in the neighborhood of the air molecules it encounters, and the probability of producing ion pairs increases. This effect accounts for the increase in specific ionization as the particle moves farther from the source.

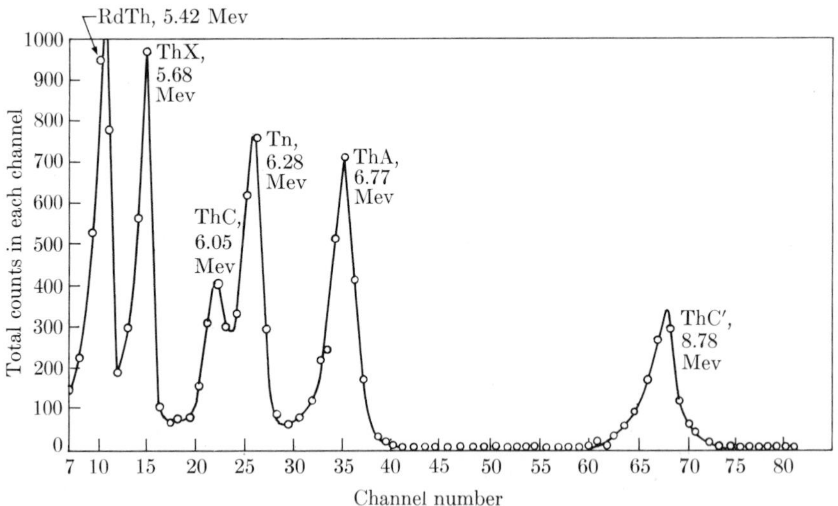

FIG. 13–7. Analysis of α-emitters in a multichannel pulse analyzer: radiothorium and its descendants. (Reprinted by permission of the United States Atomic Energy Commission.[4])

Finally, electrons are captured by the α-particle and a neutral helium atom is formed which can no longer cause ionization.

The total ionization caused by an α-particle can also be used to determine the energy of the particle. If the entire path of an α-particle is contained in an ion chamber and the total number of ion pairs formed is measured, the energy of the particle can be determined, since the average number of electron volts per ion pair is known for various gases.[7,8,11] Instead of measuring the amount of charge collected as a result of the ionization caused by each α-particle, the voltage pulses formed are analyzed by means of ingeniously designed electronic circuits called *multichannel, differential pulse height selectors.*[9,10] The output of the ionization chamber is amplified and then fed into a number of different pulse-selector circuits. Each circuit is set to select pulses of a different height, so that pulses of a number of different sizes are recorded simultaneously. If there are enough of these channels, the whole range of pulse heights may be recorded at once. A plot of energy against channel number is made by using samples of α-emitters whose energies are known from magnetic deflection methods, and this plot serves as a calibration. An instrument of this kind,[10] with 48 channels, has been very useful in a great deal of the work on artificial α-emitting radionuclides discussed in the last chapter. An example of the analysis with this instrument of the ionization produced in argon by a sample containing RdTh (Th^{228}) and its descendants is shown in Fig. 13-7. The channel number represents the size of the pulse (which depends on the energy); the curve was taken in two stages, RdTh to ThA in the first stage and ThC′ in the second stage. The number of α-particles emitted by each nuclide is proportional to the area under the peak.

The discussion of the absorption of α-particles has so far been limited to absorption by air. Other gases can be used and, in fact, studies of the ionization produced in argon have recently[11] helped to clarify some problems. Photographic emulsions have always been useful in the study of the radiations from radioactive substances. Until the development of electrical counting instruments, the emulsion was one of the most important tools in the field of radioactivity. Rapid advances in the design of electronic instruments pushed photographic emulsions into a secondary position, but recent improvements in the preparation of emulsions have made them once again important tools in radioactive measurements.[12] Measurements of track lengths in emulsions can be correlated with particle energies, giving range-energy curves for different charged particles.[13] Because of their usefulness in nuclear physics, these emulsions are called *nuclear emulsions.*

Range measurements can also be made with solid foils as absorbers. Thin, uniform foils are placed over an α-emitting sample, and the thickness needed to absorb the α-particles completely is a measure of the range.

Mica, aluminum, or gold foils are used commonly because they can be prepared easily in different thicknesses with reasonable uniformity. The range of α-particles in such materials is very small; the particles from RaC′ which have an extrapolated range of 6.953 cm in air have extrapolated ranges of 0.0036 cm in mica, 0.00406 cm in aluminum, and 0.00140 cm in gold.

Another quantity of great value in the treatment of the absorption of charged particles by matter is the *stopping power*, defined as the energy lost by the particle per unit path in the substance,

$$S(E) = -\frac{dE}{dx}, \tag{13-7}$$

where E is the classical kinetic energy. The stopping power varies with the energy of the particle, and the range of the particle is given by

$$R = \int_0^R dx = \int_0^{E_0} \frac{dE}{S(E)}, \tag{13-8}$$

where E_0 is the initial kinetic energy.

The stopping power $S(E)$ of a substance can be determined experimentally by measuring (e.g., by magnetic deflection) the energy of the particles which have gone through a certain thickness of the substance. When the energy loss is found in this way for different initial velocities, the range in the substance can be deduced from Eq. (13–8) as a function of the initial energy. If the range is known as a function of energy, the stopping power can be obtained from the relation

$$\frac{dR}{dE} = \frac{1}{S(E)}. \tag{13-9}$$

One of the reasons for the importance of the stopping power is that it can be calculated theoretically both from classical mechanics and quantum mechanics.(14,15) The theory of the stopping power depends on knowledge of the behavior of electrons in atoms. It makes predictions which are in good agreement with the results of experiments on the ionization produced by fast charged particles, on the energy loss, and on the range. The theory provides an understanding of the empirical information concerning the absorption of α-particles by matter, and there is a sound theoretical, as well as empirical, basis for the determination of charged particle energies from range and ionization measurements.

Although a detailed treatment of the stopping power is beyond the scope of this book, the usefulness of the concept can be illustrated by discussing one of its applications. The energy lost by a nonrelativistic charged particle per unit length of its path in a given substance can be expressed in the following form derived from theory,

$$S = -\frac{dE}{dx} = \frac{4\pi z^2 e^4 N}{m_e v^2} Z \ln\left(\frac{2m_e v^2}{I}\right), \tag{13–10}$$

where ze and v are the charge and the speed of the particle, respectively,
m_e is the mass of an electron,
Z is the atomic number of the substance,
N is the number of atoms per cubic centimeter of the substance, and
I is a quantity, called the average excitation potential of an atom of the substance, which must be obtained experimentally.

For an α-particle, $z = 2$, and Eq. (13–10) may be rewritten

$$S' = -\frac{m_e v^2}{16\pi e^4 N}\frac{dE}{dx} = Z \ln\left(\frac{2m_e v^2}{I}\right). \tag{13–11}$$

The quantity on the left, called S' for convenience, is proportional to the stopping power; it can be calculated and measured for different values of the atomic number Z of the absorbing substance. The quantity I depends, among other things, on Z, so that a graph of S' against Z should give a curve which is not quite a straight line. The agreement between theory and experiment is very good, as can be seen from Fig. 13–8; the solid curve represents the theoretical prediction, the circles represent experimental results. Calculations of ranges from Eqs. (13–8) and (13–10) give values in good agreement with experiment.

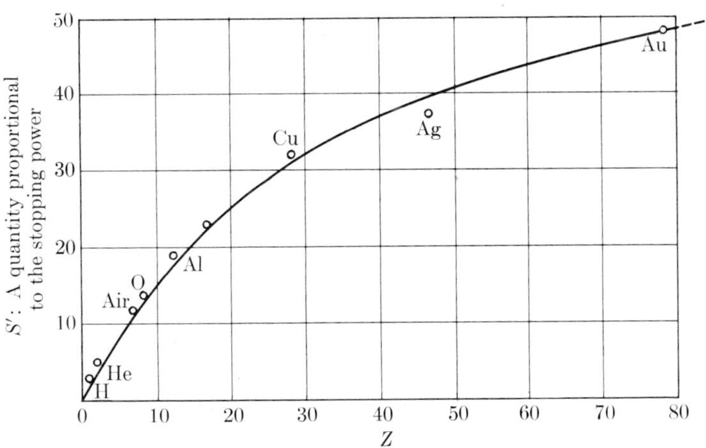

FIG. 13–8. The variation of the stopping power with atomic number.

Since the stopping power varies with the velocity of the particle, it is convenient to consider the ratio of the stopping power to that of a standard substance, a ratio approximately independent of the velocity. Air at 15°C and 760 mm Hg is usually taken as the standard. The *relative stopping power* of a substance, defined as the ratio of the range in air to the range in the substance, is often used,

$$\text{Relative stopping power} = \frac{\text{Range of } \alpha\text{-particle in air}}{\text{Range of } \alpha\text{-particle in substance}}. \tag{13-12}$$

For some purposes it is useful to express the stopping power of a substance in an alternative form, the *equivalent thickness in units of mg/cm²* defined by the relationship

$$\text{Equivalent thickness in mg/cm}^2 = \text{Range} \times \text{Density} \times 1000. \tag{13-13}$$

This quantity gives the mass per unit area, or the thickness, of material needed to absorb the α-particles. The thickness that is equivalent in stopping power to one centimeter of air is obtained by dividing the equivalent thickness, defined by Eq. (13–13), by the range of the α-particles in air. Then, from Eqs. (13–10) and (13–11),

$$\text{Thickness in mg/cm}^2 \text{ equivalent to 1 cm of air} = \frac{\text{Density} \times 1000}{\text{Relative stopping power}}. \tag{13-14}$$

Values of the quantities discussed are listed in Table 13–2 for some of the most frequently used foil materials. The values given are for α-particles from RaC′, which have an extrapolated range of 6.953 cm in air at 15°C and 760 mm Hg. The extrapolated range is used here because this is the quantity for which experimental values are given in the table.

TABLE 13–2

RANGE AND STOPPING POWER FOR RaC′ ALPHA-PARTICLES IN VARIOUS SUBSTANCES

Substance	Extrapolated range, cm	Relative stopping power	Density, gm/cm³	Equivalent thickness, mg/cm²	Thickness in mg/cm² equivalent to 1 cm of air
Mica	0.0036	1930	2.8	10.1	1.45
Aluminum	0.00406	1700	2.70	11.0	1.57
Copper	0.00183	3800	8.93	16.3	2.35
Gold	0.00140	4950	19.33	27.1	3.89

13–3 Range-energy curves. The ranges in air of the α-particles from some natural α-emitters are listed in Table 13–3 together with the energies measured by the magnetic deflection method. The ranges are quoted from Reference 5 and the energies from Reference 2b. The values show clearly the differences between the mean and extrapolated ranges as well as the general relation between range and energy. The range increases monotonically with the energy, and for mean ranges between 3 and 7 cm, the empirical equation $\overline{R} = 0.318E^{3/2}$, with $\overline{R}$ in centimeters of air at 15°C and 760 mm Hg, and E in Mev, holds fairly well. The extrapolated range, as obtained from the number-distance curve, is greater than the mean range; the difference depends on the magnitude of the mean range and varies from about 0.05 cm for particles with a mean range of 4 cm to about 0.15 cm for particles with a mean range of 11 cm. It is easier to determine the extrapolated range experimentally, and the mean range can then be calculated quite well[15] from the extrapolated range. The mean range has the advantage of being somewhat less dependent on the particular experi-

Table 13–3

Ranges and Energies of Some Natural Alpha-Emitters

Radionuclide	Mean range in air, cm	Range from extrapolated number-distance distribution, cm	Range from extrapolated ionization, cm	Energy, Mev
$_{86}Em^{219}$ (An)	5.240 ± 0.015	5.312	5.272	6.542 ± 0.006
	5.692 ± 0.015	5.769	5.727	6.807 ± 0.006
$_{84}Po^{215}$ (AcA)	6.457 ± 0.008	6.542	6.496	7.383 ± 0.012
$_{83}Bi^{211}$ (AcC)	4.984 ± 0.015	5.053	5.015	6.273 ± 0.013
	5.429 ± 0.015	5.503	5.462	6.620 ± 0.013
$_{84}Po^{211}$ (AcC′)	6.555 ± 0.015	6.641	6.595	7.442 ± 0.015
$_{86}Em^{220}$ (Tn)	5.004 ± 0.018	5.073	5.035	6.2823 ± 0.0013
$_{84}Po^{216}$ (ThA)	5.638 ± 0.008	5.714	5.672	6.7746 ± 0.0013
$_{83}Bi^{212}$ (ThC)	4.730 ± 0.008	4.796	4.778	6.0466 ± 0.0027
$_{84}Po^{212}$ (ThC′)	8.570 ± 0.007	8.676	8.616	8.7801 ± 0.004
	9.724 ± 0.008	9.841	9.780	9.4923 ± 0.004
	11.580 ± 0.008	11.713	11.643	10.5432 ± 0.004
$_{86}Em^{222}$ (Rn)	4.051 ± 0.008	4.109	4.076	5.4861 ± 0.007
$_{84}Po^{218}$ (RaA)	4.657 ± 0.008	4.722	4.685	5.9982 ± 0.0008
$_{84}Po^{214}$ (RaC′)	6.907 ± 0.006	6.997	6.953	7.6804 ± 0.0009
	7.793 ± 0.015	7.891	7.839	8.2771 ± 0.0036
	9.04 ± 0.02	9.15	9.09	9.0649 ± 0.0040
	11.51 ± 0.02	11.64	11.57	10.5058 ± 0.0045
$_{84}Po^{210}$ (RaF)	3.842 ± 0.006	3.897	3.870	5.3007 ± 0.0026

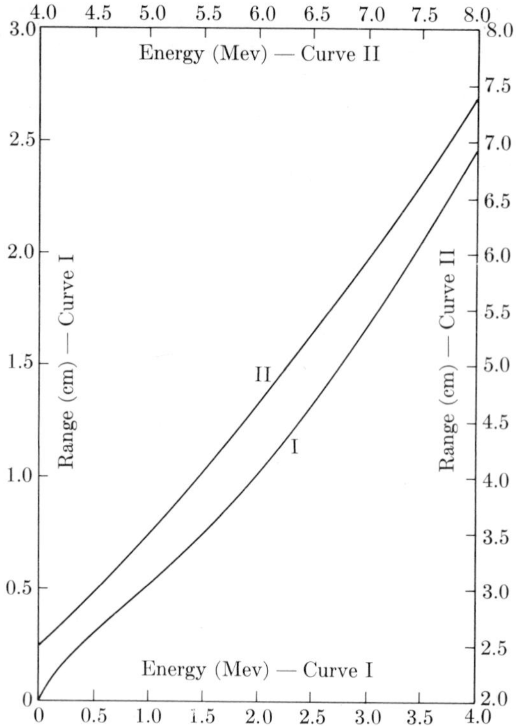

FIG. 13–9. Range-energy curves for α-particles (Bethe[16]).

mental setup, and is now generally used in range tables and for range-energy curves.

When the mean range is plotted against the energy, the resulting range-energy curve can be used to find the energy of α-particles whose range in air is measured. The data of Table 13–3 would give a range-energy curve down to an energy of only about 5 Mev, which would limit the usefulness of the curve. Information at lower energies can be obtained either by studying the range and energy of particles which have been partially slowed down, or of particles produced in nuclear reactions. Two reactions producing α-particles have been found to be valuable for the range-energy relation. The first of these is

$$\mathrm{Li}^6 + \mathrm{n} \rightarrow \mathrm{He}^4 + \mathrm{H}^3 + Q,$$

with a Q-value of 4.788 ± 0.023 Mev. The energy of the α-particle is 2.057 ± 0.010 Mev and the range is 1.04 ± 0.02 cm. The second reaction is

$$\mathrm{B}^{10} + \mathrm{n} \rightarrow \mathrm{Li}^7 + \mathrm{He}^4 + Q,$$

with a Q-value of 2.316 ± 0.006 Mev. The energy of the α-particle is 1.474 ± 0.004 Mev and its range is 0.720 ± 0.015 cm. With the points provided by these nuclear reactions, and with recent improvements in the knowledge of the relationship between ionization and energy loss, it has been possible to prepare more extensive and more accurate range-energy curves.(16,17) A curve for α-particles is shown in Fig. 13–9; similar curves are available for protons and deuterons.

The use of the range-energy curve may be illustrated by some simple examples. A recent measurement of the mean range of α-particles from U^{238} (UI) gave a value of 2.70 ± 0.02 cm. According to curve II of Fig. 13–9, this range corresponds to an energy of 4.25 Mev. The energy as determined from total ionization measurements is 4.180 ± 0.015 Mev, and the agreement is not bad. For Th^{232}, the range was found to be 2.49 ± 0.02 cm, and the energy from the curve is 4.02 Mev; the energy determined from the total ionization is 3.99 Mev, and the agreement is very good.

13–4 Alpha-particle spectra. Long-range particles and fine structure. In the treatment of the energy and range of α-particles, it has been implied that all α-particles from a given active nuclide have the same initial velocity and energy. Careful experiments have shown, however, that a given nuclide often emits particles with a number of different energies. Radium C′ and ThC′ emit a few *long-range particles* having energies considerably greater than that of the main group of particles. The existence of long-range α-particles was first observed by Rutherford and Wood,(18) who studied the absorption of the α-rays from a sample of ThC. Thorium C emits α-particles with a mean range of 4.73 cm; it also decays by β-emission to ThC′ which, in turn, emits α-particles with a mean range of 8.57 cm. A sample of ThC contains ThC′ and acts as a source of α-particles with both of the ranges cited. Rutherford and Wood found that a few particles from such a source were able to pass through an absorbing screen sufficiently thick to stop all particles with a range of 8.6 cm. When more absorbing material was used the penetrating particles were found to have a range of about 11 cm in air, and it was clear that they were different from the normal particles emitted by ThC and ThC′.

Later studies with the cloud chamber, ionization chamber, and magnetic spectrograph indicated that ThC′ emits two groups of long-range particles with energies of 9.492 Mev and 10.543 Mev, respectively, as compared with an energy of 8.780 Mev for the particles in the main group. The most recent magnetic spectrograph measurements show the presence of a third group with an energy of 10.422 Mev. For each million particles of the main group there are about 40 particles in the 9.492 Mev group, 20 in the 10.422 Mev group, and 170 in the 10.543 Mev group.

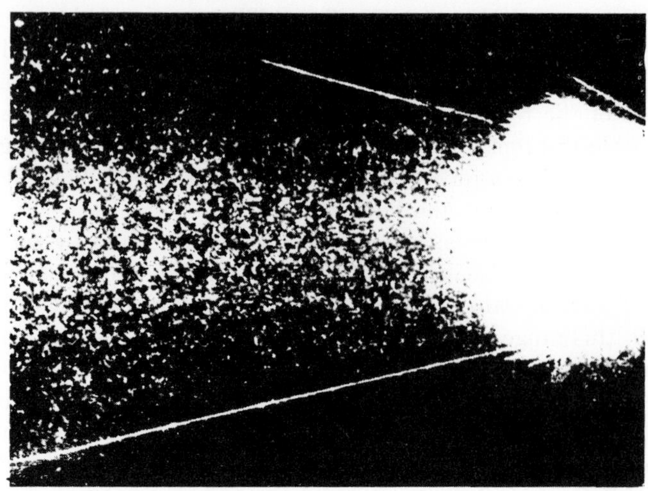

FIG. 13–10. A cloud chamber photograph of α-particles from RaC′ showing two long-range particles. (Nimmo and Feather, *Proc. Roy. Soc.* (London) **A122,** 668 (1929).)

TABLE 13–4

LONG-RANGE ALPHA-PARTICLES FROM RaC′

Mean range, cm	Alpha-particle energy, Mev	Relative number of particles	Alpha-disintegration energy, Mev	Difference of alpha-disintegration energy from that of main group, Mev
6.870	7.680	10^6	7.827	0.000
7.755	8.277	0.43	8.435	0.608
....	8.938	0.45	9.108	1.281
9.00	9.065	22	9.238	1.411
....	9.313	0.38	9.490	1.663
....	9.489	1.35	9.669	1.842
....	9.657	0.35	9.841	2.014
....	9.779	1.06	9.965	2.138
....	9.905	0.36	10.093	2.266
....	10.074	1.67	10.265	2.438
....	10.146	0.38	10.339	2.512
....	10.326	1.12	10.523	2.696
11.47	10.506	0.23	10.706	2.879

Measurements with the magnetic spectrograph have also shown that RaC′, whose main group of α-particles has an energy of 7.680 Mev, emits 12 groups of long-range particles with energies ranging from 8.277 Mev to 10.506 Mev.[19] Altogether, about 30 long-range particles are emitted per million of normal range. A cloud-chamber picture of typical long-range particles from RaC′ is shown in Fig. 13–10. The long-range α-particles from RaC′ are listed in Table 13–4. Another quantity, the *alpha disintegration energy*, is also listed; it will be discussed in Section 13–5.

Until 1930 it was thought that with the exception of the rare long-range particles, the α-particles emitted by an active nuclide all had the same initial energy. In that year it was shown by Rosenblum[20] in very careful magnetic deflection experiments that the normal α-particles emitted by some active nuclides fall into several closely spaced velocity groups. The velocities and energies of the different groups differ so little that the ranges of all the particles lie within the region of straggling. For this reason the different groups could be separated only in a magnetic spectrograph of high resolving power. The discrete, closely spaced components of the α-rays are said to form a *spectrum*, or to show *fine structure*. A great deal of research has been done on α-particle spectra since Rosenblum's discovery and the spectra of many natural and artificial α-emitters are now known. Thus, Th^{232} has two groups, with energies 3.994 Mev(76%) and 3.936 Mev(24%). Ionium (Th^{230}) has at least four groups, with energies of 4.682 Mev(76.3%), 4.615 Mev(23.4%), 4.471 Mev(0.2%) and 4.436 Mev(0.07%); it may have several other groups with very small abundances. Radium (Ra^{226}) has three groups with energies of 4.777 Mev(94.3%), 4.593 Mev(5.7%), and 4.34 Mev(0.01%), while RaAc(Th^{227}) has 14 groups with energies between 6.030 Mev and 5.661 Mev and relative abundances varying from 25 to 0.1%.

TABLE 13–5

THE ALPHA-PARTICLE SPECTRUM OF ThC

Group	Alpha-particle energy, Mev	Relative number of particles, %	Alpha-disintegration energy, Mev	Difference of alpha-disintegration energy from that of α_0 group, Mev
α_0	6.086	27.2	6.203	0.000
α_1	6.047	69.9	6.163	0.040
α_2	5.765	1.7	5.874	0.329
α_3	5.622	0.15	5.730	0.473
α_4	5.603	1.1	5.711	0.492
α_5	5.478	0.016	5.584	0.619

The α-particle spectrum of ThC (Bi^{212}) is given in detail in Table 13–5. The values of the energy used are averages of two sets of measurements given in the isotope table of Hollander, Perlman, and Seaborg.

The information now available about α-particle spectra leads to the conclusion that these spectra may be divided into three groups:

1. Spectra consisting of a single group or "line," for example, Rn, RaA, RaF.

2. Spectra consisting of two or more discrete, closely spaced (in velocity or energy) components with intensities of the same, or of only a slightly different, order of magnitude, for example, ThC, An, AcX, ThX, Pa, RdAc.

3. Spectra consisting of a main group and groups of much higher energy (long-range) particles, the latter containing, however, only a very small fraction (10^{-4} to 10^{-7}) of the number of particles in the main group. The third kind of spectrum occurs only in two extremely short-lived nuclides RaC′ and ThC′.

13–5 Nuclear energy levels. The discovery of α-spectra gave rise to the problem of why α-particles should be emitted by a given nucleus only with certain discrete energies. In view of the successful interpretation of atomic spectra in terms of discrete electronic energy levels, as in the Bohr theory of the atom and in wave mechanics, it seemed reasonable to try to account for discrete α-spectra in terms of nuclear energy levels. It is supposed that there is a number of discrete energy levels in the nucleus, and that a nucleus is normally in its lowest energy state, but under certain conditions it may exist for short times in excited states, i.e., in configurations having more than the normal amount of energy. A nucleus in such an excited state would be expected to give up its energy by some emission process, for example, by the emission of radiation. To apply these ideas, it must be possible to determine energy levels in the nucleus experimentally. As in the case of atoms, nuclei must be excited to emit radiation so that the energy levels in the nucleus may be deduced from the frequencies of this radiation. It was known that those α-emitters that have discrete α-particle spectra also emit γ-rays. Careful measurements of the α-particle energies and the γ-ray energies led to the conclusion that the γ-rays are emitted by the product nucleus that has been left in an excited state after the emission of an α-particle. Thus, the process of α-emission sometimes leaves excited nuclei which then emit radiation, and the way in which the energies of the γ-rays and the α-particles can be correlated in terms of nuclear energy levels was shown by Gamow.[(21)]

Before proceeding further with the discussion, it is necessary to distinguish between the kinetic energy of an α-particle and the total energy change in an α-decay process; the latter quantity has been called the *α-disintegration energy.* When an α-particle is emitted, the product, or residual, nucleus recoils, carrying with it a certain amount of energy. The α-disintegration energy is the sum of the kinetic energies of the α-particle and the product nucleus, and is found as follows. The principle of conservation of momentum requires that

$$Mv = M_r v_r, \tag{13–15}$$

where M_r is the mass of the product nucleus, v_r is its velocity, and the quantities without subscripts refer to the α-particle. Where relativistic effects are neglected, the α-disintegration energy E_α is given by

$$E_\alpha = \tfrac{1}{2}Mv^2 + \tfrac{1}{2}M_r v_r^2.$$

From Eq. (13–15), $v_r = (M/M_r)v$, and

$$E_\alpha = \frac{1}{2}\,Mv^2\left(1 + \frac{M}{M_r}\right). \tag{13–16}$$

The α-disintegration energy is obtained, therefore, simply by multiplying the kinetic energy of the α-particle by the quantity

$$1 + \frac{\text{mass of } \alpha\text{-particle}}{\text{mass of product nucleus}}.$$

The α-disintegration energy is listed in the fifth column of Tables 13–4 and 13–5 for the α-particle groups from RaC′ and ThC, respectively.

The interpretation of complex α-spectra in terms of nuclear energy levels can be illustrated by the case of ThC(Bi^{212}), which emits six groups of α-particles with energies listed in Table 13–5. The most energetic particles have a kinetic energy of 6.086 Mev corresponding to an α-disintegration energy of 6.203 Mev. It is supposed that a ThC nucleus always releases 6.203 Mev of energy when it decays by α-emission to form a nucleus of ThC″(Tl^{208}), and that this amount of energy is associated with one of the most energetic α-particles. The emission of one of these particles is assumed to leave the product nucleus in its lowest energy state, or *ground* state. All of the disintegration energy has gone into the kinetic energy of the α-particle and the recoil energy of the ThC″ nucleus. Suppose now that a ThC nucleus emits an α-particle of the α_1-group for which the disintegration energy is 6.163 Mev or 0.040 Mev less than the total available α-disintegration energy. The ThC″ nucleus, which retains the 0.040 Mev of energy, should be left in an excited state; it might be expected to undergo a transition to the ground state by emitting electromagnetic radiation (in analogy with the transitions undergone by atoms which have electrons in excited states). If this is the case, the radiation should appear in the form of a γ-ray with an energy of 0.040 Mev. Gamma

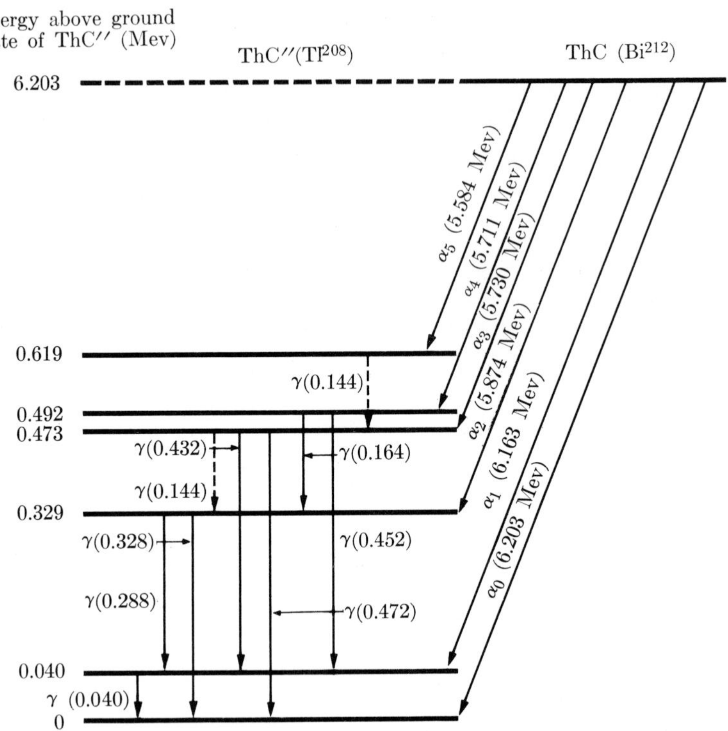

FIG. 13–11. Decay scheme for the α-decay of ThC to ThC″.

rays of this energy are, in fact, found experimentally when ThC undergoes α-decay. By extending this analysis we can postulate the existence of at least five excited states of the ThC″ nucleus and we would expect to find additional transitions from more excited states to less excited states. These transitions should involve the emission of γ-rays. Experimentally, eight γ-rays have so far been identified, with energies of 0.040, 0.144, 0.164, 0.288, 0.328, 0.432, 0.452, and 0.472 Mev, respectively. Each of these γ-rays can be matched (within the uncertainties of the measurements) to an energy difference between two excited states, and the energies of the α-particle groups and the γ-rays form a consistent scheme.

As a result of the correlation between the energies of the α-particles and the γ-rays, it is possible to construct nuclear energy level diagrams analogous to the atomic energy level diagrams shown in Chapter 7. The level diagram for the daughter nucleus is usually shown together with the decay data for the parent nucleus and the result is called a *decay scheme*. The scheme for the decay of ThC to ThC″ is shown in Fig. 13–11. The diagonal lines represent the different α-particle groups observed in the decay of ThC; the disintegration energies are also shown in this example. The

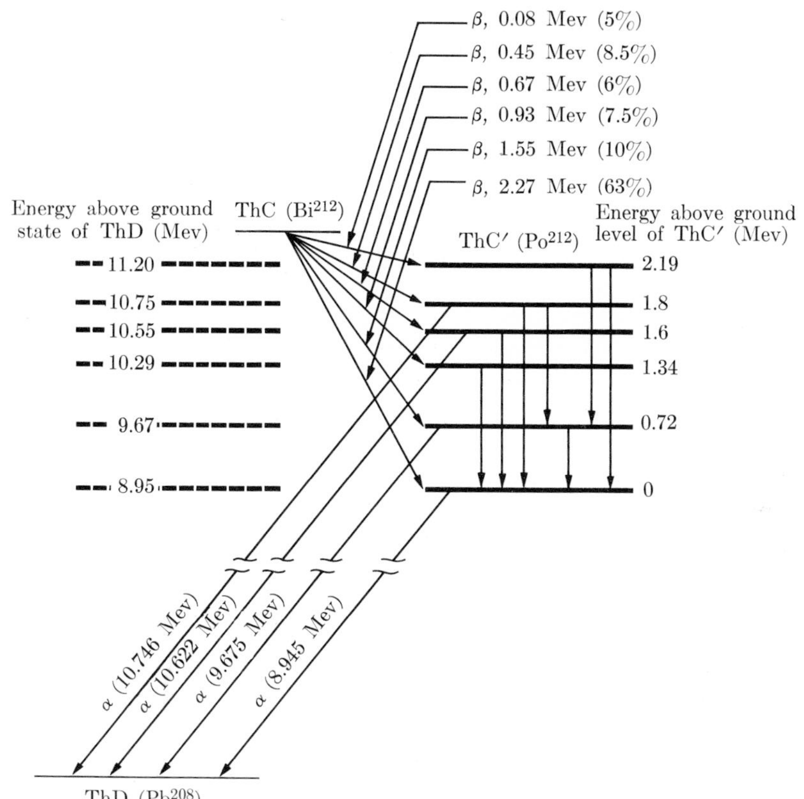

FIG. 13–12. Energy level diagram for the ThC → ThC′ → ThD decays, showing the origin of the long-range α-particles of ThC′.

vertical lines represent the γ-rays found experimentally. Each γ-ray corresponds to the energy difference between two states, although there is some ambiguity in the case of the 0.144-Mev γ-ray which may correspond, within the limits of experimental error, to either of two possible transitions. It is evident from the diagram that γ-rays are not observed for all transitions which seem to be possible. The absence of these γ-rays may be caused either by their small intensity, which may make them hard to detect, or by nuclear selection rules which may make the probability of these transitions very small. There is a theory which accounts in a satisfactory way for the radiations from excited nuclei and the selection rules involve the angular momentum and parity of the initial and final states.

The α-spectra of a large number of natural and artificial α-emitters have now been correlated with the γ-ray spectra of the decay products, and energy level schemes have been worked out. This procedure represents a start on the general problem of nuclear energy levels because it provides one of the methods by which levels can be determined. Other methods will be discussed in later chapters.

The origin of the long-range particles emitted by RaC′ and ThC′ can also be interpreted in terms of energy levels. It is possible for a nucleus, before an α-disintegration, to be in an excited state; the excitation may result from a previous disintegration. It has been found in the study of the emission of β- and γ-rays from nuclei that the emission of a nuclear β-ray sometimes leaves the newly formed nucleus in an excited state. In most cases the excited nucleus goes to its ground state by emitting a γ-ray of the proper energy. This process will be discussed in greater detail in the next chapter. In some cases, however, the newly formed excited nucleus is an α-emitter, and gets rid of its excess energy by emitting α-particles with greater energy than that of the normal particles. The existence of long range α-particles is thus explained as being caused by the decay of an excited nucleus, and the extra energy of the α-particle measures the excitation energy of the initial nucleus. An energy level diagram for the ThC-ThC′ decays showing the origin of the long range α-particles of ThC′ is given in Fig. 13–12.

13–6 The theory of alpha-decay. When the range, energy, half-life, and disintegration constant of the natural α-emitters are compared, certain remarkable facts stand out, which are illustrated by the data listed in Table 13–6. With the exception of the rare long-range particles, the ranges lie between 2.5 cm and 8.6 cm, and the α-disintegration energies vary from 4.05 Mev to 8.95 Mev. The ratio of the longest to the shortest range is about 3.5 and the ratio of the greatest to the least energy is between 2 and 2.5. But the half-lives vary from 1.39×10^{10} years for the longest-lived nuclide to 3.0×10^{-7} sec for the shortest, and the disintegration constants vary from 1.58×10^{-18} sec^{-1} to 2.31×10^{6} sec^{-1}. A factor of two or three in the energy corresponds to a factor of 10^{24} in the half-life and disintegration constant. In other words, the disintegration constant varies extremely rapidly with small changes in energy. The longest-lived nuclides emit the least energetic α-particles, while the shortest-lived nuclides emit the most energetic particles. These empirical facts were first correlated in the Geiger-Nuttall[(22)] rule, which stated that when the logarithm of the disintegration constant (in sec^{-1}) was plotted against the logarithm of the range (in cm of standard air) for the α-emitting members in the same radioactive series, an approximately straight line was obtained. The lines for the three natural series turned out to be practically parallel. The rule may be represented by the formula

$$\log \lambda = A \log R + B. \tag{13–17}$$

Table 13–6

Range, Energy, Half-Life, and Disintegration Constant of Some Alpha-Emitters

Nuclide	Mean range, cm of standard air	Alpha-disintegration energy, Mev	Half-life	Disintegration constant, sec^{-1}
Th^{232}	2.49	4.06	1.39×10^{10} y	1.58×10^{-18}
Ra^{226} (Ra)	3.30	4.86	1.62×10^{3} y	1.36×10^{-11}
Th^{228} (RdTh)	3.98	5.52	1.9 y	1.16×10^{-8}
Em^{222} (Rn)	4.05	5.59	3.83 d	2.10×10^{-6}
Po^{218} (RaA)	4.66	6.11	3.05 m	3.78×10^{-3}
Po^{216} (ThA)	5.64	6.90	0.16 s	4.33
Po^{214} (RaC′)	6.91	7.83	1.64×10^{-4} s	4.23×10^{3}
Po^{212} (ThC′)	8.57	8.95	3.0×10^{-7} s	2.31×10^{6}

The constant A is the slope of the line and has nearly the same value for the three series, but B has different values. The Geiger-Nuttall rule is illustrated in Fig. 13–13. In Section 13–3 it was noted that the mean range was found empirically to be proportional to $E^{3/2}$, where E is the kinetic energy of the α-particle. The Geiger-Nuttall rule may then be written in the form

$$\log \lambda = A' \log E + B', \tag{13–18}$$

which expresses the relationship between the logarithm of the disintegration constant and that of the energy.

The facts of α-particle emission presented a difficult theoretical problem which is brought out sharply when the spontaneous emission of α-particles by a nucleus is compared with the scattering of α-particles by the same nucleus. Consider the case of emission and scattering by U^{238}. In experiments on the scattering of α-particles, it was found that the fastest natural α-particles available, those of ThC′ with energies of nearly 9 Mev, are unable to penetrate close enough to the nucleus to show departures from the Coulomb law. It follows that at least up to a distance of about 3×10^{-12} cm from the nucleus, the potential energy of the α-particles in the field of the nucleus is still expressed by the Coulomb formula

$$U(r) = \frac{2Ze^2}{r}.$$

For $r = 3 \times 10^{-12}$ cm, the potential energy is about 9 Mev, and increases at least up to some maximum value for smaller values of r, until the

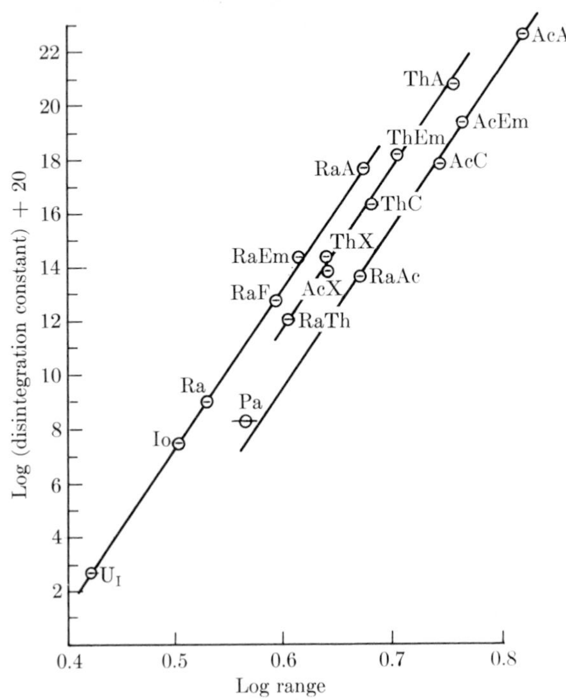

FIG. 13–13. The Geiger-Nuttall rule.

Coulomb law breaks down. This repulsive potential prevents the incident α-particle from entering the nucleus, and may be considered to form a *potential barrier*.

Since radioactive uranium nuclei emit α-particles, the latter may be supposed to exist, at least for a short time before emission, within these nuclei. The interactions between a radioactive nucleus and an α-particle inside and outside the nucleus may then be represented by a potential energy curve such as that in Fig. 13–14. The rising portion of the curve from r_2 to r_1 indicates increasing repulsion of an α-particle as it approaches the nucleus. Close to the nucleus and inside it, the shape of the potential energy curve is not known with certainty, but the Coulomb potential must break down and be replaced by an attractive potential. The interaction between the nucleus and the α-particle in the region of uncertainty may be represented by a constant attractive potential U_0, exerted over a distance r_0 called the *effective radius of the nucleus*. This type of potential is spoken of as a potential *well* of depth U_0 and width or *range* r_0. Under classical mechanics, the α-particle could exist inside the potential well, with a kinetic energy equal to $E_\alpha - U$. (The horizontal line E_α in the figure represents the α-disintegration energy when the particle is far from the

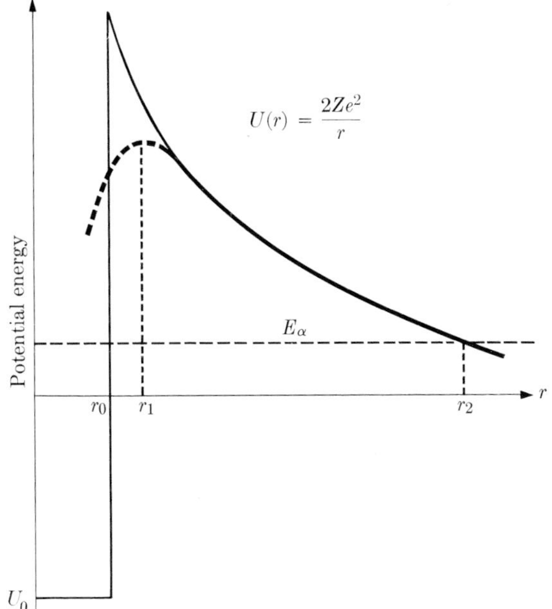

FIG. 13-14. Alpha-particle emission: potential energy of a nucleus and an α-particle as a function of their separation.

nucleus.) To escape from the nucleus, the particle would have to have a kinetic energy at least as great as the value of the energy at the maximum of the potential energy curve. Similarly, an α-particle approaching the nucleus from outside could penetrate the nucleus only if it had enough kinetic energy to get over the potential energy barrier.

Thus, the maximum value of the potential energy curve corresponds to the value of the kinetic energy which an α-particle must have according to classical mechanics to get into the nucleus from outside, or to escape from inside the nucleus. This maximum value must be greater than about 9 Mev, since α-particles from ThC′ are scattered by uranium. The uranium nucleus, however, emits an α-particle with an energy of about 4 Mev, and it is very difficult to understand how the particles contained in the inside of the nucleus can go over a potential barrier which is more than twice as high as their total energy. Classical physics provides no solution to this problem, nor can it account for the extremely large variations in half-life corresponding to small changes in energy, i.e., the Geiger-Nuttall rule.

The paradox was resolved when the problem of a potential barrier such as that involved in α-emission or α-scattering was analyzed by the methods of wave mechanics. The mathematical treatment will not be given, but the results and their implications will be discussed. If the motion of a particle in the neighborhood of a potential barrier is treated wave-mechanically, i.e., if the Schroedinger equation for an α-particle is solved, it is found that there is a finite probability that the particle can leak through the barrier even though its kinetic energy is less than the potential energy represented by the height of the barrier. In other words, a particle impinging on the barrier will not necessarily be reflected, but may pass through the barrier and continue its forward motion. Gamow[(23)] and, independently, Condon and Gurney,[(24)] in 1928, showed how this comes about. The probability that an α-particle can leak through the barrier (the "tunnel effect") can be calculated; the *permeability* of a simple barrier is given mathematically by an expression of the type

$$P = \exp\left[-\frac{2\sqrt{2M}}{\hbar}\int (U - E_\alpha)^{1/2}\,dr\right], \tag{13–19}$$

where $\hbar = h/2\pi$, M is the mass of the α-particle, and the integral is taken over the entire region in which $U(r) > E_\alpha$. The wave-mechanical treatment of α-decay yields an expression for the disintegration constant which depends on the permeability or *transparency* of the barrier. An approximate form of the result is

$$\lambda = \frac{v_\alpha}{r_0}\exp\left[-\frac{8(Z-2)e^2}{\hbar v_\alpha}(\alpha_0 - \sin\alpha_0\cos\alpha_0)\right], \tag{13–20}$$

where

$$\alpha_0 = \text{arc cos}\left[\frac{Mv_\alpha^2 r_0}{4e^2(Z-2)}\right]^{1/2};$$

r_0 is the radius of the nucleus or, more strictly, the "effective radius of the nucleus for α-decay"; v_α is the velocity of the α-particle relative to the nucleus, and is equal to $v\,[1 + (M/M_r)]$, where v is the measured velocity of the α-particle and M_r is the mass of the recoil nucleus. The quantity E_α is the alpha-disintegration energy given by

$$E_\alpha = E\,[1 + M/M_r)],$$

and $(Z - 2)$ is the charge on the product nucleus.

The expression for the disintegration constant may be considered to consist of two factors. One, v_α/r_0, gives in a rough way the number of times per second an α-particle inside the nucleus collides with the potential barrier; the exponential factor gives the probability that the α-particle will leak through the barrier. Equation (13–20) has some interesting properties. Since λ is proportional to an exponential function, a relatively small change in the argument of the exponential may cause a large change in the value of λ. The argument of the exponential depends on the velocity of the α-particle and a small change in the velocity affects the value of λ strongly. This effect may be seen more directly when some numerical values are inserted. Consider the factor $8e^2(Z - 2)/\hbar v_\alpha$ in the exponential (the function $\alpha_0 - \sin\alpha_0 \cos\alpha_0$ is of the order unity and varies slowly with v_α). The value of e is 4.80×10^{-10} esu, so that $e^2 = 2.304 \times 10^{-19}$, $\hbar = 1.054 \times 10^{-27}$. For $Z - 2 = 90$, the factor is

$$\frac{8(2.304)10^{-19}(90)}{(1.054)10^{-27}v_\alpha} = \frac{157.5 \times 10^9}{v_\alpha}.$$

Values for the velocities of α-particles range from 1.4 to 2.2×10^9 cm/sec, so that the value of the factor is of the order of magnitude 100. A small change in the value of v_α or v will, therefore, have a large effect on the value of the disintegration constant. It is also seen that the greater the velocity of the α-particle, the greater is the value of the disintegration constant, and the theoretical expression agrees in a general way with the observed facts. Equation (13–20) represents a highly important result of the application of wave mechanics to nuclear physics.

The theory also offers a clue to the meaning of the Geiger-Nuttall rule. Equation (13–20) becomes, when the common logarithm of both sides is taken,

$$\log\lambda = \log\frac{v_\alpha}{r_0} - \frac{8e^2(Z-2)}{2.303\hbar v_\alpha}(\alpha_0 - \sin\alpha_0\cos\alpha_0); \qquad (13\text{–}21)$$

the factor 2.303 enters because of the use of common logarithms. It can be shown that

$$\alpha_0 - \sin\alpha_0\cos\alpha_0 \approx \frac{\pi}{2} + \text{terms of higher order},$$

and, to a good approximation,

$$\log\lambda = \log\frac{v_\alpha}{r_0} - \frac{4\pi e^2(Z-2)}{2.303\hbar v_\alpha} + \frac{8e}{2.303\hbar}[(Z-2)r_0M]^{1/2} + \cdots. \qquad (13\text{–}22)$$

The variation of $\log v_\alpha/r_0$ is small because both r_0 and v_α vary relatively little, and the higher order terms in Eq. (13–22) are small. All of these terms may be lumped into one term which is practically a constant. Then

$$\log\lambda \approx a - b\,\frac{(Z-2)}{v_\alpha}, \qquad (13\text{–}23)$$

where a and b are constants. According to this equation, if the logarithm of the disintegration constant is plotted against the reciprocal of the

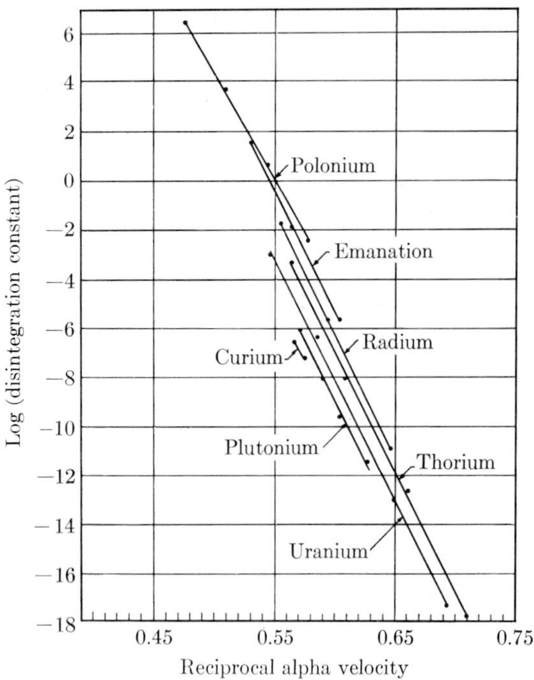

FIG. 13–15. Logarithm of the disintegration constant *vs.* reciprocal of the velocity, for constant Z. Disintegration constant λ in $\sec^{-1}$; velocity in units of 10^9 cm/sec[25].

α-velocity for the α-emitting isotopes of a radioactive element (Z constant), the result should be a straight line. The equation may be regarded as a theoretical form of the Geiger-Nuttall rule and may conveniently be compared with experiment. A comparison has been made by Kaplan,[25] who showed that α-emitters with even values of A and Z give results which are in good agreement with the theory. It can be seen from Fig. 13–15 that when $\log \lambda$ is plotted against $1/v_\alpha$, for even Z and A, straight lines are indeed obtained; the slopes of the lines are in good agreement with the values predicted by the theory. The results can also be shown in a graph like that of Fig. 13–16 in which the logarithm of the half-life is plotted against the "effective total decay energy," Q_{eff}[26]; the latter is just the α-disintegration energy with a small correction for the screening effect of the atomic electrons. A square-root scale is used for Q_{eff}; the points are experimental, and the last digit in the mass number of the α-emitter is given beside each point. The figure contains information more recent than that of Fig. 13–15 so that there are more experimental data; again, very good straight lines are obtained. The relationships between the

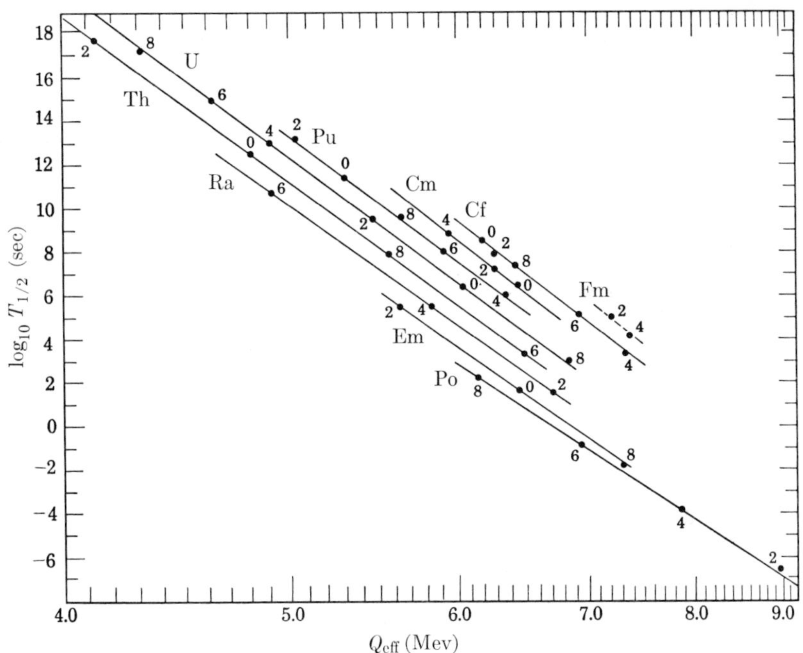

FIG. 13–16. Half-life for alpha-decay as a function of the square root of the effective total decay energy, Q_{eff}. The energy scale is linear in $Q_{\text{eff}}^{-1/2}$; Q_{eff} includes the electron screening correction. The numbers give the last digit of the mass number of the alpha-emitter; for example, 4 on the uranium scale refers to U^{234}.[(26)]

logarithm of the half-life and the α-disintegration energy of the elements with even values of Z are shown in Fig. 13–17. There are analogous curves for even-Z-odd-A nuclides, and for odd-Z nuclides.

The above discussion has been limited to the case of the even-even α-emitting nuclides in order to avoid complications which are beyond the scope of this book. The application of the theory to α-emitters with odd Z and even A, even Z and odd A, and odd Z and odd A is less straightforward, in part because the probability of α-emission depends on the difference in the angular momenta of the parent and daughter nuclei, and most of the values of the momenta are not yet known when either A or Z is odd. For this reason and others, the theory of α-decay cannot yet be considered a closed subject, although it has achieved an important degree of success.

If the energy and the disintegration constant are known for a given nucleus, Eq. (13–20) can be used to calculate the value of r_0, the effective radius of the nucleus for α-emission. It must be pointed out that the

TABLE 13–7

NUCLEAR RADII FROM ALPHA-DECAY: EVEN-EVEN ALPHA-EMITTERS

Parent nucleus	Product nucleus	Radius of product nucleus, cm × 10^{13}	$R_0 = r_0 A^{-1/3}$, cm × 10^{13}
Cm^{242}	Pu^{238}	9.32	1.50
Cm^{240}	Pu^{236}	9.29	1.50
Pu^{240}	U^{236}	9.40	1.52
Pu^{238}	U^{234}	9.30	1.51
Pu^{236}	U^{232}	9.30	1.51
U^{238}	Th^{234}	9.54	1.55
U^{236}	Th^{232}	9.36	1.52
U^{234}	Th^{230}	9.39	1.53
U^{232}	Th^{228}	9.33	1.53
U^{230}	Th^{226}	9.33	1.53
Th^{232}	Ra^{228}	9.48	1.55
Th^{230}	Ra^{226}	9.34	1.53
Th^{228}	Ra^{224}	9.33	1.54
Th^{226}	Ra^{222}	9.34	1.54
Ra^{226}	Em^{222}	9.35	1.54
Ra^{224}	Em^{220}	9.33	1.55
Ra^{222}	Em^{218}	9.30	1.55
Em^{222}	Po^{218}	9.34	1.55
Em^{220}	Po^{216}	9.36	1.57
Em^{218}	Po^{214}	9.48	1.58
Po^{218}	Pb^{214}	9.23	1.54
Po^{216}	Pb^{212}	9.19	1.54
Po^{214}	Pb^{210}	9.18	1.55
Po^{212}	Pb^{208}	9.05	1.53

radius of the nucleus is a quantity which cannot be measured directly and is obtained by calculation from some formula which also contains measured quantities. The particular formula used depends on the experimental process under consideration, and the value of the radius obtained must be regarded as an *effective* value for the given process. Fortunately, the values obtained from different processes agree quite well, and lie between 10^{-13} and 10^{-12} cm. The heavy α-emitting nuclei have radii close to 10^{-12} cm, as can be seen from Table 13–7. The values of the radii given are those of the product nuclei, and were obtained from a more rigorous form of the theory. The radii are found to conform to the formula

$$r_0 = R_0 A^{1/3} \times 10^{-13} \text{ cm},$$

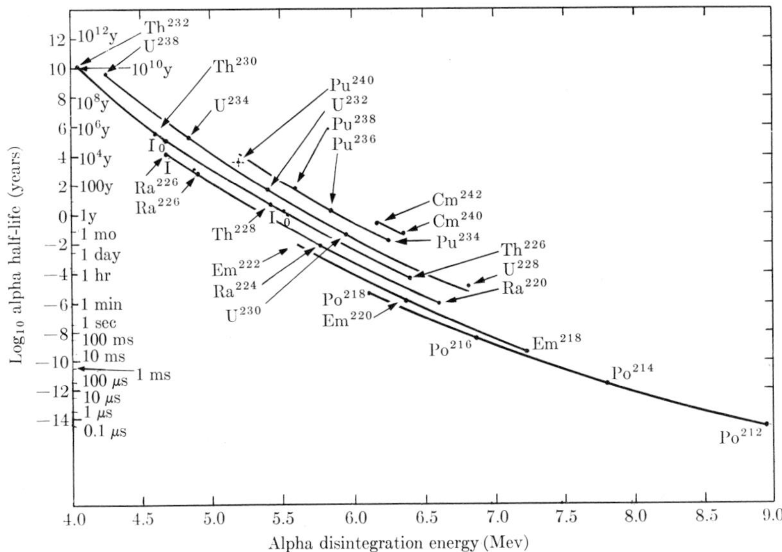

FIG. 13–17. Relationship between half-life and α-disintegration energy for some even-even α-emitters (Perlman, Ghiorso, and Seaborg(27)).

where R_0 is a constant closely equal to 1.53 and A is the atomic weight. The radius turns out to be proportional to the cube root of the atomic weight, a fact which will be useful in later work.

There is, finally, a considerable amount of work, which has only been touched upon, on the regularities among the properties of α-emitters. This work, sometimes referred to as the "systematics of α-emitters" has for its object the correlation of the available experimental information about α-emission. It is well summarized in the general references by Perlman and Rasmussen and by Hanna.

References

GENERAL

RUTHERFORD, CHADWICK, and ELLIS, *Radiations from Radioactive Substances.* New York: Macmillan, 1930, Chapters 2, 3, 4, 5, 7.

R. D. EVANS, *The Atomic Nucleus.* New York: McGraw-Hill, 1955, Chapters 16, 22.

I. PERLMAN and J. O. RASMUSSEN, "Alpha Radioactivity," *Handbuch der Physik,* Vol. 42. Berlin: Springer Verlag, 1957, pp. 109–204.

G. C. HANNA, "Alpha Radioactivity," *Experimental Nuclear Physics,* ed. by E. Segrè. New York: Wiley, 1959, Vol. III, Part IX.

F. ASARO and I. PERLMAN, "Table of Alpha-Disintegration Energies of the Heavy Elements," *Revs. Mod. Phys.* **26,** 456, 1954; **29,** 831, 1957.

H. A. BETHE and J. ASHKIN, "Passage of Radiation Through Matter," *Experimental Nuclear Physics,* E. Segrè, ed. New York: Wiley, 1953, Vol. I, Part II.

D. STROMINGER, J. M. HOLLANDER and G. T. SEABORG, "Table of Isotopes," *Revs. Mod. Phys.* **30,** 585–904, 1958.

PARTICULAR

1. RUTHERFORD, WYNN-WILLIAMS, LEWIS, and BOWDEN, "Analysis of α-Rays by an Annular Magnetic Field," *Proc. Roy. Soc.* (London) **A139,** 617 (1933).

2. (a) G. H. BRIGGS, "A Determination of the Absolute Velocity of the α-Particles from Radium C′," *Proc. Roy. Soc.* (London) **A157,** 183 (1936); (b) "Energies of Natural α-Particles," *Revs. Mod. Phys.* **26,** 1 (1954).

3. S. ROSENBLUM and G. DUPOUY, "Mésure Absolue des Vitesses des Principaux Groupes des Rayons α," *J. phys. et radium* **4,** 262 (1933).

4. A. H. JAFFEY, "Radiochemical Assay by α- and Fission Measurements," Manhattan District Declassified Document MDDC-1336, Feb. 6, 1947; also in *National Nuclear Energy Series, Plutonium Project Record,* Vol. 14A, *The Actinide Elements,* G. T. Seaborg and J. J. Katz, eds. New York: McGraw-Hill, 1954, Chapter 16.

5. M. G. HOLLOWAY and M. S. LIVINGSTON, "Range and Specific Ionization of Alpha-Particles," *Phys. Rev.* **54,** 18 (1938).

6. N. FEATHER and R. R. NIMMO, "The Ionization Curve of an Average α-Particle," *Proc. Cambridge Phil. Soc.* **24,** 139 (1928).

7. RUTHERFORD, WARD, and WYNN-WILLIAMS, "The Ranges of the α-Particles from the Radioactive Emanations and 'A' Products and from Polonium," *Proc. Roy. Soc.* (London) **A136,** 349 (1932).

8. I. CURIE and S. T. TSIEN, "Parcours des Rayons α de l'Ionium", *J. phys. et radium* **6,** 162 (1945).

9. W. E. GLENN, JR., "Pulse Height Distribution Analyzer," *Nucleonics* **4,** 50 (June 1949).

10. GHIORSO, JAFFEY, ROBINSON, and WEISSBORD, "A 48-Channel Pulse-Height Analyzer for Alpha-Energy Measurements," *The Transuranium Elements,* ed. by Seaborg, Katz, and Manning. *National Nuclear Energy Series,* Vol. 14B, Book 2, p. 1226. New York: McGraw-Hill, 1949.

11. JESSE, FORSTAT, and SADAUSKIS, "The Ionization in Argon and in Air by Single α-Particles as a Function of their Energy," *Phys. Rev.* **77,** 782 (1950).

12. H. YAGODA, *Radioactive Measurements with Nuclear Emulsions.* New York: Wiley, 1949.

13. GAILAR, SEIDLITZ, BLEULER, and TENDAM, "Range-Energy Relations for Alpha-Particles and Deuterons in the Kodak NTB Emulsion," *Rev. Sci. Instr.* **24,** 126 (1953).

14. F. BLOCH, "The Theory of Stopping Power," in Lecture Series in Nuclear Physics, Manhattan District Declassified Document MDDC-1175, Lectures 11, 12.

15. M. S. LIVINGSTON and H. A. BETHE, "Nuclear Physics, Part C. Nuclear Dynamics, Experimental," *Revs. Mod. Phys.* **9,** 245 (1937), Section XVI.

16. H. A. BETHE, "The Range-Energy Relations for Slow Alpha-Particles and Protons in Air," *Revs. Mod. Phys.* **22,** 213 (1950).

17. W. P. JESSE and J. SADAUSKIS, "Range-Energy Curves for Alpha-Particles and Protons," *Phys. Rev.* **78,** 1 (1950).

18. E. RUTHERFORD and A. B. WOOD, "Long-Range Alpha-Particles from Thorium," *Phil. Mag.* **31,** 379 (1916).

19. W. B. LEWIS and B. V. BOWDEN, "An Analysis of the Fine Structure of the α-Particle Groups from ThC and of the Long-Range Groups from ThC′," *Proc. Roy. Soc.* (London) **A145,** 235 (1934).

20. S. ROSENBLUM, "Progrès Récents dans l'Etude du Spectre Magnétique des Rayons α," *J. phys. et radium* **1,** 438 (1930).

21. G. GAMOW, "Fine Structure of α-Rays," *Nature* **126,** 397 (1930).

22. H. GEIGER and J. M. NUTTALL, "The Ranges of the α-Particles from Various Radioactive Substances and a Relation Between Range and Period of Transformation," *Phil. Mag.* **22,** 613 (1911); **23,** 439 (1912); **24,** 647 (1912).

23. G. GAMOW, "Zur Quantentheorie des Atomkernes," *Z. Phys.* **51,** 204 (1928).

24. R. W. GURNEY and E. U. CONDON, "Quantum Mechanics and Radioactive Disintegration," *Nature* **122,** 439 (1928); *Phys. Rev.* **33,** 127 (1929).

25. I. KAPLAN, "The Systematics of Even-Even Alpha-Emitters," *Phys. Rev.* **81,** 962 (1951).

26. C. J. GALLAGHER, JR., and J. O. RASMUSSEN, "Alpha-Decay Hindrance Factor Calculations," *J. Inorg. Nucl. Chem.* **3,** 333 (1957).

27. PERLMAN, GHIORSO, and SEABORG, "Systematics of Alpha-Radioactivity," *Phys. Rev.* **77,** 26 (1950).

Problems

1. Calculate the mass, relative to that of the rest mass, the kinetic energy in ergs and Mev, and the value of Hr for α-particles moving at each of the following speeds (in cm/sec): 1×10^8, 5×10^8, 7.5×10^8, 1×10^9, 1.25×10^9, 1.50×10^9, 1.75×10^9, 2.0×10^9, 2.5×10^9, 3.0×10^9, 5.0×10^9, 1.0×10^{10}, and 2.0×10^{10}. Plot the energy in Mev against the value of Hr in gauss-cm.

2. Find the speeds of α-particles with energies of 1 Mev, 2 Mev, 4 Mev, 5 Mev, 6 Mev, 8 Mev, 10 Mev, respectively.

3. Find the speed and Hr-value of the α-particles emitted by each of the isotopes of polonium listed in Table 13–3.

4. In an investigation of the $Al^{27}(d,\alpha)Mg^{25}$ reaction, an aluminum target was bombarded with 2.10-Mev deuterons. The α-particles coming off at 90° were analyzed in a 180° magnetic spectrograph, and 10 different groups of α-particles, each with a different Hr-value, were observed. The Hr-values were, in kilogauss-cm: 393, 379, 369, 354, 345, 329, 325, 323, 305, and 290. Find the Q-value corresponding to each group of α-particles. (The interpretation of the occurrence of the different α-particle groups is treated in Chapter 16.)

5. Plot the values of the kinetic energy listed in Table 13–3 against those of the mean range in air. Use the resulting curve and those of Fig. 13–9 to find the mean ranges in air of the α-particles of Problems 1 and 2, which fall within the energy limits covered by the curves.

6. If the range in air is known, the range in another substance, denoted by the subscript x, may be obtained, to a first approximation, from the empirical formula

$$R_x = R_{\text{air}} \times \frac{N_{\text{air}}}{N_x} \times \left[0.563 \frac{Z_x}{(Z_x + 10)^{1/2}}\right]^{-1},$$

where the N's represent the number of atoms per cubic centimeter and Z_x is the atomic number of the substance x. The density of air may be taken as 0.001226 gm/cm^3 and the "atomic weight of air" as 14.4. Calculate (a) the range in aluminum of the α-particles from each isotope of polonium listed in Table 13–3 (use the mean range in air), (b) the relative stopping power in aluminum, (c) the equivalent thickness in mg/cm^2, (d) the thickness in mg/cm^2 equivalent to one cm of air. Compare the results with the experimental values given in Table 13–2.

7. The nuclide Rn^{211} emits three groups of α-particles, with kinetic energies of 5.847 Mev, 5.779 Mev, and 5.613 Mev, respectively. Associated with the α-particles are γ-rays with energies of 0.0687 Mev, 0.169 Mev, and 0.238 Mev. Construct a decay scheme based on these data. Compare it with the scheme given by Strominger, Hollander, and Seaborg (gen. ref.).

8. The nuclide U^{233} emits six groups of α-particles, with kinetic energies of 4.816 Mev, 4.773 Mev, 4.717 Mev, 4.655 Mev, 4.582 Mev, and 4.489 Mev, respectively. Gamma-rays with energies 0.0428 Mev, 0.0561 Mev, and 0.099 Mev have also been reported. Construct a decay scheme based on these data.

9. The nuclide Am^{241} emits six groups of α-particles, with kinetic energies of 5.534 Mev, 5.500 Mev, 5.477 Mev, 5.435 Mev, 5.378 Mev, and 5.311 Mev,

respectively. Gamma-rays are found, with energies of 0.0264 Mev, 0.0332 Mev, 0.0435 Mev, 0.0555 Mev, 0.0596 Mev, 0.103 Mev, 0.159 Mev. Construct a decay scheme based on these data.

10. The height U of the Coulomb barrier around a nucleus of charge Z_1e and radius r for a particle of positive charge Z_2 may be found to a first approximation by calculating the energy of Coulomb repulsion at a distance equal to the radius of the nucleus,

$$U = Z_1Z_2e^2/r.$$

Assume that the nuclear radius is given by the formula $r = 1.5 \times 10^{-13}A^{1/3}$ cm, where A is the mass number of the nucleus. Calculate the height of the barrier for α-particles and the nuclei Ne^{20}, Ca^{40}, Zn^{66}, Sn^{112}, Yb^{174}, and Th^{232}. Repeat the calculation for protons and deuterons.

11. The artificially produced radioactive nuclide Pa^{230} may be considered the parent of a collateral (4n + 2) series. Find its radioactive properties from the nuclide chart or the "Table of Isotopes" of Strominger, Hollander, and Seaborg. Trace the series by means of a graph of A against Z (cf. Fig. 10–10), and show its relationship to the uranium series.

12. It has been shown [(25)] that the α-emission of the thorium isotopes can be described by the relation

$$\log \lambda = 56.13 - 105.07/v_\alpha.$$

The α-energy in the case of Th^{224} is 7.33 Mev; what is the half-life of Th^{224} as estimated from the above relation?

13. What is the Q-value of the reaction $Po^{210} \rightarrow Pb^{206} + He^4$? If the atomic mass of Pb^{206} is 206.0386, what is the atomic mass of Po^{210}?

CHAPTER 14

BETA-DECAY

The study of the properties of the β-particles emitted by the natural and artificial radioactive nuclides has added greatly to our knowledge of the structure and properties of atomic nuclei. As in the case of α-particles, the early interest was centered on the series of transformations of uranium, thorium, and actinium. At present, the main interest is in the information that β-particle emission gives about nuclear energy levels and decay schemes, and in the theory of β-decay and its relation to other fundamental nuclear problems. Many more nuclides decay by electron emission, positron emission, and orbital electron capture than by α-particle emission, and the β-decay processes yield information about hundreds of nuclear species not limited to those with large masses. Consequently, information about the energy levels and decay schemes of light and intermediate weight nuclides as well as those in the region of the natural radioactive elements can be obtained by studying their β-radiations. This study is an important part of nuclear spectroscopy.

The emission of β-particles differs from that of α-particles in respect to the spectrum of the energies of the emitted particles. The most characteristic feature of the spontaneous β-disintegration of a nucleus is the continuous distribution in energy of the emitted electrons, which is in sharp contrast to the line spectra observed for α-particles. The continuous energy distribution led, as will be seen, to serious theoretical problems. These problems have been treated with considerable success, but only by means of a theory involving new and radical ideas. Beta-decay is important, therefore, not only because of its relationship to the practical problems of nuclear physics, but also because of the conceptual problems involved.

14–1 The velocity and energy of beta-particles. The velocity, or momentum, of β-particles can be measured by means of the deflection of the path of the particles in a magnetic field. One kind of instrument which is often used is the semicircular-focusing, or 180°, magnetic spectrograph similar in principle to the instrument used for α-particles (Fig. 13–1). The details of the design of the two instruments are different because of the differences between the properties of α- and β-particles. For example, the value of the charge-to-mass ratio is much greater for electrons than for α-particles and much smaller magnetic fields may therefore be used to deflect β-particles. Instead of fields of about 10,000 gauss as in the case of the α-particle spectrograph, fields of 1000 gauss or less are strong enough. In the earlier instruments, the magnetic field was usually uniform,[1] but in the more recent designs the focusing of the particles has been improved by using a suitably inhomogeneous, or "shaped," field.[2] The velocities, or momenta, of the β-particles emitted by a radioactive source can be measured with high precision with these instruments, and when the velocities are known, the energies can be computed.

It was found in early experiments that β-particles emitted by naturally radioactive nuclides may have velocities up to about 0.99 that of light. In general, the energies of the β-particles, both positive and negative, are smaller than those of the α-particles emitted by radioactive nuclides. Most of the β-particles have energies smaller than 4 Mev, while nearly all of the α-particles have energies greater than 4 Mev. At the same kinetic energy, the β-particle, because of its much smaller mass, travels much faster than the α-particle. An α-particle with an energy of 4 Mev has a velocity about 1/20 that of light, but a 4-Mev electron would have a velocity close to 0.995 that of light.

The β-particles must be treated relativistically because of their large velocities. The formulas for the velocity and kinetic energy, analogous to Eqs. (13–5) and (13–6) for α-particles, are

$$v = Hr\frac{e}{m_0}\sqrt{1-\frac{v^2}{c^2}}, \tag{14–1}$$

$$T = m_0c^2\left[\frac{1}{\sqrt{1-\frac{v^2}{c^2}}}-1\right], \tag{14–2}$$

where m_0 is the rest mass of the electron and e is the magnitude of the electronic charge. It is useful to express the quantities Hr and T as functions of the ratio v/c and Eq. (14–1) may be written

$$\frac{v}{c} = Hr\frac{e}{m_0c}\sqrt{1-\frac{v^2}{c^2}}. \tag{14–3}$$

The specific charge of the electron, in electromagnetic units, is $e/m_0 = 1.75888 \times 10^7$ emu/gm and the velocity of light is 2.99793×10^{10} cm/sec. Then

$$\begin{aligned} Hr\text{ (gauss-cm)} &= \frac{m_0c}{e}\frac{v}{c}\left(1-\frac{v^2}{c^2}\right)^{-1/2} \\ &= 1704.5\frac{v}{c}\left(1-\frac{v^2}{c^2}\right)^{-1/2}. \end{aligned} \tag{14–4}$$

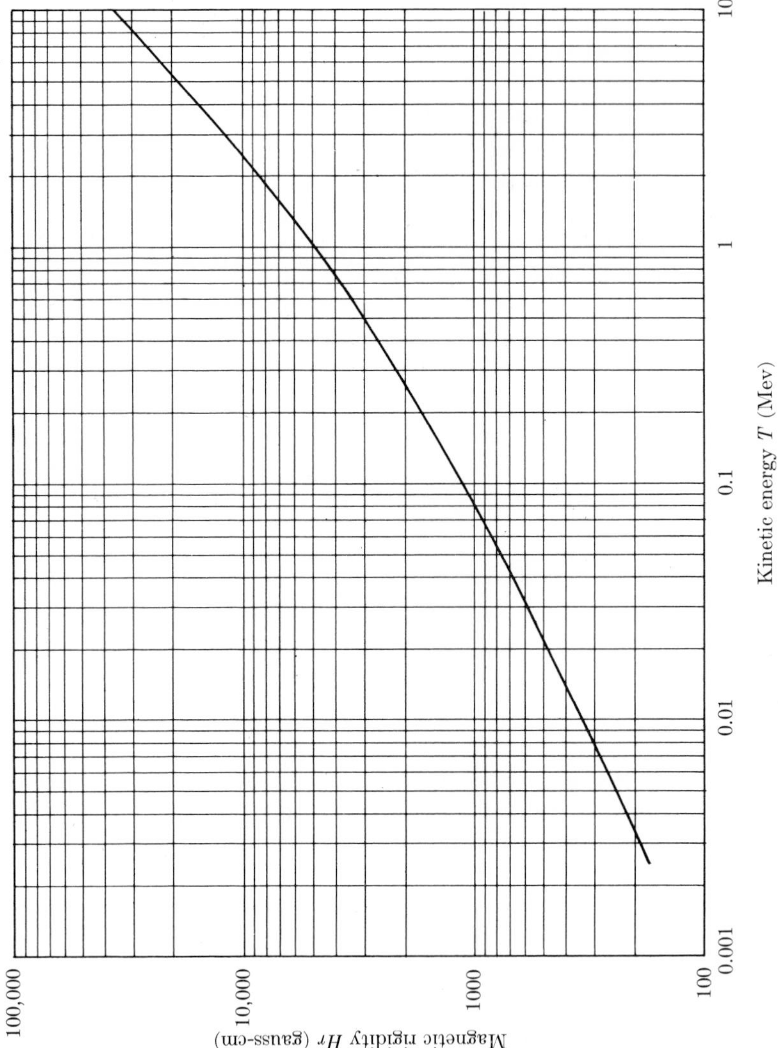

FIG. 14–1. Kinetic energy of β-particles as a function of the magnetic rigidity Hr.

TABLE 14-1

MAGNETIC RIGIDITY (Hr) AND KINETIC ENERGY OF BETA-PARTICLES AS FUNCTIONS OF VELOCITY

Velocity, v/c	Magnetic rigidity, Hr: gauss-cm	Kinetic energy, T: Mev
0	0	0
0.10	171.3	0.00258
0.15	258.6	0.00585
0.20	347.9	0.01054
0.25	440.1	0.01676
0.30	536.0	0.02468
0.35	636.9	0.03450
0.40	743.9	0.04655
0.45	858.9	0.06121
0.50	984.1	0.07905
0.55	1122	0.10085
0.60	1278	0.1278
0.65	1458	0.1614
0.70	1671	0.2045
0.75	1933	0.2616
0.80	2273	0.3407
0.85	2750	0.4590
0.90	3519	0.6613
0.92	4001	0.7928
0.94	4696	0.9868
0.96	5844	1.314
0.98	8394	2.057
0.99	11,960	3.111
0.995	16,980	4.605
0.996	19,000	5.208
0.997	21,960	6.901
0.998	26,910	7.573
0.999	38,085	10.92

The quantity m_0c^2 is the energy associated with the rest mass of the electron and is 0.5110 Mev. Hence, the kinetic energy is

$$T = 0.511 \left[\frac{1}{\sqrt{1 - \frac{v^2}{c^2}}} - 1 \right] \cdot \tag{14-5}$$

Values of Hr and T are listed in Table 14-1 for different values of v/c, and the kinetic energy T is plotted as a function of Hr in Fig. 14-1.

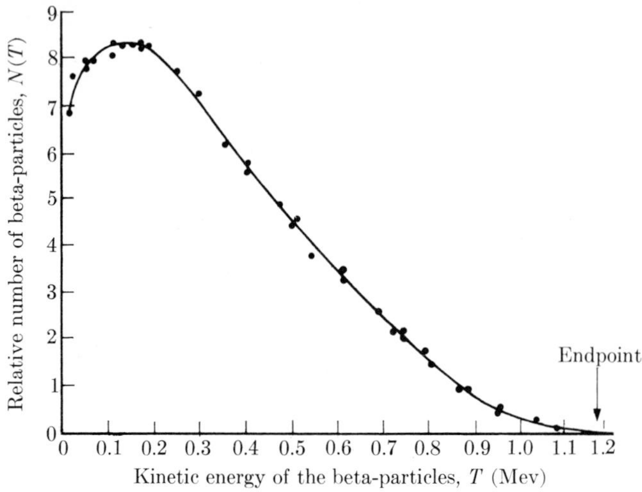

FIG. 14–2. The β-particle spectrum of RaE (Neary[(9)]).

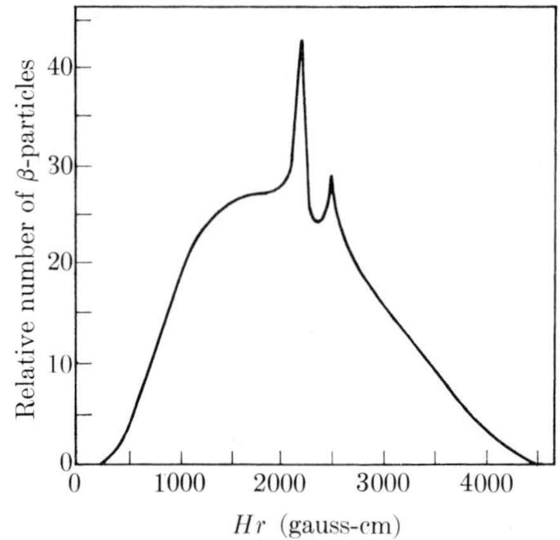

FIG. 14–3. Beta-particle spectrum of Au^{198}, showing the line spectrum. [Fan, *Phys. Rev.* **87,** 258 (1952).]

The results obtained when the β-particles (electrons) from a radioactive nuclide are studied with a magnetic spectrograph are more complex than those obtained with α-particles. If the β-particles, after being deflected, are allowed to fall on a photographic plate, a fogging of the plate is observed which extends for a definite distance and then ceases at a point characteristic of the β-emitting nuclide. This fogging is continuous, but of nonuniform intensity along the plate. In addition to this *continuous primary spectrum,* several sharp lines are sometimes seen on the plate at positions characteristic of the emitting substance. These lines form the so-called secondary, or *line spectrum*, observed for many, but not all, β-emitters. It is difficult, because of the continuous spectrum, to determine the numbers of β-particles with different energies from the darkening of the photographic plate. A velocity distribution curve (or an energy distribution curve) can be obtained by using a Geiger counter as the β-particle detector instead of a photographic plate; when this is done, the instrument is more properly called a spectrometer. The counter is placed in a fixed position (i.e., the value of r is fixed), the magnetic field is varied, and the number of β-particles reaching the counter per unit time is obtained for different values of H. Since each value of H corresponds to a different value of Hr and hence to a different value of the kinetic energy, the numbers of particles corresponding to different energy values are obtained. The continuous spectrum of β-particle energies found for RaE is shown in Fig. 14–2. In this case no line spectrum is found and the upper limit or maximum energy is at 1.17 Mev. When a line spectrum is also present, the lines appear as distinct peaks superimposed upon the continuous distribution curve, as in Fig. 14–3.

Beta-spectra will be discussed in greater detail in Section 14–4. For the present it will be noted that the spectrum of RaE shows the most important features of continuous spectra. Every continuous β-spectrum has a definite maximum, the height and position of which depend on the nucleus emitting the particles. There is also a definite upper limit of energy for the particles emitted by a nuclide; this upper limit, or *endpoint*, of the continuous spectrum is different for different nuclides. The shape of the curve at the lower energies is not known with certainty because it is hard to make accurate measurement on the low-energy particles. These properties of continuous spectra are observed for both natural and artificial β-emitters, and when the emitted particles are electrons or positrons.

The semicircular-focusing magnetic spectrometer is just one of several types of instruments for the analysis of beta-rays. Another type, often used, is the *solenoidal magnetic spectrometer* which is a variety of *magnetic lens spectrometer;* an example[3,18] is shown in Fig. 14–4. It consists of a long solenoid with the source of β-radiation on the axis near one end and a thin-walled Geiger-Mueller counter at the other end, about 90 cm from the source. The solenoid produces a homogeneous magnetic field whose direction is parallel to the axis of the cylinder. The field forces the particles to move along helical paths defined by a series of disks and rings, which act as baffles. By adjusting the current in the solenoid, and hence the magnetic field, β-particles from the source can be focused so that they make a single turn of the helix and just reach the counter. For a given value of the magnetic field only those particles which have a particular velocity will reach the counter. The velocity can be calculated from the

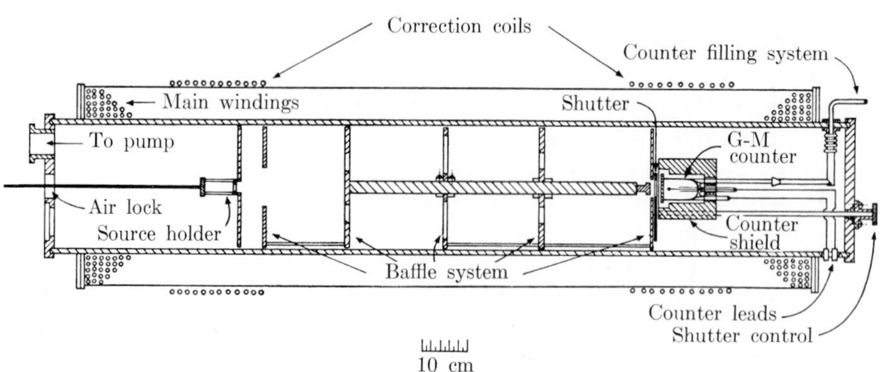

FIG. 14–4. Schematic diagram of a solenoidal magnetic spectrometer.[3,18]

magnetic field and the dimensions of the apparatus. In practice, however, instruments of this type are usually calibrated by means of electrons of known speed, and the velocity can then be determined directly from the current through the solenoid. In this way, the relative numbers of β-particles with different velocities can be determined.

The instrument just described has a relatively long, uniform magnetic field and is called a *long lens* magnetic spectrometer. Because of the difficulty of producing a really uniform magnetic field, the so-called *short lens* magnetic spectrometer is often used. In this case, the magnetic field is not uniform, but may be stronger in the region around the axis than at the axis, or may vary along the axis. The analysis of the motion of the β-particles is somewhat more complicated than in the case of the uniform field, but velocity and energy distributions can again be obtained with the help of suitable calibration. For further details about these and other instruments, the reader is referred to a recent review.[4]

14–2 The absorption of beta-particles. Range, ionization, and energy loss. The energies of β-particles can be determined from measurements of their absorption in matter,[5,6] as was the case with α-particles. Beta-particles are much more penetrating than α-particles and there are marked differences in the methods used for measuring the absorption in the two cases. An α-particle with a kinetic energy of 3 Mev has a range in standard air of about 2.8 cm and produces about 4000 ion pairs/mm of path. A 3-Mev β-particle has a range in air of over 1000 cm and produces only about 4 ion pairs/mm of path. Although the use of air as an absorbing medium for β-particles is therefore impractical, the particles are penetrating enough so that solid absorbers can be used.

Aluminum foils are the most frequently used absorbers, though gold and mica are often used. In absorption experiments, the absorber foils are

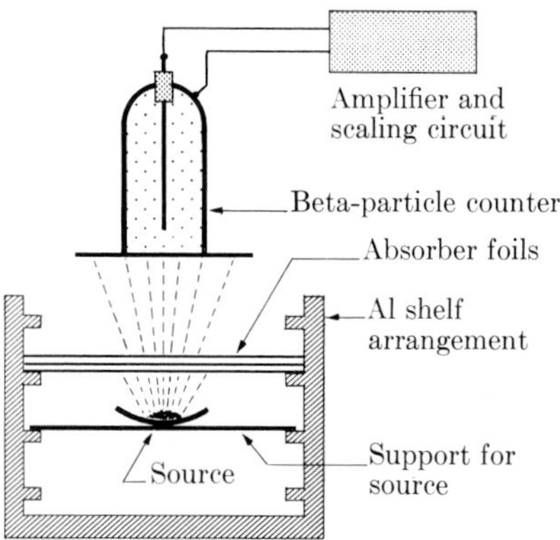

FIG. 14-5. Diagram of an experimental arrangement for measuring the absorption of β-particles (Glendenin[(5)]).

placed between the active source and a thin-windowed detector, usually a Geiger counter or an ionization chamber, and the counting rate is determined as a function of the absorber thickness. A convenient experimental arrangement is shown schematically in Fig. 14-5. A typical absorption curve is shown in Fig. 14-6 for the particles which form the continuous β-spectrum. The counting rate, plotted on a logarithmic scale, decreases linearly, or very nearly so, over a large fraction of the absorber thickness. In other words, the number of β-particles decreases exponentially with absorber thickness, to a good approximation, and the absorption may be represented by the formula

$$A(x) = A_0 e^{-\mu x}, \tag{14-6}$$

where A_0 is the counting rate, or activity, without absorber, $A(x)$ is the activity observed through a thickness x, and μ is the *absorption coefficient.* The activity does not decrease to zero as the absorber becomes very thick but becomes practically constant at a value which represents the so-called "background." There is always some radiation present which contributes to the counting rate even though it does not represent β-particles from the source. Near its end, the absorption curve deviates from the exponential form, and the point at which it meets the background is called the range R_β of the β-particles.

The exponential form of the curve is accidental, since it also includes the effects of the continuous energy distribution of the β-particles, and of the scattering of the particles by the absorber. The range R_β is the distance traversed by the most energetic particles emitted, and corresponds to the energy at the endpoint of the continuous spectrum. It is possible to obtain range-energy curves for electrons in different absorbers, and the

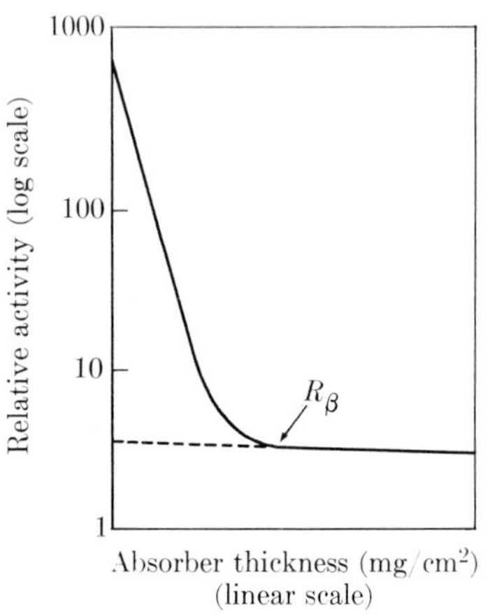

FIG. 14–6. A typical absorption curve for β-particles in aluminum. The β-particles are accompanied by γ-rays which form part of the background.

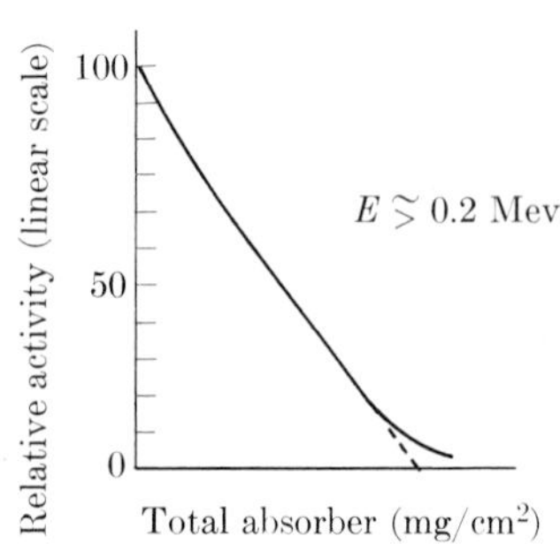

FIG. 14–7. A typical absorption curve for monoenergetic β-particles (Glendenin[(5)]).

endpoint energies of β-spectra can then be determined by absorption measurements. The absorption method does not give results comparable in accuracy to those obtained with the magnetic spectrometer, but it gives results of fair accuracy and has the advantages of simplicity and speed.

Thicknesses in absorption measurements are often given, as indicated in Fig. 14–6, in units of grams per square centimeter or milligrams per square centimeter of absorber. The actual thickness is multiplied by the density, and the resulting quantity, which may be called the *superficial density*, is used as a measure of the thickness. This usage has certain practical advantages. It is often inconvenient to measure the thickness of very thin foils, but their weight and cross section can be measured easily, and the weight divided by the cross section is equal to the product of the density and the thickness. Furthermore, it turns out that if the amount of absorber is expressed as the product of the density and thickness, the range is nearly independent of the nature of the absorber. This result follows from the fact that the range of a given β-particle depends not only on its initial energy, but also on the number of electrons with which it collides in passing through the absorber. The latter number depends on the density of electrons in the absorber or on the number of electrons per unit mass. The ability of an element to stop β-particles depends, therefore, on the ratio of the atomic number to the mass number, i.e., on Z/A. Although this ratio decreases from the light to the heavy elements, the effect of this variation on the thickness needed to stop β-particles is not large. In the case of aluminum, with $Z/A = 13/27 = 0.48$, the range of a 1-Mev β-particle is about 400 mg/cm^2; in gold, with $Z/A = 79/197 = 0.40$, the range is about 500 mg/cm^2, which is not very different from the range in aluminum.

The range may be obtained from the absorption curve simply by inspecting the curve, a method that is called the *visual* or *inspection* method. To find where the curve meets the background many accurate experimental points are needed close to the range thickness. Although this method is the simplest way of analyzing the absorption curve, it is probably the least reliable. In any case in which the curve approaches the background very slowly it is difficult to obtain good accuracy, and this kind of absorption curve is met often. To avoid the difficulties of the visual method, comparison methods have been developed[(6)] in which the range of the β-particles from a given nuclide is measured in terms of the range of a standard emitter. The standard emitter is one with an absorption curve that is particularly favorable for the visual method, and the absorption curve of the nuclide whose range is sought is compared in a detailed way with the curve of the standard substance. Still other methods of analyzing absorption curves have been developed;[(6)] these depend on fitting the experimental absorption curve with an empirical formula which includes either the range or the endpoint energy as a parameter.

The discussion so far may be applied to both β^--particles (electrons) and β^+-particles (positrons). In the case of β^+-radiation the shape of the absorption curve is much the same as for β^--particles with the exception of certain differences in the background. It is possible to correct for these differences, and ranges of β^+-particles can be determined with the methods used for β^--particles.

The absorption of monoenergetic β-particles is somewhat different from that of the particles which constitute the continuous spectrum. The particles which form a line in the discrete line spectrum often seen in β-decay are monoenergetic β^--particles and their absorption has practical importance. The absorption of monoenergetic electrons with energies greater than about 0.2 Mev is practically linear, as in Fig. 14-7. If the counting rate is plotted against the absorber thickness on ordinary graph paper, the curve closely approximates a straight line over most of the absorber thickness, with a small tail at the end of the curve. The linear portion of the curve can be extrapolated to cut the thickness axis at a value called the *extrapolated* or *effective range*. For monoenergetic electrons with energies less than 0.2 Mev, the absorption curve deviates more and more from a straight line as the energy decreases, and it becomes increasingly difficult to determine the effective range by absorption methods.

Quantitative results for ranges of β-particles in aluminum will be given in Section 14-3 in the course of the treatment of the range-energy curve for electrons.

The details of the passage of β-particles through matter are more complicated than is the case with α-particles. The differences are caused, among other things, by the much smaller mass of the β-particle and its greater speed. A β-particle may lose a large fraction of its energy in a single collision with an atomic electron, with the result that straggling is much more marked in the case of electrons than with heavier particles. Beta-particles are also scattered much more easily by nuclei than are α-particles, so that their paths are usually not straight. Consequently, even if a beam of β-particles is initially monoenergetic, the straggling and scattering make possible widely different path lengths for particles passing through the same thicknesses of absorber. The range is, therefore, a less precisely defined quantity for electrons than for α-particles or protons.

The energy lost by an electron per unit path length can be calculated from theory, and a formula has been derived for the energy lost by the particles because of ionization of absorber atoms. It is more complicated than the corresponding formula given for α-particles, Eq. (13–10), because relativistic effects must be taken into account. The theory predicts in a satisfactory way the dependence of the energy loss on the energy of the β-particle and on the atomic number of the absorbing element.

For high energy β-particles, an additional mechanism for losing energy must be taken into account. When an electron passes through the electric (Coulomb) field of a nucleus, it loses energy by radiation. This energy appears as a continuous x-ray spectrum called *bremsstrahlung* or *braking radiation*. The energy loss per unit path length because of radiation may be denoted by $(dT/dx)_{\text{rad}}$ as distinct from $(dT/dx)_{\text{ioniz}}$, the energy loss by ionization. The ratio of the two losses is given approximately by

$$\frac{(dT/dx)_{\text{rad}}}{(dT/dx)_{\text{ioniz}}} = \frac{TZ}{800},$$

where T is the β-particle energy in Mev and Z is the atomic number of the absorber. In heavy elements such as lead ($Z = 82$), the energy loss by radiation is significant even for a 1-Mev particle; in aluminum ($Z = 13$) the radiation loss amounts to only a few percent for the energies available from β-emitters.

In general, it can be said that even though the interaction between β-particles and matter is considerably more complicated than is the case for α-particles, the processes are well understood and there is good agreement between theory and experiment.

14–3 Range-energy relations for beta-particles. When the ranges of β-particles are known from absorption experiments, a range-energy relation can be used to obtain the maximum energies of β-spectra. The range-energy relation for aluminum has been determined and can be repre-

TABLE 14–2*

ENERGIES AND RANGES OF BETA-PARTICLES

Nuclide	Endpoint energy, Mev	Range, mg/cm^2
Ra^{228}	0.053	6
Rb^{87}	0.13	20
Nb^{95}	0.146	30
Lu^{176}	0.22	48
Co^{60}	0.31	81
Zr^{95}	0.400	122
Be^{10}	0.555	181
I^{131}	0.600	213
Sb^{124}	0.65	254
Mn^{56}	0.73	277
Au^{198}	0.97	399
C^{11}	0.98	447
Ba^{140}	1.022	426
Mn^{56}	1.05	462
Cd^{115}	1.13	527
Bi^{210} (RaE)	1.17	508
N^{13}	1.24	557
Na^{24}	1.39	601
Na^{24}	1.39	621
Sr^{89}	1.50	741
P^{32}	1.71	810
Te^{129}	1.80	812
Mg^{27}	1.80	821
Mg^{27}	1.80	885
Homogeneous rays	2.00	966
Y^{90}	2.18	1065
Bi^{212} (ThC)	2.25	1023
Rh^{106}	2.30	1080
Pa^{234}	2.32	1105
Sb^{124}	2.37	1220
As^{76}	2.56	1384
Rh^{104}	2.6	1198
Mn^{56}	2.86	1440
Cu^{62}	2.92	1440
Homogeneous rays	3.00	1540
Pr^{144}	3.07	1575
As^{76}	3.12	1454
Rh^{106}	3.55	1770

*Katz and Penfold: *Revs. Modern Phys.* **24,** 28 (1952).

sented either by a graph or by empirical equations.[6] The results of accurate determinations of some energies and ranges in aluminum are listed in Table 14–2, and the corresponding range-energy curve is shown in Fig. 14–8. The differently marked points represent ranges obtained by different workers who used various methods, and it is clear that there is good agreement among them. The values of the energy were determined with magnetic spectrometers and are taken from N.B.S. circular 499: *Nuclear Data*. The relation between range and energy can also be expressed by the empirical formulas

$$R = 412T_0^{1.265-0.0954 \ln T_0}; \qquad T_0 < 2.5 \text{ Mev}, \tag{14–7}$$

where T_0 is the endpoint energy, and

$$R = 530T_0 - 106; \qquad T_0 > 2.5 \text{ Mev}. \tag{14–8}$$

In Eqs. (14–7) and (14–8), T_0 is in Mev and R in mg/cm^2. As noted, Eq. (14–7) represents a good fit to the experimental results for endpoint energies less than 2.5 Mev, and Eq. (14–8) represents a good fit above 2.5 Mev. When a range is determined by an absorption measurement, the energy can be obtained either from the graph or from the appropriate empirical formula.

14–4 Beta-particle spectra. The continuous spectrum. The problems set by β-particle spectra are among the most important in nuclear physics, and the main features of these spectra will therefore be discussed in some detail. The discussion will be limited to the continuous spectrum because the β-particles which form this spectrum are the primary disintegration particles. The line spectrum, or secondary spectrum, is now known to consist of extranuclear electrons ejected from the atom by a process called *internal conversion*. A nucleus in an excited state can pass spontaneously to a state of the same nucleus, but of lower energy, either by emitting a γ-ray with an energy $h\nu$ equal to the difference between the energies of the two nuclear states, or by giving the energy to an electron in the K-, L-, . . . , shell of the same atom. The electron is ejected with kinetic energy $h\nu - E_K$, $h\nu - E_L, \ldots$, where E_K, E_L, . . . are the binding energies of the electron in the K-, L-, . . . , shells, respectively. This process is called *internal conversion* by analogy with the ordinary photoelectric conversion of the energy of the γ-ray in an atom other than the one from which the γ-ray was emitted. It was thought for some time that the energy of the nuclear transition was first emitted as γ-radiation, the γ-ray then ejecting a photoelectron just as it would from any atom through which it might pass. It has been shown that this explanation is incorrect and that the process of internal conversion is caused by the direct inter-

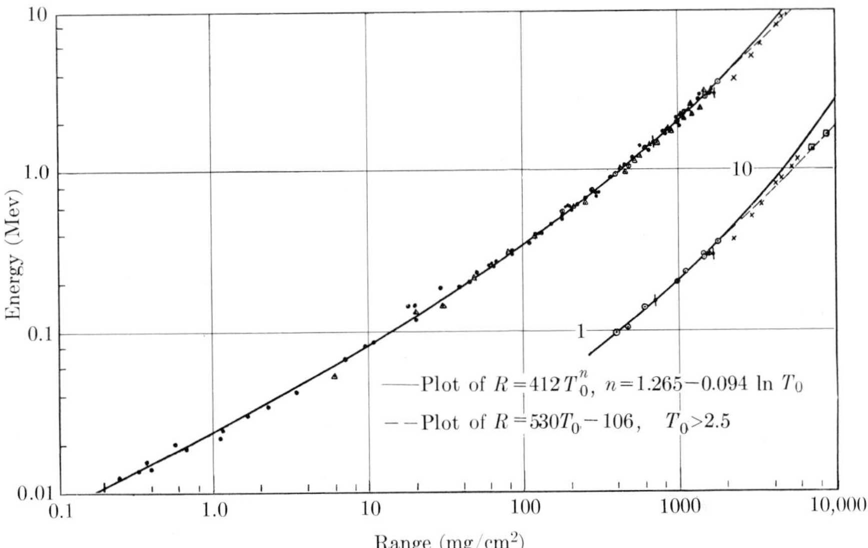

FIG. 14–8. Range-energy curve for β-particles (Katz and Penfold[(6)]).

action of the nucleus with its surrounding electrons. In any case, the nuclear charge does not change in internal conversion, so that this process is not a form of β-decay. The electron line spectrum is, therefore, related to the emission of γ-radiation rather than to the process of β-decay and will be treated more fully in the next chapter.

In view of the above discussion, the terms "β-spectrum" and "continuous β-spectrum" are synonymous, and the monoenergetic electrons which make up the lines found with the magnetic spectrograph will be disregarded in the rest of this chapter. Some typical β-spectra are shown in Fig. 14–9. In each case, the ordinate is proportional to the number of β-particles detected per unit range of Hr, and the abscissa is Hr. Since Hr is proportional to the momentum, the curves are momentum-distribution curves. They can be converted to energy distribution curves by converting the values of Hr to energy values with the aid of Fig. 14–1. The momentum-distribution curve has the same properties as the energy distribution curve and is used often because it represents the actual experimental results more directly. The relationship between the two distribution curves may be seen by comparing Fig. 14–9(e) for RaE with Fig. 14–2. In each case, the curve passes through a maximum and then decreases to zero at a value of the abscissa which represents the endpoint energy.

In some cases of β-emission, a single continuous spectrum is found, and therefore a single endpoint energy; such a spectrum is called *simple* and

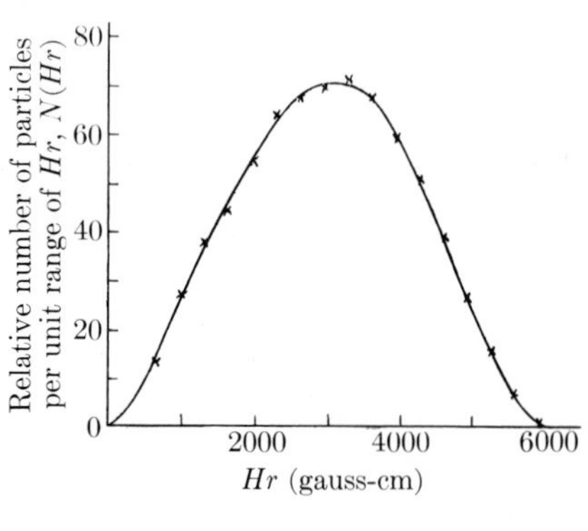

(a) Na^{24}

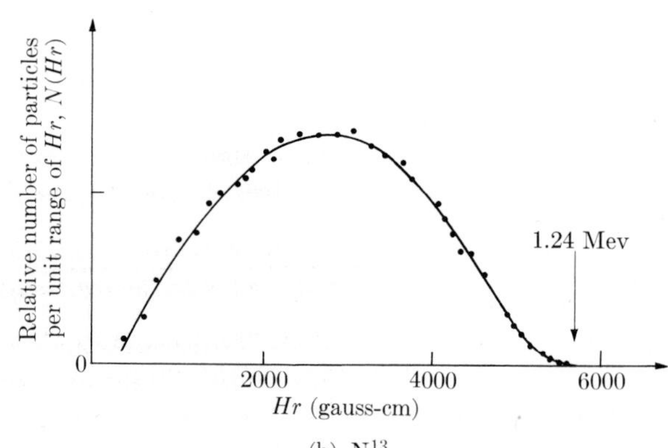

(b) N^{13}

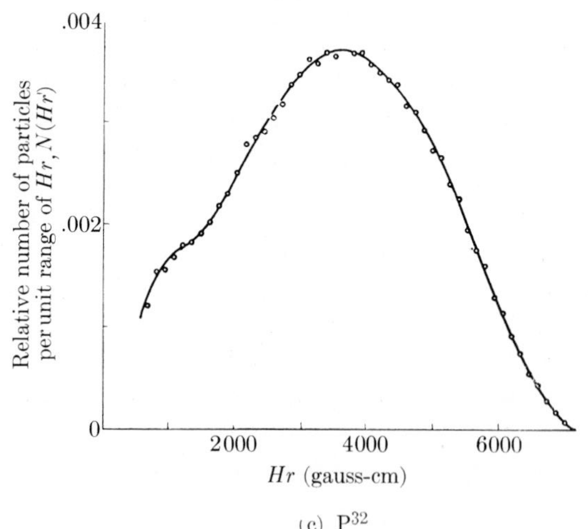

(c) P^{32}

FIG. 14–9. (*continued on next page*)

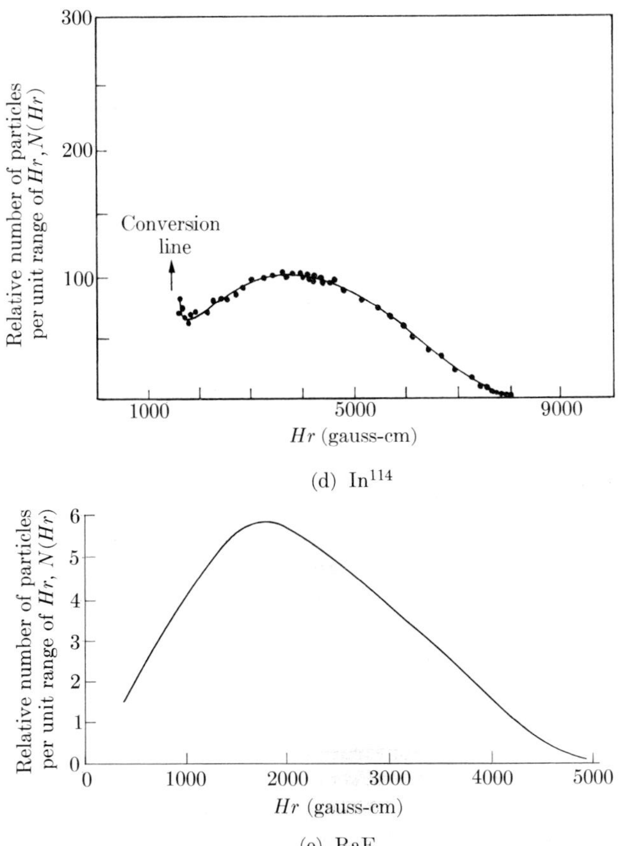

FIG. 14–9. Continuous β-particle spectra. (a) Na^{24}[Siegbahn, *Phys. Rev.* **70,** 127 (1946)]. (b) N^{13} [Siegbahn and Slatis, *Ark. Ast. Math. Fysik* **32A,** No. 9 (1945)]. (c) P^{32} [Warshaw, Chen, and Appleton, *Phys. Rev.* **80,** 288 (1950)]. (d) In^{114} (Lawson and Cork(17).) (e) RaE (Neary(9).)

is analogous to the emission of a single group of α-particles by an α-emitter. All of the spectra shown in Fig. 14–9 are simple. The emission of the β-particles may or may not be associated with the emission of γ-rays. Thus Na^{24}, Fig. 14–9(a), emits γ-rays as well as β-particles, while the other nuclides do not emit γ-rays. The nuclide N^{13}, Fig. 14–9(b), is a positron emitter and, as seen from the figure, its continuous spectrum is similar in its general features to the other spectra. Radium E, Fig. 14–9(e), is a naturally occurring radioactive nuclide, while the other nuclides are produced artificially. Phosphorus 32, Fig. 14–9(c), is especially interesting because of its use as a biological tracer material, and In^{114}, Fig. 14–9(d), is interesting for historical reasons which will be

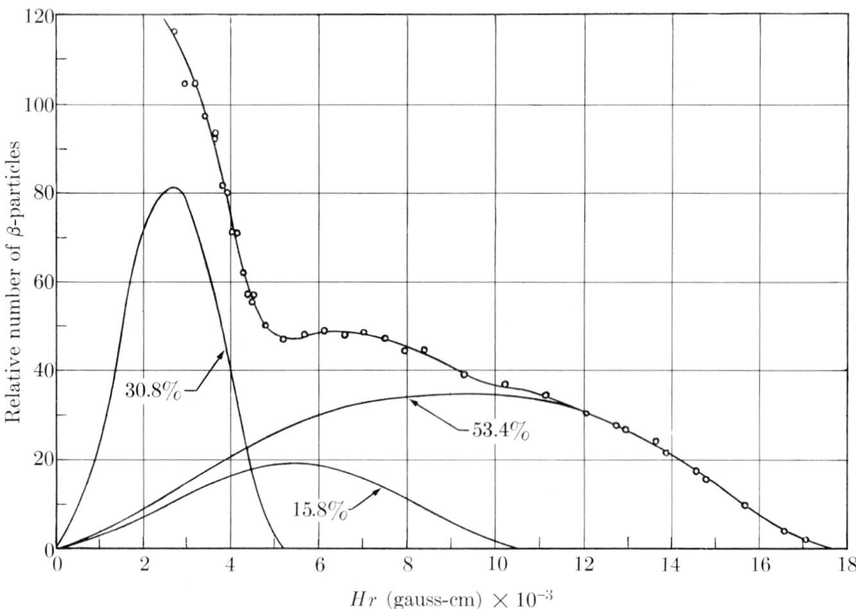

FIG. 14–10. The β-particle spectrum of Cl^{38} [Langer, *Phys. Rev.* **77,** 50 (1950)].

mentioned later. The endpoint energies for Na^{24}, Na^{13}, P^{32}, In^{114}, and RaE are 1.39, 1.20, 1.70, 1.98, and 1.17 Mev, respectively.

Many β-emitters have *complex spectra*, which means that there are two or more spectra with different endpoint energies and intensities. These spectra can be separated by analyzing the distribution curve obtained with the magnetic spectrometer. The spectrum of Cl^{38} is shown in Fig. 14–10, together with the three simple spectra of which it is composed. The curves are again momentum-distribution curves rather than energy-distribution curves. The endpoint energies are 4.81 Mev, 2.77 Mev, and 1.11 Mev, respectively, and the relative intensities are 53%, 16%, and 31%, respectively. Two γ-rays have been observed along with the β-particles, one with an energy of 2.15 Mev, the other with an energy of 1.60 Mev. One of the isotopes of antimony (Sb^{124}) has an even more complex spectrum made up of *five* simple spectra with endpoints of 2.291 Mev (21%), 1.69 Mev (7%), 0.95 Mev (7%), 0.68 Mev (26%), and 0.50 Mev (39%); at least six different γ-rays are observed.

The average energy of the β-particles which form the continuous spectrum can be calculated from the energy distribution curve. The total energy is found by integrating the product of the number of particles with a given energy and that energy. The number of particles is determined by integrating the number of particles with a given energy with respect to the energy; this number is just the area under the distribution curve. Expressed mathematically, the average energy is

$$\overline{T} = \frac{\int_0^{T_0} N(T) T\, dT}{\int_0^{T_0} N(T)\, dT}, \tag{14–9}$$

where T_0 is the endpoint energy, T is the energy, and $N(T)\,dT$ is the number of particles with energies between T and $T + dT$. For RaE, with an endpoint energy of 1.17 Mev, the average energy found in this way is 0.34 Mev. The average energy associated with a disintegration can be determined directly by placing a β-emitter inside a calorimeter designed to absorb all the known disintegration products, and by measuring the total heat energy produced by a known number of disintegrations. Experiments of this kind have been performed with RaE, and three independent observers[7,8,9] have obtained values of 0.35 ± 0.04 Mev, 0.337 ± 0.020 Mev, and 0.340 Mev, respectively. These calorimetric results are in excellent agreement with the value 0.34 Mev obtained from the distribution curve, so that the average energy is only about one-third of the maximum energy, 1.17 Mev.

It can be shown that when a β-transformation takes place, the energy of the nucleus decreases by an amount just equal to the maximum or endpoint energy of the emitted spectrum. Consider the decay of ThC to the ground state of ThD. This process can take place in either of two ways, as will now be shown. Thorium C (Bi^{212}) can decay by electron emission (maximum energy, 2.250 Mev), to ThC′, which decays by α-emission (α-disintegration energy, 8.946 Mev) to the ground state of ThD(Pb^{208}). The total energy change is $2.250 + 8.946 = 11.196$ Mev. On the other hand, ThC can decay to ThC″ by α-emission (α-disintegration energy, 6.203 Mev); ThC″ can then decay by β-emission (maximum energy, 1.792 Mev) to an excited state of ThD. The latter then decays by two successive γ-ray emissions with a total energy of 3.20 Mev to the ground state of ThD. The total energy change in the second mode of decay is $6.203 + 1.792 + 3.20 = 11.195$ Mev. The total energy change must be the same in the two cases, and it is, provided that the maximum β-decay energy is used in the calculations. If the average energy were used, there would be no such agreement.

Another example is the decay of C^{14} to N^{14} by electron emission. According to the result just obtained, the difference between the masses of C^{14} and N^{14} should be just equal to the mass corresponding to the endpoint energy of the β-spectrum of C^{14}, since no γ-ray is emitted. The latter energy is 0.155 Mev or $0.155/931.15 = 0.000166$ amu. The difference in mass between C^{14} and N^{14} can be obtained independently by considering the threshold reaction

$$C^{14} + p \rightarrow N^{14} + n + Q,$$

for which the value of Q has been found experimentally to be -0.628 ± 0.004 Mev. From Eq. (11–6), the mass difference is given by

$$M(C^{14}) - M(N^{14}) = m\,(\text{neutron}) - m\,(\text{proton}) - 0.628/931.15$$
$$= 1.008983 - 1.008144 - 0.000674 = 0.000165\ \text{amu},$$

which is just the same as the mass difference given by the maximum energy of the β-spectrum. Many other similar cases have been studied, and in each case the same result is obtained. There is, therefore, no doubt that the maximum energy of the spectrum is the one which enters into the balance of mass and energy in nuclear reactions.

The above results lead to the following paradox: a parent nucleus in a definite state of energy emits an electron and leaves a product nucleus which is also in a definite energy state. The energy of the emitted electron, however, is not equal to the difference between the energy of the nucleus before and after emission, but may have any value between zero and the maximum energy of the continuous spectrum; the average energy of the emitted particle is only about one-third of the maximum energy.[(10)] Part of the energy, approximately two-thirds in fact, seems to have disappeared, and the principle of conservation of energy seems to be violated. In view of the fact that this principle is basic to atomic and nuclear physics, as well as to all other branches of physics and chemistry, the possibility that it might not hold for β-decay presented a serious challenge to physical theory.

14–5 The theory of beta-decay. Basis of the theory. In 1934, there seemed to be only two ways of accounting for the difficulties raised by the continuous β-spectrum. One way was to give up the idea that energy is conserved during a β-decay process. This solution was not regarded with favor because the principle of conservation of energy had been uniformly successful in all of its previous applications and it was difficult to accept the idea that it should fail in β-decay. Furthermore, the concept of conservation is practically an article of faith to physical scientists and almost any other way out of a difficulty would be preferable to giving up this concept. The second solution, which was applied successfully by Fermi in his theory of β-decay, is based on the *neutrino hypothesis* first suggested by Pauli. It was postulated that an additional particle, the neutrino, is produced in β-decay and carries away the missing energy. Neutrinos had not been detected experimentally, and their properties must be such as to make them very hard to detect. It was, therefore, postulated further that the neutrino is electrically neutral and has a very small mass, i.e., very small compared with that of the electron; the neutrino mass may even be zero. The absence of charge is in accord with the requirement of conservation of charge during β-decay, so that this property is not an arbitrary one.

Under the neutrino hypothesis, each β-decay process is accompanied by the release of an amount of energy given by the endpoint of the spectrum. This disintegration energy is shared among the β-particle, the neutrino, and the recoil nucleus. It has been proved in analytical mechanics that when energy is divided among three particles, the principles of conservation of energy and momentum do not determine uniquely the energy and momentum of each particle, as they do in two-body problems. When this result is applied to β-decay it means that different nuclei of the same species distribute their equal available energies among the product particles in a continuous range of different ways. Thus, the hypothesis that a neutrino is also emitted makes it possible to account in a general way for the continuous spectrum of β-particle energies. To be acceptable, however, a theory of β-decay must accomplish more than this; it must account for the shape of the distribution curve, the existence of an endpoint energy, and the existence of a maximum in the curve. The theory must also give the correct relationship between the average energy and the endpoint energy. Finally, it must account for certain correlations, which will be discussed later in this section, between the endpoint energies and the decay constants in β-disintegrations. The Fermi theory of β-decay and its extensions, based on the neutrino hypothesis, have succeeded in accounting for all of the features of β-decay just listed. Although it may, at first, seem strange and arbitrary to postulate the existence of an undetected particle, this hypothesis has led to a highly successful theory of β-decay. The success of the theory may be regarded as presumptive evidence for the existence of the neutrino, and has led to a great deal of experimental and theoretical work on attempts to detect neutrinos directly. This work has finally succeeded, and the successful experiments will be discussed in Section 14–8.

The Fermi theory is based on the following ideas. When a nucleus emits a β-particle its charge changes by one unit, while its mass is practically unchanged. When the ejected β-particle is an electron, the number of protons in the nucleus is increased by one, and the number of neutrons is decreased by one. In positron emission the number of protons decreases by one and the number of neutrons increases by one. Beta-transformations may then be represented by the following processes:

$$\beta^{-}\text{-emission: } {}_0\text{n}^1 \rightarrow {}_1\text{H}^1 + {}_{-1}\text{e}^0 + \nu, \tag{14–10}$$

$$\beta^{+}\text{-emission: } {}_1\text{H}^1 \rightarrow {}_0\text{n}^1 + {}_1\text{e}^0 + \nu, \tag{14–11}$$

where ν represents the neutrino. The neutron is not to be regarded as composed of a proton, an electron, and a neutrino, but is considered to be transformed into those three particles at the instant of β-emission. Similarly, the proton is transformed at the instant of β-emission. The neutron or proton that is transformed is not, of course, a free particle but is bound in the nucleus by the nuclear forces. It has been found[11] that the free neutron undergoes β-decay according to Eq. (14–10), and has a half-life of about 12 min, and this phenomenon will be discussed in more detail later on. In a more complex radioactive nucleus, the neutrons cannot act as free particles; their decay is a property of the nucleus as a whole and is not to be confused with the decay of the free neutron.

Another property of the neutrino can be deduced from Eq. (14–10) or Eq. (14–11); this property is the intrinsic angular momentum or spin. It was seen in Chapter 8 that the measured values of the total angular momentum of different nuclei can be accounted for on the hypothesis that nuclei contain only protons and neutrons. A difficulty with the angular momentum arises, however, when one nucleus changes to another by emitting a β-particle. Suppose that a nucleus A emits a β-particle β, leaving a product nucleus B, and that no neutrino is emitted. The angular momentum of the system $(B + \beta)$ consists of three parts: the angular momentum of B, the rotational angular momentum of B and β about their common center of mass, and the spin of β. For the total angular momentum to be conserved, the resultant of these three contributions must be equal to the angular momentum of the initial nucleus A. Since the mass number of the nucleus is not changed by the decay process, the first contribution (angular momentum of B) must be equal to the angular momentum of A, or differ from it by an integral multiple of $h/2\pi$. It is shown in wave mechanics that the second contribution, which is a rotational angular momentum, can only be zero, or an integral multiple of $h/2\pi$. Finally, the spin of the β-particle (electron or positron) is known to be $\frac{1}{2}(h/2\pi)$. According to wave mechanics, when two or more angular momenta are added, the resultant is either equal to the sum of the components, or is less than the sum by an integral multiple of $h/2\pi$. When the three contributions are added in accordance with this rule, it is found that the total angular momentum of the system $(B + \beta)$ must differ from that of A by an odd multiple of $\frac{1}{2}(h/2\pi)$. The total angular momentum of a system of even mass number would be an integral multiple of $h/2\pi$ for the parent nucleus and a half-integral multiple of $h/2\pi$ for the product nucleus and β-particle together. For a system of odd mass number, the total angular momentum would be half-integral for the parent nucleus and integral for the product nucleus and β-particle together. If, however, it is assumed that a neutrino with spin $\frac{1}{2}(h/2\pi)$ is emitted, the resultant of the spins of the electron and the neutrino is integral (1 or 0, in units of $h/2\pi$). The resultant angular momentum of *all* particles left after the decay, product nucleus, electron, and neutrino, is then integral or half-integral according to whether that of the parent nucleus is integral or half-integral, and the total angular momentum of the system can be conserved.

An analogous argument can be made for the statistics of a system undergoing β-decay. Without the neutrino, the number of particles in the system differs by one unit before and after the β-transformation. This difference implies that the statistics would change, either from Fermi-Dirac to Bose-Einstein or vice versa. This kind of nonconservation is also unpleasant for the physicist to contemplate, and is avoided by assuming that the neutrino obeys the Fermi-Dirac statistics. The number of particles then changes by 2 units and, since neutron, proton, electron and neutrino are Fermi-Dirac particles, the initial and final nuclei obey the same kind of statistics.

The properties assumed for the neutrino may then by summarized as follows: no charge; very small mass, possibly zero, at least very small compared with the electron mass; spin equal to $\frac{1}{2}(h/2\pi)$; Fermi-Dirac statistics.

The neutrino hypothesis and the theory developed by Fermi apply also to the process of orbital electron capture, which may be represented by the equation

$$_1H^1 + {}_{-1}e^0 \rightarrow {}_0n^1 + \nu. \qquad (14\text{-}12)$$

It was seen in Section 12-2 that orbital electron capture can occur provided that the mass of the initial atom is greater than that of the final atom. The energy corresponding to this mass difference is released. In the process, the atomic number of the nucleus is changed from Z to $Z - 1$; the outer electrons rearrange themselves, and x-rays are emitted which are characteristic of the product nucleus. The total energy released, ΔE, must be the sum of the binding energy, E_e, of the orbital electron, the recoil energy, E_r, of the nucleus, and the kinetic energy of the neutrino. But ΔE as measured may be of the order of Mev, while the binding energy of the electron and the recoil energy of the nucleus are very much smaller. Nearly all of the available energy must be carried off by the neutrino, since no β-particle or γ-ray is observed; only in this way can energy be conserved. The neutrinos emitted during orbital electron capture all have the same energy, which is equal to $\Delta E - E_e - E_r$. The last result follows from the fact that the energy $\Delta E - E_e$ is divided between only two particles, whose energy and momentum are uniquely determined from simple kinematics. It also follows from Eq. (14-12) that the total angular momentum and statistics are conserved only if the spin of the neutrino is $\frac{1}{2}(h/2\pi)$ and if the neutrino obeys the Fermi-Dirac statistics.

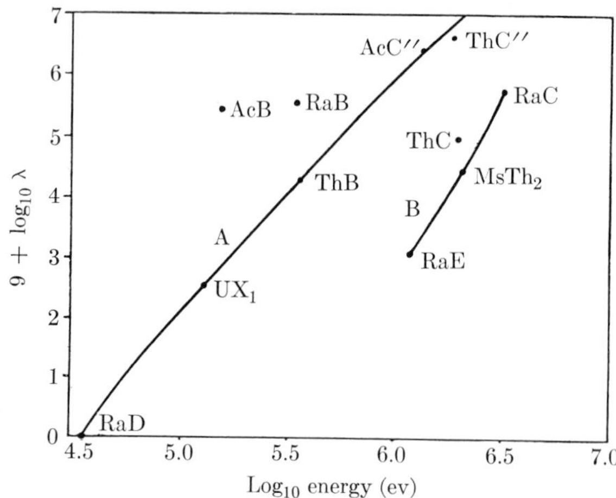

FIG. 14–11. Sargent diagram for natural β-emitters. The log of the decay constant is plotted against the log of the maximum energy of the β-particles (Sargent[12]).

Before the Fermi theory is discussed, it will help to consider certain correlations which have been observed between the endpoint energy and the half-life or disintegration constant. These correlations have an important place in the systematics of β-decay. Sargent[12] found that when the logarithm of the disintegration constant was plotted against the logarithm of the endpoint energy for the naturally occurring β-emitters, most of the points fell on or near two straight lines, as shown in Fig. 14–11. These lines, called *Sargent curves*, represent empirical rules which are analogous to the Geiger-Nuttall rule in α-decay. In contrast with the results for α-decay, in β-decay the disintegration constant does not vary so rapidly with the energy, and the two curves do not correspond each to one radioactive series. For a given value of the endpoint energy, the upper curve in Fig. 14–11 gives a value of λ which is about 100 times as great as the corresponding value on the lower curve. For a given value of the energy, a transformation on the lower curve is, therefore, about 100 times less likely than one on the upper curve. The upper curve is said to represent *allowed transitions*, and the lower curve is said to represent *forbidden transitions*. The terms *allowed* and *forbidden* are to be considered in a relative sense, and stand for different degrees of probability of spontaneous disintegration.

Sargent's idea of plotting log λ against log T_0 has been extended to artificially made β-emitters and a large number of Sargent curves has been obtained.[13] It is necessary to consider separately the curves for nuclides of small, intermediate, and large atomic numbers. In each case several curves are obtained, one of which is considered to represent allowed transitions; the others represent transitions with different degrees of forbiddenness. These regularities in the experimental results mean that the theory of β-decay must account for the different relative probabilities of β-emission as well as for the properties of the energy spectrum.

14–6 The theory of beta-decay. Results and comparison with experiment. The problem of the theory of β-decay is to calculate the probability of the processes of Eqs. (14–10), (14–11), and (14–12). To make the calculations it is necessary to introduce a force which can produce the changes expressed by those equations, i.e., convert a neutron into a proton, or vice versa, and at the same time produce an electron or positron and a neutrino. Fermi introduced such a force and treated its operation mathematically by methods analogous to those used in the quantum mechanical treatment of the emission of electromagnetic radiation. The Fermi theory[14,15] yields a formula for the energy distribution of the emitted β-particles, i.e., for the fraction of the nuclei which disintegrate per unit time by emitting a β-particle with kinetic energy between T and $T + dT$. This fraction is

$$P(T)\,dT = G^2|M|^2F(Z,T)\,(T + m_0c^2)\,(T^2 + 2m_0c^2T)^{1/2}\,(T_0 - T)^2\,dT. \tag{14–13}$$

Equation (14–13) will be discussed in some detail, and its predictions compared with experimental results. The quantity G is a natural constant which represents the strength of the Fermi force; its actual magnitude will not be important in the following discussion. $F(Z,T)$ is a complicated function which describes the effect of the Coulomb field of the nucleus on the emitted β-particle. For $Z = 0$, F is unity; it changes very slowly as Z increases up to about 20, after which it varies more rapidly. Electron and positron emission are treated by taking Z positive in the formula for F when electrons are emitted, and negative when positrons are emitted. The factor $|M|^2$ is the square of the absolute magnitude of a wave mechanical quantity M, called a *matrix element*, which is a measure of the relative probability that the nucleus will undergo a β-transformation. In most cases, M cannot be calculated exactly, but general considerations about it make possible theoretical distinctions between allowed and forbidden transitions. The other quantities on the right side of Eq. (14–13) form a statistical factor which describes how the energy T_0 is shared among the electron, the neutrino, and the product nucleus. The statistical factor is responsible for the general properties of the continuous β-spectrum; T_0 is the energy available for the disintegration and m_0 is the rest mass of the electron.

The structure of Eq. (14–13) may then be described as follows. The factor $G^2|M|^2$ is a measure of the relative probability that the nucleus may emit a β-particle and a neutrino; $F(Z,T)$ shows the effect of the Coulomb field of the nucleus on the emission of the β-particle, and the statistical factor shows what fraction of the available disintegration energy is carried off by the β-particle. All of the factors together give the required probability that a β-particle will be emitted with the appropriate energy.

For transitions of the allowed type in nuclides with low values of Z, Eq. (14–13) may be simplified; F may be taken equal to unity and M may be taken to be independent of the energy. The distinction between allowed and forbidden transitions will be made clearer later in this section. Until then, an allowed transition should be regarded simply as one with a relatively high probability, i.e., relatively short half-life and large disintegration constant. With the above approximation, the theoretical decay probability may be written

$$P(T)\,dT = (G')^2(T + m_0c^2)\,(T^2 + 2m_0c^2T)^{1/2}(T_0 - T)^2\,dT, \tag{14–14}$$

where G' is a new constant. Equation (14–14) holds for such β-emitters as the neutron, H^3, He^6, C^{11}, N^{13}, O^{15}, and F^{17}, and it is easier to discuss than Eq. (14–13). It shows that for $T = 0$ and $T = T_0$ the probability of β-decay vanishes, while for values of T between 0 and T_0, $P(T)$ is positive. When $P(T)$ is plotted against T, the value is zero when $T = 0$, increases as T increases, passes through a maximum, and then decreases to zero at $T = T_0$. For a given value of T_0, the average value of the energy of the emitted β-particle can be found by integration and is in good agreement with measured values of the average energy.[10] The theoretical expression for the spectrum in an allowed transition thus gives correctly the most important features of experimental β-spectra.

The theoretical values of the disintegration constant λ and the mean life τ are obtained by integrating Eq. (14–13) or Eq. (14–14) over the possible values of the kinetic energy:

$$\lambda = \frac{1}{\tau} = \int_0^{T_0} P(T)\,dT. \tag{14–15}$$

In the case of the simple allowed spectra represented by Eq. (14–14), the integration can be done analytically and λ can be expressed directly in terms of the endpoint energy T_0. It is found that, to a first approximation, the disintegration constant is proportional to the fifth power of the endpoint energy

$$\lambda \approx kT_0^5 \ldots,$$

where k is a constant. The last equation may be written

$$\log \lambda \approx \log k + 5 \log T_0,$$

and should represent, at least roughly, the Sargent curves. It is found that this is the case, since the slopes of the Sargent lines are approximately 5. When the effects of $F(Z,T)$ and M must be taken into account, as in allowed transitions of nuclei with greater values of Z, or in forbidden transitions, the comparison is less direct and will be discussed later.

In practice, the quantitative comparison between the theoretical and experimental spectra is made in an ingenious way involving the so-called *Kurie plot,* or *Fermi plot.* It is convenient to rewrite Eq. (14–13) before discussing this method. It will be assumed that M is independent of the energy and can be regarded as a constant. The expression $(T + m_0c^2)$ is just the total energy E of the emitted electron, since $m_0c^2(= 0.511 \text{ Mev})$ is the rest energy of the electron. It can be shown that in relativistic mechanics, the expression $(T^2 + 2m_0c^2T)^{1/2}$ is equal to the "momentum" pc. The difference $T_0 - T$ is equal to $E_0 - E$. The fraction $P(T)\,dT$ is proportional to the number $N(T)\,dT$ of β-particles with energies between T and $T + dT$. Equation (14–13) may then be replaced by

$$N(T) \sim FEp(T_0 - T)^2 = FEp(E_0 - E)^2, \tag{14–16}$$

or

$$[N(T)/FEp]^{1/2} \sim T_0 - T = E_0 - E. \tag{14–17}$$

The Fermi theory predicts, therefore, that if the quantity $[N(T)/FEp]^{1/2}$ is plotted against T or against E, the result should be a straight line which intersects the energy axis at $T = T_0$, or at $E = E_0$. If $N(T)$ is taken to be the number of β-particles observed experimentally, then the plot should be a test of the Fermi theory. This method of comparing theory and experiment, suggested by Kurie, Richardson, and Paxton,[(16)] has the advantage that it is independent of the actual measurement of T_0. If a straight line is obtained, it is good evidence for the validity of the theory, and the Kurie plot then provides a straight-line extrapolation method for finding the endpoint energy.

In practice, the momentum distribution of the emitted β-particles is often used. A function $P(p)$ can be defined so that $P(p)\,dp$ is the fraction of disintegrations that give β-particles with momenta between p and $p + dp$. To each energy interval dT there corresponds a momentum interval dp. The number of processes is the same, by definition, for corresponding intervals, so that $P(T)\,dT = P(p)\,dp$, or $N(T)\,dT = N(p)\,dp$. It is then found that the energy distribution, Eq. (14–13), is replaced by the momentum distribution

$$P(p)\,dp = cG^2|M|^2F(Z,p)p^2(T_0 - T)^2\,dp. \tag{14–18}$$

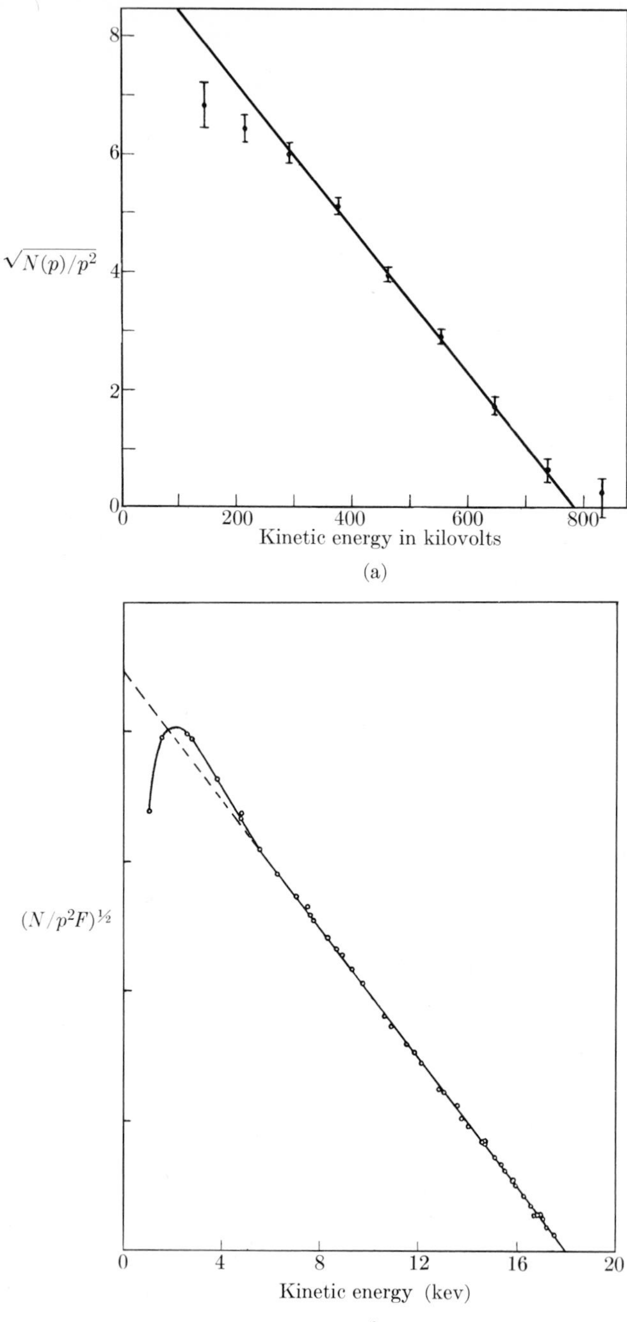

FIG. 14–12. (*continued on next page*)

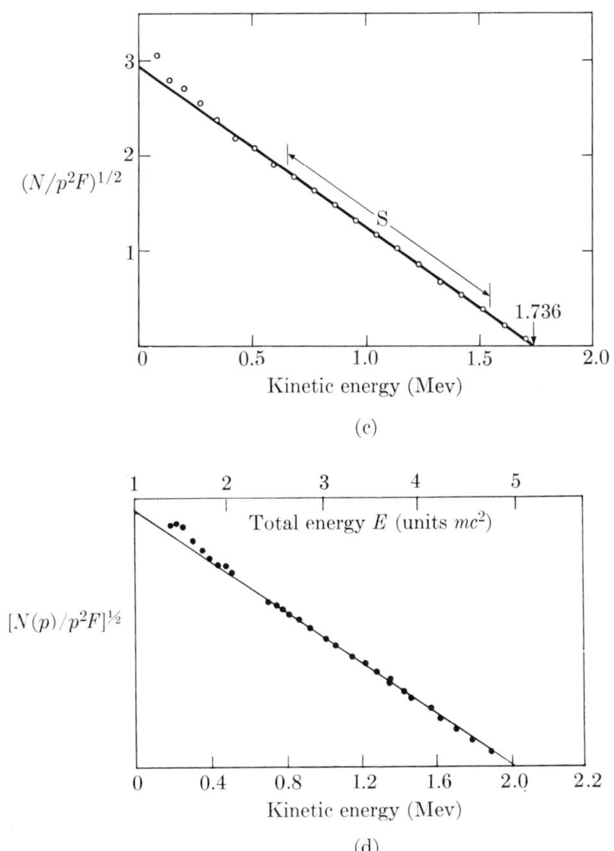

FIG. 14–12. Kurie, or Fermi, plots for allowed β-transformations. (a) The neutron (Robson[11]). (b) Tritium: H^3 (Langer and Moffat[27]). (c) O^{15} [Kistner et al., *Phys. Rev.* **105,** 1339 (1957)]. (d) In^{114} (Lawson and Cork[17]).

In the Kurie plot, the function $[N(p)/Fp^2]^{1/2}$ can be plotted against T, where $N(p)$ is the number of β-particles counted which have a momentum p. The result should again be a straight line.

Kurie plots are shown in Fig. 14–12 for some β-emitters of special interest, the neutron, tritium (H^3), O^{15}, and In^{114}. A straight line is obtained in each case, as well as an accurate value of the endpoint energy. These results, together with many others like them, provide strong evidence for the validity of the Fermi theory of allowed spectra. The Kurie plot for the neutron, Fig. 14–12(a), yields an endpoint energy of 0.782 $\pm$ 0.013 Mev. This value is just the one predicted from the difference between the masses of the neutron and the proton, which is 0.00084 amu or 0.782 Mev. Tritium is the simplest electron-emitting nuclide, apart from the neutron; its Kurie plot, Fig. 14–12(b), gives an endpoint energy of 17.95 ± 0.10 kev. This value and the plot shown are important because, as will be discussed later, they make possible an estimate of the mass of the neutrino. The plot for O^{15}, Fig. 14–12(c), is an example of the spectrum of an allowed positron transformation; the endpoint energy is 1.736 ± 0.010 Mev. Finally, the Kurie plot for In^{114}, Fig. 14–12(d), was the first case[(17)] for which unambiguous confirmation of the Fermi theory of allowed transformations was obtained.

For certain highly forbidden transformations, for which the half-life is relatively long and the disintegration constant small, the shape of the spectrum departs considerably from that predicted by the simpler forms of the theory. These deviations have been accounted for in a satisfactory way as a result of further development of the theory and refinements in the experimental methods. It turns out that in these cases, the matrix element M also depends on the energy and affects the shape of the spectrum. When this additional energy dependence is taken into account, good agreement is obtained between theory and experiment.[(18,19,20)]

The effect of $|M|^2$ on the disintegration constant and the half-life, and hence its relation to the problem of allowed and forbidden transformations, will be discussed next. Although the magnitude of $|M|^2$ cannot be calculated accurately in most cases, part of its effect on λ may be estimated without detailed calculations. The theory makes it possible to express $|M|^2$ as an infinite series, and to estimate in a rough way the relative magnitudes of the successive terms of the series. It is found that the second term is smaller than the first by a factor of about 100, the third term is smaller than the second by about the same factor, and so on. The magnitude of $|M|^2$ is, therefore, practically the same as that of the first nonvanishing term. If the first term is not zero, the succeeding terms may be neglected; $|M|^2$ then has a certain value, as do $P(T)$ and λ, and the transformation is called *allowed.* If the first term of the series vanishes, and the second term does not, the third and succeeding terms may be neglected. The value of $|M|^2$ is given by the second term of the series, and is smaller than in the allowed case by a factor of about 100. The values of $P(T)$ and λ are also correspondingly smaller than in the allowed case, and the transformation is called *first forbidden* or *once forbidden.* Similarly, if the third term of the series is the first nonvanishing one, $|M|^2$, P, and λ are still smaller, by another factor of about 100, and the transformation is called *second forbidden* or *twice forbidden.* Higher degrees of forbiddenness may be arrived at by continuing this argument.

The question immediately arises as to why the first term, or the first and second, or any number of leading terms in the expansion for $|M|^2$ should vanish. The theory again provides an answer. In a nuclear transformation, the magnitude I of the total angular momentum of the nucleus may change by an integral multiple of $h/2\pi$, with the difference ΔI appearing, for example, as orbital angular momentum of the electron and the neutrino. As in the case of atomic transitions (cf. Sections 7–3, 4), there is a selection rule which defines the values that ΔI may have. If a particular transformation obeys the selection rule, the first term in the series for $|M|^2$ does not vanish, and the transition is allowed. This result follows from the way in which the angular momenta of the initial and final nuclei are worked into the expression for $|M|^2$ by the theory. If the selection rule is not obeyed, the first term in the series for $|M|^2$ vanishes and the transition can be, at most, first forbidden. There is also a selection rule for first forbidden transitions. If it is obeyed, the transition actually is a first forbidden one; if not, the transition can be, at most, second forbidden. Depending on which selection rule is obeyed, a β-transformation may be an allowed one, or may have a particular degree of forbiddenness.

The selection rules derived from the theory depend on the way in which the special interaction introduced into the theory is defined. This interaction can be defined in several different ways, and there is no *a priori* physical reason for choosing one definition over another. There is, therefore, some arbitrariness in the theory which still remains to be removed. The first, and simplest, definition of the interaction was suggested by Fermi and led to the selection rule $\Delta I = 0$ for allowed transformations. Inconsistencies were found between theory and experiment, and Gamow and Teller used a different definition of the interaction, which led to the selection rule $\Delta I = 0, \pm 1$. The most recent work in this branch of the field of β-decay indicates that the actual interaction must be a combination of the Fermi and Gamow-Teller types. The problem of the exact definition of the interaction that leads to β-decay has not yet been completely solved, but nearly all of the experimental facts of β-decay can be explained by the theory, and the problem mentioned seems to be well on the way to solution.[20]

In spite of the fact that the matrix element $|M|^2$ cannot be calculated accurately in most cases, it is possible to treat differences between allowed and forbidden transitions quantitatively, or at least semiquantitatively. The mean life of an allowed transition, for which $|M|^2$ is independent of the energy, is, from Eqs. (14–13) and (14–15),

$$\frac{1}{\tau} = \lambda = G^2|M|^2 \int_0^{T_0} F(Z, T)(T + m_0c^2)(T^2 + 2m_0c^2T)^{1/2}(T_0 - T)^2\, dT. \tag{14–19}$$

The integral in the last equation is a function only of Z and T_0. Its value may be computed by analytical or numerical integration depending on whether or not approximate expressions for F may be used. In either case, the integral may be represented by a function $f(Z,T_0)$, and Eq. (14–19) may be written

$$1/\tau = G^2|M|^2 f(Z,T_0), \tag{14–20}$$

or

$$f\tau = 1/G^2|M|^2. \tag{14–21}$$

The product $f\tau$ is a quantity whose numerical value depends only on the atomic number and the measured values of the endpoint energy and the mean life. If the dependence of $|M|^2$ on energy is neglected in the case of forbidden transitions, the last equation may be applied also to those transitions. Although some error is involved because $|M|^2$ for forbidden transitions does depend to some extent on the energy, the product $f\tau$ may be used as an indication of the degree of forbiddenness. Since the value of $|M|^2$ for a first forbidden transition is smaller by a factor of about 100 than the value for an allowed transition, it would be expected that the value of $f\tau$ would be about 100 times as large for a first forbidden transformation as for an allowed one.

In practice, the product of the function f and the half-life is usually calculated. The half-life has been denoted by $T_{1/2}$, so that the values of the product $fT_{1/2}$ should be considered, with $fT_{1/2} = 0.693f\tau$. The half-life, however, is often denoted by t, and the product ft is called the *comparative half-life*, or the ft-value. The ft-values of several hundred β-emitters have now been determined[(21)] and analyzed.[(20,22)] They range from about 1000 sec to about 10^{18} sec and, for convenience, the logarithm of the ft-value is usually considered rather than the ft-value itself. The smallest values of the comparative half-life are found for a group of light nuclei for which $\log ft = 2.7$ to 3.7. This group includes the lightest β-emitters, such as the neutron ($\log ft = 3.21$), H^3 ($\log ft = 3.06$), He^6 ($\log ft = 2.74$). It also includes the positron emitters: ${}_5C^{11}$, ${}_7N^{13}$, ${}_8O^{15}$, ${}_9F^{17}$, ${}_{10}Ne^{19}$, ${}_{11}Na^{21}$, ${}_{12}Mg^{23}$, ${}_{13}Al^{25}$, ${}_{14}Si^{27}$, ${}_{15}P^{29}$, ${}_{16}S^{31}$, ${}_{17}Cl^{33}$, ${}_{18}A^{35}$, ${}_{19}K^{37}$, ${}_{20}Ca^{39}$, ${}_{21}Sc^{41}$, and ${}_{22}Ti^{43}$. In the latter nuclides, the mass number A is equal to $2Z - 1$, and the number of protons exceeds the number of neutrons by one. The emission of a positron makes the product nucleus have one more neutron than proton. Thus, in the reaction

$${}_6C^{11} \rightarrow {}_5B^{11} + {}_1e^0 + \nu,$$

the parent nucleus ${}_6C^{11}$ has six protons and five neutrons, while the product nucleus ${}_5B^{11}$ has five protons and six neutrons. In these cases, the initial and product nuclides are called *mirror nuclides*, because of the symmetrical relationships between the numbers of protons and neutrons. The emission of a positron has a particularly high probability in the mirror nuclides and the ft-values are especially low. The transformations of this group with $\log ft$ between 2.7 and 3.7 are described as *favored* allowed transitions.

Another group of nuclides is found for which the values of log ft lie between 4 and 5.8. The transformations are described as *normal* allowed transitions, and include such cases as B^{12}, N^{12}, S^{35}, Cu^{64}, and In^{114}. The nuclide Cu^{64} is particularly interesting because it undergoes electron emission, positron emission, and K-electron capture; the value of log ft is 5.29 for electron emission and 4.94 for positron emission, and both transitions are allowed.

Values of log ft in the range 6–9 are characteristic of first forbidden or once-forbidden transitions. Nuclides such as Zr^{97}, Cd^{115}, Ba^{140}, and W^{187} are included in this group; RaE is a borderline case which is probably first forbidden. There is a relatively small number of second forbidden transitions, for which the value of log ft is in the range 12.2–13.5; Cl^{36}, Te^{99}, Cs^{135}, and Cs^{137} are examples of this kind of transformation. Finally, Be^{10}, with a value of log ft equal to 13.7, and K^{40} with log ft = 17.6, are cases of nuclides whose β-transformations are even more highly forbidden. The degree of forbiddenness in nearly all of the above cases can be correlated in a satisfactory way with the shape of the spectrum and with the known or assumed angular momenta of parent and daughter nuclei.

14–7 Energy levels and decay schemes. Beta-transformations often yield information about the energy levels of the product nuclei and about decay schemes. These transformations are sometimes accompanied by γ-rays, and the presence of the γ-radiation means that the product nucleus is formed in an excited state and passes to its ground state by emitting one or more γ-rays. If no γ-ray is emitted, the β-transition is directly to the ground state of the product nucleus. In the case of O^{15} a positron is emitted, and no γ-ray is observed. The endpoint energy is 1.73 Mev, and the β-disintegration energy is 1.73 Mev + 1.02 Mev = 2.75 Mev. The latter result follows from the fact, discussed in Section 12–2, that two extra electron masses are needed for positron emission to be energetically possible. It follows that the difference between the ground state energies of O^{15} and N^{15} is 2.75 Mev, which is also the energy equivalent to the difference between the masses of these two nuclides. The nuclide F^{20} (Fig. 14–13a) emits electrons with an endpoint energy of 5.41 Mev, and also 1.63-Mev γ-rays. The total disintegration energy is 7.04 Mev, and the product nucleus Ne^{20} has an excited state 1.63 Mev above the ground state.

The beta-spectra of O^{15} and F^{20} are simple and the decay schemes are also simple. Sometimes, however, two or more groups of β-particles are emitted. The spectrum is then complex; it can be broken down into two or more simple spectra, and γ-rays are observed. In the case of O^{14}, in more than 99% of the disintegrations, positrons are emitted with an endpoint energy of 1.84 Mev; 2.30-Mev γ-rays are also observed. The total

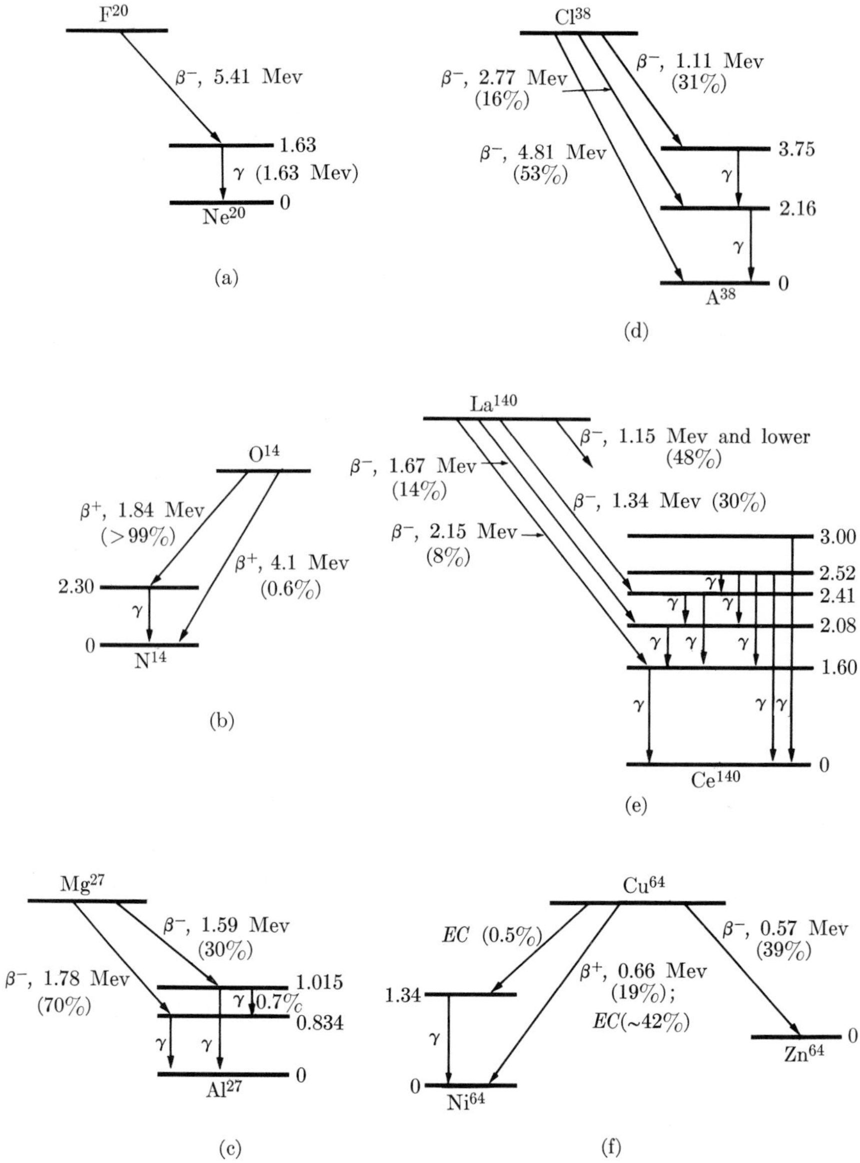

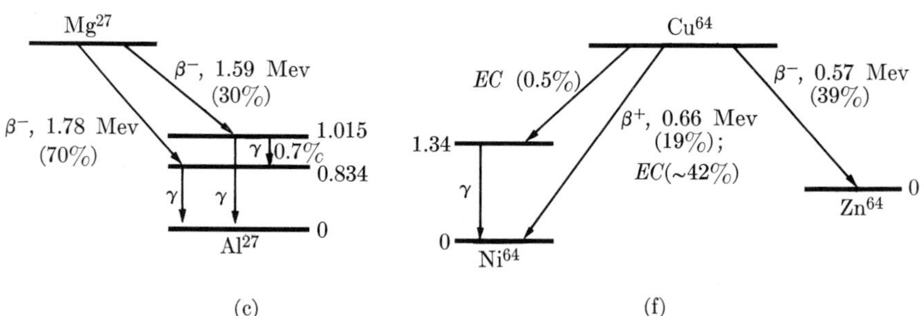

FIG. 14–13. Decay schemes for β-emitters. (a) F^{20}. (b) O^{14}. (c) Mg^{27}. (d) Cl^{38}. (e) La^{140}. (f) Cu^{64}. [Strominger, Hollander, and Seaborg, *Revs. Mod. Phys.* **30,** 585 (1958).]

disintegration energy is 1.84 Mev + 1.02 Mev + 2.30 Mev = 5.16 Mev, of which 2.86 Mev is the difference in energy between the ground state of O^{14} and the excited state of the product nucleus N^{14}. The N^{14} nucleus passes to its ground state by emitting a 2.30-Mev γ-ray. In about 0.6% of the disintegrations O^{14} undergoes a transition directly to the ground state of N^{14} by emitting 4.1-Mev positrons. The decay scheme is shown in Fig. 14–13(b). The electron-decay of Mg^{27} is more complicated: about 70% of the disintegrations correspond to an endpoint energy of 1.78 Mev and about 30% to an endpoint energy of 1.59 Mev; γ-rays are observed with energies of 0.834 Mev and 1.015 Mev, respectively, and in less than one percent of the disintegrations, a γ-ray with an energy of 0.18 Mev is observed. Coincidence experiments show that the 1.78-Mev β-ray and the 0.834-Mev γ-ray belong to the same transition, (i.e., they are *coincident*), and that the 1.59-Mev β-ray and the 1.015-Mev γ-ray belong to the same transition. A decay scheme consistent with all of these data is shown in Fig. 14–13(c). The direct transition from the ground state of Mg^{27} to the ground state of Al^{27} by electron-emission is evidently highly forbidden. The electron decay of Cl^{38} has three groups of electrons, as discussed in Section 14–4; two γ-rays have been observed, and the decay scheme is shown in Fig. 14–13(d). A still more complicated scheme is that of $_{57}La^{140}$, shown in Fig. 14–13(e); at least four groups of electrons are involved and at least nine γ-rays have been observed.

The nuclide Cu^{64} is a particularly interesting case of β-decay because it emits both electrons and positrons and also undergoes orbital electron capture. In 39% of the disintegrations, an electron is emitted; the β^--spectrum is simple and the endpoint energy is 0.57 Mev. The product nucleus Zn^{64} is formed in its ground state. In 19% of the disintegrations, a positron is emitted with an endpoint energy of 0.66 Mev; the product nucleus Ni^{64} is formed in its ground state. In 42% of the disintegrations, a K-electron is captured. In nearly all of the captures, the product nucleus Ni^{64} is formed in its ground state, but in a small fraction of the K-captures, a γ-ray is observed with an energy of 1.34 Mev. There is, therefore, an excited level of Ni^{64} 1.34 Mev above the ground state. It has been shown that the γ-ray is observed *only* in coincidence with the orbital electron capture, and is not associated with the emission of either the electron or the positron. The decay scheme of Cu^{64} is shown in Fig. 14–13(f).

The examples of decay schemes that have just been discussed represent a small fraction of the schemes that have been analyzed. Several hundred decay schemes have now been determined, varying in complexity and in the detail with which they have been worked out. Compilations of these schemes are useful in many phases of nuclear physics, and the reader is referred to that by Strominger et al., listed among the general references at the end of the chapter.

14–8 The neutrino. The success of the Fermi theory of β-decay in accounting for the properties of β-transformations provides indirect evidence of the existence of neutrinos. This success encouraged attempts to show experimentally that neutrinos exist.(23) But the properties which the neutrino must have to make it fit the requirements of the theory, in particular the zero charge and very small mass, imply that the interaction between neutrinos and matter is extremely small. The smallness of this interaction is emphasized by the fact that the discovery of the neutron with a mass very much greater than that assumed for the neutrino presented some difficulty, and it is not surprising that attempts to detect ionization caused by neutrinos failed. Additional indirect evidence for the existence of neutrinos has, therefore, been sought from studies of the momentum and energy balances in β-decay. It has also been possible to deduce, from nuclear data, an upper limit for the rest mass of the neutrino. Finally, in 1956, a nuclear reaction induced by neutrinos was observed, an inverse β-decay, and the experimental search for the neutrino has ended successfully.

We consider first the problem of the mass of the neutrino. Estimates of this mass can be obtained from two types of experiments, comparison of observed maximum energies of β-ray spectra with known available decay energies and the shape of β-ray spectra near the endpoint.

The greatest kinetic energy with which a β-particle can be emitted is, for electrons,

$$T_0 = (\Delta M - \mu)c^2, \tag{14–22a}$$

where ΔM is the difference in mass between parent and daughter atoms and μ is the rest mass of the neutrino. For positron emission the relation is

$$T_0 = (\Delta M - 2m_0 - \mu)c^2, \tag{14–22b}$$

where m_0 is the rest mass of the electron or positron. Comparison of the observed endpoints, T_0, with atomic mass differences ΔM, determined from Q-values for nuclear reactions or from atomic masses, can be used to estimate the neutrino mass. The most direct comparison is that with a reaction which is just the reverse of a β-decay: a (p,n) reaction is combined with a β^--emission, or an (n,p) reaction is combined with a β^+-emission. For example, the reaction H^3(p,n)He^3 with $Q = -0.764 \pm 0.001$ Mev can be combined with the β^--decay of H^3 with $T_0 = 0.0181 \pm 0.0002$ Mev. We have

$$Q = [M(H^3) - M(He^3) - M_n + M_p]c^2,$$

$$\mu c^2 = T_0 - [M(H^3) - M(He^3)]c^2;$$

elimination of the difference $M(H^3) - M(He^3)$ between the two equations and insertion of the values of Q, T_0 and the neutron-proton mass difference gives $\mu c^2 = 0 \pm 0.0015$ Mev, or $\mu = 0 \pm 0.003m_0$. In this case the upper limit for the rest mass of the neutrino is $0.003m_0$ or about 1.5 kev, and comes from the uncertainty in the Q-value. About a dozen similar examples involving light nuclei can be used, all of which yield limits below $\mu = 0.01m_0$, with the limits depending on the uncertainty in the Q-values and the endpoint energies.

In Fermi's formulation of the theory of β-decay, the only condition on the rest mass of the neutrino was that it be small compared to the rest mass of the electron. The precise theoretical shape of an allowed β-spectrum in the neighborhood of the endpoint depends on the actual value of the rest mass of the neutrino, and comparison of theoretical and experimental spectra near the endpoint can be used to obtain estimates of the rest mass of the neutrino. Analyses of this kind[(24)] yield $\mu \leq 5 \times 10^{-4}m_0$, or a rest energy of 250 ev or less.

There are a number of reactions in which indirect evidence of the presence of the neutrino has been obtained from the momentum given to a recoiling nucleus. One of the objects of recoil experiments has been to test the hypothesis that a single neutrino is emitted in a β-decay process. The simplest reaction in which to study the nuclear recoil is orbital electron capture. With the assumption that a single neutrino is emitted the reaction is

$$_{Z+1}\mathrm{X}^A + {}_{-1}\mathrm{e}^0 \rightarrow {}_Z\mathrm{Y}^A + \nu. \tag{14-23}$$

If the initial nucleus is at rest, conservation of momentum requires that

$$p(\mathrm{Y}) = Mv = \sqrt{2ME_r} = p_\nu, \tag{14-24}$$

where M, v, and E_r are the mass, velocity, and kinetic energy of the recoiling nucleus, respectively; p represents momentum, p_ν being that of the neutrino. The velocity with which the nucleus recoils is small because of its enormous mass relative to that of the neutrino, and its kinetic energy is just $\frac{1}{2}Mv^2$. The neutrino, however, must be treated relativistically because its very small mass must be associated with high speed to give kinetic energies of the order of Mev; its total energy E_ν can be then written [see Problem 9(d) of Chapter 6]

$$E_\nu = [(\mu c^2)^2 + p_\nu^2 c^2]^{1/2}. \tag{14-25}$$

By Eq. (14-24), $p_\nu = \sqrt{2ME_r}$ so that $E_\nu = [(\mu c^2)^2 + 2c^2ME_r]^{1/2}$, and the kinetic energy of the recoil nucleus is given by

$$E_r = \frac{E_\nu^2 - (\mu c^2)^2}{2Mc^2}. \tag{14-26}$$

Conservation of energy in orbital electron capture requires that

$$E_\nu = [M(\mathrm{X}) - M(\mathrm{Y})]c^2 - E_r - B_{K,L}, \tag{14–27}$$

where $B_{K,L}$ is the binding energy of the electron in the K- or L-shell. In most neutrino experiments, $E_\nu \gg E_r + B_{K,L}$ so that, to a good approximation,

$$E_\nu = M(\mathrm{X}) - M(\mathrm{Y}). \tag{14–28}$$

The mass difference can be obtained from Q-values for nuclear reactions or mass spectroscopic data, and E_ν is then known; μc^2 is generally much smaller than E_ν, and Eq. (14–26) becomes

$$E_r = \frac{E_\nu^2}{2Mc^2}\cdot \tag{14–29}$$

It is convenient to express E_ν in units of $m_0c^2 = 0.511$ Mev, and the last equation may be written

$$E_r = \frac{m_0c^2}{2(Mc^2/m_0c^2)}\left(\frac{E_\nu}{m_0c^2}\right)^2\cdot \tag{14–30}$$

If we now express M in atomic mass units, and note that $m_0 = 5.488 \times 10^{-4}$ amu, Eq. (14–28) becomes

$$E_r = \frac{140.2(E_\nu/m_0c^2)^2}{M}\ \mathrm{ev}, \tag{14–31}$$

where M is the atomic mass of the recoiling nucleus in amu.

If the assumption that a single neutrino is emitted is correct, the recoiling atom should have a single energy, since the entire energy of the transformation is carried off by the nucleus and the neutrino. If, on the other hand, two or more neutrinos are emitted, there should be a continuous distribution of recoil energies. A number of orbital electron capture experiments have been made in which the energy spectrum of the recoiling nucleus has been measured;[(25,26)] the nuclides studied were Be^7, A^{37} and Cd^{107}. The simplest of these cases is that of A^{37}, which decays exclusively to the ground state of Cl^{37}. The energy E_ν is given by the $\mathrm{A}^{37} - \mathrm{Cl}^{37}$ mass difference; a value of 0.816 ± 0.004 Mev for this mass difference has been obtained from the Q-value of the $\mathrm{Cl}^{37}(\mathrm{p,n})\mathrm{A}^{37}$ reaction and the neutron-proton mass difference. Equation (14–31) then gives $E_r = 9.67 \pm 0.08$ ev. The recoil energy has been measured[(24)] in several independent experiments with results of 9.7 ± 0.8 ev,[(27)] 9.6 ± 0.2 ev[(28)] and 9.65 ± 0.05 ev,[(29)] respectively. These results show clearly that monoenergetic single neutrinos are emitted. The recoil experiments with Be^7 and Cd^{107} are complicated by the emission of γ-rays; e.g., Li^7 is left in an excited state in about 10% of the disintegrations and a 0.478-Mev γ-ray is emitted. The $Be^7 - Li^7$ mass difference is 0.864 ± 0.003 Mev; in 90% of the decays the recoils should be monoenergetic with 57.3 ± 0.5 ev energy if a single neutrino is emitted. The 10% branch going to the excited state of Li^7 should show a continuous recoil spectrum extending from nearly zero to 57.3 ev and resulting from the nearly simultaneous emission of a neutrino and a γ-ray. Experimentally, recoil energy spectra were found with maxima of 56.6 ± 1.0 ev[30,31] and 55.9 ± 1.0 ev,[32] in good agreement with 57.3 ± 0.5 ev. The interpretation of the continuous spectrum, however, was complicated by experimental difficulties; although it did not prove conclusively that only one neutrino is emitted in orbital capture in Be^7, the results were consistent with that hypothesis.

In β-particle emission the momentum conditions are more complicated than in electron capture, because the recoil momentum is determined by the proportion in which the decay energy is shared between the electron and the neutrino and by the angle between the two. It is possible to measure the angle between the β-particle and the recoil nucleus and, from conservation relations, to deduce the angle between electron and neutrino. These electron-neutrino angular correlation experiments yield information concerning the form of the interaction involved in β-decay, i.e., the force which converts the neutron into a proton and at the same time produces an electron or positron and neutrino.[33] This problem is beyond the scope of this book and will not be discussed further.

The successful detection of neutrinos was made possible by the availability of a source of enormous numbers of neutrinos (a nuclear reactor) and of extremely sensitive radiation detectors. The details of the experiment to be described involve some properties of neutrons, positrons, and electrons which have not yet been discussed and it will be necessary to use some information from later chapters.

The nuclear reactions which appeared most likely to permit the detection of neutrinos are the inverse β-decay processes. The inverse of the β^--emission process, Eq. (14–10), is obtained from Eq. (14–10) by transposing the electron $({}_{-1}e^0)$ to the left side of equation, noting that $-({}_{-1}e^0) = {}_1e^0$, and reversing the direction of the reaction; the result is

$$\nu + {}_1H^1 \rightarrow {}_0n^1 + {}_1e^0. \qquad (14\text{–}32a)$$

The inverse of the β^+-emission process, Eq. (14–11), can be written

$$\nu + {}_0n^1 \rightarrow {}_1H^1 + {}_{-1}e^0, \qquad (14\text{–}32b)$$

and the inverse of orbital electron capture can be written

$$_ZY^A + \nu \rightarrow {}_{Z+1}X^A + {}_{-1}e^0,$$

or (14–32c)

$$_0n^1 + \nu \rightarrow {}_1H^1 + {}_{-1}e^0.$$

Before proceeding with the discussion of the neutrino detection experiments we must consider an additional complication. It was suggested, in accord with relativistic quantum mechanics, that the neutrinos involved in β^--decay are different from those involved in β^+-decay and orbital electron capture. Recent theoretical and experimental[34,35,36,37] work which will be discussed in the next section has shown that a distinction must indeed be made depending on the relation between the spin and momentum of the neutrino. The *neutrino* (ν) is now defined as a particle with its spin vector parallel to its momentum vector; and the *antineutrino* (ν^*) as a particle with spin opposite in direction, or antiparallel, to its momentum. The spin and momentum together define the sense of a screw, the neutrino acting like a right-handed screw and the antineutrino like a left-handed screw. The choice of which particle is called neutrino or antineutrino is an arbitrary one. The choice shown in the following equations for the direct β-processes is the one which has been generally adopted; the equations are written in simpler notation than that of Eqs. (14–10), (14–11), and (14–32)

β^--emission: $$n \rightarrow p + \beta^- + \nu^*, \qquad (14\text{–}33a)$$

β^+-emission: $$p \rightarrow n + \beta^+ + \nu, \qquad (14\text{–}33b)$$

orbital electron capture: $$p + \beta^- \rightarrow n + \nu. \qquad (14\text{–}33c)$$

The antineutrino is emitted in β^--decay; the neutrino is emitted in β^+-decay and orbital electron capture. The inverse reactions are

inverse β^-: $$\nu^* + p \rightarrow n + \beta^+, \qquad (14\text{–}34a)$$

inverse β^+: $$\nu + n \rightarrow p + \beta^-, \qquad (14\text{–}34b)$$

inverse orbital electron capture: $$\nu + n \rightarrow p + \beta^-. \qquad (14\text{–}34c)$$

In a nuclear reactor, nuclei of a fissionable material, for example, U^{235}, are split into two fragments of intermediate atomic mass (Chapter 19). The fragments are highly unstable, containing too many neutrons for the number of protons. They therefore decay, by several successive β^--emissions, to stable or long-lived fission products, liberating antineutrinos in the process. If these antineutrinos are permitted to bombard a hydrogenous material, reactions of the form of Eq. (14–34a) should occur, and the problem is to detect the simultaneous production of a neutron and a positron. This detection is possible because of certain properties of neutrons and positrons. When a positron approaches an electron, the two interact in a very special way, they disappear and two γ-rays (photons) appear in their place (Chapter 15). This mutual annihilation takes place in a very short time in matter and *prompt* γ-rays are formed, which can be observed with a scintillation detector. A neutron produced in a hydrogenous material undergoes elastic collisions with protons and loses energy (Chapter 18). When it is slowed down it can be captured in an (n,γ) reaction, especially if a relatively heavy material is present. Cadmium is such a heavy material with a very high probability for capturing slow neutrons in (n,γ) reactions. The experiments of Reines, Cowan, and co-workers(38,39) for the detection of antineutrinos were based on the use of a large, liquid, hydrogenous scintillation counter (terphenyl dissolved in triethylbenzine) containing cadmium. When an antineutrino interacts with a proton, the annihilation of the positron produced is signalled by a prompt γ-ray pulse. A delayed pulse should appear several microseconds after the prompt pulse because of the capture of the neutron in the cadmium after being slowed down. The time required for the slowing down of the neutron can be calculated and the pulses due to the prompt disappearance of the positron and the delayed capture of the neutron are counted by a delayed coincidence method. In the later experiment(39) the detector consisted of a multiple-layered arrangement of three scintillation counters and two target tanks with one of the latter consisting of a polyethylene box containing a water solution of cadmium chloride. A schematic diagram of the detector is shown in Fig. 14-14. The liquid scintillator was surrounded with 110 photomultiplier tubes connected in parallel in each tank. A typical event is shown in the figure. An antineutrino induces an inverse β^--decay at point 1. A prompt pulse appears in both the top and central scintillation counters because of the two γ-rays from the annihilation of the positron at point 2, which is close to point 1. The neutron, after undergoing several collisions, is captured in cadmium at point 3 with the appearance of several capture γ-rays and a delayed pulse. To show that the prompt and delayed pulses were due to a reaction induced by antineutrinos it was necessary to relate the pulse rate to the power level of the reactor and to the number of protons in a target box; to show that the prompt pulses were due only to the annihilation of positrons, and that the delayed pulses were due only to the capture of neutrons in cadmium. The effect of fast neutrons from the reactor had to be ruled out along with other possible spurious effects. The final result was a maximum neutrino signal rate of 2.88 ± 0.22 counts/hr with a signal-to-background ratio of three to one; this result has been accepted as convincing evidence of a nuclear reaction induced by antineutrinos.

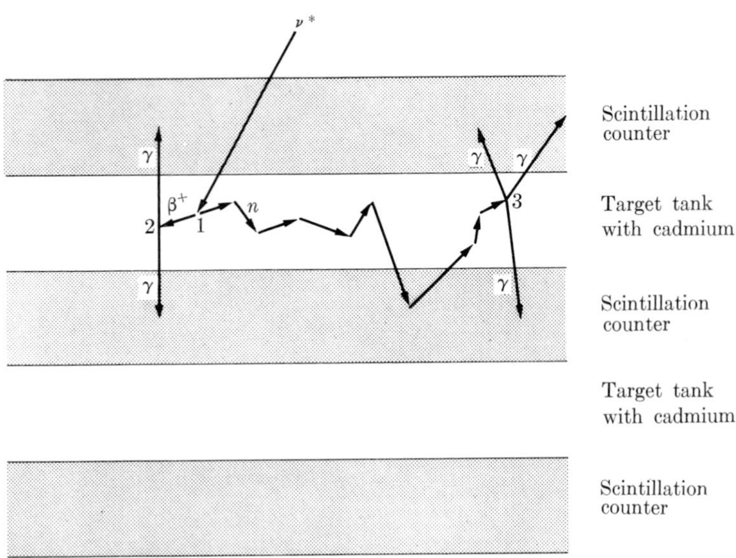

FIG. 14–14. Schematic diagram of the antineutrino detector used in the experiment of Ref. 38.

An experiment complementary to that on the detection of antineutrinos has been in progress for some time[40,41] in which an attempt has been made to detect the reaction

$$\nu^* + \text{Cl}^{37} \rightarrow \text{A}^{37} + \beta^-. \tag{14–35}$$

Comparison with Eq. (14–34c) shows that this reaction would be the inverse of orbital electron capture in A^{37} if the neutrino and antineutrino were identical. According to theory, however, these particles are different and the antineutrinos from a nuclear reactor should not induce the reaction (14–35). In the most recent version of the experiment, 3000 gallons of carbon tetrachloride, free of argon, were exposed to antineutrinos from the same reactor used in the experiment of Reines and Cowan. At the end of the irradiation the CCl_4 was swept with 12,000 liters of helium gas to remove argon. A small measured volume of argon gas was introduced into the tanks before sweeping to serve as a carrier and to measure the recovery of argon. The argon was then separated from the helium in a charcoal trap cooled with liquid nitrogen, and its β-activity was counted in a small Geiger counter. The results showed that the induction of inverse β-decay in Cl^{37} by antineutrinos, the reaction (14–35), is much less probable than the reaction $\nu^*(\text{p,n})\beta^+$. The quantitative measure of nuclear reactions is the *cross section* (see Section 16–3) which has values of the order of 10^{-24} cm^2 for most nuclear reactions. The value found[42] for the

$\nu^*(\text{p,n})\beta^+$ reaction is $(11 \pm 2.5) \times 10^{-44}$ cm^2; for the reaction (14–35), the value of the cross section was found to have an upper limit of 0.2×10^{-45} cm^2. These results show that the probability of antineutrino-induced inverse orbital electron capture is much smaller than that of the absorption of antineutrinos by protons. They show also that the probability for antineutrino-induced reactions is about twenty orders of magnitude smaller than that for ordinary nuclear reactions.

14–9 Symmetry laws and the nonconservation of parity in β-decay. The problem of the conservation of momentum and energy in β-decay led, as we have seen, to the neutrino hypothesis and eventually to the detection of the neutrino. In recent years it has been discovered that an important symmetry law, that of right-left symmetry, breaks down in certain reactions, the so-called *weak interactions*, of which β-decay is an example. To see how this comes about it is necessary that we consider the relationship between symmetry or invariance principles and conservation laws. We shall follow the discussion by Yang.(43)

During the first half of the 20th century it was recognized(44) that, in general, a symmetry or invariance principle (see Sections 6–4 and 6–5) generates a conservation law. For example, the invariance of physical laws under a displacement in space has as a consequence the law of conservation of momentum; the invariance under a rotation in space has as a consequence the law of conservation of angular momentum. In the theory of relativity the concept of invariance was extended to invariance under the Lorentz transformation, and the requirement that the equations of a physical theory satisfy this type of symmetry principle became a fundamental condition on a physical theory. This new type of invariance is related to the impossibility of detecting absolute motion, to the variation of mass with velocity, and to the relationship between mass and energy.

The development of quantum mechanics extended the use of symmetry principles still further, into the very details and language of atomic physics, and the consequences of these principles appear explicitly in the methods and results of quantum mechanics.(45) Thus, the quantum numbers which designate the states of an atomic system are often closely related to the numbers which represent the symmetries of the system. One example is that of the total orbital angular momentum quantum number L of the electrons in an atom. The atom is symmetrical with respect to a three-dimensional rotation about the nucleus, and orbital angular momentum is conserved; the Schroedinger equation for the atom must be invariant under such rotations and the detailed solutions of that equation, which embody the conservation of angular momentum, depend explicitly on L. A second example involves a kind of symmetry typical of quantum mechanics. Since one electron cannot be distinguished from another, an atom is symmetrical with respect to the interchange of electrons. The Schroedinger equation which describes an n-electron atom must therefore be invariant under a transformation which involves any permutation of its n electrons. This invariance turns out to be related to the multiplicity of the spectral terms of the atom. In more formal terms, "the Schroedinger equation for an n-electron atom is invariant under the symmetric group of order n," and the degree or order n of this invariance is closely related to the multiplicity of the spectral lines. In fact, the general structure of the periodic system (because of the way it is built up in terms of quantum numbers) may be said to be a consequence of the spatial symmetry of Coulomb's law, the law of force between the nucleus and the electrons. The combination of relativity and quantum mechanics leads to an especially striking example of the fundamental role of symmetry laws. The existence of the positron was anticipated theoretically by the requirement that the quantum mechanical equation for the electron, the Dirac equation, must be invariant under the Lorentz transformation.[(46)] The existence of other antiparticles,[(47,48)] e.g., antiproton, antineutron, antineutrino, has a similar theoretical interpretation, and their discovery testifies to the power and beauty of the invariance principles.

Another basic type of symmetry is that between right and left, or symmetry under a reflection. The principle of invariance involved may be stated in the following way: Any process which occurs in nature can also occur as it is seen reflected in a mirror; the mirror image of any object is also a possible object in nature; the motion of any object as seen in a mirror is also a motion which would be permitted by the laws of nature; an experiment made in a laboratory can also be made in the way it appears as seen in a mirror, and any resulting effect will be the mirror image of the actual effect. More precisely, we expect that the laws of nature are invariant under reflection, and experience seems to support this idea.

In quantum mechanics it has been shown[(49)] that the consequence of the right-left symmetry (invariance under reflection) of the electromagnetic forces in the atom appears as the conservation of parity. It will be recalled from Section 7–9 that parity is a quantum number which takes on the value $+1$ or -1 depending on whether the Schroedinger wave function is unchanged or changes sign under a reflection through the origin of the spatial coordinates. Experimentally (in the study of atomic spectra) it has been found[(50)] that in an atomic transition with the emission of a photon the parity of the initial state is equal to the total parity of the final state, i.e., to the product of the parities of the final atomic state and the photon emitted. Since right-left symmetry was unquestioned in interactions other than electromagnetic, the concept of parity and its conservation was taken over into nuclear physics, in particular into nuclear reactions and β-decay, and into the analysis of reactions involving new particles such as mesons and hyperons (which we do not treat in this book). In these developments, the concept of parity and the law of conservation of parity proved to be very fruitful, and the success accompanying their use was taken as support for the validity of right-left symmetry.

It has been mentioned that β-decay is an example of a weak interaction and we must define what is meant by this description. To do so, we must compare the forces involved in β-decay with those involved in other reactions. The interactions with which we are most familiar so far are those in which electromagnetic radiation is emitted (see Sections 7–2 and 7–8). The force involved is that between charged particles and is proportional to e^2. It is convenient to use a dimensionless quantity to compare the magnitudes of forces, and the one which enters naturally into atomic theory is $e^2/\hbar c = 1/137.04$, (where $\hbar = h/2\pi$), or approximately 10^{-2}. This quantity is spoken of as the "coupling constant for electromagnetic interactions." The forces binding neutrons and protons together in the nucleus are much stronger than the electromagnetic forces and also much more complicated (they will be treated in Chapter 17). The nuclear forces may be characterized by a coupling constant $f^2/\hbar c \approx 1$, where f is a constant analogous to e; these forces are referred to as *strong interactions*. The forces in β-decay, those linking the neutron, proton, electron, and neutrino (or antineutrino), may be characterized by the coupling constant $g^2/\hbar c \approx 10^{-13}$, where g is a constant related to the constant G of Section 14–6. They are very small compared to the nuclear and electromagnetic forces, and it is for this reason that they are called weak interactions. For comparison, the analogous coupling constant of Newtonian gravitational forces is of the order 10^{-39}.

Weak interactions occur not only in β-decay but also in the decay of certain particles found in cosmic rays and nuclear reactions at high energies (several hundred Mev and higher). Among these particles are the τ and θ mesons which have been discovered in recent years, and whose properties have been the subject of intensive study. Results were obtained which seemed paradoxical in that all the physical properties of the parent particle in τ-decay and θ-decay seemed to be identical, but detailed analyses indicated that the final states had opposite parities. Lee and Yang[(34)] showed that the results would fall into a consistent scheme only if parity were not conserved in these decay processes, i.e., only if these processes were not invariant under reflection. This surprising possibility led them to investigate the experimental information concerning parity conservation. They found impressive evidence for the validity of this law in atomic physics, the domain of electromagnetic interactions, and in nuclear physics when strong interactions were concerned. There were, however, no experimental data relevant to the question of the conservation of parity in

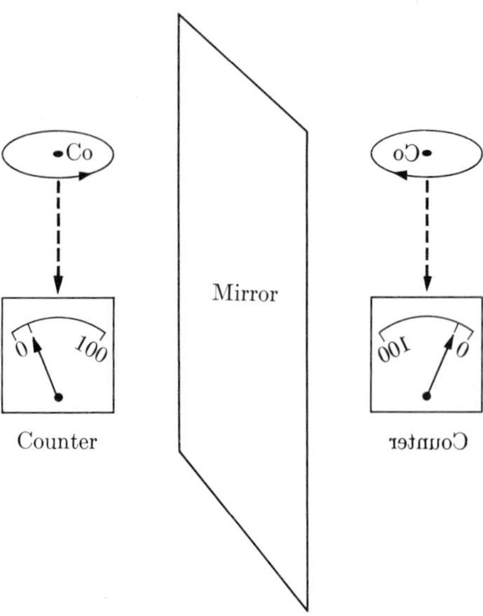

FIG. 14–15. Schematic diagram of an experiment to test the conservation of parity in β-decay (Yang[(43)]).

weak interactions; no experiment had ever been designed specificially to test invariance under reflection in the β-decay of nuclei or elementary particles.

Lee and Yang proposed some experiments, one of which involved the β-decay of Co^{60}, to test invariance under reflection. The basic idea of this experiment is to build two sets of experimental arrangements which are mirror images of each other, and which contain weak interactions, as shown in Fig. 14–15. The experiment then consists in determining whether the two arrangements always give the same results in terms of the readings of their meters (or counters). The practical feasibility of such an experiment depends on the following facts. The cobalt nucleus has a spin angular momentum; it rotates with a well-defined angular momentum when it is in its normal state. In a piece of Co^{60} under normal conditions, the nuclear spins are oriented in all directions because of the thermal motion and the electrons are emitted in all directions. Under certain conditions, however, the nuclei can be oriented in a particular direction; all the nuclei can be forced to align their axes of rotation parallel to a given direction so that they all rotate in the same sense. This can be done because the nuclear spin gives rise to a magnetic moment which can be acted upon by a magnetic field external to the nucleus. At very low temperatures, less than 0.1°K, the external field can force the spins into a given direction,[(50)] and the nuclei are said to be polarized. The low temperature

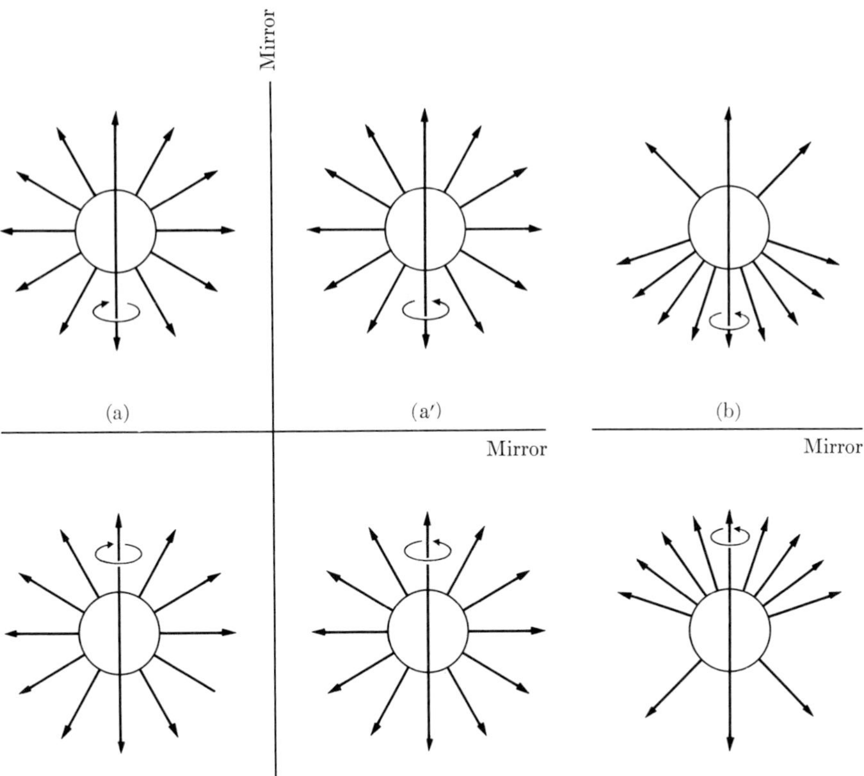

FIG. 14–16. Cobalt β-decay: possible electron emission patterns and their mirror images.

is needed to reduce the thermal motion of the nuclei to a minimum and make the orderly alignment possible. The experiment proposed by Lee and Yang was simply to line up the spins of the electron-emitting nuclei of Co^{60} along the same axis, and then see whether the β-particles were emitted preferentially in one direction or the other along the axis. If there is preferential emission in one direction, the mirror image of the experiment would give preferential emission in the opposite direction; if there is no preferential emission of β-particles, the mirror image would give the same result. Possible emission patterns are shown in Fig. 14–16. The circles represent samples of Co^{60} and the arrows represent electrons, with the arrowheads showing the direction of emission. The nuclei in a given sample are aligned so that they spin in the sense indicated by the curved arrow. In case (a) there is no preferred direction of emission; the electrons are emitted isotropically. In the image in the horizontal mirror, the sense of rotation is the same and the electron distribution is

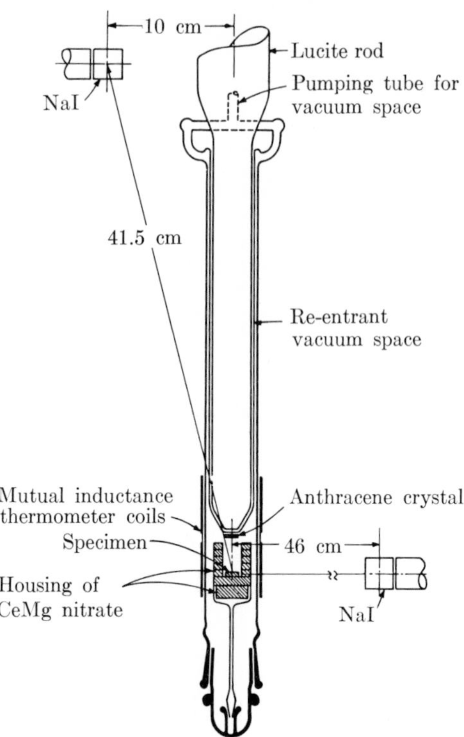

FIG. 14–17. The experimental arrangement used by Wu, Ambler, Hayward, Hoppes, and Hudson in their demonstration of the asymmetric β-particle distribution from polarized cobalt nuclei.[36]

again isotropic. In (a′), the direction in which the nuclei are aligned is rotated through 180°; electron emission is again isotropic and the mirror image of the experiment is identical with the experiment itself. The image in the vertical mirror looks just like the original turned upside down. Cases (a) and (a′) show the results expected with right-left symmetry and conservation of parity. In case (b), more electrons are emitted in the direction *opposite* to that in which the nuclei are aligned. The electron emission is anisotropic and the mirror image shows more electrons being emitted *in the direction* of the nuclear spins. The mirror image experiment would then give results different from those of the original experiment. This kind of result would be expected if parity is not conserved.

The Co^{60} decay experiment was made by Wu and co-workers.[36] Cobalt-60 emits an electron of energy 0.312 Mev in over 99% of its decays and two other electrons in the remaining decays; several γ-rays are also emitted. It was known[51] that Co^{60} nuclei can be polarized in cerium magnesium (cobalt) nitrate and the degree of polarization determined by

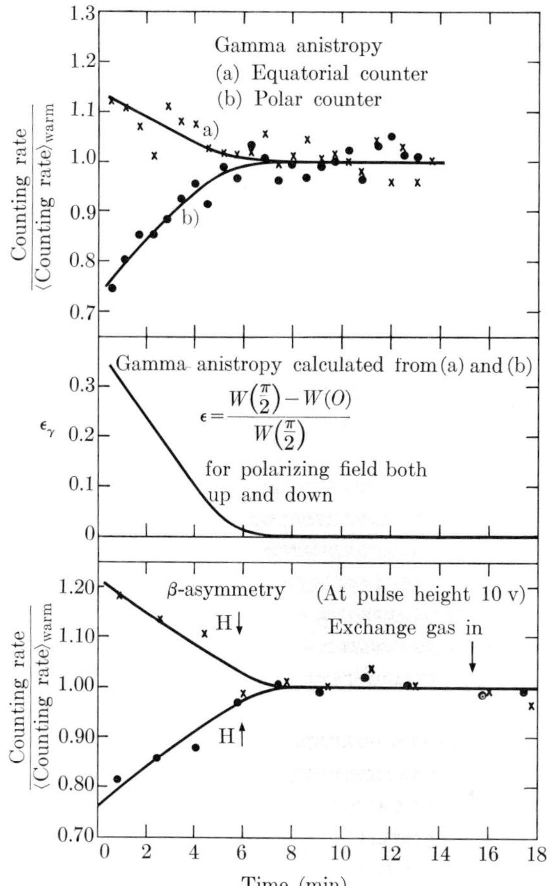

FIG. 14–18. The results of the experiment of Wu et al. on the asymmetric distribution of the β-particles from polarized cobalt nuclei.(36)

measuring the anisotropy of the γ-rays. In the β-decay experiment, additional difficulties were introduced by the need to put a β-particle counter into the cryostat in which the sample was to be cooled, and to locate the nuclei to be polarized in a thin surface layer. The cooling is accomplished by a special method of low-temperature physics called adiabatic demagnetization. A schematic diagram of the part of the cryostat of interest is shown in Fig. 14–17.

To detect the β-particles, a thin anthracene crystal was located inside the vacuum chamber near the cobalt source. The scintillations were transmitted to a photomultiplier located at the top of the cryostat. To measure the extent of the polarization of the Co^{60} nuclei, two NaI γ-ray scintillation counters were used, one in the equatorial plane and one near the polar position. The observed γ-ray anisotropy provided a measure of the polarization and hence of the temperature of the sample. Samples were made by taking good single crystals of cerium magnesium nitrate and growing on the upper surface an additional crystalline layer about 0.002 in thick, containing a few microcuries of Co^{60}. The cerium magnesium sulfate is needed because the field produced by an external magnet cannot, by itself, align the nuclear magnetic moments due to the spin; the external field is used to line up atoms (provided by the cerium magnesium sulfate), with special properties, and the magnetic field produced by these atoms in turn, lines up the nuclei. After the material in the cryostat was cooled to about 0.01°K, the magnet used in the cooling process was turned off; a vertical solenoid was raised around the lower part of the cryostat within 20 sec after the demagnetization, and the counting was started.

The results of the experiment are shown in Fig. 14–18. The time scale is actually a temperature scale measuring time from the instant when cooling is stopped. The curve labelled "gamma-anisotropy" is a measure of the extent of polarization; after about 8 min the nuclei have become sufficiently warm so that their thermal motion causes the alignment to become random, and the γ-anisotropy disappears. The curve labelled "β asymmetry" is the significant one; it indicates the number of electrons emerging in the direction of the magnetic field (the direction in which the nuclei are aligned), and the number emerging in the opposite direction. It is evident from the figure that more electrons are emitted in the direction opposite that of the magnetic field, i.e., in the direction opposite to that in which the nuclei are aligned. In other words, the experimental result corresponds to case (b) of Fig. 14–16; the electrons are emitted in a preferred direction, the principle of right-left symmetry is violated, and parity is not conserved. The size of the effect is remarkable; the electron intensity in one direction along the axis of rotation was found to be 40% greater than in the opposite direction, and there was no doubt about the result.

Many experiments have now been made on β-decay and on the decay of mesons and other fundamental particles(37) with results similar to that of the Co^{60} experiment. The breakdown of the law of conservation of parity has led to a modified theory of the neutrino(35,52) and to extensive research into some of the basic concepts of physics,(49) especially into the concepts of symmetry and invariance and their relation to conservation laws.

References

GENERAL

Rutherford, Chadwick, and Ellis, *Radiations from Radioactive Substances.* New York: The Macmillan Co., 1930, Chapters 12, 13, 14.

R. D. Evans, *The Atomic Nucleus.* New York: McGraw-Hill, 1955, Chapters 6, 17, 18, 19, 20, 21.

K. Siegbahn, ed., *Beta- and Gamma-Ray Spectroscopy.* New York: Interscience Publishers Inc.; Amsterdam: North-Holland Publishing Co., 1955.

M. Deutsch and O. Kofoed-Hansen, "Beta-Rays," *Experimental Nuclear Physics,* E. Segrè, ed. New York: Wiley, 1959, Vol. III, Part XI.

J. S. Allen, *The Neutrino.* Princeton, New Jersey: Princeton University Press, 1958.

C. S. Wu, "The Neutrino", *Theoretical Physics in the Twentieth Century. A Memorial Volume to Wolfgang Pauli,* M. Fierz and V. F. Weisskopt, eds. New York: Interscience, 1960.

R. W. King, "Table of Total Beta-Disintegration Energies," *Revs. Mod. Phys.* **26,** 327 (1954).

L. J. Lidofsky, "Table of Total Beta-Disintegration Energies," *Revs. Mod. Phys.* **29,** 773 (1957).

Strominger, Hollander, and Seaborg, "Table of Isotopes," *Revs. Mod. Phys.* **30,** 585 (1958).

PARTICULAR

1. J. L. Lawson and A. W. Tyler, "The Design of a Magnetic Beta-Ray Spectrometer," *Rev. Sci. Instr.* **11,** 6 (1940).

2. L. M. Langer and C. S. Cook, "A High Resolution Nuclear Spectrometer," *Rev. Sci. Instr.* **19,** 257 (1948).

3. C. M. Witcher, "An Electron Lens Type of Beta-Ray Spectrometer," *Phys. Rev.* **60,** 32 (1941).

4. R. W. Hayward, "Beta-Ray Spectrometers," *Advances in Electronics,* Vol. 5, p. 97. New York: Academic Press (1953).

5. L. E. Glendenin, "Determination of the Energy of Beta-Particles and Photons by Absorption," *Nucleonics* **2,** 12 (Jan. 1948).

6. L. Katz and A. S. Penfold, "Range-Energy Relations for Electrons and the Determination of Beta-Ray End-Point Energies by Absorption," *Revs. Mod. Phys.* **24,** 28 (1952).

7. C. D. Ellis and W. A. Wooster, "The Average Energy of Disintegration of Radium E," *Proc. Roy. Soc.* (London) **A117,** 109 (1927).

8. L. Meitner and W. Orthmann, "Uber eine Absolute Bestimmung der Energie der primaren β-Strahlen von Radium E," *Z. Physik.* **60,** 143 (1930).

9. G. J. Neary, "The β-Ray Spectrum of Radium E," *Proc. Roy. Soc.* (London) **A175,** 71 (1940).

10. Marinelli, Brinckerhoff, and Hine, "Average Energy of Beta-Rays Emitted by Radioactive Isotopes," *Revs. Mod. Phys.* **19,** 25 (1947).

11. J. M. Robson, "The Radioactive Decay of the Neutron," *Phys. Rev.* **83,** 349 (1951).

12. B. W. SARGENT, "The Maximum Energy of the β-Rays from Uranium X and other Bodies," *Proc. Roy. Soc.* (London) **A139,** 659 (1933).

13. N. FEATHER, *Nuclear Stability Rules.* Cambridge: University Press, 1952, Chapter 3.

14. E. FERMI, "Versuch einer Theorie der β-Strahlen," *Z. Physik.* **88,** 161 (1934).

15. E. J. KONOPINSKI, "Beta-Decay," *Revs. Mod. Phys.* **15,** 209 (1943).

16. KURIE, RICHARDSON, and PAXTON, "The Radiations Emitted from Artificially Produced Radioactive Substances. I. The Upper Limit and Shapes of the β-Ray Spectra from Several Elements," *Phys. Rev.* **49,** 368 (1936).

17. A. W. LAWSON and J. M. CORK, "The Radioactive Isotopes of Indium," *Phys. Rev.* **57,** 982 (1940).

18. C. S. WU, "Recent Investigation of the Shapes of β-Ray Spectra," *Revs. Mod. Phys.* **22,** 386 (1950)

19. T. H. R. SKYRME, "Theory of Beta-Decay," *Progress in Nuclear Physics.* Vol. 1, p. 115. New York: Academic Press, 1950.

20. E. J. KONOPINSKI and L. M. LANGER, "The Experimental Clarification of the theory of Beta-Decay," *Ann. Rev. of Nuc. Sci.,* Vol. 2. Stanford: Annual Reviews, Inc., 1953; see also, C. S. Wu, "The Neutrino," op. cit. gen. ref.

21. A. M. FEINGOLD, "Table of *ft*-Values in Beta-Decay," *Revs. Mod. Phys.* **23,** 10 (1951)

22. E. FEENBERG and G. TRIGG, "The Interpretation of Comparative Half-Lives in the Fermi Theory of Beta-Decay," *Revs. Mod. Phys.* **22,** 399 (1950).

23. H. R. CRANE, "The Energy and Momentum Relations in Beta-Decay and the search for the Neutrino," *Revs. Mod. Phys.* **20,** 278 (1948).

24. J. S. ALLEN, *The Neutrino,* see gen. ref., Chapter 2.

25. J. S. ALLEN, *The Neutrino,* see gen. ref., Chapter 3.

26. O. KOFOED-HANSEN, "Neutrino Recoil Experiments," *Beta-and Gamma-Ray Spectroscopy,* K. Siegbahn, ed., see gen. ref., p. 357.

27. G. W. RODEBACK and J. S. ALLEN, "Neutrino Recoils following the Capture of Orbital Electrons in A^{37}," *Phys. Rev.* **86,** 446 (1952).

28. O. KOFOED-HANSEN, "Neutrino Recoil Spectrometer. Investigation of A^{37}," *Phys. Rev.* **96,** 1045 (1954).

29. A. H. SNELL and F. PLEASONTON, "Spectrometry of Recoils from Neutrino Emission in Argon-37," *Phys. Rev.* **97,** 246 (1955); **100,** 1396 (1955).

30. J. S. ALLEN, "Experimental Evidence for the Existence of a Neutrino," *Phys. Rev.* **61,** 692 (1942).

31. P. B. SMITH and J. S. ALLEN, "Nuclear Recoils Resulting from the Decay of Be^7," *Phys. Rev.* **81,** 381 (1951).

32. R. DAVIS, JR., "Nuclear Recoil Following Neutrino Emission from Beryllium 7," *Phys. Rev.* **86,** 967 (1952).

33. J. S. ALLEN, *The Neutrino,* see gen. ref., Chapters 4, 5.

34. T. D. LEE and C. N. YANG, "Question of Parity Conservation in Weak Interactions," *Phys. Rev.* **104,** 254 (1956).

35. T. D. LEE and C. N. YANG, "Parity Nonconservation and a Two-Component Theory of the Neutrino," *Phys. Rev.* **105,** 1671 (1957).

36. Wu, Ambler, Hayward, Hoppes, and Hudson, "Experimental Test of Parity Conservation in Beta Decay," *Phys. Rev.* **105,** 1413 (1957).

37. C. S. Wu, "Parity Experiments in Beta Decays," *Revs. Mod. Phys.* **31,** 783 (1959).

38. Cowan, Reines, Harrison, Kruse, and McGuire, "Detection of the Free Neutrino; a Confirmation," *Science* **124,** 103 (1956).

39. Reines, Cowan, Harrison, McGuire, and Kruse, "Detection of the Free Antineutrino," *Phys. Rev.* **117,** 159 (1960).

40. R. Davis, "An Attempt to Observe the Capture of Reactor Neutrinos in Cl^{37}," (a) *Phys. Rev.* **97,** 766 (1955); (b) *Radioisotopes in Scientific Research,* Vol. 1 (Proc. First UNESCO Int. Conf. Paris, 1957). London: Pergamon Press, 1958, p. 728.

41. R. Davis, Jr. and D. S. Harmer, "An attempt to Observe the Cl^{37} $(\nu^*,e^-)A^{37}$ Reaction Induced by Reactor Antineutrinos," unpublished.

42. F. Reines and C. L. Cowan, Jr., *The Free Antineutrino Absorption Cross Section.* Part 1. Measurement of the Cross Section by Protons. Proceedings of the Second International Conference on the Peaceful Uses of Atomic Energy. Geneva: United Nations, 1958. Vol. 30, p. 253.

43. C. N. Yang, "Law of Parity Conservation and other Symmetry Laws," *Science* **127,** 565 (1958).

44. E. P. Wigner, "Invariance in Physical Theory," *Proc. Am. Phil. Soc.* **93,** 521 (1949).

45. E. P. Wigner, *Group Theory and its Application to the Quantum Mechanics of Atomic Spectra.* New York: Academic Press, 1959, Chapters 17, 18.

46. P. A. M. Dirac, *The Principles of Quantum Mechanics,* 4th ed. Oxford: Clarendon Press, 1958, Chapter 12.

47. E. Segrè, "Antinucleons," *Ann. Rev. of Nuc. Sci.* **8,** 127 (1958).

48. A. M. Shapiro, "Table of Properties of the 'Elementary' Particles," *Revs. Mod. Phys.* **28,** 164 (1956).

49. G. C. Wick, "Invariance Principles of Nuclear Physics," *Ann. Rev. of Nuc. Sci.* **8,** 1 (1958).

50. Blin-Stoyle, Grace, and Halban, "Oriented Nuclear Systems," *Prog. Nuc. Phys.,* O. R. Frisch, ed. London: Pergamon Press, 1953. Vol. 3, p. 63.

51. Ambler, Grace, Halban, Kurti, Durand, and Johnson, "Nuclear Polarization of Co^{60}," *Phil. Mag.* (7) **44,** 216 (1953).

52. T. D. Lee, "Weak Interactions and Nonconservation of Parity," *Science* **127,** 569 (1958).

Problems

1. Show that the magnetic rigidity for electrons may be written in the form

$$Hr = A(T^2 + 1.022T)^{1/2},$$

where $A = 3335.8$ gauss-cm/Mev, and T is the kinetic energy of the electron. [*Hint:* Use the result of Problem 12 of Chapter 6.]

2. Show that the kinetic energy may be written in the form

$$T = [a^2 + b(Hr)^2]^{1/2} - a,$$

where $a = 0.511$ Mev, and $b = 8.989 \times 10^{-8}$ Mev^2/(gauss-cm)^2. [*Hint:* Use the result of Problem 11 of Chapter 6.]

3. An absorption curve of a sample emitting β^--particles was taken with aluminum as the absorber and the following data were obtained.

Absorber thickness, mg/cm^2	Activity, counts/min	Absorber thickness, mg/cm^2	Activity, counts/min
0.375	16000	4.875	36
0.750	6100	7.375	13
1.75	2100	10.00	11
2.75	520	13.00	10
3.75	170		

Find, by inspection, the range of the β^--particles. What is the maximum energy of the β-particles? What is the half-thickness of Al for these particles?

4. The absorption of the β-rays from P^{32} was measured with aluminum and the following results were obtained.

Absorber thickness, mg/cm^2	Relative intensity	Absorber thickness, mg/cm^2	Relative intensity
0	1000	550	8
100	600	600	3.5
200	375	650	1.5
250	250	700	0.75
300	165	750	0.50
350	110	800	0.40
400	65	850	0.35
450	37	900	0.33
500	18	950	0.32

Find the range by inspection; then find the maximum energy from (a) Eq. (14–7), (b) Feather's empirical formula $R = 543T - 160$, where R is in mg/cm^2 and T is in Mev.

5. In a spectrometer study of the β-spectrum of P^{32}, the following data were obtained.

Hr, gauss-cm	Intensity, relative units	Hr, gauss-cm	Intensity, relative units
500	14	4000	91
1000	27	4500	82
1500	42	5000	68
2000	58	5500	49
2500	76	6000	31
3000	91	6500	12
3500	94	7000	3

The intensity is proportional to the number of emitted electrons per unit Hr interval. Plot (a) the momentum distribution of the electrons, (b) the energy distribution of the electrons, (c) a Kurie plot of the momentum distribution, (d) a Kurie plot of the energy distribution. In (c) and (d), assume that the Coulomb factor is constant. What is the extrapolated endpoint energy in each of the graphs?

6. In the following examples of β-decay, the endpoint energy of the transition is given together with the mass of the initial or final atom. Calculate the unknown mass.

Transition	Endpoint energy, Mev	Mass of initial atom, amu	Mass of final atom, amu
$C^{14}(\beta)N^{14}$	0.156	14.007685	?
$F^{17}(\beta)O^{17}$	1.748	17.007499	?
$P^{29}(\beta)Si^{29}$	3.94	?	28.985661
$S^{35}(\beta)Cl^{35}$	0.167	?	34.97996

7. The reactions $P^{31}(d,\alpha)Si^{29}$, $Si^{29}(d,p)Si^{30}$, and $Si^{30}(d,p)$ Si^{31} have the Q-values 8.158, 8.388, and 4.364 Mev, respectively. Calculate the energy available for the β-decay of Si^{31}.

8. Find the difference in mass (in Mev) between a neutron and a hydrogen atom from (a) the masses, as given in Table 11–1; (b) the endpoint of the β-spectrum of the neutron; (c) the reactions $O^{18}(p,n)F^{18}(Q = -2.447$ Mev), and $F^{18}(\beta)O^{18}$ with an endpoint energy of 0.645 Mev; (d) the reactions O^{16} $(d,p)O^{17}$ ($Q = 1.917$ Mev), $O^{16}(d,n)F^{17}$ ($Q = -1.624$ Mev), and $F^{17}(\beta)O^{17}$ with endpoint energy of 1.748 Mev.

9. The atomic masses of Ni^{64}, Cu^{64}, Zn^{64}, and Ga^{64} are 63.94813, 63.94994, 63.94932, and 63.95710 amu, respectively. (a) Which of these nuclides are stable? Which are radioactive? (b) For those nuclides which can decay by

β-emission, what are the endpoint energies of the spectra and the Q-values of the decay reactions?

10. Show that the ratio of the average energy to the maximum energy of a weak β-emitter ($T_0 \ll m_0c^2$) is $\frac{1}{3}$, when the energy distribution is given by the Fermi formula of Eq. (14–14).

11. A sample of hydrogen containing 2.57 cm^3 of tritium at N.T.P. was found to produce 0.1909 cal/hr of heat. The half-life of tritium is 12.46 years. Find (a) the disintegration rate, (b) the average energy of the β-particles emitted, (c) the ratio of average energy to maximum energy.

12. A sample of RaE contains 4.00 mg. If the half-life is 5.0 days and the average energy of the β-particles emitted is 0.34 Mev, at what rate in watts does the sample emit energy?

13. In the nuclear fission of U^{235}, about 9% of the total energy liberated comes from the β-decay of fission products. How much power is carried away by neutrinos from a nuclear power plant which liberates heat at the rate of 100,000 kw?

14. Plot log λ against log T_0 for the positron emitters listed below. The result is a Sargent curve for allowed, favored β-decay. From the curve, get the maximum energy of the particles emitted by Al^{25} (half-life of 7.3 sec) and Cl^{33} (half-life of 2.0 sec).

Nuclide	Half-life, sec	Endpoint energy, Mev	Nuclide	Half-life, sec	Endpoint energy, Mev
C^{11}	1200	0.98	Mg^{23}	11.9	2.99
N^{13}	606	1.24	Si^{27}	4.9	3.48
O^{15}	122	1.68	P^{29}	4.6	3.6
F^{17}	70	1.72	Si^{31}	3.2	3.86
Ne^{19}	18.4	2.18	A^{35}	1.86	4.40
Na^{21}	22.8	2.52	Sc^{41}	0.87	4.94

15. The nuclide Sm^{153} emits four groups of β-rays, with endpoint energies of 0.83, 0.72, 0.65, and 0.13 Mev, respectively; γ-rays with energies of 0.1032, 0.0697, 0.172, 0.545, and 0.615 Mev, respectively, are also emitted. Construct a decay scheme to fit these data. (Note that the γ-ray energies are known with higher precision than the β-ray energies.)

16. The nuclide As^{76} emits four groups of β-rays with endpoint energies of 2.97, 2.41, 1.76, and 0.36 Mev, respectively, and γ-rays with energies 0.561, 0.643, 1.200, 1.40, and 2.05 Mev. Devise a decay scheme to fit these data.

17. The nuclide Tc^{94} decays by orbital electron capture and β^+-emission; γ-rays are found, with energies 3.27, 2.73, 1.85, and 0.874 Mev, respectively, as well as positrons with an endpoint energy of 2.41 Mev. Devise a decay scheme to fit these data. Compare your scheme with that given in the "Table of Isotopes" of Strominger, Hollander, and Seaborg (gen. ref.).

CHAPTER 15

GAMMA-RAYS AND GAMMA-DECAY

The importance of γ-rays as a source of information about nuclear energy levels has been shown in Chapters 13 and 14 in connection with α-decay and β-decay. In those chapters, the emission of γ-rays was regarded simply as a means by which a nucleus can pass from an excited state to a less excited one. Gamma-decay is in itself, however, a subject of great theoretical and practical importance and will therefore be discussed in greater detail in this chapter. The study of γ-decay depends on the ability to measure the energy of γ-rays with high precision, but the problem of measuring these energies is different from that presented by charged particles. Since γ-rays are electromagnetic radiations and have no electric charge, they cannot be deflected by magnetic or electric fields. Consequently, direct measurements of their energies with a magnetic spectrometer are not possible. The mechanism of the absorption of γ-rays by matter is also different from that of charged particles, as is indicated by the very much greater penetrating power of γ-rays. Because of these properties of γ-rays, it will be necessary to examine first the interactions between γ-rays and matter. The methods used for measuring γ-ray energies will then be treated because they are based on these interactions, and then the subject of γ-decay will be taken up.

15–1 The absorption of gamma-rays by matter: experimental data. The basic property of the absorption of γ-rays is the exponential decrease in the intensity of radiation as a homogeneous beam of γ-rays passes through a thin slab of matter. When a beam of γ-rays of intensity I is incident on a slab of thickness Δx, the change in intensity of the beam as it passes through the slab is proportional to the thickness and to the incident intensity

$$\Delta I = -\mu I \, \Delta x, \tag{15–1}$$

where the proportionality constant μ is called the *absorption coefficient.* If the γ-ray photons all have the same energy, μ is independent of x, and the integration of Eq. (15–1) yields

$$I/I_0 = e^{-\mu x}. \tag{15–2}$$

Equation (15–2) gives the intensity of radiation I after a beam of initial

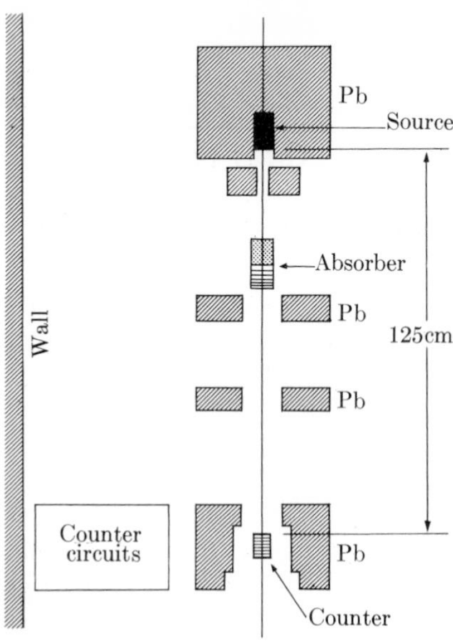

FIG. 15–1. Schematic plan view of the apparatus used for the measurement of total γ-ray absorption coefficients (Davisson and Evans[(1)]).

intensity I_0 has traversed a thickness x of a given material. The intensity may be written

$$I = Bh\nu, \tag{15–3}$$

where B is the number of photons crossing unit area in unit time, and $h\nu$ is the energy per photon; B is often called the *flux*, defined as the number of photons per square centimeter per second, and I is the corresponding energy flux. Equation (15–2) may then be written

$$B/B_0 = e^{-\mu x}. \tag{15–4}$$

The equations which describe the absorption of γ-rays are the same as those for x-rays as discussed in Section 4–1. This result is not surprising because both kinds of rays are electromagnetic radiations. In fact, it is not possible to differentiate between γ-rays and x-rays on the grounds of differences in their properties, since no differences have been found. The terms γ-rays and x-rays are used now chiefly to distinguish between the sources of the rays; γ-rays come from nuclei, while x-rays are the high energy radiations resulting from jumps of extranuclear electrons in atoms, or produced by artificial sources such as a Coolidge tube or a betatron. The discussion of absorption in this section applies to both γ-rays and x-rays.

It is necessary to define precisely the conditions under which Equations (15–1), (15–2), and (15–4) are valid. These conditions are (1) the γ-rays

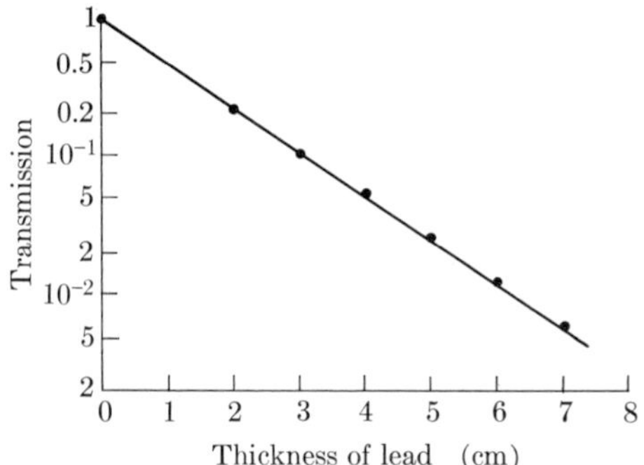

FIG. 15–2. Experimental verification of the narrow-beam attenuation formula for γ-rays, and measurement of the absorption coefficient in lead. Source, 1.14 Mev γ-rays from Zn^{65}.

are monoenergetic, i.e., the beam is homogeneous; (2) the beam is collimated and of small solid angle; (3) the absorber is "thin." An experimental arrangement for studying the decrease in intensity, or attenuation, of a γ-ray beam under these conditions is shown in Fig. 15–1. This arrangement provides a narrow collimated beam and is spoken of as "good geometry." The thick lead shielding blocks, separated by a narrow space, act as the collimator. The counter used to detect the γ-rays is surrounded by a lead shield to absorb extraneous γ-rays and electrons which might otherwise be scattered into it and counted. In a "good geometry" experiment, photons are removed from the beam either by absorption or by scattering. A photon which undergoes a scattering collision with an atom in the absorber slab may be deflected enough so that it does not reach the detector and is lost from the beam just as though it had actually been absorbed. The decrease in the intensity of the beam in passing through the slab is a measure of the combined effects of outright absorption and deflection. The term *absorption* is often used as synonymous with *attenuation*, but this practice causes less confusion than might be expected because the meaning is usually clear from the context.

The validity of Eqs. (15–2) and (15–3) can be shown by a "good geometry" experiment. The results of such an experiment are given in Fig. 15–2; the absorber is lead, and the source is Zn^{65}, which emits γ-rays with an energy of 1.14 Mev. When the fractional transmission is plotted on a logarithmic scale against the absorber thickness on a linear scale, the result is a straight line in agreement with the equations. The slope of the line gives the value μ (cm^{-1}) $= 0.7068 \pm 0.0051$. Many experimental determinations of absorption coefficients have been made in this way and it has been found that the value of the absorption coefficient depends on the nature of the absorber and on the initial energy of the γ-rays. Some values obtained by Davisson and Evans[(1)] are listed in the third column of Table 15–1. It is evident that for a given element, the absorption coefficient decreases as the energy of the γ-rays increases. The value of μ also varies from element to element and is, in general, greater for heavy elements than for light elements. Theoretical values are listed for future reference.

The thickness x may be expressed in cm or in gm/cm^2, as was done in Chapter 14 in the discussion of the absorption of β-particles. Since the product μx must be dimensionless, μ may be correspondingly expressed in cm^{-1} or cm^2/gm. It is sometimes convenient to express the absorber thickness in atoms/cm^2 or electrons/cm^2 rather than in gm/cm^2; the absorption coefficient then has the units cm^2/atom or cm^2/electron. The absorption coefficient is usually denoted by μ when the units are cm^{-1}, by μ/ρ for cm^2/gm, by ${}_e\mu$ for cm^2/electron, and by ${}_a\mu$ for cm^2/atom. The subscripts e and a are placed at the lower left in order to avoid confusion later on. In terms of ${}_e\mu$ and ${}_a\mu$, the other coefficients are

$$ {}_a\mu = Z\,{}_e\mu, \tag{15–5}$$

$$ \frac{\mu}{\rho} = N\left(\frac{Z}{A}\right){}_e\mu = \frac{N}{A}\,{}_a\mu, \tag{15–6}$$

$$ \mu = \rho N\left(\frac{Z}{A}\right){}_e\mu = \frac{\rho N}{A}\,{}_a\mu, \tag{15–7}$$

where Z is the atomic number, A is the atomic weight, N is Avogadro's number, and ρ is the density in gm/cm^3. In Eq. (15–7), the product $\rho N/A \cdot Z$ is the number of electrons/cm^3 of absorber and, since ${}_e\mu$ has the units cm^2/electron, μ comes out in cm^{-1}. The coefficient ${}_a\mu$ has the units cm^2/atom, and when it is multiplied by $\rho N/A$, the number of atoms/cm^3, the result is again μ in cm^{-1}. The coefficients ${}_e\mu$ and ${}_a\mu$ are often called the "cross section per electron" and the "cross section per atom," respectively, because of their units.

The attenuation of a γ-ray beam may also be expressed in terms of a quantity called the *half-thickness*, i.e., the thickness of absorber needed to reduce the intensity to half its initial value. Equation (15–2) may be written in terms of the common logarithm

$$ \log\,(I/I_0) = -0.4343\mu x. \tag{15–8}$$

When $I/I_0 = \frac{1}{2}$,

$$ \log\,(\tfrac{1}{2}) = -0.4343\mu x_{1/2}, $$

and

$$ \mu = \frac{0.693}{x_{1/2}}, \tag{15–9}$$

Table 15–1

Measured Values of Some Gamma-Ray Absorption Coefficients

Source	Absorber	Experimental absorption coefficient, μ: cm^{-1}	Theoretical absorption coefficient, μ: cm^{-1}
Mn^{54} (0.835 Mev)	Al	0.1823 ± 0.0003	0.1820
	Cu	0.5782 ± 0.0013	0.5718
	Sn	0.4683 ± 0.0014	0.4628
	Ta	1.210 ± 0.004	1.228
	Pb	0.9368 ± 0.0041	0.9256
Zn^{65} (1.14 Mev)	Al	0.1571 ± 0.0022	0.1559
	Cu	0.4862 ± 0.0070	0.4914
	Sn	0.3923 ± 0.0054	0.3858
	Ta	0.9127 ± 0.0100	0.9536
	Pb	0.7068 ± 0.0051	0.7057
Na^{24} (2.76 Mev)	Al	0.0956 ± 0.0026	0.1001
	Cu	0.3164 ± 0.0080	0.3273
	Sn	0.2668 ± 0.0045	0.2692
	Ta	0.6433 ± 0.0055	0.6467
	Pb	0.4776 ± 0.0045	0.4644

where $x_{1/2}$ is the half-thickness. In the example of Fig. 15–2, the half-thickness of lead for 1.14 Mev γ-rays is 0.98 cm.

A useful rule of thumb may be deduced from the absorption coefficients. In Eqs. (15–6) and (15–7), the ratio Z/A changes very slowly as Z increases and, as will be shown later, ${}_e\mu$ is approximately the same for all elements in a certain energy region. Hence, the absorption coefficient shows the smallest variation from element to element when expressed as ${}_e\mu$ or μ/ρ. From Eq. (15–9),

$$\frac{\mu}{\rho} = \frac{0.693}{(x_{1/2})\rho}, \tag{15–10}$$

and, since μ/ρ varies slowly with Z, the product $(x_{1/2})\rho$ also varies slowly from element to element. It follows from this result that the greater the density the smaller the thickness needed of a given material to decrease the γ-ray intensity to a specified extent. For this reason, heavy metals such as iron, and especially lead, are used for shielding against γ-rays and x-rays. The approximate constancy of μ/ρ and $(x_{1/2})\rho$ means that the weights of different materials needed to decrease the intensity of the radiation by a certain fraction are very nearly the same. But for substances of higher density, the volume, and hence the thickness, will be less than for materials of lower density.

15–2 The interaction of gamma-rays with matter. The interaction of γ-rays with matter is markedly different from that of charged particles such as α- or β-particles. The difference is apparent in the much greater penetrating power of γ-rays and in the absorption laws. Gamma-rays and x-rays, which are both electromagnetic radiations, show a characteristic exponential absorption in matter, and have no definite range such as is found for charged particles. Charged particles, especially heavy ones, lose their energy during the course of a large number of collisions with atomic electrons. The energy loss occurs in many small steps and the particle gradually slows down until it is stopped altogether and absorbed. When a beam of γ-ray photons is incident on a thin absorber, however, each photon that is removed from the beam is removed individually in a single event. The event may be an actual absorption process, in which case the photon disappears, or the photon may be scattered out of the beam. The *one shot* nature of the removal process is responsible for the exponential absorption. For, the number of photons that can be removed in passing through a thickness Δx of absorber is proportional to Δx and to the number of photons reaching Δx; this kind of dependence leads directly to the exponential absorption law, as shown by Eqs. (15–1) and (15–2).

Three processes are mainly responsible for the absorption of γ-rays. These are (1) photoelectric absorption, (2) Compton scattering by the electrons in the atoms, and (3) production of electron-positron pairs as a result of the interaction between γ-rays and the electric fields of atomic nuclei. Quantum mechanics makes it possible to derive formulas for the probability of each process and the probability can be expressed as an absorption coefficient or as a cross section. The total absorption coefficient, which appears in Eqs. (15–1) and (15–2), is the sum of the absorption coefficients for the three different processes. The problem is further complicated by the fact that the absorption coefficient depends on the energy of the incident γ-rays as well as on the nature of the absorbing material. Consequently, the absorption of γ-rays cannot be described by a single formula, or by a range-energy curve. Each partial cross section or absorption coefficient must be evaluated as a function of energy for a given material, and tables or sets of curves can then be prepared from which values of the total absorption coefficient of a given material can be obtained. Thus,

$$\mu(E) = \tau(E) + \sigma(E) + \kappa(E), \tag{15–11}$$

where τ, σ, and κ denote the photoelectric, Compton, and pair formation coefficients, respectively, and each absorption coefficient is related to the corresponding cross section by Eq. (15–7). In the following sections, formulas for the cross sections for the three processes will be written down, discussed, and illustrated by means of curves and tables. Before the detailed treatment is begun, the processes will be discussed in a general way.

In the photoelectric process, all the energy $h\nu$ of the incident photon is transferred to a bound electron which is ejected from the atom with a kinetic energy $T = h\nu - I$, where I is the ionization potential of the electron. The electron may be ejected from the absorber or, if the absorber is not too thin, is more likely to be reabsorbed almost immediately because of the short range of an electron in a solid. At low photon energies, below 50 kev for aluminum and 500 kev for lead, the photoelectric effect gives the chief contribution to the absorption coefficient.

As the energy of the radiation increases, Compton scattering replaces the photoelectric effect as the chief means of removing photons from the initial beam. It was seen in Section 6–7 that in Compton scattering the incident photon is scattered by one of the atomic electrons and the latter is separated from its atom. The photon moves off at an angle with its original direction and with less energy than it had initially. The change of direction serves to remove the photon from the incident γ-ray beam. Compton scattering gives the main contribution to the absorption coefficient between 50 kev (0.050 Mev) and 15 Mev for aluminum, and between 0.5 and 5 Mev for lead. In these energy ranges, the energy of the incident γ-ray is much greater than the binding energy of the atomic electrons. The process may therefore be regarded as the scattering of a photon by a free electron initially at rest and, since each electron in the atom scatters independently, the Compton absorption coefficient per atom is proportional to the atomic number Z.

At high enough energies, both the photoelectric absorption and the Compton scattering become unimportant compared with pair formation. In the latter process, in the Coulomb field of an atomic nucleus, a γ-ray with sufficient energy disappears and an electron and positron are created. The *total* energy of the pair is equal to the energy $h\nu$ of the incident γ-ray; the kinetic energy T of the pair is

$$T = h\nu - 2m_0c^2, \tag{15–12}$$

if the small recoil energy of the nucleus is neglected. For pair production to occur, $h\nu$ must be greater than $2m_0c^2$ or 1.02 Mev. Pair production cannot occur for $h\nu < 2m_0c^2$ because this amount of energy is needed to supply the rest energy of the two particles. A *pair* of particles, with opposite charge, must be formed if charge is to be conserved. At photon energies greater than 5 Mev for lead and 15 Mev for aluminum, the probability of pair production is greater than that for Compton scattering and continues to increase with increasing energy.

In addition to the three effects discussed, there are some minor effects which may contribute to the attenuation of a γ-ray beam. The most important of these is coherent scattering by whole atoms or molecules, which may add several percent to the absorption coefficients of high-Z materials at low γ-ray energies. Among the other effects are (1) the nuclear photoelectric effect, in which high-energy photons eject neutrons from the nuclei of high-Z materials, and (2) Thomson and Compton scattering by nuclei rather than by electrons. For most practical purposes, all of these effects may be neglected.

The results of the theory of the interaction between γ-rays and matter have been collected and compared with experiment in the excellent review article(2) by C. M. Davisson and R. D. Evans. Many tables and curves are given there which present much of the presently available information in useful form.

15–3 Photoelectric absorption. Formulas for the probability that a photon of energy $h\nu$ will undergo photoelectric absorption have been derived by quantum mechanical methods. Several different formulas must be used if the probability, expressed as an absorption coefficient, is to be calculated over the range of photon energies from 0.1 Mev to 5 or 10 Mev. The reason for this is that the theory is complicated and different assumptions are made in different energy ranges in order to ease the mathematical difficulties. The main features of the dependence of the absorption coefficient on the atomic number of the absorber and on the energy of the photon are shown by the simplest of the formulas. If the energy of the photon is sufficiently small so that relativistic effects are not important, but large enough so that the binding energy of the electrons in the K-shell may be neglected, the cross section per atom ${}_a\tau$ for photoelectric absorption is

$$ {}_a\tau = \phi_0 Z^5 \left(\frac{1}{137}\right)^4 4\sqrt{2}\left(\frac{m_0c^2}{h\nu}\right)^{7/2}, \tag{15–13}$$

where

$$\phi_0 = \frac{8\pi}{3}\left(\frac{e^2}{m_0c^2}\right)^2 = 6.651 \times 10^{-25}\ \text{cm}^2. \tag{15–14}$$

In Eq. (15–13), $h\nu$ is the energy of the incident photon, m_0c^2 is the rest energy of the electron, and Z is the atomic number of the absorbing material; ϕ_0 is a convenient unit for measuring the cross section and represents the cross section for the scattering of low energy photons by a free electron at rest. This kind of scattering, called *Thomson scattering*, was discussed in Section 4–2. The simple equation (15–13) applies only to the ejection of electrons from the K-shell of the atom, which accounts for about 80% of the photoelectric effect.

The most important property of ${}_a\tau$ is the strong dependence on the atomic number and on the energy of the incident photon; ${}_a\tau$ is proportional to Z^5 and inversely proportional to $(h\nu)^{7/2}$. The Z^5 dependence means that, for a given photon energy, the process of photoelectric absorption is much more important in heavy metals such as lead than in light ones such as aluminum. The energy dependence shows that for a given element, the effect is much greater at small photon energies than at large ones. It follows that the photoelectric effect is especially important in the absorption of low energy photons by heavy elements.

In the more rigorous formulas, the dependence of ${}_a\tau$ on Z and $h\nu$ is not quite the same as that shown in Eq. (15–13). Thus, the variation of ${}_a\tau$ with Z does not follow the simple fifth power law, although for practical purposes the deviation from a power law is small. The exponent giving the best agreement with theory increases with γ-ray energy and is between 4 and 5 for energies above 0.35 Mev, being approximately 4.5 at 1.13 Mev and 4.6 at 2.62 Mev. The rigorous dependence of ${}_a\tau$ on the energy of the γ-rays also cannot be described by a simple power law. The cross section decreases rapidly with increasing energy, approximately as $(h\nu)^{-3}$ for energies less than 0.5 Mev, and as $(h\nu)^{-1}$ for energies greater than 0.5 Mev. Values obtained from the more rigorous theory are given in Table 15–2. For purposes of comparison and ease of use, the values of the quantity

TABLE 15–2

THE PHOTOELECTRIC EFFECT*

Energy of γ-rays		$Z = 0$	$Z = 13$ Al	$Z = 26$ Fe	$Z = 38$	$Z = 50$ Sn	$Z = 65$	$Z = 82$ Pb
$n = m_0c^2/h\nu$	Mev							
0		0.353	0.275	0.228	0.203	0.188	0.166	0.159
0.1	5.108	0.436	0.343	0.289	0.255	0.230	0.206	0.182
0.125	4.086	0.453	0.362	0.306	0.272	0.247	0.216	0.188
0.194	2.633	0.524	0.412	0.347	0.302	0.270	0.237	0.205
0.25	2.043	0.581	0.447	0.375	0.331	0.290	0.259	0.225
0.375	1.362	0.740	0.556	0.462	0.403	0.366	0.319	0.278
0.50	1.022	0.956	0.694	0.575	0.500	0.440	0.397	0.341
0.75	0.6711	1.528	1.162	0.947	0.781	0.690	0.600	0.522
1.0	0.5108	2.375	1.862	1.506	1.237	1.038	0.878	0.747
1.25	0.4086	3.578	2.756	2.188	1.815	1.512	1.262	1.031
1.5	0.3405	5.3	3.834	3.062	2.528	2.116	1.750	1.362
2.0	0.2554	9.48	7.46	5.34	4.24	3.55	2.96	2.17
3.0	0.1703	24.3	17.61	12.44	8.81	7.88	6.25	4.18
4.0	0.1277	46.8	33.1	22.8	14.8	14.2	10.5	6.64
5.0	0.1022	79.9	53.4	34.6	21.8	21.6	15.6	9.61

*Numerical values of ${}_a\tau/Z^5n$ in units of 10^{-32} cm^2.

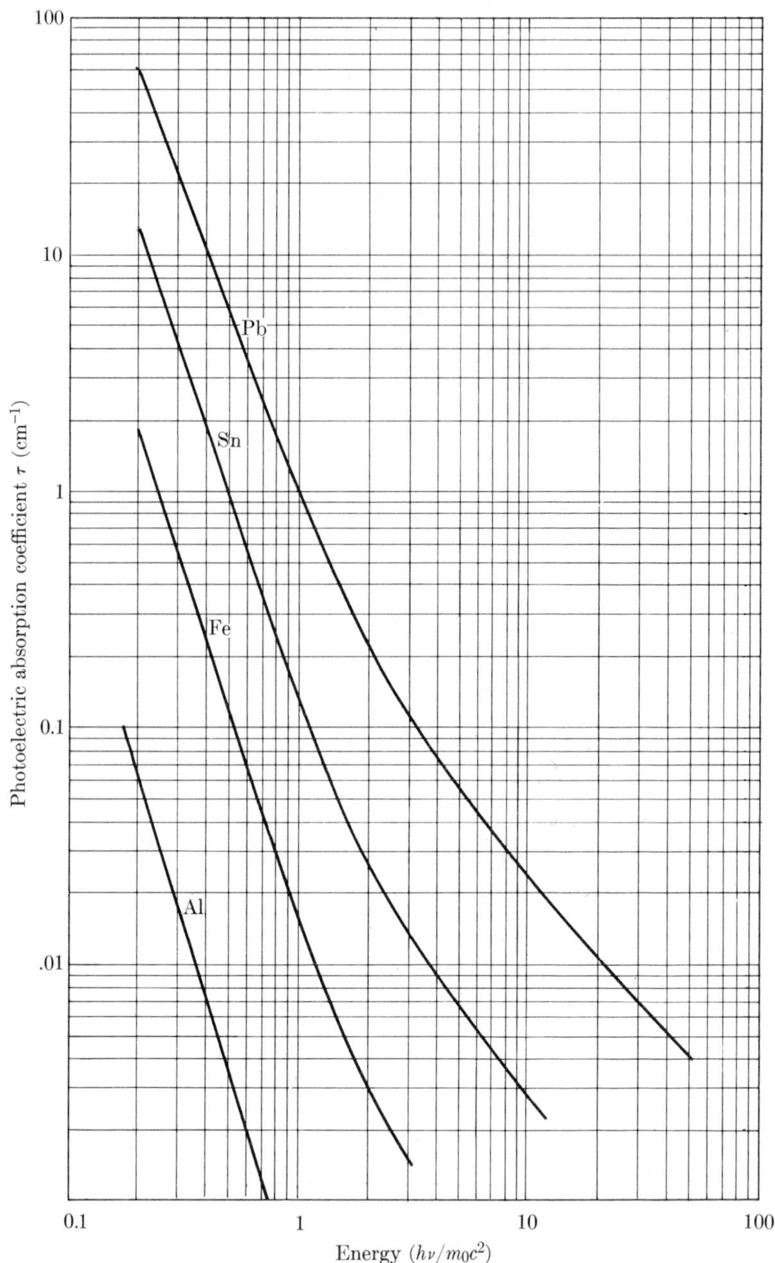

Fig. 15–3. The absorption coefficient per centimeter as a function of the photon energy for photoelectric absorption in aluminum, iron, tin, and lead.

${}_a\tau/Z^5 n$ are listed, where n is the ratio $m_0c^2/h\nu$. Increasing values of n correspond to decreasing γ-ray energies. For a particular absorber, say lead, the values of ${}_a\tau$ are obtained for different γ-ray energies by multiplying the values listed in the table first by $Z^5 = (82)^5$, and then by the appropriate value of n. Curves obtained in this way are shown in Fig. 15–3 for aluminum ($Z = 13$), iron ($Z = 26$), tin ($Z = 50$), and lead ($Z = 82$). It will be noted that the values of ${}_a\tau/Z^5 n$ listed in the table are in units of 10^{-32} cm^2. Values of ${}_a\tau$ for energies and atomic numbers other than those listed can be obtained by plotting the appropriate curves and interpolating. The values for $Z = 0$ are included to allow interpolation for elements with $Z < 13$.

15–4 Compton scattering. The main features of Compton scattering were discussed in Section 6–7, but it will be useful to repeat some of the results obtained there in somewhat different form. The energy lost by a photon in a single Compton scattering process can be obtained from Eq. (6–52), which gives the change in wavelength of the photon,

$$\lambda - \lambda_0 = \frac{h}{m_0c}(1 - \cos\phi). \qquad (15\text{–}15)$$

In Eq. (15–15), λ_0 is the wavelength of the photon before collision, λ is the wavelength after the collision, and ϕ is the angle between the initial and final directions of the photon as shown in Fig. 6–7. Since $\lambda = c/\nu$, Eq. (15–15) may be written

$$\frac{c}{\nu} - \frac{c}{\nu_0} = \frac{h}{m_0c}(1 - \cos\phi),$$

or

$$\frac{1}{\nu} = \frac{1}{\nu_0} + \frac{h}{m_0c^2}(1 - \cos\phi).$$

Inversion of both sides of the last equation gives

$$\nu = \frac{1}{\dfrac{1}{\nu_0} + \dfrac{h}{m_0c^2}(1 - \cos\phi)}.$$

If the top and bottom of the right side are multiplied by ν_0, and if both sides of the resulting equation are multiplied by h, the result is

$$h\nu = \frac{h\nu_0}{1 + \dfrac{h\nu_0}{m_0c^2}(1 - \cos\phi)}. \qquad (15\text{–}16)$$

Equation (15–16) gives the energy of the scattered photon in terms of the initial energy and the scattering angle. The recoiling electron has a kinetic energy given by

$$T = h\nu_0 - h\nu$$
$$= \frac{(1 - \cos\phi)\,\dfrac{h\nu_0}{m_0c^2}}{1 + \dfrac{h\nu_0}{m_0c^2}(1 - \cos\phi)}. \qquad (15\text{–}17)$$

The kinetic energy of the electron has its maximum value when $\cos\phi = -1$ or $\phi = 180°$, and the photon is scattered directly backward. The electron energy, in this case, is

$$T_{\max} = \frac{h\nu_0}{1 + (m_0c^2/2h\nu_0)}. \qquad (15\text{–}18)$$

The electron receives the least energy in a grazing collision in which the photon continues with its initial frequency in the forward direction, and the electron is ejected with very nearly zero velocity in a direction perpendicular to that of the photon path.

The discussion so far refers only to a single Compton scattering process. In order to treat the contribution of the Compton effect to the attenuation of a beam of γ-rays in matter, it is necessary to calculate the probability that such a scattering process will occur. This probability was calculated on the basis of relativistic quantum mechanics by Klein and Nishina. Although the details of the theory are complicated, the results may be expressed in straightforward formulas with which calculations can be made quite easily. A formula was obtained for the cross section per electron ${}_e\sigma$ for the removal of photons from the incident beam by scattering:

$${}_e\sigma = \tfrac{3}{4}\phi_0$$
$$\times\left\{\frac{1+\alpha}{\alpha^2}\left[\frac{2(1+\alpha)}{1+2\alpha} - \frac{1}{\alpha}\ln(1+2\alpha)\right] + \frac{1}{2\alpha}\ln(1+2\alpha) - \frac{1+3\alpha}{(1+2\alpha)^2}\right\}, \qquad (15\text{–}19)$$

where α has been defined as

$$\alpha = \frac{h\nu_0}{m_0c^2}, \qquad (15\text{–}20)$$

and ϕ_0 is given by Eq. (15–14). When ${}_e\sigma$ is multiplied by $\rho N(Z/A)$, the result is the Compton absorption coefficient σ (cm^{-1}),

$$\sigma\,(\text{cm}^{-1}) = \rho N \frac{Z}{A}\,{}_e\sigma. \qquad (15\text{–}21)$$

The coefficient σ is a measure of the probability that a photon is scattered out of the beam per centimeter of absorber, and since the beam is initially homogeneous, σ is also a measure of the total amount of energy removed from the beam per centimeter of absorber. It is sometimes necessary to know the amount of energy carried off by the scattered photons or the amount of energy absorbed by the recoil electrons. These amounts may be expressed in terms of two quantities denoted by ${}_e\sigma_s$ and ${}_e\sigma_a$. The former, ${}_e\sigma_s$, may be called the "Compton scattering cross section per electron for the energy of the scattered photon." It is given by

$$ {}_e\sigma_s = \frac{3}{8}\,\phi_0\left[\frac{1}{\alpha^3}\ln(1+2\alpha) + \frac{2(1+\alpha)(2\alpha^2-2\alpha-1)}{\alpha^2(1+2\alpha)^2} + \frac{8\alpha^2}{3(1+2\alpha)^3}\right]. \tag{15-22} $$

The quantity ${}_e\sigma_a$ is the "Compton cross section per electron for the energy absorbed." The coefficient ${}_e\sigma$ is simply the sum of ${}_e\sigma_s$ and ${}_e\sigma_a$:

$$ {}_e\sigma = {}_e\sigma_s + {}_e\sigma_a, \tag{15-23} $$

so that when ${}_e\sigma$ and ${}_e\sigma_s$ are known, ${}_e\sigma_a$ is obtained by subtraction. In analogy with Eq. (15-21),

$$ \begin{aligned} \sigma_s &= \rho N \frac{Z}{A}\,{}_e\sigma_s, \\ \sigma_a &= \rho N \frac{Z}{A}\,{}_e\sigma_a. \end{aligned} \tag{15-24} $$

Finally, the sum

$$ \sigma = \sigma_s + \sigma_a, \tag{15-25} $$

shows how the total Compton coefficient is broken down into a coefficient for the energy scattered out of the beam and a coefficient for the energy removed by the recoil electrons. The electrons, because of their very short range, are often absorbed and their energy may appear in the form of heat. In some practical problems it is necessary to know how much heat is produced as a result of the attenuation of a beam of γ-rays in passing through a given absorber. The contribution of Compton scattering to this heat can be obtained when ${}_e\sigma_a$ is known for the particular γ-ray energy.

It is convenient to tabulate the values of ${}_e\sigma$, ${}_e\sigma_s$, and ${}_e\sigma_a$, i.e., the values in cm^2/electron, because these are independent of the properties of the absorber. The values for a particular absorber are then obtained by using Eqs. (15-21) and (15-24). Table 15-3 gives the cross sections for the Compton effect ${}_e\sigma$, ${}_e\sigma_s$, and ${}_e\sigma_a$ in units of 10^{-25} cm^2/electron. The values of the absorption coefficients per centimeter, σ_s, σ_a, and σ, for the special case of lead ($Z = 82$) are plotted in Fig. 15-4.

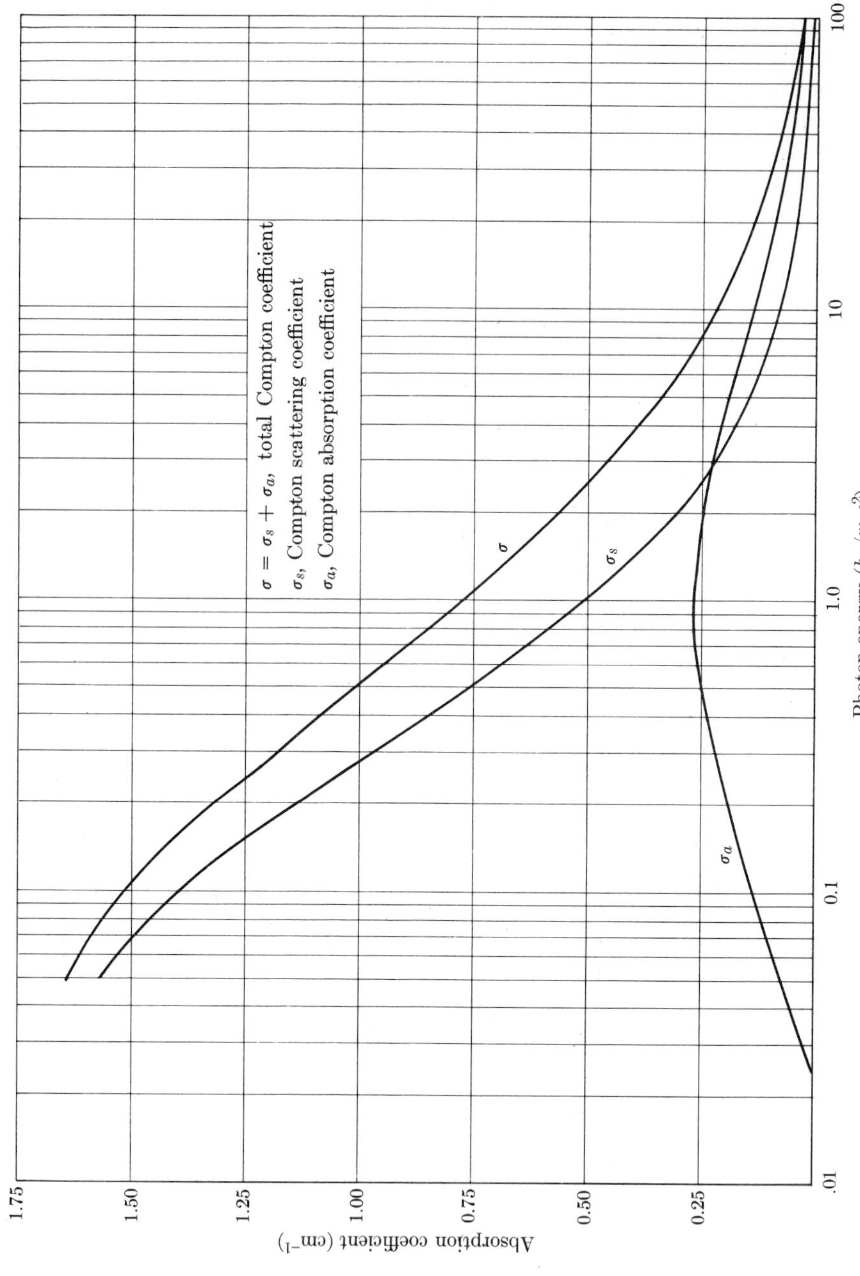

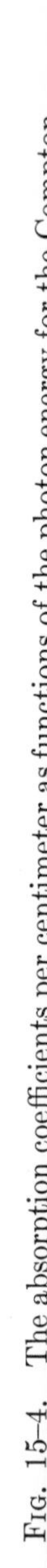

FIG. 15–4. The absorption coefficients per centimeter as functions of the photon energy for the Compton

TABLE 15–3

CROSS SECTIONS FOR THE COMPTON EFFECT*

Initial energy		$_e\sigma$	$_e\sigma_s$	$_e\sigma_a$
$\alpha = h\nu/m_0c^2$	Mev			
0.025	0.0128	6.31	6.31	0.00
0.05	0.0256	6.07	5.79	0.28
0.075	0.0383	5.83		
0.10	0.0511	5.599	5.138	0.461
0.15	0.0767	5.243		
0.20	0.102	4.900	4.217	0.683
0.25	0.128	4.636		
0.30	0.153	4.410	3.597	0.813
0.40	0.204	4.032	3.152	0.880
0.50	0.255	3.744	2.818	0.928
0.60	0.307	3.507	2.551	0.956
0.70	0.358	3.309	2.337	0.972
0.80	0.409	3.140	2.158	0.982
0.90	0.460	2.994	2.008	0.986
1.0	0.511	2.866	1.879	0.987
1.5	0.767	2.397	1.432	0.965
2.0	1.022	2.090	1.164	0.926
2.5	1.278	1.868	0.983	0.884
3.0	1.533	1.696	0.852	0.844
3.5	1.789	1.559	0.753	0.806
4.0	2.044	1.446	0.674	0.772
4.5	2.300	1.351	0.611	0.739
5.0	2.555	1.269	0.559	0.710
6.0	3.066	1.136	0.477	0.659
7.0	3.577	1.031	0.417	0.614
8.0	4.088	0.9465	0.370	0.5765
9.0	4.599	0.876	0.333	0.543
10.0	5.108	0.8168	0.3023	0.5145
12.0	6.132	0.722	0.256	0.466
20.0	10.22	0.502	0.158	0.344
30.0	15.33	0.371	0.107	0.264
50.0	25.55	0.250	0.066	0.184
70.0	35.77	0.191	0.047	0.144
100.0	51.10	0.143	0.033	0.110

*In units of 10^{-25} cm^2/electron.

Equations (15–19) and (15–22) show that the Compton scattering per electron is independent of Z, so that the scattering per atom is proportional to Z. The mass scattering coefficient σ/ρ is given by

$$\frac{\sigma}{\rho} = N \frac{Z}{A} {}_e\sigma, \tag{15–26}$$

and varies only slowly with Z. For the light elements Z/A is closely equal to $\frac{1}{2}$, so that for a given photon energy σ/ρ is practically constant for these elements. The total scattering coefficient per electron, ${}_e\sigma$, decreases with increasing photon energy, as can be seen from the values in Table 15–3. The decrease is quite slow at low values of the energy, and for energies above 0.5 Mev, ${}_e\sigma$ is roughly proportional to $(h\nu)^{-1}$. Thus Compton scattering decreases much more slowly with increasing energy than does photoelectric absorption; even in heavy elements, it is the most important process in the energy range from about 0.6 to 2.5 Mev.

In the treatment so far, it has been assumed that a singly-scattered photon was removed from the γ-ray beam. This assumption may be considered to define a *thin* absorber. In a *thick* absorber some of the singly-scattered photons which have been considered lost will be scattered and rescattered and may eventually reach the detector. The problem of repeated, or multiple, Compton scattering of γ-rays is important in connection with the shielding of personnel against the γ-radiation emitted by accelerators or nuclear reactors. Considerable progress has been made in the theoretical and experimental treatment of this problem, and the reader is referred to the book by H. Goldstein listed in the general references at the end of the chapter.

15–5 Electron-positron pair formation. The third mechanism by which electromagnetic radiation can be absorbed is the production of electron-positron pairs. This process, which has no analogy in classical physics, may be accepted as a strictly experimental phenomenon. The discovery of the positron and of pair formation, however, represents one of the great triumphs of modern physical theory, and for this reason a short digression seems worth while. It was mentioned very briefly at the end of Section 7–5 that the Schroedinger equation, the fundamental equation of wave mechanics, did not satisfy the requirement of invariance postulated by the thoery of special relativity. When the Bohr theory of the atom was replaced by wave mechanics, the problem of the relativistic treatment of the electron still remained. This problem was solved by Dirac, who derived an equation, the Dirac electron equation, which is one of the most important developments in modern physics. Dirac required that the equation which represents the motion of an electron be invariant under the Lorentz transformation (see Sections 6–4 and 6–5).

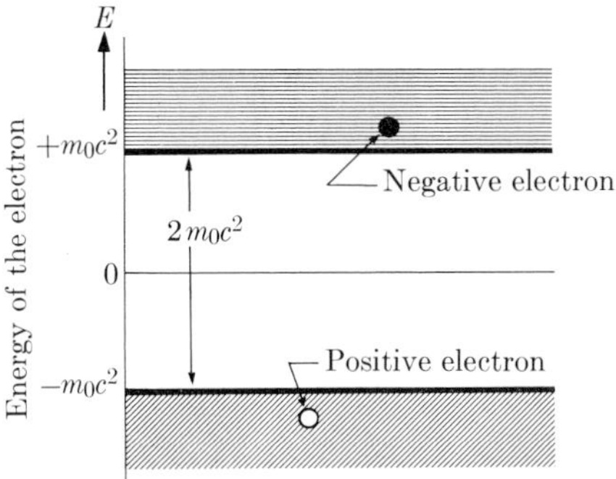

FIG. 15–5. Possible values of the energy of an electron according to the Dirac theory.

The solution of the Dirac equation led in a natural way to the formula for the fine structure of spectral lines and to the spin of the electron. But to solve the Dirac equation it was necessary to assume that the electron can exist in two sets of quantum states, one of positive energy (including the rest energy), and the other of negative energy. It was found that the possible values of the energy of a free electron are either greater than $+m_0c^2$ or smaller than $-m_0c^2$, and that no possible energies for the electron exist between these two limits. This state of affairs is shown in Fig. 15–5, where the shaded regions are those in which values of the energy exist. Electrons in states of positive energy behave in the usual manner of electrons that are ordinarily observed, while electrons in states of negative energy should have properties which have no classical analogy. These electrons could simply be neglected if they were considered from a classical viewpoint. For, in that case, the value of the energy of an electron could change only in a continuous way, and an electron in a positive energy state could not bridge the discontinuous gap between a state of energy $+m_0c^2$ and one of energy $-m_0c^2$. In other words, a state of negative energy would have no real physical meaning. According to quantum theory, however, an electron can make a discontinuous transition from one energy state to another, so that there is no way of ruling out a jump from a positive energy state to a negative energy state.

Dirac avoided the difficulty by assuming that the states of negative energy are real, but that they are all usually occupied. The electrons which are ordinarily observed are those in positive energy states. Suppose that one electron is missing in the distribution of negative energy states. The empty state, according to Dirac, would appear as a particle with positive energy and positive charge, since a particle of negative energy and negative charge is absent. This empty place, or "Dirac hole," would therefore behave like a positively charged particle. Dirac first assumed

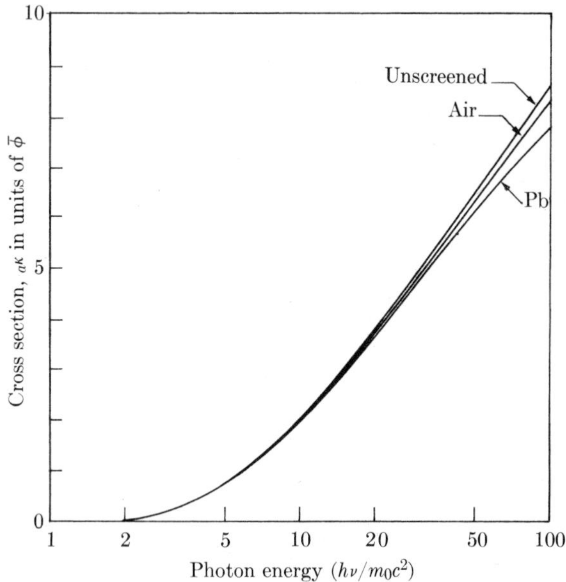

Fig. 15–6. Pair-formation cross section per atom as a function of the photon energy; $_a\kappa$ in units of $\bar{\phi}$. (By permission, from *Experimental Nuclear Physics*, Vol. I., E. Segrè, ed., New York: Wiley, 1953.)

TABLE 15–4

PAIR-FORMATION CROSS SECTIONS

Photon energy, Mev	$_a\kappa/\bar{\phi}$		$_a\kappa/Z^2$, in units of 10^{-27} cm²/atom	
1.533	0.086		0.050	
2.043	0.327		0.189	
3.065	0.905		0.524	
5.108	1.98		1.15	
	Al	Pb	Al	Pb
7.662	2.93	2.86	1.70	1.66
10.22	3.66	3.60	2.12	2.08
17.01	5.07	4.90	2.94	2.84
25.54	6.14	5.96	3.56	3.46

that it represented a proton, but this assumption had to be discarded because it could explain neither the difference in mass between a proton and an electron nor the stable existence of the proton.

The dilemma was resolved in 1933 when Anderson, during the course of a study of cosmic rays by means of a cloud chamber, observed particles with the same mass as that of the electron and with an electric charge equal in magnitude but opposite in sign to that of the electron. This particle was given the name *positron* and identified with the Dirac hole. In terms of Dirac's theory, the production of a positron is interpreted as follows. A photon of energy greater than $2m_0c^2$ can raise an electron from a state of negative energy to a state of positive energy. The disappearance of an electron from a negative energy state leaves a hole, which means the appearance of a positron; the appearance of an electron in a positive energy state means the appearance of an ordinary electron. Thus, a pair of particles is created. Since each particle must have a rest mass equal to m_0c^2, the photon must have an energy of at least $2m_0c^2$, or 1.02 Mev. The creation process usually takes place in the electric field in the neighborhood of a nucleus, because some body must be given recoil energy and momentum in order that energy and momentum be conserved in the system.

The formulas for the probability of pair formation are more complicated than those for photoelectric absorption and Compton scattering, and will not be written down here. Values of the cross section per atom, ${}_a\kappa$, for pair production can be listed for certain photon energies and curves for the variation of ${}_a\kappa$ with energy can be given. The cross section is zero for photon energies less than 1.02 Mev; for greater energies, it increases at first slowly, then more rapidly. It is proportional to Z^2, so that for a given photon energy, pair formation increases quite rapidly with atomic number. Pair production cross sections are usually expressed in units of the quantity

$$\bar{\phi} = \frac{Z^2}{137}\left(\frac{e^2}{m_0c^2}\right)^2 = Z^2 \times 5.796 \times 10^{-28}\,\mathrm{cm}^2. \qquad (15\text{–}27)$$

Curves of ${}_a\kappa/\bar{\phi}$ as a function of $\alpha = h\nu/m_0c^2$ are shown in Fig. 15–6 for air and lead; values of ${}_a\kappa/\bar{\phi}$ and ${}_a\kappa/Z^2$ for Al and Pb are listed at several photon energies in Table 15–4. At photon energies above about $10m_0c^2$ or about 5 Mev, it is necessary to take into account an effect called *screening*. At these energies the electron-positron pair is probably formed some distance from the nucleus. When this distance lies outside some of the electron shells, the field in which the pairs are created is smaller than that of the nucleus alone and the probability of pair formation is reduced somewhat. Consequently, at higher energies (up to 25 Mev) the values for lead are somewhat smaller than those for air or aluminum; at still greater energies, the screening effect becomes more marked.

It follows from the way in which the cross section for pair formation depends on Z and on the energy that this process is important for high energies and for heavy elements. The contribution of pair formation to the total absorption coefficient μ for lead is equal to that of the Compton effect at about 4.75 Mev. Above this energy, pair formation predominates.

The process of pair production is closely related to the reverse process, that of electron-positron annihilation. A positron, after being formed, is slowed down by collisions with atoms until it is practically at rest. It then interacts with an electron which is also practically at rest. The two particles disappear, and two photons appear moving in opposite directions, each with an energy of 0.511 Mev, equal to the rest energy of an electron. Two photons, rather than one, are needed to allow momentum to be conserved if the annihilation occurs away from nuclei. The photons which appear on the annihilation of an electron-positron pair are called *annihilation radiation*, and the absorption of γ-rays by the pair-production process is always complicated by the appearance of this low-energy secondary radiation.

15–6 The absorption of gamma-rays by matter. Comparison of experimental and theoretical results. Theoretical values of the absorption coefficient ${}_a\mu$ or μ can now be obtained from the values of the partial coefficients given in Tables 15–2, 15–3, and 15–4 and from graphs of those data. The following procedure is used:

1. Values of ${}_a\tau/Z^5n$ for different values of n are found from Table 15–2 or from plots of the data in the table. Values of the photoelectric cross section per atom ${}_a\tau$ are obtained on multiplying by Z^5n.
2. Values of the Compton cross section per atom ${}_a\sigma$ are obtained from the data of Table 15–3 on multiplying the values of ${}_e\sigma$ by Z.
3. Values of ${}_a\kappa/\bar{\phi}$ are obtained from Table 15–4 or Fig. 15–6 for the same values of $n(= 1/\alpha)$ used in the first two steps, and multiplied by $\bar{\phi} = 5.796 \times 10^{-28}Z^2$ to give the pair production cross section per atom ${}_a\kappa$.

The total absorption coefficient per atom ${}_a\mu$ is then

$$ {}_a\mu = {}_a\tau + {}_a\sigma + {}_a\kappa. \tag{15–28}$$

The mass absorption coefficient μ/ρ is

$$\frac{\mu}{\rho} = \frac{N}{A}\,({}_a\tau + {}_a\sigma + {}_a\kappa). \tag{15–29}$$

The absorption coefficient μ is

$$\mu = \frac{\rho N}{A}\,({}_a\tau + {}_a\sigma + {}_a\kappa). \tag{15–30}$$

Tables of values of ${}_a\tau$, ${}_a\sigma$, ${}_a\kappa$, and ${}_a\mu$ have been prepared by Davisson and Evans[(2)] for 24 elements ranging from hydrogen ($Z = 1$) to bismuth ($Z = 83$). Their values for lead are listed in Table 15–5 together with the values of μ/ρ and μ. The values of the cross section per atom are in units of 10^{-24} cm^2/atom. Mass absorption coefficients μ/ρ (cm^2/gm) are listed in Table 15–6 for four commonly used absorbers (water, aluminum, iron, and lead).

The total and partial absorption cross sections per centimeter for lead are plotted in Fig. 15–7, and the general features of the absorption of γ-rays can be seen by considering these curves. At very low energies, absorption by photoelectrons predominates, but decreases rapidly with increasing energy. As it decreases, attenuation by the Compton effect becomes relatively more important, until the two effects are equal at about 0.5 Mev. At energies slightly less than 1 Mev most of the attenuation is

Table 15–5

Values of the Absorption Coefficients for Lead*

Photon energy, Mev	Photo-electric ${}_a\tau$	Compton ${}_a\sigma$	Pair formation ${}_a\kappa$	Total ${}_a\mu$	Coefficient per cm, μ, cm^{-1}	Mass coefficient μ/ρ, cm^2/gm
0.1022	1782	40.18		1822	59.9	5.30
0.1277	985	38.01		1023	33.6	2.97
0.1703	465	35.04		500	16.4	1.45
0.2554	161	30.70		192	6.31	0.558
0.3405	75.7	27.63		103.3	3.39	0.300
0.4086	47.8	25.74		73.5	2.42	0.214
0.5108	27.7	23.50		51.2	1.68	0.149
0.6811	14.5	20.73		35.2	1.16	0.102
1.022	6.31	17.14		23.45	0.771	0.0682
1.362	3.86	14.81	0.1948	18.87	0.620	0.0549
1.533		13.91	0.3313			
2.043	2.08	11.86	1.247	15.19	0.499	0.0442
2.633						
3.065		9.313	3.507			
4.086	0.869	7.761	5.651	14.28	0.469	0.0415
5.108	0.675	6.698	7.560	14.93	0.491	0.0434
6.130		5.917	9.119			
10.22	0.316	4.115	14.04	18.47	0.607	0.0537
15.32	0.206	3.042	18.00	21.25	0.698	0.0618
25.54	0.122	2.044	23.24	25.41	0.835	0.0739

*In units of 10^{-24} cm^2/atom.

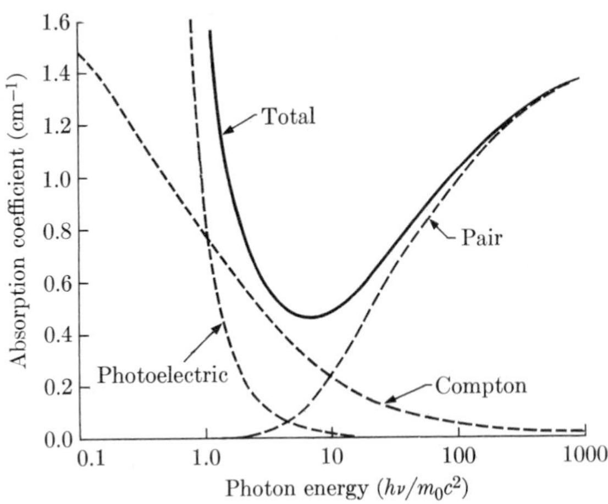

FIG. 15–7. Total absorption coefficient of lead, showing the contributions of photoelectric absorption, Compton scattering, and pair formation.

TABLE 15–6

MASS ABSORPTION COEFFICIENTS*

Photon energy, Mev	Water	Aluminum	Iron	Lead
0.1	0.167	0.160	0.342	5.29
0.15	0.149	0.133	0.182	1.84
0.2	0.136	0.120	0.138	0.895
0.3	0.118	0.103	0.106	0.335
0.4	0.106	0.0922	0.0918	0.208
0.5	0.0967	0.0840	0.0828	0.145
0.6	0.0894	0.0777	0.0761	0.114
0.8	0.0786	0.0682	0.0668	0.0837
1.0	0.0706	0.0614	0.0595	0.0683
1.5	0.0576	0.0500	0.0484	0.0514
2.0	0.0493	0.0431	0.0422	0.0451
3.0	0.0396	0.0353	0.0359	0.0410
4.0	0.0339	0.0310	0.0330	0.0416
5.0	0.0302	0.0284	0.0314	0.0430
6.0	0.0277	0.0266	0.0305	0.0455
8.0	0.0242	0.0243	0.0298	0.0471
10.0	0.0221	0.0232	0.0300	0.0503

*μ/ρ, cm^2/gm.

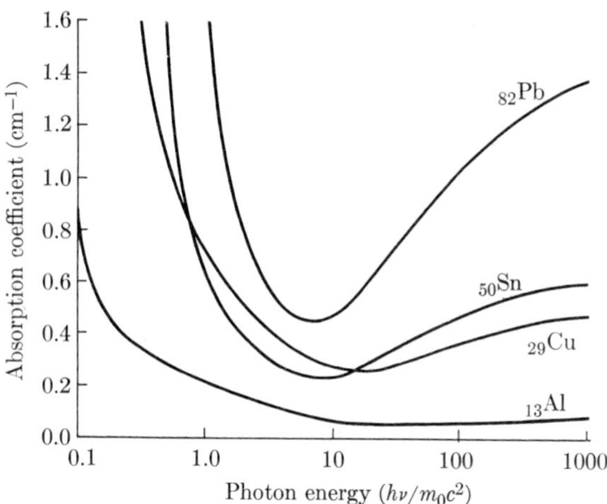

FIG. 15-8. The total absorption coefficients of aluminum, copper, tin, and lead.

caused by the Compton effect. Absorption by pair formation starts at about 1 Mev and increases while the other effects decrease, until at high energies the absorption is almost completely by pair production. The curve of total absorption coefficient *vs.* energy has a minimum at energies where the Compton effect and pair production become comparable. While these features are characteristic of the absorption curves for all elements, the particular energy at which one process or another is important varies from element to element. The following table indicates the energy ranges in which the various effects make their greatest contributions in a light element (aluminum) and in a heavy element (lead).

	Photoelectric effect	Compton effect	Pair formation
Aluminum	<0.05 Mev	0.05–15 Mev	>15 Mev
Lead	<0.5 Mev	0.5 –5 Mev	>5 Mev

The variation of the absorption coefficient *vs.* energy curve from element to element is shown by the curves for aluminum, copper, tin, and lead in Fig. 15-8.

The simplest method of testing the theory of the absorption of γ-rays is to measure the total absorption coefficient by the method discussed in Section 15-1. Although measurements of this kind are basically simple, the experimental results depend on the knowledge of the energy of the photons emitted by the source, on the purity of the absorber, and on many other experimental details. The production of artificial radionuclides has provided sources of monoenergetic γ-rays with energies in the range from

0.5 to 2.8 Mev, and monoenergetic γ-rays beams of higher energy have been obtained as products of nuclear reactions. It has been found that in the energy range from 0.1 Mev to 6 Mev the theoretical values of the absorption coefficient as calculated from the present theories of the photoelectric effect, Compton effect, and pair production, are in good agreement with the experimentally measured values. The agreement has been found for a large number of elements in the periodic table from carbon to lead. The reader is again referred to the review article by Davisson and Evans for a detailed discussion, including many experimental curves, of the experimental results. A partial comparison of theory and experiment is given in Table 15–1.

The partial absorption coefficients have also been measured experimentally, although measurements of this kind are more difficult, and hence less extensive, than those of the total absorption coefficient. The theory of the individual effects has been found to give cross sections or partial absorption coefficients in good agreement with the experimental results. Thus, the theory of photoelectric absorption, the Klein-Nishina formulas for the Compton effect, and the theory of pair formation have all been shown to predict the correct results. The theory of these processes and the experimental verification are treated in detail in Heitler's book, *The Quantum Theory of Radiation.*[(3)]

15–7 The measurement of gamma-ray energies. Several general methods can be used to measure the energies of γ-rays. Since γ-rays are electromagnetic radiations, the most direct method is to determine the wavelength and hence the energy by using a crystal as a diffraction grating (Section 4–3). A high-precision instrument called a *curved-crystal focusing-type spectrometer* has been used by DuMond,[(4,5)] and the principle on which it is designed is shown schematically in Fig. 15–9. The quartz diffracting crystal C is bent and clamped so that the diffracting planes meet, when extended, in a line at β, normal to the plane of the figure. The radius of curvature of the crystal is then equal to the diameter of the focusing circle F shown (2.0 m). If the source of γ-rays is at the real focus R, and if the Bragg condition, Eq. (4–10), is satisfied, there is a diffracted beam which enters the detector D as if it came from a virtual source at V; the Bragg angle is denoted by θ. For each different wavelength there is a particular position of the source somewhere on the focusing circle which yields a diffracted beam. The counting rate of the detector is determined as a function of the source position as the source is moved around the focusing circle; the rate has a sharp maximum at the point for which a strong diffracted beam occurs. The position of this point determines the wavelength and hence the frequency and energy of the γ-rays. This method can give results with very high precision, e.g., the energy of

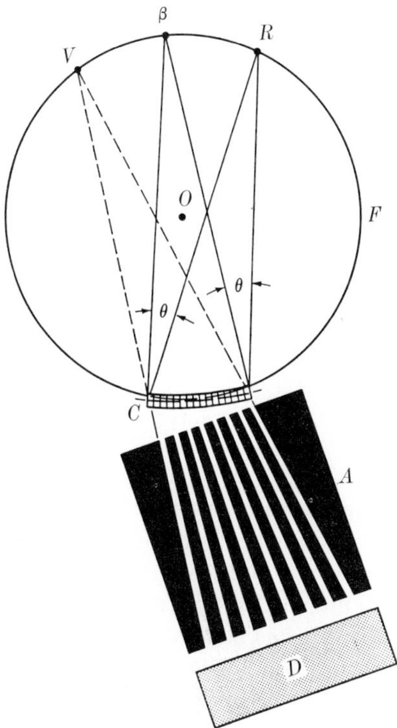

FIG. 15–9. Schematic diagram of a crystal γ-ray spectrometer (DuMond[4,5]).

the γ-ray associated with the β-decay of Au^{198} was determined as 411.770 ± 0.036 kev, and results with similar precision have been obtained for the energies of many other nuclear γ-rays.[5] Crystal diffraction methods, however, have two disadvantages. First, the measurements become more difficult and less precise as the energy of the γ-rays increases and the wavelength decreases, with the result that they cannot be carried much above 1 Mev. Second, they require highly active sources, and these are often not available.

DuMond's γ-ray spectrometer was also used to measure the energy of the annihilation radiation.[5] The photons from the annihilation of positrons from Cu^{64} were found to have an energy of 510.941 ± 0.067 kev. The "best value" for the electron rest energy at the time of the experiment was 510.969 ± 0.015 kev, so that the measured value of the energy of the annihilation radiation agreed very well with the value predicted by the theory of pair formation and annihilation. The two numbers quoted can be used to calculate the possible difference in mass between the positron and the electron, and it was found that to a part in 10^4 there is no evidence for any difference in mass between positive and negative electrons. This result has special interest because, together with a recent direct comparison of the ratio m/e for the electron and positron, it provides independent experimental proof that the electron and positron are identical except for the sign of the electric charge. In the comparison experiment,[6] electrons from a hot filament and positrons from Na^{22} were analyzed in a mass spectrometer with a reversible electric field and a fixed magnetic field. The particles followed identical trajectories in opposite directions in the spectrometer. The magnitude of the electric field needed for focusing was taken as a measure of m/e for the particle involved. The result obtained was

$$\frac{(m/e)^{-} - (m/e)^{+}}{m/e} = (26 \pm 71) \times 10^{-6},$$

where the ratios with the minus and plus signs refer to the values obtained by focusing the particles in the experiment, and the denominator is the accepted value of the ratio for the electron. Thus, both the mass and the specific charge of the positron and electron have been shown to be equal to within very small amounts. Although the comparison experiment is not directly related to γ-rays, it has been included here because of the way in which it supplements the annihilation radiation experiment.

The curved crystal spectrometer has been improved and simplified,[30,31] and applied to problems such as the study of the γ-rays from neutron capture. As an example of the precision that can be obtained, a recent measurement[32] of the deuteron binding energy gave the result 2.2255 ± 0.0015 Mev.

Gamma-rays of moderate energy are most often studied by means of the photoelectrons and Compton recoil electrons which they eject from a suitable material called a *radiator*. The energies of the ejected electrons are measured with a magnetic spectrometer, as described in Section 14–1.

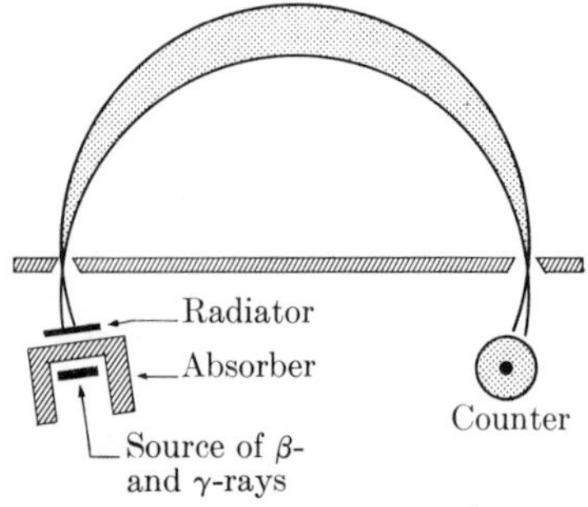

FIG. 15–10. Diagram of an arrangement for measuring the energies of γ-rays from a source of β- and γ-rays.

When the Compton effect is used, the source is enclosed in an absorber of material of low atomic number such as aluminum, thick enough to stop all the primary electrons or other charged particles but not the γ-rays. The Compton electrons ejected from the radiator, which is usually a thin foil, are focused in the spectrometer and form a continuous spectrum with a fairly sharply defined upper limit. The upper limit corresponds to electrons ejected in the forward direction from the surface of the absorber, and is related to the γ-ray energy by Eq. (15–18). If T_m is the maximum electron energy, and $h\nu$ is the desired γ-ray energy [this amounts simply to dropping the subscript zero in Eq. (15–18)], then

$$h\nu = \tfrac{1}{2}[T_m + (T_m^2 + 2T_m m_0 c^2)^{1/2}]. \tag{15–31}$$

A schematic diagram of an arrangement for measuring the energies of γ-rays from a source which emits both β-particles and γ-rays is shown in Fig. 15–10.

If the radiator is a thin foil of material of intermediate or high atomic number, the photoelectrons ejected from it form line spectra. Lines appear corresponding to electrons from the K-shell and the L-shell, and for spectrometers of very high resolution even M electrons may be resolved. The energy of the electrons in a given line can be determined from the position of the line (i.e., from the Hr value) and the γ-ray energy is obtained by adding the binding energy of the electron in the shell from which it was ejected to the measured electron energy.

Some of the β-particle spectrometers discussed in Section 14–1 have been used to determine γ-ray energies by measuring the energies of Compton

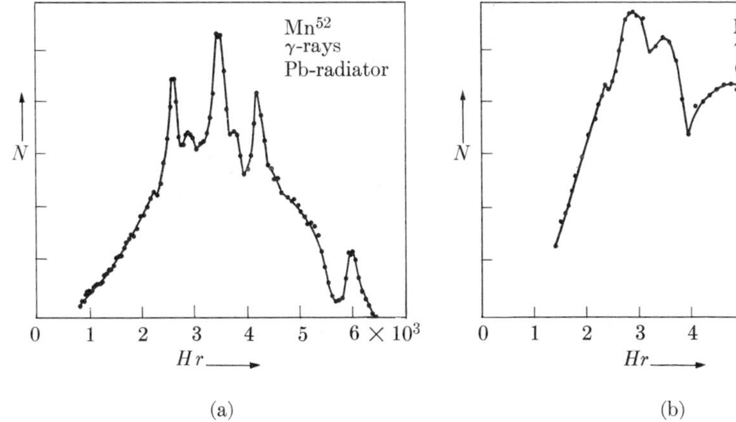

Fig. 15–11. (a) Photoelectrons and Compton recoil electrons ejected from a lead radiator by γ-rays from Mn^{52}; the peaks are caused by photoelectrons. (b) Compton recoil electrons ejected from a copper radiator by γ-rays from Mn^{52} (Peacock and Deutsch[(8)]).

electrons and photoelectrons. Deutsch, Elliott, and Evans[7] have treated the theory, design, and use of a short magnetic lens spectrometer for both β- and γ-rays. The spectra of β-particles from two radiators, produced by γ-rays from Mn^{52}, and measured with that instrument,[8] are shown in Fig. 15–11. Figure 15–11(a) shows the photoelectrons and Compton recoil electrons ejected from a lead radiator by γ-rays from Mn^{52}. There are four pronounced peaks caused by γ-rays of energies 0.510 ± 0.01 Mev, 0.734 ± 0.015 Mev, 0.94 ± 0.02 Mev, and 1.46 ± 0.03 Mev. Figure 15–11(b) shows the spectrum of Compton recoil electrons produced in a copper radiator indicating the same four γ-rays. The lowest energy γ-ray is doubtless annihilation radiation, since Mn^{52} is a positron emitter.

At higher energies, above 2 or 3 Mev, γ-rays are best studied with a pair spectrometer, because the probability of pair formation increases with increasing photon energy, while the Compton and photoelectric cross sections decrease. A collimated beam of γ-rays falls on a radiator and

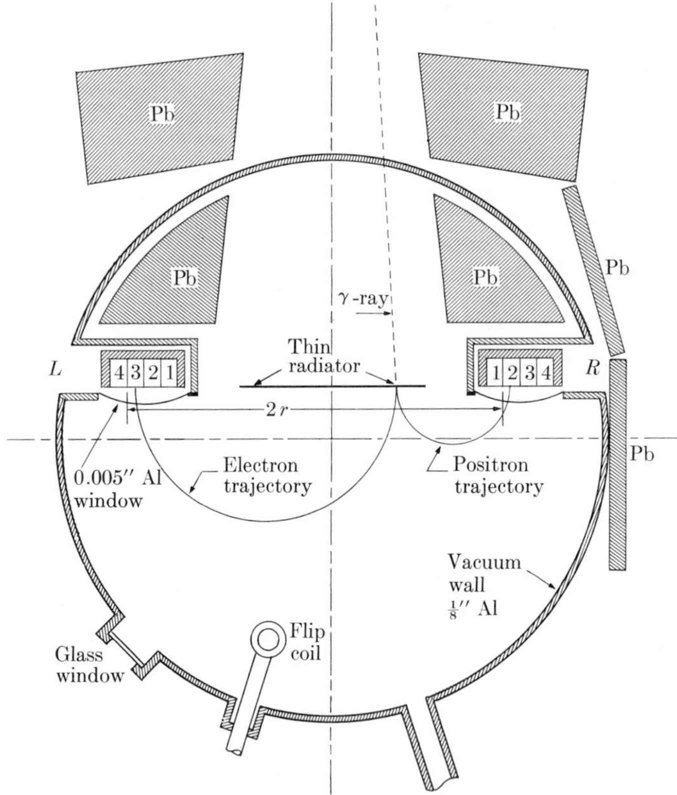

FIG. 15–12. Schematic diagram of a pair spectrometer (Walker and McDaniel[11]).

ejects electron-positron pairs from it. The electrons and positrons are focused separately in a uniform magnetic field into a number of counters. It is necessary to detect simultaneously the electron and positron produced by one γ-ray, and this can be done by using coincidence techniques.[9,10]

In the coincidence method only the *simultaneous* detection of *two* events in different counters is recorded. It is possible, with suitable electronic circuits, to achieve a "resolving time" of a fraction of a microsecond, that is, less than 10^{-6} sec, with Geiger counters. With ionization chambers and electron multipliers, smaller resolving times ($\sim 10^{-8}$ sec) can be achieved. Two events occurring within a time interval less than the resolving time and detected by separate counters are said to be *in coincidence*. The energies of the paired electron and positron are determined from their Hr values, and the sum of the energies together with the rest energies (1.02 Mev) gives the energy of the γ-ray.

A schematic diagram of a pair spectrometer used by Walker and McDaniel[11] is shown in Fig. 15–12. A uniform magnetic field is applied perpendicular to the plane of the paper. A parallel beam of γ-rays falls on a thin radiator foil placed perpendicular to the beam. Electron-positron pairs are produced which, for γ-rays of fairly high energy (>3 Mev), go very nearly in the forward direction. The electron and the positron may each fall into one of four counters placed in the plane of the radiator foil. The counters are connected into coincidence circuits in pairs, in such a way that pairs produced by γ-rays of the same energy are counted and recorded.

Energies of γ-rays can also be obtained from measurements of the energies of internal conversion electrons. The process of internal conversion has been discussed very briefly in Section 14–4; in this process the secondary electrons come, not from a radiator, but from the radioactive atom itself. The electron energies are determined by magnetic analysis and appear as the secondary or line spectrum of β-emitters.

Absorption methods can also be used to determine the energies of γ-rays. In principle, a measurement of the attenuation of a γ-ray beam by a known absorber in a "good geometry" experiment would give a value of the absorption coefficient. For a given absorber, the absorption coefficient is known as a function of energy either from theory, or as a result of measurements with γ-rays of known energy, and the unknown energy can then be found from the measured value of the absorption coefficient. A minor complication is introduced by the fact that a given absorption coefficient may correspond to two different photon energies because of the minimum in the absorption coefficient-energy curve, as shown in Figs. 15–7 and 15–8. The correct energy can be found by obtaining absorption curves in two different materials.

Unfortunately, the "good geometry" method requires highly active sources because only a small fraction of the γ-rays emitted by the source

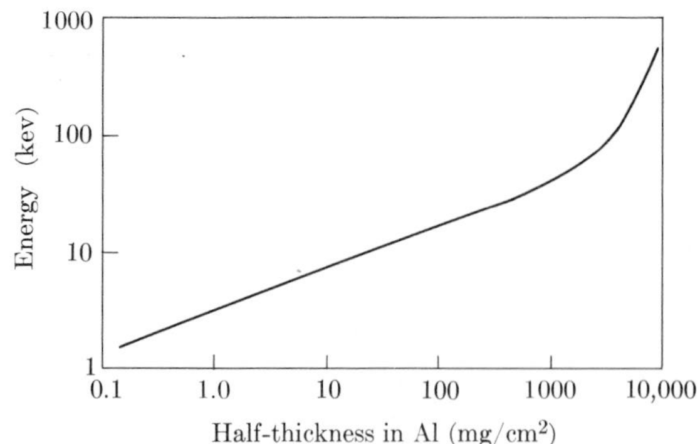

FIG. 15–13. Half-thickness values in aluminum for low-energy γ-rays [L. E. Glendenin, *Nucleonics*, **2**, No. 1 (1948)].

is used in the experiment. When weak sources only are available, absorption experiments with "poor geometry" may be used. An arrangement similar to that of Fig. 14–5 is feasible, but with the absorbers placed against the counter and with the source placed as far away from the counter as the intensity of the source allows. The β-particle counter of the figure is, of course, replaced by a γ-counter. Under these conditions and if the γ-radiation is homogeneous, an exponential curve (straight line on semilogarithmic graph paper) is obtained over a range of at least 3 or 4 half-thicknesses. The energy may be obtained from a half thickness curve of known energy such as that for aluminum shown in Fig. 15–13, and this method yields fairly accurate energy values below 1 Mev.

It is also possible to determine γ-ray energies by measuring the maximum range of Compton electrons ejected from a radiator by the γ-rays. The range measurements may be made in a cloud chamber or with coincidence methods. In the latter method, the Compton electrons pass through two thin-walled Geiger counters connected to a coincidence circuit, and absorbers are placed between the two counters. The coincidence rate is plotted against the absorber thickness; it decreases rapidly with increasing absorber thickness and becomes zero when the endpoint of the Compton electrons is reached. This method can be used for energies up to about 5 Mev. It has been discussed in detail and extended to higher energies by Fowler, Lauritsen, and Lauritsen.[(12)]

The last method to be mentioned is that of the scintillation spectrometer,[(13,14)] the basis of which was discussed in Section 2–8. In this method, a scintillation counter acts as a proportional counter for measuring the energies of the electrons produced by the γ-radiation, and hence the energies of the γ-rays. Thallium-activated sodium iodide is most often used as the phosphor because a large sodium iodide crystal has a high efficiency for the absorption of γ-rays. The high atomic number of iodine permits the measurement of "lines" due mainly to photoelectrons and electron-positron pairs. The secondary electrons produce pulses which can be displayed on an oscilloscope screen and photographed. The amount of light resulting from the capture of a γ-ray, as well as the pulse height, is a nearly linear function of the energy of the γ-ray, and the counting rate can be determined as a function of pulse height or energy. The result is a spectrum whose maxima give the energies of the γ-rays.

15-8 Gamma-decay: internal conversion. It was supposed in Sections 13-5 and 14-7 that a nucleus which has been left in an excited state following α-decay or β-decay can pass to a less excited state either by emitting a γ-ray or by internal conversion. It was shown that with this supposition it is possible to account for the energies of the γ-rays associated with α-decay and β-decay, and to write down consistent decay schemes.

It was not always clear that the emission of the γ-rays follows that of the α- or β-particles, and considerable work was done in the 1920's to prove this. Several different, independent proofs were developed, two of which will be described briefly. One method, used by Rutherford and Wooster,[15] depended on the measurement of the secondary x-ray lines which accompany the decay of RaB. Radium B, with a half-life of 26.8 minutes, emits electrons and γ-rays, and the β-particle spectrum shows a number of lines corresponding to internal conversion in various electron shells. After the internal conversion electrons are emitted, the vacancies in the electron shells are filled (see Section 7-7) with the emission of characteristic x-rays. Rutherford and Wooster determined the wavelengths of some of the L x-ray lines which accompany the β-decay of RaB and found that the wavelengths agreed excellently with the accepted values for atomic number 83. Since the decay of RaB causes a change of atomic number from 82 to 83, the result showed that internal conversion takes place after the change in atomic number.

The emission of an internal conversion electron after a radioactive decay implies that a reorganization of the product nucleus takes place after the decay process has occurred. It was proved unambiguously that the amount of the energy released in internal conversion is the same as that released in γ-ray emission, so that both processes represent the same reorganization of the product nucleus, and the emission of the γ-ray must therefore follow that of the α- or β-particle. Table 15-7 gives the energies of a group of β-ray lines of RaB, as measured by Ellis.[16] The electron shell in which the conversion took place was identified from the characteristic x-radiation, and the binding energies of the electrons in the different shells were known. The sum of the kinetic energy of the electron and its

TABLE 15–7

CONVERSION OF A GAMMA-RAY IN VARIOUS ELECTRON SHELLS

β-Ray energy, kev	Binding energy for $Z = 83$	Conversion level	γ-Ray energy, kev
36.74	16.34	L_{I}	53.08
37.37	15.67	L_{II}	53.04
39.63	13.38	L_{III}	53.01
48.85	3.99	M_{I}	52.84
49.10	3.68	M_{II}	52.78
49.66	3.17	M_{III}	52.83
51.90	0.93	N_{I}	52.83
52.64	0.20	O	52.84
		Average	52.91

binding energy was taken to be the energy of the converted γ-ray, and in the example shown, the same value, 52.91 kev, for the γ-ray energy was inferred from all of the γ-ray lines, within the experimental error.

The energies of the γ-rays emitted by a given nuclide can also be measured directly with a crystal spectrometer, as described in Section 15–7. Independent measurements of this type gave the value 53.3 kev for the energy of a γ-ray associated with the decay of RaB, in good agreement with the value of 52.9 kev inferred from the β-ray lines. Similar agreement was obtained for many γ-rays between direct measurements of their energies and inferences from the β-ray lines, proving that the emission of a γ-ray and of an internal conversion electron represent the same nuclear reorganization. Since the emission of the internal conversion electron follows the radioactive disintegration, the emission of the γ-ray must also follow the particle decay process.

Another ingenious experiment on the time of emission of the conversion electrons was that of Ellis and Wooster.[17] They placed a source of RaB inside a thick platinum tube, on the outside of which was placed a thin coating of RaB. The γ-rays from the inside source ejected photoelectrons from the platinum, with energies $h\nu - K_{\text{Pt}}$, $h\nu - L_{\text{Pt}}, \ldots,$ where K_{Pt} is the binding energy of the electron in the K-shell of platinum, and so on. At the same time the γ-rays from the source on the outside of the tube were accompanied by internal conversion electrons with energies $h\nu - K_{\text{RaB}}$, $h\nu - L_{\text{RaB}} \ldots .$ The electrons were studied by magnetic deflection, and photographs were taken simultaneously, and on the same photographic plate, of the photoelectric spectrum and of the internal conversion spectrum. The difference in energy of corresponding groups of electrons is $K_{\text{RaB}} - K_{\text{Pt}}$, $L_{\text{RaB}} - L_{\text{Pt}} \ldots$, where K_{RaB} is the binding energy of an electron in the K-shell of RaB, etc. The energy difference for the K-lines was calculated theoretically as $K_{82} - K_{78}$ or $K_{83} - K_{78}$, depending on whether the internal conversion electrons and γ-rays were assumed to come from the parent nucleus RaB ($_{82}Pb^{214}$) or from the product nucleus RaC ($_{83}Bi^{214}$), and compared with experimental results. The binding energies of the electrons in the K- and L-shells of platinum ($Z = 78$) and of the elements with $Z = 82$ and 83 were well known, and the experimental values of the energy differences could be obtained from the spacing between lines (Hr values) as seen on the photographic plate. This procedure was used for several different lines and in each case the experimental result agreed with the value calculated for $Z = 83$ and not with the value calculated for $Z = 82$. These results showed again that the internal conversion electron and, therefore, the γ-ray, is emitted after the β-particle. Other experiments were done with both α-emitters and β-emitters, and in each case the γ-rays were shown to be emitted after the particle disintegration.

The proof that the γ-rays are emitted by the product nucleus and not the parent nucleus must now be followed by a brief discussion of a generally accepted confusion in terminology. The γ-rays which have been considered up to now are all emitted immediately after the α- or β-particle, and it has not been possible to measure the time interval between the two decay processes, because it is too short. According to theory, the γ-ray lifetime for these nuclei is of the order of magnitude of 10^{-13} to 10^{-11} sec. Consequently, the γ-ray intensity decays with the same half-life as that of the parent nucleus. In the case of RaB, with a half-life of 26.8 min, the γ-ray intensity also decays with a half-life of 26.8 min, and the practice arose and has been continued of calling these γ-rays the γ-rays of RaB rather than those of RaC. In another case, ThC goes by α-decay to ThC″ and has a spectrum with fine structure, emitting six groups of α-particles. The product nucleus ThC″ emits at least eight different γ-rays which can be fitted into a consistent decay scheme as in Fig. 13–11. Nevertheless, these γ-rays are usually spoken of as the γ-rays of ThC and are so listed in tables and charts of nuclides.

The relative frequency with which γ-emission and internal conversion occur in a given nuclide is a matter of great interest from the standpoint of theory and is expressed by a quantity called the *internal conversion coefficient.* This coefficient is defined as the ratio of the number of conversion electrons to the number of γ-ray photons emitted and there are different coefficients for the different electron shells. Values of the internal conversion coefficients can be calculated from quantum mechanics, but rigorous calculations are very laborious, requiring the use of large electronic computing machines. The coefficient should increase rapidly with the atomic number, approximately as Z^3 over part of the range, and decreases rapidly as the γ-ray energy increases. The conversion coefficient is a difficult quantity to determine experimentally with any accuracy, but some of the detailed predictions of the theory have been verified by means of ingenious experiments.[18] In general, internal conversion is found most often in heavy elements when low-energy (<0.5 Mev) γ-rays are emitted. In the lighter elements and with higher energy γ-rays, internal conversion is relatively infrequent.

15–9 Gamma-decay and nuclear energy levels: theory. The study of the γ-rays and internal conversion electrons emitted in transitions from excited levels of nuclei to less excited or stable levels is one of the main sources of our knowledge of low energy nuclear levels. These levels are characterized by their stationary properties, energy, angular momentum (spin), and parity. The energies of the γ-rays or conversion electrons determine the energies of the transitions which give, in turn, the energy differences between levels. Some information about the angular momenta and parities of the levels can be obtained by combining experimental results with theoretical relationships between the decay probability and the energy of the γ-rays. More precise information can be obtained with the aid of certain specialized methods. One such method is the quantitative study of internal conversion coefficients; another method depends on the angular correlation between successive γ-rays emitted by an excited nucleus; a third depends on the angular distribution of the γ-rays emitted by nuclei aligned in a magnetic field (see Section 14–9).

A complete theory of γ-decay would require the use of the quantum theory of radiation together with a detailed description of the quantum states of the nucleus. A knowledge of nuclear forces would be needed to formulate the problem, but these forces are not yet understood. Consequently, the theory must be based on a rough model of the nucleus and on a method of treating the interaction between the nucleus and radiation. In the present state of our knowledge of nuclear structure, therefore, the properties of nuclear levels and the transitions between levels cannot be predicted accurately and in detail. Nevertheless, the combination of experimental results with the available theory makes possible the determination of angular momenta and parities and the construction of energy level and decay schemes. We shall, therefore, discuss the theory briefly with the object of arriving at equations relating the disintegration constant for γ-decay and the energy of the γ-rays. Equations (15–40) and (15–41) below, will be seen to be analogous to Eqs. (13–20) and (13–22) for α-decay and to Eqs. (14–13) and (14–14) for β-decay. Even an elementary discussion of γ-decay is complicated, however, by the need to show explicitly the effects of angular momentum and parity, effects which did not appear in the relatively simple cases of α-decay and β-decay considered earlier.

The derivation of Eqs. (15–40) and (15–41) is beyond the scope of this book, but the discussion which follows should lead to some familiarity with the nomenclature of γ-decay and may serve to introduce the reader to some of the ideas underlying the theory.

It has proved to be very useful to express the interaction (force or potential) between the nucleus and electromagnetic radiation in terms of an infinite series of powers of the ratio $R/\lambdabar$, where R is a linear dimension which characterizes the nucleus and is taken as the radius of the nucleus, while $\lambdabar$ is a quantity which characterizes the radiation and is the wavelength divided by 2π. The nuclear radius can be written in the form (cf. Section 13–6)

$$R = R_0 A^{1/3} \times 10^{-13} \text{ cm}, \tag{15–32}$$

where R_0 is a constant with a value of about 1.5, and A is the mass number of the nucleus. The wavelength of the γ-radiation is given by

$$\lambdabar = \frac{\hbar}{p} = \frac{\hbar}{h\nu/c} = \frac{\hbar c}{E}.$$

When E is in Mev and $\lambdabar$ in cm,

$$\lambdabar = \frac{197}{E \text{ (Mev)}} \times 10^{-13} \text{ cm}, \tag{15–33}$$

and

$$\frac{R}{\lambdabar} = \frac{R_0 A^{1/3} E \text{ (Mev)}}{197}. \tag{15–34}$$

Now, $A^{1/3}$ has values up to about 6, and E is generally about 1 or 2 Mev, so that the ratio $R/\lambdabar$ is nearly always less than 0.1. The expression for the decay probability (disintegration constant) is given by theory as an infinite series in powers of $(R/\lambdabar)^2$. The successive terms in the series decrease very rapidly and the decay probability is given almost completely by the first nonvanishing term in the series, as in the case of β-decay (Section 14–6). This term gives rise to an allowed transition; the succeeding terms, which are much smaller, give rise to forbidden transitions. Although the ratio $(R/\lambdabar)^2$ is small for most of the interesting nuclear transitions, the forbidden transitions are still important in γ-decay. One reason is that the quantity observed in experiments is the transition rate and, at nuclear energies, even very low transition rates can be measured. This situation is in strong contrast with that of atomic transitions where light is emitted or absorbed. For atoms, the value of $R/\lambdabar$ is of the order $10^{-8}/10^{-5} = 10^{-3}$, which is much smaller than the values for nuclei, so that the forbidden atomic transitions are relatively more forbidden than the nuclear ones. In addition, the experimental material in atomic transitions consists of spectral lines, and the intensities of the forbidden lines are so small as to make these lines very difficult to observe. Since the forbidden transitions are important in γ-decay, each term in the expansion of the decay probability must be examined to see what it means physically and under what conditions it becomes important.

The expansion procedure, which is adapted from classical electromagnetics, separates the radiation from a nucleus into distinct types called *multipole radiations.* This separation corresponds to sorting the emitted γ-rays (photons) according to the amount of angular momentum $L\hbar$ carried off by each photon. In other words, for a given γ-ray energy, photons can be emitted which differ in their angular momentum. The decay probability is the sum of the partial probabilities for the emission of these different types of γ-rays; but in this sum one term, the first nonvanishing term, predominates. The partial probabilities will be seen to depend very strongly on the angular momentum carried off by the photons, and this is one reason for the usefulness of the procedure. Photons can have only integral values of L; the value $L = 0$ is ruled out as a consequence of the fact that electromagnetic waves are transverse in nature. The multipole radiation is characterized by its *order*, given by 2^L. For $L = 1$, the radiation has been shown by theory to correspond to that emitted by a vibrating dipole; for $L = 2$, it corresponds to the radiation from a vibrating quadrupole, for $L = 3$ to that from an octupole, and so on. For each value of L there are two classes of radiation, called *electric* and *magnetic* multipole radiations, which differ in their parity. An electric multipole has even parity when L is even, and odd parity when L is odd. Magnetic multipole radiation has odd parity when L is even, and even parity when L is odd; parity of electric multipole $= (-1)^L$, parity of magnetic multipole $= -(-1)^L$, where $+1$ means even parity and -1 means odd parity. A given γ-decay process may leave the parity of the wave function of the nucleus unchanged, or there may be a change in the parity; in either case, the parity of the wave function of the entire system, nucleus and radiation, is conserved.

As in atomic transitions, there are selection rules which must be obeyed; these are shown in Table 15–8 together with the symbols used for the different types of radiation. The selection rules for emission of γ-rays (or for their absorption) define those combinations of L and parity which give nonvanishing values of the transition probability.

In the case of electric multipole transitions of order 2^L it has been shown[19] that the decay probability can be written in the following form when the independent particle or *shell* model of the nucleus (see Chapter 17) is used:

$$\lambda_{EL} = 2\pi\nu \frac{e^2}{\hbar c} S \left(\frac{R}{\lambdabar}\right)^{2L}; \tag{15–35}$$

TABLE 15-8

SELECTION RULES AND SYMBOLS FOR GAMMA RADIATION

Classification	Change in angular momentum of nucleus, in units of $\hbar$	Symbol	Parity change of nuclear wave function
Electric 2^L-pole	L	EL	No for L even Yes for L odd
Magnetic 2^L-pole	L	ML	Yes for L even No for L odd
Electric dipole	1	$E1$	Yes
Magnetic dipole	1	$M1$	No
Electric quadrupole	2	$E2$	No
Magnetic quadrupole	2	$M2$	Yes
Electric octupole	3	$E3$	Yes
Magnetic octupole	3	$M3$	No
Electric 16-pole	4	$E4$	No
Magnetic 16-pole	4	$M4$	Yes

here λ_{EL} is the disintegration constant (or reciprocal mean life) for γ-decay with the emission of EL radiation, and will not be confused with the wavelength λ; ν is the frequency of the emitted radiation. The quantity S depends on L and is given by

$$S = \frac{2(L+1)}{L[1 \times 3 \times 5 \ldots (2L+1)]^2} \left(\frac{3}{L+3}\right)^2; \qquad (15\text{–}36)$$

it has the following values.

L	S
1	2.5×10^{-1}
2	4.8×10^{-3}
3	6.25×10^{-5}
4	5.3×10^{-7}
5	3.1×10^{-9}

The probability of emission of magnetic multipole radiation is given by

$$\lambda_{ML} = 2\pi\nu \frac{e^2}{\hbar c} (10S) \left(\frac{\hbar}{McR}\right)^2 \left(\frac{R}{\lambdabar}\right)^{2L}, \qquad (15\text{–}37)$$

where M is the mass of a nucleon. For a given value of L, electric multipole radiation is more probable than magnetic multipole radiation, with

$$\frac{\lambda_{EL}}{\lambda_{ML}} = \frac{1}{10}\left(\frac{R}{\hbar/Mc}\right)^2 = 4.4A^{2/3}; \qquad (15\text{–}38)$$

we have used $\hbar/Mc = 0.211 \times 10^{-13}$ cm and $R = 1.4\,A^{1/3} \times 10^{-13}$ cm, which is considered to be suitable for γ-decay calculations. The disintegration constants can be expressed in terms of the γ-ray energy and the atomic mass of the nucleus if we use Eqs. (15–32) and (15–33) and note that $e^2/\hbar c = 1/137$; for example,

$$\lambda_{EL} = \frac{2\pi\nu}{137}\,S(R_0A^{1/3})^{2L}\left(\frac{E\text{ (Mev)}}{197}\right)^{2L}. \qquad (15\text{–}39)$$

The frequency ν is proportional to the energy so that, on multiplying the numerator and denominator of the right side of the last equation by h, and inserting numerical values, we get

$$\lambda_{EL} = 2.4S(R_0A^{1/3})^{2L}\left(\frac{E\text{ (Mev)}}{197}\right)^{2L+1} 10^{21}\text{ sec}^{-1}; \qquad (15\text{–}40)$$

where $R_0A^{1/3}$ is just the radius of the nucleus in units of 10^{-13} cm. Equation (15–37) becomes

$$\lambda_{ML} = 0.55SA^{-2/3}(R_0A^{1/3})^{2L}\left(\frac{E\text{ (Mev)}}{197}\right)^{2L+1} 10^{21}\text{ sec}^{-1}. \qquad (15\text{–}41)$$

Equations (15–40) and (15–41) are convenient forms for calculation and for comparison with experimental results.

It is evident from Eqs. (15–35) and (15–37) or from Eqs. (15–40) and (15–41) together with the values of S, that the probability of γ-decay should decrease very rapidly with increasing values of L. Since the energy of the γ-rays is at most a few Mev, Eqs. (15–40) and (15–41) show that, for a given value of L, the probability of γ-decay should increase with increasing values of the energy of the γ-rays. The theoretical results are illustrated by the numerical examples given in Table 15–9; values of the mean life (the reciprocal of the disintegration constant) for γ-decay with emission of different types of multipole radiation are given for a nucleus with a radius of 6×10^{-13} cm ($A \approx 80$). The results show that when the angular momentum carried off by the γ-ray is small ($L = 1$ or 2), that is, when the change in the angular momentum of the nucleus is small, the decay probability is large and the mean life is very short. But when the change in the angular momentum of the nucleus is large (3 or more units), the decay probability may become very small and the mean life may become long.

TABLE 15-9

APPROXIMATE THEORETICAL VALUES OF THE MEAN LIFE FOR GAMMA-DECAY

Type of radiation	Change in angular momentum of nucleus, L: in units of $\hbar$	Mean life, sec		
		Gamma-ray energy		
		1.00 Mev	0.20 Mev	0.05 Mev
$E1$	1	3×10^{-16}	3×10^{-14}	2×10^{-12}
$M1$	1	3×10^{-14}	3×10^{-12}	2×10^{-10}
$E2$	2	7×10^{-12}	3×10^{-8}	3×10^{-5}
$M2$	2	8×10^{-10}	3×10^{-6}	4×10^{-3}
$E3$	3	6×10^{-7}	5×10^{-2}	8×10^{2}
$M3$	3	7×10^{-5}	6	9×10^{4}
$E4$	4	6×10^{-2}	1×10^{5}	3×10^{10}
$M4$	4	7	1×10^{7}	3×10^{12}
$E5$	5	8×10^{4}	4×10^{12}	2×10^{19}
$M5$	5	9×10^{6}	4×10^{14}	2×10^{21}

The values given by Eqs. (15-40) and (15-41) are rough estimates because of the approximate nature of the theory; for example, a factor which depends on the actual value of the angular momentum of the initial nucleus is not included. The use of a particular nuclear model, the independent particle model, may be a more serious limitation. But the theory should give the general trends of the decay probability with angular momentum, energy, and mass number, and its usefulness will depend on how well the theoretical predictions agree with experiment.

The theory of the more specialized methods mentioned at the beginning of this section (internal conversion, angular correlations, and angular distributions) is still more complicated than that relating decay constant and energy and will not be discussed.

15-10 Gamma-decay and nuclear energy levels: experimental results and nuclear isomerism. 1. *Measured lifetimes and nuclear isomerism.* Most known γ-decay rates have been determined by the direct measurement of the lifetimes τ of the excited states involved.[20] The total decay rate λ of an excited state is given by

$$\frac{1}{\tau} = \lambda = \lambda_\gamma + \lambda_e, \tag{15-42}$$

where λ_γ is the rate of emission of photons and λ_e is the rate of emission of conversion electrons. The internal conversion coefficient $\alpha = \lambda_e/\lambda_\gamma$ can be measured,(21) or it can be calculated theoretically(22) by means of a method which is independent of the calculation of λ_γ. The transition rate by γ-decay alone is then

$$\lambda_\gamma = \frac{1}{\tau(1+\alpha)} = \frac{1}{\tau_\gamma}, \tag{15–43}$$

and it is this experimental value which is to be compared with theory. But, before we make this comparison, we must look more closely at the experimental lifetimes.

The theoretical values of lifetimes shown in Table 15–9 indicate that some mean lives for γ-decay may be long enough to be measurable. Although most γ-ray lifetimes have been found to be very short, more than 250 cases are known in which the lifetime for γ-decay is measurable, with observed half-lives from 10^{-10} sec to many years. These "delayed" transitions are called *isomeric transitions* and the states from which they originate are called *isomeric states* or *isomeric levels.* Because of these delayed transitions, pairs of nuclear species exist which have the same atomic and mass numbers, i.e., they are isotopic and isobaric, but have different radioactive properties; these nuclides are called *nuclear isomers* and their existence is referred to as *nuclear isomerism.*(23) The existence of excited states with long lifetimes makes possible the direct study of such stationary properties as the angular momentum and magnetic moment. It must be noted, however, that the isomeric property of a state is not absolute but is only relative to properties of states of lower energy. An isomeric state is usually denoted by the letter *m* attached to the mass number of the nuclide; thus, In^{113m} represents an isomeric level of In^{113}.

The first case of nuclear isomerism to be discovered was that of UX_2 and UZ, which have the same atomic number and the same mass number but have different half-lives and emit different radiations. Both UX_2 and UZ grow out of UX_1 ($_{90}Th^{234}$) by β^--decay, which gives them the same mass number 234, and both have the atomic number $90 + 1 = 91$. Uranium X_2 has a half-life of 1.18 minutes; it emits three groups of electrons with endpoint energies of 2.31 Mev (90%), 1.50 Mev (9%), and 0.58 Mev (1%). On the other hand, UZ emits four groups of β-particles with endpoint energies of 0.16 Mev (28%), 0.32 Mev (32%), 0.53 Mev (27%), and 1.13 Mev (13%), and has a half-life of 6.7 hours. Uranium X_2 is more highly excited than UZ, and about 0.15% of the UX_2 nuclei decay to the ground state UZ by emitting a γ-ray with an energy of 0.394 Mev. Uranium *Z*, the ground state of $_{91}Pa^{234}$, then decays to UII ($_{92}U^{234}$) by electron emission. Nearly 100% of the UX_2 atoms decay directly to UII by electron emission.

This example of isomerism was unique for many years, until the discovery of artificial radioactivity, and especially the production of many new radio-nuclides by neutron bombardment, led to the discovery of many isomeric pairs. The first well-established case of artificially produced isomers was found in bromine. When a target containing bromine was bombarded with slow neutrons, the product was found to show three different half-lives for β-decay: 18 min, 4.9 hr, and 34 hr. Chemical analysis identified the carrier of each of these activities as bromine. This result was surprising because there are only two stable isotopes of bromine, Br^{79} and Br^{81}, and in this region of the periodic table the usual reaction with slow neutrons is the (n, γ) process with the formation of the next higher isotope of the element in question. With bromine, therefore, only two radioactive products would be expected

$$_{35}Br^{79} + {_0n^1} \rightarrow {_{35}Br^{80}} + \gamma,$$

$$_{35}Br^{81} + {_0n^1} \rightarrow {_{35}Br^{82}} + \gamma,$$

and one of the new bromine isotopes seemed to have two half-lives for β-decay. The identity of this isotope was determined by bombarding bromine with 17-Mev γ-rays, with the following reactions:

$$_{35}Br^{79} + \gamma \rightarrow {_{35}Br^{78}} + {_0n^1},$$

$$_{35}Br^{81} + \gamma \rightarrow {_{35}Br^{80}} + {_0n^1}.$$

Two products were obtained, again with three half-lives, 6.4 min, 18 min, and 4.4 hr. Since Br^{80} was formed in both cases, and the 18 min and 4.4 hr half-lives were observed in both cases, the two half-lives were attributed to two isomeric states of Br^{80}.

It has been found possible to separate the bromine isomers chemically, and it has been shown that the 4.4 hr isomer decays into the 18 min isomer by emitting two γ-rays in cascade, with energies of 0.049 Mev and 0.037 Mev, respectively. The 18 min isomer, which is $_{35}Br^{80}$ in its ground state, then decays by electron emission (92%) to $_{36}Kr^{80}$, which is stable, and by positron emission (3%) and orbital electron capture (5%) to the ground state of $_{34}Se^{80}$.

An explanation of isomerism was proposed by von Weizsäcker in 1936[(24)]. He suggested that an isomer is an atom whose nucleus is in an excited state and has an angular momentum which differs by several units from that of any lower energy level including the ground state. Von Weizsäcker showed that while γ-ray transitions requiring only one or two units of spin change could be expected to occur with half-lives of the order of 10^{-13} sec, those with larger spin change might result in longer and measurable half-lives; each additional unit of spin change could increase the half-life of a transition of given energy by a factor of the order of 10^6. Later, more detailed calculations, such as those discussed in Section 15–9, as well as experimental results have verified von Weizsäcker's hypothesis.

2. *Experimental relations between lifetime and energy.* The measured quantities in studies of γ-decay are λ_γ (or τ_γ) and E; A is a parameter which varies from nuclide to nuclide. Equation (15–40) for electric multipole radiation may be written

$$\tau_\gamma = \text{constant} \times A^{-2L/3}\, E^{-(2L+1)}, \qquad (15\text{–}44)$$

or

$$\log\,(\tau_\gamma A^{2L/3}) = \text{constant} - (2L+1)\log E. \qquad (15\text{–}45)$$

Hence, for a given type of electric multipole radiation, a graph of log $(\tau_\gamma A^{2L/3})$ against log E should be a straight line with slope $-(2L+1)$. Similarly, for magnetic multipole transitions, we have

$$\log\,(\tau_\gamma A^{(2L-2)/3}) = \text{constant} - (2L+1)\log E. \qquad (15\text{–}46)$$

The comparison of experimental and theoretical results can be discussed more conveniently if we write out Eqs. (15–40) and (15–41) explicitly for the different types of γ-radiation. These equations are listed in Table 15–10; the values of the constants were actually obtained from an improved version[(25)] of Eqs. (15–40) and (15–41). It has been found experimentally[(26)] that the measured lifetimes for $L = 2$, 3, and 4 fit the following equations.

$$L = 2\colon \log \tau_\gamma\ (\text{sec}) = 4 - 5\log E\ (\text{kev}) \qquad (15\text{–}47\text{a})$$

$$L = 3\colon \log \tau_\gamma\ (\text{sec}) = 17.5 - 7\log E\ (\text{kev}) \qquad (15\text{–}47\text{b})$$

$$L = 4\colon \log \tau_\gamma\ (\text{sec}) = 27.7 - 9\log E\ (\text{kev}) \qquad (15\text{–}47\text{c})$$

The general form of the dependence of the experimental lifetime on the γ-ray energies is, therefore, that predicted by theory, although the results, as expressed by Eqs. (15–47), are not detailed enough to test the dependence of lifetime on mass number. The best quantitative agreement between theory and experiment is obtained for $M4$ transitions for which the experimental data conform to the relation

$$\tau_\gamma(\text{sec}) = 1.0 \times 10^4 (2I_i + 1) A^{-2} E^{-9}; \qquad (15\text{–}48)$$

in this case the dependence of lifetime on both energy and mass number is just as predicted by theory. The factor $(2I_i + 1)$, where I_i is the angular momentum (in units of $\hbar$) of the isomeric state is a statistical weight factor; with this factor the mean deviation of the experimental lifetime from the value given by Eq. (15–48) is estimated to be less than 30%,

Table 15-10

Theoretical Expressions for the Decay Constant

Type of gamma-radiation	Disintegration constant for gamma-decay, λ_γ: cm^{-1}
$E1$	$1.5 \times 10^{14} A^{2/3} E^3$
$M1$	$2.8 \times 10^{13} E^3$
$E2$	$1.6 \times 10^{8} A^{4/3} E^5$
$M2$	$1.2 \times 10^{8} A^{2/3} E^5$
$E3$	$1.1 \times 10^{2} A^{2} E^7$
$M3$	$1.8 \times 10^{2} A^{4/3} E^7$
$E4$	$5.0 \times 10^{-5} A^{8/3} E^9$
$M4$	$1.5 \times 10^{-4} A^{2} E^9$
$E5$	$1.6 \times 10^{-11} A^{10/3} E^{11}$
$M5$	$7.5 \times 10^{-11} A^{8/3} E^{11}$

which represents good agreement for this kind of nuclear data. This agreement is especially important because most long-lived isomers emit γ-radiation of the $M4$ type.

The experimental lifetimes for $M1$, $M2$, and $M3$ transitions also depend on energy and mass number in the general way predicted on the basis of the independent particle model; but the actual numerical values of the lifetimes may be two or three orders of magnitude greater than the theoretical values. Similar results have been found for the electric multipole transitions; the general form of the dependence of lifetime on energy and mass number agrees with theory. For $E3$, $E4$, and $E5$ transitions, the decay rates are generally smaller than those predicted by theory, but there are many $E2$ transitions, especially in certain even-even nuclei, which have considerably higher transition probabilities than predicted by theory.

It will be remembered that the theory mentioned refers to a particular model of the nucleus, the independent particle, or shell, model. The experimental data have been interpreted as showing that this model provides a useful basis for the prediction of electromagnetic transitions in nuclei, although it is not satisfactory in all its details. The discrepancies between theory and experiment have, therefore, been used to improve the model; modifications have been made in it and their validity tested by comparison with experiment. The study of isomers has been especially useful in this way and, during the last few years, there has been a gradual shift of emphasis. Isomers, which were at first of interest as a semi-empirical branch of radioactivity, are now playing an increasing role in the investi-

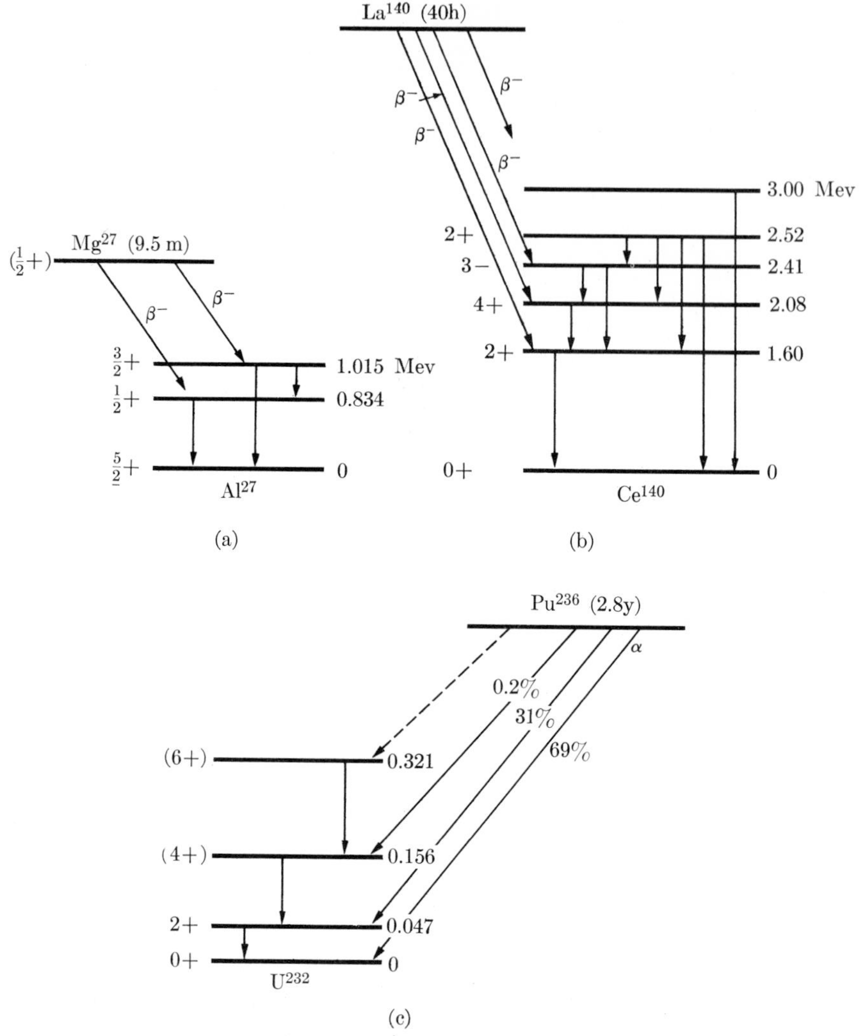

FIG. 15–14. Energy level and decay schemes with emission of γ-rays from short-lived excited states. (a) Mg^{27}. (b) La^{140}. (c) Pu^{236}. [Strominger, Hollander, and Seaborg, *Revs. Mod. Phys.* **30,** 585 (1958).]

gation of nuclear structure. The interest has shifted from the general relations between lifetimes and energies to details of the nuclear structure which determine the multipole character and probability of γ-ray transitions. One result[(27,28)] has been the accumulation of evidence in favor of a more highly developed model of the nucleus, the *collective* model which will be discussed in Chapter 17.

3. *Energy levels and decay schemes.* Some decay schemes involving the emission of γ-rays by short-lived excited states are shown in Fig. 15–14. Values of the angular momentum and parity assigned to a level are indicated at the left end of the horizontal line representing the level; the energy above that of the ground level is given at the right end of the lines. When the angular momentum of a state has been measured directly, it is denoted by underlining that quantum number in the decay scheme, for example, $\underline{\frac{5}{2}}+$, for the ground state of Al^{27} in (a). If the angular momentum is determined uniquely by other methods, it is designated by the quantum number without any other modification, e.g., $\frac{1}{2}+$, for the first excited state of Al^{27}. Ground states of even-even nuclei are known to have zero angular momentum and often fall into this category. Probable values of the angular momentum are indicated with parentheses, as in the case of the ground state of Mg^{27} in (a). Parity assignments, indicated by a plus sign (even parity) or minus sign (odd parity), depend heavily on theory and the underlining and parentheses do not apply to them. Figures 15–14 (a) and (b) contain information not given for Al^{27} and Ce^{140} in Fig. 14–13. For some states, it has not yet been possible to assign any values of angular momentum and parity, e.g., the 3.00 Mev level of Ce^{140}. Some decay schemes are very complicated, and extensive and careful analysis is needed to work them out. Thus, Bi^{206} decays by orbital electron capture to Pb^{206}, and 12 excited levels and 28 γ-rays have been found so far. Values of the energy, angular momentum, and parity have been established for 10 of the levels, and probable values assigned to the other two; the 28 observed γ-rays have been fitted into a decay scheme which meets all the experimental and theoretical requirements.[(29)]

Examples of energy level and decay schemes for nuclear isomers are shown in Fig. 15–15; these schemes show that members of isomeric pairs can decay in different ways. In the simplest mode of decay the isomeric level decays by γ-emission to the ground level, which is stable: In^{113m}, with a half-life of 104 min, emits a γ-ray with an energy of 0.392 Mev and becomes stable In^{113}, as shown in Fig. 15–15(a). The angular momenta of the stable and isomeric levels have both been measured in this case; they differ by four units and the transition has been shown to be of the $M4$ type. In some isomeric pairs the ground state, instead of being stable, may be radioactive and decay by β^--emission. Thus, 14 hr Zn^{69m} emits a 0.438 Mev γ-ray and goes to the ground state of Zn^{69}, which then decays by electron emission to the ground state of Ga^{69}; the decay scheme is shown in Fig. 15–15(b). The decay of both members of an isomeric pair may be more complex, as in the case of Br^{80}, Fig. 15–15(c); Br^{80m} decays to Br^{80} by emitting two γ-rays in cascade; Br^{80} then decays by electron emission to Kr^{80}, and by positron emission and orbital electron capture to Se^{80}. When the excited level decays only by γ-emission to the ground

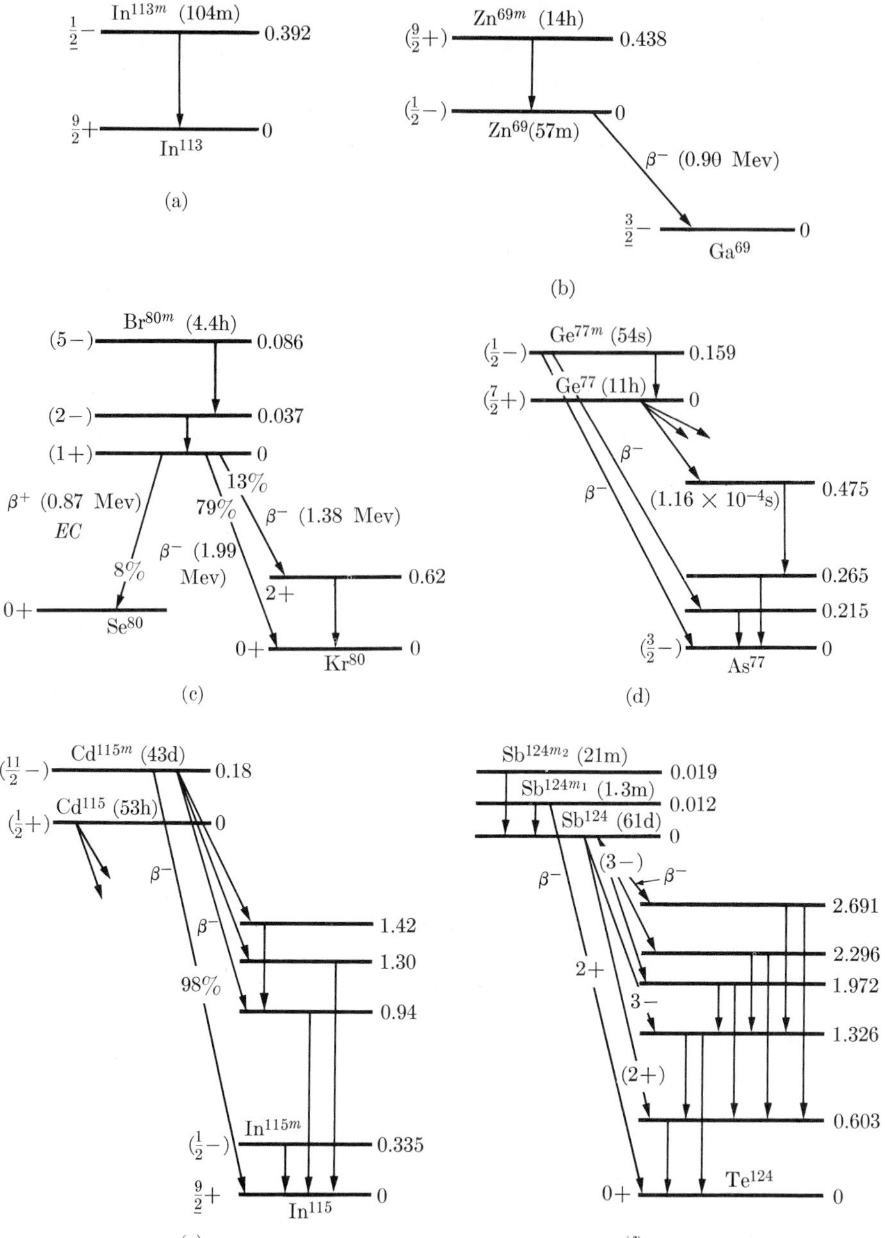

FIG. 15–15. Decay schemes of nuclear isomers. (a) In^{113m}. (b) Zn^{69m}. (c) Br^{80m}. (d) Ge^{77m}. (e) Cd^{115m}. (f) Sb^{124m_1} and Sb^{124m_2}. [Strominger, Hollander, and Seaborg, *Revs. Mod. Phys.* **30,** 585 (1958).]

level, as in the above cases, the members of the isomeric pair are said to be *genetically related.* The more excited member of an isomeric pair may decay in two ways: Ge^{77m} decays to Ge^{77} by emission of a γ-ray; it can also decay by electron emission to the ground level of As^{77} or to an excited level; the ground level of Ge^{77} is unstable and decays to excited levels of As^{77} by electron emission; the decay scheme for Ge^{77m} and Ge^{77} is shown in Fig. 15–15(d). The members of an isomeric pair may decay independently by β^--emission: Cd^{115m} and Cd^{115} are not genetically related, and both have complex β-spectra. The decay scheme is shown in Fig. 15–15(e); it is especially interesting because two of the energy levels of the product nucleus, In^{115}, form an isomeric pair. Triple isomerism has also been observed and Fig. 15–15(f) shows an example; Sb^{124m_2}, with a half-life of 21 min, decays by γ-emission to the ground state, while Sb^{124m_1}, with a half-life of 1.3 min, decays both by γ-emission and β^--emission. The ground state of Sb^{124} is unstable and also decays by β^--emission.

For a detailed summary of isomeric decay schemes the reader is referred to a recent review by Alburger (see gen. ref.).

References

GENERAL

RUTHERFORD, CHADWICK, and ELLIS, *Radiations from Radioactive Substances.* New York: The Macmillan Co., 1930, Chapters 12, 15.

R. D. EVANS, *The Atomic Nucleus.* New York: McGraw-Hill, 1955, Chapters 5, 23, 24, 25.

K. SIEGBAHN, ed., *Beta- and Gamma-Ray Spectroscopy.* New York: Interscience; Amsterdam: North-Holland Publishing Co., 1955.

M. DEUTSCH and O. KOFOED-HANSEN, "Gamma-Rays," *Experimental Nuclear Physics,* E. Segrè, ed., New York: Wiley, 1959. Vol. III, Part X.

D. E. ALBURGER, "Nuclear Isomerism," *Encyclopedia of Physics.* Vol. 42 (Nuclear Reactions III), pp. 1–108. Berlin: Springer, 1957.

F. AJZENBERG-SELOVE, ed., *Nuclear Spectroscopy.* New York: Academic Press, 1960. Parts A, B.

G. W. GRODSTEIN, *X-Ray Attenuation Coefficients from 10 kev to 100 Mev,* National Bureau of Standards Circular 583. Washington, D.C.: U.S. Government Printing Office, 1957.

H. A. BETHE and J. ASHKIN, "Passage of Radiations Through Matter," *Experimental Nuclear Physics,* E. Segrè, ed., New York: Wiley, 1953 (Vol. I. Part II).

STROMINGER, HOLLANDER, and SEABORG, "Table of Isotopes," *Revs. Mod. Phys.* **30,** 585 (1958).

WAY, KING, MCGINNIS and VAN LIESHOUT, *Nuclear Level Schemes, A = 40 to A = 92,* U.S. Atomic Energy Commission Document TID-5300, Sept. 1955.

H. GOLDSTEIN, *Fundamental Aspects of Reactor Shielding.* Reading, Mass., U.S.A.: Addison-Wesley, 1959.

PARTICULAR

1. C. M. Davisson and R. D. Evans, "Measurements of Gamma-Ray Absorption Coefficients," *Phys. Rev.* **81,** 404 (1951).

2. C. M. Davisson and R. D. Evans, "Gamma-Ray Absorption Coefficients," *Revs. Mod. Phys.* **24,** 79–107, 1952.

3. W. Heitler, *The Quantum Theory of Radiation,* 3rd ed. Oxford University Press, 1954.

4. J. W. M. DuMond, "High Resolving Power, Curved-Crystal Focussing Spectrometer for Short Wave-Length X-Rays and Gamma-Rays," *Rev. Sci. Instr.* **18,** 626 (1947); also, "Gamma-Ray Spectroscopy by Direct Crystal Diffraction," *Ann. Revs. Nucl. Sci.* **8,** 163 (1958).

5. Muller, Hoyt, Klein, and DuMond, "Precision Measurements of Nuclear Gamma-Ray Wavelengths of Ir^{192}, Ta^{182}, RaTh, Rn, W^{187}, Cs^{137}, Au^{198}, and Annihilation Radiation," *Phys. Rev.* **88,** 775 (1952).

6. Page, Stehle, and Gunst, "Direct Comparison of m/e for the Positron and the Electron," *Phys. Rev.* **89,** 1273 (1953).

7. Deutsch, Elliott, and Evans, "Theory, Design and Applications of a Short Magnetic Lens Electron Spectrometer," *Rev. Sci. Instr.* **15,** 178 (1944).

8. W. C. Peacock and M. Deutsch, "Disintegration Schemes of Radioactive Substances, IX. Mn^{52} and V^{48}," *Phys. Rev.* **69,** 306 (1946).

9. J. W. Dunworth, "Coincidence Methods in Nuclear Physics," *Rev. Sci. Instr.* **11,** 167 (1940).

10. A. C. G. Mitchell, "The Use of Coincidence Counting Methods in Determining Nuclear Disintegration Schemes," *Revs. Mod. Phys.* **20,** 296 (1948).

11. R. L. Walker and B. D. McDaniel, "Gamma-Ray Spectrometer Measurements of Fluorine and Lithium Under Proton Bombardment," *Phys. Rev.* **74,** 315 (1948).

12. Fowler, Lauritsen, and Lauritsen, "Gamma-Radiation from Excited States of Light Nuclei," *Revs. Mod. Phys.* **20,** 236–277 (1948).

13. P. R. Bell, "The Scintillation Method," *Beta- and Gamma-Ray Spectroscopy,* K. Siegbahn, ed. (see gen. ref.), p. 133.

14. Eriksen, Jenssen, and Sunde, "Scintillation Counter as Gamma-Ray Spectrometer," *Physica* **18,** 9 (1952).

15. E. Rutherford and W. A. Wooster, "The Natural X-Ray Spectrum of RaB," *Proc. Cambridge Phil. Soc.* **22,** 834 (1925).

16. C. D. Ellis, "The γ-Rays of Radium (B + C) and of Thorium (C + C′)," *Proc. Roy. Soc.* (London) **A143,** 350 (1934).

17. C. D. Ellis and W. A. Wooster, "The Atomic Number of a Radioactive Element at the Moment of the Emission of the γ-Rays," *Proc. Cambridge Phil. Soc.* **22,** 844 (1925).

18. G. D. Latyshev, "Interaction of γ-Rays with Matter and the Spectroscopy of γ-Radiation," *Revs. Mod. Phys.* **19,** 132 (1947).

19. J. M. Blatt and V. F. Weisskopf, *Theoretical Nuclear Physics.* New York: Wiley, 1952, Chapter 12.

20. R. E. Bell, "Measurement of Short Lifetimes of Excited States: Measurement by Delayed Coincidence; Comparison Methods," *Beta- and Gamma-Ray Spectroscopy,* K. Siegbahn, ed. (see gen. ref.), p. 494.

21. M. Deutsch and O. Kofoed-Hansen, "Gamma-Rays," *Experimental Nuclear Physics*, E. Segrè, ed. (see gen. ref.), p. 360.

22. M. E. Rose, "Theory of Internal Conversion," *Beta- and Gamma-Ray Spectroscopy*, K. Siegbahn, ed. (see gen. ref.), p. 396.

23. E. Segrè and A. C. Helmholz, "Nuclear Isomerism," *Revs. Mod. Phys.* **21**, 271 (1949).

24. C. F. von Weizsäcker, "Metastabile Zustande der Atomkerne," *Naturwiss.* **24**, 813 (1936).

25. S. A. Moszkowski, "Theory of Multiple Radiation," *Beta- and Gamma-Ray Spectroscopy*, K. Siegbahn, ed. (see gen. ref.), p. 373.

26. M. Goldhaber and A. W. Sunyar, "Classification of Nuclear Isomers," *Phys. Rev.* **83**, 906 (1951).

27. M. Goldhaber and A. W. Sunyar, "Classification of Nuclear Isomers," *Beta- and Gamma-Ray Spectroscopy* (see gen. ref.) p. 453.

28. L. I. Rusinov and D. A. Varshalovich, "Electromagnetic Transitions in Isomeric Nuclei," *J. Nucl. Energy*, Part A: Reactor Science, **10**, 170 (1959). Translated from *Atomnaya Energiya* **5**, 432 (1958).

29. D. E. Alburger and M. H. L. Pryce, "Energy Levels in Pb^{206} from the Decay of Bi^{206}," *Phys. Rev.* **95**, 1482 (1954).

30. Chupp, DuMond, Gordon, Jopson, and Mark, "Precise Determination of Nuclear Energy Levels in Heavy Elements," *Phys. Rev.* **112**, 518 (1958).

31. Kazi, Rasmussen, and Mark, "Six-Meter Radius Bent-Crystal Spectrograph for Nuclear Gamma Rays," *Rev. Sci. Instr.*, **31**, 983 (1960).

32. Kazi, Rasmussen, and Mark, "Measurement of the Deuteron Binding Energy using a Bent-Crystal Spectrograph," *Phys. Rev.* **123**, 1310 (1961).

Problems

1. When a beam of γ-rays passes through aluminum under "good geometry" conditions, the intensity varies with the energy of the rays and the thickness of the aluminum, as shown in the following table.

Thickness of aluminum, cm	Relative intensity		
	Energy = 0.835 Mev	Energy = 1.14 Mev	Energy = 2.75 Mev
0	1.00	1.00	1.00
2.5	0.634	0.675	0.788
5	0.401	0.456	0.622
7.5	0.254	0.308	0.490
10	0.161	0.208	0.387
12.5	0.102	0.141	0.305
15	0.065	0.095	0.240
20	0.026	0.043	0.150
25	0.010	0.020	0.093
30	0.0042	0.0090	0.058
35	0.0017	0.0041	0.036
40	.....	0.0019	0.022

Find, graphically, the value of the absorption coefficient $\mu(\text{cm}^{-1})$ in each case. Compare the results with the values given in Table 15–1.

2. When γ-rays of energy 2.76 Mev pass through varying thicknesses of aluminum, tin, and lead, the intensity varies with the thickness of each substance as shown in the table on the next page. Find, graphically, the value of $\mu(\text{cm}^{-1})$ in each case and compare the results with the values given in Table 15–1.

3. Calculate the values of the cross section ${}_a\tau$ for photoelectric absorption in aluminum and lead at the photon energies listed in Table 15–2. Plot the values obtained against the photon energy in Mev.

4. Calculate the values of the cross sections σ, σ_s, and σ_a for Compton scattering in aluminum at 10 values of α which cover the energy range of Table 15–3. Plot the resulting values against the photon energy in Mev.

5. From the curve of Fig. 15–6, calculate the values of the cross section ${}_a\kappa$ for pair formation in aluminum and lead at photon energies of 1, 5, 10, 15, 20, and 25 Mev.

6. Calculate, from theory, the values of $\mu(\text{cm}^{-1})$ for copper for γ-rays of energy 0.835 Mev, 1.14 Mev, and 2.76 Mev. Compare the results with the theoretical and experimental values given in Table 15–1.

7. A heterogeneous beam of γ-rays from a radioactive source is allowed to fall on an aluminum radiator and the electrons ejected are analyzed in a magnetic spectrograph. The electrons fall into three groups with maxima at 11960, 5845, and 2750 gauss-cm, respectively. What are the energies of the three γ-rays?

Aluminum		Tin		Lead	
Thickness, cm	Intensity	Thickness, cm	Intensity	Thickness, cm	Intensity
0	1.00	0	1.00	0	1.00
2.5	0.788	2	0.586	1	0.620
5.0	0.622	3	0.449	2	0.384
7.5	0.490	5	0.263	3	0.238
10.0	0.387	7	0.154	4	0.147
12.5	0.305	8	0.118	6	0.057
15	0.240	10	0.069	8	0.022
20	0.150	12	0.041	10	0.008
25	0.093	16	0.014	12	0.003
30	0.058	18	0.008		
35	0.036				
40	0.022				

8. In an experiment on the γ-rays from Ba^{131}, the rays are allowed to fall on a lead radiator. Four groups of photoelectrons are observed, corresponding to Hr values of 1250, 1445, 2050, and 2520 gauss-cm, respectively. The binding energy of the K-shell electrons in lead is 0.0891 Mev. What are the energies of the γ-rays?

9. In an early experiment on the γ-rays from RaB, Ellis measured the energies of the photoelectrons ejected from radiators of wolfram, platinum, lead, and uranium. In each case, three groups of K-electrons were observed with the following energies, in Mev.

	W	Pt	Pb	U
a.	0.166	0.158	0.149	0.122
b.	0.220	0.212	0.203	0.174
c.	0.276	0.269	0.260	0.231

The binding energies of the K-electrons in the four radiators are 0.0693, 0.0782, 0.0891, and 0.1178 Mev, respectively. Find the mean energy and the wavelength of each of the three γ-rays.

10. Terbium-160 decays by β-emission to excited states of dysprosium-160. Two of the γ-rays emitted by the latter nuclide give rise to an internal conversion line spectrum with the following energies: 32.9, 78.0, 84.7, 86.1, 143.0, 187.0, and 194.7 kev. The binding energies for K-, L-, M-, and N-shell electrons in dysprosium are 53.4, 8.6, 1.9, and 0.4 kev, respectively. Find the energies of the two γ-rays, and the shell and γ-ray with which each conversion line is associated.

11. Derive an expression for the kinetic energy (in ev) with which the nucleus recoils when a γ-ray is emitted; express the energy of the γ-ray in kev, and the mass of the recoil nucleus in amu.

12. The reactions Cr(p,n)Mn, and Fe(d,α)Mn both lead to activities of 21.3 min and 5.8 days, while the (α,n) reaction on vanadium gives a product with a half-life of 310 days. To what nuclide, or nuclides, do the two shorter activities belong?

13. The nuclide Cr^{49} decays by positron emission, emitting three groups of particles with endpoint energies of 1.54 Mev, 1.39 Mev, and 0.73 Mev, respectively. Gamma-rays with energies of 0.15 Mev, 0.061 Mev, and 0.089 Mev, are observed. Construct a decay scheme to fit these data.

14. The nuclide Cr^{51} decays by orbital electron capture to V^{51} and the Q-value for the ground state transition is 0.75 Mev. Three γ-rays are observed, with energies of 0.325 Mev, 0.320 Mev, and 0.65 Mev respectively, the first two of which are in cascade. Construct an energy level and decay scheme to fit these data.

15. The irradiation of Co^{59} with neutrons produces a pair of isomers of Co^{60}, one with a half-life of 10.4 min, the other with a half-life of 5.2 years. The 10.7-min level has an excitation energy of 0.059 Mev and transforms mainly to the ground state of Co^{60} by internal conversion; it also decays by emitting β^--particles with an endpoint energy of 1.54 Mev. The ground state level of Co^{60} decays by β^--emission with an endpoint energy of 0.31 Mev. Two γ-rays are observed, in cascade, with energies of 1.17 Mev and 1.33 Mev, respectively. The Q-value for the transition from the isomeric state of Co^{60} to the ground state of Ni^{60} is 2.876 Mev. Construct an energy level and decay scheme based on these data.

CHAPTER 16

NUCLEAR REACTIONS

In the discussion of nuclear reactions in Chapter 11, the emphasis was on the production of new nuclear species and the interest was mainly in the change of one nucleus into another. The quantitative aspects stressed were those of the mass and energy balance. The study of nuclear reactions is also important for other reasons. Information about the relative probabilities of different reactions provides clues to the problem of nuclear structure and offers a testing ground for ideas about nuclear forces. Any nuclear transmutation process may lead to an excited state of the product nucleus, and the decay of the excited state then gives information about energy levels and decay schemes. The process may be caused by any kind of incident particle and may result in the emission of any kind of particle (proton, neutron, deuteron, α-particle, or γ-ray). The incident and emitted particle may even be of the same kind, as in scattering processes. The great number of nuclear reactions which are possible provide, therefore, a wealth of experimental data for the field of nuclear spectroscopy and for the theory of nuclear structure. It is from this standpoint that nuclear reactions will be considered in the present chapter.

16–1 Nuclear reactions and excited states of nuclei. In Chapter 11, the nuclear reactions considered were of the type

$$x + X \rightarrow Y + y, \tag{16–1}$$

or, in more compact notation, $X(x, y)Y$. The equation and the notation both mean that particle x strikes nucleus X to produce nucleus Y and particle y. The particles x and y may be elementary particles or γ-rays, or they may themselves be nuclei, e.g., α-particles or deuterons. The transmutation represented by Eq. (16–1) does not cover all the reactions of interest. In the general case, more than one particle may emerge, or the outgoing particle may be the same as the incident particle. For the purpose of this chapter, the nuclear reactions to be considered may be represented by the set of equations

$$x + X \rightarrow \begin{cases} X + x, \\ X^* + x, \\ Y + y, \\ Z + z, \text{ etc.} \end{cases} \tag{16–2}$$

In the first two reactions of the set, the outgoing particle is of the same kind as the incident particle, and the process is called *scattering*. The first reaction represents *elastic scattering*, in which the total kinetic energy of the system, projectile plus target, is the same before the collision as after. Some kinetic energy is usually transferred from the projectile to the target nucleus, but the latter is left in the same internal, or nuclear, state as before the collision. A collision between two billiard balls, in which neither ball is damaged or otherwise changed, is an example of an elastic collision. The second reaction represents *inelastic scattering*, in which the target nucleus X is raised into an excited state X^*, and the total kinetic energy of the system is decreased by the amount of the excitation energy given to the target nucleus. The other reactions of the set represent different possible nuclear transmutations in which the product nuclei may be formed in their ground states or, more often, in excited states. The excited product nucleus usually decays very quickly to the ground state with the emission of γ-rays.

The fact that the product nucleus in a transmutation process can be left in an excited state was discovered by measuring the energies of the protons emitted in (α, p) reactions on light elements, the first nuclear reactions studied in detail. When a given light element was bombarded with monoenergetic natural α-particles, one or more groups of protons, each containing particles of the same energy, was observed for a particular direction of emission.[1,2] The existence of the distinct proton groups was demonstrated by their different ranges. When boron was bombarded with α-particles from polonium with a range of 3.8 cm, or an energy of 5.30 Mev, two groups of emergent protons were found at right angles to the incident α-particle beam; these protons had ranges of 20 cm and 50 cm, respectively. Similar results were obtained with other light elements such as fluorine and aluminum; when the latter was bombarded with α-particles from RaC′, four groups of protons were observed[3] for each of several different energies of the incident α-particles.

The energy balance of the reactions, or Q-value, can be calculated for each energy group by the methods of Section 11-2. The proton group with the greatest energy gives the greatest Q-value, and this value is supposed to correspond to the ground state of the product nucleus. A proton group of lower energy gives a lower Q-value, and the difference between the greatest Q and a smaller one gives the excitation energy of the product nucleus after emission of the lower energy proton. These ideas and their application involve some of the basic procedures of nuclear physics and will be illustrated by an analysis of two experiments, on the reaction

$$_{13}\text{Al}^{27} + {}_{2}\text{He}^{4} \rightarrow [{}_{15}\text{P}^{31}] \rightarrow {}_{14}\text{Si}^{30} + {}_{1}\text{H}^{1} + Q,$$

which are representative of work relating nuclear reactions to energy levels. In one experiment,[(4)] an aluminum foil was bombarded with α-particles accelerated in a cyclotron to an energy of 7.3 Mev. The ranges of the protons emerging at an angle of 0° with the direction of the incident beam were measured by counting coincidences between protons and γ-rays. Four groups of protons were observed with ranges in air of 101.6 cm, 60.8 cm, 40.8 cm, and 25–30 cm, respectively. The corresponding energies, obtained from range-energy curves for protons, are 9.34, 6.98, 5.55, and 4.2–4.65 Mev. The energies of the incoming α-particle and the outgoing proton are known for each group, as are the mass numbers of the initial and final nuclei and, since the angle at which the protons are observed is 0°, Q-values can be calculated from Eq. (11–10). The greatest proton energy gives $Q = 2.22$ Mev, which corresponds to the formation of Si^{30} in its ground state; the three lower-energy proton groups give Q-values of -0.06, -1.44, and about -2.4 Mev, respectively. There are, therefore, three excited states of Si^{30} with energies 2.28, 3.66, and about 4.6 Mev above that of the ground state. The ground state energy is taken as the zero point of the energy scale for the various states. The results, which are collected in the first part of Table 16–1, are in good agreement with those of earlier experiments.[(3,5)] It was also found in this experiment

TABLE 16–1

EXCITED STATES OF Si^{30} FROM THE BOMBARDMENT OF ALUMINUM BY ALPHA-PARTICLES

Experiment 1[(4)] Energy of α-particles = 7.3 Mev			*Experiment 2*[(6)] Energy of α-particles = 21.54 Mev		
Energy of protons emitted at 0°, Mev	Q-value, Mev	Energy of excited state, Mev	Energy of protons emitted at 90°, Mev	Q-value, Mev	Energy of excited state, Mev
9.34	2.22	0	.....	2.22	0
6.98	−0.06	2.28	.....	....	....
5.55	−1.44	3.66	16.85	−1.27	3.49
4.2–4.65	−2.4	4.6			
			14.96	−3.22	5.44
			13.28	−4.96	7.18
			12.29	−5.98	8.20
			11.27	−7.04	9.26
			10.68	−7.65	9.87
			9.72	−8.64	10.86

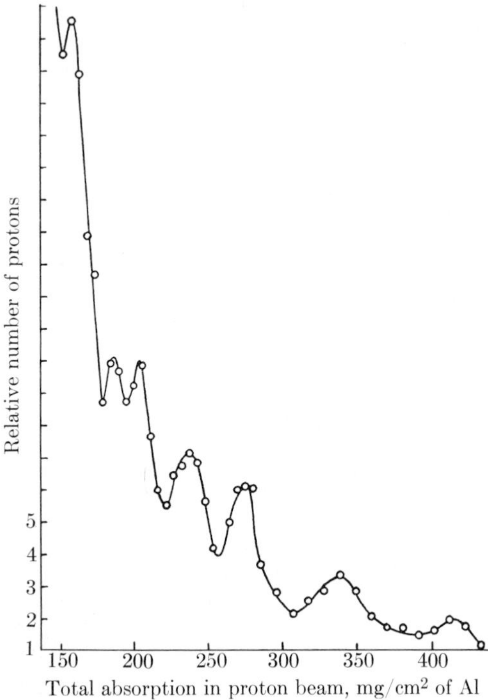

FIG. 16–1. Proton groups from the reaction $Al^{27}(\alpha, p)Si^{30}$ (Brolley *et al.*[6]).

that no γ-rays accompanied the highest-energy group of protons, while the lower-energy protons did show coincidences with γ-rays; the lower-energy protons are indeed associated with excited states of the Si^{30} nucleus, while the highest-energy group is associated with the ground state.

In another experiment,[6] a thin aluminum foil was bombarded in vacuum with α-particles accelerated in a cyclotron to an energy of 21.54 Mev. The protons emerging at an angle of 90° with the direction of the incident beam were counted with two proportional counters placed in series and arranged for coincidence counting. The ranges of the proton groups in aluminum were measured by allowing the protons to pass through aluminum foils before entering the counting system. Under the conditions of the counting procedure, the number of protons counted in a particular group showed a peak at the end of the range so that when the number of counts is plotted against the thickness of aluminum, each peak in the curve corresponds to the range or energy of a different proton group. The change of the counting rate with the thickness of the aluminum absorber is shown in Fig. 16–1. The energies of the proton groups were obtained from the ranges in aluminum by means of the known theoretical relationship between range and energy in that material, and are listed in the second part of Table 16–1. The Q-values are obtained from Eq. (11–9). The range of the most energetic proton group could not be determined well in this experiment because the counting rates became too low, so the value $Q = 2.22$ Mev was used for the reaction leading to the ground state of Si^{30}, and gave the energy levels listed in the table. The energy level values at 3.66 Mev and 3.49 Mev correspond to the same excited state, the discrepancy being experimental. The level at about 4.6 Mev observed in the first experiment was not observed in the second experiment, but other work has established the existence of a level at 4.75 Mev. There are, therefore, at least 9 excited energy levels of Si^{30}, as shown in the table, and it is probable that there are more which have not been excited by the (α, p) reaction.

The emission of groups of particles is not limited to (α, p) reactions, but is also found in many other reactions. In each case, the different energy groups correspond to different states of the product nucleus, the lower-energy groups corresponding to excited states which decay to the ground state by emitting γ-rays.

Another important feature of many nuclear reactions was discovered during the investigation of (α, p) reactions when the energy of the incident α-particles was varied. It was found in the early experiments that when the energy of the incident α-particles was increased the yield of protons did not increase monotonically, as might be expected on intuitive grounds, but showed sharp maxima at certain discrete values of the energy.[(3)] In the bombardment of Al by α-particles, the yield of protons had peaks at α-particle energies of 4.0, 4.49, 4.86, 5.25, 5.75, and 6.61 Mev, and was markedly lower in the energy range between any two of these peaks. The occurrence of maxima in the reaction rate at different energies is called *resonance*, and the particular energies at which the maxima occur are called *resonance energies*. The reason for the use of these terms will be discussed later. At each resonance energy of the incident α-particles the different proton groups are observed, and Q-values can be calculated from the known resonance energy and the corresponding energies of the proton groups. These Q-values are the same as those obtained for nonresonance α-particle energies, and lead to the same excited states of the product nucleus. The experimental results obtained at the resonance energies are often more useful than those obtained at other energies because more events occur at the resonance energies and more precise information can be obtained about the Q-values and excited levels in the *compound nucleus*. The resonance phenomenon does not, however, add anything new about the energy levels of the product nucleus.

Resonance is important in nuclear reactions because knowledge of the resonance energies in a reaction yields information about certain energy levels of the nucleus referred to in Chapter 11 as the *compound nucleus*.

The transmutations discussed in Chapter 11 were written in a way which indicates that the incident particle combines with the target nucleus to form an intermediate compound nucleus. When Al is bombarded with α-particles, the compound nucleus is $[P^{31}]$, and it turns out that the existence of resonance energies is related to the energy levels of P^{31}. Resonance is not limited to (α, p) reactions but is also found in other nuclear reactions.

16–2 The compound nucleus. In 1936, Bohr proposed his theory[7,8] of the compound nucleus, which has been extremely useful in the correlation and interpretation of nuclear reactions. The basic ideas of this theory will therefore be considered before a detailed discussion of nuclear reactions is undertaken. Bohr assumed that a nuclear reaction takes place in two steps:

(1) The incident particle is absorbed by the initial, or target, nucleus to form a *compound nucleus.*

(2) The compound nucleus disintegrates by ejecting a particle (proton, neutron, α-particle, etc.) or a γ-ray, leaving the final, or product, nucleus.

Bohr assumed also that the mode of disintegration of the compound nucleus is independent of the way in which the latter is formed, and depends only on the properties of the compound nucleus itself, such as its energy and angular momentum. The two steps of the reaction can then be considered separate processes:

1. Incident particle + initial nucleus → compound nucleus.
2. Compound nucleus → product nucleus + outgoing particle.

Bohr's assumptions are in accord with many of the facts of nuclear transmutation. It was shown in Section 11–9 that when a given nuclide is bombarded with particles of a single type, several different new nuclides can be formed. When Al^{27} is bombarded with protons, the new nuclide may be Mg^{24}, Si^{27}, Si^{28}, or Na^{24}. According to Bohr, the interaction between Al^{27} and a proton is assumed to give the compound nucleus $[{}_{14}Si^{28}]$, which can then disintegrate in any one of several ways: into Mg^{24} and an α-particle; or into Si^{27} and a neutron; or into Si^{28} and a γ-ray; or even into Na^{24}, 3 protons, and a neutron. Bohr's assumption is also in accord with the picture of the nucleus as a system of particles held together by very strong short-range forces. When the incident particle enters the nucleus, its energy is quickly shared among the nuclear particles before re-emission can occur, and the state of the compound nucleus is then independent of the way it was formed. That this conclusion is reasonable can be shown by the following arguments. On being captured, the incident particle makes available a certain amount of *excitation energy*, which is nearly equal to the kinetic energy of the captured particle plus its binding energy in the compound nucleus. The magnitude of the excitation energy can be calculated from the masses of the incident particle, target nucleus, and compound nucleus, and the kinetic energy of the incident particle. Consider, for example, the capture of a neutron by Al^{27} to form the compound nucleus $[Al^{28}]$,

$$_{13}Al^{27} + {}_0n^1 \rightarrow [_{13}Al^{28}].$$

The masses of the interacting neutron and nucleus are 1.00898 and 26.99008 amu, or a total of 27.99906 amu; that of the compound nucleus is 27.99077 amu. The mass excess is 0.00829 amu, corresponding to 7.72 Mev, to which must be added the kinetic energy of the incident neutron. In the case of a slow neutron, the kinetic energy may be neglected. If the incident neutron has a kinetic energy of 1 Mev, the excitation energy is nearly $7.72 + 1 = 8.72$ Mev and the energy of the compound nucleus $[_{13}Al^{28}]$ is greater than the energy of the ground state of Al^{28} by this amount. Immediately after the formation of the compound nucleus, the excitation energy may be considered to be concentrated on the captured particle, but this additional energy is rapidly distributed among the other particles in the compound nucleus as a result of interactions among the nuclear particles. The distribution presumably takes place in a random way. At a given instant, the excitation energy may be shared among several nucleons; at a later time it may be shared by other nucleons, or it may eventually again become concentrated on one nucleon or combination of nucleons. In the latter case, if the excitation energy is large enough, one nucleon, or a combination of nucleons, may escape, and the compound nucleus disintegrates into the product nucleus and outgoing particle. The energy that must be concentrated on a single nuclear particle or group of particles in order to separate it from the compound nucleus is called the *separation* or *dissociation energy*, and is usually about 8 Mev.

As a result of the random way in which the excitation energy is distributed in the compound nucleus, the latter has a lifetime which is relatively long compared with the time that would be required for a particle to travel across the nucleus. The latter time interval, sometimes called the *natural nuclear time*, is of the order of magnitude of the diameter of the nucleus divided by the speed of the incident particle. If the incident particle is a 1-Mev neutron, its speed is about 10^9 cm/sec. Since the diameter of the nucleus is of the order of 10^{-12} cm, the time required for a 1-Mev neutron to cross the nucleus is of the order of 10^{-21} sec. Even a slow neutron with a velocity of 10^5 cm/sec would need only about 10^{-17} sec to cross the nucleus. It will be seen below that the lifetime of a compound nucleus may be as long as 10^{-15} or 10^{-14} sec, which is long compared with the natural nuclear time. During its relatively long lifetime, the compound nucleus "forgets" how it was formed, and the disintegration is independent of the mode of formation. The compound nucleus may be said to exist in a "quasi-stationary" state, which means that although it exists for a time interval which is very long compared with the natural nuclear time, it can still disintegrate by ejecting one or more nucleons. These quasi-stationary states are usually called *virtual states* or *virtual levels* in contrast to *bound states* or *bound levels,* which can decay only by emitting γ-radiation.

There are many ways in which the excitation energy of the compound nucleus can be divided among the nuclear particles and, since each distribution corresponds to a virtual level, there are many possible virtual levels of the compound nucleus. These levels are closely related to the phenomenon of resonance. It is reasonable to assume that if the energy of the incident particle is such that the total energy of the system, incident particle plus target nucleus, is equal to the excitation energy of one of the levels of the compound nucleus, the probability that the compound nucleus will be formed is much greater than if the energy falls in the region between two levels. The system is analogous to that of a radio wave and a tuned receiver circuit. When the frequency (energy) of the incoming wave is equal to that of the circuit, the wave and circuit are in resonance and the reception is good; when the two frequencies (energies) are not equal, the reception is poor. The occurrence of a resonance peak in the rate of a nuclear reaction when the energy of the incoming particles is varied shows that the compound nucleus has an energy level whose excitation energy is very nearly the sum of the binding energy of that particle and its kinetic energy. In the $Al^{27}(\alpha, p)Si^{30}$ reaction discussed above, 18 resonances have been observed at α-particle energies from 3.95 Mev to 8.62 Mev; the binding energy of the α-particle in the compound nucleus $[P^{31}]$ is 9.68 Mev. The excitation energy of a level cannot be obtained just by adding the resonance energy to the binding energy, because that would neglect the motion of the compound nucleus. Some of the energy of the incident particle is used in supplying kinetic energy to the compound nucleus and is not available for excitation.

The fraction of the particle energy that must be added to the binding energy in order to get the excitation energy can be derived as follows. If the target nucleus is assumed to be at rest before the collision, the momentum of the incident particle must be equal to that of the compound nuecleus,

$$m_x v = M_{CN} V,$$

where v, V, and M_{CN} are the speed of the incident particle, and the speed and mass of the compound nucleus, respectively. Then

$$V = v \frac{m_x}{M_{CN}}.$$

The portion of the energy of motion of the incident particle that goes into exciting the initial nucleus will be called E'_x, and is given by

$$E'_x = \frac{1}{2} m_x v^2 - \frac{1}{2} M_{CN} V^2 = \frac{1}{2} m_x v^2 - \frac{1}{2} M_{CN} v^2 \frac{m_x^2}{M_{CN}^2}$$
$$= \frac{1}{2} m_x v^2 \left(1 - \frac{m_x}{M_{CN}}\right) = E_x \left(1 - \frac{m_x}{M_{CN}}\right). \tag{16-3}$$

In calculations of this kind, the masses may be replaced by mass numbers without introducing any serious errors. Equation (16–3) may also be written

$$E'_x = E_x \left(\frac{M_X}{m_x + M_X}\right), \tag{16-4}$$

since $M_{CN} = M_X + m_x$.

The energies of the excited levels of $[P^{31}]$ which correspond to α-particle resonances are given in Table 16–2. The excitation energies of these levels are considerably greater than the dissociation energy (~8 Mev) for a particle and the compound nucleus can disintegrate by particle emission so that the levels are virtual. The reactions $Al^{27}(\alpha, p)Si^{30}$ and $Al^{27}(\alpha, n)P^{30}$ have been observed, a result which is consistent with the idea that the compound nucleus $[P^{31}]$ formed by the bombardment of aluminum with α-particles can disintegrate in more than one way.

Each excited state of the compound nucleus, whether bound or virtual, has a certain mean lifetime τ; there is a certain period of time, on the average, during which the nucleus remains in a given excited state before decaying by emission of either a particle or a γ-ray. The reciprocal of the mean life is the disintegration constant, which gives the probability per unit time of the emission of a particle or γ-ray. In the discussion of energy states excited by nuclear reactions it is customary to use, instead of the disintegration constant, a quantity proportional to it, called the *level width*, and defined by the relation

$$\Gamma = \frac{h}{2\pi\tau}. \tag{16-5}$$

The level width has the units of energy, and its use is based on an application of the Heisenberg uncertainty principle (Section 7–8). According to this principle, the accuracy with which the energy and time can be determined for a quantized system such as an atomic nucleus is limited by the relationship

$$(\Delta E)\,(\Delta t) \sim \frac{h}{2\pi}, \tag{16-6}$$

TABLE 16–2

LEVELS OF THE COMPOUND NUCLEUS [P^{31}] EXCITED IN THE BOMBARDMENT OF ALUMINUM BY α-PARTICLES

α-particle resonance energy, Mev	Energy to be added to the binding energy, Mev	Excited level of [P^{31}], Mev
3.95	3.44	13.12
4.53	3.95	13.63
4.70	4.09	13.77
4.84	4.22	13.90
5.12	4.46	14.14
5.29	4.61	14.29
5.64	4.91	14.59
6.01	5.23	14.91
6.38	5.56	15.24
6.57	5.72	15.40
7.00	6.10	15.78
7.20	6.27	15.95
7.34	6.39	16.07
7.60	6.62	16.30
8.04	7.00	16.68
8.24	7.18	16.86
8.42	7.33	17.01
8.62	7.51	17.19

where ΔE is the uncertainty in the energy and Δt is the uncertainty in the time. The mean lifetime of an excited state may be identified with the uncertainty Δt corresponding to an uncertainty ΔE in the energy. The latter is defined as the width Γ, in energy, of the excited level, giving Eq. (16–5). A state with a very short mean lifetime is poorly defined in energy and the width Γ is relatively large, while a long-lived state is sharply defined in energy and the width is relatively small. For each possible mode of decay, there is a different probability of decay and, therefore, a different *partial width* Γ_y for each decay product. The *total width* Γ of an energy level is then the sum of the individual partial widths;

$$\Gamma = \Gamma_\gamma + \Gamma_\alpha + \Gamma_p + \cdots, \tag{16–7}$$

where Γ_γ, the *radiation width*, is a measure of the probability per unit time for emission of a γ-ray, Γ_α is the *α-particle width*, Γ_p is the *proton width*, and so on.

The concept of level widths is useful because values of these widths can often be obtained from measurements of resonances, as will be shown later. When the total width is known, the value of the mean lifetime is given by

$$\tau\text{ (sec)} = \frac{h\text{ (erg-sec)}}{2\pi\Gamma\text{ (erg)}}$$
$$= \frac{1.06 \times 10^{-27}}{\Gamma\text{ (ev)} \times 1.60 \times 10^{-12}}$$
$$= \frac{6.6 \times 10^{-16}}{\Gamma\text{ (ev)}}. \tag{16-8}$$

Level widths are usually given in ev; for a wide level, Γ may be of the order of 10^4 ev, as in the case of light nuclei, and the lifetime is 6.6×10^{-20} sec, while a sharp level with a Γ of 0.1 ev has the relatively long mean life of 6.6×10^{-15} sec. When the narrow level is that of a compound nucleus, the lifetime of that level can be as long as 10^{-15} sec or 10^{-14} sec, as mentioned earlier in this section. Along with the level width, the level spacing D, or mean distance between levels, can be obtained from resonance measurements and is an important quantity in nuclear spectroscopy. The level width and level spacing are useful not only for characterizing compound nuclei, but for any excited nuclear states, and can be applied to both bound and virtual levels.

The study of resonance phenomena in nuclear reactions can now be seen to have great importance because it provides information about the width and spacing of the energy levels of the compound nucleus. The dependence of the width and spacing on the mass number and the excitation energy of the nucleus provides a test of theories of nuclear reactions and nuclear models. The partial widths of a level of the compound nucleus give the relative probabilities for different modes of disintegration, and these probabilities also yield information about nuclear structure.

16-3 Cross sections for nuclear reactions. To study nuclear reactions in detail, it is necessary to have a quantitative measure of the probability of a given nuclear reaction. This quantity must be one which can be measured experimentally and calculated in such a way that the theoretical and experimental values can be compared readily. The quantity that is most often used for this purpose is the *cross section* of a nucleus for a particular reaction, usually denoted by σ with an appropriate subscript. The term "cross section" has already been met in connection with the attenuation of a γ-ray beam, when it was used as a measure of the probability that a photon is removed from the beam; the cross section was regarded there as a form of absorption coefficient. In the discussion of nuclear re-

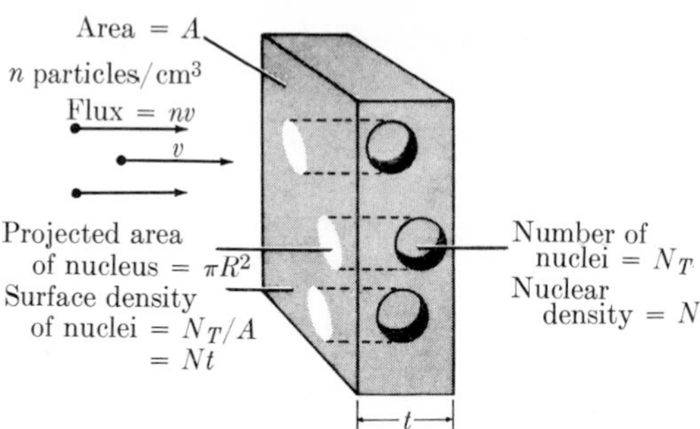

FIG. 16–2. Cross sections for nuclear reactions as geometrical areas (Hughes[(12)]).

actions, a more detailed consideration of the idea of cross sections will prove useful. The concept of a nuclear cross section can be most easily visualized as the cross-sectional area, or target area, presented by a nucleus to an incident particle. If the nuclei are considered as spheres of radius R cm and the incident particles as point projectiles, then the target area, or cross section σ, of each nucleus is given by

$$\sigma = \pi R^2 \text{ cm}^2. \tag{16–9}$$

A particle which passes normally through a thin sheet of material of area A containing N_T nuclei has a probability $N_T\sigma/A$ of colliding with a nucleus provided that there is no "overlapping" of the nuclei, i.e., that $N_T\sigma/A$ is small (Fig. 16–2). The quantity N_T/A, which is the number of nuclei per square centimeter or the surface density of nuclei, is equal to Nt, where N is the number of nuclei per cubic centimeter and t is the thickness of the sheet. For an incident beam containing n particles per cubic centimeter, moving with velocity v, the number of particles passing through the sheet is nv per square centimeter per second, and the collision rate can be expressed as

$$\text{collisions per square centimeter per second} = nv \cdot \frac{N_T\sigma}{A} = nv\sigma Nt;$$

the collision cross section is then

$$\sigma = \frac{\text{collisions per square centimeter per second}}{nvNt}. \tag{16–10}$$

The quantity nv, which is the number of particles in the incident beam crossing one square centimeter of area each second, is called the *particle flux*. Equation (16–10) then shows that the cross section for collision is the number of collisions per unit volume per second for unit incident flux and unit nuclear density.

The above discussion can be applied to any nuclear reaction, and the term "collision" can be replaced by the appropriate term for any nuclear process. Thus, σ_s may denote the cross section for scattering of a given kind of particle, and may consist of two parts, σ_e, the cross section for elastic scattering, and σ_i, the cross section for inelastic scattering. It is possible to speak of a cross section σ_a for the absorption of a particular kind of particle, defined as the number of particles that are absorbed, or disappear, per cubic centimeter per second for unit incident flux and unit nuclear density. There are also cross sections for individual reactions, such as $\sigma(\alpha, \text{p})$, $\sigma(\alpha, \text{n})$, and $\sigma(\text{p}, \alpha)$. The number of particles removed from the beam is obtained by adding the numbers removed by all the processes which can take place; particles can be removed by being absorbed in nuclear reactions or by being scattered out of the beam. In an experiment, the number of events of a given kind (collisions, scatterings, absorptions, or other processes) is counted, and the quantities nv and N are measured; the cross section for the particular process is then given by a relationship of the type of Eq. (16–10). The collision cross section, which corresponds to the effect of all possible processes, is usually called the *total cross section*, σ_t and is the sum of the cross sections for the individual reactions. There is a direct analogy between σ_t and the total absorption coefficient for γ-rays, expressed in units of square centimeter per atom, which is the sum of a cross section for photoelectric absorption, a cross section for Compton scattering, and a cross section for pair formation. As in the case of γ-radiation, the total cross section can be found by measuring the transmission of a particle beam in a "good geometry" experiment.

The expression for the number of processes of a given kind, say the ith kind, contains the product $N\sigma_i$. This product represents the cross section, for the ith process, of all the atoms in a cubic centimeter of material. It is sometimes called the *macroscopic* cross section and denoted by

$$\Sigma_i = N\sigma_i;$$

it has the unit cm^{-1}.

A rough idea of the magnitude of cross sections for nuclear reactions can be obtained from Eq. (16–9) for the geometrical cross section of a nucleus. In the discussion of the theory of α-decay, it was found that the radii of α-emitting nuclei may be represented by the formula

$$R = 1.5A^{1/3} \times 10^{-13} \text{ cm},$$

where A is the atomic weight. If it is assumed that this formula may be applied to all nuclei, which will be shown to be a reasonably good approximation, then

$$\sigma = \pi R^2 = \pi(1.5)^2 A^{2/3} \times 10^{-26} \text{ cm}^2.$$

For a nucleus of intermediate mass, say $A = 125$,

$$\sigma = \pi(2.25)(25) \times 10^{-26}\,\text{cm}^2 = 1.8 \times 10^{-24}\,\text{cm}^2,$$

and the geometrical cross section of a nucleus is of the order of 10^{-24} cm^2. Cross sections for nuclear reactions are often expressed in terms of the unit 10^{-24} cm^2, and this unit is called the *barn*, abbreviated as b.

Although it is convenient and simple to introduce the cross section as a target area, and to get a rough idea of its magnitude by calculating the geometrical cross section, this procedure must not be taken seriously. The experimental meaning of the cross section comes from its use as a measure of the number of nuclear events which occur under a given set of experimental conditions, as expressed by Eq. (16–10) and the subsequent discussion. Nuclear cross sections are found to have values ranging from small fractions of a barn to hundreds of thousands of barns, and these values often differ greatly from the geometrical cross section. A given nucleus can have widely different cross sections for different nuclear reactions and the values represent the relative probabilities of those reactions. Under certain special conditions scattering cross sections and geometrical cross sections can be related directly, and the measured values of the scattering cross sections are then used to determine the values of the nuclear radii. The elastic scattering of neutrons with energy greater than 10 Mev is a case in point, and will be discussed in Section 16–5E.

The results of an experimental study of a nuclear reaction can be expressed in terms of the number of processes which take place under the conditions of the experiment or in terms of the cross section for the reaction. The advantage of using the cross section lies in the fact that its value is independent of the flux of incident particles and the density of the material used for the target, and these quantities are not essential to the nuclear reaction itself. When the cross section for a particular reaction between a given particle and a given nuclide is known, it is possible to predict the number of reactions that will take place when a sample of the nuclide is exposed to a known flux of those particles for a given length of time. This kind of information is needed in many practical problems, as in the manufacture of artificial radionuclides, and cross sections are a valuable form of nuclear data. In many cases, the results of the study of a nuclear reaction are expressed directly in terms of the number of processes that take place, or of the number of outgoing particles. If the main interest is in the determination of the energies of the different groups of particles emitted, it is enough to measure the energies by measuring the ranges or by magnetic deflection, without knowing exactly how many processes have taken place. The relative intensities of different groups of product particles can also be measured by methods which are independent of the number of processes, as can the energies of resonances in nuclear reactions.

The concepts of cross section and level width can be applied to resonances in a quantitative way. The probability of the reaction $X(x, y)Y$ may be denoted by the cross section $\sigma(x, y)$. According to the two-step view of nuclear reactions,

$$\sigma(x, y) = \sigma_C(x) \cdot (\text{relative probability of the emission of } y), \qquad (16\text{–}11)$$

where $\sigma_C(x)$ is the cross section for the formation of the compound nucleus in a collision between the particle x and the target nucleus X. The relative probability of the emission of the particle (or γ-ray) y is just Γ_y/Γ, where Γ_y is the partial level width for y and Γ is the total level width. Then

$$\sigma(x, y) = \sigma_C(x) \frac{\Gamma_y}{\Gamma}. \qquad (16\text{–}12)$$

In general, the values of the cross sections and level widths depend on the energy of the incident particle, and on the charge and mass of the target nucleus. One of the problems of nuclear theory is the calculation of $\sigma_C(x)$, Γ, Γ_y, and therefore of $\sigma(x, y)$. In the particularly important case of resonance processes, a theoretical formula for the cross section was derived by Breit and Wigner.[9,10] The rigorous derivation of the Breit-Wigner formula is a difficult problem, but the physical meaning of the formula can be discussed qualitatively. In its simplest form, the Breit-Wigner formula gives the value of the cross section in the neighborhood of a single resonance level formed by an incident particle with zero angular momentum. Under these conditions, the formula is

$$\sigma(x, y) = \frac{\lambda^2}{4\pi} \frac{\Gamma_x \Gamma_y}{(E - E_0)^2 + (\Gamma/2)^2}, \qquad (16\text{–}13)$$

where λ is the de Broglie wavelength of the incident particle defined in Section 7–8, E is the energy, E_0 is the energy at the peak of the resonance, and the Γ's are the widths already defined. Equation (16–13) may be regarded as containing three factors. The first is a measure of the probability of forming a compound nucleus and, according to wave mechanics, is proportional to λ^2. The second factor,

$$\frac{1}{(E - E_0)^2 + (\Gamma/2)^2},$$

is the mathematical expression for the resonance property and may be called the *resonance factor*. The denominator of this factor has its smallest value when $E = E_0$, and the cross section then has its greatest value. As the value of E departs from E_0, the denominator increases in value, and the resonance factor and cross section decrease. The third factor is the probability for definite types of disintegration of the compound nucleus and is expressed by the partial widths Γ_x and Γ_y.

The Breit-Wigner formula is an example of the application of wave mechanics to nuclear physics, and it is typical of such applications that the wavelength of the incident particle appears in the formula. The de Broglie wavelength of a particle is defined as

$$\lambda = \frac{h}{mv}, \tag{16–14}$$

where h is Planck's constant and mv is the momentum of the particle. It is convenient to express λ in terms of the kinetic energy rather than the momentum and, since $E = \frac{1}{2}mv^2$,

$$\lambda = \frac{h}{\sqrt{2mE}}. \tag{16–15}$$

An idea of the magnitudes of λ and σ may be obtained by considering the incident particle to be a neutron; when the numerical values of h and m are inserted, and E is expressed in electron volts, the result is

$$\lambda\ (\text{cm}) = \frac{2.87 \times 10^{-9}}{\sqrt{E\ (\text{ev})}}. \tag{16–16}$$

A neutron with a kinetic energy of 10^4 ev has a wavelength of 2.87×10^{-11} cm, while a 1-Mev neutron has a wavelength of 2.87×10^{-12} cm. Since $\sigma(x, y)$ is proportional to λ^2, its value should be of an order of magnitude similar to that obtained from the geometrical estimate in this section. Its actual value in any particular case will, of course, depend also on the other quantities in Eq. (16–13). It is found that the values of resonance cross sections vary over a wide range, from fractions of a barn to the order of 10^6 barns.

Values of $\sigma_0(x, y)$ (the cross section at the peak of the resonance), E_0, and Γ can be obtained from experiments in which the particle transmission through the target is measured as a function of the energy of the incident particles in the neighborhood of the resonance energy. The values of σ_0 and E_0 can be deduced from the transmission curve. It can be shown from Eq. (16–13) that the "full width at half-maximum," that is, the energy interval between the two points at which the cross section is $\sigma_0/2$, is just the total width Γ. The total width can thus be determined directly from the actual observed width of the cross section resonance. When the total width is known, the values of the partial widths can be obtained by further experiments, or may sometimes be inferred with the aid of certain theoretical considerations.

The rigorous form of the Breit-Wigner formula is considerably more complicated than Eq. (16–13). It contains an additional factor, of order unity, which depends on the angular momenta of the particles involved in the reaction. It also takes into account the fact that when two or more resonances are close together, the simple "one-level formula" (16–13) is no longer correct because the resonances may "overlap" and affect each other. Fortunately, for many practical purposes, the one-level formula is adequate.

16–4 Experimental results. General considerations. To discuss the experimental information about nuclear reactions in an orderly way, it is necessary to classify the reactions in some consistent scheme. Blatt and Weisskopf[11] have shown that nuclear reactions can be classified into different groups characterized by the nature of the incident particle, its energy, and the atomic weight of the target nucleus. Their classification and general treatment have proved successful, and will be followed during the remainder of this chapter.

The incident particles to be considered include neutrons, protons, α-particles, and deuterons; reactions initiated by γ-rays will also be treated.

The range of energy E of the incident particle may be divided roughly into five regions:

I. Low energies: $0 < E < 1000$ ev.
II. Intermediate energies: 1 kev $< E <$ 500 kev.
III. High energies: 0.5 Mev $< E <$ 10 Mev.
IV. Very high energies: 10 Mev $< E <$ 50 Mev.
V. Ultrahigh energies: $E >$ 50 Mev.

The target nuclei may be divided into three groups according to the values of the mass number A:

A. Light nuclei: $1 < A < 25$.
B. Intermediate nuclei: $25 < A < 80$.
C. Heavy nuclei: $80 < A < 250$.

This classification of nuclear reactions serves to separate them according to their general character, the particles emitted, and their energy distribution, as well as other properties. Although the distinctions between the different groups are not clear-cut, they are sufficiently marked to make the experimental results fall into definite patterns.

It is not surprising that reactions initiated by neutrons have, in general, different characteristics from those initiated by charged particles. The neutron, because of its lack of charge, can penetrate relatively easily into a nucleus, and the probability of forming a compound nucleus is relatively

Table 16–3

Reactions with Intermediate and Heavy Nuclei

(Reprinted by permission from *Theoretical Nuclear Physics* by J. M. Blatt and V. F. Weisskopf. 1952. John Wiley and Sons, Inc.)

This table lists the nuclear reactions occurring in each group. The symbols listed refer to the emerging particle in a reaction characterized by the type of target, the type of incident particle (columns), and the energy range (rows). The order of symbols in each group corresponds roughly to the order of the yields of the corresponding reactions. Reactions whose yield is usually less than about 10^{-2} of the leading one are omitted.

Abbreviations: el = elastic, inel = inelastic, res = resonances. The abbreviation (res) refers to all reactions listed in the box. The elastic scattering of charged particles is omitted, since it cannot easily be separated from the non-nuclear Coulomb scattering. Fission is also omitted, since it occurs only with a few of the heaviest elements.

	Intermediate nuclei				Heavy nuclei			
Incident particle	n	p	α	d	n	p	α	d
Energy of incident particle								
I. Low: 0–1 kev	n (el) γ (res)				γ n (el) (res)			
II. Intermediate: 1–500 kev	n (el) γ (res)	n γ α (res)	n γ p (res)	p n	n (el) γ (res)			
III. High: 0.5–10 Mev	n (el) n (inel) α p (res for lower energies)	n p (inel) α (res for lower energies)	n p α (inel) (res for lower energies)	p n pn 2n	n (el) n (inel) p γ	n p (inel) γ	n p γ	p n pn 2n
IV. Very high: 10–50 Mev	2n n (inel) n (el) p np 2p α three or more particles	2n n p (inel) np 2p α three or more particles	2n n p np 2p α (inel) three or more particles	p 2n pn 3n d (inel) tritons three or more particles	2n n (inel) n (el) p pn 2p α three or more particles	2n n p (inel) np 2p α three or more particles	2n n p np 2p α (inel) three or more particles	p 2n np 3n d (inel) tritons three or more particles

high. Positively charged particles must overcome the repulsive Coulomb forces exerted by the positively charged nucleus; they must either penetrate the Coulomb barrier, as discussed in Chapter 13, or they must have enough energy to go over the top of the barrier. Consequently, a neutron can cause a reaction when it has a very low kinetic energy, while charged particles need much higher energies, and many more neutron reactions have been observed than charged particle reactions.

The light nuclei (group A) are characterized by the fact that it is extremely difficult to describe nuclear reactions in that group in terms of general rules. The character and yield of most of the reactions seem to depend strongly on the structure of the individual nucleus; and separation energies and Q-values differ by much larger amounts among the light elements than within any other group. The reactions with intermediate or heavy nuclei are of different character in the different energy regions. For example, charged particles with energies less than 500 kev (corresponding to the energy groups I and II) cannot penetrate to the nucleus in groups B and C, and neutron reactions are practically the only ones observed. In region I (low energy) resonance capture of neutrons by heavy nuclei predominates, while in energy region II the most important reaction is the resonance elastic scattering of neutrons. In energy region III charged particle reactions become important; the residual nucleus can be left in several excited states and the emitted particles can form groups with different energies. In region IV the energy is high enough for reactions in which more than one particle is emitted for each incident particle [(p, 2n), (p, np), etc.]. Finally, in region V, the energy of the incident particle is so high that many nucleons can be ejected by a single incident particle, and the theory of the compound nucleus breaks down.

In Table 16–3, taken from Blatt and Weisskopf,[(10)] the different reaction types are arranged according to the groups in which they appear. Reactions with light nuclei are not listed because, as indicated above, they cannot be fitted into any general scheme. The reactions in each group are listed in the order of their importance. In general, the reaction which has the larger cross section precedes the one with lower yield and, although there are exceptions to this scheme, it is a useful one.

In the following sections of this chapter, the different groups of reactions will be discussed in a general way and illustrated by specific examples. The emphasis will be on the use of nuclear reactions as sources of information about nuclei rather than on the experimental techniques or the theoretical details. The latter can be found in the literature references.

16–5 Neutron-induced reactions. Reactions induced by neutrons will be discussed first because of their great importance in both "pure" and "applied" nuclear physics. Experimental methods have been devel-

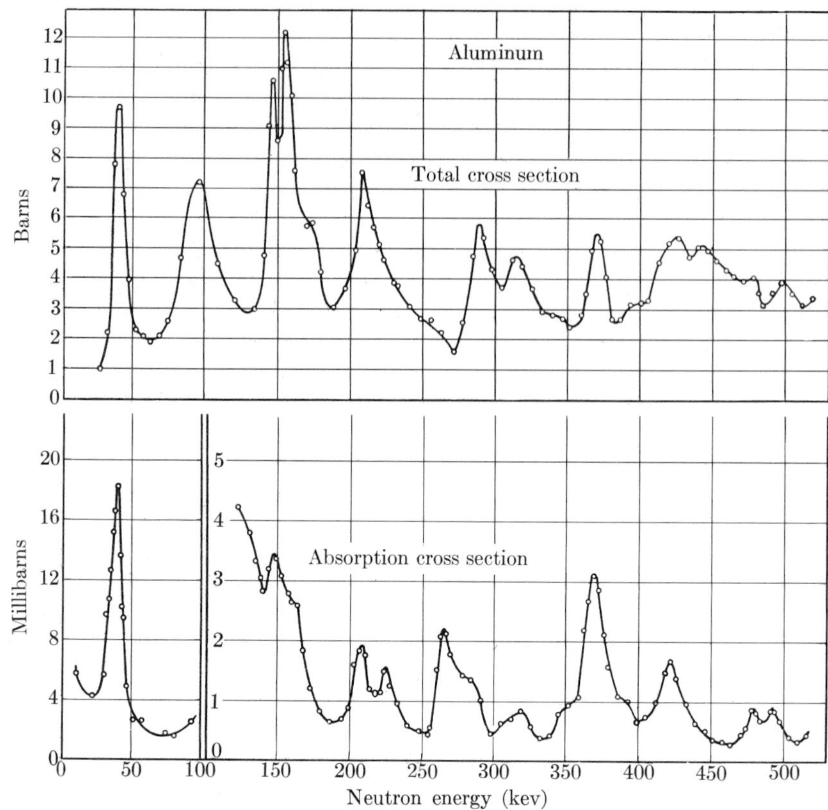

FIG. 16–3. Total and capture cross sections of aluminum for neutrons in the intermediate energy region (Henkel and Barschall(13)).

oped for obtaining neutrons of different energies, for detecting them, and for measuring the cross sections of the different neutron-induced reactions. These methods form a highly specialized branch of nuclear physics and have been treated in detail by Hughes;(12) they will be discussed in Chapter 18, which is devoted to the subject of neutron physics. For the purposes of the present chapter, it will simply be stated that beams of monoenergetic neutrons can be obtained at energies ranging from small fractions ($\sim 10^{-4}$) of an electron volt to many millions of electron volts, and that the cross sections or yields of the different reactions induced by these neutrons can be measured by means of appropriate devices.

A. *Low and intermediate energy, intermediate nuclei.* The most important reactions between neutrons of low or intermediate energy and nuclei of intermediate atomic weight are elastic scattering (n, n), and radiative capture (n, γ). The bombardment of Al^{27} by neutrons provides an interesting example of these reactions. The total cross section σ_t and the radiative capture cross section $\sigma(n, \gamma)$ have been measured as functions of the incident neutron energy. The total cross section was obtained from transmission measurements. The capture cross section was obtained by making use of the fact that the product Al^{28} is an electron emitter and its yield can be determined by measuring the induced radioactivity. Some of the results[13] are shown in Fig. 16–3, in which the total and radiative capture cross sections are plotted as functions of energy in the range from 10 to about 500 kev. The values of the total cross section lie between 1 and 12×10^{-24} cm^2, while the capture cross sections are of the order of 10^{-27} cm^2, or three orders of magnitude smaller. Since the only reactions that have been observed when Al^{27} is bombarded with intermediate-energy neutrons are elastic scattering and radiative capture, the total cross section is practically equal to the elastic scattering cross section. The capture cross section shows resonances at the same energies as the total cross section. This result would be expected according to the compound nucleus picture and serves as a confirmation of that picture. It follows from the values of the cross section, and from the Breit-Wigner formula that the scattering or neutron width Γ_n is about 1000 times as great as the radiation width Γ_γ. The experimental results show that the neutron widths are in the range 5–20 kev, so that the radiation widths are of the order of 10 ev.

Over twenty resonances have been found for neutron energies in the range from 0.009 Mev to 0.8 Mev and, since Al has only one stable isotope, they must correspond to excited levels of Al^{28}. They are listed in Table 16–4, together with the corresponding excited levels of the compound nucleus $[Al^{28}]$ computed as shown in Section 16–2. The binding energy of the neutron in this nucleus is 7.724 Mev, as calculated from the difference

$$\text{mass of Al}^{27} + \text{mass of neutron} - \text{mass of Al}^{28}.$$

The observed level spacing for Al^{28}, the distance between neighboring resonances, varies from 10 kev to 90 kev, with an average value of about 40 kev.

The scattering cross section has an appreciable value between resonances, being still of the order of 10^{-24} cm^2. At neutron energies where there is no resonance, this cross section represents the probability that the neutrons are scattered without the formation of a compound nucleus. In this case, the nucleus acts like a hard sphere of radius R, and it has been shown that for neutrons with energies up to about 1 Mev the scattering cross section is

$$\sigma_s = 4\pi R^2.$$

This type of scattering is called *potential scattering*, as distinct from resonance scattering, because the force between the nucleus and the neutrons

TABLE 16–4

LEVELS OF THE COMPOUND NUCLEUS [Al^{28}] EXCITED IN THE BOMBARDMENT OF ALUMINUM BY NEUTRONS.

Neutron resonance energy, Mev	Energy to be added to the binding energy, Mev	Excited level of [Al^{28}] = 7.724 + energy in second column, Mev
0.0091	0.0088	7.733
0.040	0.039	7.763
0.095	0.092	7.816
0.145	0.140	7.864
0.155	0.149	7.873
0.210	0.203	7.927
0.225	0.217	7.941
0.265	0.256	7.980
0.290	0.280	8.004
0.315	0.304	8.028
0.370	0.357	8.081
0.425	0.410	8.134
0.445	0.429	8.153
0.480	0.463	8.187
0.500	0.482	8.206
0.530	0.511	8.235
0.570	0.550	8.274
0.620	0.598	8.322
0.650	0.627	8.351
0.705	0.680	8.404
0.795	0.767	8.491

can be described mathematically by a potential function analogous to the Coulomb potential in the scattering of α-particles by nuclei. The potential which describes the neutron scattering is not electrical in nature, but is a characteristically nuclear potential.

One part of the low energy region I, the so-called *thermal* energy region, has a special property in the case of the (n, γ) reaction. A beam of thermal neutrons can be obtained from a neutron-producing apparatus in which the neutrons are slowed down in some medium and emerge with energies in equilibrium with those of the thermal motion of the atoms of the medium. At room temperature, the thermal neutron energy region is in the neighborhood of 0.025 ev. Although the Breit-Wigner formula is strictly valid only near a resonance, it may be applied in the thermal energy region in the absence of a resonance if it is assumed that the energy

E_0 is the resonance energy nearest to the thermal region. It turns out that all of the factors in the Breit-Wigner formula are constant compared with Γ_n, which is proportional to the neutron velocity v, and λ, which is proportional to $1/v$. The formula for the (n, γ) reaction then reduces to

$$\sigma(\text{n}, \gamma) = \frac{\text{constant}}{v}, \tag{16–17}$$

and the cross section for radiative capture is inversely proportional to the neutron velocity. In the thermal energy region, the radiative capture of neutrons by aluminum is described by the "$1/v$ law"; at a neutron energy of 0.025 ev, the value of the (n, γ) cross section is 0.24×10^{-24} cm^2, and is much greater than the values in the resonance region.

B. *Low-energy, heavy nuclei.* The only reactions possible for low-energy neutrons on heavy nuclei are, with a few exceptions, elastic scattering and radiative capture. When the positive charge of the nucleus is large, the effect of the Coulomb barrier prohibiting the emission of charged particles of low energy is even greater than with intermediate nuclei. In reactions between low-energy neutrons and heavy nuclei, the cross sections often show resonances very close together in energy. The level distances are often of the order of 10 to 100 ev and the excited states of the compound nuclei in this region are usually close together. There are some important exceptions to this rule. In certain nuclides, no resonances have been found, and in others, especially lead and bismuth, the spacing between resonances is very large (many kev). These nuclides are now thought to have a special structure interpreted in terms of the shell model of nuclear structure, which will be discussed in the next chapter.

An example of a heavy nucleus with resonances very close together is given in Fig. 16–4, which shows the variation of the total cross section of Ag^{109} in the energy range from one ev to about 5000 ev.[(14)] At the higher energies, the resonances cannot be resolved well because of experimental difficulties; the energy resolution which can be achieved is shown by the triangles. In heavy elements such as silver, the resonances are mainly capture rather than scattering resonances, and the neutron width Γ_n is small compared with the radiation width Γ_γ. The total cross section is practically equal to the capture cross section σ(n, γ) near resonance; between resonances, the total cross section is approximately equal to the (potential) scattering cross section $4\pi R^2$. The total cross section[(14)] of an exceptional nucleus Bi^{209}, between one ev and 10,000 ev, is shown in Fig. 16–5; only two widely spaced resonances are seen, and the curve is very different in character from the curve for Ag^{109}.

Neutron capture resonances have been analyzed extensively, and have been used to test the Breit-Wigner formula. Figure 16–6 shows a low-energy capture resonance in cadmium;[(15)] the solid line was calculated

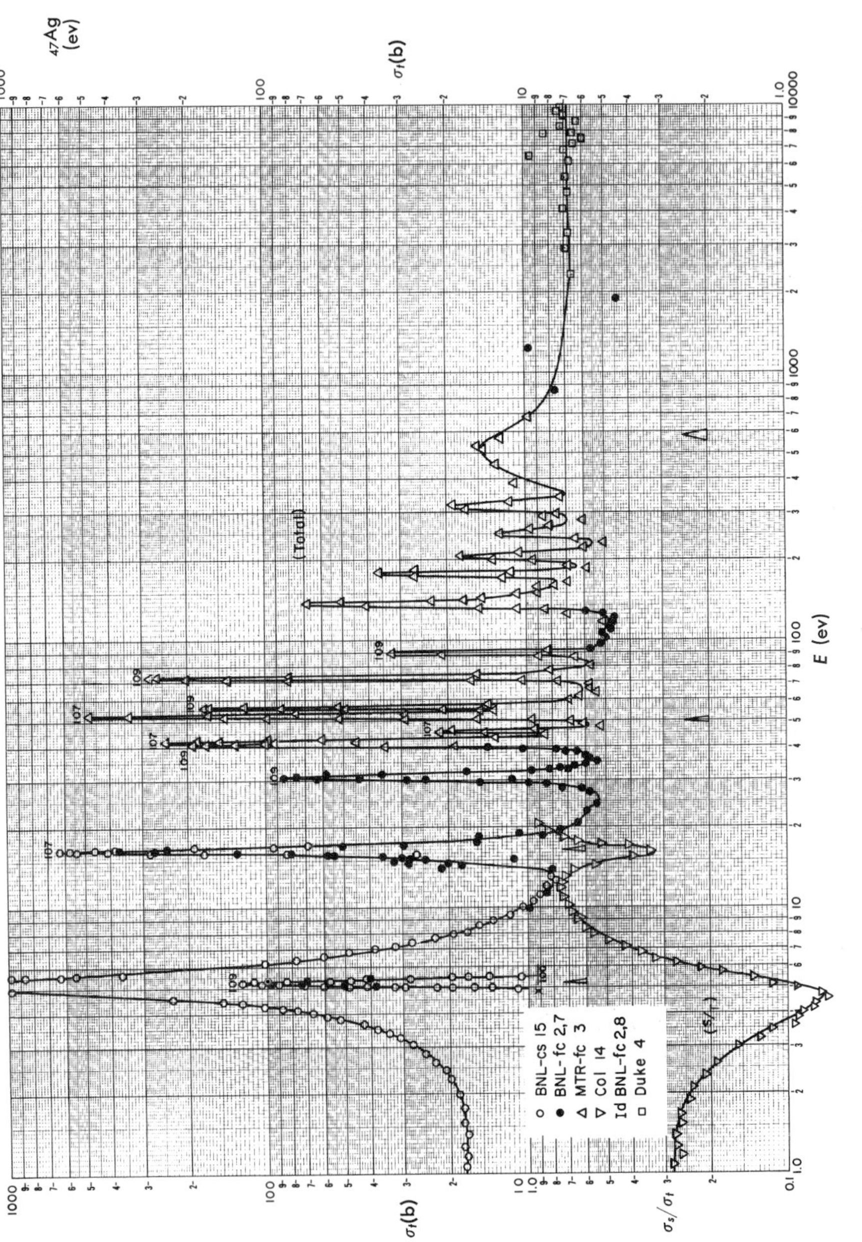

FIG. 16–4. The total neutron cross section of silver in the low-energy region.(14)

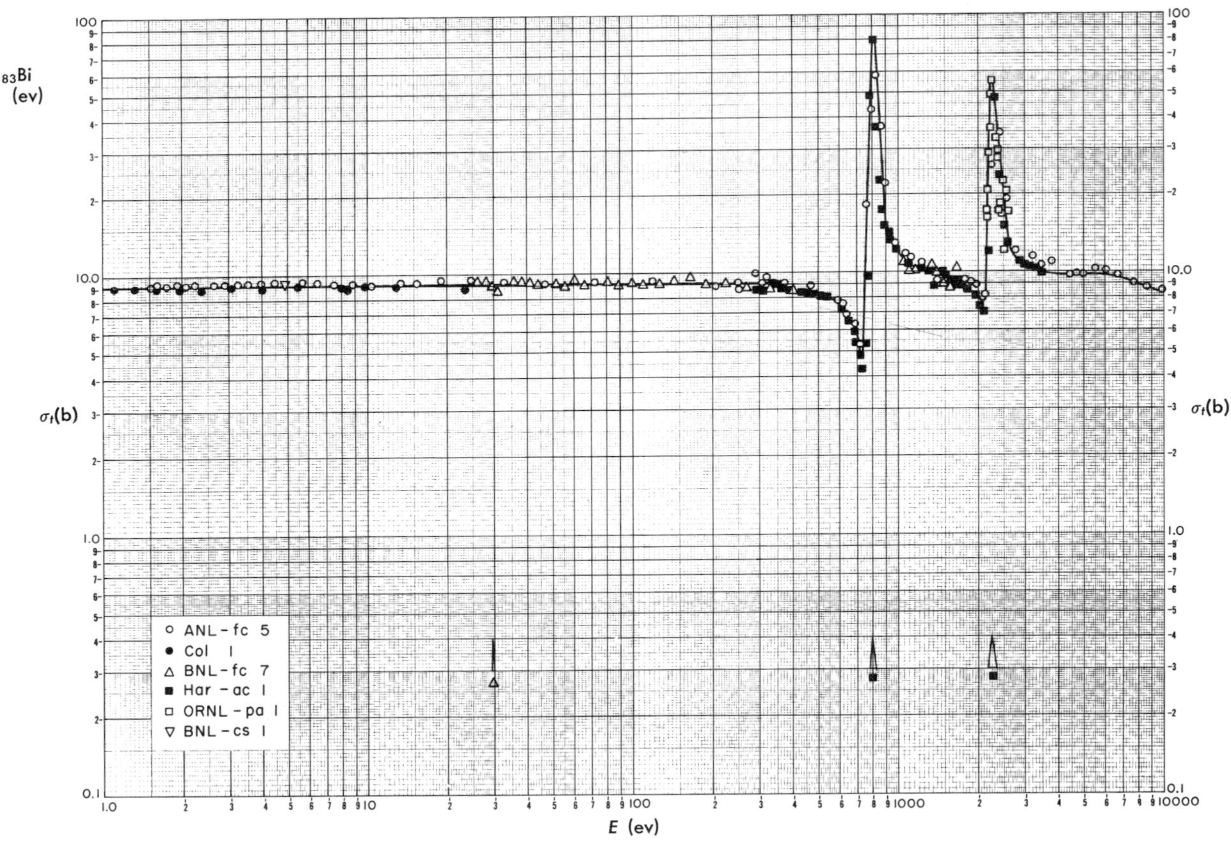

FIG. 16–5. The total neutron cross section of bismuth in the low-energy region.(14)

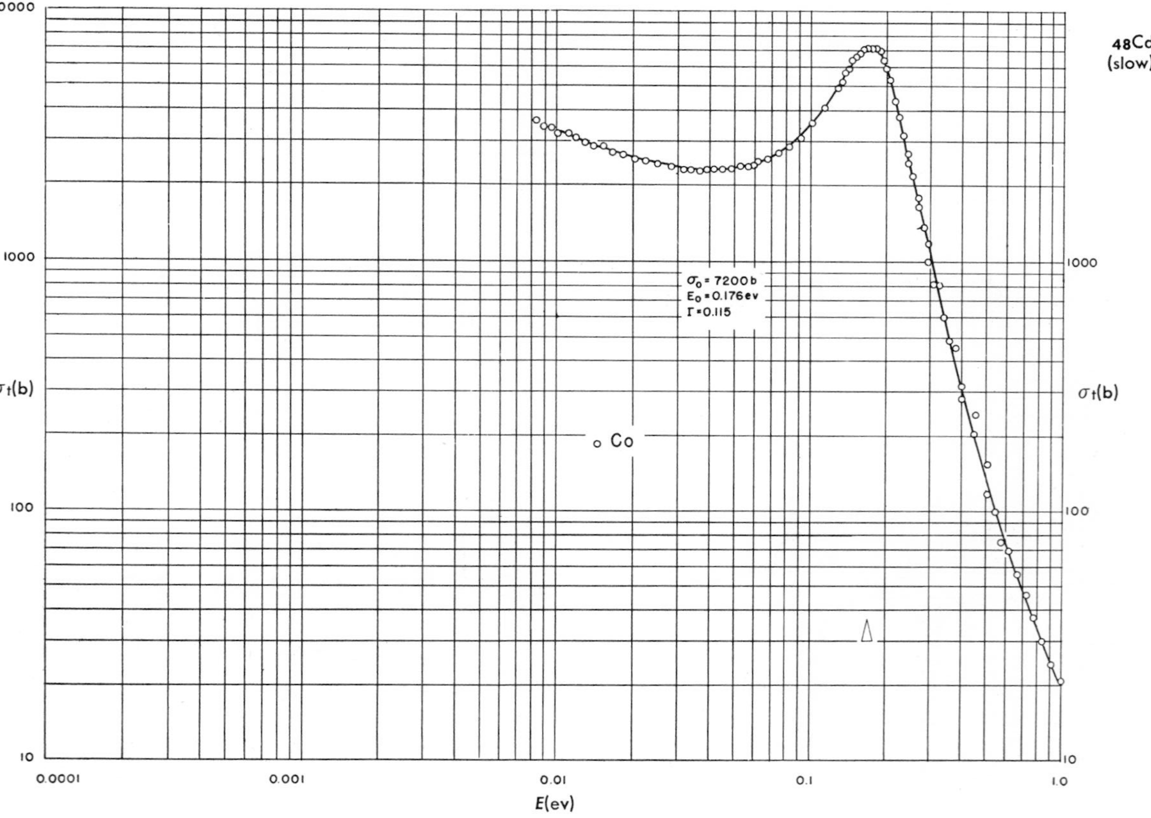

Fig. 16–6. The low-energy resonance in cadmium (Rainwater *et al.*[15]).

from the one-level formula (16–13) with the constants $\sigma_0 = 7200$ b, $E_0 = 0.176$ ev, and $\Gamma = 0.115$ ev, and the circles represent the experimental points. The agreement between the formula and the experimental results is excellent and provides a quantitative proof of the validity of the formula.

At thermal energies, in the absence of resonances, $\sigma(\mathrm{n}, \gamma)$ follows the $1/v$ law, as was the case with intermediate nuclei.

C. *Intermediate-energy, heavy nuclei.* In the intermediate-energy region, the reactions are similar in nature to those with intermediate nuclei. Resonance scattering is more important than resonance capture and $\Gamma_{\mathrm{n}} > \Gamma_\gamma$. The spacing between resonances, however, is smaller than for intermediate nuclei. It is hard to explore the intermediate-energy region because of experimental difficulties with neutrons of these energies. Individual resonances cannot be resolved and the experiments give information about cross sections averaged over many resonances. These cross sections in turn yield information about the average level spacing. When the Breit-Wigner formula (16–13) is averaged over many resonance levels and applied to (n, γ) reactions at neutron energies of about 1 Mev, it is found that the averaged (n, γ) cross section is given by

$$\overline{\sigma(\mathrm{n}, \gamma)} = (\lambda^2/2)(\Gamma_\gamma/D), \tag{16–18}$$

where D is the level spacing. If $\overline{\sigma(\mathrm{n}, \gamma)}$ is measured, and if Γ_γ is calculated theoretically, values of the level spacing can be obtained. The radiative capture cross section has been measured for a large number of nuclides by Hughes[(16)] for 1-Mev neutrons. In the scheme of Section 16–4, this energy lies in region III (high energy) rather than in region II (intermediate). These reactions will be discussed here because they are a borderline or overlapping case which could be discussed in either place, but it is more convenient to discuss them here. The measured values of the capture cross section are plotted against the mass number of the target nucleus in Fig. 16–7; the cross sections are given in units of millibarns (thousandths of a barn). As the mass number increases, the capture cross section increases until A is about 100. The radiation width Γ_γ decreases, according to theory, but less rapidly than the cross section increases. The result is that the level spacing D, from Eq. (16–18), decreases quite rapidly as the number of particles in the nucleus increases. At higher values of the mass number, the cross section does not seem to behave regularly; but if the cross section is plotted against the number of neutrons in the nucleus, as in Fig. 16–8, it is evident that the irregularities are caused by sharp drops in the value of the cross section that occur at neutron numbers of 50, 82, and 126. These exceptionally low values of the cross section correspond to unusually large level spacings such as those mentioned previously for lead and bismuth, both of which

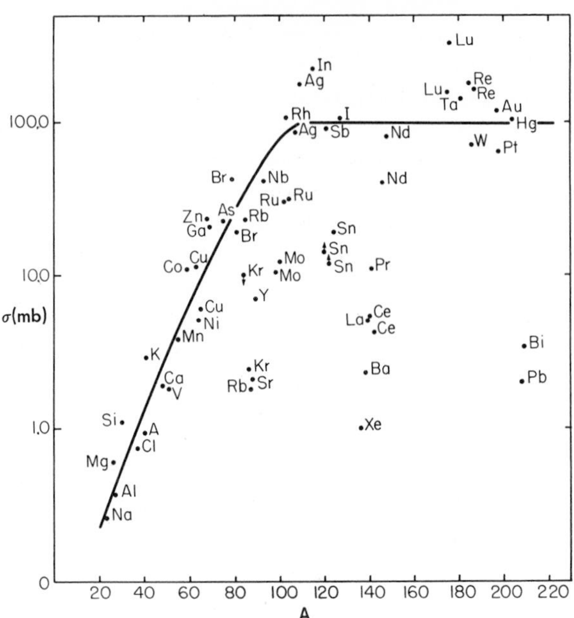

FIG. 16–7. Neutron capture cross sections at an effective energy of 1 Mev (Hughes[12]).

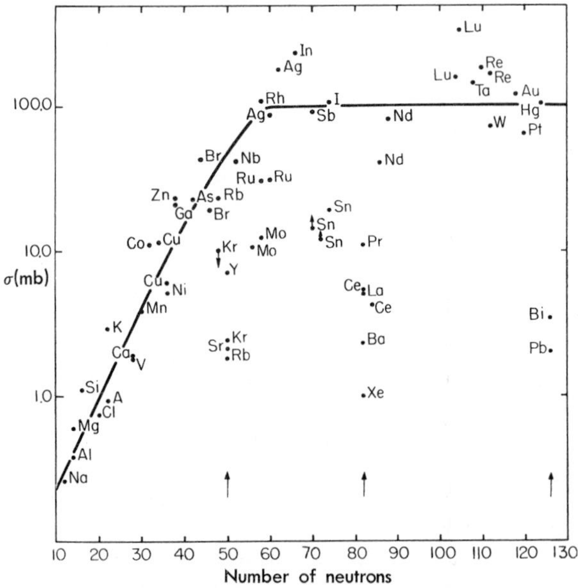

FIG. 16–8. Neutron capture cross sections at 1 Mev as a function of neutron number (Hughes[12]).

have 126 neutrons. The values of the level spacing obtained in this way vary from about 5×10^4 ev for Na^{23} to about 1 ev for the ordinary heavy nuclides. The nuclides with 50, 82, or 126 neutrons have much greater level spacings than neighboring nuclides, and it will be seen in the next chapter that these neutron numbers seem to be associated with particular nuclear structures.

D. *High-energy, heavy and intermediate nuclei.* When the energy of the incident neutron is about 1 Mev or higher, new types of processes can occur, such as inelastic scattering and the emission of charged particles. The emission of neutrons is more probable than that of charged particles because of the Coulomb barrier. The emitted neutrons can leave the residual nucleus in excited states so that some of the scattering is inelastic rather than elastic. The excited states, which are bound states, decay by γ-emission, and information about them can be obtained from the energies of the γ-rays and from the energy distribution of the scattered neutrons. The determination of inelastic neutron scattering cross sections is a difficult experimental problem, and only a relatively small amount of information has been obtained in this way. The cross sections for the reactions in which charged particles are emitted are much smaller than those for inelastic scattering and radiative capture, especially when the target nucleus has a high value of Z, because of the effect of the Coulomb barrier.

E. *Very high energy, heavy and intermediate nuclei.* The scattering of neutrons with energies greater than 10 Mev has been used to determine values of the nuclear radius.[17,18] According to theory, the total cross section at high energies approaches the value

$$\sigma_t = 2\pi R^2, \tag{16–19}$$

so that if σ_t is determined by transmission measurements, R can be obtained. The rigorous relationship between σ_t and R is more complicated than the last equation.[19] The formula for σ_t actually includes a term which depends on the energy of the neutrons, and the value of the cross section given by Eq. (16–19) is an asymptotic one which is reached only for extremely high energies. The results of two sets of experiments, one with 14-Mev neutrons, the other with 25-Mev neutrons, are listed in Table 16–5; the values of R were obtained from the rigorous formula. The quantity R_0 is defined by the relationship

$$R = R_0 A^{1/3} \times 10^{-13} \text{ cm}, \tag{16–20}$$

and should be a constant if the nuclear radius is proportional to the cube root of the nuclear mass. The results show that R_0 is indeed practically constant, and the values of nuclear radii conform with the formula (16–20)

TABLE 16–5

NUCLEAR RADII FROM NEUTRON CROSS SECTIONS AT HIGH ENERGIES

Element	Energy, Mev	Observed σ_t, 10^{-24} cm^2	Nuclear radius, R, 10^{-13} cm	R_0
Be	14	0.65	2.4	1.17
B	14	1.16	3.4	1.54
C	25	1.29	3.8	1.65
O	25	1.60	4.3	1.71
Al	14	1.92	4.6	1.53
	25	1.85	4.6	1.52
S	14	1.58	4.1	1.30
Cl	25	1.88	4.7	1.44
Fe	14	2.75	5.6	1.46
Cu	25	2.50	5.5	1.38
Zn	14	3.03	5.9	1.48
Se	14	3.35	6.3	1.46
Ag	14	3.82	6.8	1.44
	25	3.70	6.9	1.46
Cd	14	4.25	7.2	1.48
Sn	14	4.52	7.4	1.52
Sb	14	4.25	7.3	1.46
Au	14	4.68	7.5	1.33
Hg	14	5.64	8.3	1.42
	25	5.25	8.4	1.44
Pb	14	5.05	7.8	1.32
Bi	14	5.17	7.9	1.34

over the entire range of the elements, with the possible exception of the very lightest ones. The average value of R_0 is close to 1.4 and agrees reasonably well with the value 1.53 obtained from α-decay data in Section 13–6. The two results should not be expected to be the same because they are not the results of direct measurements of the nuclear radius. The quantity R, as it appears in the theory, is the distance at which nuclear forces begin to act. These forces should act differently on a neutron approaching the nucleus and an α-particle leaving it, and it is not surprising that different values are obtained for nuclear radii from the analysis of different kinds of experiments. The fact that the values obtained by different indirect methods for the dimensions of atomic nuclei agree as well as they do represents a triumph for nuclear theory.

Values of nuclear radii have also been obtained recently from studies of the scattering of neutrons with energies in the neighborhood of 1 kev.[20]

The interpretation of the experimental results is more complicated than at the higher energies, but the results are similar, with $R_0 = 1.35$.

In the very high energy region, the incident particle has enough energy to cause the emission of more than one particle from the compound nucleus, and (n, 2n), (n, np), (n, 3n), etc., reactions, may be expected. Again, because of the Coulomb barrier, the emission of neutrons is more likely than that of charged particles. The (n, 2n) reaction has been studied in a number of cases,[(21)] and its threshold is usually in the neighborhood of 10 Mev. Above the threshold, the cross section increases with energy,[(22)] reaching a maximum somewhere near 20 Mev, and then decreases with increasing energy of the incident neutrons.

16–6 Reactions induced by protons and alpha-particles. A. *High energy.* The cross sections of nuclear reactions induced by protons or α-particles are extremely small for particle energies below 0.1 Mev except for a few cases among the very light nuclei. At low or intermediate energies, the Coulomb barrier prevents any appreciable interaction between positively charged particles and nuclei. When the energy of the incident particles is high enough, protons and α-particles may be scattered elastically or inelastically; they may undergo radiative capture, or they may cause the emission of a charged particle. The (α, p) reaction, the first one studied in detail, has already been discussed in Section 16–1. The properties of reactions induced by protons and α-particles can be illustrated by the reactions of protons with aluminum. The reaction $Al^{27}(p, \gamma)Si^{28}$ yields information about the excited levels of the compound nucleus [Si^{28}]. A large number of very sharp resonances has been found,[(23)] some of which are shown in Fig. 16–9. In these experiments, an aluminum target was bombarded with protons accelerated in a Van de Graaff electrostatic generator, and the γ-ray yield was measured as a function of the proton energy. The sharpness of the resonances (the low value of the total width) shows that the lifetimes of the excited states are relatively long. About 100 resonances have been found at proton energies between 0.2 Mev and 4 Mev.[(24)] The binding energy of the proton as obtained from the mass difference, $Al^{27} + p - Si^{28}$, is 11.59 Mev, and the excited levels lie between 11.7 and 15.6 Mev.

The reaction $Al^{27}(p, \alpha)Mg^{24}$ has also been studied and more than 30 resonances have been observed[(24,25)] in the energy range from 1.4 Mev to 4 Mev. These resonances also give information about the excited levels of [Si^{28}] and should occur at the same energies as those of the (p, γ) reaction. The elastic scattering of protons by aluminum also shows resonances and, with few exceptions, at every resonance obtained with the (p, γ) or (p, α) reactions, there is an elastic scattering resonance. This result is expected because, according to the compound nucleus theory,

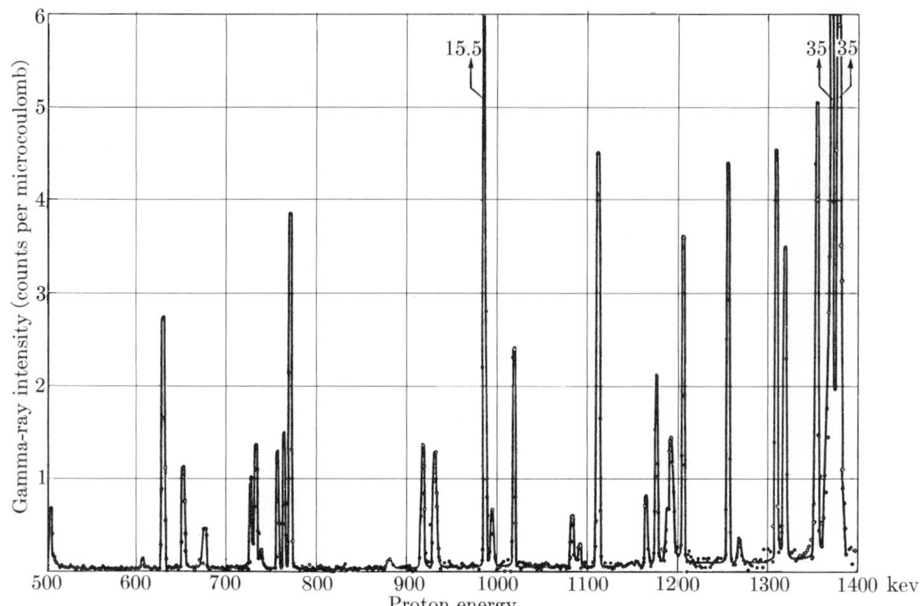

FIG. 16–9. Yield of the (p, γ) reaction in aluminum, as a function of the proton energy (Brostrom, *et al.*[23]).

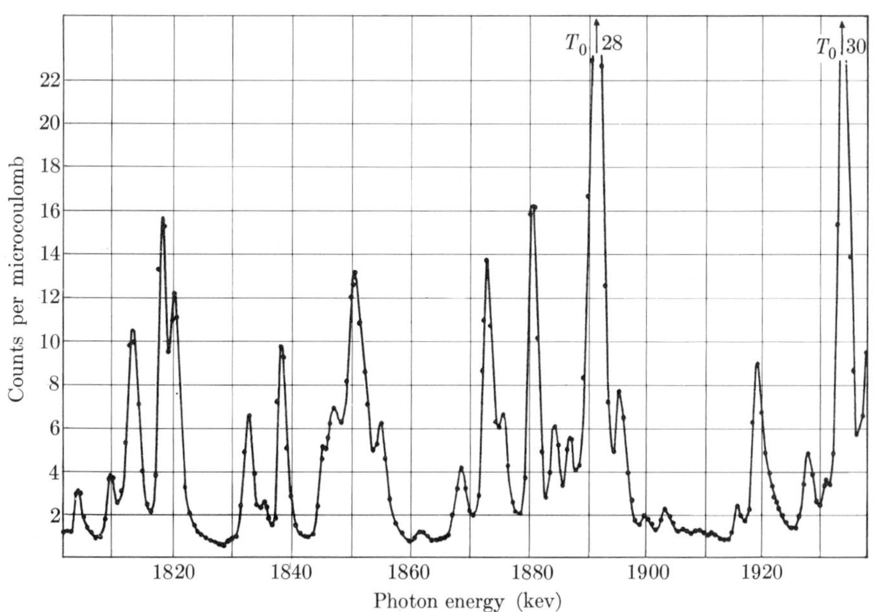

FIG. 16–10. Yield of the (p, n) reaction in Cl^{37}, as a function of the proton energy (Schoenfeld *et al.*[28]).

any state of the compound nucleus formed by proton capture must be capable of decay by emission of a proton with the full energy available.

The inelastic scattering of protons by aluminum shows resonances and gives information about excited states of the compound nucleus $[Si^{28}]$; it also provides an important method for determining the level spectrum of the *target* nucleus Al^{27}. This reaction may be described by the equations

$$_{13}Al^{27} + {}_1H^1 \rightarrow [{}_{14}Si^{28}] \rightarrow {}_{13}Al^{27*} + {}_1H^1,$$
$$_{13}Al^{27*} \rightarrow {}_{13}Al^{27} + \gamma,$$

where the asterisk denotes an excited state of Al^{27}. The method generally used for studying the inelastic scattering of protons is to bombard the scatterer with a collimated beam of monoenergetic protons, and then, by means of range or magnetic deflection measurements, to determine the energy distribution of the protons scattered through a known angle. The scattered protons usually appear in several groups with different energies, and the corresponding nuclear energy levels can be obtained from the equations of conservation of energy and momentum. In this way, Al^{27} has been found to have more than 20 excited levels at energies between 0.84 Mev and 5.95 Mev above the ground state.(26) They are bound levels and decay by γ-emission.

When the energy of the incident protons or α-particles is high enough, a neutron may be emitted. The (p, n) and (α, n) reactions are threshold reactions; if the target nucleus is stable, the threshold proton energy for neutron emission is always greater than 782 kev, the neutron-proton mass difference. One effect of the Coulomb barrier is to increase the threshold to still higher values. Below the threshold for neutron emission, radiative capture, scattering, and the emission of charged particles are important. Above the threshold, the relative importance of neutron emission increases with increasing energy of the bombarding particle, and when the compound nucleus has enough excitation energy to emit a neutron of 1 Mev or more, neutron emission becomes the dominant reaction. Threshold energies have been determined(27) for a number of (p, n) reactions among the light and intermediate nuclides. In these experiments, protons were usually accelerated in an electrostatic generator and the neutron yield was measured. In the case of Cl^{37}, the threshold for the reaction $Cl^{37}(p, n)A^{37}$ is at 1.64 Mev. Between this energy and 2.51 Mev over 100 resonances have been found(28) corresponding to excited states of the compound nucleus $[A^{38}]$. A sample of the experimental results in the range from 1.80 Mev to 1.94 Mev is shown in Fig. 16–10. The resonances were resolved well enough so that an average separation between levels of about 5 kev was observed. Similarly, in the reaction $Cr^{53}(p, n)Mn^{53}$, 261 resonances have been observed,(29) corresponding to excited levels of the compound nucleus [Mn^{54}], for proton energies between 1.42 Mev and 2.47 Mev, with an average spacing of 3.9 kev.

These results are an indication of the complexity of nuclear spectra and of the great difficulty of accounting in detail for these spectra on the basis of a theory of nuclear structure. One method of avoiding this problem is to develop an approximate theory which does not attempt to reproduce individual resonances but simply tries to describe the general trend of the magnitude and energy dependence of nuclear cross sections on the basis of a very rough picture of nuclear structure. A schematic theory of this kind has been proposed by Feshbach and Weisskopf[(19)] and has been applied to the (p, n) and (α, n) reactions, among others. When the energy of the incident particle is high enough, the emission of a neutron is the dominant reaction, and the other reactions are negligible in comparison. The approximate theory can then be applied to the calculation of the cross sections averaged over many resonances, giving the values for the cross sections for the formation of the compound nuclei. If the latter are denoted by $\sigma_C(\mathrm{p})$ and $\sigma_C(\alpha)$, then

$$\sigma(\mathrm{p, n}) \cong \sigma_C(\mathrm{p}); \quad \sigma(\alpha, \mathrm{n}) \cong \sigma_C(\alpha).$$

When the energy levels are very close together, as in the heavier-intermediate and heavy nuclei, it is very difficult to resolve the resonances, and experimental measurements give values of $\sigma(\mathrm{p, n})$ and $\sigma(\alpha, \mathrm{n})$ averaged over many resonances. If the resolution is sufficiently poor, the detailed resonance structure is not seen, and curves such as those in Fig. 16–11 are obtained. The circles represent experimental points for the reactions $Ag^{107}(\mathrm{p, n})Cd^{107}$ and $Ag^{109}(\mathrm{p, n})Cd^{109}$, and the dashed curves are theoretical results; the latter depend on the value chosen for the nuclear radius. Good agreement was obtained when the formula (16–20) was used for the radius with $R_0 = 1.5$. This agreement is typical of results obtained for (p, n) and (α, n) reactions on a series of nuclides; in nearly every case, the experimental results could be accounted for by the theory provided that a value of R_0 between 1.3 and 1.5 was used.[(30)] These values of R_0 are in good agreement with the values obtained from α-decay and neutron scattering so that the application of the approximate theory of Feshbach and Weisskopf to (p, n) and (α, n) reactions seems to be successful.

B. *Very high energy.* When the energy of the incident particles is greater than 10 Mev, more than one particle can be emitted. The compound nucleus decays most often by emitting a neutron, but the residual nucleus may be left in a sufficiently excited state to emit a second neutron; the most common reactions of this type are the (p, 2n) or (α, 2n) reactions. At still higher energies, three neutrons may be emitted. When bismuth

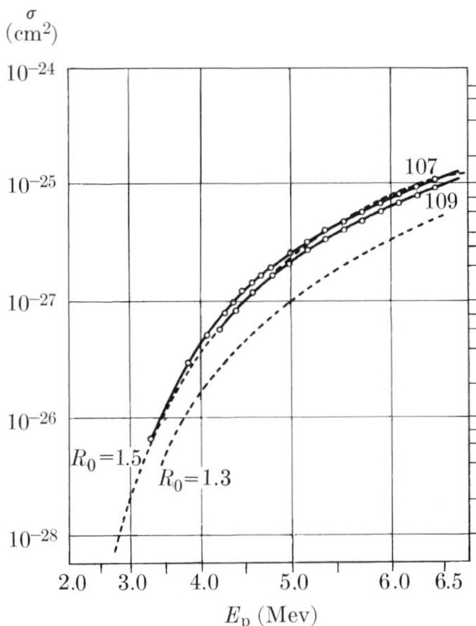

FIG. 16–11. Cross sections for the reactions Ag^{107}(p, n)Cd^{107} and Ag^{109}-(p, n)Cd^{109}. The circles are experimental points and the dashed curves are the theoretical results of Shapiro.[30]

is bombarded by α-particles,[31] the (α, 2n) reaction is first observed when the α-particle energy is about 20 Mev. The cross section increases with increasing energy, and reaches its maximum value of about one barn at about 30 Mev. At this energy, the (α, 3n) reaction is first observed. As the energy is increased above 30 Mev, $\sigma(\alpha$, 2n) decreases, and reaches a value of one-quarter of a barn at about 40 Mev, while $\sigma(\alpha$, 3n) increases to $1\frac{1}{4}$ barns.

In this energy region, Ghoshal[32] has made a set of measurements which serves as an interesting test of the theory of the compound nucleus. The idea underlying the experiment can be developed from Eq. (16–12). For a given excited state of a compound nucleus Γ is a constant and the ratio Γ_y/Γ may be replaced, for convenience, by γ_y. The cross section for a reaction of the type $A + a \rightarrow [C^*] \rightarrow B + b$ may be written

$$\sigma(a, b) = \sigma_a(\epsilon)\gamma_b(E), \tag{16–21}$$

where $\sigma_a(\epsilon)$ is the cross section for the absorption of the particle a with kinetic energy ϵ by the target nucleus A to form a particular state $[C^*]$ of the compound nucleus; $\gamma_b(E)$ is the probability of disintegration of $[C^*]$ into the final nucleus B and the particle b. The excitation energy E of the compound state $[C^*]$ is $E = \epsilon + \beta_a$, where β_a is the binding energy of the particle a to the target nucleus A. If the compound nucleus $[C^*]$ is formed in the same state of excitation, but by the reaction between a different initial particle and target nucleus, then the cross section for the reaction $A' + a' \rightarrow [C^*] \rightarrow B + b$ is

$$\sigma(a', b) = \sigma_{a'}(\epsilon')\gamma_b(E), \tag{16–22}$$

where ϵ' is the kinetic energy of the incident particle a'. Since the excitation energy is assumed to be the same in the two cases, but the binding energies of the incident particles are different, ϵ' is different from ϵ; but $\gamma_b(E)$ is the same in the two cases because of Bohr's basic assumption that the mode of decay of the compound nucleus $[C^*]$ is independent of the way in which it is formed.

If $[C^*]$ disintegrates in a different way to give the particle d and the residual nucleus D, Eqs. (16–21) and (16–22) are replaced by

$$\sigma(a, d) = \sigma_a(\epsilon)\gamma_d(E), \tag{16–23}$$

and

$$\sigma(a', d) = \sigma_{a'}(\epsilon')\gamma_d(E). \tag{16–24}$$

It follows from the last four equations that

$$\frac{\sigma(a, b)}{\sigma(a, d)} = \frac{\gamma_b(E)}{\gamma_d(E)} = \frac{\sigma(a', b)}{\sigma(a', d)}. \tag{16–25}$$

An experimental test of Eq. (16–25) provides a direct test of the validity of Bohr's theory of the compound nucleus.

Ghoshal bombarded Cu^{63} with protons and Ni^{60} with α-particles, producing the compound nucleus $[Zn^{64}]$ in both cases. The energy (mass) difference between the two target nuclei is such that a proton energy ϵ and an α-particle energy $\epsilon' = \epsilon + 7$ Mev produce the compound nucleus $[Zn^{64}]$ in the same state of excitation. Ghoshal measured the cross sections of the reactions

$$Ni^{60}(\alpha, n)Zn^{63}, \quad Ni^{60}(\alpha, 2n)Zn^{62}, \quad Ni^{60}(\alpha, pn)Cu^{62};$$

$$Cu^{63}(p, n)Zn^{63}, \quad Cu^{63}(p, 2n)Zn^{62}, \quad Cu^{63}(p, pn)Cu^{62}.$$

According to the analysis leading to Eq. (16–25), the ratios $\sigma(\alpha, n)$: $\sigma(\alpha, 2n)$:$\sigma(\alpha, pn)$ for Ni^{60} should agree, within the limits of experimental errors, with the ratios $\sigma(\alpha, n)$:$\sigma(\alpha, 2n)$:$\sigma(\alpha, pn)$ for Cu^{63}. The experimental results are shown in Fig. 16–12, where the six cross sections are plotted as functions of the kinetic energy of the α-particles and protons. The proton energy scale has been shifted by 7 Mev with respect to the α-particle energy scale in order to make the peaks of the proton curves correspond with those of the α-particle curves. It can be seen from the

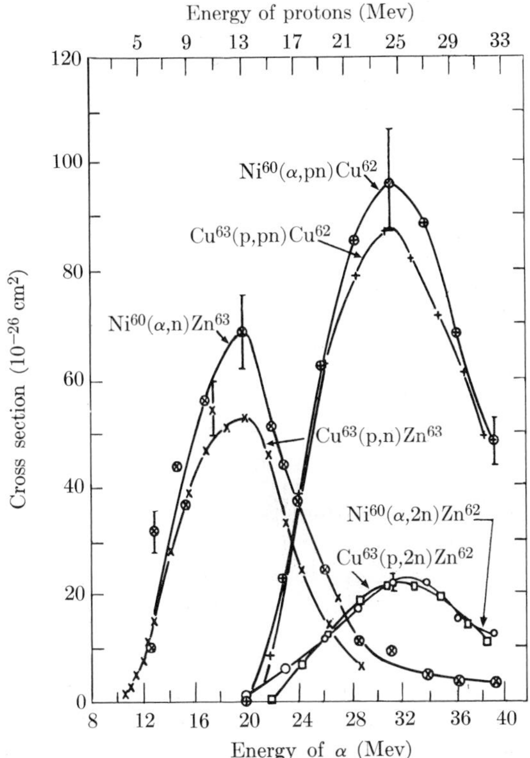

FIG. 16–12. Experimental cross sections for the (p, n), (p, 2n), and (p, pn) reactions on Cu^{63} and for the (α, n), (α, 2n), and (α, pn) reactions on Ni^{60} as functions of the proton and α-particle energies, respectively (Ghoshal[32]).

figure that the ratios agree quite well, providing evidence for the validity of the compound nucleus hypothesis.

C. *Coulomb Excitation.* A reaction different in nature from those discussed so far can occur when a charged particle interacts with a nucleus without specific nuclear forces coming into play; there can be inelastic scattering and excitation of the nucleus because of the Coulomb forces.[33,34] When a charged particle (α-particle or proton) with a few Mev of energy moves past a target nucleus, the latter is subjected to a rapidly varying electric field. This field may give rise to transitions from the ground state to excited states of the target nucleus. An excited state produced in this way then decays by emission of γ-rays or conversion electrons, and the decay can be described in terms of appropriate multipole radiation. A cross section for the production of the γ-ray can be measured; it is directly related to the transition probability, and is determined from the particle beam current, the thickness of the target, and the efficiency of the detector.

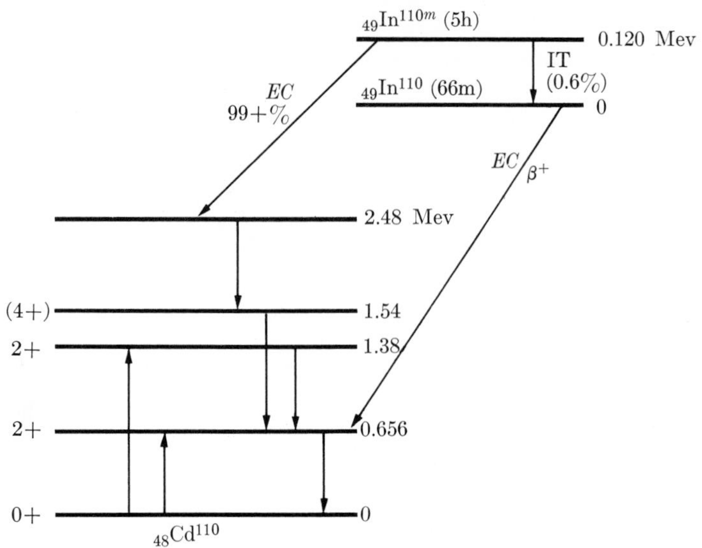

FIG. 16–13. Low-lying energy states of In^{110m} showing data obtained by Coulomb excitation. [Strominger, Hollander, and Seaborg, *Revs. Mod. Phys.* **30,** 585 (1958).]

The cross section for electric excitation is small (about 10^{-29} or 10^{-28} cm^2) compared with those for direct nuclear transformations, and can be observed only when the latter are extremely improbable, as in collisions of protons or α-particles of intermediate energy with intermediate or heavy nuclei. Because of the large Coulomb forces, the classical distance of closest approach (cf. Chapter 3) is then large compared to the nuclear radius and the probability of direct nuclear reaction is very small.

Coulomb excitation has been studied in many nuclei, and the theory of the process has been developed extensively.(35) The nuclear states most strongly produced in these reactions are the low-lying states (1 Mev or lower) induced by the electric quadrupole field of the incident particles, and the γ-rays emitted in this case are of the $E2$ type. The Coulomb excitation reaction is, therefore, a useful tool for the study of low-lying states; new information has been obtained for previously known levels, and many new levels have been identified. A decay scheme including data obtained by means of Coulomb excitation is shown in Fig. 16–13; the arrows pointing upward represent the transitions excited in the reaction.

16–7 Limitations of the compound nucleus theory. The information concerning nuclear reactions discussed in Chapter 11 and so far in this chapter is qualitatively in harmony with the theory of the compound nucleus. The existence of sharply defined resonances (in agreement with theory based on Bohr's ideas) shows that the assumption that the mode of disintegration of the compound nucleus is independent of the way in which the latter is formed is valid within a certain limited region of excitation energy. When the energy of this incident particle is within the region of sharp and well-defined resonances and if the energy coincides with or is near to a resonance, the independence assumption is justified. The nuclear reaction then produces only one quantum state of the compound nucleus, and the properties of a given quantum state are independent of the way it is produced. But, when the energy of the incident particle is high, several states of the compound nucleus may be excited. The uncertainty ΔE [cf. Eq. (16–6)] may be large enough to overlap several states, and their relative excitation may depend on how the nucleus is excited. The decay of these states may then depend on the mode of excitation, and the independence assumption requires further testing.

At these higher energies the independence hypothesis is hard to test quantitatively, in a direct and general way, because it is not easy to reach the same range of excitation energy in the compound nucleus with different reactions. Ghoshal's experiments, discussed in the last section, form one of a few such tests. The results show that the independence assumption seems to be valid for a compound nucleus with an excitation energy of 15 to 40 Mev when produced by protons on Cu^{63} or by α-particles on Ni^{60}. There are, however, reactions for which the independence hypothesis seems to break down in that the decay probabilities of the compound nucleus depend upon the way in which the compound nucleus is formed. For example, a comparison has been made[36] of the relative probabilities of emission of protons and of neutrons from compound nuclei formed either with protons or with neutrons. When nuclei with mass numbers in the range 48–71 were bombarded with 21-Mev protons and 14-Mev neutrons, proton emission was found to be more probable than neutron emission when the reaction was initiated by protons.

There is also indirect evidence against the independence assumption.[8] According to Bohr's picture of nuclear reactions, independence is related to the sharing of the excitation energy by all the particles in the compound nucleus; the direction of incidence or the position in the nucleus of the particle initiating the reaction should have no effect on the decay of the compound nucleus. But reaction products are sometimes emitted with much greater energy than would be expected if the energy of the incoming particle were shared among all the constituents of the nucleus; and the reaction products often show an angular distribution with more of them in the direction of the incident particle than would be expected. In these  cases the compound nucleus seems to have more memory of the initial process than is compatible with the independence hypothesis. Certain experimental results have also led to a further examination of the assump-

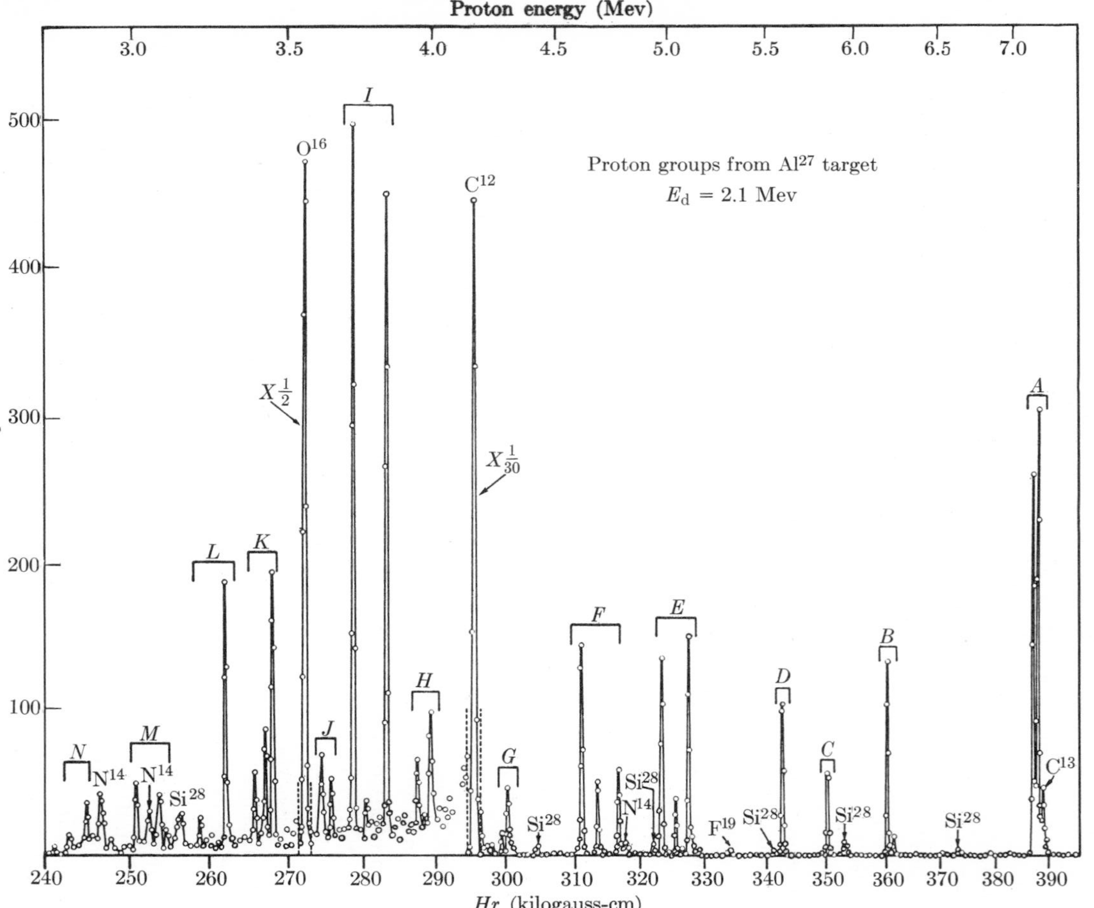

FIG. 16–14. Proton groups from the reaction $Al^{27}(d, p)Al^{28}$, in which aluminum was bom-

tion that the compound nucleus is formed *immediately* after the incident particle enters the nucleus, and it has been found that this assumption must be changed. At excitation energies above the region of sharply defined resonances, it has therefore become necessary to introduce a modified picture of nuclear reactions. This treatment will be discussed in Chapter 17 after the subject of nuclear models has been introduced.

There are additional limitations of the compound nucleus hypothesis. Bohr, in his original statement of the hypothesis, pointed out that his two-step analysis of nuclear reactions would not apply to light nuclei; there are too few particles in such nuclei to make possible the large number of energy exchanges needed for the independence hypothesis to be valid. It will also be seen in later sections of this chapter that certain reactions initiated by deuterons are not adequately described in terms of Bohr's hypothesis, and that some photon-induced reactions may occur through a mechanism other than the formation of a compound nucleus.

16-8 Deuteron-induced reactions: intermediate nuclei, high energy. Reactions induced by deuterons have been a fruitful source of information about the energy levels of nuclei. The product particles, protons, neutrons, or α-particles, often appear in groups with different energies and are related to the low-lying excited levels of the product nucleus. Up to now, the most important results from the standpoint of nuclear spectroscopy have been obtained with the light and lighter-intermediate nuclides in the high-energy region. Because of the special nature of the light nuclei, the discussion of deuteron-induced reactions will be limited to intermediate nuclei and the high energy region III.

The reaction $Al^{27}(d, p)Al^{28}$ is an example of the way in which deuterons have been used in the study of nuclear levels. In one experiment,[(37)] an aluminum target was bombarded with deuterons accelerated in an electrostatic generator. The target was placed between the pole pieces of a 180° magnetic spectrograph, and charged particles leaving the target at 90° with respect to the incident beam were analyzed in the spectrograph. Nuclear emulsion plates were used for recording the charged particles, and proton, deuteron, and α-particle tracks were recognized by the track lengths. The proton tracks were counted under a microscope in small strips of plate material and the number of tracks in each strip was plotted against the product Hr, where H is the magnetic field strength and r is the radius of the path of the protons in the magnetic field. A sample of the results is shown in Fig. 16–14 for incident deuterons with an energy of 2.1 Mev; the proton peaks that have been assigned to the $Al^{27}(d, p)Al^{28}$ reaction are designated by capital letters. Other peaks caused by the presence of impurities were also observed. These impurities were known and their peaks were identified by comparison with the results obtained with other target materials. Fifty different proton groups were found corresponding to the ground state and 49 excited states of Al^{28}. An energy level diagram of these states is shown in Fig. 16–15. These states decay by γ-emission and are bound states of Al^{28}; together with the virtual levels listed in Table 16–3, they indicate how the spectrum of a lighter-intermediate nucleus looks. In a later experiment,[(38)] 7.0-Mev deuterons were used as well as a magnetic spectrograph with greater resolution; 100 proton groups were resolved with nearly all of the new groups having energies lower than those shown in Fig. 16–14. The new proton groups correspond to energy levels of Al^{28} about 6 Mev or more above the ground state.

The γ-rays resulting from the decay of excited states of Al^{28} could not be studied under the conditions of the Al^{27}(d, p)Al^{28} reaction discussed above. They have been observed in studies of the γ-rays from excited states of Al^{28} formed as a result of neutron capture in Al^{27}. When the captured neutron is a slow one, its kinetic energy is negligible compared with its binding energy in the compound nucleus, and the binding and excitation energies are practically identical. The compound nucleus, in this case [Al^{28}], decays by γ-emission through excited states of the product nucleus Al^{28}, and the energies of the γ-rays give information about these excited states. About 30 γ-rays have been observed, and their energies measured with a pair spectrometer.[(39)]

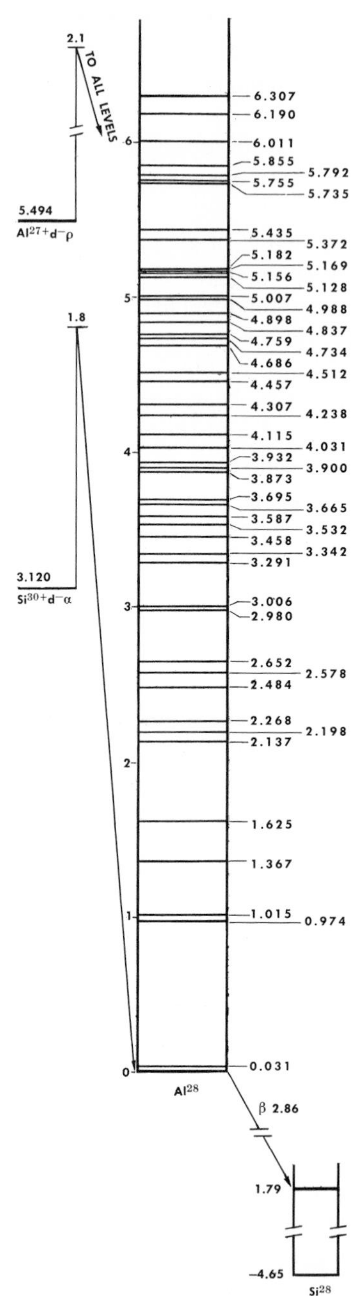

Fig. 16–15. Energy-level diagram for Al^{28} (Enge *et al.*[(37)]).

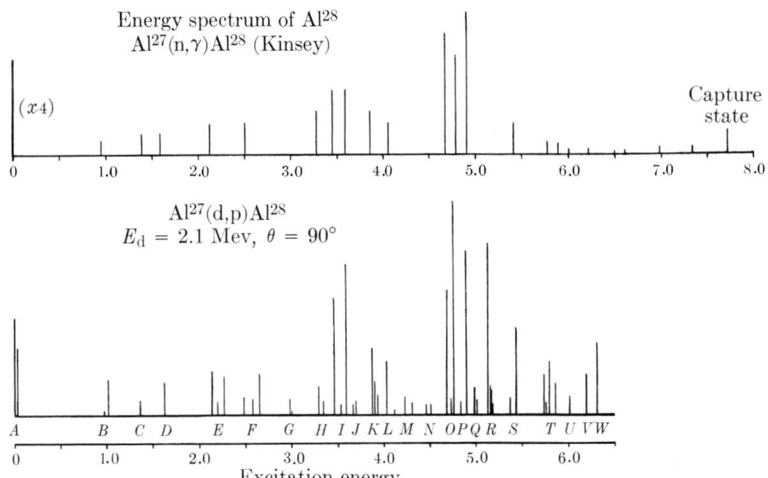

FIG. 16–16. Energy-level diagram for Al^{28} with relative intensities (Enge *et al.*(37)).

The most energetic γ-ray had an energy of 7.72 Mev, which is just the binding energy of the captured neutron, and there can be little doubt that this γ-ray represents the direct transition from the capture state to the ground state. The differences in energy between this γ-ray and those with less energy can be related to bound states of Al^{28}. The states inferred in this way are compared in Fig. 16–16 with those given by the (d, p) reaction. The upper part of the diagram was constructed from (n, γ) studies, and the height of the lines represents the relative intensities of the γ-rays. The lower diagram shows the states as deduced from the (d, p) studies and indicates the relative proton yields. The levels given by the two methods are in good agreement, and the maxima and minima in the relative intensities have the same locations in the two diagrams.

The (d, α) reaction has been used in a similar way as, for example, in the reaction Al^{27}(d, α)Mg^{25}. Excited states of Mg^{25} can also be obtained from the reaction Mg^{24}(d, p)Mg^{25} and the results of a study(40) of the two reactions are listed in Table 16–6. The levels of Mg^{25} obtained from the two reactions agree very well.

The (d, n) reaction often yields groups of neutrons whose energies lead to levels of the product nucleus. An example(41) is the reaction Mg^{26}(d, n)Al^{27}, which gives excited states of Al^{27} at 0.88 ± 0.07, 1.92 ± 0.07, 2.75 ± 0.07, 3.65 ± 0.07, 4.33 ± 0.07, 5.32 ± 0.07, and 5.81 ± 0.07 Mev. These levels are in good agreement with levels found from the inelastic scattering of protons. The inelastic scattering of deuterons, like that of protons, gives information about the levels of the target

TABLE 16–6

ENERGY LEVELS IN Mg^{25}*

Mg^{24}(d, p)Mg^{25}	Al^{27}(d, α)Mg^{25}
0.582 ± 0.006	0.584 ± 0.006
0.976 ± 0.006	0.977 ± 0.010
1.612 ± 0.006	1.610 ± 0.010
1.957 ± 0.006	1.958 ± 0.010
2.565 ± 0.006	2.558 ± 0.010
2.742 ± 0.008	2.729 ± 0.010
2.806 ± 0.007	2.791 ± 0.015
3.405 + 0.007	3.404 ± 0.012
3.899 + 0.008	3.896 ± 0.015
3.972 ± 0.010	3.965 ± 0.015

*Excitation energy in Mev.

nucleus. Levels of Al^{27} have been found in this way[(42)] with excitation energies of 0.97, 2.39, 3.17, 4.74, and 5.76 Mev. The experimental error in this case was about 0.2 Mev, and within the limits of this error, the levels agree with some of those obtained from the inelastic scattering of protons by Al^{27}, and from the (d, p) reaction on Mg^{26}. Finally, the reaction Si^{21}(d, α)Al^{27} gives levels at 0.84 and 1.01 Mev,[(43)] in agreement with the two lowest levels obtained by the three other methods.

The theoretical treatment of deuteron-induced reactions is more difficult than the treatment of the reactions discussed earlier in this chapter. One reason is that new types of processes occur with deuterons which cannot be accounted for on the basis of the compound nucleus picture. These processes result from the following properties of the deuteron: (a) the deuteron is a very loosely bound structure with a binding energy of only 2.23 Mev as compared with 28.3 Mev for the α-particle and 7 or 8 Mev for the average particle in a nucleus; (b) its charge distribution is very unsymmetric, i.e., the center of mass and the center of charge (the proton) do not coincide as they do in the α-particle.

According to the compound nucleus theory, a nucleus $X(Z, A)$ bombarded by deuterons should form the compound nucleus $[C(Z + 1, A + 2)]$, which can then decay in different ways:

$$X(Z, A) + \mathrm{d} \rightarrow [C(Z + 1, A + 2)] \rightarrow \begin{cases} Y(Z + 1, A + 1) + \mathrm{n}, \\ Y(Z, A + 1) + \mathrm{p}, \\ Y(Z - 1, A - 2) + \alpha. \end{cases} \tag{16–26}$$

The special properties of the deuteron make it possible for other processes to take place in addition to those of the scheme (16–26). Because of the finite distance between the proton and neutron in the deuteron one of these particles may reach the nuclear surface before the other. The nuclear interaction energies (average binding energy per nucleon) are much higher than the binding energy of the deuteron, and the constituent of the deuteron that arrives first at the nuclear surface is quickly separated from its partner. If the kinetic energy of the deuteron is not very high, the Coulomb field of the nucleus keeps the proton away and the neutron is more likely to enter the nucleus than the proton. If the second constituent of the deuteron hits the nucleus an instant later, the compound nucleus $[C(Z + 1, A + 2)]$ is formed just as in the scheme (16–26). But if the second nucleon (proton) misses the nucleus, the process that takes place is

$$X(Z, A) + \mathrm{d} \rightarrow [C'(Z, A + 1)] + \mathrm{p}. \tag{16–27}$$

The compound nucleus $[\mathrm{C}'(Z, A + 1)]$ may decay by emission of some other particles, or by γ-decay. If the proton penetrates the Coulomb barrier and hits the target nucleus first, the process may be

$$X(Z, A) + \mathrm{d} \rightarrow [C''(Z + 1, A + 1)] + \mathrm{n}. \tag{16–28}$$

The processes (16–27) and (16–28) are called *stripping reactions* because a nucleon is stripped out of the deuteron by the target nucleus. At relatively low energies, the reaction (16–27) is called the "Oppenheimer-Phillips process" after the two workers who first described it in 1935.

The additional processes which can take place with deuterons make the cross sections for these reactions greater than those for reactions with other charged particles of similar energy. As a result, deuteron reactions are of great practical importance in the production of radioactive nuclides and in the study of nuclear reactions. The characteristics of stripping reactions are closely related to the properties of the state of the product nucleus, and the investigation of these reactions has been of special use in nuclear spectroscopy.[44,59]

16–9 Reactions induced by gamma-rays. The disintegration of nuclei by γ-rays occurs when a photon is absorbed whose energy $h\nu$ is greater than the separation energy of a particle (proton, neutron, or α-particle). When this condition is satisfied, a (γ, p), (γ, n), or (γ, α) reaction is observed. A process of this type is often called a *photodisintegration;* the term *nuclear photoeffect* is sometimes used in analogy with the atomic photoelectric effect.

The simplest example of this type of reaction is the photodisintegration of the deuteron. This reaction has a threshold at a γ-ray energy just equal to the binding energy of the deuteron, and very careful determinations of the threshold energy have been made because of the great importance of the binding energy of the deuteron in the theory of the nucleus. In a highly precise measurement[(45)] of the threshold, deuterium was bombarded with x-rays from the action of electrons on a gold target; the electrons from a cathode-ray "gun" were accelerated in an electrostatic generator. In the neighborhood of 2 Mev, the energy of the electrons is practically identical with that of the most energetic x-rays, and if the electron energy is measured very carefully, that of the x-rays is known with high precision. The neutrons resulting from the reaction $d + \gamma \rightarrow p + n$ were counted as a function of the electron energy, and the *photoneutron* threshold was obtained by extrapolating the neutron yield versus energy curve to zero yield. The results of the experiment are shown in Fig. 16–17. The open circles give the neutron count; the dots represent the square root of the difference between the total neutron count and the background count. According to theory, this quantity should be proportional to the energy above threshold. The straight lines show that this is the case and allow an accurate extrapolation to zero yield. The binding energy of the deuteron was determined as 2.226 ± 0.003 Mev. The shape of the yield curve is typical for photodisintegration reactions at energies just above the threshold.

In the experiment just described, the threshold for the photodisintegration of beryllium was also determined. The reaction in this case is

$$_4\text{Be}^9 + \gamma \rightarrow {}_2\text{He}^4 + {}_2\text{He}^4 + \text{n},$$

and the threshold energy gives the binding energy of the neutron in the Be^9 nucleus as 1.666 ± 0.002 Mev.

Photoneutron thresholds have been measured for many (γ, n) reactions[(46)] over the entire range of elements, chiefly to get neutron binding energies. The measurement of neutron binding energies is one method of establishing mass differences with great accuracy and also provides information about nuclear structure. The neutron whose binding energy is determined is the "last" neutron, that is, the one removed from a stable nucleus or added to a stable nucleus. The energy needed to remove a neutron from a stable nucleus is given by the threshold of the (γ, n) or (n, 2n) reaction, or by the Q-value of the ground-state group from a (p, d) reaction or a (d, H^3) reaction. These reactions furnish neutron binding data for the target nucleus. The energy released on the addition of a neutron to a stable nucleus is given by the γ-rays from the capture of slow neutrons (n, γ) or can be obtained from the Q-value of the ground-state proton group from a (d, p) reaction.[(47)]

With the exception of deuterium and beryllium, the binding energy of the last neutron lies between about 5 and 13 Mev. The experimental values of the binding energy have been tabulated and analyzed as a function of the number of neutrons in the nucleus.[(47)] It was found that

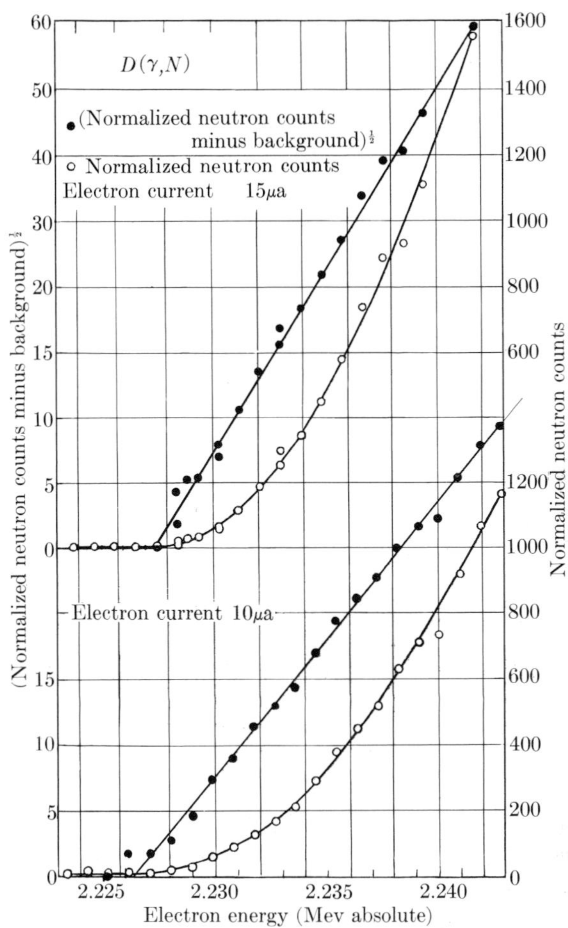

Fig. 16–17. The photoneutron yield from deuterium near threshold (Mobley and Laubenstein[45]).

nuclei with 28, 50, 82, and 126 neutrons have special properties; in each case there is a sharp break in the binding energy of the neutron added to the nucleus. Thus the binding energy of the 127th neutron is about 2.2 Mev less than that of the 126th neutron. The binding energy of the 50th neutron is about 2 Mev greater than that of the 51st neutron; similarly, there is a break at 82 neutrons and probably one at 28 neutrons. These results indicate that the nuclei with 28, 50, 82, and 126 neutrons are markedly more stable than their neighbors. In Section 16–5C, the special nature of nuclei with 50, 82, and 126 neutrons was shown by the low values of their capture cross sections for 1-Mev neutrons. A theory of nuclear structure which accounts for the particularly stable nature of these nuclei will be discussed in the next chapter. These results provide another example of the way in which the study of nuclear reactions is related to the problem of nuclear structure.

The (γ, n) cross section has been measured as a function of the energy of the γ-rays[(48)] for a series of elements covering the periodic table. In the intermediate and heavy nuclei, the cross section increases quite rapidly above the threshold energy, reaching a maximum somewhere near 20 Mev. The maximum value of the cross section increases with increasing mass number; it is of the order of 10^{-26} cm^2 for the lighter-intermediate nuclei such as aluminum, and reaches a value of nearly 10^{-24} cm^2 in bismuth.

Studies of (γ, p) reactions[(49)] show that they are similar in nature to the (γ, n) reactions in that the cross section versus energy curves have the same general shape. An interesting theoretical problem arises because of the relative magnitudes of (γ, n) and (γ, p) cross sections. In reactions involving the intermediate and heavy nuclei, it seems natural to assume that the Bohr theory of the compound nucleus should apply, and that the process is divided into two parts: (1) the absorption of the γ-ray photon forming a compound nucleus with the excitation energy $E = h\nu$, and (2) the decay of the compound nucleus by the emission of any one of several different particles. Because of the blocking effect of the Coulomb barrier, the most probable process is the emission of a neutron leading to a (γ, n) reaction; protons are expected to be emitted much more rarely. In many cases, the relative yield of (γ, n) and (γ, p) reactions does not agree with these ideas, and proton emission has been found to be much stronger than expected. These unexpectedly large (γ, p) cross sections have been interpreted in terms of a direct photodisintegration process.[(50)] It is supposed that the energy of the γ-ray photon is absorbed by a single proton in the nucleus which is then emitted before the energy can be shared with the other nucleons. Theoretical calculations show that only a small number of protons would have to be ejected by this direct process to give rise to a (γ, p) cross section appreciably larger than the one expected from the compound nucleus alone, and there seems to be evidence supporting the idea of a direct photoproton effect.

At γ-ray energies in the neighborhood of 20 Mev, the (γ, 2n) and (γ, np) reactions are observed and their cross sections increase with increasing energy, while the (γ, n) and (γ, p) cross sections decrease.

In contrast with (γ, p) reactions, (γ, α) reactions can be interpreted adequately at present in terms of the compound nucleus theory, as is shown, for example, by recent experiments[(51)] on the reaction $V^{51}(\gamma, \alpha)Sc^{47}$. A sample of vanadium was bombarded with γ-rays with energies which could be varied from 10.5 to 25 Mev. The yield was determined as a function of γ-ray energy by measuring the activity of the Sc^{47} produced; the ntensity of the 160-kev γ-ray characteristic of Sc^{47} was determined with

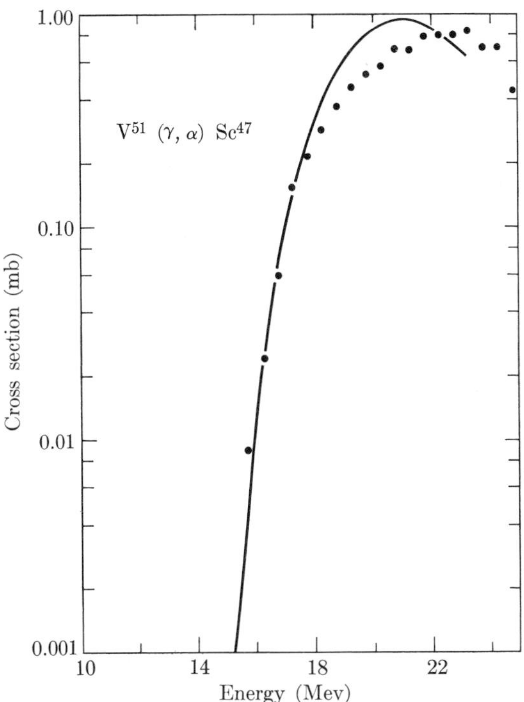

FIG. 16–18. Excitation function for the reaction $V^{51}(\gamma, \alpha)Sc^{47}$.(51)

a NaI(Tl) scintillator and a pulse-height analyzer. Vanadium consists almost entirely (99.75%) of V^{51}, and the products of the (γ, n) and (γ, p) reactions on that nuclide are stable. Hence those reactions, which are more probable than the (γ, α) reaction, do not interfere with the analysis of the latter. The (γ, α) reaction was not observed below 15.5 Mev because of the effect of the Coulomb barrier on the emission of α-particles. The excitation function (curve of yield or cross section against energy) has a maximum of 0.8 millibarns at 23 Mev and agrees with the theoretical curve based on the compound nucleus theory, as shown in Fig. 16–18. This result seems to be a general property of (γ, α) reactions in medium-weight nuclei.

It is not surprising that the idea of the compound nucleus applies to these reactions; α-particles probably do not exist as such in the nucleus so that direct ejection of an α-particle is very unlikely. It is reasonable to suppose that the excitation energy brought by the incident γ-ray is shared among many particles and that, under favorable conditions, four nucleons *fuse* to form an α-particle in a kind of internal conversion process. The emission of the α-particle from the excited compound nucleus is then analogous to the emission of α-particles from radioactive nuclei.

16–10 Reactions at ultra-high energies. At energies of about 50 Mev and up, nuclear reactions are quite different in character from those at lower energies. When a particle with ultra-high energy strikes a nucleus, it may be absorbed to form a very highly excited compound nucleus. It is more likely, however, that the particle will collide with one or more nucleons and then escape with part of its original kinetic energy. By means of this scattering process, one or a few nucleons may get a large fraction of the energy of the incident particle. The latter nucleon or group of nucleons may be ejected from the nucleus without further collision, or they may share their energy with a few other nucleons. In most collisions with ultra-high energy particles, the target nuclei are excited to energies far greater than the binding energies of individual nucleons. Nucleons or groups of nucleons such as α-particles or lithium nuclei escape until the residual nucleus is left in a relatively stable state. At energies of 100 Mev or more, a dozen or more nucleons may escape and many combinations of protons and neutrons have been observed. Reactions of this kind are described by the term *spallation*. The theory of the compound nucleus does not apply to these reactions because there is not enough time for the excitation energy to be shared among all the nucleons before particles start being emitted. It is no longer possible to separate the reaction into two independent phases and it is necessary to treat separately the interaction of the incoming particle with the individual nucleons of the target nucleus. The region of ultra-high energy also includes reactions induced by cosmic rays, including the production of new kinds of particles called *mesons*. The complexity of both the experimental information and the theoretical ideas is such that further discussion of ultra-high energy nuclear physics is beyond the scope of this book. The interested reader is referred to the literature(52,53) for relatively easy treatments of these subjects.

16–11 Reactions with light nuclei: energy levels of light nuclei. A great deal of information about the excited states of light nuclei has now been accumulated from nuclear reactions. Some of the most recent compilations(54,57) include more than 400 energy levels of the 40-odd nuclei of atomic number less than 11; another(25) includes many levels of nuclei with Z between 11 and 20. The progress made in recent years in the methods of studying excited states, together with the establishment of an accurate and consistent set of nuclear masses (Table 11–1), has made it possible to locate these states with high precision. In the majority of the nuclei in this region, the first 10 Mev of excitation have been studied thoroughly enough so that most of the stationary states are probably known. Although the reactions do not fall into the kind of patterns that were found for intermediate and heavy nuclei, the energy levels provide a body of experimental data in which important clues to the problems of nuclear forces and nuclear structure can be found. In the discussion of light nuclei, therefore, the emphasis will be on the levels rather than on the reactions. With a few exceptions, the reactions are of the types which have already been discussed, and the treatment of the light nuclei in this section can be regarded as a further application of the methods and ideas which have been developed in the earlier sections of this chapter.

The energy levels of the light nuclei are sufficiently well defined and separated so that it is possible to present the level structure for each nucleus in the form of a diagram showing the known levels and indicating the nuclear reactions in which they are involved. In addition to the position of the level, other information is also desired in nuclear spectroscopy, such as the width or lifetime of an excited state. Two additional parameters that are important in the description of a nuclear state are the angular momentum quantum number and the parity. The latter quantities can really be understood only in terms of quantum mechanics, and their effect on nuclear reactions has therefore not been mentioned in this chapter. They influence the angular distribution of the reaction products, and information about them can be obtained from the way in which the yield of the product particle varies with the angle between the direction of emission of the product particle and the direction of the incident particle. Information about the angular momentum and parity can also be obtained from the probability of β-decay and that of γ-decay. When the values of these parameters for a state are known, they are also shown in the level diagram.

The way in which the energy-level diagram is constructed can be illustrated by treating a typical nucleus in some detail. The nucleus N^{15} has been chosen because of the large number of reactions in which it is either the compound nucleus or the product nucleus. All of the data which will be considered have been taken from the compilation by Ajzenberg and Lauritsen referred to above,[(57)] where the references to the original sources are given. The energy level diagram for N^{15} is shown in Fig. 16–19. Although it appears formidable at first sight, it will be seen to provide a concise and easily interpreted summary of the available data.

Altogether, 25 reactions among various light nuclei provide information about the excited states of N^{15}. These are: $B^{11}(\alpha, n)N^{14}$, $B^{11}(\alpha, p)C^{14}$, $C^{12}(\alpha, p)N^{15}$, $C^{12}(t, p)C^{14}$, $C^{13}(d, n)N^{14}$, $C^{13}(d, p)C^{14}$, $C^{13}(d, \alpha)B^{11}$, $C^{13}(d, t)C^{12}$, $C^{14}(d, n)N^{15}$, $C^{14}(p, n)N^{14}$, $C^{15}(\beta^-)N^{15}$, $N^{14}(n, n)N^{14}$, $N^{14}(n, \alpha)B^{11}$, $N^{14}(n, p)C^{14}$, $N^{14}(n, \gamma)N^{15}$, $N^{14}(n, 2n)N^{13}$, $N^{14}(n, t)C^{12}$, $N^{14}(d, p)N^{15}$, $N^{14}(t, pn)N^{15}$, $O^{15}(\beta^+)N^{15}$, $O^{18}(p, \alpha)N^{15}$, $C^{14}(p, \gamma)N^{15}$, $O^{16}(\gamma, p)N^{15}$, $O^{16}(n, d)N^{15}$, and $O^{17}(d, \alpha)N^{15}$. In 12 reactions, N^{15} is the product nucleus, and in 13 it is the compound nucleus. Low-lying levels (bound states) are given by the proton groups from the reactions

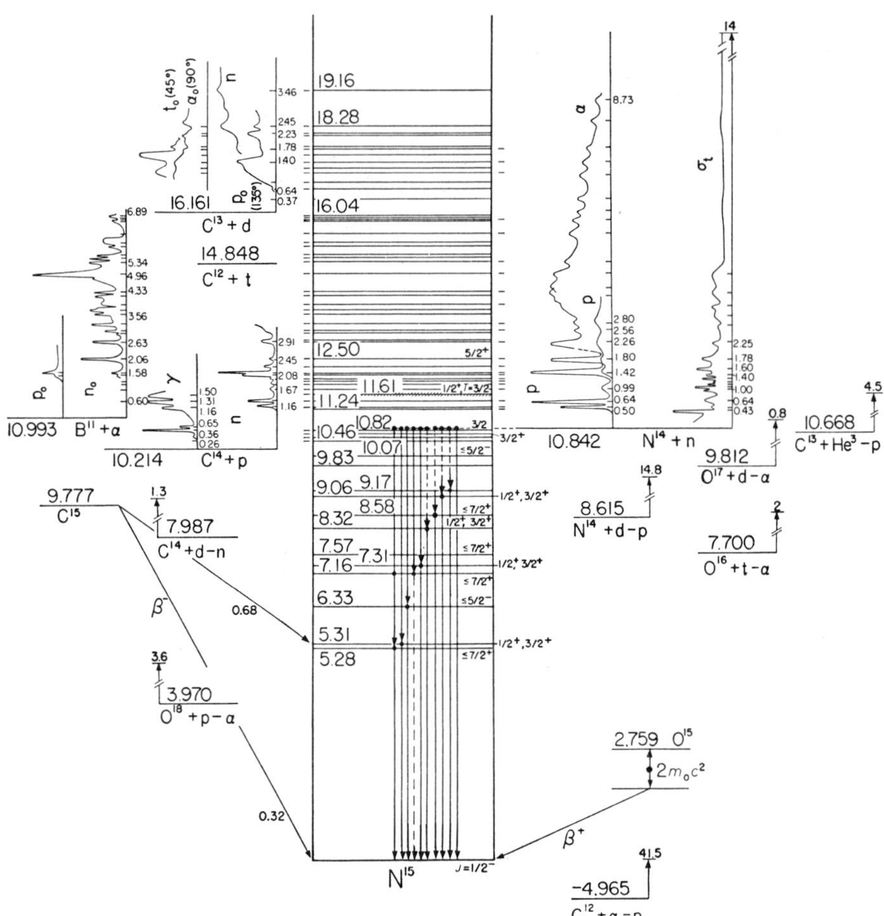

Fig. 16–19. Energy levels of N^{15} (Ajzenberg-Selove and Lauritsen[(57)]).

N^{14}(d, p)N^{15} and $C^{12}(\alpha$, p)N^{15}, by the neutron groups from the reaction C^{14}(d, n)N^{15}, and by the γ-rays from the reaction $N^{14}(n, \gamma)N^{15}$. These levels are listed in Table 16–7. Although no single reaction gives all the levels, there is good agreement among the values obtained from the different reactions. The middle column of Fig. 16–19 shows the energy states of N^{15}. The numbers indicate the energy in Mev above the ground state of N^{15}, which serves as the zero of the energy scale for the whole diagram. The N^{15} spectrum is remarkable for the rather great excitation (5.28 Mev) of the lowest excited level and for the fact that this level is a doublet with a spacing of 30 kev. The angular momenta are indicated by the J-values as obtained from the analysis of the angular distribution of

Table 16–7

Low-Lying Levels of N^{15} from Various Reactions*

$N^{14}(d, p)N^{15}$ proton groups	$N^{14}(n, \gamma)N^{15}$ γ-rays	$C^{12}(\alpha, p)N^{15}$ proton groups	$C^{14}(d, n)N^{15}$ neutron groups
5.276	5.275		
5.305		5.3	5.34
6.328	6.325	6.5	6.32
7.164	7.164		
7.309	7.356		7.46
8.315	8.278		
	9.156		

* Excitation energy in Mev.

Table 16–8

Resonance Levels in N^{15} in Various Reactions*

$C^{14}(p, n)N^{14}$	$N^{14}(n, p)C^{14}$	$N^{14}(n, n)N^{14}$	$N^{14}(n, \alpha)B^{11}$	$C^{13}(d, n)N^{14}$	$B^{11}(\alpha, n)N^{14}$
		11.235			
11.294	11.30	11.293			
11.430	11.431	11.430			
11.760	11.761	11.765			
11.877		11.879			
		11.943			
11.964		11.964			
12.096		12.094			
12.147	12.155	12.142	12.16		
12.328		12.323			12.38
12.495	12.47	12.494			
	12.94		12.94		12.95
			13.22		
	13.45		13.40		13.34
					13.63
					14.05
			14.26		14.29
					14.50
			14.63		14.71
			14.90		
			15.62		
			15.89		
			16.13		
			16.47	16.65	
			17.04	16.89	
			17.30	17.49	
			17.78	17.69	
			18.43		
			18.97		

* Excitation energy in Mev.

the protons from the reaction $N^{14}(d, p)N^{15}$. When two numbers are given for the angular momentum, it means that the assignment is not yet certain but the choice has been narrowed down to the values given. Even parity is indicated by a plus sign, odd parity by a minus sign. The reactions which give the low-lying levels are indicated at either side of the central column, together with the binding energy of the incident particle in the compound nucleus. For example, the reaction $N^{14}(d, p)N^{15}$ is shown to the left of the central column together with the Q-value for the reaction, which is 8.609 Mev. The formation of N^{15} by the β^+-decay of O^{15} and the β^--decay of C^{15} are also shown.

Virtual levels of N^{15}, as given by six different reactions, are listed in Table 16–8. About 30 levels, obtained from resonances in these reactions, are shown in the upper part of the central column of Fig. 16–19. The reactions which give rise to these resonances are shown on the sides of the diagram, together with the binding energy of the incident particle in the compound nucleus and a sketch of the cross section curve for each reaction. Thus, the reactions $N^{14}(n, n)N^{14}$, $N^{14}(n, p)C^{14}$, and $N^{14}(n, \alpha)B^{11}$ are indicated to the left of the upper part of the central column. The binding energy of the last neutron in $[N^{15}]$ is 10.834 Mev and this value serves as

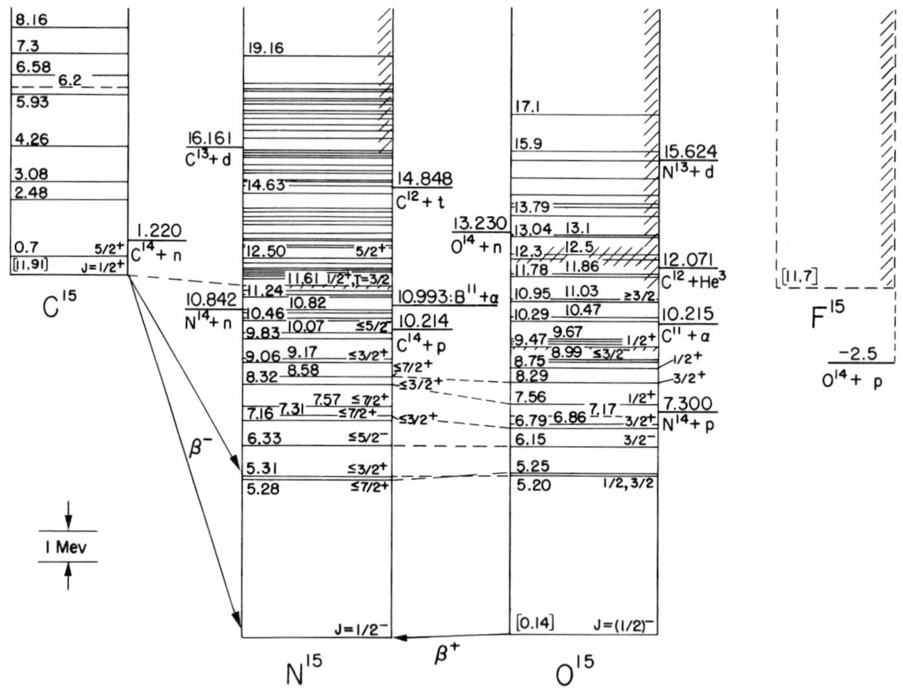

FIG. 16–20. Energy levels of N^{15} and O^{15}.(58)

a zero point for the energies of the particles in the various neutron-induced reactions. The energy of the excited level is obtained from the binding energy and the resonance energy as shown in Section 16–2. The values of the resonance levels as given by the different reactions are in excellent agreement. The levels which are cross-hatched are known to be particularly broad, that is, to have a particularly large Γ.

The comparison of the energy level diagrams of the light nuclei shows that there are similarities in some of the level spectra which have a bearing on the problem of nuclear forces.[55,56,58] It has been found, for example, that the distribution of low-lying levels in mirror nuclei is similar. In Fig. 16–20, these levels[51] are shown for N^{15} and O^{15}. The former nucleus has 7 protons and 8 neutrons, while the latter has 8 protons and 7 neutrons. Results like these provide evidence on the relative magnitudes of the force between two protons and that between two neutrons and, in fact, support the hypothesis that these two forces are equal.

References

GENERAL

R. D. Evans, *The Atomic Nucleus.* New York: McGraw-Hill, Chapters 12, 13, 14.

P. Morrison, "A Survey of Nuclear Reactions," *Experimental Nuclear Physics,* E. Segrè, ed. New York: Wiley, 1953, Vol. II, Part VI.

L. Eisenbud and E. P. Wigner, *Nuclear Structure.* Princeton, New Jersey: Princeton University Press, 1958, Chapters 4, 9, 10.

J. M. Blatt and V. F. Weisskopf, *Theoretical Nuclear Physics.* New York: Wiley, 1952, Chapters 8, 9, 10.

P. M. Endt and M. Demeur, eds. *Nuclear Reactions.* Amsterdam: North-Holland Publishing Co.; New York: Interscience Publishers, 1959.

F. Ajzenberg-Selove, ed., *Nuclear Spectroscopy.* New York: Academic Press, 1960.

ARTICLES IN "*The Encyclopedia of Physics*" (*Handbuch der Physik*) Berlin: Springer, 1957–1959.

1. W. E. Burcham, "Nuclear Reactions, Levels, and Spectra of Light Nuclei," Vol. 40, *Nuclear Reactions I,* pp. 1–201.
2. B. B. Kinsey, "Nuclear Reactions, Levels, and Spectra of Heavy Nuclei," Vol. 40, *Nuclear Reactions I,* pp. 202–372.
3. J. Rainwater, "Resonance Processes by Neutrons," Vol. 40, *Nuclear Reactions I,* pp. 373–409.
4. A. Wattenberg, "Nuclear Reactions at High Energies," Vol. 40, *Nuclear Reactions I,* pp. 450–537.
5. G. Breit, "Theory of Resonance Reactions and Allied Topics," Vol. 41, *Nuclear Reactions II,* pp. 1–407.
6. G. R. Bishop and R. Wilson, "The Nuclear Photoeffect," Vol. 42, *Nuclear Reactions III,* pp. 309–361.

COLLECTIONS OF ENERGY LEVEL DATA

1. F. AJZENBERG and T. LAURITSEN, "Energy Levels of Light Nuclei, IV," *Revs. Mod. Phys.* **24,** 321 (1952).

2. F. AJZENBERG and T. LAURITSEN, "Energy Levels of Light Nuclei, V," *Revs. Mod. Phys.* **27,** 77 (1955).

3. F. AJZENBERG-SELOVE and T. LAURITSEN, "Energy Levels of Light Nuclei," *Nuclear Physics* **11,** 1–340 (1959).

4. P. M. ENDT and J. C. KLYVER, "Energy Levels of Light Nuclei (Z = 11 to Z = 20)," *Revs. Mod. Phys.* **26,** 95 (1954).

5. P. M. ENDT and C. M. BRAAMS, "Energy Levels of Light Nuclei (Z = 11 to Z = 20), II," *Revs. Mod. Phys.* **29,** 683 (1957).

6. WAY, KING, MCGINNIS, and VAN LIESHOUT, *Nuclear Energy Levels (A = 40 to A = 92; Calcium to Zirconium).* U.S. Atomic Energy Commission Document TID-5300. Washington, D.C.: U.S. Government Printing Office, 1955.

7. STROMINGER, HOLLANDER and SEABORG, "Table of Isotopes," *Revs. Mod. Phys.* **30,** 585 (1958).

PARTICULAR

1. CHADWICK, CONSTABLE, and POLLARD, "Artificial Disintegration by α-Particles," *Proc. Roy. Soc.* (London) **A130,** 463 (1931).

2. J. CHADWICK and J. E. R. CONSTABLE, "Artificial Disintegration by α-Particles. Part II, Fluorine and Aluminum," *Proc. Roy. Soc.* (London) **A135,** 48 (1932).

3. W. E. DUNCANSON and H. MILLER, "Artificial Disintegration by RaC′ α-Particles: Aluminum and Magnesium," *Proc. Roy. Soc.* (London) **A146,** 396 (1934).

4. B. B. BENSON, "Proton-Gamma Ray Coincidence Counting with Cyclotron Bombardment," *Phys. Rev.* **73,** 7 (1948).

5. M. S. LIVINGSTON and H. A. BETHE, "Nuclear Physics. C. Nuclear Dynamics, Experimental," *Revs. Mod. Phys.* **9,** 245 (1937).

6. BROLLEY, SAMPSON, and MITCHELL, "Energy Levels of Si^{30}," *Phys. Rev.* **76,** 624 (1949).

7. N. BOHR, "Neutron Capture and Nuclear Constitution," *Nature* **137,** 344 (1936).

8. F. L. FRIEDMAN and V. F. WEISSKOPF, "The Compound Nucleus," *Niels Bohr and The Development of Physics,* W. Pauli, ed. London: Pergamon Press, 1955, pp. 134–161.

9. G. BREIT and E. WIGNER, "Capture of Slow Neutrons," *Phys. Rev.* **49,** 519 (1936).

10. J. M. BLATT and V. F. WEISSKOPF, *Theoretical Nuclear Physics.* New York: Wiley, 1952, Chapter 8.

11. J. M. BLATT and V. F. WEISSKOPF, Ref. 10 above, Chapter 9.

12. D. J. HUGHES, *Pile Neutron Research.* Reading, Mass.: Addison-Wesley, 1953.

13. R. L. HENKEL and H. H. BARSCHALL, "Capture Cross Sections for Fast Neutrons," *Phys. Rev.* **80,** 145 (1950).

14. D. J. Hughes and R. B. Schwartz, "Neutron Cross Sections," *Brookhaven National Laboratory Report BNL-325*, 2nd ed. Washington, D.C.: U.S. Government Printing Office, July 1958.

15. Rainwater, Havens, Wu, and Dunning, "Slow Neutron Velocity Spectrometer Studies, I. Cd, Ag, Sb, Ir, Mn," *Phys. Rev.* **71,** 65 (1947).

16. D. J. Hughes, Ref. 12 above, pp. 109–114.

17. R. Sherr, "Collision Cross Sections for 25-Mev Neutrons," *Phys. Rev.* **68,** 240 (1945).

18. J. M. Blatt and V. F. Weisskopf, Ref. 10 above, pp. 481–483.

19. H. Feshbach and V. F. Weisskopf, "A Schematic Theory of Nuclear Cross Sections," *Phys. Rev.* **76,** 1550 (1949).

20. Seth, Hughes, Zimmerman, and Garth, "Nuclear Radii by Scattering of Low-Energy Neutrons," *Phys. Rev.* **110,** 692 (1958).

21. B. L. Cohen, "(n, 2n) and (n, p) Cross Sections," *Phys. Rev.* **81,** 184 (1951).

22. Brolley, Fowler, and Schlacks, "(n, 2n) Reactions in C^{12}, Cu^{63} and Mo^{92}," *Phys. Rev.* **88,** 618 (1952).

23. Brostrom, Huus, and Tangen, "Gamma-Ray Yield Curve of Aluminum Bombarded with Protons," *Phys. Rev.* **71,** 661 (1947).

24. P. M. Endt and J. C. Klyver, "Energy Levels of Light Nuclei, ($Z = 11$ to $Z = 20$)," *Revs. Mod. Phys.* **26,** 95 (1954).

25. P. M. Endt and C. M. Braams, "Energy Levels of Light Nuclei. II." *Revs. Mod. Phys.* **29,** 683 (1957).

26. Browne, Zimmerman, and Buechner, "Energy Levels in Al^{27}," *Phys. Rev.* **96,** 725 (1954).

27. Richards, Smith, and Browne, "Proton-Neutron Reactions and Thresholds," *Phys. Rev.* **80,** 524 (1950).

28. Schoenfeld, Duborg, Preston, and Goodman, "The Reaction $Cl^{37}(p, n)A^{37}$; Excited States in A^{38}," *Phys. Rev.* **85,** 873 (1952).

29. Lovington, McCue, and Preston, "The (p, n) Reaction on Separated Isotopes of Chromium," *Phys. Rev.* **85,** 585 (1952).

30. M. M. Shapiro, "Cross Sections for the Formation of the Compound Nucleus by Charged Particles," *Phys. Rev.* **90,** 171 (1953).

31. E. L. Kelly and E. Segrè, "Some Excitation Functions of Bismuth," *Phys. Rev.* **75,** 999 (1949).

32. S. N. Ghoshal, "An Experimental Verification of the Theory of the Compound Nucleus," *Phys. Rev.* **80,** 939 (1950).

33. C. L. McClelland and C. Goodman, "Excitation of Heavy Nuclei by the Electric Field of Low-Energy Protons," *Phys. Rev.* **91,** 760 (1953).

34. T. Huus and C. Zupancic, "Excitation of Nuclear Rotational States by the Electric Field of Impinging Particles," *Kgl. Danske Videnskab, Selskab. Mat. fys. Medd.* **28,** No. 1 (1953).

35. Alder, Bohr, Huus, Mottelson, and Winther, "Study of Electromagnetic Excitation with Accelerated Ions," *Revs. Mod. Phys.* **28,** 432 (1956).

36. B. L. Cohen and E. Newman, "(p, pn) and (p, 2n) Cross Sections in Medium Weight Elements," *Phys. Rev.* **99,** 718 (1955).

37. Enge, Buechner, and Sperduto, "Magnetic Analysis of the $Al^{27}(d, p)Al^{28}$ Reaction," *Phys. Rev.* **88,** 961 (1952).

38. BUECHNER, MAZARI, and SPERDUTO, "Magnetic Spectograph Measurements on the Al^{27}(d, p)Al^{28} Reaction," *Phys. Rev.* **101,** 188 (1956).

39. KINSEY, BARTHOLOMEW, and WALKER, "Neutron Capture of γ-Rays from F, Na, Mg, Al, and Si," *Phys. Rev.* **83,** 519 (1951).

40. ENDT, ENGE, HAFFNER, and BUECHNER, "Excited States of Mg^{25} from the Al^{27}(d, α)Mg^{25} and Mg^{24}(d, p)Mg^{25} Reactions," *Phys. Rev.* **87,** 27 (1952).

41. SWANN, MANDEVILLE, and WHITEHEAD, "The Neutrons from the Disintegration of the Separated Isotopes of Magnesium by Deuterons," *Phys. Rev.* **79,** 598 (1950).

42. K. K. KELLER, "The Al^{27}(d, p)Al^{28} and Al^{27}(d, d′)Al^{27} Reactions," *Phys. Rev.* **84,** 884 (1951).

43. D. M. VAN PATTER and W. W. BUECHNER, "Investigation of the (d, p) and (d, α) Reactions of the Silicon Isotopes," *Phys. Rev.* **87,** 51 (1952).

44. S. T. BUTLER, "Angular Distributions from (d, p) and (d, n) Nuclear Reactions," *Proc. Roy. Soc.* (London) **A208,** 559 (1951).

45. R. C. MOBLEY and R. A. LAUBENSTEIN, "Photon-Neutron Thresholds of Beryllium and Deuterium," *Phys. Rev.* **80,** 309 (1950).

46. SHER, HALPERN, and MANN, "Photo-Neutron Thresholds," *Phys. Rev.* **84,** 387 (1951).

47. J. A. HARVEY, "Neutron Binding Energies from (d, p) Reactions and Nuclear Shell Structure," *Phys. Rev.* **81,** 353 (1951).

48. MONTALBETTI, KATZ, and GOLDEMBERG, "Photoneutron Cross Sections," *Phys. Rev.* **91,** 659 (1953).

49. J. HALPERN and A. K. MANN, "Cross Sections of Gamma-Proton Reactions," *Phys. Rev.* **83,** 370 (1951).

50. E. D. COURANT, "Direct Photodisintegration Processes in Nuclei," *Phys. Rev.* **82,** 703 (1951).

51. P. DYAL and J. P. HUMMEL, "Excitation Function for the $V^{51}(\gamma, \alpha)Sc^{47}$ Reaction," *Phys. Rev.* **115,** 1264 (1959).

52. R. B. LEIGHTON, *Principles of Modern Physics.* New York: McGraw-Hill, 1959, Chapters 20, 21.

53. RICHTMYER, KENNARD, and LAURITSEN, *Introduction to Modern Physics,* 5th ed. New York: McGraw-Hill, 1955, Chapter 11.

54. F. AJZENBERG and T. LAURITSEN, "Energy Levels of Light Nuclei. V," *Revs. Mod. Phys.* **27,** 77 (1955).

55. W. E. BURCHAM, "The Low-Lying Excited States of Light Nuclei," *Progress in Nuclear Physics,* O. R. Frisch, ed. New York: Academic Press, Vol. 2, pp. 174–234 (1952).

56. T. LAURITSEN, "Energy Levels of Light Nuclei," *Annual Review of Nuclear Science,* Vol. I. Stanford: Annual Reviews Inc., 1952.

57. F. AJZENBERG-SELOVE and T. LAURITSEN, "Energy Levels of Light Nuclei," *Nuclear Physics* **11,** 1 (1959).

58. T. LAURITSEN and F. AJZENBERG-SELOVE, "Energy Levels of the Light Nuclei," *Annual Reviews of Nuclear Science* **10,** 409 (1961).

59. M. K. BANERJEE, "The Theory of Stripping and Pickup Reactions," in *Nuclear Spectroscopy,* F. Ajzenberg-Selove, ed. New York: Academic Press, 1960. Part B, p. 695.

Problems

1. To which states of the nucleus Mg^{25} do the different groups of α-particles considered in Problem 4 of Chapter 13 belong? Draw a diagram to scale showing the location of these states.

2. A beryllium target was bombarded with α-particles accelerated to a kinetic energy of 21.7 Mev in a cyclotron. It was reported that four groups of protons were observed corresponding to Q-values of -6.92, -7.87, -8.57, and -10.74 Mev, respectively. (a) What were the energies of the proton groups observed at an angle of 90° with the incident beam? (b) Which of the Q-values corresponds to the ground state of the product nucleus, and how does it compare with the value calculated from the atomic masses? (c) To which levels of the product nucleus do the different proton groups correspond? (d) What are the threshold energies for the excitation of the different states of the product nucleus?

3. The reaction $C^{13}(d, p)C^{14}$ has a resonance at a deuteron energy of 2.45 Mev. At what α-particle energy does this datum lead you to predict a resonance level for the reaction $B^{11}(\alpha, n)N^{14}$?

4. The nucleus N^{14} has an excited state at an energy 12.80 Mev above the ground level. At what energy of the incident particle would you expect a resonance to occur in each of the following reactions: (a) $B^{10}(\alpha, n)N^{13}$; (b) $C^{12}(d, p)C^{13}$?

5. A target of boron enriched in the isotope B^{10} was bombarded with deuterons with an energy of 1.510 Mev, and the protons from the reaction $B^{10}(d, p)B^{11}$ were analyzed with a magnetic spectrograph. Eleven groups of protons were observed with the following Hr values, in kilogauss-centimeters: 448, 399, 338, 322, 266, 264, 245, 190, 171, 156, and 151. Calculate (a) the energy in Mev of the protons in each group, (b) the Q-value corresponding to each group of protons, (c) the energy level of the product nucleus corresponding to each proton group.

6. The nuclide F^{19} was bombarded with protons of varying energy and the yields of neutrons from the (p, n) reaction were measured. Resonances were observed at the following proton energies, in Mev: 4.29, 4.46, 4.49, 4.57, 4.62, 4.71, 4.78, 4.99, 5.07, and 5.20. The threshold energy for the reaction was found experimentally to be 4.253 Mev. Calculate from the data (a) the mass difference $Ne^{19} - F^{19}$, (b) the endpoint energy of the β-decay of Ne^{19}, (c) the excitation energies of the states of Ne^{20} which correspond to the resonances.

7. The nuclide Na^{23} was bombarded with deuterons in a cyclotron and the yield of the (d, p) reaction was determined in terms of the deuteron current in microampere hours (μah) and the radioactivity of the product nucleus. For each μah, 1500 microcuries of activity were produced owing to the formation of Na^{24}, which has a half-life of 15.06 hr. (a) How many particles are there in 1 μah of deuterons? (b) What was the yield of the reaction in protons per million deuterons?

8. A gold foil 0.02 cm thick is irradiated for 5 min with a beam of thermal neutrons with flux of 10^{12} neutrons/cm^2-sec. The nuclide Au^{198}, with a half-life of 2.7 days, is produced by the reaction $Au^{197}(n, \gamma)Au^{198}$. The density of gold is 19.3 gm/cm^3 and the cross section for the reaction is 98.7

$\times 10^{-24}$ cm^2. (a) How many atoms of Au^{198} are produced per square centimeter of foil? (b) What is the activity of the foil in millicuries per square centimeter?

9. A tantalum foil 0.02 cm thick and with a density of 16.6 gm/cm^3 is irradiated for 2 hours with a beam of thermal neutrons of flux 10^{12} neutrons/cm^2-sec. The radionuclide Ta^{182}, with a half-life of 114 days, is formed as a result of the reaction $Ta^{181}(n, \gamma)Ta^{182}$, and the foil has an activity of 12.3 rd/cm^2 immediately after the irradiation. Find (a) the number of atoms of Ta^{182} produced, (b) the cross section for the reaction $Ta^{181}(n, \gamma)Ta^{182}$.

10. When the nuclide C^{12} is bombarded with slow neutrons, γ-rays with an energy of 4.947 Mev are produced by the reaction $C^{12}(n, \gamma)C^{13}$. Compare this energy with the binding energy of the last neutron in C^{13} (a) as calculated from the atomic masses of the nuclei involved, (b) as deduced from the Q-value, 2.723 Mev, of the reaction $C^{12}(d, p)C^{13}$, (c) as deduced from the Q-value, -0.281 Mev, of the reaction $C^{12}(d, n)N^{13}$ and the endpoint energy 1.200 Mev of the positron decay of N^{13}.

11. The effect of a (d, p) reaction is to add a neutron to the target nucleus. Show that the binding energy of the last neutron in the product nucleus is given by the sum of the Q-value for the (d, p) reaction and the binding energy of the deuteron. Q-values of 4.48 Mev, 5.14 Mev, and 1.64 Mev have been obtained for the reactions $Pb^{206}(d, p)Pb^{207}$, $Pb^{207}(d, p)Pb^{208}$, and $Pb^{208}(d, p)Pb^{209}$, respectively. What are the binding energies of the last neutron in Pb^{207}, Pb^{208}, and Pb^{209}?

12. A sample of lead was bombarded with thermal neutrons and the capture γ-rays were analyzed with a pair spectrometer. Two γ-rays were observed, one with an energy of 6.734 ± 0.008 Mev, the second with an energy of 7.380 $\pm$ 0.008 Mev. What reactions were responsible for these rays?

13. The difference in mass between Fe^{54} and Fe^{56} has been shown mass-spectroscopically to be 1.99632 amu. What is the binding energy of the last two neutrons in Fe^{56}? How does this value compare with the value obtained from the following set of nuclear reactions?

$$Fe^{54}(n, \gamma)Fe^{55}; \quad Q = 9.28 \text{ Mev},$$
$$Fe^{55}(K \text{ capture})Mn^{55}; \quad Q = 0.21 \text{ Mev},$$
$$Mn^{55}(n, \gamma)Mn^{56}; \quad Q = 7.25 \text{ Mev},$$
$$Mn^{56}(\beta^-)Fe^{56}; \quad Q = 3.63.$$

14. The reactions $N^{14}(n, p)C^{14}$ and $C^{14}(p, n)N^{14}$ show resonances at the energies listed below.

$N^{14}(n, p)C^{14}$	$C^{14}(p, n)N^{14}$
495 kev	1163 kev
639 kev	1312 kev
998 kev	1668 kev
1120 kev	1788 kev
1211 kev	1884 kev

(a) Show that each pair of resonances corresponds to the same excited level of N^{15}. (b) Why is each resonance in the reaction $C^{14}(p, n)N^{14}$ higher by the same amount (close to 670 kev) than the corresponding resonance in the $N^{14}(n, p)C^{14}$ reaction?

15. Is it ever possible to use neutrons from the reaction $C^{14}(p, n)N^{14}$ to produce the inverse reaction $N^{14}(n, p)C^{14}$ in another target? Explain.

CHAPTER 17

NUCLEAR FORCES AND NUCLEAR STRUCTURE

A large amount of experimental information about atomic nuclei is now available, examples of which have been given in preceding chapters. Although the data and their interpretation are consistent with the idea that nuclei are built up of protons and neutrons, there is as yet no satisfactory explanation of how nuclei are held together. It has not yet been possible to analyze the forces which hold neutrons and protons together and the structure of nuclei is not understood. There is no picture or theory of the nucleus which can unite the available information about nuclei into a consistent body of knowledge like that of the quantum-mechanical theory of atomic structure. It is possible only to show how the problem of nuclear forces and nuclear structure has been attacked, to point to some successes, and to indicate some limitations and dilemmas.

Information about the nature of nuclear forces has been sought in two ways. The first is through the study of the interaction between two nucleons in order to discover the nature of the forces between nuclear particles. The properties of the deuteron and the scattering of protons and neutrons by protons have been analyzed with this objective. It is hoped that when the properties of the interactions between nucleons are known, it will be possible to derive from them the properties of complex nuclei; this program has so far been only partially successful. The second method is the study of the properties of complex nuclei and the attempt to deduce from these properties the nature of the interactions between the particles which make up the nuclei; this method also has been only partially successful. Between the two methods, however, something has been learned about nuclear forces, and this information can be used to suggest models of the nucleus. Although no single model can account for the complete range of nuclear properties, there are several which are useful.

In this chapter, the following procedure will be adopted. It will first be shown how some qualitative and semiquantitative ideas about nuclear forces and structure can be deduced from the properties of complex nuclei. The analysis of the interactions between nucleons will then be discussed in order to get some quantitative information about nuclear forces. Finally, some nuclear models and their applications will be discussed.

17–1 Nuclear binding energies and the saturation of nuclear forces. It has been shown (Table 9–3 and Fig. 9–11) that the average binding energy per particle in a nucleus has approximately a constant value for all but the lightest nuclei. In other words, the total binding energy of a nucleus is nearly proportional to the number of particles in the nucleus. Now, if every particle in the nucleus is supposed to interact with every other particle, the interaction energy, and therefore the binding energy, should be approximately proportional to the number of interacting pairs. Since each of the A particles in the nucleus would interact with $(A - 1)$ other particles, the number of interacting pairs would be $A(A - 1)/2$, and the binding energy would be proportional to this quantity. In heavier nuclei, A may be neglected in comparison with A^2 and the binding energy would be proportional to the square of the number of particles in the nucleus. This calculated result is in sharp contrast with the experimental result that the total binding energy is nearly proportional to A. To get out of this difficulty, it is necessary to assume that a nuclear particle does not interact with all the other particles in the nucleus, but only with a limited number of them. This situation is analogous to that in a liquid or solid, in which each atom is linked by chemical bonds to a number of nearest neighbors rather than to all the other atoms, and the chemical binding energy is practically proportional to the number of atoms present.

The consideration of different types of chemical binding leads to the conclusion that the best analogy to nuclear forces is represented by homopolar binding like that of the hydrogen molecule. There is a strong attraction between two hydrogen atoms to form the molecule H_2, but a third hydrogen atom cannot be bound strongly to a hydrogen molecule. The hydrogen molecule is said to be *saturated*. An assembly of many hydrogen atoms, as in a drop of liquid hydrogen, has a chemical binding energy approximately equal to that of the corresponding number of hydrogen molecules and is, therefore, approximately proportional to the number of hydrogen atoms present.

The correct dependence of nuclear binding energies on the number of particles in the nucleus can be obtained, then, if it is assumed that the forces between nucleons show saturation in much the same way as the forces of homopolar chemical binding. Furthermore, if a particular nuclear particle does not interact with all the other particles in the nucleus but only with some of its neighbors, then nuclear forces must have a *short range*. The force, or interaction energy, between two nucleons must fall off rapidly as the particles are separated; molecular forces also behave in this way. In a nucleus with many particles, a given nucleon can interact only with some of its neighbors and not with nucleons which are relatively far from it. The range of the specifically nuclear forces must consequently be smaller than the radius of any but the lightest nuclei.

The above discussion shows how certain properties of nuclear forces, namely, their saturation property and their short range, can be deduced from the variation of the binding energy per particle with the number of particles in the nucleus. In this discussion, the repulsive electrostatic forces between the protons have been neglected. The magnitude of the Coulomb energy between two protons is simply e^2/r, where e is the charge on a proton, 4.8×10^{-10} esu, and r is the distance between the protons. A rough idea of the magnitude of this energy can be found by taking $r = 3 \times 10^{-13}$ cm. Then

$$\frac{e^2}{r} = \frac{(4.8)^2 \times 10^{-20}}{(3 \times 10^{-13})(1.60 \times 10^{-6})} \cong 0.5 \text{ Mev}.$$

The Coulomb energy between two protons is therefore small compared with the average binding energy per particle, which is about 8 Mev. Despite its smallness, the Coulomb repulsion becomes important for heavier nuclei because of the saturation of the attractive nuclear forces. For the Coulomb force shows no saturation, and the total energy of the Coulomb interaction is proportional to the number of proton pairs in the nucleus, $\frac{1}{2}Z(Z - 1)$. The total Coulomb energy has been shown[1] to be

$$\text{Total Coulomb energy} = \frac{3}{5} Z(Z - 1) \frac{e^2}{R}, \tag{17-1}$$

where R is the nuclear radius. Since $R \sim A^{1/3}$ and Z is roughly proportional to A, the Coulomb energy is roughly proportional to $A^{5/3}$. But the total binding energy is proportional to A, so that the relative importance of the repulsive electrostatic energy increases with increasing mass number, roughly as $A^{2/3}$.

One result of the increased importance of the Coulomb energy for nuclei with greater Z (or A) is to decrease the binding energy per particle as the mass number increases. Figure 9–11 shows that in the neighborhood of $Z = 50$ or $A = 135$, the binding energy per particle starts to decrease. For very heavy nuclei, with A greater than 200, the effect is strong enough so that some of these nuclei are unstable with respect to α-disintegration.

17–2 Nuclear stability and the forces between nucleons. The information about stable nuclides contained in Table 9–1 and Fig. 9–9 allows certain deductions to be made about the forces between pairs of nucleons. The mass number A is approximately equal to twice the nuclear charge Z. This relationship is particularly accurate for the light nuclei; for heavier nuclei, the value of A increases more rapidly than $2Z$. Since A is the total number of particles in the nucleus and Z is the number of protons, the number of neutrons is very nearly equal to the number of protons in the light nuclei, but in the heavier nuclei the number of neutrons increases more rapidly than the number of protons. The equality between the numbers of protons and neutrons in the light nuclei can be interpreted as showing that there is a strong attractive force between a neutron and a proton. The stability of the deuteron, which is made up of a neutron and a proton, supports this conclusion.

If there is a strong specifically nuclear attractive force between a neutron and a proton, it is reasonable to suppose that there are similar forces between two neutrons and between two protons. If so, the force between two neutrons must be very nearly equal to that between two protons, apart from the electrostatic force between two protons. For, if the attraction between two neutrons were greater than that between a neutron and a proton or between two protons, the most abundant stable light nuclei would contain more neutrons than protons rather than equal numbers of these particles. A similar argument shows that the attractive force between two protons cannot be greater than that between a neutron and a proton, or between two neutrons. The existence of a force between two neutrons may be inferred from the way in which the neutron number increases with the proton number. As the Coulomb energy increases, the ratio of the number of neutrons to that of protons increases gradually from 1 for the light nuclei to 1.6 for uranium, with $Z = 92$. This increase may be interpreted as a necessary increase in the total attractive forces in the nucleus needed to overcome the effects of the electrostatic repulsion. Since each nuclear particle interacts only with a limited number of other nucleons, it is reasonable to suppose that some of the neutrons interact with other neutrons.

It will be assumed, in view of the preceding discussion, that there are specifically nuclear, attractive forces between a neutron and a proton, between two neutrons, and between two protons. These forces are abbreviated as n-p, n-n, and p-p, respectively. The n-n and p-p forces are assumed to be equal or very nearly equal in magnitude and to be not greater than the n-p force; they may, however, be much smaller than the n-p force, or they may have the same, or very nearly the same, magnitude as the n-p force. Either possibility would be consistent with the observed mass-to-charge ratio of the stable nuclides. The two possibilities can be expressed in the form

$$\text{(a)}\quad \text{n-p} \approx \text{n-n} \approx \text{p-p},$$

or

$$\text{(b)}\quad \text{n-p} \gg \text{n-n};\ \text{n-p} \gg \text{p-p};\ \text{n-n} \approx \text{p-p}.$$

Wigner,[(2,3,4)] in a general treatment of nuclear energy levels, made the assumption that the relationship described by case (a) holds, that is, that the specifically nuclear n-p, n-n, and p-p forces are equal. This assumption is known as the hypothesis of the *charge independence of nuclear forces.* The alternative assumption that the n-n and p-p forces are equal, but not necessarily equal to the p-n force, is known as the hypothesis of the *charge symmetry of nuclear forces.* It is less restrictive than the hypothesis of charge independence.

17–3 Energy levels of light nuclei and the hypothesis of the charge independence of nuclear forces. The hypothesis of the charge symmetry of nuclear forces can be tested by studying the energy levels, especially the ground states, of certain light nuclei. According to this hypothesis, the difference in the total energy between the ground states of a pair of mirror nuclei (isobars with a neutron excess of ± 1) should be accounted for solely by Coulomb repulsion and the difference in mass between neutron and proton. The simplest mirror nuclei are H^3 and He^3. The former contains 1 proton and 2 neutrons, or two n-p bonds and one n-n bond; He^3 contains 2 protons and 1 neutron, or two n-p bonds and one p-p bond. The binding energies of H^3 and He^3, calculated from the data of Table 11–1, are

$$\text{B.E.}(\text{H}^3) = 8.482 \text{ Mev}; \qquad \text{B.E.}(\text{He}^3) = 7.711 \text{ Mev}.$$

The difference in binding energies, 0.771 Mev, is small compared with either binding energy, so that the n-n force in H^3 is not very different from the p-p force in He^3. The small difference between the binding energies is attributed to the Coulomb repulsion between the two protons in He^3, since there is no Coulomb force in H^3. The effect is in the right direction, since He^3 is less strongly bound than H^3. If He^3 is assumed to be a sphere of radius R_c, its Coulomb energy[(1)] is given by

$$\text{Coulomb energy of He}^3 = \frac{6}{5}\frac{e^2}{R_c} = 0.771 \text{ Mev}. \tag{17–2}$$

This equation can be solved for R_c, with the result $\mathrm{R}_c = 2.24 \times 10^{-13}$ cm, which is a reasonable value for the "radius" of a light nucleus such as He^3. The hypothesis that the n-n force in H^3 is equal to the p-p force in He^3 is, therefore, consistent with the known information about these nuclei.

The positron-emitters, ${}_ZX^A$, with $A = 2Z - 1$, for which the parent and daughter nuclei are mirror nuclei, provide more evidence for the equality of the n-n and p-p forces. In each pair of mirror nuclei, the parent has one more proton than neutron, and the daughter has one more neutron than proton. The number of n-p bonds is presumably the same in parent and daughter, but the parent nucleus should have one more p-p bond and one less n-n bond than the daughter. If the n-n and p-p forces are equal, it should be possible to calculate the difference in the ground-state energy (atomic or nuclear mass) from the simple formula

$$\Delta E\text{ (calc)} = \Delta E\text{ (Coulomb)} - (\text{n} - \text{H}^1), \tag{17–3}$$

where ΔE (Coulomb) is the difference between the Coulomb energies of the pair, and $(\mathrm{n} - \mathrm{H}^1)$ represents the energy equivalent of the difference in mass between a neutron and a proton. If Z is the nuclear charge of the parent nucleus, then the difference in energy is given by

$$\Delta E\ (\text{calc}) = \frac{3}{5}\frac{e^2}{R}[Z(Z-1) - (Z-1)(Z-2)] - (\mathrm{n} - \mathrm{H}^1). \quad (17\text{–}4)$$

The radius R is taken to be the same for the parent and daughter nuclei because the mass number is the same and the difference in the nuclear mass should have a negligible effect on the radius. The value of R used in calculating ΔE comes from the formula

$$R = 1.42A^{1/3} \times 10^{-13}\ \text{cm}. \quad (17\text{–}5)$$

The constant 1.42, obtained from measurements of the scattering of fast neutrons by nuclei, is used because those measurements included work on light nuclei; it is, therefore, more suitable than the value obtained from the decay of heavy radionuclides. The value of ΔE (calc) should be the same as the experimental value of the positron decay energy if the n-n and p-p forces are equal as assumed.

Equation (17–4) can be written

$$\Delta E\ (\text{calc}) = \frac{6}{5}\frac{e^2}{R}(Z-1) - (\mathrm{n} - \mathrm{H}^1)$$

$$= \frac{6}{5}\frac{e^2(Z-1)}{(1.42)A^{1/3}10^{-13}} - (\mathrm{n} - \mathrm{H}^1). \quad (17\text{–}6)$$

When numerical values for e^2 and $(\mathrm{n} - \mathrm{H}^1)$ are inserted, and ΔE is expressed in Mev, the result is

$$\Delta E\ (\text{calc}) = \frac{6}{5}\frac{(4.80)^2 \times 10^{-20}}{(1.42)10^{-13}(1.60)10^{-6}}\frac{(Z-1)}{A^{1/3}} - 0.78\ \text{Mev},$$

or

$$\Delta E\ (\text{calc}) = 1.22\frac{(Z-1)}{A^{1/3}} - 0.78\ \text{Mev}. \quad (17\text{–}7)$$

The results of the calculations are given in Table 17–1, together with the experimental values of ΔE; the agreement is good, and the results are consistent with the assumption that the n-n and p-p forces are equal.

Although the mirror nuclei provide evidence in support of the hypothesis of the charge symmetry of nuclear forces, they tell nothing about the magnitude of the n-p force as compared with the n-n and p-p forces. There are light nuclei whose energy levels do yield such information. Consider, for example, the three isobars Be^{10}, B^{10}, and C^{10}, which contain 4 protons

TABLE 17–1

ENERGY DIFFERENCE BETWEEN MIRROR NUCLEI

Parent nucleus	Daughter nucleus	Charge of parent nucleus, Z	Cube root of mass number	Energy difference	
				ΔE (calc), Mev	ΔE (exp), Mev[73, 74]
C^{11}	B^{11}	6	2.22	1.95	1.98
N^{13}	C^{13}	7	2.35	2.33	2.22
O^{15}	N^{15}	8	2.47	2.68	2.76
F^{17}	O^{17}	9	2.57	3.02	2.77
Ne^{19}	F^{19}	10	2.67	3.33	3.26
Na^{21}	Ne^{21}	11	2.76	3.64	3.52
Mg^{23}	Na^{23}	12	2.84	3.94	4.06
Al^{25}	Mg^{25}	13	2.92	4.24	4.26
Si^{27}	Al^{27}	14	3.00	4.51	4.78
P^{29}	Si^{29}	15	3.07	4.77	4.98
S^{31}	P^{31}	16	3.14	5.05	5.41
Cl^{33}	S^{33}	17	3.21	5.30	5.53
A^{35}	Cl^{35}	18	3.27	5.57	5.95
Ca^{39}	K^{39}	20	3.39	6.07	6.45
Sc^{41}	Ca^{41}	21	3.45	6.29	5.96

and 6 neutrons, 5 protons and 5 neutrons, and 6 protons and 4 neutrons, respectively. Because of the equality of p-p and n-n forces, Be^{10} and C^{10} should correspond in much the same way that mirror nuclei do; their ground-state energy difference should be obtained from the difference in Coulomb energy and the energy equivalent of the difference in mass between two neutrons and two protons. The calculated and observed energy differences do indeed agree, as has been shown by a calculation similar to that made for the mirror nuclei.[5] This kind of argument can be extended to B^{10}, but only if the n-p force is equal to the n-n and p-p forces, since the number of n-p bonds in B^{10} should differ from the number of n-p bonds in Be^{10} and C^{10}. The theoretical treatment is more complicated than that for the mirror nuclei and will not be given here. It turns out that a certain excited state of B^{10} should bear a close relation to the ground state of Be^{10} and C^{10}; there is reasonably good evidence that this excited state does exist. An analogous result is obtained for the triad C^{14}-N^{14}-O^{14}, and these results provide evidence for the charge independence of nuclear forces, that is, that p-p = n-n = n-p. Although the evidence from energy levels is not as clear as that for the equality of p-p and n-n forces, a more detailed survey[6] of the known energy levels of light isobars indicates that the assumption of the charge independence of the specifically nuclear forces is a good approximation to the true state of affairs within the nucleus.

The hypothesis of the charge independence of nuclear forces has proved increasingly fruitful: It has led, for example, to the introduction of a new quantum number T, the isotopic or *isobaric* spin quantum number, which has helped in the understanding of nuclear energy levels and nuclear reactions.[2,4,8] If nuclear forces are, in fact, charge independent, that is, if the forces between nucleons are the same for neutrons and protons, we may regard a nucleon as a single entity having two states, the proton and the neutron. The charge of a nucleon may then be treated as a variable and a new variable, the isotopic or isobaric spin τ, may be introduced for nucleons. The adjective *isotopic* or *isobaric* comes from the fact that the variable refers to protons and neutrons, and "isobaric" is preferable because the proton and neutron are isobars, not isotopes; the term "*i*-spin" is often used for convenience. The new variable has a role similar to that of the electron spin variable σ, which is the reason for the "spin" part of the name. The isobaric spin variable τ distinguishes between protons and neutrons just as the electron spin variable σ is used to distinguish electrons with spins parallel to a given axis from electrons with spins in the opposite direction. Thus, τ may be assigned the value $+\frac{1}{2}$ for a neutron and $-\frac{1}{2}$ for a proton, and the charge of a nucleon may be expressed by $(\frac{1}{2} - \tau)$. The quantum number T bears the same relation to the isobaric spin variable τ that the spin quantum number S does to the electron spin variable σ. In the atomic case there are $2S + 1$ independent states belonging to the same quantum number S, with each state having a different value of the z-component, S_z, of the spin angular momentum. Similarly, there are $2T + 1$ independent nuclear states associated with the isobaric spin quantum number T, each state belonging to a different value of the quantity T_ζ, where $T_\zeta = T, \mathrm{T} - 1, \ldots, -T$. The quantity T_ζ may be called the "component of T in the direction of positive charge." In analogy with S_z, which is given for an atom by the sum of the S_z's of the individual electrons, T_ζ is given for a nucleus by

$$T_\zeta = \sum_i \tau_i,$$

where the sum is over all the nucleons. Hence, if the nucleus has N neutrons and Z protons,

$$T_\zeta = \tfrac{1}{2}(N - Z), \tag{17-8}$$

or just half the neutron excess of the nucleus. The values of T_ζ associated with a particular T describe states of several isobaric nuclei, and all of these states depend in the same way on the space and ordinary spin variables of the nucleons. If the forces are the same for neutrons and pro-

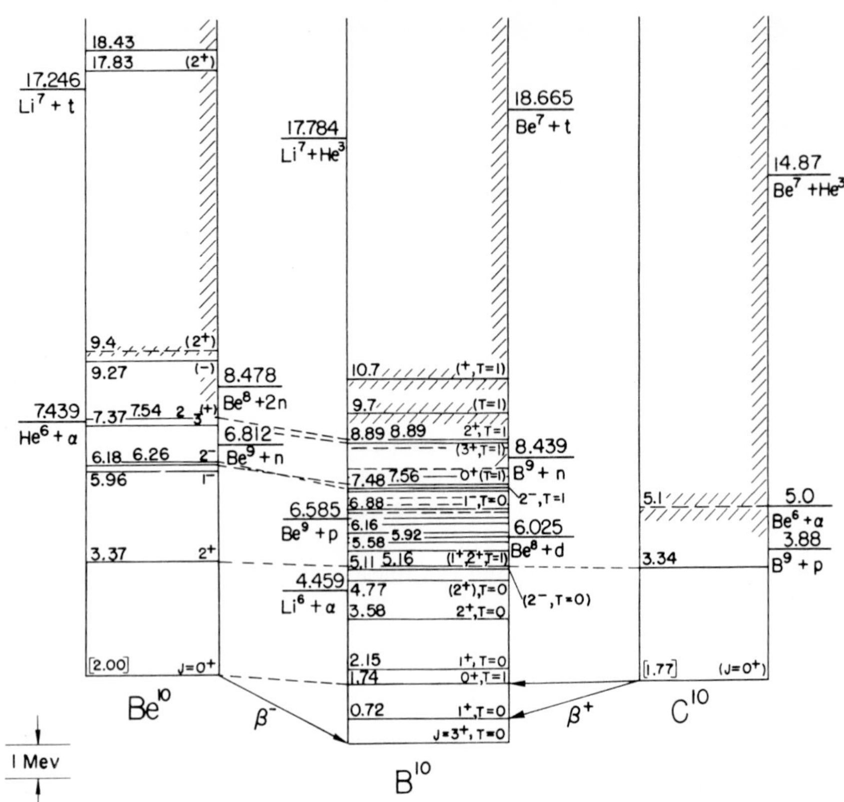

FIG. 17–1. Energy levels of the isobars Be^{10}, B^{10} and C^{10}.[76]

tons, these states should have the same energy (apart from their Coulomb energy and neutron-proton mass difference) and are said to form a T multiplet. Two states of a T multiplet differ only in that neutrons are substituted in one of these states for some of the protons in the other state, or vice-versa; and the total angular momentum is the same for all states of a T multiplet, as is the parity.

An example of the application of the concepts of charge independence and isobaric spin to the energy levels of light nuclei[7] is shown in Fig. 17–1. The level diagrams for the nuclides Be^{10}, B^{10}, and C^{10} are grouped into an isobaric set to show the correspondence of levels comprising isobaric spin multiplets. The relative positions of the ground states in each set have been adjusted to the extent that the first order Coulomb energy differences and the n-H mass differences have been removed. Levels for which the correspondence seems well established are connected by dashed lines.

The concept of the T multiplet has been applied to β-decay, γ-decay, and to nuclear reactions. There are selection rules involving the permissible change in T in β-decay, in γ-decay, and in nuclear reactions, and there is increasing evidence that total isobaric spin is conserved in nuclear reactions(8) in a manner analogous to the conservation of total nuclear angular momentum. The success of these applications supplies additional support for the hypothesis of the charge independence of nuclear forces.

17–4 The nuclear radius. The treatment of the Coulomb energy in the last section is an approximate one, based on elementary ideas about the size and shape of the atomic nucleus. These ideas have had to be modified, because of both theoretical and experimental work,(9,10) although the conclusions concerning the charge independence of nuclear forces are not affected. It was assumed in the last section that the nuclear charge is spread uniformly over a sphere of radius R, with R obtained from experiments on nuclear reactions; certain quantum-mechanical effects were also neglected. More accurate calculations have been made in which the experimental data on mirror nuclei were used to determine the nuclear radius.(11) The radius may still be expressed by the relation

$$R = R_0 A^{1/3}, \tag{17–9}$$

but with $R_0 = (1.28 \pm 0.05) \times 10^{-13}$ cm. These results are in good agreement with those of precise calculations of nuclear Coulomb energies.(12) Nuclear radii have also been determined experimentally with methods which yield information about the electric charge distribution. For example, electrons interact very little with nuclei through specifically nuclear forces and are affected only by the electric charge distribution of the nucleus. Experiments on the scattering of fast electrons by nuclei can consequently be interpreted(13) to give the radius of the equivalent sphere of electric charge with the result

$$R = (1.2 \pm 0.1) \times 10^{-13} A^{1/3}\ \text{cm}.$$

Similar results have been obtained(14) from experiments on the interaction between nuclei and μ-mesons (see Section 21–4).

The values for nuclear radii obtained by means of these methods are somewhat smaller than those obtained from experiments involving nuclear forces, e.g., from cross sections for the scattering of neutrons or protons or from α-radioactivity. The former methods give information about the size and shape of the charge distribution while the latter give information about the nuclear potential energy distribution. The two distributions are not necessarily the same, and should not be confused in precise work.

In particular, the assumption of a uniform spherical charge distribution seems to be an oversimplification.

The extent of recent work in this field and the scope of the results is indicated by the material in References 15 and 75.

17–5 The interaction between two nucleons. It has been shown in the earlier sections of this chapter how qualitative information about nuclear forces can be obtained from the properties of complex nuclei. Quantitative information can best be obtained from the experimental and theoretical investigation of phenomena involving only two nucleons, and *two-nucleon problems* are, therefore, the basic problems of nuclear physics. These problems include the scattering of protons by protons, which gives information about the strength and the range of the p-p forces, the scattering of neutrons by protons, the properties of the deuteron, and the capture of neutrons by protons, all of which give information about the strength and range of the n-p force. The force between two neutrons cannot be studied directly because there is no stable nucleus which consists only of neutrons, and because experiments on the scattering of neutrons by neutrons are not practical. The properties of complex nuclei are the only source of information about the n-n force.

The analysis of the experimental results and the theory are both complicated, involving the quantum-mechanical theory of collision processes and of the deuteron. These problems are treated in some of the books listed at the end of the chapter and in review articles(16,17,18) and will not be discussed in detail here. Instead, the ideas underlying the experiments and theory will be sketched and some of the results will be indicated. We start with the problem of the deuteron.

The successful application of wave mechanics to α-radioactivity showed that nuclear problems as well as atomic problems should be understandable in terms of wave mechanics. There is, however, a serious difference between atomic and nuclear problems; in the former the force between the positively charged nucleus and an electron is known to be derived from the Coulomb electrostatic potential; in the latter the details of the force law are unknown. In the case of the deuteron, the simplest nucleus with more than one nucleon, certain facts are known, and it is necessary to seek a force law, or potential between the proton and neutron, which will account for the facts. The most important experimental basis for the theory of the deuteron is its binding energy, 2.23 Mev; in addition, the spin, magnetic moment, and quadrupole moment are known. The Schroedinger equation for the deuteron is

$$\nabla^2\psi + \frac{2m}{\hbar^2}[E - V]\psi = 0, \qquad (17\text{–}10)$$

where ∇^2 is the Laplacian second order differential operator which is, in rectangular cartesian coordinates,

$$\nabla^2 = \frac{\partial^2}{\partial x^2} + \frac{\partial^2}{\partial y^2} + \frac{\partial^2}{\partial z^2};$$

m is the reduced mass,

$$m = \frac{M_n M_p}{M_n + M_p} \approx \frac{1}{2} M \quad \text{(of proton or neutron).} \tag{17-11}$$

The quantity E is known; for the ground state of the deuteron it is negative and numerically equal to the binding energy. The potential V is not known and the procedure used is to start with some reasonable assumed function and modify it as necessity requires, keeping it as simple as the experimental facts permit. The simplest assumption that can be made about the force between the neutron and proton is that it is a *central* force, acting along a line joining the two particles, and that it can be derived from a potential $V(r)$, where r is the distance between the particles. Since the force is attractive, $V(r)$ is negative and decreases with decreasing r. The wave function is then a function of r only, and may be written as $\psi(r)$. It is convenient to use spherical coordinates in the Schroedinger equation. The lowest, most stable, quantum state is that for which the angular momentum $l = 0$ (the S-state). There is then spherical symmetry and terms which involve the polar and azimuthal angles drop out so that the Schroedinger equation for the deuteron becomes

$$\frac{d^2\psi(r)}{dr^2} + \frac{2}{r}\frac{d\psi(r)}{dr} + \frac{2m}{\hbar^2}[E - V(r)]\psi(r) = 0. \tag{17-12}$$

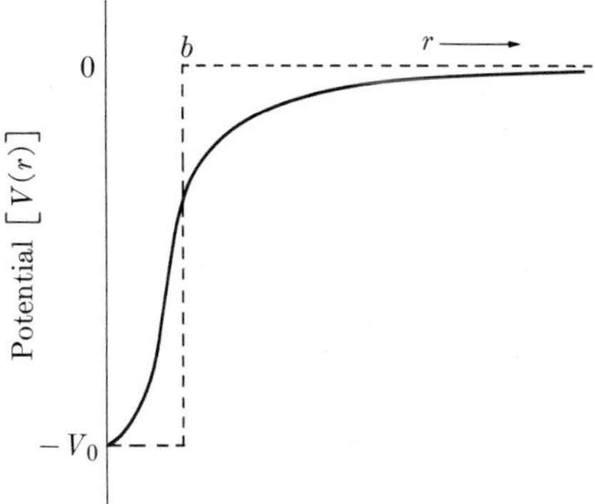

FIG. 17-2. Attractive potential between a neutron and a proton as a function of the distance between the two particles. The dotted lines represent the square-well potential.

Equation (17–12) can be simplified by making the substitution

$$\psi(r) = \frac{u(r)}{r}, \tag{17–13}$$

with the result

$$\frac{d^2u(r)}{dr^2} + \frac{M}{\hbar^2}[E - V(r)]u(r) = 0. \tag{17–14}$$

Various potentials can be used subject, of course, to the requirement that the force has short range. The potentials which have been used most frequently are

(a) square well: $V(r) = -V_0 \quad (r < b)$

$\qquad\qquad\qquad\quad = 0 \quad (r > b);$

(b) exponential: $V(r) = -V_0 e^{-\mu r};$

(c) Gaussian: $V(r) = -V_0 e^{-\mu^2 r^2};$

(d) Yukawa: $V(r) = -V_0(e^{-\mu r}/\mu r).$

The simplest of these is the square well shown in Fig. 17–2; the magnitude of the potential is assumed to be significantly large only within the short range $0 < r < b$, and to be zero for $r > b$. This kind of potential was used in the theory of α-decay (Section 13–6) to describe the interaction between the α-particle and the residual nucleus except that a repulsive (positive) potential was used there. The binding energy alone does not permit the determination of both the magnitude and the range, but gives only a relationship between these two quantities. The result of the analysis may be written

$$b^2V_0 = 1.48 \times 10^{-24} \text{ Mev}\cdot\text{cm}^2. \tag{17–15}$$

If the range b is assumed to be 2×10^{-13} cm, a reasonable value for the deuteron, then V_0 would be 37 Mev. The other short-range potential functions give about the same results as the square well.

The exact solution of the problem of the deuteron is complicated by several factors. The n-p potential does not depend on r alone, since the deuteron is known to have a quadrupole moment (see Section 8–5) and is, therefore, not spherically symmetrical. The potential depends also on the relative orientation of the spins of the neutron and the proton. In fact, a detailed analysis of the ground state of the deuteron shows that a neutron and a proton can form a stable deuteron only if their spins are parallel. This stable state is known as the *triplet state,* in contrast to the unstable *singlet state* in which the spins are antiparallel. To account for the shape (quadrupole moment) of the deuteron, it is necessary to postulate that the n-p potential depends on the angle between the spins and the line of connection of the two particles. The n-p force is consequently noncentral or, more precisely, not completely central. Finally, the analysis of the deuteron alone does not give all of the desired information about the n-p force, although it does provide useful clues to the solution of the problem.

More detailed information about the n-p force is obtained from the analysis of the scattering of neutrons by protons. In some typical scattering experiments,(19,20,21) cyclohexane, C_6H_{12}, was bombarded with neutrons of known energy, and the transmission was measured in a "good geometry" experiment. The neutron transmission is given by

$$I/I_0 = e^{-n\sigma_t x},$$

where σ_t is the total cross section in cm^2, n is the number of scattering nuclei per cubic centimeter, and x is the thickness of the scatterer in cm. Cyclohexane contains carbon in addition to hydrogen, and the effect of the carbon was determined by doing separate transmission measurements on a carbon target. The total cross section per atom of hydrogen is then

$$\sigma_t(\mathrm{H}) = \tfrac{1}{12}[\sigma_t(\mathrm{C_6H_{12}}) - 6\sigma_t(\mathrm{C})]. \tag{17-16}$$

The absorption cross sections of hydrogen and carbon are negligible compared with the respective scattering cross sections in the neutron energy ranges studied (0.35 to 6.0 Mev), so that the total cross section can be taken equal to the scattering cross section.

The measured cross sections can be analyzed by methods which lead in a not very direct way to information about the properties of the n-p potential. The difficulties mentioned in connection with the problem of the deuteron are met also in the scattering problem. At neutron energies less than 10 Mev, which are relatively low for scattering experiments of this kind, the range of the n-p potential turns out to be close to 2×10^{-13} cm for the triplet potential. For a square-well potential, the corresponding depth is 35 Mev. The results for the singlet potential are 2.8×10^{-13} cm for the range, and 11.8 Mev for the depth. It has been found, however, that for neutron energies less than 10 Mev, the details of the shape of the potential cannot be obtained from neutron-proton scattering experiments, in spite of the difficult and complicated analytical methods used. The results of the experiments are now analyzed by means of a theory called the *effective range theory* which yields two quantities called the *effective range* and the *scattering length.* These quantities represent average properties of the potential and determine the scattering cross section; they are related to both the range and magnitude of the potential, and are used because they make the analysis of experimental data easier. In particular, they allow calculations to be made on the basis of very general assumptions about the nature of nuclear potentials. Both quantities have different values in the triplet and singlet cases. Recent results for the effective range are, for the n-p forces,

$$\begin{aligned} \text{triplet effective range} &= (1.70 \pm 0.03) \times 10^{-13} \text{ cm}, \\ \text{singlet effective range} &= (2.4 \pm 0.4) \times 10^{-13} \text{ cm}. \end{aligned} \tag{17–17}$$

Values for these quantities can also be obtained from other measurements. An estimate of the triplet effective range based on measurements of the cross section for the photodisintegration of the deuteron[22,23] gives a result in good agreement with the value 1.70×10^{-13} cm. The cross section for the capture of thermal neutrons by protons[24] leads to a value of 2.5×10^{-13} cm for the singlet effective range.

Recent values for the scattering lengths are

$$\begin{aligned} \text{triplet scattering length} &= 5.38 \times 10^{-13} \text{ cm}, \\ \text{singlet scattering length} &= -2.37 \times 10^{-12} \text{ cm}. \end{aligned} \tag{17–18}$$

There is only one proton-proton force, or potential, corresponding to antiparallel spins. This *singlet* interaction is a consequence of the Pauli exclusion principle (Section 7–6) according to which two similar particles cannot have the same quantum numbers. Two protons can interact strongly only when their quantum states are the same except for the spins, which must be antiparallel. The hypothesis of the charge independence of nuclear forces would then require that the values of the effective range and scattering length for the p-p potential be the same as those for the singlet n-p potential.

Many proton-proton scattering experiments have been done with protons of relatively low energy (10 Mev or less).[16] Well-collimated, monoenergetic beams of protons can be obtained from a Van de Graaff electrostatic generator. The protons are scattered by the protons in a sample of hydrogen gas, and the scattered particles are detected at different angles by means of an ionization chamber. The cross section for the proton-proton scattering process can be determined as a function of the energy of the incident particles. The differences between p-p scattering and the Rutherford scattering of α-particles by nuclei are interesting. In the latter case, the measured scattering is caused by the Coulomb repulsion between the α-particle and the nucleus. The repulsive forces are very large and deviations from the Coulomb force law are observed only for highly energetic particles and light nuclei. The occurrence of deviations points to the existence of specifically nuclear forces. In p-p scattering, the Coulomb potential is relatively small, and the bombarding particles are scattered in a way which depends on the combined effect of the specifically nuclear forces and the Coulomb forces. The effect of the Coulomb forces must be subtracted from the total effect to obtain the desired information about the nuclear p-p force.

The results of the scattering experiments can be analyzed by methods similar to those used for n-p scattering. Additional complications are introduced into the theory by the Coulomb repulsive force between the protons and by the fact that the bombarding and target particles are of the same kind. When the experimental data are analyzed in terms of the range and depth of the potential which fit the measured values of the cross sections, the results agree quite well with those obtained for n-p scattering. Thus, a square-well potential with a range of 2.8×10^{-13} cm would have a depth of 11.35 Mev as compared with a singlet n-p depth of 11.8 Mev for the same range. If other shapes of the potential well are used, equally good agreement is obtained between the strengths of the potential needed for n-p and p-p forces. These results, together with those of Section 17–3 on the equality of the n-n and p-p forces (apart from the Coulomb effect), provide experimental evidence for the validity of the assumption of the charge independence of nuclear forces.

Just as in n-p scattering, p-p scattering results below 10 Mev do not give details of the shape of the potential. The effective range theory has also been applied to p-p scattering, and gives the results

$$\begin{aligned} \text{effective p-p range} &= (2.65 \pm 0.07) \times 10^{-13} \text{ cm}, \\ \text{p-p scattering length} &= (-7.67 \pm 0.5) \times 10^{-13} \text{ cm}. \end{aligned} \tag{17–19}$$

The value for the effective range agrees, within the stated limits of error, with the value for the singlet n-p effective range [Eq. (17–17)]. The scattering lengths are quite different in the two cases, a result which seems to imply that the depths of the potentials are significantly different in the two cases. It has been shown, however, that there are magnetic interactions between the particles, and that when these are taken into account, the p-p and singlet n-p scattering lengths agree.

The results of the scattering work can be summarized, then, by the statement that at relatively low energies (less than 10 Mev) the nuclear potentials for the singlet n-p interaction and the p-p interaction are identical, in agreement with the hypothesis of the charge independence of nuclear forces.

17–6 The status of the problem of nuclear forces. The information collected in the last section about the interaction between two nuclear particles at low energy can be summarized as follows: there exists a force between nuclear particles which is attractive (except for the electrostatic repulsion between two protons), and which has a range of about 2×10^{-13} cm. The force is spin dependent, being different in the parallel spin (triplet) and antiparallel spin (singlet) cases. It is not completely central, but depends to some extent on the angle between the axis of the spins and the line between the two particles. The nuclear force does not seem to depend on the type of nucleon; the same force can be assumed between any pair of nucleons, two protons, two neutrons, or a neutron and a proton, without coming into conflict with any low-energy result. In other words, the nuclear forces seem to be charge independent at low energies.

Unfortunately, the information just listed is incomplete and leads to serious problems, some of which have not yet been resolved. One of these problems has to do with the saturation of nuclear forces. The study of complex nuclei has shown that the binding energy per nucleon is approximately constant and it has been deduced from this fact (Section 17–1) that nuclear forces must have the property of saturation. It has also been found that the nuclear radius is proportional to the cube root of the nuclear mass, that is, the nuclear volume is proportional to the number of particles in the nucleus, a result which also implies the saturation of nuclear forces. It turns out, however, that if the nuclear forces are attractive, short-range forces, they cannot have the saturation property required by the binding energies and nuclear radii. A study of the potential and kinetic energy of the nucleus as a function of the radius has shown[25] that under the action of purely attractive forces, the nucleus would be stable only if its radius were about half as great as the range of nuclear forces. Each nucleon would then interact with every other nucleon and the binding energy would be proportional to A^2, in contradiction with the observed fact that the binding energy is proportional to A. Nuclear radii would be smaller than the values obtained from experiment.

The difficulty has been partially resolved by assuming that the force, or part of the force, between two nucleons is sometimes attractive, sometimes repulsive, depending on the state of the two nucleons with respect to each other. A force of this kind is of a more general type than those used in ordinary atomic physics, and it has been necessary to broaden the concepts of potential energy and force in order to apply them to the nucleus. A clue to the kind of generalization needed was supplied by the treatment of the molecular ion H_2^+, which consists of two protons and one electron. To account for the properties of H_2^+, a force was needed which is a mixture of *ordinary* and *exchange* terms. The exchange terms arose from the possibility that the electron could exchange positions between the protons. It was assumed that the electron could jump from one proton to the other, and that the H_2^+ ion could be regarded as consisting of a hydrogen atom and a proton, which can change their positions. The possibility of this exchange introduced new terms into the mathematical expressions for the potential and the force, and the new contribution to the force is known as an *exchange force*. Exchange forces have the property of being attractive or repulsive depending on the states of the particles involved, and it was postulated that forces of this kind exist between nucleons. Although there is not yet a satisfactory explanation for the existence of nuclear exchange forces, these forces can be treated mathematically and lead to the correct saturation properties. The introduction of exchange forces does not affect the low-energy interaction between two nucleons, and removes the contradiction with the saturation requirement which follows from the assumption of purely attractive nuclear forces.

Although exchange forces cannot be described accurately in terms of elementary ideas, a simple, if not entirely correct, picture can be given of how they lead to saturation. Immediately after the discovery of the neutron (1932), Heisenberg suggested that when a neutron interacts with a proton the single electric charge jumps from one nucleon to the other so that, in the first such jump, the original proton changes into a neutron and the neutron into a proton. In this way, a neutron may interact with only one proton at a time, and the n-p force has the property of saturation. After Fermi proposed his theory of β-decay, these ideas were extended. According to this theory, a neutron can change into a proton by emitting an electron and a neutrino, and a proton can change into a neutron by emitting a positron and a neutrino. Heisenberg suggested that nuclear forces are exchange forces in which electrons (or positrons) and neutrinos are exchanged between two nucleons. The strength of these nuclear forces could then be calculated from the observed probability of β-decay. Unfortunately, when these calculations were made, the nuclear forces that should result from the electron-neutrino interchange were weaker by a factor of about 10^{-14} than those required by a square-well potential with a range of the order of 10^{-13} cm. Although the theory was pretty, it did not work.

After the failure of this theory, Yukawa (1935) predicted the existence of a particle, now called a *meson*, having a rest mass between the electron and nucleon masses. He proposed a theory of nuclear forces which involves these mesons and is referred to as the *meson theory of nuclear forces*. The discovery of such a particle in the cosmic radiation stimulated a great deal of experimental and theoretical research in the fields of nuclear theory, cosmic rays, and high-energy particle physics. Several different mesons with different charges and rest masses have been discovered, and many attempts have been made to fit them into a consistent scheme of nuclear forces. These attempts, however, have not yet succeeded in reproducing quantitatively the known properties of nuclear forces.

Another serious problem has to do with the properties of nuclear forces deduced from nucleon scattering experiments at high energy. It has been emphasized that the hypothesis of the charge independence of nuclear forces seems to be valid at *low* energies. It has also been mentioned that the low-energy data do not give definite information about the detailed properties of the n-p and p-p potentials. It was expected on theoretical grounds that the results of scattering experiments at high energies would give results which are very sensitive to the details of the interaction, its dependence on the distance between the particles, and to its exchange properties. The development of high-energy particle accelerators has made it possible to extend the energy range of neutron-proton and proton-proton scattering to energies up to 400 Mev.

Several difficulties have appeared as a result of the work on high-energy scattering, and the data available at present have led to confusion rather than to clarity. The proton-proton scattering at high energies shows some features which are quite different from the neutron-proton scattering. For example, the p-p scattering cross section at energies greater than 100 Mev is almost independent of the angle of scattering, in strong contrast to the n-p scattering. Although certain differences between the p-p and n-p scattering are to be expected even if the nuclear force law is the same, the observed differences are such that they cannot be explained in terms of the theory and forces that account for the low-energy results. The charge-independent forces which account for the low-energy results seem to fail when applied to the high-energy scattering data. This dilemma has not yet been resolved; it is receiving a great deal of attention and it may still be possible to find a force law which can account for both types of scattering.

Thus, in spite of the existence of a large amount of experimental material, it has not yet been possible to analyze the forces which hold neutrons and protons together. Although considerable progress has been made in recent years, knowledge of the nature of nuclear forces is still in an early, if not rudimentary, stage. It is, therefore, not possible to develop a theory of nuclear structure based on the present knowledge of nuclear forces.

17–7 Nuclear models. The shell, or independent particle, model. In the absence of a detailed theory of nuclear structure, attempts have been made to correlate nuclear data in terms of various models. Several models have been proposed, each based on a set of simplifying assumptions and useful in a limited way. Each model serves to correlate a portion of our experimental knowledge about nuclei, usually within a more or less narrow range of phenomena, but fails when applied to data outside of this range.

The nuclear shell model is one of the most important and useful models of nuclear structure. Its conception and name come from the results of empirical correlations of certain nuclear data. It has become apparent in recent years that many nuclear properties vary periodically in a sense similar to that of the periodic system of the elements. Most of these properties show marked discontinuities near certain even values of the proton or neutron number. The experimental facts indicate that especially stable nuclei result when either the number of protons Z or the number of neutrons $N = A - Z$ is equal to one of the numbers[26,27,28] 2, 8, 20, 50, 82, 126. These numbers are commonly referred to as *magic numbers.* Although this name is not a strictly scientific one, it serves the purpose of identifying the numbers simply and conveniently. The magic numbers of neutrons and protons have been interpreted as forming closed shells of neutrons or protons in analogy with the filling of electron shells in atoms, and the neutron and proton shells appear to be independent of each other. It has been shown that the existence of the closed neutron and proton shells corresponding to the magic numbers can be described theoretically in terms of a particular nuclear model. The remainder of this section will be devoted to a summary of some of the evidence which led to the nuclear shell model, and to a brief discussion of the ideas involved in the theory.

One of the most important pieces of evidence for the magic numbers comes from the study of the stable nuclides. The nuclei for which Z and N are 2 or 8, He^4 and O^{16} respectively, are more stable than their neighbors, as can be seen from the binding energies (Table 9–3). Evidence for some of the other magic numbers has been obtained from very careful studies[29,30] of the details of the binding energy curve in the region of intermediate and heavy nuclei. When the average binding energy per nucleon (obtained from mass spectrographic measurements and nuclear disintegration data) is plotted against the mass number, the curve is not really smooth as in Fig. 9–11, but has several kinks in it. These kinks or breaks correspond to sudden increases in the value of the binding energy per nucleon. They have been found to occur at $Pb^{208}(Z = 82, N = 126)$, $Ce^{140}(Z = 58, N = 82)$, $Sn^{120}(Z = 50, N = 70)$, and $Sr^{88}(Z = 38, N = 50)$. In other studies,[31,32] masses have been determined in the neighborhoods of $Z = 20$ and 28, and $N = 20$ and 28, and maxima have been found at these values for the binding energy per nucleon. As in other properties, the evidence for the special stability of nuclei with N or Z equal to 28 is less marked than that for the magic numbers. Along with the numbers 14 and 40, 28 is sometimes called a *semimagic* number.

Stability is usually related not only to high binding energy, but also to high natural abundance; in all cases where the binding energies are known, a definite correlation is found between the binding energies and the abundances. A table of the relative abundances of nuclei compiled from data on the composition of the earth, sun, stars, and meteorites has been published by Brown;[33] pronounced peaks are found at $O^{16}(N, Z = 8)$, $Ca^{40}(N, Z = 20)$, $Sn^{118}(Z = 50)$; Sr^{88}, Y^{89}, $Zr^{89}(N = 50)$; Ba^{138}, La^{139}, $Ce^{140}(N = 82)$; and at $Pb^{208}(Z = 82, N = 126)$.

The relative stability of different elements is also indicated by the number of stable isotopes per element. A graph of this number is shown

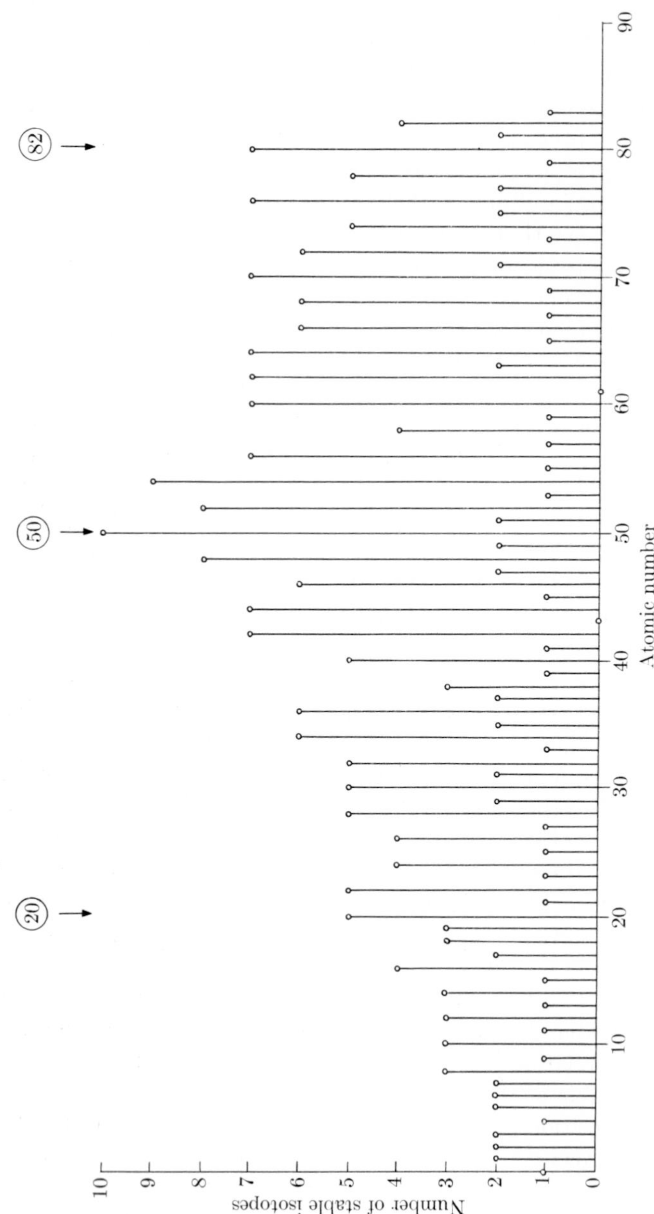

FIG. 17–3. The number of stable isotopes per element as a function of the atomic number (Flowers[27]).

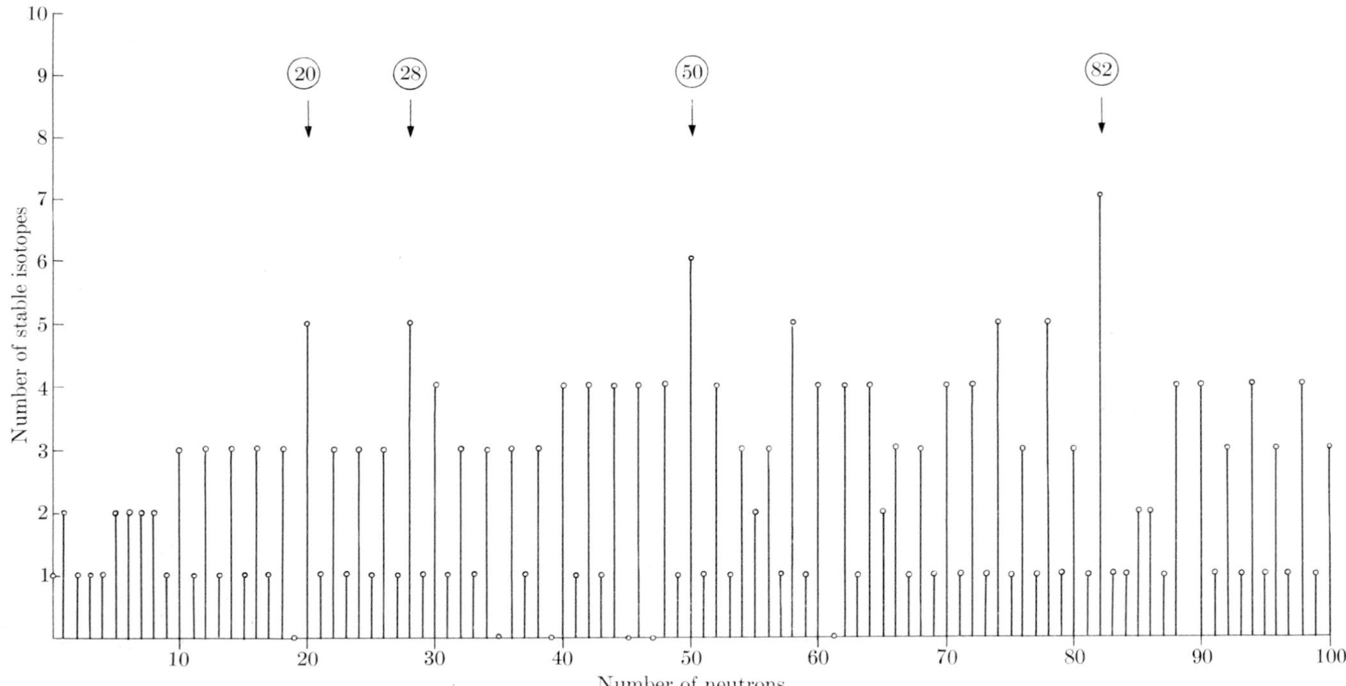

FIG. 17–4. The number of stable nuclides as a function of the neutron number (Flowers[27]).

in Fig. 17–3 as a function of Z. The lightest element with more than two stable isotopes is oxygen ($Z = 8$); the first element with four stable isotopes is sulfur, whose heaviest isotope is S^{36} with 20 neutrons. Five isotopes occur first in calcium ($Z = 20$); seven appear in molybdenum, of which the lightest is Mo^{92} with 50 neutrons. Ten isotopes occur only once—in tin with 50 protons. Figure 17–4 represents the number of stable nuclides having a given number of neutrons; such nuclei are sometimes called *isotones*. The peaks at 20, 28, 50, and 82 neutrons are clear.

The energies of the α-particles emitted by heavy radioactive nuclei provide strong evidence[34] for the magic numbers $Z = 82$ and $N = 126$. The fact that lead ($Z = 82$) is the stable end product of the three natural radioactive series points to the special nature of nuclei with $Z = 82$. The nuclides ${}_{85}At^{213}$ and ${}_{84}Po^{212}$, which have 128 neutrons, emit exceptionally energetic α-particles, 9.2 Mev and 8.78 Mev, respectively; Po^{212} has the extremely short half-life of 3×10^{-7} sec, while that of At^{213} has not yet been measured. In both of these cases, the daughter nucleus has 126 neutrons; there seems to be a strong tendency for nuclei with 128 neutrons to emit an α-particle and change into much more stable nuclei with 126 neutrons. When the *parent* nucleus has 126 neutrons, the α-energies should be particularly low, since the decay starts from a nucleus with low energy (high binding energy). These expectations are borne out by the experimental data; Po^{210} and At^{211}, with 126 neutrons, have relatively long half-lives and low α-energies; $Bi^{209}(N = 126)$ is stable, whereas the neutron-richer isotopes Bi^{210} and Bi^{211} are α-radioactive.

Similar relations have been found in β-decay.[35] The energies of the emitted β-particles are especially large when the number of neutrons or protons in the product nucleus corresponds to a magic number. The results of the investigations on α-decay and β-decay show that there are energy discontinuities of about 2 Mev in the neighborhood of the magic numbers 28, 50, 82, and 126. Since the average binding energy per particle in a nucleus is about 8 Mev, the effect in energy is about 25%.

It has already been noted in Section 16–5C that the absorption cross sections for neutrons of about 1 Mev in energy are unusually low for nuclei containing 50, 82, or 126 neutrons.[36] The cross sections for these nuclei are lower by a factor of about fifty than those of neighboring nuclei; the low cross section is an indication of large level spacing in the compound nucleus formed by the neutron capture. Similar evidence has been obtained from thermal neutron capture cross sections.[37] The observed values are unusually small for nuclei with magic neutron numbers, and nuclei with 50 protons (tin) and 82 protons (lead) have smaller capture cross sections than their neighbors. Figure 17–5 is a graph of the thermal neutron cross sections of nuclides with even mass number as a function of the number of neutrons. The solid lines join the largest values of the cross

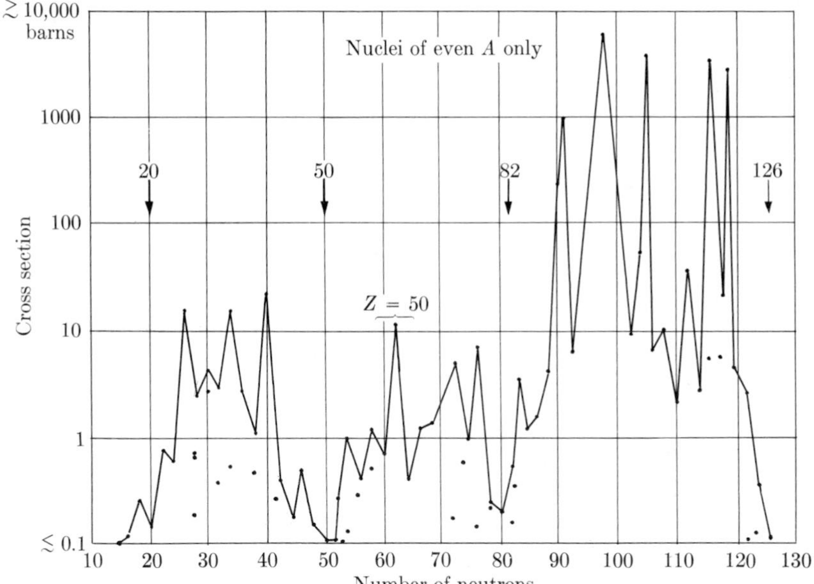

FIG. 17–5. Thermal neutron cross section as a function of the neutron number (Flowers [27]).

sections when two or more nuclei exist, for a given value of N, whose cross sections have been measured. This procedure emphasizes the larger values and tends to mask the appearance of the low values associated with the magic number nuclei. Nevertheless, real dips in the curve appear at or near the neutron numbers 20, 50, 82, and 126. A similar trend is shown by a graph for nuclei with odd values of A.

The binding energy of the last neutron added to a nucleus has been determined from measurements of the Q-values of (d, p) reactions, (n, γ) reactions, (γ, n) reactions, and (d, t) reactions. These reactions and some of the results have been mentioned in Section 16–9. Values of this binding energy can be calculated from the semiempirical binding energy formula (17–30) below. The difference between the observed and calculated values was studied as a function of the number of neutrons,[38] and the results are shown in Fig. 17–6; ΔE is the difference between the observed neutron binding energy for a nucleus with $N + 1$ neutrons and the neutron binding energy calculated from the semiempirical mass formula. The figure shows sharp discontinuities in the value of ΔE at 50 and 126 neutrons, and another drop in the region of 82 neutrons. There is a break near, but not exactly at, 28 neutrons; the evidence for the semimagic neutron number 28 is not as clear in this study as in some of the others discussed above.

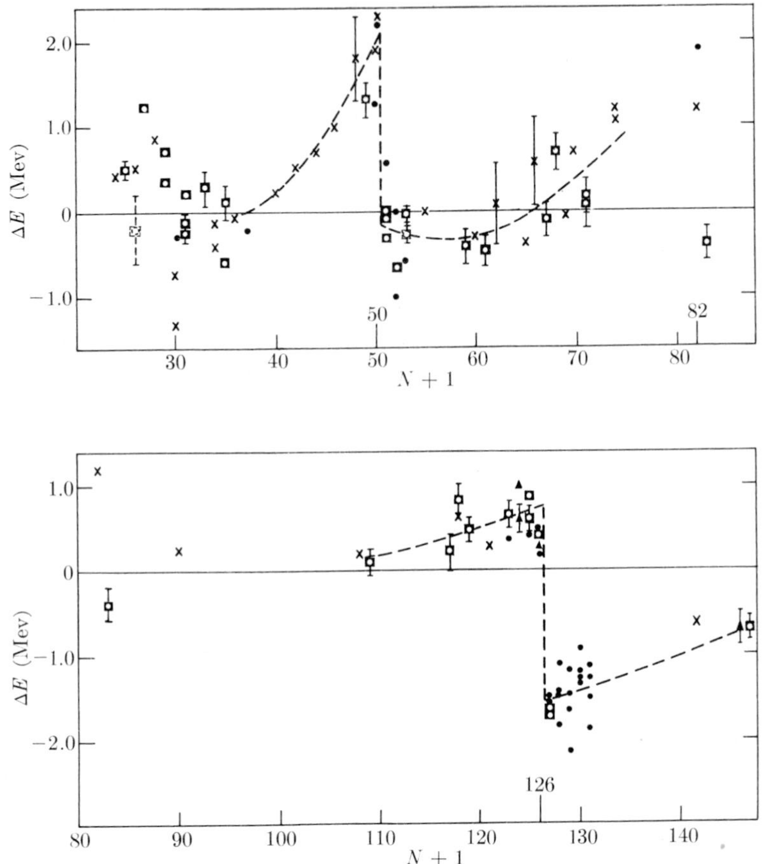

17–6. Binding energy of the last neutron added to a nucleus as a function of the neutron number. ΔE is the difference between the theoretical and experimental values (Harvey[38]).

The correlations discussed are examples of the simpler kinds of evidence for the special properties of nuclei containing magic numbers of neutrons or protons. The results have been interpreted as an indication that neutrons and protons within the nucleus are arranged into shells within the nucleus like electrons in atoms. Each shell is limited to a certain maximum number of protons or neutrons. When a shell is filled, the resulting configuration is particularly stable and has an unusually low energy.

Several theories have been proposed to account for the neutron and proton shells.[39,40,41,42] It is assumed that each nucleon moves in its orbit within the nucleus, independently of all other nucleons. The orbit is determined by a potential energy function $V(r)$ which represents the average effect of all interactions with other nucleons, and is the same for each particle. Each nucleon is regarded as an independent particle and the interaction between nucleons is considered to be a small perturbation on the interaction between a nucleon and the potential field. This model is often called the *independent-particle model* as well as the nuclear-shell model. There is a direct analogy between the theoretical treatment of a nucleon in a nucleus and an electron in an atom. The potential energy $V(r)$ is analogous to the Coulomb energy, and the orbit (or quantum state) of a nucleon is analogous to an orbit (or quantum) state of an electron. The theoretical treatment of the nucleon parallels that of the electron as described in Chapter 7, and closed nucleon shells are found analogous to the closed electron shells in the atom. The analogy cannot be carried too far, however, for several reasons. The potential which describes the nuclear attractions is quite different from the Coulomb potential; it has a form between the square-well potential $V = -V_0$, and the so-called *oscillator potential* $V = -V_0 + ar^2$, where r is the distance between the nucleon and the center of force and a is a constant; the Coulomb potential is proportional to $1/r$. Furthermore, two different kinds of particles must be considered, protons and neutrons, rather than a single kind, electrons. The Pauli exclusion principle (Section 7–6) must be applied to both protons and neutrons; it excludes two protons from occupying the same quantum state, and two neutrons from having the same quantum numbers.

The theoretical results are shown in Fig. 17–7. The energy levels for a three-dimensional rectangular well potential are shown on the right (the well has actually been assumed to be infinitely deep). The levels for the simple harmonic oscillator potential are shown on the left. The numbers in parentheses are the total number of electrons in the indicated shell and the shells below. For the oscillator potential there would be closed shells at 2, 8, 20, 40, 70, 112 and 168 nucleons. Although closed shells are predicted at 2, 18, and 20 nucleons, the numbers 50, 82 and 126 are missing and closed shell numbers appear which have no apparent connection with experimental relations. The infinitely deep rectangular well potential also predicts closed shells at 2, 8 and 20 nucleons but, in addition, at many other *nonmagic* numbers, with 50, 82 and 126 again missing. It has been shown that to get the correct shell numbers, i.e., those deduced from the experimental data, another assumption must be made. For reasons which are not yet fully understood, it is necessary to assume that there is a marked difference in the energy of a nucleon according to whether its spin angular momentum is parallel or antiparallel to its orbital angular momentum. When this assumption of *strong spin-orbit coupling* is made and its effects are taken into account,[(41,43)] the theory gives closed nucleon shells at 2, 8, 20, 50, 82, and 126 particles, in agreement with the experimental data. The occurrence of magic numbers of protons and

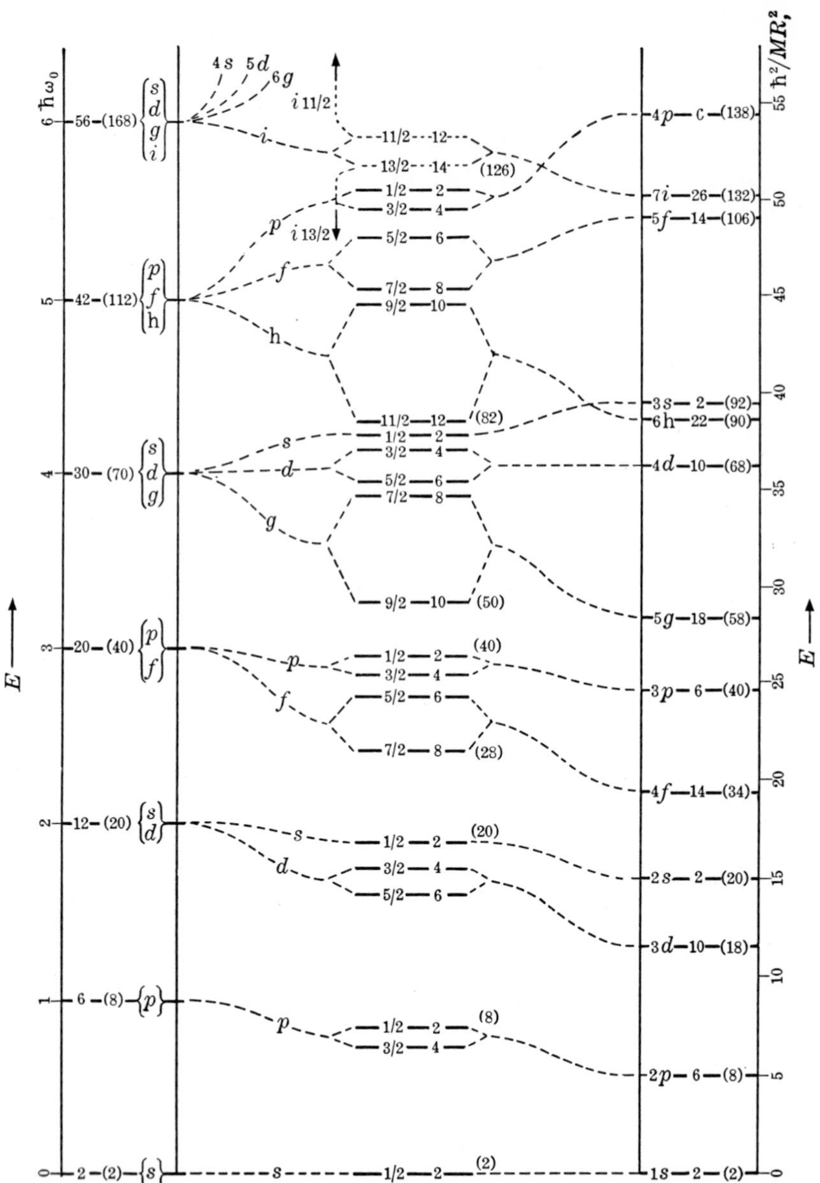

FIG. 17–7. Nuclear shells. The energy levels of an infinitely deep rectangular well are shown on the right, and those of a simple harmonic oscillator potential on the left. The numbers in parentheses give the total of electrons in the indicated and lower shells. The levels in the center are at energies intermediate between those of a square-well potential and the simple harmonic oscillator; in addition a strong spin-orbit interaction is assumed for these levels.[41]

neutrons can then be said to be understood in the sense that a model has been devised which gives the correct numbers when treated mathematically.

The nuclear-shell model has been applied successfully to a variety of nuclear problems. A detailed discussion of these applications cannot be given here, but some of the successes of the shell theory will be mentioned. The model makes it possible to predict the total angular momenta of nuclei and the results are in good agreement with experiment. It is then possible to assign values of the total angular momentum to nuclei for which this quantity has not been measured, in particular to β-radioactive nuclei. This assignment is very useful in the study of β-decay. It has been mentioned (Section 14–6) that the probability of β-decay depends on the difference between the total angular momenta of the initial and product nuclei. The theoretical predictions with regard to the allowed or forbidden nature of many β-decay processes, based on shell model assignments of the angular momentum, have been compared with the experimental results, and reasonably good agreement has been obtained.[(44,45,46)]

The subject of nuclear isomerism has been studied from the viewpoint of the nuclear-shell model,[(47,48,49)] and a correlation has been found between the distribution of isomers and the magic numbers. When the known long-lived isomers ($T_{1/2} \geq 1$ sec) with odd mass number A are plotted against their odd proton or odd neutron number, as in Fig. 17–8, it is clear that there are groupings or islands of isomers just below the magic numbers 50, 82, and 126. There is a sharp break at each of these numbers; isomerism disappears when a shell is filled and does not appear again until the next shell is about half full. An example of the discontinuity in the distribution of isomers is given by the odd-A isotopes of xenon: $_{54}\text{Xe}^{127}$, $_{54}\text{Xe}^{129}$, $_{54}\text{Xe}^{131}$, $_{54}\text{Xe}^{133}$, and $_{54}\text{Xe}^{135}$ with 73, 75, 77, 79, and 81 neutrons, respectively, all show isomerism, but no isomerism has been found in $_{54}\text{Xe}^{137}$ and $_{54}\text{Xe}^{139}$ with 83 and 85 neutrons, respectively. Isomerism (the existence of long-lived excited states) occurs when neighboring energy levels of a nucleus have a large difference in total angular momentum, so that the transition is highly forbidden. The shell model can be used to predict the total angular momenta of low-lying excited levels. It predicts that the conditions for isomerism should exist below the magic numbers 50, 82, and 126, but not immediately above them; in the latter regions, neighboring levels are expected to have small angular momentum differences. The model also predicts correctly when isomerism should appear in the unfilled shells. The classification of isomers according to the type of transitions (Section 15–10) is also in excellent agreement with level assignments based on the shell model.

The experimental data on magnetic moments and electric quadrupole moments have also been interpreted with some success in terms of the nuclear shell model. As an example, at the proton numbers 2, 8, 20, 50,

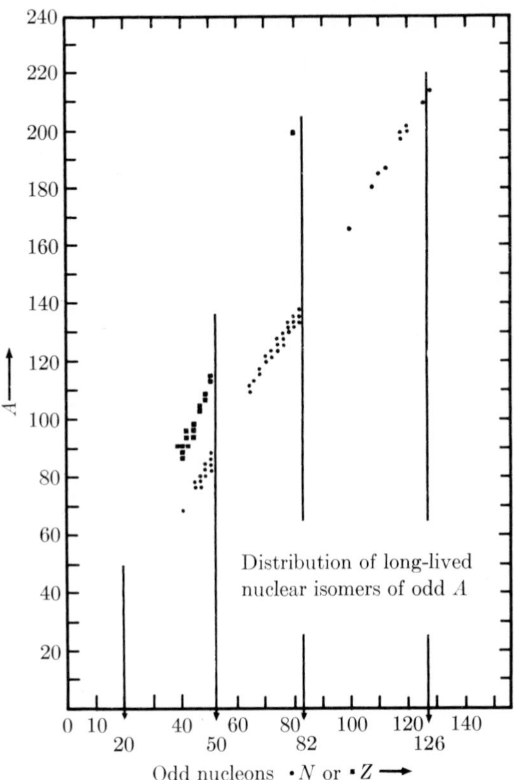

Fig. 17–8. Distribution of long-lived isomers of odd A as a function of either atomic number or neutron number (adapted from Goldhaber and Hill[47]).

and 82, the quadrupole moment is zero or small. When a new shell begins to form, the quadrupole moment is negative; as the number of protons in the unfilled shell is increased, Q becomes positive and increases until it reaches a maximum when the shell is about $\frac{2}{3}$ filled; Q then decreases to zero at the magic proton numbers,[50] after which it becomes negative. This behavior is in accord with the predictions of shell theory. In some cases, however, the quadrupole moment is much larger than expected from the independent-particle model.[51] The magnitude of the quadrupole moment is a measure of the deviation of a nucleus from spherical shape, and the shell model seems to underestimate this deviation. The large values of the moment indicate that in some cases the nucleus is far from spherical. These large values can be explained if it is assumed that the part of the nucleus consisting of the filled shells forms a core which can be deformed by the nucleons in the unfilled shell. When a spheroidal, rather than a spherical, core is treated mathematically, the quadrupole moments come out closer to the experimental values.[52] This modification leads to an extension of the shell model which will be discussed very briefly in Section 17-9.

17-8 The liquid drop model and the semiempirical binding energy formula. A great deal of importance has been attached to the mass and binding energy of the nucleus, and a formula which would allow the calculation of nuclear masses or binding energies would be very useful. Such a formula has been developed and is called the *semiempirical mass formula*, or the *semiempirical binding energy formula*;[53,54] it is related to the liquid-drop model of the nucleus. Some of the properties of nuclear forces (saturation, short-range), which have been deduced from the approximately linear dependence of the binding energy and the volume on the number of particles in the nucleus, are analogous to the properties of the forces which hold a liquid drop together. Hence, a nucleus may be considered to be analogous to a drop of incompressible fluid of very high density ($\sim 10^{14}$ gm/cm^3). This idea has been used, together with other classical ideas such as electrostatic repulsion and surface tension, to set up a semi-empirical formula for the mass or binding energy of a nucleus in its ground state. The formula has been developed by considering the different factors which affect the nuclear binding, and weighting these factors with constants derived from theory where possible and from experimental data where theory cannot help.

The main contribution to the binding energy of the nucleus comes from a term proportional to the mass number A; since the volume of the nucleus is also proportional to A, this term may be regarded as a *volume energy:*

$$E_v = a_v A. \tag{17-20}$$

The Coulomb energy between the protons tends to lower the binding energy and its effect appears as a term with a minus sign. The total Coulomb energy of a nucleus of charge Z is given by Eq. (17-1), and its effect on the binding energy is represented by the term

$$\begin{aligned} E_c &= -a'_c \frac{3}{5} \frac{Z(Z-1)e^2}{R} \\ &= -4a_c \frac{Z(Z-1)}{A^{1/3}}, \end{aligned} \tag{17-21}$$

where

$$a_c = \frac{3}{20} \frac{e^2}{R_0 \times 10^{-13}}. \tag{17-22}$$

In Eq. (17-21), the radius R has been expressed in terms of $A^{1/3}$, and the factor 4 has been introduced in order to conform to the notation often found in the literature.

The binding energy is also reduced because the nucleus has a surface. Particles at the surface interact, on the average, only with half as many other particles as do particles in the interior of the nucleus. In the attractive term (17–20), it was assumed that every nucleon has the same access to other nucleons and it is necessary, therefore, to subtract a term proportional to the surface area of the nucleus. This *surface energy* is represented by

$$E_s = -a_s A^{2/3}, \tag{17–23}$$

and is analogous to the surface tension of a liquid.

The binding energy formula needs another term to represent the so-called *symmetry effect*. For a given value of A, there is a particular value of Z which corresponds to the most stable nuclide. For light nuclides, where the Coulomb effect is small, this value is $A/2$, as is seen from the fact that the numbers of protons and neutrons are equal in the most abundant light nuclides. In the absence of the Coulomb effect, a departure from the condition $Z = A/2$ would tend to lead to instability and a smaller value of the binding energy. A term proportional to some power of the *neutron excess* $(A - 2Z)$ would represent the magnitude of this effect. The second power is chosen because the term then vanishes for $Z = A/2$, as does its derivative with respect to Z. The latter condition corresponds to the maximum value of the binding energy in the absence of the Coulomb energy. A detailed study of the symmetry effect shows that it is also inversely proportional to A, with the result that the symmetry energy can be written

$$E_\tau = -a_\tau \frac{(A - 2Z)^2}{A}. \tag{17–24}$$

It was noted in Section 9–4, during the discussion of the stable nuclides, that those with even numbers of protons and neutrons are the most abundant and presumably the most stable. Nuclei with odd numbers of both protons and neutrons are the least stable, while nuclei for which either the proton or neutron number is even are intermediate in stability. It has been shown that this "odd-even" effect can be represented by a term E_δ whose value depends in the following way on the numbers of protons Z and neutrons N in the nucleus:

A	Z	N	E_δ	
even	even	even	$+\delta/2A$	
odd	even	odd	0	(17–25)
odd	odd	even	0	
even	odd	odd	$-\delta/2A$	

The formula for the binding energy is obtained by combining all of the terms just discussed into the equation

$$\text{B.E.} = a_v A - 4a_c \frac{Z(Z-1)}{A^{1/3}} - a_s A^{2/3} - a_\tau \frac{(A-2Z)^2}{A} + E_\delta. \quad (17\text{–}26)$$

Equation (17–26) can also be rewritten to give the nuclear (or atomic) mass, since the mass and binding energy are related by the equation

$$M = ZM_p + (A - Z)M_n - \frac{\text{B.E.}}{c^2},$$

where M_p, M_n, and M are the masses of the proton, neutron, and nucleus, respectively. The semiempirical mass formula is then

$$M = ZM_p + (A - Z)M_n - \frac{a_v}{c^2} A + \frac{4a_c}{c^2} \frac{Z(Z-1)}{A^{1/3}} + \frac{a_s}{c^2} A^{2/3} + \frac{a_\tau}{c^2} \frac{(A-2Z)^2}{A} - \frac{\delta}{2c^2 A}. \quad (17\text{–}27)$$

The equations (17–26) and (17–27) express the binding energy and the atomic mass of a nuclide as a function of A and Z. The values of the constants can be determined by a combination of theoretical calculations and adjustments to fit experimental values of the masses (or binding energies). The values of the constants a_v, a_c, a_τ, a_s, and δ have been determined, with the results

$$a_v = 14.0 \text{ Mev}, \quad a_c = 0.146 \text{ Mev},$$
$$a_\tau = 19.4 \text{ Mev}, \quad a_s = 13.1 \text{ Mev};$$

the quantity δ is 270 Mev for nuclides with even values of A and Z, zero for odd A nuclides, and -270 Mev for even A, odd Z nuclides. In practice, masses of atoms are usually measured, and Eq. (17–27) is used to calculate atomic masses by using the mass of the hydrogen atom instead of the mass of the proton. When the values $M_n = 1.008982$ amu, and M_{H} (instead of M_p) $= 1.008142$ amu, are inserted, Eq. (17–27) can be written

$$M = 0.99395A - 0.00084Z + 0.0141A^{2/3} + 0.021 \frac{(A-2Z)^2}{A} + \frac{0.00063Z(Z-1)}{A^{1/3}} - \frac{\delta'}{A}, \quad (17\text{–}28)$$

where

$$\delta' = \begin{cases} 0.145 \text{ amu, even } A, \text{ even } Z, \\ 0, \text{ odd } A, \text{ even } Z; \text{ odd } A, \text{ odd } Z, \\ -0.145 \text{ amu, even } A, \text{ odd } Z. \end{cases} \quad (17\text{–}29)$$

Equation (17–26) becomes

$$\text{B.E. (Mev)} = 14.0A - 0.584\frac{Z(Z-1)}{A^{1/3}} - 13.1A^{2/3} - 19.4\,\frac{(A-2Z)^2}{A} + E_\delta, \tag{17-30}$$

where

$$E_\delta = \begin{cases} \dfrac{135}{A}, & \text{even } A,\ \text{even } Z, \\ 0, & \text{odd } A,\ \text{even } Z;\ \text{odd } A,\ \text{odd } Z, \\ \dfrac{-135}{A}, & \text{even } A,\ \text{odd } Z. \end{cases} \tag{17-31}$$

Several formulas similar to Eq. (17–28) have been developed with somewhat different values for the various constants depending on the amount of effort put into the determination of those values. One of these formulas has been used to calculate the mass values corresponding to about 5000 pairs of Z, A values.(55) The semiempirical mass or binding energy formula reproduces quite accurately the values for many nuclei but does not account for all of the features of the nuclear binding energy, as will be shown later. It does give the main features of the dependence of binding energies on mass number and charge number.

The binding energy formula helps account for some of the important features of the stability properties of nuclei, in particular the β-activity and stability properties of isobars. This application of the binding energy formula will now be illustrated. An examination of the nuclide chart shows that isobars (nuclides with the same mass number) are subject to interesting stability rules.(56) One of these rules is that pairs of stable isobaric nuclides differ in charge number much more frequently by two units than by one. More than 60 examples of isobaric pairs are known for which $\Delta Z = 2$, but only 2 pairs are known for which $\Delta Z = 1$; the latter are ${}_{48}\text{Cd}^{113}$ and ${}_{49}\text{In}^{113}$, and ${}_{51}\text{Sb}^{123}$ and ${}_{52}\text{Te}^{123}$. The values $A = 113, 123$ are the only odd values of A for which there are more than one stable nuclide. For even values of A, there may be two stable isobars, and in a few cases ($A = 96, 124, 130, 136$) even three. In accordance with the rule stated above, the isobars differ in atomic number by two units, as in ${}_{50}\text{Sn}^{124}$, ${}_{52}\text{Te}^{124}$, and ${}_{54}\text{Xe}^{124}$; the isobars with odd Z, ${}_{51}\text{Sb}^{124}$ and ${}_{53}\text{I}^{124}$, are unstable.

Consider now the binding energy formula (17–30) for a given odd value of A. The only variable terms are those which contain Z, and these depend on Z^2; the odd-even term vanishes. If the binding energies for a series of

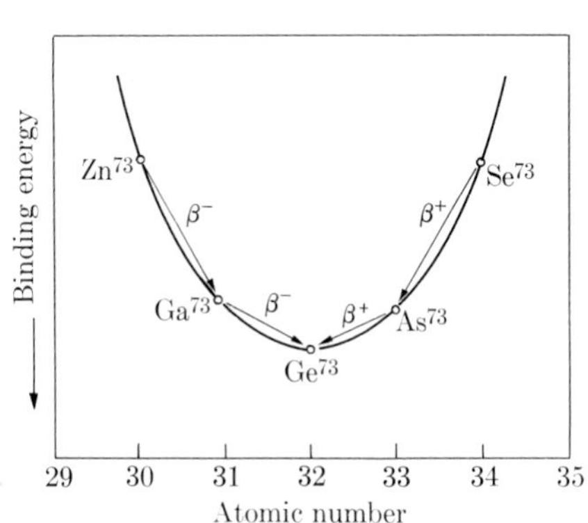

FIG. 17-9. Binding energies and decay properties of nuclides of odd mass number ($A = 73$).

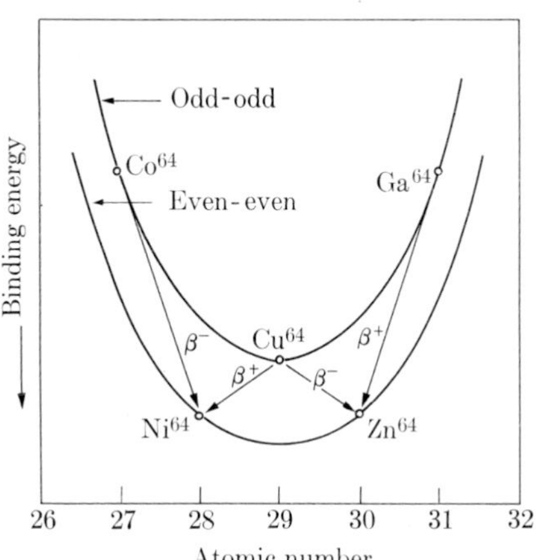

FIG. 17-10. Binding energies and decay properties of nuclides of even mass number ($A = 64$).

nuclides with constant odd A and varying Z are plotted against Z, the result is a parabola-like curve such as that shown in Fig. 17-9, in which A has been taken as 73. The binding energies decrease with increasing values of the ordinate, so that the bottom of the curve represents the most stable nucleus of the series, with the greatest binding energy. In general, there will be one isobar for which the binding energy is at or near the bottom of the curve, and this will be the only stable member of the series. All isobars whose binding energies are less than that of the stable one will lie on the two arms of the curve. Their masses will be greater than that of the stable isobar and they will decay by the emission of an electron or a positron or by K-capture, subject to the conditions expressed in Eqs. (12-4), (12-7), and (12-10). Isobars to the left of the stable one have fewer protons than the stable one and decay by electron emission,

$$_{30}\text{Zn}^{73} \xrightarrow{\beta^-} {}_{31}\text{Ga}^{73} \xrightarrow{\beta^-} {}_{32}\text{Ge}^{73} \text{ (stable)}.$$

Isobars which lie to the right of the lowest point in the binding energy curve contain an excess of protons and decay by positron emission, or by K-capture, or by both, e.g.,

$$_{34}\text{Se}^{73} \xrightarrow{\beta^+,K} {}_{33}\text{As}^{73} \xrightarrow{K} {}_{32}\text{Ge}^{73} \text{ (stable)}.$$

The results are different for isobars with even A because of the odd-even effect. The binding energy curve, considered as a function of Z, will still be parabolic in shape. But nuclides with even numbers of protons and neutrons have an additional positive energy contribution (135 Mev/A), while nuclei with odd numbers of protons and neutrons will have binding energies decreased by this amount. Hence, there are two curves, one for even Z-even A nuclides and another for odd Z-even A nuclides, as in Fig. 17–10. The odd-even nuclides lie on the upper curve and are therefore unstable with respect to those on the lower curve, with the result that no stable odd Z-even A nuclides should exist. The only exceptions to this rule are the light nuclides H^2, Li^6, B^{10}, and N^{14}. As in other respects, light nuclei must be treated separately; a model like the liquid-drop model would not be expected to apply to these nuclei, which have too few particles to be treated as a fluid drop with a continuous distribution of material. It can be seen from the curve for even Z-even A nuclides, that two isobars with Z-values differing by two units can lie close to the bottom of the lower curve; these constitute stable isobaric pairs. Nuclides to the left of the stable nuclides decay by electron emission, and those to the right by positron emission or K-capture or both. The isobar at the bottom of the odd-even curve may decay by either positron or electron emission.

The binding energy or mass formula does not include closed shell effects; but it can be used to provide a base line from which shell effects can be calculated. Studies of the energetics of β-decay(35) and mass-spectroscopic measurements(31,32) show that the mass of any isobar which has Z or $N = 20$, 28, 50, 82 or 126 lies about 1 or 2 Mev below the value predicted by the smoothly varying mass formula. There are, therefore, discontinuities in the atomic masses at the magic numbers and a shell-structure term can be added to the semiempirical mass or binding energy formula to take the effect into account.(57)

The liquid-drop model has other important applications; the theory of the compound nucleus is based on this model, and the phenomenon of nuclear fission, which will be discussed in Chapter 19, can be described in terms of the distortion and division of a liquid drop.

17–9 The collective nuclear model. The successes of the liquid-drop and nuclear-shell models seem to lead to a serious dilemma. The liquid-drop model can account for the behavior of the nucleus as a whole, as in nuclear reactions and nuclear fission. In the latter, certain nuclei actually divide into two smaller nuclei and the division can be described in terms of the deformation of a drop; an explanation in terms of the motion of a single nucleon, or any number of nucleons, seems impossible. The liquid-drop model is said to describe the *collective* behavior of the nucleus, and the excitation of the nucleus is treated as surface oscillations, elastic vibrations, and other such "collective" modes of motion. On the other hand, many nuclear phenomena seem to show that nucleons behave as individual, and nearly independent, particles. Hence there are two entirely different ways of regarding nuclei, with a basic contradiction between them. The reasonable conclusion is that the two different models or pictures are incomplete parts of a larger or more general model. The problem of the electric quadrupole moments has led to such a model, which includes both the liquid-drop and independent-particle aspects; this new model is called the *collective model.*[58,59,60] It is assumed, in this model, that the particles within the nucleus exert a centrifugal pressure on the surface of the nucleus as a result of which the nucleus may be deformed into a permanently nonspherical shape; the surface may undergo oscillations (liquid-drop aspect). The particles within the nucleus then move in a nonspherical potential like that assumed in order to account for the quadrupole moments. Thus, the nuclear distortion reacts on the particles and modifies somewhat the independent-particle aspect. The nucleus is regarded as a shell structure capable of performing oscillations in shape and size. The result is that the collective model can be made to describe such drop-like properties as nuclear fission[61,62] while at the same time preserving the shell model characteristics and, in fact, improving on the earlier shell model.

The simplest type of collective motion which has been identified experimentally is connected with rotations of deformed nuclei.[63,64] If the rotational collective motion is sufficiently slow, it will not affect the internal structure of the nucleus. According to classical physics, the rotational energy is proportional to the square of the angular velocity ω, so that

$$E_{\text{rot}} = \tfrac{1}{2}I\omega^2. \tag{17–32}$$

The parameter I denotes an effective moment of inertia of the nucleus which can be calculated if a specific model is assumed for the internal structure. Rotational energy levels are obtained when the angular momentum is quantized and the result is, for even-even nuclei,

$$E_{\text{rot}} = \frac{\hbar^2}{2I} J(J + 1), \tag{17–33}$$

where J is the total angular momentum quantum number of the nucleus. In the case of a spheroidal nucleus, the deformation is symmetric with respect to reflection in the nuclear center. As a result, J is restricted to even values, 0, 2, 4, 6, $\cdots$, and the parity should be even. According to the theory, the deformation (deviation from spherical shape) should be greatest and the rotational levels most easily observed for nuclei with numbers of nucleons far from closed shells. The first excited state should be a 2+ state; the second excited state should be a 4+ state with energy $5 \cdot 4/3 \cdot 2 = 3\frac{1}{3}$ times that of the first excited state. The third and fourth states should have $I = 6+$ and $8+$, respectively, and energies 7 and 12 times that of the 2+ state. The experimental data agree quite well with these predictions, as is evident from Fig. 17–11, which shows the experi-

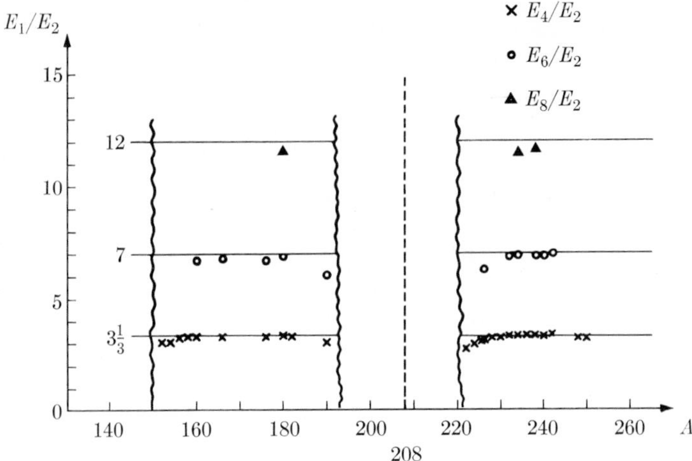

FIG. 17–11. Experimentally observed energy ratios for rotational levels in even-even nuclei.(74)

mentally observed energy ratios for excited rotational states in even-even nuclei with mass numbers in the regions $150 < A < 190$ and $A > 220$. These are just the regions farthest from closed shells. The theory also predicts correctly the properties of the low-lying levels of nearly spherical nuclei close to closed shells.(64) Another successful application is to the problem of the cross sections for Coulomb excitation and for the probability of $E2$ transitions.(65) Finally, the theory has been applied successfully to problems of magnetic moments, quadrupole moments, and isomeric transitions.(60,66)

17–10 The optical model for nuclear reactions. In the study of nuclear reactions in Chapter 16, it became evident that the theory of nuclear reactions based on the concept of the compound nucleus (the *statistical model* for nuclear reactions) is not always reliable. The statistical nature of the compound nucleus theory implies that its predictions are at best averages, and do not take into account the differences between specific nuclei. It is not surprising, therefore, that a more detailed model is needed for the description of nuclear reactions.

One of the most serious failures of the compound nucleus theory is its inability to account for the large-scale energy dependence of total neutron cross sections. The statistical model assumes that the compound nucleus is formed immediately when the incident neutron reaches the nuclear surface. The cross section for reaching the surface turns out to be a monotonically decreasing function of the energy, varying as $E^{-1/2}$ for small energies and reaching the asymptotic value $2\pi R^2$ for large energies.(67)

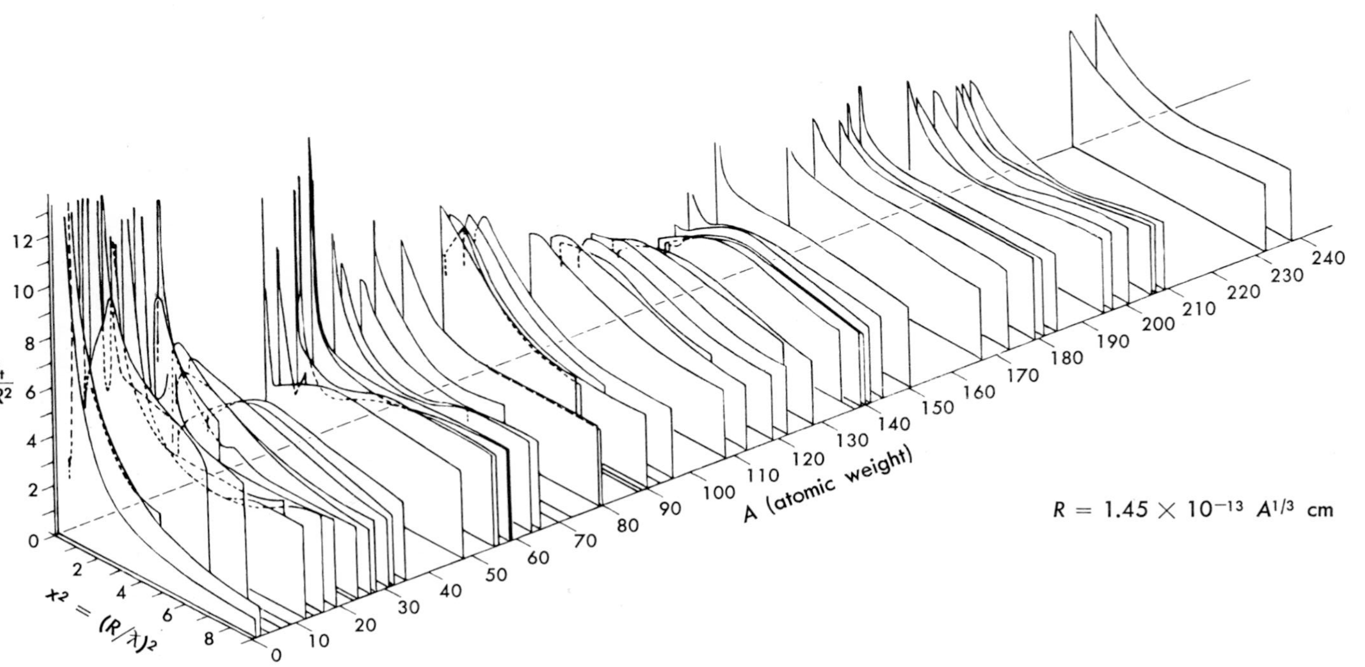

FIG. 17-12. Observed total neutron cross sections as a function of energy and mass number.(69,71) The energy is expressed as $x^2 = (R/\lambda\!\!\!^-)^2$, where $\lambda\!\!\!^-$ is the de Broglie wavelength of the neutrons.

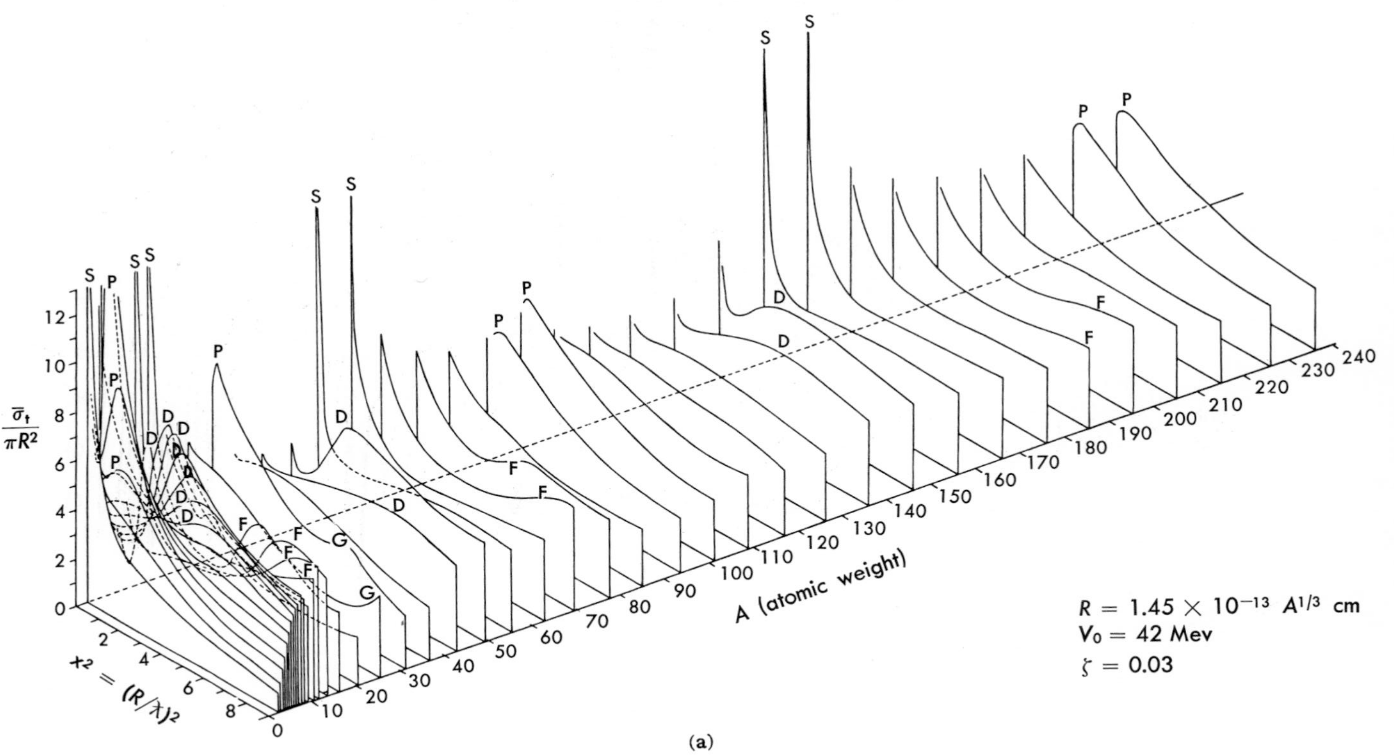

$\frac{\bar{\sigma}_t}{\pi R^2}$
$x^2 = (R/ƛ)^2$
A (atomic weight)
$R = 1.45 \times 10^{-13} A^{1/3}$ cm
$V_0 = 42$ Mev
$\zeta = 0.03$

(a)

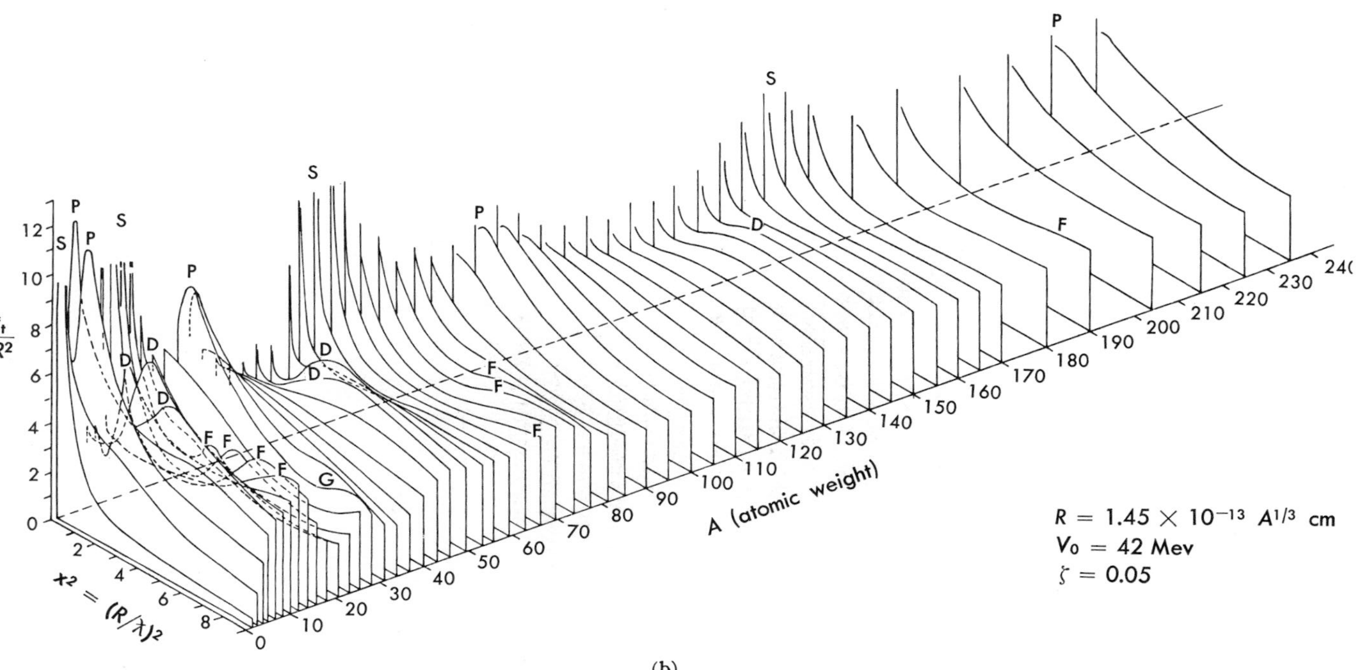

FIG. 17–13. (a) Calculated neutron total cross sections as a function of energy and mass number, for a well depth $V_0 = 42$ Mev, radius $R = 1.45 \times 10^{-13} A^{1/3}$ cm, $\zeta = 0.03$. (b) The same for $\zeta = 0.05$.(71)

The experimental results do not agree; at excitation energies above the region of sharply defined resonances, the observed total cross sections show a more complicated behavior which seems to point to a combination of single particle and compound nucleus properties. When the cross sections are plotted against energy E and atomic number A in a three-dimensional graph, as in Fig. 17–12, certain regularities are observed, with maxima and minima at values of E which shift in a systematic way with increasing A.(68,69) At the neutron energies involved, between 0.1 Mev and several Mev, and for intermediate or heavy nuclei, individual resonances cannot be resolved, and the measured cross sections are averages over many levels. The problem is that of accounting for this *gross structure* of the total neutron cross section.

The large-scale maxima are similar to those predicted by the theory of a single particle moving in an attractive potential $V(r)$, e.g., a square well,(70) but such a potential does not permit absorption of the particle as required by the compound nucleus concept. The single particle and compound nucleus pictures have been combined successfully in the *optical model*.(71,72) This model describes the effect of the nucleus on the incident particle by a potential well $-V_0(r)$, but allows for the possibility of compound nucleus formation by adding to the potential a negative imaginary part, $-iV_1(r)$. This part produces an absorption of the incident particle within the nucleus, and this absorption is supposed to represent the formation of the compound nucleus. On the optical model, compound nucleus formation does not occur "immediately" nor with complete certainty. Even if the incident particle has entered the nucleus, it is removed from its free particle state only with some delay and with a certain probability. If $V_0(r)$ and $V_1(r)$ are reasonably constant over the nucleus, it is possible to define a *coalescence coefficient* which is the probability per unit length for the incident particle in nuclear matter to form the compound nucleus. In this picture, if an incident particle gets into the nucleus, it is reflected back and forth before escaping or being absorbed. The theory accounts very well for the behavior of the neutron cross sections, as is evident in Fig. 17–13, with the rather simple potential,

$$V_0 = -U, \qquad \text{for } r < R,$$

$$V_0 = 0, \qquad \text{for } r > R,$$

$$V_1 = \zeta V_0,$$

and the values

$$V_0 = 42 \text{ Mev}, \qquad \zeta = 0.03 \text{ to } 0.05,$$

$$R = 1.45 \times 10^{-13} A^{1/3} \text{ cm}.$$

In the optical model, the nucleus is not "black" to the wave representing the incident particle; instead it acts like a gray sphere, partly absorbing and partly refracting the incoming wave. There is an analogy with physical optics in that the nucleus acts like a spherical region with a given refractive index (attractive potential well) and opacity. It is this analogy which is responsible for the adjective *optical*, and for the name *cloudy crystal ball* model, which is sometimes used. The model has also been extended and applied with more complicated potential functions to other nuclear reactions and cross sections.[72]

References

GENERAL

R. D. Evans, *The Atomic Nucleus.* New York: McGraw-Hill, 1955, Chapters 2, 10, 11, Appendix C2.

H. A. Bethe and P. Morrison, *Elementary Nuclear Theory,* 2nd ed. New York: Wiley, 1956.

L. R. B. Elton, *Introductory Nuclear Theory.* New York: Interscience Publishers, Inc., 1959, Chapters 3, 4, 5.

L. Eisenbud and E. P. Wigner, *Nuclear Structure.* Princeton, New Jersey: Princeton University Press, 1958, Chapters 5, 6, 7, 8.

L. Landau and Ya. Smorodinsky, *Lectures on Nuclear Theory* (translated from the Russian). New York: Plenum Press, 1959.

E. Fermi, *Nuclear Physics.* Notes Compiled by J. Orear, A. H. Rosenfeld, and R. A. Schluter. Revised Edition. Chicago: University of Chicago Press, 1950.

J. M. Blatt and V. F. Weisskopf, *Theoretical Nuclear Physics.* New York: Wiley, 1952.

R. G. Sachs, *Nuclear Theory.* Reading, Mass.: Addison-Wesley, 1953.

N. F. Ramsey, "Nuclear Two-Body Problems and Elements of Nuclear Forces," *Experimental Nuclear Physics,* E. Segrè, ed., Vol. I, Part IV. New York: Wiley, 1952.

K. B. Mather and P. Swan, *Nuclear Scattering.* Cambridge: Cambridge University Press, 1958.

M. G. Mayer and J. H. D. Jensen, *Elementary Theory of Nuclear Shell Structure.* New York: Wiley, 1955.

E. Feenberg, *Shell Theory of the Nucleus.* Princeton, New Jersey: Princeton University Press, 1955.

ARTICLES IN THE *Encyclopedia of Physics* (*Handbuch derPhysik*) Vol. 39. Berlin: Springer Verlag, 1957.

1. L. Hulthén and M. Sugawara, "The Two-Nucleon Problem," pp. 1–143.
2. D. L. Hill, "Matter and Charge Distribution within Atomic Nuclei," pp. 178–240.
3. J. P. Elliott and A. M. Lane, "The Nuclear Shell Model," pp. 241–410.
4. S. A. Moszkowski, "Models of Nuclear Structure," pp. 411–559.

PARTICULAR

1. H. A. Bethe, "Nuclear Physics A. Stationary States of Nuclei," *Revs. Mod. Phys.* **8,** 82 (1936); esp. p. 96.

2. E. Wigner, "On the Consequences of the Symmetry of the Nuclear Hamiltonian on the Spectroscopy of Nuclei," *Phys. Rev.* **51,** 107 (1937).

3. E. Feenberg and E. Wigner, "On the Structure of the Nuclei Between Helium and Oxygen," *Phys. Rev.* **51,** 95 (1937).

4. E. Wigner and E. Feenberg, "Symmetry Properties of Nuclear Levels," *Rep. Prog. Phys.* London: Physical Society. Vol. 8 (1941), pp. 274–317.

5. D. R. Inglis, "The Energy Levels and the Structure of Light Nuclei," *Revs. Mod. Phys.* **25,** 390 (1953).

6. T. Lauritsen, "Energy Levels of Light Nuclei," *Ann. Revs. Nuc. Sci.* **1,** 67 (1952).

7. F. Ajzenberg-Selove and T. Lauritsen, "Energy Levels of Light Nuclei," (a) *Nuclear Physics* **11,** 1–340 (1959); (b) *Ann. Revs. Nuc. Sci.* **10,** 409 (1961).

8. W. E. Burcham, "Isotopic Spin and Nuclear Reactions," *Progress in Nuclear Physics,* O. R. Frisch, ed. London: Pergamon Press, **4,** 171 (1955).

9. K. W. Ford and D. L. Hill, "The Distribution of Charge in the Nucleus," *Ann. Revs. Nuc. Sci.* **5,** 25 (1955).

10. J. M. C. Scott, "The Radius of a Nucleus," *Progress in Nuclear Physics,* O. R. Frisch, ed. London: Pergamon Press, **5,** 157 (1956).

11. O. Kofoed-Hansen, "Mirror Nuclei Determinations of Nuclear Size," *Revs. Mod. Phys.* **30,** 449 (1958).

12. A. E. S. Green, "Nuclear Sizes and the Weiszacker Mass Formula," *Revs. Mod. Phys.* **30,** 569 (1958).

13. R. Hofstadter: (a) "Electron Scattering and Nuclear Structure," *Revs. Mod. Phys.* **28,** 214 (1956); (b) "Nuclear and Nucleon Scattering of High-Energy Electrons," *Ann. Revs. Nuc. Sci.* **7,** 231 (1957).

14. E. M. Henley, "Nuclear Radii from Mesonic Atoms," *Revs. Mod. Phys.* **30,** 438 (1958).

15. "Papers from the International Congress on Nuclear Sizes and Density Distributions held at Stanford University, Dec. 17–19, 1957," *Revs. Mod. Phys.* **30,** 412–584 (1958).

16. J. D. Jackson and J. M. Blatt, "The Interpretation of Low-Energy Proton-Proton Scattering," *Revs. Mod. Phys.* **22,** 77 (1950).

17. G. L. Squires, "The Neutron-Proton Interaction," *Progress in Nuclear Physics,* O. R. Frisch, ed. London: Pergamon Press, **2,** 89 (1952).

18. G. Breit and M. H. Hull, Jr., "Advances in Knowledge of Nuclear Forces," *Am. J. Phys.* **21,** 184–220 (1953).

19. Bailey, Bennet, Bergstrahl, Nuckolls, Richards, and Williams, "The Neutron-Proton and Neutron-Carbon Scattering Cross Sections for Fast Neutrons," *Phys. Rev.* **70,** 583 (1946).

20. D. H. Frisch, "The Total Cross Sections of Carbon and Hydrogen for Neutrons of Energies from 35 to 490 kev," *Phys. Rev.* **70,** 589 (1946).

21. Lampi, Freier, and Williams, "The Total Scattering Cross Section of Neutrons by Hydrogen and Carbon," *Phys. Rev.* **80,** 853 (1950).

22. Barnes, Carver, Stafford, and Wilkinson, "The Photodisintegration of the Deuteron at Intermediate Energies, I," *Phys. Rev.* **86,** 359 (1952).

23. D. H. Wilkinson, "The Photodisintegration of the Deuteron at Intermediate Energies, II," *Phys. Rev.* **86,** 373 (1952).

24. Harris, Muehlhause, Rose, Schroeder, Thomas, and Wexler, "Thermal Neutron-Proton Capture," *Phys. Rev.* **91,** 125 (1953).

25. J. M. Blatt and V. F. Weisskopf, *Theoretical Nuclear Physics.* New York: Wiley, 1952, Chapter III.

26. M. G. Mayer, "On Closed Shells in Nuclei," *Phys. Rev.* **74,** 235 (1948).

27. B. H. Flowers, "The Nuclear Shell Model," *Progress in Nuclear Physics,* O. R. Frisch, ed. London: Pergamon Press, **2,** 235 (1952).

28. M. H. L. Pryce, "Nuclear Shell Structure," *Reports on Progress in Physics.* London: Physical Society, **17,** 1 (1954).

29. H. E. Duckworth and R. S. Preston, "Some New Atomic Mass Measurements and Remarks on the Mass Evidence for Magic Numbers," *Phys. Rev.* **82,** 468 (1951).

30. H. E. Duckworth, "Evidence for Nuclear Shells from Atomic Mass Measurements," *Nature* **170,** 158 (1952).

31. Collins, Nier, and Johnson, "Atomic Masses from Titanium through Zinc," *Phys. Rev.* **86,** 408 (1952).

32. B. G. Hogg and H. E. Duckworth, "An Atomic Mass Study of Nuclear Shell Structure in the Region $28 \leq N \leq 50, 28 \leq Z \leq 40$," *Can. J. Phys.* **31,** 942 (1953).

33. H. Brown, "Table of Relative Abundances of Nuclear Species," *Revs. Mod. Phys.* **21,** 625 (1949).

34. Perlman, Ghiorso, and Seaborg, "Systematics of Alpha-Radioactivity," *Phys. Rev.* **77,** 26 (1950).

35. C. D. Coryell, "Beta-Decay Energetics," *Ann. Revs. Nuc. Sci.,* Vol 2. Stanford: Annual Reviews, Inc., 1953.

36. Hughes, Garth, and Levin, "Fast Neutron Cross Sections and Level Density," *Phys. Rev.* **91,** 1423 (1953).

37. B. H. Flowers, *op. cit.* Ref. 27, p. 241.

38. J. A. Harvey, "Neutron Binding Energies from (d, p) Reactions and Nuclear Shell Structure," *Phys. Rev.* **81,** 353 (1951).

39. E. Feenberg and K. C. Hammack, "Nuclear Shell Structure," *Phys. Rev.* **75,** 1877 (1949).

40. L. W. Nordheim, "On Spins, Moments and Shells in Nuclei," *Phys. Rev.* **75,** 1894 (1949).

41. Haxel, Jensen, and Seuss, "Modellmässige Deutung der ausgezeichneten Nukleonzahlen in Kernbau," *Z. Physik.* **128,** 295 (1950).

42. M. G. Mayer, "On Closed Shells in Nuclei," *Phys. Rev.* **75,** 1969 (1949).

43. M. G. Mayer, "Nuclear Configuration in the Spin-Orbit Coupling Model. I. Empirical Evidence," *Phys. Rev.* **78,** 16 (1950).

44. Mayer, Moszkowski, and Nordheim, "Nuclear Shell Structure and Beta-Decay. I. Odd-A Nuclei," *Revs. Mod. Phys.* **23,** 315 (1951).

45. L. W. Nordheim, "Nuclear Shell Structure and Beta-Decay. II. Even-A Nuclei," *Revs. Mod. Phys.* **23,** 322 (1951).

46. M. G. MAYER, "Nuclear Shell Systematics: Classification of β-Transitions," *Beta- and Gamma-Ray Spectroscopy,* K. Siegbahn, ed. Amsterdam: North-Holland Publishing Co.; New York: Interscience. 1955, pp. 433–452.

47. M. GOLDHABER and R. D. HILL, "Nuclear Isomerism and Shell Structure," *Revs. Mod. Phys.* **24,** 179–239 (1952).

48. M. GOLDHABER and A. W. SUNYAR, "Classification of Nuclear Isomers," *Phys. Rev.* **83,** 906 (1951).

49. M. GOLDHABER and A. W. SUNYAR, "Classification of Nuclear Isomers," *Beta- and Gamma-Ray Spectroscopy, op. cit.* Ref. 46, pp. 453–467.

50. W. GORDY, "Relation of Nuclear Quadrupole Moment to Nuclear Structure Theory," *Phys. Rev.* **76,** 139 (1949).

51. TOWNES, FOLEY, and LOW, "Nuclear Quadrupole Moments and Shell Structure," *Phys. Rev.* **76,** 1415 (1949).

52. J. RAINWATER, "Nuclear Energy Level Argument for a Spheroidal Nuclear Model," *Phys. Rev.* **79,** 432 (1950).

53. E. FEENBERG, "Semi-empirical Theory of the Nuclear Energy Surface," *Revs. Mod. Phys.* **19,** 233 (1947).

54. A. E. S. GREEN and N. A. ENGLER, "Mass Surfaces," *Phys. Rev.* **91,** 40 (1953).

55. N. METROPOLIS, *Table of Atomic Masses.* Chicago: Institute for Nuclear Studies, 1948.

56. N. FEATHER, *Nuclear Stability Rules.* Cambridge: University Press, 1952, Chapter 1.

57. A. E. S. GREEN and D. F. EDWARDS, "Discontinuities in the Nuclear Mass Surface," *Phys. Rev.* **91,** 46 (1953).

58. A. BOHR, "The Coupling of Nuclear Surface Oscillations to the Motion of Individual Particles," *Kgl. Danske Videnskab. Selskab, Mat-fys. Medd.* **26,** No. 14 (1952).

59. A. BOHR and B. R. MOTTELSON, "Collective and Individual-Particle Aspects of Nuclear Structure," *Kgl. Danske Videnskab. Selskab, Mat-fys. Medd.* **27,** No. 16 (1953).

60. A. BOHR and B. R. MOTTELSON, "Collective Nuclear Motion and the Unified Model," *Beta- and Gamma-Ray Spectroscopy,* K. Siegbahn, ed., *op. cit.* Ref. 46, pp. 469–493.

61. D. L. HILL and J. A. WHEELER, "Nuclear Constitution and the Interpretation of Fission Phenomena," *Phys. Rev.* **89,** 1102 (1953).

62. J. A. WHEELER, "Nuclear Fission and Nuclear Stability," *Niels Bohr and the Development of Physics,* W. Pauli, ed. London: Pergamon Press, 1955, pp. 163–184.

63. A. BOHR, "Rotational States of Atomic Nuclei." Thesis, Copenhagen: E. Munksgaards Forlag, 1955, 53 pp.

64. B. R. MOTTELSON, "Collective Motion in the Nucleus." *Revs. Mod. Phys.* **29,** 186 (1957).

65. ALDER, BOHR, HUUS, MOTTELSON, and WINTHER, "Study of Electromagnetic Excitation with Accelerated Ions," *Revs. Mod. Phys.* **28,** 432 (1956).

66. F. VILLARS, "The Collective Model of Nuclei," *Ann. Revs. Nuc. Sci.* **7,** 185 (1957).

67. H. Feshbach and V. F. Weisskopf, "A Schematic Theory of Nuclear Cross Sections," *Phys. Rev.* **76,** 1550 (1949).

68. H. H. Barschall, "Regularities in the Total Cross Sections for Fast Neutrons," *Phys. Rev.* **86,** 431 (1952).

69. Okazaki, Darden and Walton, "Total Cross Sections of Rare Earths for Fast Neutrons," *Phys. Rev.* **93,** 461 (1954).

70. F. L. Friedman and V. F. Weisskopf, "The Compound Nucleus," *Niels Bohr and the Development of Physics,* W. Pauli, ed. London: Pergamon Press, 1955, pp. 134–162.

71. Feshbach, Porter, and Weisskopf, "Model for Nuclear Reactions with Neutrons," *Phys. Rev.* **96,** 448 (1954).

72. H. Feshbach, "The Optical Model and Its Justification," *Ann. Revs. Nuc. Sci.* **8,** 49 (1958).

73. Strominger, Hollander, and Seaborg, "Table of Isotopes," *Revs. Mod. Phys.* **30,** 585 (1958).

74. R. Wallace and J. A. Welch, Jr., "Beta Spectra of the Mirror Nuclei," *Phys. Rev.* **117,** 1297 (1960).

75. L. R. B. Elton, *Nuclear Sizes.* Oxford University Press, 1961.

Problems

1. The total binding energies (obtained from the measured atomic masses) of the nuclei listed below are given in Table 9–3. Compute the Coulomb energy of each of these nuclei and the ratio of Coulomb energy to total binding energy. Plot all three quantities against the atomic number and the mass number. The nuclei are Be^9, Al^{27}, Cu^{63}, Mo^{98}, Xe^{130}, W^{184}, U^{238}.

2. Calculate the binding energies of the nuclei of Problem 1 from the semiempirical binding energy formula (17–29), and compare the results with the experimental values.

3. Calculate the atomic masses of the nuclides of Problem 1 from the semiempirical mass formula (17–28) and compare the results with the experimental values of Table 9–3.

4. Repeat the calculations of Problems 1, 2, and 3 for the nuclides Ca^{40}, Sn^{120}, and Pb^{208}. What differences are shown by these nuclides from the behavior of the nuclides of Problem 1? How would you account for these differences?

5. From the semiempirical mass formula, Eq. (17–28), derive a condition for the most stable nucleus at a given mass number. [*Hint:* The most stable nucleus with a given mass number should occur for that value of Z, denoted by Z_0, for which the nuclear mass, as given by Eq. (17–28), is a minimum.] What is the value of Z_0 for $A = 27, 64, 125$, and 216? Compare the results with the table of stable nuclides, Table 9–1.

6. Plot $A - Z_0$ against Z_0, where Z_0 is determined from the equation of Problem 5. Compare the resulting curve with the plot of the known nuclides, Fig. 12–1.

7. Calculate, from the semiempirical mass formula, the energy available for the following decay processes, and compare the results with the experimental values. (a) the α-decay of radium, (b) the α-decay of Po^{214}(RaC′), (c) the β-decay of Hf^{170} (experimental endpoint energy = 2.4 Mev), (d) the β (plus γ)-decay of Au^{198} (experimental decay energy = 1.374 Mev).

8. Calculate the Q-value of the reaction $Fe^{54}(n, \gamma)Fe^{55}$ from the semiempirical mass formula, and compare the result with the experimental value given in Problem 13 of Chapter 16.

9. Calculate, from the semiempirical binding energy formula, the binding energies of the last neutron in Pb^{207}, Pb^{208}, and Pb^{209}. Compare the results with the experimental values given in Problem 11 of Chapter 16.

10. Calculate the approximate kinetic energy of a nucleon inside a nucleus under the following assumptions: (1) Because of the short-range interaction with neighboring nucleons, a nucleon occupies a volume, the radius of which is roughly half the range of nuclear forces, with the range taken as $b \simeq 2$ to 3×10^{-13} cm. (2) This radius is equal to $\lambda\!\!\!^{-} = \lambda/2\pi$, where λ is the de Broglie wavelength of a nucleon inside the nucleus.

11. When a thermal neutron (a neutron with practically zero kinetic energy) is captured in a nucleus with Z protons and N neutrons, the excitation energy in the compound nucleus is given by

$$S_n = M(Z, N) + M_n - M(Z, N + 1).$$

(a) What is the difference in the excitation energy of the compound nucleus when a thermal neutron is captured by an odd-N target and by an even-N target of approximately the same mass number? (b) Calculate the difference in excitation energy for the cases

$$_{92}\mathrm{U}^{235} + \mathrm{n} \rightarrow [\mathrm{U}^{236}],$$

and

$$_{92}\mathrm{U}^{238} + \mathrm{n} \rightarrow [\mathrm{U}^{239}].$$

12. Show that the average binding energy per nucleon can be represented by

$$\frac{\text{B.E.}}{A} = a_v - \frac{a_s}{A^{1/3}} - a_c \frac{Z^2}{A^{4/3}} - a_\tau \left(1 - \frac{2Z}{A}\right)^2 + \frac{E_\delta}{A}.$$

Calculate the value of the average binding energy per particle in Xe^{124} and compare the result with that listed in Table 9–3.

13. Show that the Q-value of a nuclear reaction may be written

$$Q = \sum_p (\text{B.E.})_p - \sum_i (\text{B.E.})_i,$$

where the subscripts i, p represent initial and product nuclei, respectively. Then show that the Q-value for a (d, α) reaction on the target nucleus with atomic number Z and mass number A is given, to a good approximation, by

$$Q = (\text{B.E.})_\alpha - (\text{B.E.})_\mathrm{d} - 2a_v + a_s[A^{2/3} - (A-2)^{2/3}]$$
$$+ 4a_c\left[\frac{Z^2}{A^{1/3}} - \frac{(Z-1)^2}{(A-2)^{1/3}}\right] - a_\tau\left[\frac{2(A-2Z)^2}{A(A-2)}\right]$$
$$+ \begin{cases} \delta/A: & A \text{ even, } Z \text{ odd} \\ -\delta/A: & A \text{ even, } Z \text{ even} \\ 0: & A \text{ odd, } Z \text{ anything.} \end{cases}$$

Part III

Special Topics and Applications

CHAPTER 18

NEUTRON PHYSICS

Since the discovery of the neutron in 1932, the scope and importance of neutron physics have grown remarkably, and there is now a wide interest in the ideas, methods, and applications which have been developed in this field. Neutrons, because they are uncharged heavy particles, have properties which make them especially interesting and important in contemporary science and technology. The many nuclear reactions induced by neutrons are a valuable source of information about the nucleus, and have produced many new nuclear species. These artificially made nuclides yield further information about nuclei, and have applications in other branches of science, such as chemistry, biology, and medicine. Neutrons have direct uses as research tools; for example, their optical properties make them more useful than x-rays for certain analytical purposes. The most striking use of neutrons is in the chain reactions involving fissile materials. These chain reactions have had epoch-making military applications ("atom bombs"), and may be developed into an important industrial source of heat and electric power. These uses of neutrons and applications of neutron physics depend on knowledge of the properties of neutrons and on an understanding of their interaction with matter. Some aspects of neutron physics and its place in nuclear physics have already been discussed; the discovery of the neutron was treated in Chapter 11, as was the production of new nuclear species by means of reactions initiated by neutrons. The β-radioactivity of the free neutron was discussed in Chapter 14. The properties of neutron-induced reactions were treated in Chapter 16, and examples were given of the ways in which these reactions give new information about nuclei. Finally, the role of neutrons as constituents of the nucleus was discussed in Chapters 8 and 17. In the present chapter the emphasis will be on the neutrons themselves, their production and detection, their interaction with matter in bulk, and on methods for measuring neutron energies and cross sections for neutron-induced reactions.

18–1 The production of neutrons. Nuclear reactions are the only source of neutrons, and the (α, n) reactions on light elements which led to the discovery of the neutron are still used to produce these particles.[1] When one gram (one curie) of radium is mixed with several grams of powdered

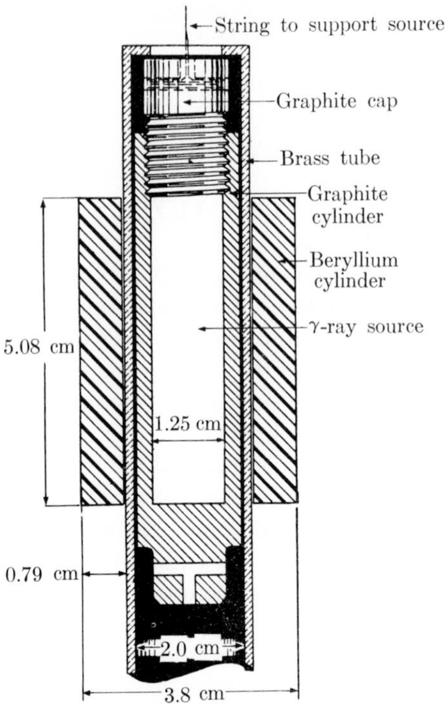

FIG. 18–1. A photoneutron source in which neutrons are produced by the disintegration of beryllium or deuterium by γ-rays from radioactive nuclides made in a chain-reacting pile (Wattenberg[2,3]).

beryllium, about 10^7 fast neutrons are emitted per second as a result of the reaction

$$\mathrm{Be}^9 + \mathrm{He}^4 \rightarrow \mathrm{C}^{12} + \mathrm{n}^1 + 5.71 \text{ Mev}.$$

The mixture can be contained in a volume of 6 or 7 cubic centimeters, and provides a convenient source of neutrons. Radium and its decay products emit α-particles with energies from 4.79 Mev to 7.68 Mev (RaC′), and the neutrons have energies from about 1 Mev to 12 or 13 Mev. The half-life of radium is long, about 1600 years, and the Ra-Be mixture provides neutrons at a sufficiently steady rate to be used as a standard of neutron emission. Radon is sometimes used to supply the α-particles instead of radium, since it is a gas and can be made into a more compact source then Ra-Be, but the Rn-Be source has the disadvantage of decaying rapidly because of the short half-life (3.8 days) of radon. Polonium-beryllium sources are used when a neutron source with relatively few γ-rays is desired; one curie of polonium mixed with powdered beryllium yields about 3×10^6 neutrons/sec. The use of polonium has several disadvantages; the neutron output is much lower than that of the Ra-Be source, special chemical facilities are needed to prepare pure polonium, and the half-life (140 days) is short compared with that of a Ra-Be source.

Photoneutron sources[2] are used frequently because the neutrons produced are practically monoenergetic, and it is not necessary to rely on natural radioactive substances. Most of these sources are based on the $Be^9(\gamma, n)Be^8$ and the $H^2(\gamma, n)H^1$ reactions. The binding energy of the last neutron is particularly low in H^2 and Be^9, and the (γ, n) reactions have low thresholds (Section 16–9), 2.23 Mev for the $H^2(\gamma, n)H^1$ reaction, and 1.67 Mev for the beryllium reaction. Very intense γ-emitters can be made cheaply in nuclear reactors, and can be made into much more intense neutron sources than are possible with natural γ-emitters. A diagram of a photoneutron source is shown in Fig. 18–1; a source of this kind can be used with Be or D_2O and a γ-emitter, and under appropriate conditions can supply up to 10^7 neutrons/sec. The properties[3,4,5] of some photoneutron sources are listed in Table 18–1; the values for the neutron yield are taken from Feld's article.[5] The yield is the number of neutrons per second from 1 gm of target at 1 cm from 1 curie of the source.

Neutrons can also be produced in particle accelerators,[6,7] and sources of this kind have been used a great deal in neutron research. The bombardment of heavy ice (D_2O), or deuterium-containing paraffin, with deuterons accelerated in a Van de Graaff electrostatic generator provides a simple and efficient source. The reaction $H^2(d, n)He^3$ is exoergic, with a Q-value of 3.28 Mev, and good neutron yields can be obtained with deuteron energies as low as 100 to 200 kev; approximately monoenergetic neutrons can be obtained if the conditions of the experiment are carefully controlled. Other reactions which can be used are

$$H^3 + H^2 \rightarrow He^4 + n^1 + 17.6 \text{ Mev},$$

$$C^{12} + H^2 \rightarrow N^{13} + n^1 - 0.26 \text{ Mev},$$

$$Li^7 + H^1 \rightarrow Be^7 + n^1 - 1.65 \text{ Mev},$$

$$H^3 + H^1 \rightarrow He^3 + n^1 - 0.764 \text{ Mev}.$$

The five reactions mentioned have proved to be the most useful sources based on bombardment with artificially accelerated charged particles, and they can provide beams of approximately monoenergetic neutrons in the range 5 kev to 20 Mev.

The neutrons produced by the methods discussed have intermediate or high energies. Slow neutrons, with energies from about 0.01 ev to a few ev, are needed for many applications; they are obtained from higher

TABLE 18–1

PROPERTIES OF PHOTONEUTRON SOURCES

Source	Half-life	Average neutron energy, kev	Neutrons per sec. per curie, 1 gm of target at 1 cm, $\times 10^{-4}$
$Na^{24} + D_2O$	14.8 hr	220 ($\pm$20)	27
$Na^{24} + Be$	14.8 hr	830 ($\pm$40)	13
$Mn^{56} + D_2O$	2.6 hr	220	0.31
$Mn^{56} + Be$	2.6 hr	100 (90%); 300 (10%)	2.9
$Ga^{72} + D_2O$	14 hr	130	6
$Ga^{72} + Be$	14 hr		5
$In^{116} + Be$	54 min	100 (40%); 300 (60%)	0.82
$Sb^{124} + Be$	60 days	30	19
$La^{140} + D_2O$	40 hr	140	0.8
$La^{140} + Be$	40 hr	620	0.3

energy neutrons by allowing the latter to move through a material in which they can lose most of their energy in scattering collisions. Neutrons lose energy through inelastic scattering collisions with medium or heavy nuclei and through elastic scattering collisions with light nuclei; the first process is more effective for neutrons with energies greater than about 1 Mev and the second for neutrons with lower energies. If a source of fast neutrons is embedded in a large mass of material containing light nuclei, the neutrons move through the material and lose energy as they undergo scattering collisions. If the absorption cross section of the material for neutrons is small compared with the scattering cross section, the neutrons continue to lose energy until their speeds are comparable to those of the thermal motion of the nuclei of the material. They are then called *thermal neutrons.* A light material with a suitably low absorption cross section for neutrons is called a *moderator* and the neutrons are said to be "slowed down" or "moderated" to thermal energies. The most frequently used moderators are water, paraffin, and graphite; heavy water and beryllium, which are expensive materials, are sometimes used. A source of fast neutrons together with a moderator, suitably arranged, provides a source of slow neutrons. Some of the details of the slowing-down process will be given in Section 18–3, and the properties of the thermal neutrons will be discussed in Section 18–4.

The most powerful sources of neutrons are those associated with nuclear reactors (chain-reacting piles). It will be shown in Chapters 19 and 20 that many neutrons are produced in a chain-reacting assembly of a fissile material and a moderator. The neutrons resulting from the fission of uranium atoms are fast, and are slowed down in the moderator. If the reactor is based on fission by slow neutrons, it contains a mixture of fast, intermediate, and slow neutrons. For some purposes, as for the production of artificial radionuclides, the mixture of neutrons with different energies can be used, but for many experiments thermal neutrons are needed. These neutrons can be obtained by the use of a *thermal column*, which is simply a large block of graphite moderator placed directly adjacent to the nuclear reactor. Neutrons escaping from the reactor enter the column and are slowed down in it; after diffusing an appropriate distance in the column, they are reduced to thermal energies. The thermal neutrons can be used inside the column to bombard targets exposed there, or they can be allowed to escape through a hole, forming a beam of particles which can be used outside the column.

18–2 The detection of neutrons. Neutrons are uncharged particles and produce a negligible amount of ionization in passing through matter, with the result that they cannot be detected directly in any instrument (Geiger counter, cloud chamber) whose action depends on the ionization caused by the particle which enters it. The detection of neutrons depends on secondary effects which result from their interactions with nuclei.[(8)] Some of the reactions are:

1. the absorption of a neutron by a nucleus with the prompt emission of a fast charged particle;
2. the absorption of a neutron with fission of the resulting compound nucleus;
3. the absorption of a neutron with the formation of a radioactive nuclide whose activity can be measured;
4. the scattering of a neutron by a light nucleus, such as a proton, as a result of which the recoiling light nucleus produces ionization.

A neutron detector based on the first type of interaction can be an ionization chamber or a proportional counter. One of the most frequently used detectors is based on the reaction $B^{10}(n, \alpha)\ Li^7$, which is exoergic, with $Q = 2.78$ Mev. The target nucleus B^{10}, which has an abundance of 18.8% in natural boron, is responsible for the high cross section (755 barns) of the latter for thermal neutrons. A chamber or counter can be filled with BF_3 gas or lined with a compound of boron; the gas detector is used more extensively. When high sensitivity is needed, natural BF_3 is replaced by $B^{10}F_3$ made from the separated isotope B^{10}. The cross section for the (n, α) reaction follows the $1/v$ law (Section 16–5A) and falls off with increasing neutron energy. The sensitivity of BF_3 counters consequently decreases with increasing neutron energy, and these counters are most useful for slow neutrons.

The fission process can be used to detect neutrons, and this method may be regarded as a special case of the first method discussed. The following reactions are typical:

$$U^{235} + n \text{ (slow or fast)} \rightarrow A + B,$$

$$U^{238} + n \text{ (fast)} \rightarrow C + D;$$

A, B, C, and D represent strongly ionizing fission fragments, which are heavily charged and highly excited nuclei. Fission fragments have kinetic energies of the order of 100 Mev and their intense ionization is easy to distinguish from that caused by protons, α-particles, or other ionizing radiations. The fissile nuclei may be introduced into an ionization chamber as a gas (UF_6) or as a wall coating. Chambers containing materials fissionable by thermal neutrons (U^{235}, U^{233}, Pu^{239}) are efficient thermal neutron detectors. Chambers containing natural uranium, or uranium from which some of the U^{235} has been removed, can be used as fast neutron detectors at energies greater than about 1.0 Mev. Thorium, protoactinium, and neptunium can also be used in this way to detect fast neutrons.

The third method is based on the fact that many nuclear reactions induced by neutrons result in radioactive product nuclei, and the neutrons are detected by means of the activity of an exposed substance. The feasibility of detecting the induced radioactivity depends on its lifetime, which cannot be appreciably shorter than the time which must elapse, for experimental reasons, between exposure to neutrons and measurement of the induced reactivity; on the other hand, the lifetime must not be so long that the decay rate is negligible. The cross section for the reaction involved must be large enough so that enough radioactive nuclei are formed during the exposure. Many substances have been found in which these requirements are met, among them being indium, gold, manganese, and dysprosium, and these can be used as detectors. Foils of the detector material are exposed to the neutron source, usually for a given length of time; they are then removed and the extent of the induced activity is determined by counting the emitted radiations with an appropriate Geiger counter, ionization chamber, scintillation counter, or other radiation detector. This *activation method* can be used for neutrons in the different energy ranges, provided the detector materials are chosen carefully.

The most common method for detecting fast neutrons is based on the observation of the ionization produced by the recoiling protons in the elastic scattering of neutrons by hydrogenous materials. An ionization chamber or counter filled with a hydrogen-containing gas, or having a window made of a solid hydrogenous material, may be used. An incident neutron can impart enough energy to a hydrogen nucleus so that the ionization caused by the moving proton can operate the detector. The hydrogenous material can also be contained in a cloud chamber or in a nuclear emulsion.

18–3 The interaction of neutrons with matter in bulk: slowing down. The interaction between neutrons and matter in bulk is very different from that of charged particles of γ-rays and is a subject that requires special treatment. It was found by Fermi (1934) that the radioactivity induced in targets bombarded with neutrons is increased when the neutrons are made to pass through a hydrogenous material placed in front of the target. Fermi and his colleagues showed that the neutrons are slowed down in the hydrogenous material, apparently without being absorbed, and the slower neutrons have a greater probability of inducing radioactivity than do more energetic neutrons. The conversion of fast neutrons into slow neutrons has been investigated in great detail both experimentally and theoretically, and the importance of slow neutrons and the slowing down process has been demonstrated beyond question. A complete and rigorous treatment of neutron slowing down involves complicated mathematics, but the underlying principles and some of the most useful results are quite simple and can be illustrated with little difficulty. For most practical purposes, the important slowing-down process is elastic scattering by light nuclei. Inelastic scattering by intermediate or heavy nuclei is important for neutrons with energies above 1 Mev, but becomes practically negligible below this energy. For simplicity, therefore, only elastic scattering by light nuclei will be considered.

The amount of energy that a neutron loses in a single collision can be calculated by solving the equations of conservation of energy and momentum for the energy of the neutron after the collision. But there is another method which is simpler and neater, and also illustrates some ideas which are very useful in the treatment of nuclear collisions. This method involves the use of two reference systems: the first is the *laboratory* or *L-system,* in which the target nucleus is assumed to be at rest before the collision, and is approached by the incident neutron; the second is the *center of mass* or *C-system,* in which the center of mass of the neutron and nucleus is considered to be at rest and both the neutron and the nucleus approach it. The collision process can be described from the viewpoint of an observer moving with the center of mass, and it turns out that the equations needed for the description are relatively simple. They can be solved quite easily and the results can then be transformed back to the L-system. In view of the importance of the slowing-down process and of the usefulness of the center of mass reference system in nuclear physics, some of the less complex aspects of the process will be treated in detail. The treatment does not require the use of quantum mechanics, as the

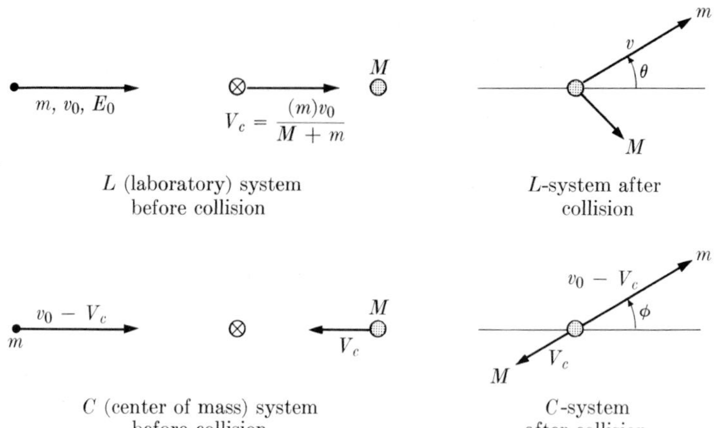

FIG. 18–2. An elastic scattering collision between a neutron and a nucleus as described in the laboratory and center of mass reference systems.

classical picture of "billiard ball" collisions is accurate enough, and the mathematics needed do not go beyond trigonometry and elementary calculus.

The relationship between the L- and C-systems is shown in Fig. 18–2. In the L-system, before the collision, the neutron of mass m moves towards the nucleus (to the right in the figure) with speed v_0, momentum mv_0, and energy E_0; the nucleus of mass M is assumed to be at rest. The speed of the center of mass V_c is given by

$$V_c = v_0 \frac{m}{M + m}. \tag{18–1}$$

After the collision, the neutron moves with speed v and energy E, at an angle θ with its original direction, and the nucleus moves off at some angle with the original direction of the neutron.

In the C-system, before the collision, the neutron moves to the right with speed

$$v_0 - V_c = v_0 \frac{M}{M + m}, \tag{18–2}$$

and the nucleus moves to the left with speed V_c. The total momentum, as measured in the C-system, is

$$m\left(\frac{Mv_0}{M + m}\right) - M\left(\frac{mv_0}{M + m}\right) = 0,$$

since momentum is a vector quantity, and the velocity of the nucleus is opposite in direction to that of the neutron. After the collision, the neutron moves at an angle ϕ with its initial direction. Since the total momentum must be conserved, its value must be zero after the collision, and the nucleus must move off at an angle $(180° + \phi)$ with the direction of the incident neutron. The fact that the momentum is zero in the C-system before and after the collision makes the arithmetic in this system simpler than in the L-system. The observer in the C-system sees only a change in the directions of the neutron and the nucleus as a result of the collision, and the two particles depart in opposite directions. In an elastic collision, in which the kinetic energy is also conserved, the speeds of the particles in the C-system must be the same as they were before the collision; otherwise there would be a change in the total kinetic energy of the two particles. The total effect in the C-system is, therefore, to change the directions of the velocities but not their magnitudes. In the L-system, in which the nucleus was originally at rest, the magnitudes of the velocities are changed, and the directions are not opposite. The neutron, which is scattered through an angle θ, has a velocity v which is the vector sum of the velocity of the neutron in the C-system and the velocity of the center of mass. The relationship between the different velocities is shown in the vector diagram of Fig. 18–3. There are two cases of special interest for which the speed of the neutron after the collision is readily obtained from the figure. In a glancing collision $\phi \approx 0$ and

$$v = v_0 \frac{M}{M + m} + v_0 \frac{m}{M + m} = v_0.$$

The amount of energy lost by the neutron is negligible, and $E = E_0$. In a head-on collision, $\phi = 180°$, and the speed of the neutron is

$$v = v_0 \frac{M}{M + m} - v_0 \frac{m}{M + m} = v_0 \left(\frac{M - m}{M + m}\right),$$

or

$$\frac{E_{\text{min}}}{E_0} = \frac{\frac{1}{2}mv^2}{\frac{1}{2}mv_0^2} = \left(\frac{M - m}{M + m}\right)^2. \tag{18–3}$$

The neutron loses the most energy in a head-on collision; when the moderator is graphite, $M = 12$, $m = 1$, and

$$\frac{E_{\text{min}}}{E_0} = \left(\frac{12 - 1}{12 + 1}\right)^2 = 0.72.$$

A neutron can therefore lose up to 28% of its energy in a collision with a carbon nucleus; a 1-Mev neutron can lose as much as 0.28 Mev per collision, and a 1-ev neutron up to 0.28 ev.

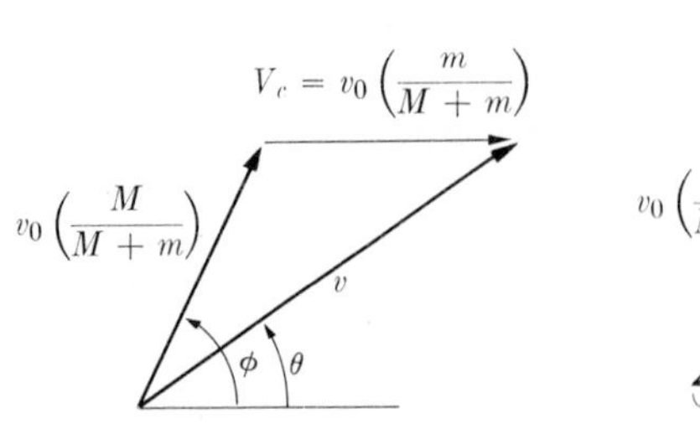

FIG. 18–3. Vector diagram for the velocity of the neutron after the collision in the laboratory and center of mass systems.

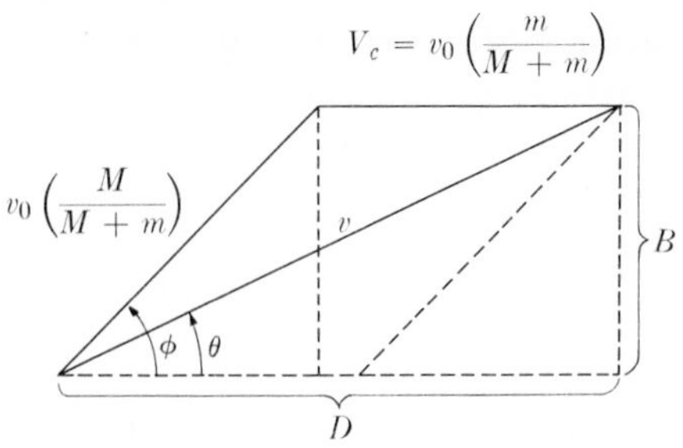

FIG. 18–4. The relationship between the scattering angle in the center of mass system and that in the laboratory system.

For intermediate values of ϕ, the neutron speed after the collision can be found as a function of ϕ by applying the trigonometric law of cosines to Fig. 18–3;

$$v^2 = v_0^2\left(\frac{M}{M+m}\right)^2 + v_0^2\left(\frac{m}{M+m}\right)^2 + 2v_0^2\left(\frac{M}{M+m}\right)\left(\frac{m}{M+m}\right)\cos\phi.$$

The ratio of the neutron energy E after collision to the initial energy E_0 is then

$$\frac{E}{E_0} = \frac{v^2}{v_0^2} = \frac{M^2 + m^2 + 2Mm\cos\phi}{(M+m)^2}. \tag{18–4}$$

If the ratio of moderator mass to neutron mass M/m is called A, the last equation becomes

$$\frac{E}{E_0} = \frac{A^2 + 1 + 2A\cos\phi}{(A+1)^2}. \tag{18–5}$$

The mass ratio A can be taken equal to the mass number of the moderator without introducing any significant error, since m is close to unity and M is very close to an integer. It is convenient to express the energy ratio in terms of the quantity

$$r = \left(\frac{A-1}{A+1}\right)^2, \tag{18–6}$$

which is a measure of the maximum energy that can be lost by the neutron in a single collision. Equation (18–5) then becomes

$$\frac{E}{E_0} = \frac{1+r}{2} + \frac{1-r}{2}\cos\phi. \tag{18–7}$$

The greatest energy loss occurs for $\phi = 180°$, when $\cos\phi = -1$, and $E = rE_0$; for $\phi = 0$, $\cos\phi = 1$, and $E = E_0$.

The scattering angle ϕ in the center of mass system can now be related to the scattering angle in the laboratory system. It is evident from Fig. 18–4 that $\cot\theta = D/B$, where

$$D = v_0 \frac{M}{M+m}\cos\phi + v_0 \frac{m}{M+m},$$

$$B = v_0 \frac{M}{M+m}\sin\phi,$$

so that

$$\cot\theta = \frac{\cos\phi + 1/A}{\sin\phi}. \tag{18–8}$$

The cosine of an angle can be obtained from the cotangent by means of the relation

$$\cos\theta = \frac{\cot\theta}{(1+\cot^2\theta)^{1/2}}.$$

When the expression (18–8) for $\cot\theta$ is inserted into the last equation, the result, after some simple arithmetic, is

$$\cos\theta = \frac{1 + A\cos\phi}{(1 + A^2 + 2A\cos\phi)^{1/2}}. \tag{18–9}$$

The quantity needed is the average value of $\cos\theta$, which can be obtained by integrating Eq. (18–9) over the possible values of ϕ, the scattering angle in the C-system. This integration depends on the probability that a neutron will be scattered through an angle between ϕ and $\phi + d\phi$, but this probability is not known *a priori*. Both experimental results and a rigorous theoretical treatment of the collision process show that the scattering is spherically symmetric (isotropic) in the C-system, provided that the initial energy of the neutrons is less than 10 Mev. This condition is satisfied in most cases of interest, in particular for the neutrons resulting from nuclear fission. Equation (18–9) can then be integrated over the element of solid angle $2\pi\sin\phi\,d\phi$, and

$$\overline{\cos\theta} = \frac{1}{4\pi}\int_0^{\pi}\cos\theta\, 2\pi\sin\phi\, d\phi$$

$$= \frac{1}{2}\int_0^{\pi}\frac{1 + A\cos\phi}{(1 + A^2 + 2A\cos\phi)^{1/2}}\sin\phi\, d\phi.$$

If $\cos\phi$ is set equal to x,

$$\overline{\cos\theta} = \frac{1}{2}\int_{-1}^{+1} \frac{1 + Ax}{(1 + A^2 + 2Ax)^{1/2}}\,dx = \frac{2}{3A}\cdot \tag{18-10}$$

When A is large, that is, for heavy scattering nuclei, $\overline{\cos\theta}$ is small and the scattering in the L-system is practically isotropic. Neutrons which collide with heavy nuclei are scattered forward as often as they are scattered backward. When A is small, as for light nuclei, more neutrons are scattered forward than backward.

The average energy loss per collision can now be calculated; the calculation involves a more useful quantity, the average decrease per collision in the logarithm of the neutron energy, denoted by ξ. This quantity is a convenient one to use in neutron slowing-down calculations because it is independent of the neutron energy. Since E/E_0, as given by Eq. (18–7), is a linear function of $\cos\phi$, and all values of $\cos\phi$ are equally probable, it follows that all values of E/E_0 are equally probable. The probability $P\,dE$ that a neutron of initial energy E_0 will have an energy, after one collision, between E and $E + dE$ is given by

$$P\,dE = \frac{dE}{E_0(1 - r)}, \tag{18-11}$$

where $E_0(1 - r)$ represents the entire range of energy values which a neutron can have after one collision. By definition,

$$\xi = \overline{\ln E_0 - \ln E} = \overline{\ln\frac{E_0}{E}}\cdot$$

Then,

$$\xi = \int_{rE_0}^{E_0} \ln\frac{E_0}{E}\,P\,dE = \int_{rE_0}^{E_0}\left(\ln\frac{E_0}{E}\right)\frac{dE}{E_0(1 - r)}\cdot$$

If x is set equal to E/E_0,

$$\xi = \frac{1}{1 - r}\int_1^r \ln x\,dx,$$

or

$$\begin{aligned}\xi &= 1 + \frac{r}{1 - r}\ln r\\ &= 1 + \frac{[(A - 1)/(A + 1)]^2 \ln[(A - 1)/(A + 1)]^2}{1 - [(A - 1)/(A + 1)]^2},\end{aligned} \tag{18-12a}$$

and ξ is independent of energy, as stated above.

Equation (18–12a) can be rewritten in the form

$$\xi = 1 - \frac{(A-1)^2}{2A} \ln\left(\frac{A+1}{A-1}\right). \tag{18–12b}$$

For $A > 10$, a convenient approximation for ξ, good to about one percent, may be used,

$$\xi \approx \frac{2}{A + \frac{2}{3}}. \tag{18–13}$$

The formulas for ξ, Eqs. (18–12a) and (18–12b), break down for two special cases, $A = 1$ (hydrogen) and $A = \infty$, because the functions on the right sides of the equations are no longer determinate. By taking the appropriate limits as $A \to 1$, and as $A \to \infty$, values of ξ can be determined. For the important case $A = 1$ (hydrogen), $\xi = 1$, and the average value of $\ln (E_0/E)$ is unity. For $A \to \infty$, $\xi \to 0$, so that a neutron loses practically no energy in an elastic collision with a heavy nucleus. This result can also be obtained from Eq. (18–6) which shows that $r \to 1$, and Eq. (18–7) which shows that $E/E_0 \to 1$. In fact, for large values of A, r can be expanded in the series

$$r = 1 - \frac{4}{A} + \frac{8}{A^2} - \frac{12}{A^3} + \cdots,$$

and for values of A greater than 50, $r \approx 1 - 4/A$. Remembering that the maximum energy loss occurs for $\cos\phi = -1$, we get $E \approx rE_0 \approx E_0(1 - 4/A)$.

It is sometimes useful to consider the average energy after a collision

$$\overline{E} = \int_{rE_0}^{E_0} EP(E)\, dE = \frac{E_0(1+r)}{2}. \tag{18–14}$$

For hydrogen ($A = 1$), $r = 0$, and the average energy of a neutron after a collision with a proton is just half of the initial energy. In a head-on collision between a neutron and a proton, which have very nearly equal masses, the former can lose all its energy; this result follows from Eq. (18–7), since $\cos\phi = -1$ and $r = 0$. This possibility distinguishes hydrogen from other moderators. For a collision with a carbon atom, $r = 0.72$, and $\overline{E} = 0.86E_0$. For $A = 200$, $r = 0.98$, $\overline{E} = 0.99E_0$, and $E_{\min} = 0.98E_0$.

When ξ is known, the average number of collisions needed to bring about a given decrease in neutron energy can easily be calculated. If the neutrons start out with an average energy of 2 Mev, and if they are to be slowed down to 0.025 ev (thermal energy at ordinary room temperature), the

TABLE 18–2

SCATTERING PROPERTIES OF SOME NUCLEI

Element	A	ξ	Number of collisions from 2 Mev to 0.025 ev
Hydrogen	1	1.00	18
Deuterium	2	0.725	25
Helium	4	0.425	43
Lithium	7	0.268	67
Beryllium	9	0.208	87
Carbon	12	0.158	114
Oxygen	16	0.120	150
Uranium	238	0.0084	2150

total logarithmic energy loss is $\ln(2 \times 10^6/0.025)$. Since the average loss per collision is ξ, the number of collisions is given by

$$\text{Average number of collisions (2 Mev to 0.025 ev)} = \frac{\ln [(2 \times 10^6)/0.025]}{\xi} = \frac{18.2}{\xi}. \tag{18–15}$$

The values of ξ and of the average number of elastic collisions needed to reduce the neutron energy from 2 Mev to 0.025 ev are listed in Table 18–2 for several nuclear species.

Although the quantity ξ is a measure of the moderating ability of a given substance, it does not tell the whole story. According to Table 18–2, hydrogen would be the best moderator, but the probability of a collision between a neutron and a hydrogen nucleus in hydrogen gas is small because of the small density of the gas. The number of atoms per unit volume must therefore be taken into account. The probability that a scattering collision will occur, the cross section for scattering, must also be considered. Finally, for a good moderator, the absorption cross section must be small, otherwise too many neutrons would be lost by absorption. As a result of these considerations, two other quantities are used to express the properties of moderators. The first of these is the *slowing-down power*, defined by

$$\text{Slowing-down power} = \xi N\sigma_s = \xi\Sigma_s = \frac{N_0\rho\xi\sigma_s}{A}, \tag{18–16}$$

where N is the number of atoms per cubic centimeter, ρ is the density, N_0 is the Avogadro number (6.02×10^{23}), and σ_s is the scattering cross section. The slowing-down power has the dimension cm^{-1}. The quantity $N\sigma_s = \Sigma_s$ is called the *macroscopic scattering cross section*, and is the

TABLE 18–3

VALUES OF THE SLOWING-DOWN POWER AND MODERATING RATIO

Moderator	Slowing-down power, cm^{-1}	Moderating ratio
H_2O	1.53	72
D_2O	0.370	12,000
Be	0.176	159
Graphite	0.064	170

probability per centimeter that a neutron will be scattered. Since ξ is the average loss in log E per collision, the slowing-down power may be interpreted as the average loss in log E per centimeter of neutron travel; it should have a relatively large value for a good moderator.

The second quantity is the *moderating ratio*, defined by

$$\text{Moderating ratio} = \frac{\xi\Sigma_s}{\Sigma_a} = \frac{\xi\sigma_s}{\sigma_a}, \tag{18–17}$$

where σ_a is the absorption cross section. The moderating ratio is a measure of the relative slowing down power and absorbing ability of a substance. Of the light elements listed in Table 18–2, lithium has by far the highest absorption cross section (about 65 b at thermal energy) and is ruled out as a moderator. Hydrogen and deuterium can be used in the form of water (ordinary and heavy, respectively), since oxygen is a good moderator, and the number of hydrogen or deuterium atoms per cubic centimeter of water is large compared with that in the respective vapors.

The properties of some good moderators are given in Table 18–3. In calculating the values of the moderating ratio, the thermal neutron absorption cross sections have been used. These values are greater than those at higher energies, so that the values of the moderating ratio are lower limits rather than accurate values, but are useful for purposes of comparison. According to these results, D_2O is the best of the moderators listed and H_2O the least effective. The relatively poor moderating ratio of H_2O is caused by the relatively high absorption cross section of hydrogen. In practice, D_2O is an extremely expensive substance and can be used only for special applications in which cost is not the primary consideration; beryllium is also expensive and is rarely used. Graphite and water are used often, as is paraffin, which is made up of hydrogen and carbon; these materials offer a satisfactory compromise between moderating ability and cost.

The above discussion of neutron moderation has touched only on some of the elementary aspects of the process. There are other problems which are important, such as the determination of the energy distribution of the neutrons during moderation, and of the distance traveled by the neutrons during the process. These problems have been treated in detail, both theoretically and experimentally, by various authors,(9,10,11) and are also discussed in the general references cited at the end of this chapter.

18–4 Thermal neutrons. As a result of the slowing-down process, neutrons reach the state in which their energies are in equilibrium with those of the atoms or molecules of the moderator in which they are moving. In a particular collision with an atom, a neutron may then gain or lose a small amount of energy; in a large number of collisions between neutrons and atoms energy gains are as probable as energy losses. The neutrons are then said to be in thermal equilibrium with the atoms or molecules of the moderator. Their behavior is similar to that of gas atoms and can be described quite accurately by the kinetic theory of gases. It is necessary, of course, that the neutrons be able to collide with the moderator atoms a sufficient number of times to reach equilibrium before they are absorbed, and this requires that the absorption cross section at low energies must be small compared with the scattering cross section. When the conditions for thermal equilibrium are satisfied, the neutrons have the well-known Maxwell distribution of velocities,(12)

$$n(v)\,dv = 4\pi n\left(\frac{m}{2\pi kT}\right)^{3/2} v^2 e^{-mv^2/2kT}\,dv, \tag{18–18}$$

where n is the total number of neutrons per unit volume, $n(v)\,dv$ is the number per unit volume with velocities between v and $v + dv$, m is the mass of the neutron, k is Boltzmann's constant, equal to 1.380×10^{-16} erg per degree absolute, and T is the absolute temperature in degrees Kelvin.

The Maxwell distribution has certain interesting properties, one of which is the value of the most probable velocity. It is evident from Eq. (18–18) that $n(v) = 0$ for $v = 0$ and for $v = \infty$, so that $n(v)$ must have a maximum value for some finite value of v. This value of v is the most probable velocity, denoted by v_0; it can be determined by the usual procedure of differentiating the right side of Eq. (18–18) with respect to v and setting the derivative equal to zero. The right side can be written as $Cf(v)$, where C represents the coefficient $4\pi n(m/2\pi kT)^{3/2}$, which is independent of v; it is also convenient to set $\alpha = (m/2kT)^{1/2}$. Then

$$n(v) = Cf(v) = Cv^2 e^{-\alpha^2 v^2},$$

and

$$\frac{1}{C}\frac{dn(v)}{dv} = 2ve^{-\alpha^2 v^2} - 2v^3\alpha^2 e^{-\alpha^2 v^2} = 0.$$

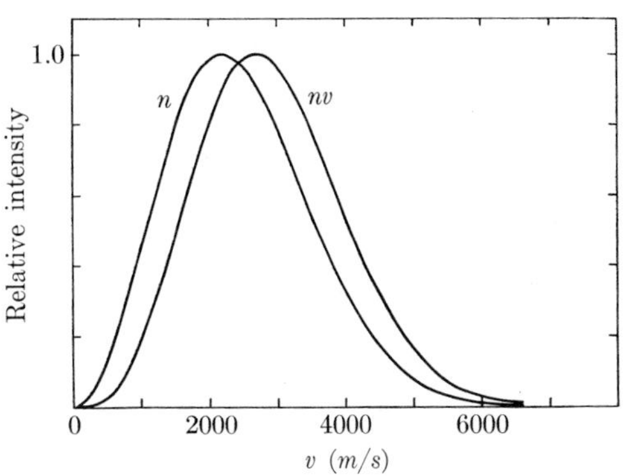

FIG. 18-5. The Maxwell distribution of thermal neutrons, showing the density n and the flux nv as functions of neutron velocity.

For the last equation to be satisfied for some value $v = v_0$, it is necessary that

$$v_0 = \frac{1}{\alpha} = \left(\frac{2kT}{m}\right)^{1/2} \tag{18-19a}$$

for each value of T. Since $n(v)$ has its maximum value when $v = v_0$, this velocity is the most probable one. The *energy corresponding to the most probable velocity* is denoted by E^0 and is

$$E^0 = \tfrac{1}{2}mv_0^2 = kT. \tag{18-19b}$$

The numerical value of v_0 is given in centimeters per second by

$$v_0 = \left(\frac{2kT}{m}\right)^{1/2} = \left(\frac{2 \times 1.380 \times 10^{-16}T}{1.675 \times 10^{-24}}\right)^{1/2} = (1.648 \times 10^8 T)^{1/2}. \tag{18-20}$$

The most probable velocity at room temperature, 20°C or 293°K, is 2200 m/sec (more precisely, 2198 m/sec), and this velocity corresponds to an energy of 0.0252 ev. The Maxwell distribution of thermal neutrons at 20°C is shown by the left-hand curve in Fig. 18-5, in which the relative neutron density $n(v)$ is plotted against the velocity in meters per second. The neutron energy corresponding to the most probable velocity is given by

$$E^0(\text{ev}) = 8.61 \times 10^{-5}T. \tag{18-21}$$

Values of v_0 and E^0 at different temperatures are listed in Table 18-4.

In addition to the most probable velocity v_0 and the corresponding energy E^0 there are several other properties of the Maxwell distribution

TABLE 18–4

MOST PROBABLE VELOCITIES OF THERMAL NEUTRONS

Temperature °C	°K	Most probable velocity, v_0: meters/sec	Energy corresponding to the most probable velocity, E^0: ev
0	273	2120	0.0235
20	293	2200	0.0252
27	300	2220	0.0258
127	400	2570	0.0344
227	500	2870	0.0431
327	600	3140	0.0517
427	700	3400	0.0603
527	800	3630	0.0689
727	1000	4060	0.0861

which are of interest. The average velocity is

$$\bar{v} = \frac{2}{\pi^{1/2}} v_0 = 1.1284 v_0, \tag{18–22}$$

and is about 13% greater than the most probable velocity. The energy distribution of the neutrons is given by

$$n(E)\, dE = \frac{2\pi n}{(\pi kT)^{3/2}} e^{-E/kT} E^{1/2}\, dE, \tag{18–23}$$

where $n(E)\, dE$ represents the number of neutrons with kinetic energy between E and $E + dE$. The average energy is

$$\overline{E} = \tfrac{3}{2}kT = \tfrac{3}{2}E^0, \tag{18–24}$$

and is half again as great as the energy corresponding to the most probable velocity. The average energy does not correspond to the average velocity, as can be verified by comparing Eqs. (18–22) and (18–24), but to the root mean square velocity v_s, given by

$$v_s^2 = \frac{2\overline{E}}{m} = \frac{2}{m}\left(\frac{3}{2}kT\right) = \frac{3}{2}\left(\frac{2kT}{m}\right) = \frac{3}{2}v_0^2,$$

or

$$v_s = 1.2248 v_0 = 1.0854 \bar{v}. \tag{18–25}$$

Although any of the quantities discussed can be used to characterize the Maxwell distribution of thermal neutrons, it is customary to describe thermal neutrons in terms of the most probable velocity and the corre-

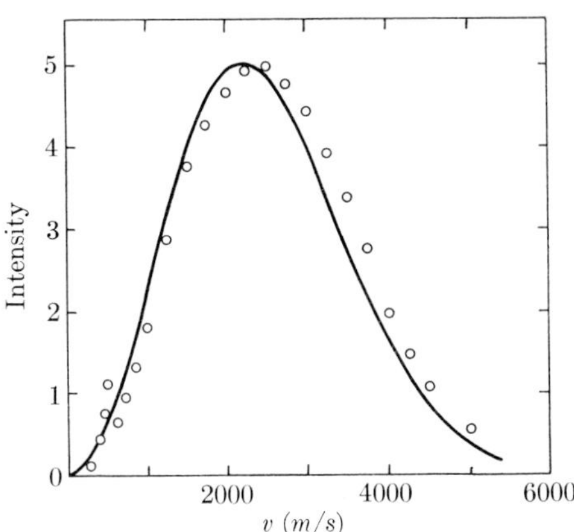

FIG. 18–6. Directly measured velocity distribution of thermal neutrons (circles) compared with the Maxwell distribution at room temperature (Ringo, cited by Hughes).

sponding energy. Thus they are usually referred to as "kT neutrons," and the cross sections quoted for thermal neutrons are usually those for a speed of 2200 m/sec, or for an energy of 0.025 ev.

The velocity distributions of thermal neutrons from different sources have been measured experimentally. It is possible to determine the relative numbers of neutrons with different speeds by means of special instruments which will be described in Section 18–6. The maximum of the measured distribution gives the most probable speed directly, and the "temperature" of the neutrons can be obtained from Eq. (18–19). For various experimental reasons, highly accurate measurements are hard to make, but the measured velocities do follow the Maxwell distribution within the experimental accuracy of about 10%. An example of a directly measured velocity distribution[(13)] is given in Fig. 18–6; the neutrons came from the thermal column of the nuclear reactor at the Argonne National Laboratory. The experimental points indicate a "neutron temperature" slightly higher than 293°K (20°C), for which the solid, theoretical curve was calculated.

Another quantity which has great importance in neutron physics is the *flux* of neutrons; it is usually defined as the product $n(v)v$ and is obtained by multiplying the neutron density $n(v)$ by the velocity. The flux distribution for thermal neutrons is given by

$$n(v)v\,dv = \frac{4n}{v_0^3\pi^{1/2}}\,v^3e^{-v^2/v_0^2}\,dv, \tag{18–26}$$

and is plotted in Fig. 18–5 for room temperature neutrons. The flux has the dimensions cm^{-2} sec^{-1} and is usually described geometrically as the number of neutrons crossing one square centimeter of area in one second. This description is correct for a unidirectional beam of neutrons of density n and velocity v crossing a plane 1 cm^2 in area normal to the beam. Under some conditions, however, this description is wrong. Suppose, for example, that the plane 1 cm^2 in area is inside a nuclear reactor or thermal column in which there is a density n of neutrons moving with velocity v, and that the neutrons can cross the area in any direction with the same probability, that is, the flux nv is isotropic. It can be shown that under these conditions the number of neutrons crossing the plane unit area per second from each side is $nv/4$. The simple geometrical description of the flux can be retained if the area considered is that of a properly chosen volume element as, for example, the surface of a sphere whose cross-sectional area is 1 cm^2. This interpretation of the flux is illustrated in Fig. 18–7, taken from Hughes' book. The concept of neutron flux is not limited to thermal neutrons but is used for neutrons with any density and velocity distribution.

The neutron density and flux, the Maxwellian distribution, and the $1/v$-law for certain neutron cross sections are linked in an instructive way in the absorption of thermal neutrons. Suppose that a sample of a certain stable material is exposed to the flux $n(v)v$ of thermal neutrons, and that a nuclear reaction occurs in which the neutron disappears, and which has a cross section σ_a. In general, σ_a depends on the energy of the neutrons, or on their velocity, and can be written as $\sigma_a(v)$. The number of reactions induced per second by neutrons of velocity v is $n(v)v\sigma_a(v)N$, where N is the number of nuclei in the target sample which can undergo the given reaction. If a new nuclide is produced in the reaction, as in the case of an (n, γ) reaction, the rate of production is

$$\frac{dN_a}{dt} = n(v)v\sigma_a(v)N, \tag{18–27}$$

where $N_a(v)$ is the number of nuclei of the new kind produced by neutrons of velocity v. If now, the cross section σ_a obeys the $1/v$-law, it can be written in the form

$$\sigma_a(v) = \frac{\sigma_{a0}v_0}{v}, \tag{18–28}$$

where v_0 represents some standard velocity and σ_{a0} is the value of the cross section at that velocity. Equation (18–27) becomes

$$\frac{dN_a}{dt} = n(v)\sigma_{a0}v_0N; \tag{18–29}$$

the v's cancel and the absorption rate depends on the neutron density n rather than on the flux nv, regardless of the neutron velocity.

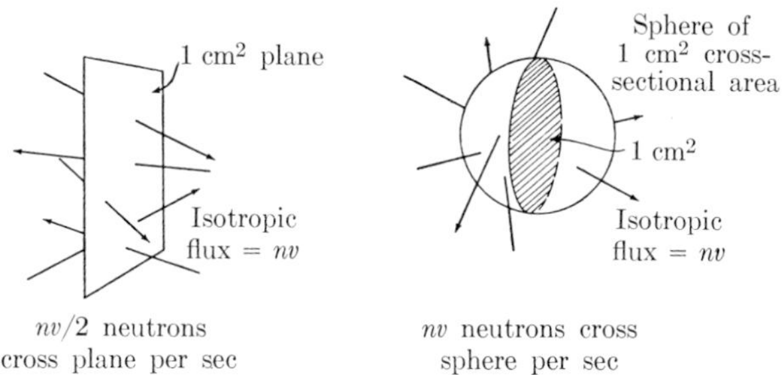

FIG. 18–7. The geometrical description of neutron flux as the number of neutrons crossing a sphere of unit cross-sectional area per second (Hughes).

If the neutron distribution is Maxwellian, $n(v)$ is given by Eq. (18–18), and the formula for the total number N_a of new nuclei formed per second is obtained by integrating Eq. (18–29) over all possible values of v,

$$\frac{dN_a}{dt} = N\sigma_{a0}v_0 \int_0^\infty 4\pi n \left(\frac{m}{2\pi kT}\right)^{3/2} v^2 e^{-mv^2/2kT}\, dv. \tag{18–30}$$

The integral on the right side of the last equation is just equal to n, the total number of neutrons per cubic centimeter, because the Maxwellian distribution just shows how the n neutrons are divided among the different possible velocities. This result can be verified by noting that

$$\int_0^\infty v^2 e^{-\beta^2 v^2}\, dv = \frac{\pi^{1/2}}{4\beta^3},$$

where β is a quantity independent of v. Then the rate at which new nuclei are formed by thermal neutrons with a Maxwellian velocity distribution, in a reaction with a $1/v$-cross section, is simply

$$\frac{dN_a}{dt} = nv_0\sigma_{a0}N. \tag{18–31}$$

The same result would be obtained for any distribution $n(v)$, provided the cross section σ_a follows the $1/v$-law, but the Maxwell distribution is the important practical one. It is customary to take v_0 as 2200 m/sec, that is, as the most probable velocity in a Maxwellian distribution at 20°C, and the thermal cross sections listed in tables and compilations are usually those measured at neutron speeds of 2200 m/sec. When the material being activated has a $1/v$-cross section, the thermal neutrons may be treated as monoenergetic with energy E^0 corresponding to the most probable speed. The thermal flux is expressed simply by multiplying the neutron density by the velocity $v_0 = 2200$ m/sec.

When the new nuclear species is radioactive, with a decay constant λ, the rate of change of the number of new radioactive nuclei is given by

$$\frac{dN_a}{dt} = nv_0\sigma_{a0}N - \lambda N_a. \tag{18–32}$$

This equation has exactly the same form as Eq. (10–6), which was met in the discussion of natural radioactivity, and can be integrated to give

$$N_a(t) = \frac{nv_0\sigma_{a0}N}{\lambda}(1 - e^{-\lambda t}), \tag{18–33}$$

where t is the length of time for the irradiation of the sample. The rate at which the radioactive nuclei decay is then

$$\lambda N_a(t) = nv_0\sigma_{a0}N(1 - e^{-\lambda t}). \tag{18–34}$$

Equation (18–34) is the basis of the method usually used to determine the *activation* cross section of a nuclide if the neutron density n is known, or for determining the neutron density when the activation cross section is known. The target, or detector, is a thin foil of a material which undergoes a reaction obeying the $1/v$-law, so that Eq. (18–34) is satisfied. The rate of decay λN_a of the induced activity is measured by standard counting procedures, and if t, λ, and σ_{a0} are known, n can be obtained. In an actual experiment, a finite time θ must elapse between the end of the irradiation period t and the time of the actual counting. The measured activity is then given by

$$A_m = nv_0\sigma_{a0}N(1 - e^{-\lambda t})e^{-\lambda\theta}. \tag{18–35}$$

The additional exponential factor corrects for the decay which occurs between the end of the irradiation period and the time of the counting procedure. This method will be discussed further in Section 18–7.

18–5 The diffusion of thermal neutrons. In work involving thermal neutrons, it is often necessary to know the spatial distribution of the neutrons, that is, the dependence of the neutron density on position. This problem can be treated with the aid of the theory of diffusion, and some of the ideas involved will be developed and then illustrated by means of an easy example. It will be assumed, for the sake of simplicity, that the neutrons are monoenergetic; in view of the results of the last section, this assumption is a reasonable one.

Suppose that there is a distribution of neutrons whose density varies only in the x-direction, and consider a plane element of area perpendicular to this direction. If the neutron density $n(x)$ were uniform (independent of x) the number of neutrons crossing this element from one side would be equal to the number crossing from the other side. But if there are more neutrons on one side of the surface element, more neutrons cross from that side, and there is a net flow of neutrons across the plane area from the side with the greater neutron density to the side with the smaller density. Neutrons are then said to *diffuse* from the region of higher neutron density to the region of lower neutron density. The motion of the neutrons may be represented by Fick's law of diffusion, according to which the net number J of neutrons flowing per unit time through a unit area normal to the direction of flow is given for the one-dimensional case by

$$J = -D\frac{dn}{dx}; \tag{18–36}$$

J is called the *current density* and D is the *diffusion coefficient.* When n has the units cm^{-3} and x is in centimeters, D has the units square centimeters per second. The fact that thermal neutrons in a moderator have the same velocity distribution as the molecules in a gas suggests that the motion of the neutrons may be treated in the same way as that of gas molecules. The assumption is made that neutrons, in a medium which does not absorb many of them, act like the molecules of a very dilute gas, and the penetration of neutrons through matter is assumed to be similar to the diffusion process in gases. With these assumptions, neutron diffusion theory, at least in its elementary form, consists of the application of the ideas and equations of the kinetic theory of gases to neutrons. In particular, the diffusion coefficient D can be interpreted in terms of concepts taken from kinetic theory. The application of these concepts is not limited to thermal neutrons but can be used for neutrons of any velocity (as, for example, in the problem of the spatial distribution of neutrons during the slowing-down process), provided that a suitable diffusion coefficient can be defined.

One of the concepts which is useful in kinetic theory is that of the *mean free path,* usually denoted by λ. Although this symbol is also used for the disintegration constant, there will be no confusion because the mean free path λ will always have a subscript denoting a particular process. The mean free path for a process is related to the cross section for that process; thus the average distance that a neutron moves between scattering collisions is called the *scattering mean free path* λ_s, defined by the relation

$$\lambda_s(\text{cm}) = 1/N\sigma_s, \tag{18–37}$$

where N is the number of nuclei per cubic centimeter and σ_s is the scattering cross section per nucleus. The absorption mean free path is given by

$$\lambda_a(\text{cm}) = 1/N\sigma_a, \tag{18–38}$$

and is the path length traveled, on the average, by thermal neutrons before being absorbed. There is also a *transport mean free path* λ_{tr} defined as

$$\lambda_{tr} = \frac{1}{N\sigma_s(1 - \overline{\cos\theta})} = \frac{1}{N\sigma_{tr}}, \tag{18–39}$$

where $\overline{\cos\theta}$ is the average value of the scattering angle in the laboratory reference system, and is given by Eq. (18–10). The transport mean free path is introduced in order to take into account the fact that the scattering is preferentially in the forward direction; it is greater than the scattering mean free path, which means that, on the average, a neutron will travel farther in a given number of collisions than if there were no preferred direction. The transport cross section $\sigma_{tr} = \sigma_s(1 - \overline{\cos\theta})$ is then a measure of the rate at which the neutron loses its forward momentum or the "memory" of its original direction. According to kinetic theory, the diffusion coefficient is given by

$$D = \tfrac{1}{3}\lambda_{tr}v. \tag{18–40}$$

Hence, if the neutron velocity is known, and if λ_{tr} is determined [by means of Eq. (18–39)] in terms of the properties N, $\overline{\cos\theta}$, and σ_s of the medium in which the neutrons are diffusing, then the rate of diffusion can be calculated from Eq. (18–36), and the current density is

$$J = -\frac{\lambda_{tr}v}{3}\frac{dn}{dx}. \tag{18–41}$$

It is now possible to show how the spatial dependence of the neutron density can be found. The example to be given is that of monoenergetic neutrons diffusing in the x-direction in a large slab of moderator with an

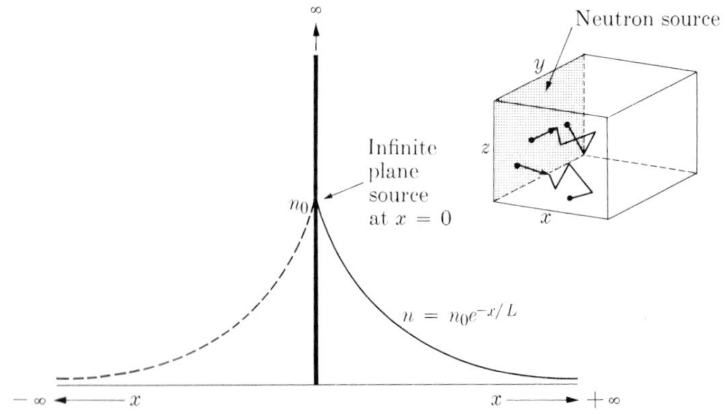

FIG. 18–8. Diffusion of thermal neutrons from an infinite plane source.

absorption cross section σ_a. This problem arises frequently in research with thermal neutrons, especially when a thermal column is used in connection with a nuclear reactor. It is assumed that there is a source which produces Q neutrons per square centimeter per second, spread uniformly over the plane boundary of the slab at $x = 0$, as shown in Fig. 18–8. If the slab is large enough (infinite), variations in the y- and z-directions can be neglected. Consider the element of the slab between x and $x + dx$. The net number of neutrons which leak into the slab per unit time is denoted by $L_x\, dx$ and is given by

$$\begin{aligned} L_x\, dx &= J(x) - J(x + dx) \\ &= -\frac{\lambda_{\text{tr}} v}{3}\left(\frac{dn}{dx}\right)_x + \frac{\lambda_{\text{tr}} v}{3}\left(\frac{dn}{dx}\right)_{x+dx} \\ &= \frac{\lambda_{\text{tr}} v}{3}\frac{d^2 n}{dx^2}\, dx. \end{aligned}$$

At equilibrium, the number of neutrons leaking into the element dx of the slab must be equal to the number of neutrons absorbed in it, which is $nv\sigma_a N\, dx$, so that

$$\frac{\lambda_{\text{tr}} v}{3}\frac{d^2 n}{dx^2} = nv\sigma_a N$$

or

$$\frac{d^2 n}{dx^2} = \frac{3N\sigma_a}{\lambda_{\text{tr}}}\, n = \frac{3}{\lambda_{\text{tr}}\lambda_a}\, n. \tag{18–42}$$

A new quantity, the *thermal diffusion length* L, is now defined by means of the relationship

$$L = +\left(\frac{\lambda_{\text{tr}}\lambda_a}{3}\right)^{1/2}, \tag{18–43}$$

so that Eq. (18–42) becomes

$$\frac{d^2 n}{dx^2} = \frac{n}{L^2}. \tag{18–44}$$

It is easily verified by direct substitution that the general solution of Eq. (18–44) is

$$n(x) = ae^{-x/L} + be^{x/L}, \tag{18–45}$$

where a and b are constants to be determined by the conditions on the problem. One condition is that n must be finite for all values of x, including $x = \infty$; it follows that since L is positive, b must be zero. The second

TABLE 18–5

DIFFUSION CONSTANTS OF MODERATORS

Material	Density, gm/cm^3	N, molecules/cm^3	L, cm	σ_{tr}, barns/molecule	σ_a, millibarns, per molecule at 2200 m/sec	λ_{tr}, cm	λ_a, cm
H_2O	1.00	0.0334×10^{24}	2.73	69.6	660	0.43	51.2
D_2O	1.10	0.0331	171	12.0	0.93	2.52	36,700
Be	1.85	0.1235	20.8	5.65	10.1	1.43	906
C	1.60	0.0803	52.0	4.54	4.5	2.74	2,900

condition is that the current density at $x = 0$ is $J(0) = Q/2$, since only half of the neutrons produced by the uniform source enter the moderator. This condition determines the value of the constant a

$$J(0) = \frac{Q}{2} = -\frac{\lambda_{tr}v}{3}\left(\frac{dn}{dx}\right)_0 = \frac{\lambda_{tr}v}{3}\frac{a}{L},$$

so that

$$a = \frac{3}{2}\frac{QL}{\lambda_{tr}v} = \frac{QL}{2D}.$$

The solution of the problem, therefore, is

$$n(x) = (QL/2D)e^{-x/L} = n_0 e^{-x/L}, \tag{18–46}$$

and the neutron density decreases exponentially with the distance from the source. The variation in $n(x)$ is shown in Fig. 18–8. The diffusion length L can be shown to be the average (air-line) distance a neutron moves from the plane $x = 0$ before absorption, in contrast to λ_a, which is the total path a neutron traces before absorption. The diffusion length is important from a practical standpoint because it can be measured experimentally. The transport and absorption cross sections can also be measured in some cases, and other constants can then be determined with the aid of the diffusion theory equations. The various diffusion constants for several moderators are listed in Table 18–5.

There are many other important applications of the theory of neutron diffusion which are beyond the scope of this book and are treated in some of the general references listed at the end of the chapter.

18–6 Cross sections for neutron-induced reactions: measurement of the total cross section. The use of the cross section for a nuclear reaction as a quantitative measure of the probability that the reaction will occur was treated in Chapter 16. The importance of neutron-induced reactions in pure and applied physics and technology makes it necessary to have a detailed knowledge of the cross sections for these reactions. The efficient production of new nuclides by neutron bombardment requires accurate information about cross sections for (n, γ) reactions. The possibility of achieving a chain reaction and the design of a nuclear reactor depend strongly on the values of the absorption and fission cross sections of fissile materials, and on the absorption and scattering cross sections of moderators and structural materials. The intelligent use of experimental data on neutron cross sections requires understanding of the methods used for measuring them, of the results of the measurements, and of the limitations of the methods.

The fact that a given nucleus generally has a different cross section for each type of neutron reaction in which it can take part, together with the fact that each cross section may vary with the neutron energy, sets obstacles in the way of building up an adequate store of cross-section values. The total cross section σ_t is the sum of the partial reaction cross sections and the partial scattering cross sections. Although the total cross section is the easiest one to measure, it does not, in general, yield enough information, and measurements of the various partial cross sections are often needed. As a result of the variation of cross section with energy it is desirable to measure values of cross sections for monoenergetic neutrons. One of the main problems which then arise is that of obtaining beams of monoenergetic neutrons of sufficient intensity to allow precise measurements. The experimental techniques involved depend on the neutron energies; they are different for slow and for fast neutrons, and it is necessary to consider measurements in different energy ranges separately.

The total cross section can be determined by measuring the neutron transmission in a "good geometry" experiment. The idea on which the determination is based is similar in principle to that discussed previously (Section 15–1) in connection with the measurement of the total attenuation coefficient or cross section for γ-rays. The neutron flux observed after a beam of neutrons has passed through a thickness x of material is given by

$$nv = (nv)_0 e^{-N\sigma_t x}, \tag{18–47}$$

where $(nv)_0$ is the flux incident on the material, N is the number of nuclei per cubic centimeter, and σ_t is the desired total cross section. Then

$$\sigma_t = \frac{1}{Nx} \ln \frac{(nv)_0}{nv}, \tag{18–48}$$

and the value of the cross section can be determined from the response of a detector in the presence and absence of the material. The geometrical

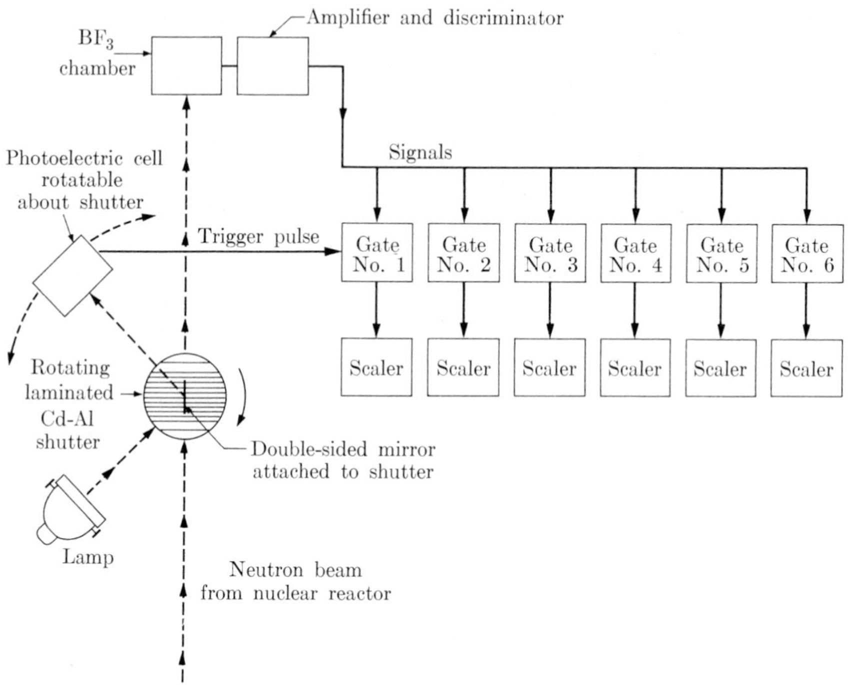

FIG. 18–9. Schematic diagram of a neutron velocity selector.[16]

requirements can usually be met without difficulty, and the emphasis is on getting sufficiently intense beams of monoenergetic neutrons, or of analyzing the effect of a nonhomogeneous beam in terms of the effects of its components. This problem has been solved for slow neutrons, with the aid of devices called *velocity selectors.*

In one type of instrument,[14,15,16] the mechanical velocity selector or *slow neutron chopper*, the design is based on the fact that cadmium is a strong absorber of neutrons with energies less than 0.3 ev, while other metals such as aluminum are weak absorbers in this energy region. A schematic diagram of a slow chopper[16] at the Harwell Laboratory (England) is shown in Fig. 18–9. A collimated beam from the thermal column of a reactor is incident on a rotating cylindrical shutter which contains a series of laminae, alternately cadmium and aluminum, mounted in a steel case. Because of the strong absorption by the cadmium, neutrons can pass through and be detected only when the laminae are parallel to the direction of the beam. Two bursts of neutrons are released at each revolution; the neutrons travel along a measured path to a detector (BF_3 counter) and are recorded, according to their time of arrival, in the following way. A double-sided mirror fixed to the end of a shutter reflects a beam of light from a fixed lamp into a photocell at an instant between successive bursts given by the angular setting of the cell. The pulse from the photocell is then fed into an electric gate unit which passes those pulses from the detector which occur during a short preset time interval immediately following the photocell impulse. As this gate closes it opens a second gate and so on through six gates, thus recording neutrons which arrive at six different times. The circuit therefore constitutes a six-channel analyzer which can be moved over a greater range of velocities by adjusting the position of the photocell. The relative transmission or neutron counting rate can then be obtained for different neutron velocities (energies) and the cross section σ_t can be determined as a function of energy. The limitations on the mechanical chopper are set by the limited speed with which the shutter can be rotated and by the use of cadmium, with its low "cut-off" energy. As a result, a slow chopper, as an instrument of this type is often called, can be used for neutrons with energies from about 10^{-4} ev to about 0.2 ev.

An example of results obtained with slow choppers is given in Fig. 18–10, in which the total cross section of Au^{197} is shown at energies between 2×10^{-4} ev and about 0.1 ev. The values denoted by black circles were obtained with the instrument described; the values denoted by open circles were obtained with a slow chopper at the Brookhaven National Laboratory. The curve is taken from a recent compilation of neutron cross sections.[(17)]

Monoenergetic beams of neutrons can be obtained through the use of a *crystal spectrometer velocity selector*, the design of which is based on the wave properties of neutrons. It can be seen from Eq. (16–16) that neutrons with energies in the range 0.01 ev to about 10 ev have wavelengths in the neighborhood of 10^{-8} or 10^{-9} cm. These wavelengths are of the same order of magnitude as the wavelengths of x-rays and of the distance between crystal planes, with the result that optical interference effects are observed when slow neutrons are reflected from crystal surfaces. When a beam of slow neutrons of wavelength λ is reflected from the surface of a crystal, reflection maxima are observed at angles given by the Bragg relation

$$n\lambda = 2d \sin \theta, \tag{18–49}$$

where n is an integer which represents the order of the reflection ($n = 1, 2, 3, \ldots$), d is the distance between crystal planes, and θ is the glancing angle for the nth order reflection. If the de Broglie relation, $\lambda = h/mv$, for the wavelength is inserted into Eq. (18–49), the velocity of the neutrons at the nth maximum is

$$v = \frac{nh}{2md \sin \theta}, \tag{18–50}$$

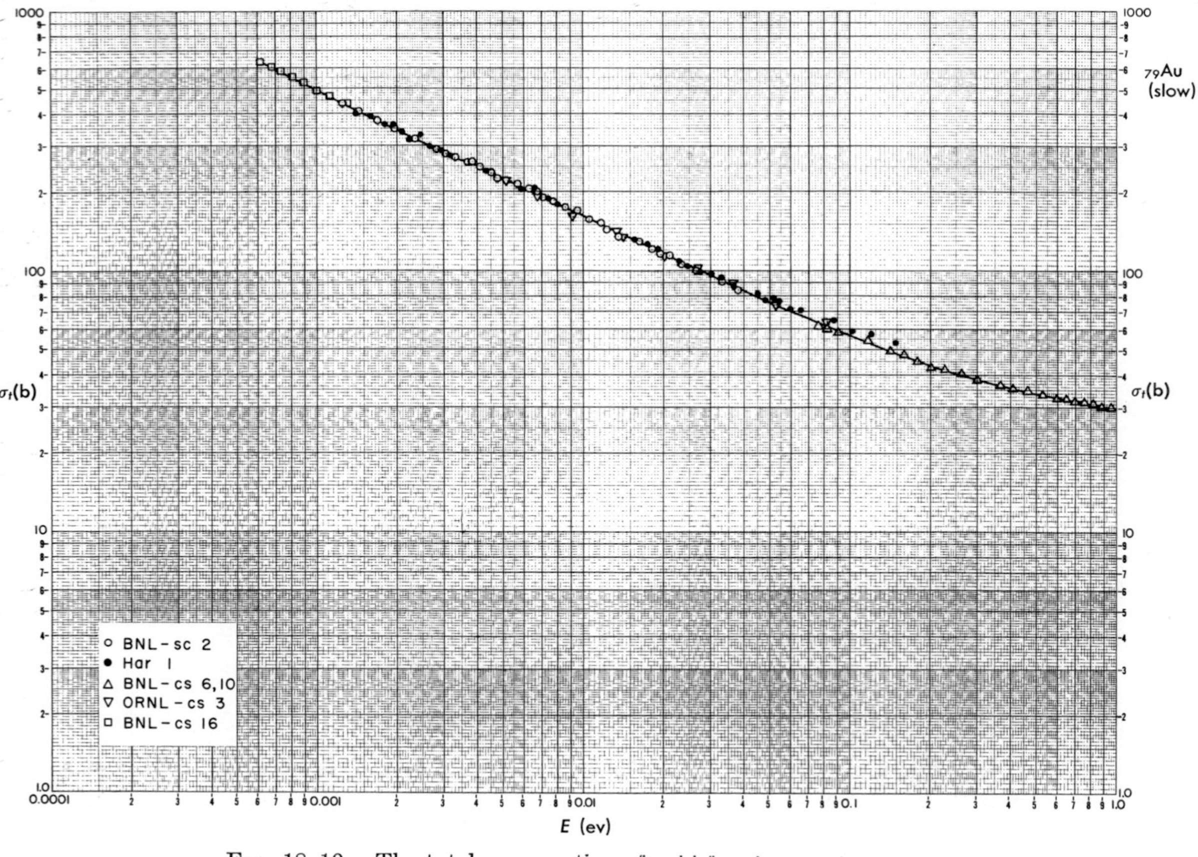

FIG. 18–10. The total cross section of gold for slow neutrons.[17]

and the corresponding energy is

$$E = \frac{1}{2} mv^2 = \frac{n^2h^2}{8md^2 \sin^2 \theta}. \tag{18–51}$$

A schematic diagram of a crystal spectrometer for neutrons would look much like the diagram of an x-ray spectrometer (Fig. 4–4), but the neutron instrument is usually larger because the source area and the detector are larger than in the case of x-rays. In the Brookhaven crystal spectrometer,[(18)] the detector is a BF_3 cylindrical proportional counter, containing $B^{10}F_3$, oriented so that neutrons travel down the axis of the counter in order to get a high detecting efficiency. In a measurement of the total cross section, a thin sheet of material is placed between the crystal and the detector, and the transmission is measured as a function of the glancing angle θ and, therefore, of the neutron energy. The values of σ_t for Au denoted by open triangles and squares in Fig. 18–10 were obtained with the Brookhaven crystal spectrometer. With this instrument, measurements can be extended into the energy range in which the slow chopper is no longer useful. Precise measurements can be made at neutron energies up to 10 ev with a beryllium crystal and useful, but less precise, results can be obtained between 10 and 50 ev. At higher energies, the glancing angles become very small and the uncertainty in the measurement of these angles, and consequently in the energies, is large. At low energies (large glancing angles) higher order reflections introduce difficulties; precise measurements can be made at energies down to about 0.05 ev with a crystal of rock salt. The energy range which can be covered by the crystal spectrometer makes this instrument especially useful for the study of the details of low-lying neutron resonances. Accurate values of the Breit-Wigner parameters (Section 16–3) can be obtained[(19)] and details of the shape of the resonance can be determined. The importance of information of this kind in connection with the understanding of nuclear reactions and of the structure of the nucleus has been discussed in Chapter 16.

It is necessary, for various reasons, to measure cross sections at neutron energies greater than those covered by the slow chopper and crystal spectrometer, and instruments have been devised for this purpose. The time-of-flight method for measuring the neutron velocity, mentioned in the discussion of the slow chopper, can be used with a cyclotron or linear accelerator, which produces a neutron beam. Fast neutrons are produced in periodic bursts lasting only a few microseconds and separated by longer intervals. They are slowed down in a paraffin block and then allowed to pass through a thin slab of material. The time required for the neutrons to reach a detector about 10 m away from the source is measured electronically. The timing is usually done by distributing the counter pulses to a series of recorders (channels) according to time and, in this way,

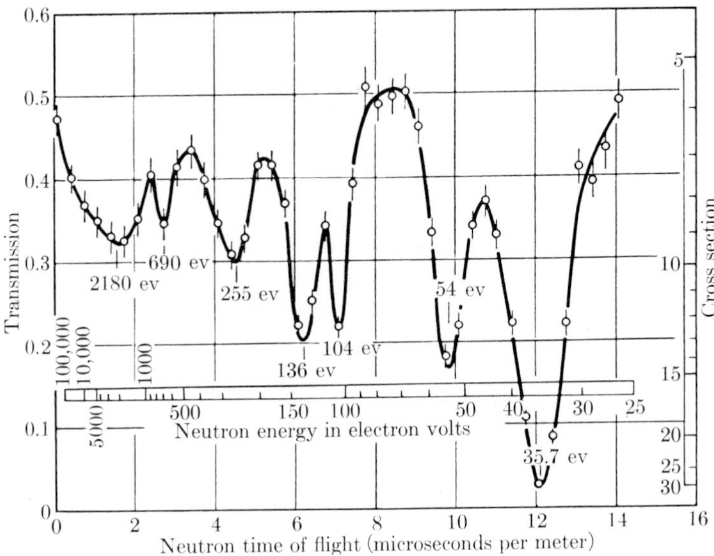

Fig. 18–11. Neutron transmission and total cross section of bromine as functions of the neutron energy, as measured with a pulsed-cyclotron, time-of-flight, velocity spectrometer (Havens and Rainwater[23]).

neutrons of certain velocities are selected from a nonhomogeneous beam. An instrument of this kind has been used[20,21,22,23] at Columbia University and a transmission curve obtained with it is shown in Fig. 18–11. The transmission and the total cross section of bromine are shown as functions of the neutron time of flight and energy; the dips in the transmission curve correspond to the resonances. It is possible with this instrument, to work with neutron energies from 0.003 ev to over 5000 ev.

The time-of-flight techniques have also been applied to fast mechanical choppers used with neutrons from a nuclear reactor. Bursts of neutrons are produced by mechanical interruptions of a beam of neutrons with energies up to several thousand electron volts. The *fast chopper* first developed at the Brookhaven National Laboratory[24,25] consists of a stationary slit system (the stator) and a moving slit system (the rotor). The stator produces two sharply defined neutron beams from a roughly collimated beam from the BNL graphite-moderated research reactor. The rotor intercepts these beams except at the instant when the slits of the rotor are aligned with those of the stator. The rotor is 30 in. in diameter, is made of aluminum and a plastic material, weighs 250 lb and can attain a speed of 12,000 revolutions per minute. Channels through the rotor permit eight bursts of neutrons to pass for each revolution. At

10,000 rev/min, the burst length is slightly less than 1 μsec; the flight path is 20 m and a scintillation neutron detector is used. Under these conditions, high precision can be obtained in transmission measurements and in the determination of total cross sections. The curve for the total cross section of silver between 1 ev and 10^4 ev (Fig. 16–4) is an example of data obtained with several instruments including a crystal spectrometer (open circles) and two fast choppers (black circles and triangles). There are many resonances which lie close together, and it was not possible to separate (*resolve*) such close levels before the development of these new, high-resolution, choppers.

The different methods used for measuring total cross sections for neutrons with energies up to 10^4 ev are compared in Reference 26.

The methods discussed so far can be used for neutrons with energies from a fraction of an electron volt to several thousand electron volts, and these energies are said to lie in the "low to intermediate" energy range. The energy range from a few kev to about 20 kev is usually referred to as the "intermediate to fast" range. Although the distinction between slow, intermediate, and fast neutrons is an arbitrary one, and depends on the particular purposes for which the neutrons are used or studied, a change in the methods of measuring cross sections occurs at energies of a few kev, and is caused by a change in the method of producing monoenergetic neutrons. At energies of a few kev and above, monoenergetic neutrons are produced directly by reactions in the target of a charged particle accelerator or by monoenergetic γ-rays producing photoneutrons in deuterium or beryllium; these methods were discussed in Section 18–1. When a beam of monoenergetic intermediate or fast neutrons has been produced in a cyclotron or Van de Graaff generator, its transmission can be determined by measuring the neutron flux[(27)] with detectors designed for the purpose, and the cross section is obtained from the transmission in the usual way. For further details about the measurement of total neutron cross sections at higher energies, the reader is referred to a review article by Barschall.[(28)] The cross section curves of Fig. 16–3 for aluminum are examples of results obtained in the energy range 10^4 ev to 1 Mev, with monoenergetic neutron beams produced with a Van de Graaff generator. Figure 18–12 shows the total cross section of bismuth[(17)] in the range 10^4 ev to 15 Mev. Figure 18–12 together with Fig. 16–5 shows how the total cross section of bismuth varies between 1 ev and 15 Mev. For completeness, Fig. 18–13 is included to show the total cross section of bismuth[(17)] between 1 ev and 10^{-4} ev. In the case of bismuth, the total cross section has been measured as a function of energy between 10^{-4} ev and 15 Mev. This element is especially interesting to physicists because it contains a magic number of neutrons and is monoisotopic, and most nuclides have not been studied so carefully. In particular, the energy

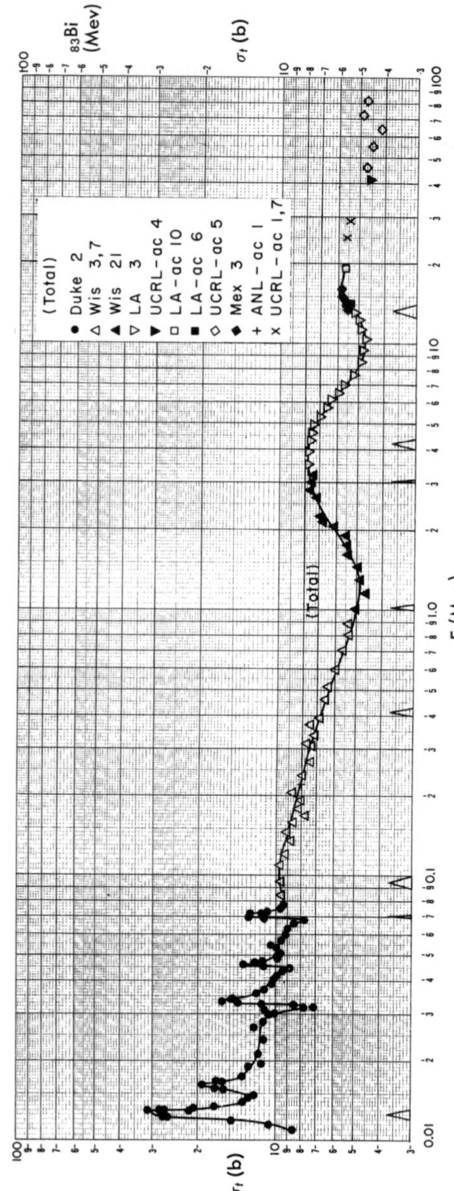

FIG. 18-12. The total cross section of bismuth as a function of neutron energy from 10^4 ev to 15 Mev.[17]

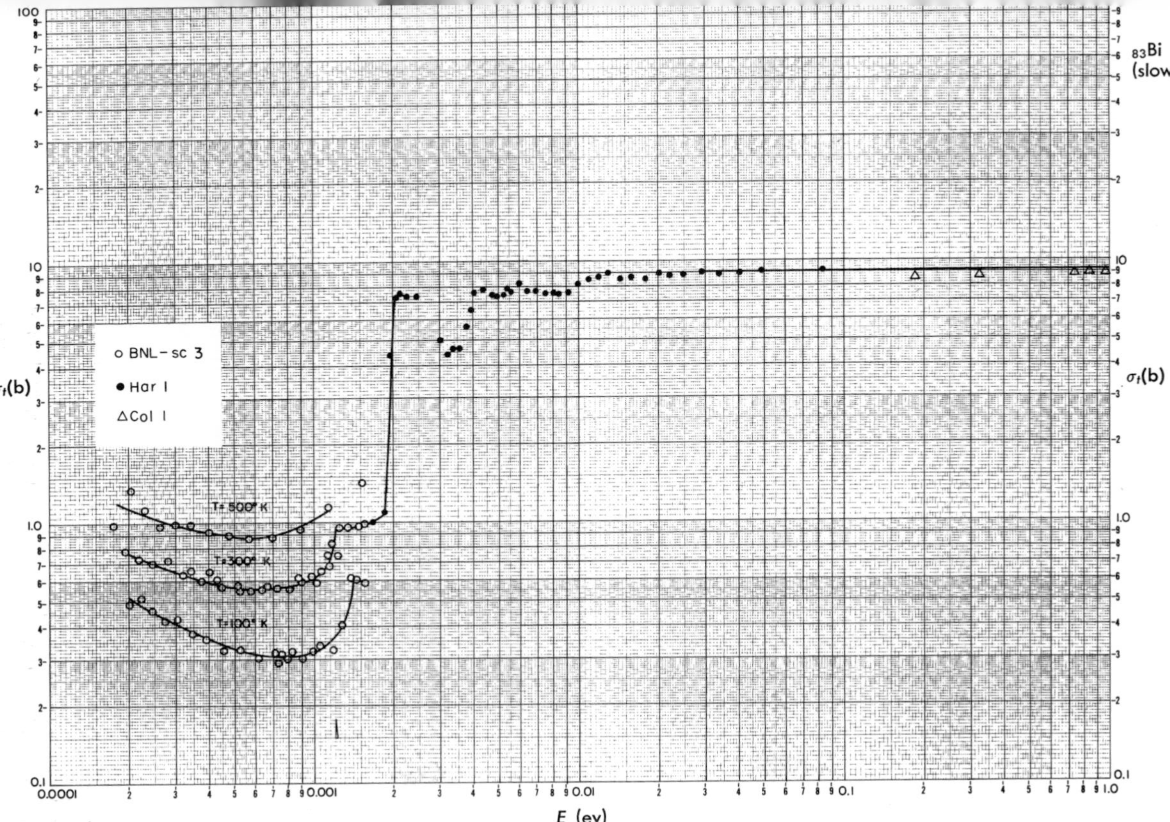

FIG. 18–13. The total cross section of bismuth as a function of neutron energy from 10^{-4} ev to 1 ev.(17)

range between 1 and 10 kev has not yet been explored well for most nuclei because it has been a region in which precise measurements could not be made; the energies were too high for time-of-flight methods and too low for direct neutron beams from accelerators. The development of the fast chopper allows measurements to be made well into this region from below, while improvements in the techniques based on the use of Van de Graaff accelerators[29] allow transmission measurements to be made with reasonably good precision at energies as low as 1 kev. Hence the gap in available cross-section values in the 1–10 kev region should be closed in the near future.

18–7 Scattering, absorption, and activation cross sections. The total cross section can be expressed as the sum of the scattering and absorption cross sections σ_s and σ_a,

$$\sigma_t = \sigma_s + \sigma_a. \tag{18–52}$$

The partial cross sections are harder to determine than the total cross section for several reasons. The total cross section can be obtained from the ratio of two counting rates, with and without the sample, and it is not necessary to know the absolute efficiency of the detector. The partial cross sections, however, cannot be expressed in terms of flux ratios, and absolute measurements are usually needed. The scattering cross section can be measured, in principle, by placing the detector outside the direct beam, say at an angle of about 90° with the direction of the beam, as shown in Fig. 18–14. If the scattering is isotropic or nearly so, the total scattering can be obtained from a single measurement in which the detector subtends a known solid angle and the number of neutrons scattered through that angle is counted. Since the scattered neutrons are spread over a 4π solid angle, the scattered intensity at any particular angle is very low. Furthermore, to avoid multiple scattering within the sample, which would increase the difficulty of interpreting the experimental results, a thin scatterer must be used and nearly all the neutrons in the beam are transmitted. Hence, only a small fraction of the neutrons is scattered and only a small fraction of these is detected. Because of these complications, scattering cross sections cannot in general be determined with the accuracy possible for total cross sections.

Scattering experiments such as that just outlined have, nevertheless, been made.[30] The result obtained for the cross section usually represents an average value over some neutron velocity distribution. This average value has no simple relationship to the cross section at any particular neutron velocity because the scattering cross section, unlike the absorption cross section, often varies in an indefinite and irregular way at thermal energies. The average scattering cross section for the Maxwell

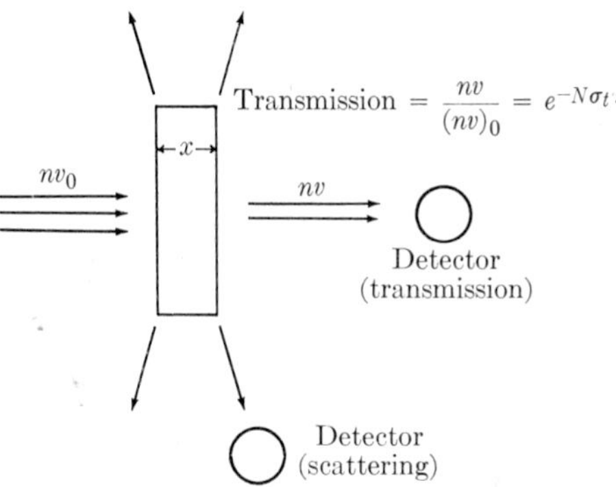

FIG. 18–14. Geometrical arrangement for measurement of total or scattering cross section.

distribution has been measured for many substances, and values are listed in the AEC Compilation[17] of cross sections for neutron reactions, because of their usefulness in certain practical applications such as design calculations for nuclear reactors. With a few exceptions these values lie between 1 and 10 b.

To overcome the difficulty of the low counting rate in thin-target scattering measurements, a thick-target method has been developed.[31,32,33] Although this method has the advantage of greater counting rates for the scattered neutrons, it has the disadvantage that the measured counting rate of scattered neutrons cannot be interpreted directly in terms of the scattering cross section. The ratio σ_s/σ_t can, however, be determined as a function of energy if the corresponding ratio for a standard material is known; the lower curve of Fig. 16–4 was determined in this way.[33] If σ_t is known, σ_s is determined. There is also a *thick-thin* target[34,35] method in which an attempt is made to combine an increased counting rate with a minimum of difficulty in the interpretation of the observed rates. A thin sample is placed in the neutron beam in a plane at a small angle to the incident beam; the sample then appears thick for the transmitted neutrons but thin for neutrons scattered at right angles to the direction of the incident beam. The scattered and transmitted neutrons are observed as functions of neutron energy for a standard target as well as for the sample whose scattering cross section is desired. The ratio of σ_s/σ_t for the sample is related to σ_s/σ_t for the standard more simply than in the thick-target method, and is determined if the latter ratio is known. The scattering cross section is again determined if σ_t is known from transmission measurements.

The thin-target method has also been improved recently.[36] Resonance scattering of neutrons from a linear accelerator was measured by comparing the number of neutrons counted in an annular detector with the number scattered by a sample of pure lead, which has a known and constant scattering cross section. Eighteen BF_3 proportional counters were arranged to form the annular detector. The comparison with the known scatterer made it unnecessary to know the absolute values of the incident flux and the counter efficiency. The detected scattered neutrons were recorded in a multi-channel time-of-flight analyzer so as to give the yield of scattered neutrons as a function of the incident neutron energy.

Measurements of the scattering cross sections for slow neutrons can also be made by studying optical interference effects.(37) Since the wavelength of slow neutrons is comparable to the interatomic distances in crystals, the scattered neutrons produce a wave capable of interfering with the incident neutron wave.(38) This coherent scattering has a probability which can be expressed in terms of the *coherent scattering cross sections* of the individual atoms. These cross sections can be determined from the analysis of the interference effects, although the theory is too involved to be discussed here. Values of the coherent scattering cross section have been determined for a large number of nuclides; again, with a few exceptions, the measured values lie between 1 and 10 b.(17) In addition to the coherent scattering there is some *incoherent*, or *diffuse*, scattering which does not show interference effects. The absence of interference between incident and scattered waves can be attributed to such factors as the presence of two or more isotopes of an element and the effect of the interaction between the spins of the neutrons and those of the scattering nuclei. When the neutron energy reaches several electron volts, the wavelength of the neutrons is too small to give coherent crystal effects and the scattering is entirely incoherent. The scattering atom acts as if it were free and not bound to other atoms; the scattering cross section measured under these conditions is called the *free atom cross section* and is the same as the total scattering cross section at these energies. It is apparent that the study of neutron scattering is complicated at low energies, but the existence of effects peculiar to neutrons helps make them valuable as research tools.(39)

The absorption cross section σ_a refers to those reactions in which a neutron is not re-emitted. The main absorption processes are the (n, γ), (n, p), and (n, α) reactions, which have already been discussed in Chapters 11 and 16. The absorption cross section of thermal neutrons can be measured most accurately with a method that is sensitive to the disappearance of a neutron, but is not affected by scattering. One such method involves the use of a chain-reacting pile.(40) If a pile is operating at a constant neutron flux level, the insertion of a substance which can absorb neutrons will result in a gradual decrease in the neutron flux. The decrease can be measured accurately and is not sensitive to the scattering properties of the absorber even when the absorption cross section is small. This method is called the *danger coefficient method* because it was first used to measure the harmful effects (with respect to the reactivity of a pile) of various materials. A modification in which a *pile oscillator* is used increases the sensitivity of this method;[41,42] the absorber is moved periodically in and out of the pile, or from one point to another inside the pile, with the result that the neutron flux in the pile oscillates; the magnitude of the flux oscillation is a measure of the absorption cross section of the sample.

The individual reactions whose cross sections make up σ_a may be measured by detecting the particles produced in the reaction, such as protons or α-particles, or by determining the activity of the product nucleus if it is radioactive. In the last case, the cross section measured is called the *activation cross section*, σ_{act}, although the particular reaction producing the activation may be an (n, γ), (n, p), (n, α), or (n, 2n) reaction. The name σ_{act} refers to the method of measuring the cross section rather than the reaction involved. The ideas on which this method are based were discussed in Section 18–4. The activation method is not restricted to thermal neutrons, but can be applied to neutrons of any energy, or to a distribution of energies; an example of the use of this method with fast neutrons was given in Section 16–5C. The activation cross section usually refers to a specific isotope of an element, since the activity of a particular isotope of the target element or of another element is usually measured.[43] In many cases, the absorption and activation cross sections are equal; for this to happen, it is necessary that one particular absorption process be much more important than the others and that the product of the reaction be radioactive. Thus, Cu^{63}, a naturally occurring isotope of copper, captures a neutron to give Cu^{64}, which has a half-life of 12.9 hr; the measured activation cross section is 4.3 ± 0.2 b; the absorption cross section of separated Cu^{63} measured by the pile oscillator method is 4.5 ± 0.1 b, so that the two cross sections agree well.

In some cases, the absorption and activation cross sections are not the same. For example, monoisotopic Bi^{209} has an absorption cross section of 0.034 ± 0.002 b as measured by the pile oscillator method. The activation cross section for the production of Bi^{210}, which is an electron emitter, is only 0.019 ± 0.002 b, or about half as great as the absorption cross section. It is now known that Bi^{209}, on capturing a neutron, can also form an isomer of Bi^{210}, which is an α-emitter with a half-life of about 10^6 years. Hence, the determination of the β-activity of the product of neutron absorption by Bi^{209} accounts for only about half of the neutrons absorbed. The cross section compilation,[17] which has now been referred to several times, consequently lists absorption and activation cross sections for thermal neutrons.

The interpretation of the total cross section is often made easy because one partial cross section is much greater than the other. In moderators (graphite, D_2O, H_2O) the absorption cross section is very small compared with the total cross section. Graphite, for example, has a total cross section of about 4.8 b in the energy range 0.02 ev to 400 ev, while its absorption cross section at 0.025 ev is only 0.0045 b and decreases according to the $1/v$-law as the energy increases. The scattering cross section is practically identical with the total cross section, and is constant in this range. Similarly, bismuth (Figs. 16–5, 18–12, and 18–13) has a total cross section of about 9 b over most of the energy range. Its thermal absorption cross section is 0.034 b and decreases according to the $1/v$-law with increasing energy, so that scattering is responsible for practically all of the total cross section. On the other hand, boron has a total cross section of 755 b at 0.025 ev, caused almost entirely by the (n, α) reaction, since the scattering cross section is only 4 b. Boron follows the $1/v$-law for absorption, and scattering does not become significant until the neutron energy reaches about 10 ev. If the absorption cross section of a nuclide is known at 0.025 ev (2200 m/sec) and there are no resonances in σ_t at low energies, it is usually safe to assume that σ_a follows the $1/v$-law (Section 16–5A) so that the absorption cross section can be calculated for other energy values. If σ_t is also known as a function of energy, σ_s can be obtained at energies other than 0.025 ev by subtracting the *calculated* values of σ_a from σ_t. This procedure saves a good deal of experimental work on the measurement of σ_s as a function of energy.

In the region of neutron energies in which the total cross section shows resonances, it is sometimes necessary to distinguish between scattering and absorption resonances,[(44)] and this can be done with the aid of activation measurements. Thus, indium and dysprosium have absorption (activation) resonances at 1.45 ev and 1.74 ev, respectively, which make these substances especially useful as neutron detectors. Manganese (Mn^{55}), Co^{59}, and W^{186} have strong scattering resonances at 337 ev, 132 ev, and 19 ev, respectively. Resonance cross sections can reach many thousands of barns, and one nuclide is known, Xe^{135}, a fission product, with an absorption cross section of about 3×10^6 b, the largest known cross section.

At high energies in the range 0.01 to 10 or 20 Mev, additional difficulties are met.[(28,45)] Scattering is no longer isotropic and the scattering cross section depends on the angle of scattering. Inelastic scattering becomes important, and is complicated by the fact that the scattered neutrons have energies which can be distributed over a wide range, and the sensitivity of the detector must be known as a function of energy. Activation measurements are harder to make at high energies than at low energies because the cross sections involved, mainly (n, γ), are relatively small and low β-activities are produced. For all of these reasons, the cross section values available in the range 0.01 Mev to about 20 Mev are mainly those for σ_t.

There are, of course, many other interesting and important properties and applications of neutrons which cannot be discussed in this book; the reader is referred for them to the books listed in the bibliography which follows.

REFERENCES

GENERAL

B. T. FELD, "The Neutron," *Experimental Nuclear Physics,* E. Segrè, ed. New York: Wiley, 1953, Vol. II, Part VII.

D. J. HUGHES, *Pile Neutron Research.* Reading, Mass.: Addison-Wesley, 1953.

E. AMALDI, "The Production and Slowing Down of Neutrons," *The Encyclopedia of Physics.* (*Handbuch der Physik*). Berlin: Springer Verlag, 1959, Vol. 38, Part 2, pp. 1–659.

L. F. CURTISS, *Introduction to Neutron Physics.* New York: Van Nostrand, 1959.

D. J. HUGHES, *Neutron Cross Sections.* London: Pergamon Press, 1957.

D. J. HUGHES, *Neutron Optics.* New York: Interscience Publishers, 1954.

F. AJZENBERG-SELOVE, ed. *Nuclear Spectroscopy.* New York: Academic Press, 1960. Part III. *Neutron Spectroscopy.* Articles by various authors, pp. 335–490.

Proceedings of the International Conference on the Peaceful Uses of Atomic Energy, Geneva, 1955, Vol. 4. "Cross Sections Important to Reactor Design." New York: United Nations, 1956 (Many articles on equipment, techniques, and results).

Proceedings of the Second International Conference on the Peaceful Uses of Atomic Energy, Geneva, 1958, Vol. 14. "Nuclear Physics and Instrumentation" (Many articles on nuclear theory, nuclear reactions, and neutron spectrometry).

PARTICULAR

1. H. L. ANDERSON, *Neutrons from Alpha-Emitters,* Preliminary Report No. 3 in Nuclear Science Series. Washington, D.C.: National Research Council, 1948.
2. A. WATTENBERG, *Photo-Neutron Sources,* Preliminary Report No. 6 in Nuclear Science Series. Washington, D.C.: National Research Council, 1949.
3. A. WATTENBERG, "Photo-Neutron Sources and the Energy of Photo-Neutrons," *Phys. Rev.* **71,** 497 (1947).
4. RUSSELL, SACHS, WATTENBERG, and FIELDS, "Yields of Neutrons from Photoneutron Sources," *Phys. Rev.* **73,** 545 (1948).
5. B. T. FELD, "The Neutron," *op. cit.* gen. ref., p. 369
6. A. O. HANSON and R. F. TASCHEK, *Monoenergetic Neutrons from Charged Particle Reactions,* Preliminary Report No. 4 in Nuclear Science Series. Washington, D.C.: National Research Council, 1948.
7. HANSON, TASCHEK, and WILLIAMS, "Monoenergetic Neutrons from Charged Particle Reactions," *Revs. Mod. Phys.* **21,** 635 (1949).

8. B. T. Feld, "The Neutron," *op. cit.*, gen. ref., pp. 404–426.

9. A. M. Weinberg and E. P. Wigner, *The Physical Theory of Neutron Chain Reactors.* University of Chicago Press, 1958, Chapters 10, 11.

10. R. E. Marshak, "Theory of the Slowing Down of Neutrons by Elastic Collisions with Atomic Nuclei," *Revs. Mod. Phys.* **19,** 185 (1947).

11. S. Glasstone and M. C. Edlund, *The Elements of Nuclear Reactor Theory.* New York: Van Nostrand, 1952, Chapter 6.

12. R. D. Present, *Kinetic Theory of Gases.* New York: McGraw-Hill, 1958, Chapter 5.

13. R. Ringo, unpublished work, cited by D. J. Hughes (*op. cit.*, p. 89).

14. Fermi, Marshall, and Marshall, "A Thermal Velocity Selector and its Application to the Measurement of the Cross Section of Boron," *Phys. Rev.* **72,** 193 (1947).

15. T. Brill and H. V. Lichtenberger, "Neutron Cross Sections with the Rotating Shutter Mechanism," *Phys. Rev.* **72,** 585 (1947).

16. P. A. Egelstaff, "The Operation of a Thermal Neutron Time-of-Flight Spectrometer," *J. Nuclear Energy,* **1,** 57 (1954).

17. D. J. Hughes and R. B. Schwartz, "Neutron Cross Sections," *Brookhaven National Laboratory Report BNL-325*, 2nd ed. July 1, 1958. Washington, D.C.: U.S. Government Printing Office.

18. (a) L. B. Borst and V. L. Sailor, "Neutron Measurements with the Brookhaven Crystal Spectrometer," *Rev. Sci. Instr.* **24,** 141 (1953); (b) Sailor, Foote, Landon, and Wood, *Rev. Sci. Instr.* **27,** 26 (1956).

19. V. L. Sailor, "The Parameters for the Slow Neutron Resonance in Rhodium," *Phys. Rev.* **91,** 53 (1953).

20. L. J. Rainwater and W. W. Havens, Jr., "Neutron Beam Spectrometer Studies of Boron, Cadmium, and the Energy Distribution from Paraffin," *Phys. Rev.* **70,** 136 (1946).

21. Rainwater, Havens, Wu, and Dunning, "Slow Neutron Velocity Spectrometer Studies, I. Cd, Ag, Sb, Ir, Mn," *Phys. Rev.* **71,** 65 (1947).

22. E. Melkonian, "Slow Neutron Velocity Spectrometer Studies of O_2, N_2, A, H_2, H_2O and Seven Hydrocarbons," *Phys. Rev.* **76,** 1750 (1949).

23. W. W. Havens, Jr., and L. J. Rainwater, "Slow Neutron Velocity Spectrometer Studies IV. Au, Ag, Br, Fe, Co, Ni, Zn," *Phys. Rev.* **83,** 1123 (1951).

24. Seidl, Hughes, Palevsky, Levin, Kato, and Sjöstrand, "Fast Chopper Time-of-Flight Measurements of Neutron Resonances," *Phys. Rev.,* **95,** 476 (1954).

25. Fluharty, Simpson, and Simpson, "Neutron Resonance Measurements of Ag, Ta, and U^{238}," *Phys. Rev.* **103,** 1778 (1956).

26. L. M. Bollinger, "Techniques of Slow Neutron Spectroscopy," *Nuclear Spectroscopy,* F. Ajzenberg-Selove, ed., *op. cit.* gen. ref., pp. 342–357.

27. Barschall, Rosen, Taschek, and Williams, "Measurements of Fast Neutron Flux," *Revs. Mod. Phys.* **24,** 1 (1952).

28. H. H. Barschall, "Method for Measuring Fast Neutron Cross Sections," *Revs. Mod. Phys.* **24,** 120 (1952).

29. Hibdon, Langsdorf, and Holland, "Neutron Transmission Cross Sections in the Kilovolt Region," *Phys. Rev.* **85,** 595 (1952).

30. M. Goldhaber and G. H. Briggs, "Scattering of Slow Neutrons," *Proc. Roy. Soc.* (London), **A162,** 127 (1937).

31. J. Tittman and C. Sheer, "The Energy Dependence of the Resonant Scattering of Slow Neutrons from Gold," *Phys. Rev.* **83,** 746 (1951).

32. B. N. Brockhouse, "Resonant Scattering of Slow Neutrons," *Can. J. Phys.* **31,** 432 (1953).

33. C. Sheer and J. Moore, "Measurement of Scattering Cross Sections for Low-Energy Neutron Resonances," *Phys. Rev.* **98,** 565 (1955).

34. J. A. Moore, "Resonance Scattering of Slow Neutrons on Indium," *Phys. Rev.* **109,** 417 (1958).

35. H. L. Foote, Jr., "Neutron Scattering Cross Section of U^{235}," *Phys. Rev.* **109,** 1641 (1958).

36. Rae, Collins, Kinsey, Lynn, and Wiblin, "An Analysis of Slow Neutron Resonances in Silver," *Nuclear Physics* **5,** 89 (1958).

37. C. G. Shull and E. O. Wollan, "Coherent Scattering Amplitudes as Determined by Neutron Diffraction," *Phys. Rev.* **81,** 527 (1951).

38. E. O. Wollan and C. G. Shull, "Neutron Diffraction and Associated Studies," *Nucleonics,* **3,** No. 1, 8–21, July 1948; **3,** No. 2, 17, August 1948.

39. G. E. Bacon, *Neutron Diffraction.* Oxford: Clarendon Press, 1955.

40. Anderson, Fermi, Wattenberg, and Zinn, "Method for Measuring Neutron Absorption Cross Sections by the Effect on the Reactivity of a Chain-Reacting Pile," *Phys. Rev.* **72,** 16 (1947).

41. Hoover, Jordan, Moak, Pardue, Pomerance, Strong, and Wollan, "Measurement of Neutron Absorption Cross Sections with a Pile Oscillator," *Phys. Rev.* **74,** 864 (1948).

42. H. Pomerance, "Thermal Neutron Capture Cross Sections," *Phys. Rev.* **83,** 64 (1951).

43. Seren, Friedlander, and Turkel, "Thermal Neutron Activation Cross Sections," *Phys. Rev.* **72,** 888 (1947).

44. B. T. Feld, *op. cit.*, gen. ref., pp. 291, 323–325.

45. J. L. Fowler and J. E. Brolley, Jr., "Monoenergetic Neutron Techniques in the 10–30 Mev Range," *Revs. Mod. Phys.* **28,** 103 (1956).

PROBLEMS

1. The Q-value for the photodisintegration of the deuteron is 2.227 ± 0.003 Mev. The mass doublet $2H^1$-H^2 is, from nuclear data, 1.5494 ± 0.0024 millimass units (mmu), and the mass of the hydrogen atom is 1.008142 ± 0.000003 amu. Find the mass of the neutron from these data.

2. It is customary to describe neutrons of different energy by the adjectives shown below:

Energy, ev	Neutron type
0.001	cold
0.025	thermal
1.0	slow (resonance)
100	slow
10^4	intermediate
10^6	fast
10^8	ultrafast
10^{10}	ultrafast (relativistic)

Compute the speed, de Broglie wavelength, and absolute temperature for each neutron type.

3. The reaction $Be^9(\alpha, n)C^{12}$ with polonium α-particles with an energy of 5.30 Mev is a useful source of neutrons. Calculate (a) the Q-value of the reaction, (b) the energy of the neutrons which emerge at angles of 0°, 90°, and 180° with the direction of the incident beam.

4. The following (d, n) reactions are good sources of fast neutrons: $H^2(d, n)He^3$, $H^3(d, n)He^4$, $C^{12}(d, n)N^{13}$, $N^{14}(d, n)O^{15}$, $Li^7(d, n)Be^8$, $Be^9(d, n)B^{10}$. Assume that the product nucleus is left in the ground state, and calculate the energy of the neutrons which would correspond to zero kinetic energy of the incident deuterons.

5. The following (p, n) reactions are useful sources of relatively low-energy neutrons: $H^3(p, n)He^3$, $Li^7(p, n)Be^7$, $Be^9(p, n)B^9$, $C^{12}(p, n)N^{12}$, $Na^{23}(p, n)Mg^{23}$, $V^{51}(p, n)Cr^{51}$. It has been shown that the minimum neutron energy is given by $E_{n,\,\min} = E_t/(A+1)^2$, where E_t is the threshold energy and A is the mass number of the target nucleus. Calculate the value of $E_{n,\,\min}$ for each reaction. The Q-value for the (p, n) reaction on vanadium is −1.532 Mev.

6. A cylindrical BF_3 counter, 10 in. long and $\frac{1}{4}$-in. in diameter, contains BF_3 at a pressure of 20 cm Hg (corrected to N.T.P.). The counter has an over-all efficiency of 2%, that is, it detects 2 out of every 100 neutrons incident on it. When the counter is exposed to neutrons with a Maxwellian velocity distribution at 20°C, the counting rate is 20,400 counts/min. (a) What is the neutron flux? (b) What would be the neutron flux if the neutrons had a Maxwellian distribution at 127°C, and the counting rate were the same? A counter of this type is usually used for relative flux measurements. (c) What would be the relative counting rates if the flux were kept constant and the temperature of the neutrons in different experiments had the values 0°C, 25°C, 50°C, 100°C, 200°C? Take the rate at 25°C equal to 100 units.

7. Compute the average logarithmic energy loss per collision, the number of collisions needed to reduce the neutron energy from 2 Mev to 0.025 ev, the slowing-down power, and the moderating ratio of fluorine, magnesium, and bismuth at room temperature. Compare the results with those for the materials listed in Tables 18–2 and 18–3. Use the following values of cross sections.

	σ_a	σ_s
F	9 mb	5 b
Mg	59 mb	6 b
Bi	30 mb	9 b

8. A gold foil 0.02 cm thick and 1 cm^2 in cross section is irradiated for one hour in a nuclear reactor. The thermal neutrons have a Maxwellian energy distribution and their flux is 10^{12} neutrons/cm^2-sec; it is assumed that all of the neutrons at the irradiation position are thermal. The gold has a density of 19.3 gm/cm^3; its thermal absorption cross section is 98.7 b and the slight deviation from the $1/v$-law may be neglected. (a) What is the activity of the foil on being removed from the reactor? (b) If the activity of the foil is measured 60 days after the irradiation, with a counter whose over-all efficiency is one percent, what is the counting rate?

9. Consider the problem of the diffusion, in a weakly absorbing medium, of monoenergetic neutrons from a point source which emits Q neutrons per second. (a) Show that the differential equation for the neutron density is

$$\frac{d^2n}{dr^2} + \frac{2}{r}\frac{dn}{dr} - \frac{n}{L^2} = 0,$$

where r is the radial distance from the source at $r = 0$, and L is the diffusion length of thermal neutrons in the medium. (b) Show that the neutron density is given by

$$n = \frac{Qe^{-r/L}}{4\pi Dr},$$

where D is the diffusion coefficient.

10. A crystal spectrometer has a beryllium crystal with a lattice spacing of 0.7323 A for the reflecting planes used for slow neutrons. What are the Bragg angles for the first order reflection of neutrons of energy 1, 3, 5, 10, 30, and 50 ev, respectively?

11. In measurements of neutron wavelength with a crystal spectrometer, there is an uncertainty $\Delta\lambda$ in the wavelength because of the uncertainty $\Delta\theta$ in the Bragg angle. Consequently, there is an uncertainty ΔE in the measured energy of the neutrons. The *resolution* of the instrument is defined as the fractional uncertainty $\Delta E/E$ in the energy. Show that (a) $\Delta E/E = 2 \cot \theta \, \Delta\theta \approx 2 \, \Delta\theta/\theta$, (b) $\Delta E/E \approx k \, \Delta\theta E^{1/2}$, where k is a certain constant. In the spectrometer of Problem 10, the uncertainty in the Bragg angle, $\Delta\theta$, is 7.8 min. Calculate the uncertainty in the energy for each of the energies of Problem 10.

12. In the time-of-flight velocity selector, the following relations hold for the energy E (ev), the velocity v (m/sec) and the time-of-flight for a 1-m path t (μsec/m):

$$v = 10^6/t, \qquad E = 51.5 \times 10^2/t^2, \qquad t = 71.5/E^{1/2}.$$

Derive the following relations.

$$\Delta v = -\frac{10^6\,\Delta t}{t^2}$$

$$\frac{\Delta v}{v} = -\frac{\Delta t}{t}, \qquad \Delta v = -10^6 v^2\,\Delta t$$

$$\frac{\Delta E}{E} = -\frac{2\,\Delta t}{t}, \qquad \Delta E = -0.028E^{3/2}\,\Delta t$$

Calculate the energy spread ΔE at energies of 1 ev, 10 ev, 100 ev, 1 kev, and 10 kev for each of the time uncertainties $\Delta t = 0.5$, 0.1, and 0.05 μsec/m.

13. A beam of thermal neutrons with an initial intensity of 10^6 neutrons/cm^2-sec is passed through a tantalum sheet 1 mm thick. The intensity of the beam emerging from the tantalum is 8.65×10^5 neutrons/cm^2-sec. The activity of the foil after irradiation for 2.0 days and a delay of 2 hr to allow for the decay of short-lived isomeric activity is 1224 counts/cm^2/sec because of the activity of Ta^{182}. The density of tantalum is 16.6 gm/cm^3 and the half-life of Ta^{182} is 114 days. What are (a) the total thermal cross section of tantalum, (b) the activation cross section for the production of Ta^{182}?

14. Arsenic has an absorption cross section for thermal neutrons of 4.1 b and a scattering cross section of 6 b; its density is 5.73 gm/cm^3. A beam of thermal neutrons is passed through a slab of arsenic 2 cm thick. By what fraction is the intensity of the beam reduced? How much of this reaction is caused by scattering? How much by absorption?

15. The total cross section of a nuclide with an isolated resonance for which the resonance scattering is negligible can be expressed in the form

$$\sigma_t(E) = \sigma_s + \left(\frac{E_0}{E}\right)^{1/2} \frac{\sigma_0 \Gamma^2}{4(E - E_0)^2 + \Gamma^2},$$

where σ_s is the scattering cross section, and σ_0 is a composite quantity involving parameters discussed in Section 16–3. The total cross section of rhodium has been measured with the crystal spectrometer of Problems 10 and 11, and the following values have been obtained for the resonance parameters: $E_0 = 1.260 \pm 0.004$ ev, $\sigma_0 = 5000 \pm 200$ b, $\Gamma = 0.156 \pm 0.005$ ev, and $\sigma_s = 5.5 \pm 1.0$ b. Plot the value of the total cross section for values of the energy between 0.2 ev and 40 ev. Calculate, from the formula, the thermal absorption cross section. How does the result compare with the measured value of 156 ± 7 b.

16. How many collisions are needed for neutrons to lose, on the average, 99% of an initial energy of 2 Mev in graphite? What does this result show, as compared with the number required to moderate the neutrons to thermal energies?

17. Show that the neutron velocity at which the maximum of the Maxwellian flux distribution occurs is given by $\sqrt{3kT/m}$.

18. The expression (18–43) for the thermal diffusion length was obtained for monoenergetic neutrons. Suppose that the thermal neutrons have a Maxwellian velocity distribution at 20°C in a very large block of graphite. Graphite has a transport cross section which may be taken to be constant and its absorption cross section varies as $1/v$. Show that the thermal diffusion length of the neutrons is given by

$$L = \frac{0.613}{N\sqrt{\sigma_{tr}\sigma_{a_0}}},$$

where the subscript zero denotes the cross section for 2200 m/sec neutrons.

19. Suppose that the only two reactions that a certain nuclide undergoes when bombarded with neutrons are elastic scattering and radiative capture. Show that the value of the total cross section at the peak of the resonance is given by

$$\sigma_{t0} = \frac{\lambda_0^2}{4}\frac{\Gamma_n}{\Gamma},$$

where the subscript zero refers to values at $E = E_0$, the energy at the peak. [*Hint:* Apply Eq. (16–13) to each of the reactions.]

20. It is known from experiment that when a given nuclide is bombarded with neutrons, Γ_γ has a constant value, while Γ_n is proportional to $E^{1/2}$. Show that the cross section for radiative capture may be written in the form

$$\sigma_r(E) = \sigma_{r0}\left(\frac{E_0}{E}\right)^{1/2}\left[\frac{1}{1 + [(E - E_0)/(\Gamma/2)]^2}\right],$$

where the subscript zero refers to values at the peak of the resonance.

CHAPTER 19

NUCLEAR FISSION

19–1 The discovery of nuclear fission. The discovery of nuclear fission was one of the results of attempts to make transuranium elements of atomic number greater than 92 by means of (n, γ) reactions followed by β-decay of the product nucleus. These attempts eventually suceeded, and some of the methods used and results obtained were discussed in Section 12–3. The interpretation of the early experiments was difficult, however, because of unexpected results which could be explained only in terms of nuclear fission.

To understand some of the difficulties that were met, it is necessary to consider briefly how radioactive elements are separated from inactive ones. A radioelement formed by a nuclear reaction is usually available only in a very small amount, possibly as small as 10^{-12} gm, and cannot be separated by means of ordinary chemical methods. But the separation can often be made with the aid of a *carrier* which is a stable substance with chemical properties similar to those of the radioelement. The element and the carrier usually belong to the same subgroup of the periodic system and can undergo similar chemical reactions. If an appreciable amount, perhaps 10 to 100 mg, of the carrier is added to a solution containing the radioelement, and if the carrier is then precipitated from the solution by the formation of an insoluble salt, the radioelement is precipitated along with the carrier. The carrier and the radioelement can then be separated by means of other chemical reactions.

In the separation of radium from other members of the uranium series, barium serves as a carrier. Radium and barium both belong to group IIA of the periodic system and form insoluble sulfates. When a solution of a barium salt is added to a solution containing a very small amount of radium and a sulfate is added, barium and radium sulfates are precipitated together. A neater method depends on the fact that barium and radium chlorides are precipitated together from concentrated solutions of hydrochloric acid. The precipitate can be dissolved in water, and the barium and radium can then be separated by repeated fractional crystallization from hydrochloric acid. In analogous ways, lanthanum acts as a carrier for actinium ($Z = 89$). Sometimes the carrier is a stable form of the radioelement, e.g., stable iodine can be used as a carrier for radioactive isotopes of iodine.

In the early experiments on the formation of transuranium elements, uranium was bombarded with neutrons and several different β-activities, distinguished by their half-lives, were detected. Carrier techniques were used to separate the elements responsible for the activities, but the number and properties of the new radioelements were such that they could not be fitted into a scheme consistent with the known properties of the heavy elements and the predicted properties of the transuranium elements. One difficulty in the analysis of the products of the bombardment of uranium with neutrons led to a remarkable conclusion. In addition to the elements which seemed to be real transuranium elements, there were four which were supposed to be β-radioactive iostopes of radium because they were precipitated with barium when the latter was used as the carrier. The decay products of these nuclides seemed to be isotopes of actinium because they were precipitated with lanthanum, the carrier for actinium. Although these results appeared to be consistent, since actinium follows radium in the periodic system, they raised two serious questions. First, the production of an isotope of radium by the neutron bombardment of uranium would require an (n, 2α) reaction, but this reaction is a very unlikely one (cf. Section 16–5), especially at low neutron energies. Second, further chemical experiments showed that the "radium" activities could not be separated from the barium carrier, and the daughter activities could not be separated from the lanthanum carrier. In 1939, Hahn and Strassmann[(1,2)] performed a beautiful and thorough set of experiments which proved beyond a doubt that the "radium" isotopes are really isotopes of barium and the "actinium" isotopes are isotopes of lanthanum. Furthermore, they showed that one of the barium isotopes resulting from the neutron bombardment of uranium could be identified, because of its half-life of 86 min, with the previously known nuclide Ba^{139}, which has the same half-life. Similarly, one of the lanthanum isotopes from the neutron bombardment of uranium was identified with the known nuclide La^{140}, which has a half-life of 40 hr.

The production of the nuclides $_{57}La^{140}$ and $_{56}Ba^{139}$ from uranium, which has the atomic number 92 and an atomic weight of nearly 240, required a hitherto unknown kind of nuclear reaction in which the uranium nucleus is split into fragments which are themselves nuclei of intermediate atomic weight. If such a process really occurs, it should also be possible to find nuclei with masses between 90 and 100 and atomic numbers of about 35. Hahn and Strassmann were able to find an active isotope of strontium ($Z = 38$) and one of yttrium ($Z = 39$) which met these requirements, as well as isotopes of krypton ($Z = 36$) and xenon ($Z = 54$). It was clear from the chemical evidence that uranium nuclei, when bombarded with neutrons, can indeed split into two nuclei of intermediate atomic weight. It was then predicted, from the systematics of stable nuclides and from the semiempirical binding-energy formula, that the product nuclei would have very great energies and would produce large numbers of ion pairs in passing through a gas. When a thin layer of uranium was put in a suitable ionization chamber connected to an amplifier and irradiated with neutrons, great bursts of ionization were observed[3,4] corresponding to energies up to 100 Mev. These pulses of ionization are extremely large compared with those of single α-particles and are easy to recognize. Further chemical work then showed that besides the reaction products mentioned above, other elements of medium mass number were formed, including bromine, molybdenum, rubidium, antimony, tellurium, iodine, and cesium. There was, therefore, ample chemical and physical evidence for the splitting of the uranium nucleus, and this process was called *fission.*[5]

It is now known that fission can be produced in various nuclides under different conditions, and some of the results will be mentioned. When small samples of the separated isotopes of uranium, prepared in a mass spectrograph, were bombarded, it was found[6] that slow neutrons cause fission of U^{235} but not of U^{238}; fast neutrons, with energies greater than one Mev, cause fission of both U^{235} and U^{238}. Thorium and Pa^{231} undergo fission, but only when bombarded with fast neutrons. Fission can also be produced in uranium and thorium by high-energy α-particles, protons, deuterons, and γ-rays. The nuclei Pu^{239} and U^{233}, formed by (n, γ) reactions on U^{238} and Th^{232}, respectively, followed by β-decay of the products, undergo fission when bombarded with slow or fast neutrons, as do other artificial heavy nuclides. Finally, some heavy nuclei have been found to undergo spontaneous fission; in this process, the nucleus divides in the ground state without bombardment by particles from outside.

In addition to the two large fission fragments, neutrons and γ-rays are emitted. Division into three fragments of comparable size (ternary fission) has been observed but is a very rare event, occurring about five times per million binary fissions. Long-range α-particles are sometimes emitted, about once in every 400 fissions. The emission of light nuclei, with masses greater than 4 and probably less than 12, is a relatively common event, occurring once in about every 80 fissions.

19–2 Fission cross sections and thresholds. The probability of fission, as compared with that of other reactions, is a matter of theoretical interest and practical importance. The U^{235} nucleus may capture a thermal neutron to form the compound nucleus [U^{236}], or the neutron may be scattered. The compound nucleus may either undergo fission, or it may emit γ-rays and decay to the ground state of U^{236}, which emits a 4.5-Mev α-particle and has a half-life of 2.4×10^7 years. The thermal cross sections[7] for the different reactions for U^{235}, U^{238}, natural uranium, and Pu^{239} are given in Table 19–1. The total cross section of U^{235} for 2200 m/sec neutrons is 698 b, the total absorption cross section is 683 b, and the fission cross section is 577 b. The ratio of the radiative capture cross section to the

TABLE 19–1

PROPERTIES OF FISSIONABLE MATERIALS[7,8]*

	U^{233}	U^{235}	Natural Uranium	Pu^{239}
σ_{abs}	578 ± 4	683 ± 3	7.68 ± 0.07	1028 ± 8
σ_f	525 ± 4	577 ± 5	4.18 ± 0.06	742 ± 4
σ_r	53 ± 2	101 ± 5	3.50	286 ± 4
σ_s		15 ± 2	8.3 ± 0.2	9.6 ± 0.5
ν (average number of neutrons per fission)	2.51 ± 0.02	2.44 ± 0.02		2.89 ± 0.03
$\alpha = \sigma_r/\sigma_f$	0.101 ± 0.004	0.18 ± 0.01		0.39 ± 0.03
$\eta = \dfrac{\nu}{1+\alpha}$	2.28 ± 0.02	2.07 ± 0.01	1.34 ± 0.02	2.08 ± 0.02

*For 2200 m/sec neutrons.

fission cross section is 0.18, so that the probability of radiative capture is about 18% that of fission. The natural abundance of U^{235} is only 0.72%, with the result that the fission and radiative capture cross sections of natural uranium are much smaller than those of the separated isotope U^{235}. In Pu^{239} the ratio of the capture cross section to that for fission is 0.39; although the fission cross section is greater than in U^{235}, the ratio of capture to fission is also greater.

Some of the artificially produced radioactive nuclides undergo fission with thermal neutrons, as do some of the shorter-lived naturally occurring radionuclides. Values of the fission cross section for 2200 m/sec neutrons are listed in Table 19–2, together with values of activation cross sections and total absorption cross sections where these are available.[7] It is evident that many heavy nuclides are fissionable, and that the probability of fission varies over a wide range of values. Only U^{235}, U^{233} and Pu^{239} have high cross sections as well as long half-lives, and either occur naturally (U^{235}) or can be produced in significant amounts in practical lengths of time (Pu^{239} and U^{233}). Hence, only these three fissionable materials are important in the large-scale applications of nuclear fission.

The fission cross section varies with energy in a complicated way, as is shown in Fig. 19–1 for U^{235}. In the thermal region σ_f varies approximately as $1/v$; starting at 0.28 ev, there are many closely spaced resonances, with at least 20 resolved resonances below 20 ev. At high energies, the fission cross section is relatively small, only about one barn in the neighborhood of 1 Mev; the fission cross sections of U^{233} and U^{235} in the

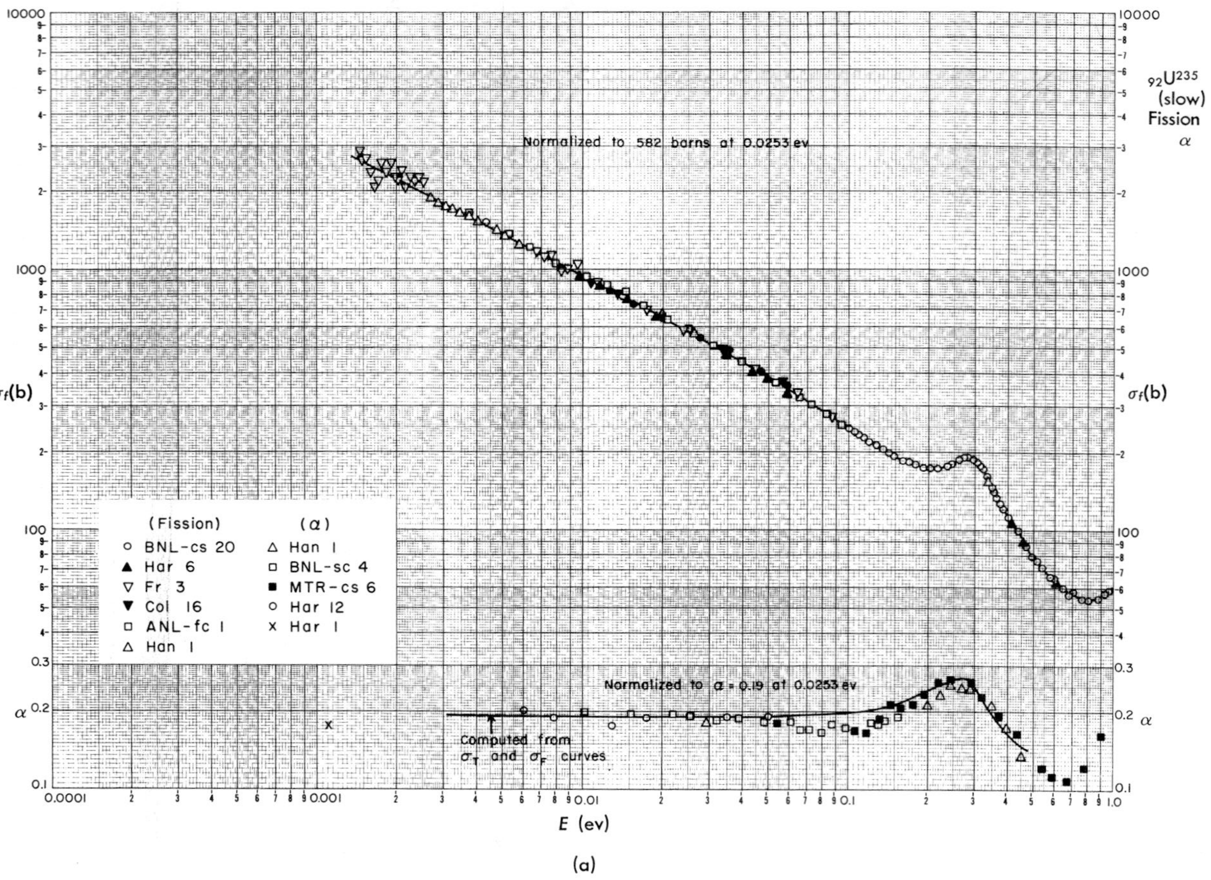
$_{92}U^{235}$
(slow)
Fission
α
σ_f(b)
α
Normalized to 582 barns at 0.0253 ev
Normalized to α = 0.19 at 0.0253 ev
Computed from
σ_T and σ_F curves
(Fission)
○ BNL-cs 20
▲ Har 6
▽ Fr 3
▼ Col 16
□ ANL-fc I
△ Han I
(α)
△ Han I
□ BNL-sc 4
■ MTR-cs 6
○ Har 12
x Har I
E (ev)
(a)

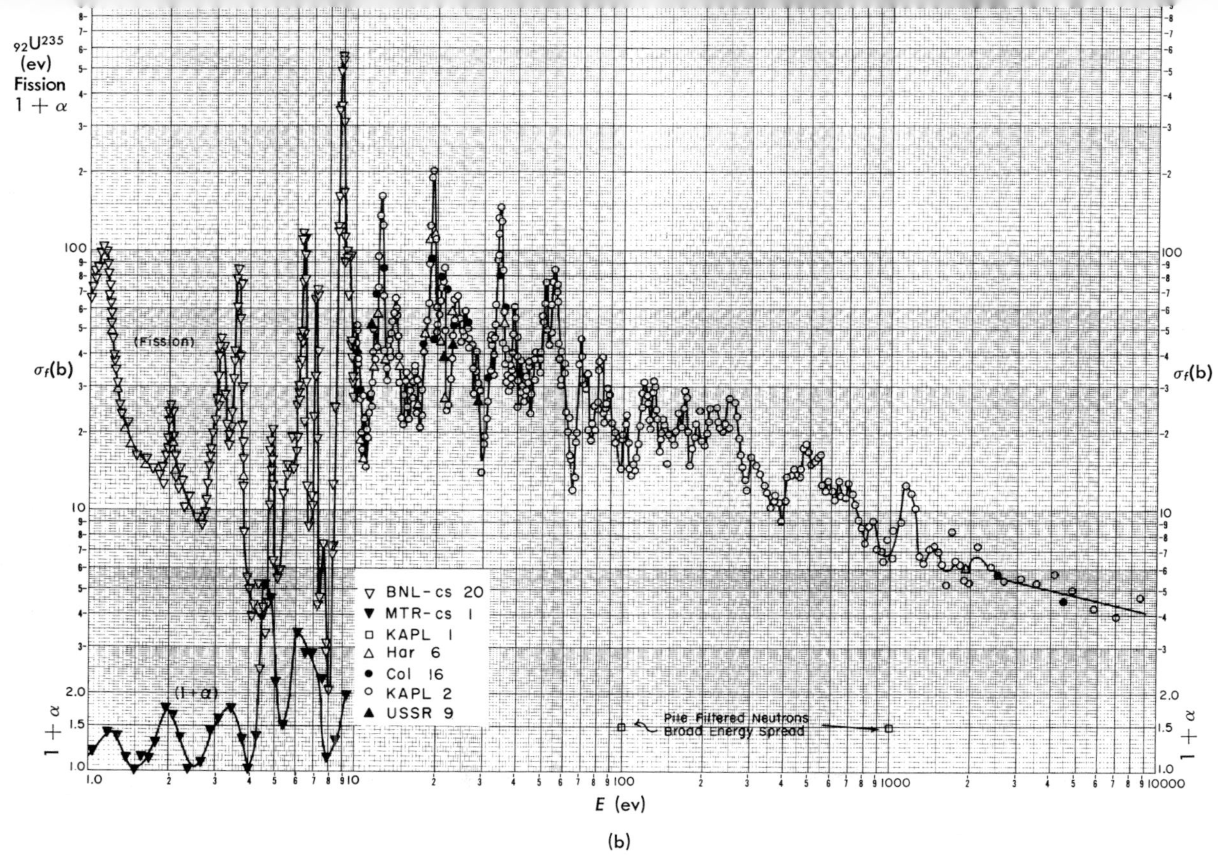

FIG. 19–1. The fission cross section of U^{235} at low energies.[7]

TABLE 19–2

THERMAL CROSS SECTIONS OF SOME HEAVY NUCLIDES[7]*

Nuclide	Half-life	Fission cross section, σ_f, b	Total absorption cross section, σ_a, b	Activation cross section, σ_{act}, b
$_{90}Th^{227}$	18.2d	1500 ± 1000		
$_{90}Th^{229}$	7.34×10^3y	45 ± 11		
$_{91}Pa^{230}$	17.3d	1500 ± 250		
$_{91}Pa^{231}$	3.4×10^4y	$(10 \pm 5) \times 10^{-3}$		200 ± 15
$_{91}Pa^{232}$	1.3d	700 ± 100		760 ± 100
$_{92}U^{230}$	20.8d	25 ± 10		
$_{92}U^{231}$	4.2d	400 ± 300		
$_{92}U^{232}$	74y	80 ± 20		300 ± 200
$_{92}U^{233}$	1.62×10^5y	525 ± 4	578 ± 4	53 ± 2
$_{92}U^{234}$	2.52×10^5y	<0.65	105 ± 4	90 ± 30
$_{92}U^{235}$	7.1×10^8y	582 ± 4	683 ± 3	101 ± 5
$_{93}Np^{234}$	4.4d	900 ± 300		
$_{93}Np^{236}$	22h	2800 ± 800		
$_{93}Np^{237}$	2.2×10^6y	$(19 \pm 3) \times 10^{-3}$	170 ± 5	169 ± 6
$_{93}Np^{238}$	2.1d	1600 ± 100		
$_{93}Np^{239}$	2.3d	<1		35 ± 10
$_{94}Pu^{238}$	86.4y	16.8 ± 0.3		400 ± 10
$_{94}Pu^{239}$	2.44×10^4y	742 ± 4	1028 ± 8	286 ± 4
$_{94}Pu^{240}$	6.6×10^3y	0.030 ± 0.045	286 ± 7	250 ± 40
$_{94}Pu^{241}$	13y	1010 ± 13	1400 ± 80	400 ± 50
$_{94}Pu^{242}$	3.75×10^5y	<0.2	30 ± 2	19 ± 1
$_{95}Am^{241}$	458y	3.2 ± 0.2	630 ± 35	750 ± 80
$_{95}Am^{242}$	100y	6400 ± 500		

*For 2200 m/sec neutrons.

neighborhood of 1 Mev are shown in Fig. 19–2a. Some heavy nuclides, for example, U^{234}, U^{236}, and U^{238}, do not undergo fission with slow neutrons, but only with fast neutrons. Fission is a threshold reaction in these nuclides, and the fission cross section varies with energy in much the same way that the cross sections for other threshold reactions do. The fast fission thresholds of U^{234}, U^{236}, and U^{238}, and the variation of the fission cross section with energy, above threshold, have been determined by bombarding the nuclides with neutrons from the reaction $H^3(p, n)He^3$. The results are shown in Fig. 19–2b.

Threshold energies and cross sections have been measured for fission processes induced by charged particles.[10] The threshold for fission by deuterons is close to 8 Mev for Th^{232}, U^{235}, and U^{238}; that for fission by

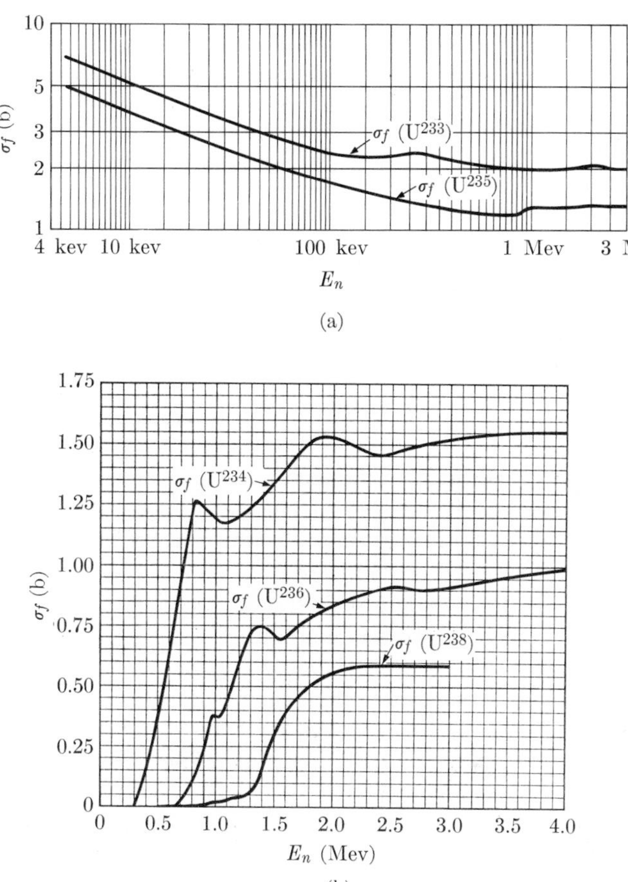

FIG. 19–2. The fission cross sections of several uranium isotopes in the Mev region.[7,9] (a) U^{233}, U^{235} (b) U^{234}, U^{236}, U^{238}.

α-particles is in the neighborhood of 21 Mev for these nuclides. In the range of energies explored so far, the value of the cross section increases with energy above the threshold and then either levels off or passes through a maximum. In the case of U^{235} bombarded with deuterons, the cross section rises to a value of one barn at about 20 Mev and then increases slowly to a value between 1.5 and 2 b at deuteron energies between 120 and 200 Mev. The cross section for the fission of U^{235} by α-particles rises to a value of 1.5 b at about 50 Mev and remains at about this value up to energies of 400 Mev.

Gamma-rays from nuclear reactions and high-energy x-rays from a betatron can cause fission, and some of the measured thresholds[11] are

TABLE 19–3

THRESHOLD ENERGIES FOR PHOTOFISSION

Nuclide	Photofission threshold, Mev
$_{92}U^{238}$	5.08 ± 0.15
$_{92}U^{235}$	5.31 ± 0.25
$_{92}U^{233}$	5.18 ± 0.27
$_{94}Pu^{239}$	5.31 ± 0.27
$_{90}Th^{232}$	5.40 ± 0.22

shown in Table 19–3. The cross sections for these reactions are generally smaller than those for neutron-induced fission. When U^{238} is bombarded by monoenergetic 6.3-Mev γ-rays from the $F^{19}(p, \gamma)Ne^{20}$ reaction, the cross section is about 3 mb; for the 17.5-Mev γ-rays from the $Li^{7}(p, \gamma)Be^{8}$ reaction, the cross section is about 30 mb.

Many heavy nuclides undergo spontaneous fission,[(12)] and this process is an alternative, less probable, method of nuclear disintegration than α-particle emission. Nuclei which undergo fission with slow neutrons have smaller spontaneous fission rates than their isotopes which undergo fission only with fast neutrons. For example, U^{235} has a half-life for spontaneous fission of about 1.8×10^{17} years, corresponding to a rate of about one fission per gram per hour while U^{238} has a fission half-life of 8.0×10^{15} years, or a rate of 25 fissions/gm/hr. Similarly, Pu^{239} has a spontaneous fission rate of 36 fissions/gm/hr, or a fission half-life of 5.5×10^{15} years, while Pu^{240} has a rate of 1.6×10^{6} fissions/gm/hr and a half-life of 1.2×10^{11} years.

Finally, fission can be produced in bismuth, lead, thallium, mercury, gold, and platinum by neutrons with very high energies, e.g., 40 Mev.

19–3 The fission products. It was indicated in Section 19–1 that a number of nuclides of intermediate charge and mass are formed when a uranium nucleus undergoes fission. The study of the nuclei formed in fission was evidently a promising source of information about the mechanism of the fission process and offered the possibility that new, hitherto unknown, nuclides might be discovered. The latter possibility became apparent when the neutron-to-proton ratios of the uranium isotopes were compared with those of some of the fission products. The compound nucleus $[_{92}U^{236}]$ has 144 neutrons and 92 protons, and the value of the ratio is $144/92 = 1.57$. The values of the ratio for the stable isotopes of some typical fission products (krypton, iodine, xenon, and cesium) vary from 1.17 to 1.52, and are appreciably lower than that for $[U^{236}]$. When that excited nucleus splits into two smaller nuclei, the neutron-to-proton ratio for at least one of them must be greater than the value compatible with stability. Such an unstable nucleus might be expected to approach stability by electron emission or, if the excitation energy is high enough, by ejection of one or more neutrons; it has been found experimentally that both of these processes occur.

The investigation of the products of the fission of U^{235} has shown that the range of mass numbers is from 72, probably an isotope of zinc with atomic number 30, to 158, thought to be an isotope of europium with atomic number 63. About 97% of the U^{235} nuclei undergoing fission yield products which fall into two groups, a "light" group with mass numbers from 85 to 104, and a "heavy" group with mass numbers from 130 to 149. The most probable type of fission, which occurs in about 7% of the total, gives products with mass numbers 95 and 139. There are 87 possible mass numbers between 72 and 158, which may represent the total number of different nuclides formed as direct fission fragments. If this were the case, the uranium nucleus should be capable of splitting in over 40 different ways. More than 60 primary products have actually been detected, so that there are at least 30 different modes of fission, a different pair of nuclei being formed in each mode.

The fission fragments have too many neutrons for stability and most of them decay by electron emission. Each fragment starts a short radioactive series, involving the successive emission of electrons. These series are called *fission decay chains*, and each chain has three members, on the average, although longer and shorter chains occur frequently. The problem of determining the masses and atomic numbers of the fission products and of identifying the members of the many decay chains is an extremely difficult one. Nevertheless, as the result of careful and persistent work, more than 60 chains have been established, and about 200 different radionuclides have been assigned to them.[13,14] An example of a long chain is

$$_{54}Xe^{140} \xrightarrow[16s]{\beta^-} {}_{55}Cs^{140} \xrightarrow[66s]{\beta^-} {}_{56}Ba^{140} \xrightarrow[12.8d]{\beta^-}$$

$$\longrightarrow {}_{57}La^{140} \xrightarrow[40h]{\beta^-} {}_{58}Ce^{140} \text{ (stable).}$$

This chain is especially interesting because it contains two of the nuclides, Ba^{140} and La^{140}, whose appearance led to the discovery of fission. The short chain

$$_{60}Nd^{147} \xrightarrow[11d]{\beta^-} {}_{61}Pm^{147} \xrightarrow[4y]{\beta^-} {}_{62}Sm^{147} \ (\sim 10^{11}\,y)$$

is important because one of its members is an isotope of the element with atomic number 61. This element had not been clearly isolated before the discovery of fission, and has now been named *promethium* (symbol, Pm). Radioactive isotopes of the element with atomic number 43, which has not been found to occur in nature, also have been identified as fission products. This element is now called *technetium* (symbol, Tc), and its longest-lived isotope occurs in the chain.

$$_{42}\text{Mo}^{99} \xrightarrow[66\text{h}]{\beta^-} {_{43}\text{Tc}^{99}} \xrightarrow[2.2\times10^5\text{y}]{\beta^-} {_{44}\text{Ru}^{99}} \text{ (stable).}$$

19–4 The mass and energy distributions of the fission products. The mass distribution of the fission products is shown most conveniently in the form of a *fission yield curve*, in which the percentage yields of the different products are plotted against mass number. The yield of any given mass can be found by measuring the abundance of a long-lived nuclide near the end of a chain, or that of the stable end product. Yield curves[15,16] for the fission of U^{235} by thermal neutrons and by 14-Mev neutrons are shown in Fig. 19–3a; curves for thermal fission of U^{233} and Pu^{239} are shown in Fig. 19–3b. The fission yield for a particular nuclide is the probability (expressed as a percentage) of forming that nuclide or the chain of which it is a member; it may also be regarded as the percentage of fissions yielding the nuclide or chain. Since two nuclei result from each fission, the total yield adds up to 200%. The yields vary from about $10^{-5}\%$ to about 7%,

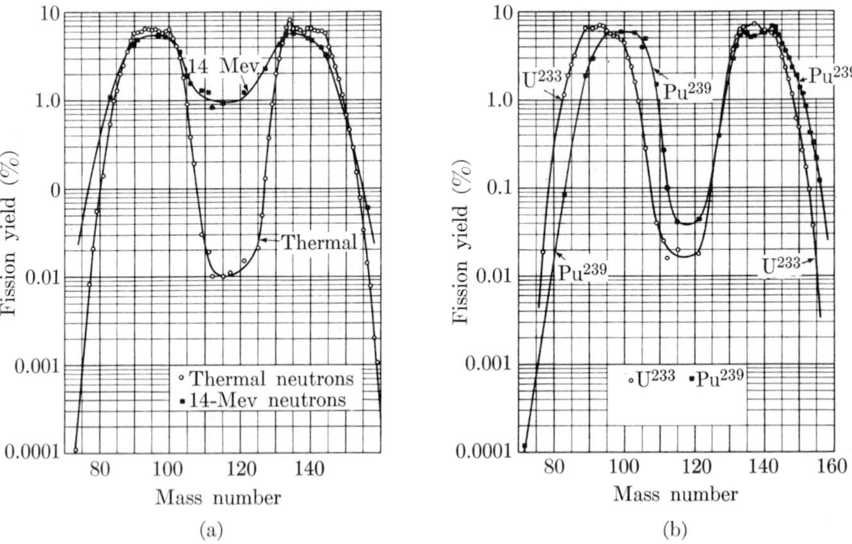

FIG. 19–3. Fission yields from U^{235}, U^{233}, and Pu^{239}: (a) thermal and fast fission of U^{235}; (b) thermal fission of U^{233} and Pu^{239}.[15]

and the range is so large that a logarithmic scale is used for the ordinate of the yield curve. Each curve shows two peaks, corresponding to the light and heavy groups of products. In the case of U^{235} the maxima lie near mass numbers 95 and 135; fission induced by slow neutrons is a highly asymmetric process, and division into two equal fragments occurs in only about 0.01% of the fissions. Small "fine-structure" peaks are evident in some of the yield curves; for U^{235}, these peaks are at masses 100 and 134. This fine structure is, for the most part, a shell effect, and can be accounted for by a detailed analysis of the neutron binding energies in nuclei with neutron numbers close to those corresponding to closed shells, i.e., 50 and 82 neutrons. The U^{235} curves illustrate the effect on the mass distribution of increasing the neutron energy above thermal. The greatest change is the increase in the probability of symmetric fission; for fission by 14-Mev neutrons, the increase is about 100-fold. The other changes are a small drop in the peak yields and a moderate increase in the most asymmetric modes of fission, i.e., a rise in the wings of the yield versus mass curve.

The mass distribution of the fission fragments can also be obtained from the distribution of their kinetic energies, which can be determined by measuring the ionization produced in an appropriate ionization chamber. In one type of chamber, the fissile material is placed on one of the electrodes and the ions which result when a fission fragment enters the region between the electrodes are collected. Since the fission fragments occur in pairs, an experiment of this kind measures the energy of one fragment only. In another type of chamber, a very thin foil is made the common cathode of two back-to-back ionization chambers. The two fragments resulting from neutron bombardment travel in opposite directions into the two chambers and the ionizations they cause are measured simultaneously. The nucleus undergoing fission can be considered to be initially at rest; and, if the neutrons emitted are neglected, the law of conservation of momentum gives

$$M_1V_1 = M_2V_2, \tag{19–1}$$

where the subscripts refer to the two fission fragments. The energies of the fragments are then in the ratio

$$\frac{E_1}{E_2} = \frac{\frac{1}{2}M_1V_1^2}{\frac{1}{2}M_2V_2^2} = \frac{M_2}{M_1}, \tag{19–2}$$

and the masses are inversely proportional to the kinetic energies; when the energy distribution has been measured, the mass distribution is obtained from Eq. (19–2).

The energy distribution of the fission products has been measured for the fission of U^{233}, U^{235}, and Pu^{239} by thermal neutrons,[17,18] and for

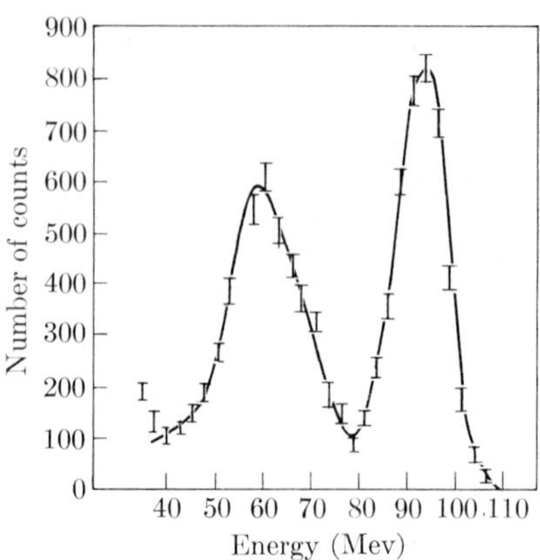

FIG. 19–4. Energy distribution of the fragments from fission of U^{235} by thermal neutrons (Brunton and Hanna[17]).

the fission of U^{235}, U^{238}, Th^{232}, and Pu^{239} by fast neutrons.[19,20] The energy distribution in the case of the thermal fission of U^{235} is shown in Fig. 19–4; this energy spectrum is typical for fission induced by slow neutrons, and the curves for U^{233} and Pu^{239} are similar to the curve shown. The peaks near 60 Mev and 95 Mev show the asymmetry of the fission process. The best ionization chamber measurements[17] give 154.7 Mev for the average total kinetic energy of the fragments from U^{235}. The energy distribution for fission induced by fast neutrons (2.5 and 14 Mev) show the same two peaks, and the maxima differ only slightly from the values for fission induced by slow neutrons. The most important difference in the shapes of the spectra is an increase in the height of the curve in the region between the two peaks when the energy of the bombarding neutrons is in the Mev range. This rise is interpreted as indicating increased probability of symmetrical fission when the neutron energy is high. In the case of U^{235}, symmetrical fission is about 100 times more probable for 14-Mev neutrons than for thermal neutrons. When the energy of the bombarding neutrons is raised to 45 Mev, there are still two peaks, but the dip between them is small, and the probability of symmetrical fission is still greater than at the lower energies. With 90-Mev neutrons, only one peak is observed, corresponding to division into two equal fragments; at these very high energies symmetrical fission is the most probable mode.[21]

The velocity distribution of the fission fragments has been studied directly by means of a time-of-flight method.[22] A thin foil with a film of U^{233}, U^{235}, or Pu^{239} is irradiated by thermal neutrons, and pairs of fragments are detected by scintillation detectors. One fragment travels only about 1 cm before striking a detector, while the other travels about

350 cm along an evacuated tube. The pulses from the two detectors are shown on the screen of a cathode-ray tube and photographed; the velocity of the fragment can be obtained from the distance between the two peaks, and the kinetic energy is then calculated from the velocity. Results obtained with this method show that the energies are greater than those given by the ionization measurements, and the average total kinetic energy of the fission fragments from U^{235} is increased by 12.4 Mev, from 154.7 Mev to 167.1 Mev. This value agrees with the result, 167.1 ± 1.6 Mev, of a calorimetric measurement.(23)

The energy distribution of the fragments from spontaneous fission has also been determined(24) and is similar to that found in neutron-induced fission.

The determination of the charge distribution in fission is more difficult than that of the energy distribution because in an ionization chamber or cloud chamber there is no obvious way of determining the nuclear charge at the instant of fission. Radiochemical methods based on attempts to measure the primary fission yield have been used;(16,25) these methods are indirect, and the problem has not yet been solved.(26)

19–5 Neutron emission in fission. The comparison of the value of the neutron-to-proton ratio of the compound nucleus [U^{236}] with the values for some of the fission products, discussed in Section 19–3, indicated also that neutrons might be emitted in fission. The fact that neutrons are emitted was shown in a simple experiment shortly after the discovery of fission.(26) A neutron source was placed at the center of a large vessel, with detectors at various distances from it to determine the neutron density in the vessel. The vessel was filled first with a uranyl sulfate solution, and then with an ammonium nitrate solution for comparison. The average neutron density was found to be greater when uranium was present, showing in a rough way that more neutrons were formed, when fission occurred, than were used up. The average value of the number of neutrons released in fission, usually denoted by ν, has been measured for various fissionable materials, and some of the results(27,28) are listed in Table 19–4. The average number of neutrons released is always greater than two and increases with the energy of the neutrons that induce fission. The number of neutrons released in any one fission process must, of course, be an integer but, since the fissionable nucleus can divide in at least 30 different ways, the *average value* ν of the number of neutrons does not have to be an integer.

In addition to ν, there is another property of fissionable materials which has practical importance, the average number of neutrons emitted per neutron absorbed by a fissionable nuclide. It is evident from the cross section values of Table 19–1 that all of the neutrons absorbed by a fis-

Table 19–4

The Average Number of Fission Neutrons as a Function of the Energy of the Neutrons Inducing Fission[27]

Neutron energy, Mev	$U^{233}+n$	$U^{235}+n$	$U^{238}+n$	$Pu^{239}+n$
Thermal	2.51 ± 0.02	2.44 ± 0.02		2.89 ± 0.03
0.08	2.58 ± 0.06	2.47 ± 0.03		3.05 ± 0.08
1.3	2.69 ± 0.05	2.61 ± 0.09		3.08 ± 0.05
1.5		2.57 ± 0.12	2.65 ± 0.09	
1.8	2.75 ± 0.06	2.72 ± 0.06		3.28 ± 0.06
2.0		2.80 ± 0.15		
4.0	3.06 ± 0.12	3.01 ± 0.12	3.11 ± 0.10	3.43 ± 0.11
14.1	3.86 ± 0.28	4.52 ± 0.32	4.13 ± 0.25	4.85 ± 0.50

sionable material do not induce fission; some absorptions result in the emission of γ-rays, i.e., radiative capture competes with fission. With U^{235}, for example, the reaction $U^{235}(n, \gamma)U^{236}$ also occurs, and U^{236} is an α-emitter with a half-life of 2.4×10^7 years. The ratio of the radiative capture cross section to the fission cross section is usually denoted by α,

$$\alpha = \sigma_r/\sigma_f, \tag{19–3}$$

and the number of fission neutrons released per neutron absorbed in a fissionable nuclide is given by

$$\eta = \nu/(1 + \alpha). \tag{19–4}$$

Values of α and η for thermal neutrons are listed in Table 19–1 for the important fissionable materials.

The average number of neutrons emitted per spontaneous fission has been determined for various heavy nuclides.[12] The rate (number of fissions per gram per second) can be measured[29] as well as the total number of neutrons.[30] For U^{238}, the rate is $(6.90 \pm 0.24) \times 10^{-3}$ fissions/gm/sec, or about 25 fissions/gm/hr, and $\nu = 2.4 \pm 0.2$, close to the value for fission induced by thermal neutrons; for Th^{232}, $\nu = 2.6 \pm 0.3$, and for Cf^{252}, $\nu = 3.53 \pm 0.15$.

The neutrons emitted as a result of the fission process can be divided into two classes, *prompt neutrons* and *delayed neutrons*. The prompt neutrons, which make up about 99% of the total fission neutrons, are emitted within an extremely short interval of time, possibly as low as 10^{-14} sec, of the fission process. It is thought that the compound nucleus $[U^{236}]$ first splits into two fragments, each of which has too many neutrons for stability, and has also the excess energy (6 Mev or more) needed to expel a neutron. The excited, unstable nucleus consequently ejects

Table 19–5

Properties of Delayed Neutrons[33]

Half-life, sec	Energy, Mev	Yield, % of total neutrons emitted in fission, Thermal and fast fission			Fast fission	
		U^{235}	U^{233}	Pu^{239}	U^{238}	Th^{232}
55	0.25	0.021	0.022	0.007	0.020	0.075
22	0.46	0.140	0.078	0.063	0.215	0.330
5.6	0.41	0.125	0.066	0.046	0.254	0.341
2.1	0.45	0.253	0.072	0.068	0.609	0.981
0.6	0.42	0.074	0.013	0.018	0.353	0.378
0.2		0.027	0.009	0.009	0.118	0.095
Total yield		0.64	0.26	0.21	1.57	2.20

one or more neutrons within a very short time after its formation; prompt γ-rays are apparently emitted at the same time. These ideas are consistent with the results of experiments on the angular correlation between the direction of the neutrons and that of the fragments,[31] the neutrons being emitted preferentially in the same direction as the fragments.

The delayed neutrons, which constitute about 0.64% of the total neutrons from the fission of U^{235}, are emitted with gradually decreasing intensity for several minutes after the actual fission process. Six well-defined groups of delayed neutrons have been observed by studying the rate of decay of the neutron intensity.[32,33] The rate of decay of each group is exponential, just as for other forms of radioactive change, and a specific half-life can be assigned to each group, as well as a mean life and decay constant. From the intensity, the fraction β_i which the group constitutes of the total (prompt and delayed) fission neutrons can be determined. The properties of the prominent groups of delayed neutrons are listed in Table 19–5. Three additional low-intensity delayed neutron groups from uranium have been reported,[34] with half-lives of 3, 12, and 125 min and yields, per fission, of 5.8×10^{-8}, 5.6×10^{-10}, and 2.9×10^{-10}, respectively.

The same groups of delayed neutrons are found in the fission of different heavy nuclides by slow or by fast neutrons, but the yields vary from nuclide to nuclide. Of the three important fissionable materials, U^{235} has the greatest yield, 0.64%, about three times that of Pu^{239}. Although the yield of delayed neutrons is less than one percent of the total number of neutrons emitted, the delayed neutrons have a strong influence on the time-dependent behavior of a chain-reacting system based on fission and play an important part in the control of the system.

The mechanism for the emission of delayed neutrons is understood. Since some of the fission products are rich in neutrons and are very unstable with respect to β-emission, a product with Z protons and N neutrons may have a β-decay energy greater than the binding energy of the last neutron in the daughter product with $Z + 1$ protons and $N - 1$ neutrons. In the β-decay, the product nucleus can be left either in the ground state or in one of the many excited states. If the excitation energy of one of the excited states of the $Z + 1$, $N - 1$ nucleus is greater than the binding energy of the last neutron, de-excitation may occur by the emission of a neutron, leaving a nucleus with $Z + 1$ protons and $N - 2$ neutrons. The neutron emission will be delayed and will appear to have the half-life of the β-decaying nuclide (Z, N), i.e., that of the delayed neutron precursor. Several of the delayed neutron groups have been related to the decay of bromine and iodine fission products.[35-39] Thus, Br^{87} is the precursor of the 55-sec delayed neutron emitter. Iodine-137, with a half-life of 24 sec, is associated with the 22-sec group. This group may consist of two or more precursors with similar half-lives, and may also contain Br^{88} which has a half-life of 16.3 sec. The 5.6-sec group includes I^{138} ($T_{1/2} =$ 6.3 sec) and possibly Br^{89}($T_{1/2} = 4.4$ sec). The decay scheme of Br^{87} is shown in Fig. 19–5. Some 70% of the β-decays of Br^{87} go to the 5.4 Mev excited level of Kr^{87}; this level lies about 0.3 Mev above the level corresponding to the binding energy of the last (51st) neutron in Kr^{87}.

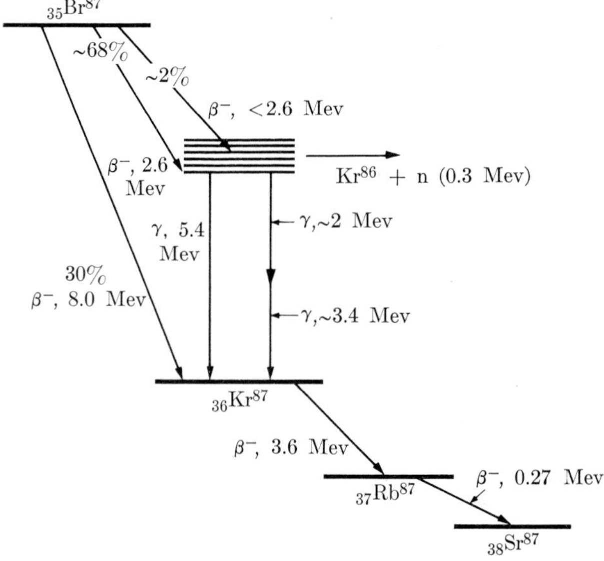

FIG. 19–5. Decay scheme of Br^{87} and its products, showing delayed neutron emission.

About 3% of the Kr^{87} nuclei in the 5.4 Mev level emit a neutron in the transition $Kr^{87} \rightarrow Kr^{86} + n + 0.3$ Mev, while the rest decay by γ-emission to the ground state of Kr^{87}. The *delayed neutrons* are delayed entirely by the 55-sec half-life of their β-ray parent Br^{87}. It should be noticed that Kr^{87} has 51 neutrons and the emission of the delayed neutron leaves a closed shell of 50 neutrons. Similarly $_{53}I^{137}$ decays by β-emission to $_{54}Xe^{137}$ which can emit a neutron, leaving $_{54}Xe^{136}$, which has a closed shell of 82 neutrons; Br^{89} and I^{139} yield, on β-decay followed by n-emission, nuclei with 50 + 2 and 82 + 2 neutrons, respectively, which are again especially stable neutron configurations.

19–6 The energy distribution of the neutrons emitted in fission. The energies of the prompt neutrons emitted in fission vary from values less than 0.05 Mev to more than 17 Mev. The determination of neutron energies over so wide a range is not an easy problem, and several methods have been used in order to cover the range. The neutrons from the thermal fission of U^{235} have been studied in great detail, and this case provides a good example of the methods and results.[40] In the energy range 0.05 to 0.7 Mev, a cloud chamber was used containing hydrogen gas and water vapor.[41] A thin foil of U^{235} was bombarded with a beam of thermal neutrons from the thermal column of a nuclear reactor. The fission neutrons were allowed to enter the cloud chamber, where they collided with protons, and the range of the recoil protons was determined by measuring the track lengths. The energy distribution of the recoil protons was obtained from the range measurements, and the energy distribution of the neutrons was obtained from that of the recoil protons. The results

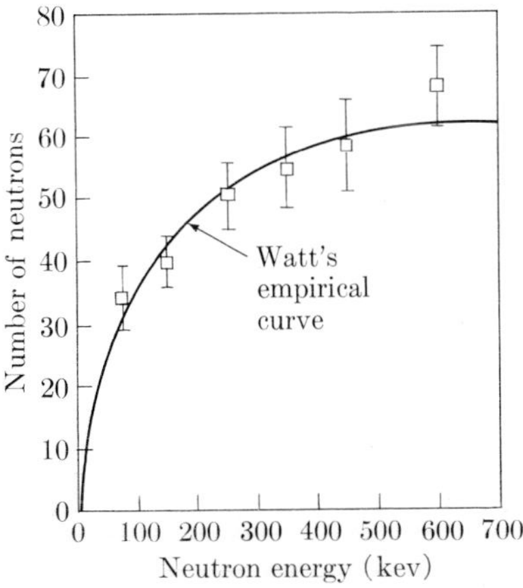

FIG. 19–6. Energies of fission neutrons: the low-energy region (Bonner *et al.*[41]).

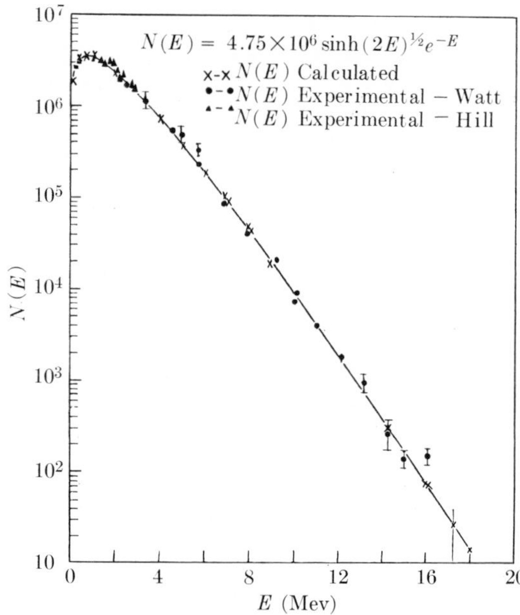

FIG. 19–7. Energies of fission neutrons: the spectrum from 0.4 Mev to 17 Mev (Watt[43]).

are shown in Fig. 19–6, where the relative number of neutrons in a given energy interval is plotted against the energy and compared with an empirical formula which will be given below.

The neutron energy spectrum between 0.4 Mev and 17 Mev was studied by measuring, with coincidence counters, the range distribution of recoil protons from hydrogenous materials.[42,43] The neutron intensity shows a broad maximum near 0.75 Mev, and decreases nearly exponentially at energies greater than 2 Mev. The energy distribution between 0.075 Mev and 17 Mev can be described by the empirical formula

$$N(E) = e^{-E} \sinh(2E)^{1/2}, \tag{19–5}$$

where $N(E)$ represents the relative number of neutrons per unit energy range as a function of the neutron energy. The experimental results between 0.4 Mev and 17 Mev are shown in Fig. 19–7, and agree with the empirical formula. The solid curve of Fig. 19–6 represents the same empirical formula, and the agreement is good also for the low-energy neutrons. The average value of the neutron energy is 2.0 ± 0.1 Mev.

The fission neutron spectra of U^{235} and Pu^{239}, in the range from 0.5 to 8 Mev, have been studied with nuclear emulsion plates,[44,45] and the results agree with those just discussed. The spectrum for the neutrons from Pu^{239} shows a maximum in the 0.6- to 0.8-Mev region and the empirical formula of Eq. (19–5) fits the experimental results; the average energy is again 2.0 Mev.

19–7 The energy release in fission. One of the most striking properties of the fission process is the magnitude of the energy released per fission, which is about 200 Mev as compared with several Mev for other nuclear reactions. An estimate of the amount of energy released per fission can be made by considering the binding energy curve, Fig. 9–11. The value of the average binding energy per particle has a broad maximum of about 8.4 Mev in the range of mass numbers from 80 to 150, and it has been shown that nearly all of the fission products have mass numbers in this range. The average binding energy per particle is about 7.5 Mev in the neighborhood of uranium. Hence, the average binding energy per particle is about 0.9 Mev greater in the fission products than in the compound nucleus $[U^{236}]$, and the excess, 0.9 Mev, is liberated in the fission process. The total amount of energy released per fission should be roughly equal to the product of the number of particles (236) multiplied by the excess binding energy per particle (0.9 Mev), or approximately 200 Mev.

The total energy release per fission can also be calculated from the nuclear masses of $[U^{236}]$ and a typical pair of fission products. It has been shown that the fission products with the greatest yields have mass numbers near $A = 95$ and $A = 139$. If $_{42}Mo^{95}$ and $_{57}La^{139}$ are taken as a pair of stable products at the ends of the chains for their respective masses, their combined mass is 138.955 amu + 94.946 amu = 233.900 amu, as given by the semiempirical mass formula. The corresponding mass number is 234, and 2 neutrons are apparently released in this particular fission process. When 2.018 amu are added for these two neutrons, the products of the reaction have a total mass of 235.918 amu. The mass of U^{235} from the semiempirical mass formula is 235.124 amu, so that the mass of the compound nucleus $[U^{236}]$ is close to 235.124 + 1.009 = 236.133 amu. The mass excess that is converted to energy is equal to 236.133 amu − 235.918 amu = 0.215 amu. Since one amu is equivalent to 931 Mev, the energy liberated in the process is $931 \times 0.215 = 198$ Mev. Although there are at least 30 different ways in which the nucleus can divide, the mass excess is approximately the same for all of these processes, and 200 Mev is a good value for the average amount of energy released per fission, as calculated in this way.

The predicted value of about 200 Mev can be compared with experimental values. The total amount of energy released per fission is the sum of the kinetic energy of the fission fragments, the kinetic energy of the emitted neutrons, the kinetic energy of the prompt γ-rays, and the

TABLE 19–6

THE ENERGY RELEASE IN THE FISSION OF U^{235} BY THERMAL NEUTRONS

Kinetic energy of the fission fragments	167 Mev
Kinetic energy of fission neutrons	5
Prompt γ-rays	7
β-decay energy	5
γ-decay energy	5
Neutrino energy	11
Total fission energy	200 Mev

total energy of the decay processes in the fission decay chains. It was shown in Section 19–4 that the average value of the total kinetic energy of the fission fragments from the thermal fission of U^{235} is 167 Mev; the uncertainty in this number is about 5 Mev. The average value of the kinetic energy carried off by the neutrons is equal to the product of the average number of neutrons emitted per fission and the average kinetic energy of the neutrons; for the thermal fission of U^{235}, this product is 2.5×2.0 Mev $= 5$ Mev. The kinetic energy of the prompt γ-rays is in the neighborhood of 5 to 8 Mev.[46,47] Finally, the average energy[48] of all radiations (β-rays, γ-rays, and neutrinos) of the fission products is 21 ± 3 Mev; about half of this energy escapes in the form of neutrinos, and the other half is divided, approximately equally, between the β-rays and the γ-rays. These results are summarized in Table 19–6; the total fission energy determined in this way is 200 Mev, with an uncertainty between 5 and 10 Mev, in good agreement with the calculated values.

The energy release in fission has also been measured calorimetrically,[49] in a way which did not include the energy carried off by γ-rays, neutrons, and neutrinos. The result obtained was 177 ± 5 Mev; if 26 Mev are added (cf. Table 19–6) for the energies missed, the total is 203 ± 8 Mev, in good agreement with the previous results. Another calorimetric measurement yielded a value of the energy released in the fission of U^{235} except for the energy carried off by neutrinos;[50] the result was 190 ± 5 Mev, which also agrees very well with the results of Table 19–6 and with the calculated values.

Results obtained with other fissile materials, such as Pu^{239} and U^{233}, are similar to those for U^{235}.

19–8 The theory of the fission process. It is not surprising, in view of the remarkable properties of the fission process, that a great deal of

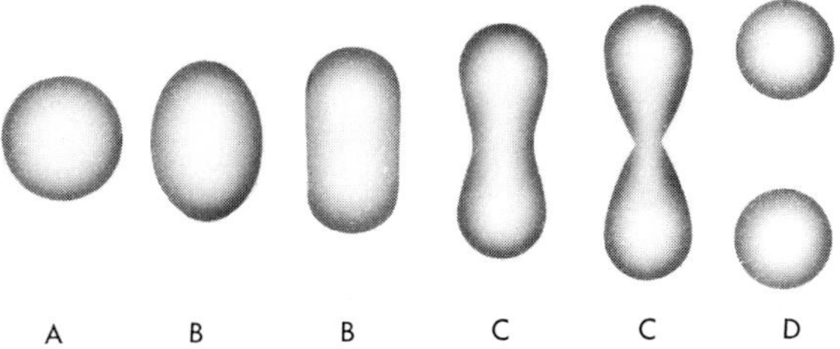

FIG. 19-8. Possible steps in the process of nuclear fission according to the liquid-drop model.

effort has been put into theoretical studies of the process. The first thorough treatment was that of Bohr and Wheeler[51] who accounted for many of the properties of fission on the basis of the liquid-drop model of the nucleus. This model was used in the last section in calculating the energy release in fission from the atomic masses of the fissile and product nuclei as given by the semiempirical mass formula; that formula was derived by considering that a charged nucleus is analogous to a liquid drop (Section 17-6). Bohr and Wheeler showed that the liquid-drop model could be used to describe the process of fission in some detail, and made successful predictions about the existence of spontaneous fission, and about the ability of various heavy nuclei to undergo fission with slow or fast neutrons.

In the spherical, liquid-drop nucleus, the shape of the drop depends on a balance involving the surface tension forces and the Coulomb repulsive forces. If energy is added to the drop, as in the form of the excitation energy resulting from the capture of a slow neutron, oscillations are set up within the drop; these tend to distort the spherical shape so that the drop may become ellipsoidal in shape. The surface tension forces tend to make the drop return to its original shape, while the excitation energy tends to distort the shape still further. If the excitation energy is sufficiently large, the drop may attain the shape of a dumbbell. The Coulomb repulsive forces may then push the two "bells" apart until the dumbbell splits into two similar drops, each of which then becomes spherical in shape. The sequence of steps leading to fission is shown in Fig. 19-8. If the excitation energy is not large enough, the ellipsoid may return to the spherical shape, with the excitation energy being liberated in the form of γ-rays, and the process is one of radiative capture rather than fission.

The potential energy of the drop in the different stages of Fig. 19-8 can be calculated under the Bohr-Wheeler theory, as a function of the degree of deformation of the drop[51,52] The form of the results is shown in Fig. 19-9, where the potential energy E is plotted against a parameter r which is a measure of the degree of deformation. At $r = 0$, which repre-

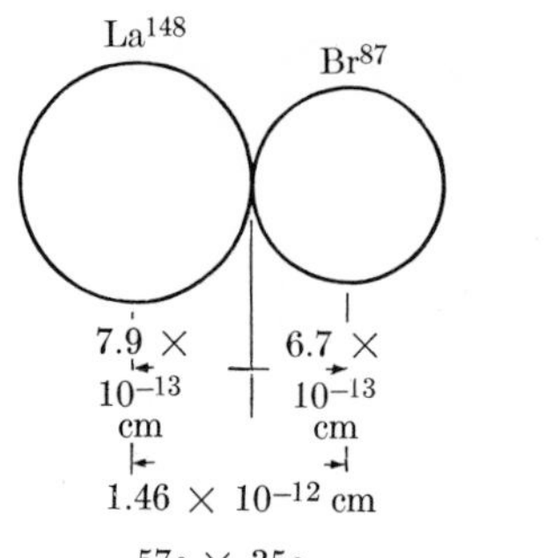

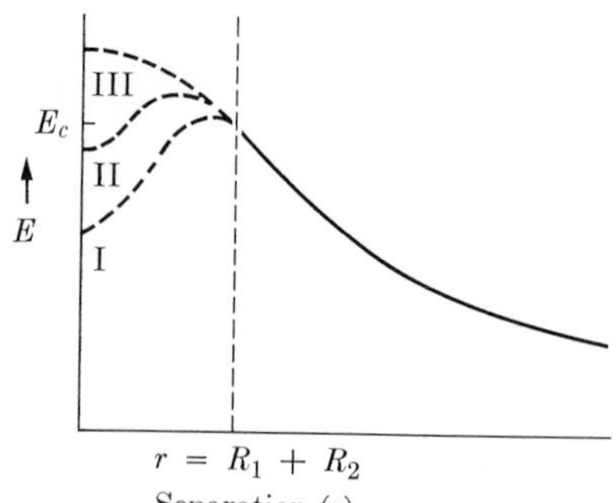

FIG. 19–9. Potential energy of two fission fragments as a function of distance between centers and the configuration at contact, where the potential energy is 197 Mev.

sents the initial spherical nuclear drop, there is available an amount of energy E_0, given by

$$E_0 = c^2[M(A, Z) - M(A_1, Z_1) - M(A_2, Z_2)], \qquad (19\text{–}6)$$

where (A, Z) represents the nucleus that may undergo fission, and the subscripts 1 and 2 stand for the possible final products. The value of E_0 corresponds to the ground state of the compound nucleus formed when the target nucleus captures a neutron, but does not include the excitation energy resulting from the neutron capture. It was shown in the last section that the value of E_0 for U^{236} is about 200 Mev.

The meaning of the graph can now be seen more easily if the process which would be the reverse of fission is considered. When the drop has split into two fragments, r is the distance between their centers; if R_1 and R_2 are the radii of the two product drops, $r = R_1 + R_2$ is the value of the deformation parameter for which the two drops just touch. For values of r smaller than $R_1 + R_2$, r represents the degree of deviation of the original drop from its spherical shape. For values of r greater than $R_1 + R_2$, the energy is just the electrostatic energy resulting from the mutual repulsion of the two positively charged nuclear fragments. The value of E in this region is $(Z_1Z_2e^2/r)$, and E increases as the distance between the two fragments is decreased; the energy at $r = \infty$ is taken equal to zero. When the two fragments just touch, at $r = R_1 + R_2$, the Coulomb energy is denoted by E_c and is

$$E_c = \frac{Z_1Z_2e^2}{R_1 + R_2}. \qquad (19\text{–}7)$$

When r is smaller than $R_1 + R_2$, the energy depends not only on the electrostatic forces, but also on the surface tension forces; the calculation of E is then complicated, and three different curves (I, II, and III) are shown for the possible variation of E in this region. The shapes of these curves and the values of E_0 are related to the mass of the nucleus, that is, to the value of A in Eq. (19–6). Stable nuclei with values of A somewhat greater than 100 are of type I, with E_0 about 50 Mev smaller than E_c. Nuclei like those of uranium, plutonium, or thorium are of type II, for which $E_c - E_0$ is about 6 Mev. For still heavier nuclei, E_0 may be greater than E_c, as in case III. Nuclei of type III should undergo fission spontaneously and would not be expected to last long.

On a classical basis, nuclei of type II should be stable with respect to fission, but according to quantum mechanics, there is a certain probability that they will undergo spontaneous fission. The fragments may "leak through" the barrier represented by E_c in the same way that α-particles do in α-decay. In U^{238}, about 25 spontaneous fissions occur per gram per hour, and the half-life for this process is about 10^{17} years.

The *activation* energy $E_c - E_0$ needed to induce fission in nuclei of type II has been calculated from the Bohr-Wheeler theory, and can be compared with the excitation energy resulting from the capture of a neutron, or other particle, or a γ-ray. The excitation energy of a compound nucleus can be computed with the aid of the semiempirical binding energy (or mass) formula, or from the masses of the nuclei involved if they are known. When the compound nucleus is formed by the capture of a slow neutron, the excitation energy differs by a negligible amount from the binding energy of the last neutron. In the case of $[U^{236}]$, the binding energy is calculated from the known masses as follows:

Mass of U^{235},	235.11704
Mass of neutron,	1.00898
	236.12602
Mass of U^{236}	236.11912
$\Delta M =$	0.00690 amu $=$ 6.4 Mev.

The values of the excitation energy calculated in this way for a number of heavy nuclei are listed in Table 19–7 and compared with the corresponding values of the activation energy obtained from the Bohr-Wheeler theory. In U^{233}, U^{235}, and Pu^{239} the excitation energy is considerably greater than the activation energy, and these nuclei would be expected to undergo fission with thermal neutrons; they do, and the thermal fission cross sections of U^{235} and Pu^{239} are large, 577 b and 742 b, respectively. In U^{238} and Th^{232}, the excitation energy is smaller than the activation energy and these nuclei do not undergo fission with thermal neutrons, as predicted. To induce fission, the neutrons must have considerable kinetic

TABLE 19–7

FISSIONABILITY OF HEAVY NUCLEI WITH THERMAL NEUTRONS

Target nucleus	Compound nucleus	Excitation energy, Mev	Activation energy, Mev
U^{233}	$[U^{234}]$	6.6	4.6
U^{235}	$[U^{236}]$	6.4	5.3
U^{238}	$[U^{239}]$	4.9	5.5
Th^{232}	$[Th^{233}]$	5.1	6.5
Pa^{231}	$[Pa^{232}]$	5.4	5.0
Np^{237}	$[Np^{238}]$	5.0	4.2
Pu^{239}	$[Pu^{240}]$	6.4	4.0

energy, and fission should be a threshold reaction in these nuclei, as it is. In Pa^{231} and Np^{237}, the excitation energy is greater than the activation energy, and these nuclides should undergo fission with thermal neutrons. Early experiments indicated that they do not; recent measurements show that there is some thermal fission, but the cross sections are very small (about 10^{-2} b) compared with those of U^{235} and Pu^{239}. The fact that the calculated values of the excitation energy are greater than those of the activation energy for Pa^{231} and Np^{237} does not agree with the very low probabilities of fission by thermal neutrons.

The values of the excitation energy listed in Table 19–7 indicate that the binding energy of the neutron added to a nucleus with an odd number of neutrons (U^{233}, U^{235}, Pu^{239}) is greater than that for a nucleus with an even number of neutrons (Th^{232}, U^{238}, Pa^{231}, Np^{237}). It is found generally that the excitation energy is greater for target nuclei with odd numbers of neutrons than for nuclei with even numbers of neutrons, and this effect is related to the presence of the "odd-even" term in the binding energy formula. Fission with thermal neutrons occurs much more often in nuclei with an odd number of neutrons than in nuclei with an even number of neutrons, as is shown by the values of the fission cross section listed in Tables 19–1 and 19–2. Nearly all of the target nuclei with large values of the cross section for fission by thermal neutrons have odd numbers of neutrons, while nearly all of the nuclei with very small cross sections have even numbers of neutrons. There is, therefore, some correlation between the values of the excitation energy and the occurrence of thermal fission, in qualitative agreement with the predictions of the Bohr-Wheeler theory. There are some obvious exceptions; ${}_{96}Cm^{240}$ has a very large cross section (~20,000 b), and ${}_{92}U^{230}$ and ${}_{92}U^{232}$ have appreciable cross sections (25 b and 80 b, respectively); each of these nuclei has an even number of neutrons. The nuclides Pa^{231} and Np^{237} have low excitation energies, corresponding to their even numbers of neutrons, but their activation energies are still lower, so that they should undergo fission with slow neutrons in spite of their odd neutron numbers. The theory based on the liquid-drop model gives incorrect values of the activation energy in these cases.

The above discussion indicates that the liquid-drop model of the nucleus has been applied with some success to the problem of nuclear fission. The theory gives a qualitative, and sometimes quantitative, picture of the process and has made possible the description of some of the properties of fission, but it also has some serious defects in addition to those already mentioned. According to the theory, the most probable mode of division of the liquid-drop nucleus is into two equal fragments, so that it fails to account for the observed asymmetry of the process; it also yields values for photofission thresholds and for spontaneous fission rates which are wrong quantitatively, and whose variations from nuclide to nuclide do not agree with experiment.

Attempts have been made to modify the liquid-drop theory of fission in order to overcome these difficulties. The application of the shell model has been suggested,[(53)] and has had some success in accounting for the asymmetry.[(54)] Another promising approach seems to be that of Hill and Wheeler,[(55)] which is based on the new collective model of the nucleus, and the reader is referred to their article for a complete and detailed discussion of the mechanism of fission. Many additional papers on the theory of fission can be found in the general references listed below.

References

GENERAL

L. A. Turner, "Nuclear Fission," *Revs. Mod. Phys.* **12,** 1 (1940).

W. J. Whitehouse, "Fission," *Progress in Nuclear Physics,* Vol. 2, O. R. Frisch, ed. New York: Academic Press, 1952.

Radiochemical Studies: The Fission Products, National Nuclear Energy Ser., Div. IV, 9, C. D. Coryell and N. Sugarman, eds. New York: McGraw-Hill, 1951.

Physics of Nuclear Fission, Supplement No. 1 of the Soviet Journal, *Atomnaya Energiya.* London: Pergamon Press, 1958.

I. Halpern, "Nuclear Fission," *Ann. Revs. Nuc. Sci.* **9,** 245 (1959).

Proceedings of the United Nations International Conference on the Peaceful Uses of Atomic Energy, Geneva, 1955, Vol. 2: Physics: Research Reactors. New York: United Nations, 1956.

Proceedings of the Second United Nations International Conference on the Peaceful Uses of Atomic Energy, Geneva, 1958, Vol. 15: Physics in Nuclear Energy. Geneva: United Nations, 1958.

Proceedings of the Symposium on the Physics of Fission, held at Chalk River, Ontario, May, 1956, Atomic Energy of Canada, Ltd, Report AECL-329, Chalk River, Ontario, July, 1956.

PARTICULAR

1. O. HAHN and F. STRASSMAN, "Uber den Nachweis und das Verhalten der bei der Bestrahlung des Urans mittels Neutronen entstehenden Erdalkalimetalle," *Naturwiss* **27,** 11, 89 (1939).

2. O. HAHN, *New Atoms.* New York: Elsevier (1950).

3. O. R. FRISCH, "Physical Evidence for the Division of Heavy Nuclei under Neutron Bombardment," *Nature* **143,** 276 (1939).

4. ANDERSON, BOOTH, DUNNING, FERMI, GLASOE, and SLACK, "The Fission of Uranium," *Phys. Rev.* **55,** 511 (1939).

5. L. MEITNER and O. R. FRISCH, "Disintegration of Uranium by Neutrons: a New Type of Nuclear Reaction," *Nature* **143,** 239 (1939).

6. NIER, BOOTH, DUNNING, and GROSSE, "Nuclear Fission of Separated Uranium Isotopes," *Phys. Rev.* **57,** 546, 748 (1940).

7. (a) D. J. HUGHES and R. B. SCHWARTZ, *Neutron Cross Sections, BNL-325,* 2nd ed. Washington, D.C.: U.S. Government Printing Office, July, 1958.

(b) D. J. HUGHES, B. A. MAGURNO, and M. K. BRUSSEL, *BNL-325,* 2nd ed. Supplement I, Jan., 1960.

8. D. J. HUGHES, "New World-Average Thermal Cross Sections," *Nucleonics,* **17,** No. 11, 132 (Nov., 1959).

9. R. W. LAMPHERE, "Fission Cross Sections of the Uranium Isotopes, 233, 234, 236, and 238, for Fast Neutrons," *Phys. Rev.* **104,** 1654 (1956).

10. J. JUNGERMAN, "Fission Excitation Functions for Charged Particles," *Phys. Rev.* **79,** 632 (1950).

11. KOCH, MCELHINNEY, and GASTEIGER, "Experimental Photo-Fission Thresholds in U^{235}, U^{238}, U^{233}, Pu^{239}, and Th^{232}," *Phys. Rev.* **77,** 329 (1950).

12. K. A. PETRZHAK, "Spontaneous Fission of Heavy Nuclei," *Physics of Nuclear Fission, op. cit.* gen. ref., pp. 129–152.

13. "Plutonium Project, Nuclei Formed in Fission: Decay Characteristics, Fission Yields, and Chain Relationships," *Revs. Mod. Phys.* **18,** 513 (1946); *J. Am. Chem. Soc.* **68,** 2411 (1946).

14. *Radiochemical Studies: The Fission Products, op. cit.* gen. ref., Book 3, Appendixes A, B, C.

15. S. KATCOFF, "Fission-Product Yields from U, Th, and Pu," *Nucleonics* **16,** No. 4, 78 (April 1958).

16. A. N. MURIN, "Charge and Mass Distributions of Fission Products," *Physics of Nuclear Fission, op. cit.* gen. ref., pp. 26–44.

17. D. C. BRUNTON and G. C. HANNA, "Energy Distribution of Fission Fragments from U^{235} and U^{233}," *Can. J. Research* **A28,** 190 (1950).

18. D. C. BRUNTON and W. B. THOMPSON, "Energy Distribution of Fission Fragments from Pu^{239}," *Can. J. Research* **A28,** 498 (1950).

19. S. S. FRIEDLAND, "Energy Distribution of Fission Fragments of U^{235} Produced by 2.5 Mev and 14 Mev Neutrons," *Phys. Rev.* **84,** 75 (1951).

20. J. S. WAHL, "Energy Distribution of Fragments from the Fission of U^{235}, U^{238}, and Pu^{239} by Fast Neutrons," *Phys. Rev.* **95,** 126 (1954).

21. J. JUNGERMAN and S. C. WRIGHT, "Kinetic Energy Release in Fission of U^{238}, U^{235}, Th^{232}, and Bi^{209} by High Energy Neutrons," *Phys. Rev.* **76,** 1112 (1949).

22. R. B. LEACHMAN, "Velocities of Fragments from Fission of U^{233}, U^{235}, and Pu^{239}," *Phys. Rev.* **87,** 444 (1952).

23. R. B. LEACHMAN and W. D. SCHAFER, "A Calorimetric Determination of the Average Kinetic Energy of the Fragments from U^{235} Fission," *Can. J. Phys.* **33,** 357 (1955).

24. W. J. WHITEHOUSE and W. GALBRAITH, "Energy Spectrum of Fragments from the Spontaneous Fission of Natural Uranium," *Phil. Mag.* **41,** 429 (1950).

25. L. E. GLENDENIN, "The Distribution of Nuclear Charge in Fission," *Technical Report No. 35,* Laboratory for Nuclear Science and Engineering, M.I.T., Cambridge, 1949.

26. VON HALBAN, JOLIOT, and KOVARSKI, "Liberation of Neutrons in the Nuclear Fission of Uranium," *Nature* **143,** 470, 680 (1939).

27. R. B. LEACHMAN, "The Fission Process—Mechanisms and Data," Paper No. 2467, *Proceedings of the Second United Nations International Conference on the Peaceful Uses of Atomic Energy, op. cit.* gen. ref., Vol. 15, 229 (1958).

28. BONDARENKO, KUSMINOV, KUTSAYEVA, PROKHOROVA, and SMIRENKIN, "Average Number and Spectrum of Prompt Neutrons in Fast-Neutron-Induced Fission," Paper No. 2187, *Proceedings of the Second United Nations International Conference on the Peaceful Uses of Atomic Energy, op. cit.* gen. ref., Vol. 15, 353 (1958).

29. E. SEGRÈ, "Spontaneous Fission," *Phys. Rev.* **86,** 21 (1952).

30. D. J. LITTLER, "A Determination of the Rate of Emission of Spontaneous Fission Neutrons by Natural Uranium," *Proc. Phys. Soc.* (London) **65A,** 203 (1952).

31. R. B. LEACHMAN, "Neutrons and Radiations from Fission," Paper No. 665, *Proceedings of the Second United Nations International Conference on the Peaceful Uses of Atomic Energy, op. cit.* gen. ref., Vol. 15, 331 (1958).

32. HUGHES, DABBS, CAHN, and HALL, "Delayed Neutrons from Fission of U^{235}," *Phys. Rev.* **73,** 111 (1948).

33. KEEPIN, WIMETT, and ZIEGLER, "Delayed Neutrons from Fissionable Isotopes of Uranium, Plutonium, and Thorium," *Phys. Rev.* **107,** 1044 (1957); *J. Nuclear Energy* **6,** 1 (1957).

34. KUNSTADTER, FLOYD, and BORST, "Long-lived Delayed Neutrons from Fission," *Phys. Rev.* **91,** 594 (1953).

35. SNELL, NEDZEL, IBSER, LEVINGER, WILKINSON, and SAMPSON, "Studies of the Delayed Neutrons. I. The Decay Curve and Intensity of the Delayed Neutrons," *Phys. Rev.* **72,** 541 (1947).

36. SNELL, LEVINGER, MEINERS, SAMPSON, and WILKINSON, "Studies of the Delayed Neutrons. II. Chemical Isolation of the 56-Second and 23-Second Activities," *Phys. Rev.* **72,** 545 (1947).

37. N. SUGARMAN, "Short-Lived Halogen Fission Products," *J. Chem. Phys.* **17,** 11 (1949).

38. (a) G. J. PERLOW and A. F. STEHNEY, "Halogen Delayed-Neutron Activities," *Phys. Rev.* **113,** 1269 (1959).

(b) A. F. STEHNEY and G. J. PERLOW, "Halogen Delayed-Neutron Activities," Paper No. 691, *Second International Conference on the Peaceful Uses of Atomic Energy, op. cit.* gen. ref., Vol. 15, 384 (1958).

39. A. C. Pappas, "The Delayed Neutron Precursors in Fission," Paper No. 583, *Second International Conference on the Peaceful Uses of Atomic Energy, op. cit.* gen. ref., Vol. 15, 373 (1958).

40. B. G. Erozolimski, "Fission Neutrons" in *Physics of Nuclear Fission, op. cit.* gen. ref., p. 64.

41. Bonner, Ferrell, and Rinehart, "A Study of the Spectrum of the Neutrons of Low Energy from the Fission of U^{235}," *Phys. Rev.* **87,** 1032 (1952).

42. D. L. Hill, "The Neutron Energy Spectrum from U^{235} Thermal Fission," *Phys. Rev.* **87,** 1034 (1952).

43. B. E. Watt, "Energy Spectrum of Neutrons from Thermal Fission of U^{235}," *Phys. Rev.* **87,** 1037 (1952).

44. N. Nereson, "Fission Neutron Spectrum of U^{235}," *Phys. Rev.* **85,** 600 (1952).

45. N. Nereson, "Fission Neutron Spectrum of Pu^{239}," *Phys. Rev.* **88,** 823 (1952).

46. Kinsey, Hanna, and Van Patter, "Gamma Rays Produced in the Fission of U^{235}," *Can. J. Research* **26A,** 79 (1948).

47. Maienschein, Peelle, Zobel, and Love, "Gamma Rays Associated with Fission," Paper No. 670, *Second International Conference on the Peaceful Applications of Atomic Energy, op. cit.* gen. ref., Vol. 15, 366 (1958).

48. K. Way and E. P. Wigner, "The Rate of Decay of Fission Products," *Phys. Rev.* **73,** 1318 (1948).

49. M. C. Henderson, "The Heat of Fission of Uranium," *Phys. Rev.* **58,** 744 (1940).

50. J. L. Meem, "Energy Release per Fission in the Bulk Shielding Reactor," *Nucleonics* **12,** No. 5, 62 (May 1954).

51. N. Bohr and J. A. Wheeler, "The Mechanism of Nuclear Fission," *Phys. Rev.* **56,** 426 (1939).

52. S. Frankel and N. Metropolis, "Calculations in the Liquid-Drop Model of the Nucleus," *Phys. Rev.* **72,** 914 (1947).

53. L. Meitner, "Fission and Nuclear Shell Model," *Nature* **165,** 561 (1950).

54. P. Fong, "Statistical Theory of Nuclear Fission: Asymmetric Fission," *Phys. Rev.* **102,** 434 (1956); also, *Phys. Rev.* **89,** 332 (1953).

55. D. L. Hill and J. A. Wheeler, "Nuclear Constitution and the Interpretation of Fission Phenomena," *Phys. Rev.* **89,** 1102 (1953).

Problems

1. The nuclides listed below are formed in fission. Indicate the fission product chain, if any, to which each nuclide gives rise: Zn^{72}, Kr^{90}, Kr^{92}, Kr^{97}, Mo^{105}, Sn^{126}, Xe^{144}, Pm^{156}.

2. Two of the stable end products of fission chains resulting from the thermal fission of U^{235} are Sr^{91} and La^{139}. What weight of each of these nuclides is formed by the complete fission of 1 gm of U^{235}?

3. Two possible sets of products of the fission of Pu^{239} by thermal neutrons are (a) Ce^{141}, Mo^{96} and 3 neutrons, (b) Pd^{108}, Xe^{129} and 3 neutrons, (c) Gd^{155}, Br^{81} and 4 neutrons. Compute the energy release in each case.

4. In a calorimetric measurement of the energy released in the fission of uranium, it was found that a 13.36 gm sample of uranium produced heat at the rate of 40 microwatts when bombarded with slow neutrons. A thin layer of uranium weighing 54 μgm, also in the calorimeter and subjected to the same neutron flux, was found to undergo fission at the rate of 340 processes/min. The fissions were counted with an ionization chamber in the calorimeter. Calculate the energy release in Mev/fission from these data.

5. Plot, against time, the total intensity of the delayed neutrons from the fission of U^{235} in a sample of uranium which has been irradiated for a time interval long compared with the period of the longest-lived delayed neutron emitter. Continue the plot until the intensity reaches 0.1% of the initial activity. Use a semilogarithmic paper.

6. It has been shown, on the basis of the liquid-drop model, that the limit of stability of nuclei against spontaneous fission is given by the quantity

$$\frac{Z^2}{A} = \frac{10}{3} a_s \frac{R_0}{e^2},$$

where a_s is the coefficient of the *surface energy* term in the semiempirical binding energy formula, R_0 is the constant coefficient in the $A^{1/3}$ formula for the nuclear radius, and e is the electronic charge.

(a) Compute the limiting value of Z^2/A.

(b) Calculate the value of Z^2/A for Ra^{226}, Th^{232}, U^{232}, U^{238}, Pu^{238}, Am^{241}, and Cu^{242}.

(c) Which of these nuclides might be expected to have significant rates of spontaneous fission? The observed rates, in fissions per gram per second, for the nuclides in (b) are $\ll 0.6$, 4.2×10^{-5}, 16, 6.9×10^{-3}, 2.1×10^3, 40, and 7.6×10^6, respectively.

7. Compute the binding energy of a thermal neutron to each of the following nuclei: Th^{227}, U^{229}, Np^{236}, Np^{237}, Np^{238}, Pu^{241}, Pu^{242}, Am^{241}, Am^{242}, and Cm^{240}. Which of these nuclides would be expected to have relatively large cross sections for fission by thermal neutrons? Compare the predictions with the values listed in Table 19–2.

[*Hint:* For Problems 3 and 7, the necessary masses can be obtained from K. T. Bainbridge, "Charged Particle Dynamics and Optics, Relative Isotopic Abundances of the Elements, Atomic Masses," *Experimental Nuclear Physics*, Vol. I, E. Segrè, ed. New York: Wiley, 1953.]

8. How would you account for the difference in the kinetic energy of the neutrons that can cause fission of U^{235} and U^{238}, respectively? [*Hint:* Use the results of Problem 11, Chapter 17.]

9. Suppose that an even-even compound nucleus (Z, A) such as U^{236} undergoes symmetric fission. (a) Show that the prompt energy release is given, to a good approximation by

$$E_f\ (\text{Mev}) = 0.22\frac{Z^2}{A^{1/3}} - 3.42A^{2/3}.$$

(b) Show that the condition for spontaneous symmetric fission to be possible energetically is $Z^2/A \geq 15$.

10. Calculate the contributions of the Coulomb and surface energies, respectively, to the energy release when the compound nucleus $[U^{236}]$ formed by the capture of a slow neutron by U^{235} undergoes symmetric fission.

11. Show that the height of the Coulomb barrier for symmetric fission is given by

$$E_c\ (\text{Mev}) = 0.15\frac{Z^2}{A^{1/3}}\cdot$$

12. Show that the condition for stability with respect to spontaneous symmetric fission is $Z^2/A \leq 49$. [*Hint:* Note that the condition is given by $E_b - E_f \geq 0$.

13. Bohr and Wheeler[(51)] showed that, for a small deformation of a liquid drop, the net change in the potential energy ΔE is given by

$$\Delta E = A\left(1 - 0.022\frac{Z^2}{A}\right)\cdot$$

For $\Delta E > 0$, the nucleus would return to its original shape; for $\Delta E < 0$, the nucleus would be unstable. Show that the condition that a nucleus be just unstable with respect to a small deformation is $Z^2/A \geq 45$. Note that

$$\frac{1}{2}\frac{a_s}{a_c} \geq 45,$$

where a_s and a_c are the coefficients which appear in the surface and Coulomb energy terms of the semiempirical binding energy formula.

14. Bohr and Wheeler[(51)] also showed that the activation energy E_a needed for nuclear drop to undergo fission is

$$E_a = A^{2/3} \times 0.89\left(1 - 0.022\frac{Z^2}{A}\right)\cdot$$

Use this formula, and also the difference $E_b - E_f$ to calculate the activation energies for the nuclei listed in Table 19–7.

CHAPTER 20

NUCLEAR ENERGY SOURCES

20–1 Nuclear fission as a source of energy. The large amount of energy released in fission, together with the emission of more than one neutron, has made it possible to use the fission process as a source of energy. The emission, on the average, of 2.5 neutrons in the fission of a U^{235} nucleus permits a chain reaction in which these neutrons produce more fissions and more neutrons, and so on. Under some conditions, the numbers of fissions and neutrons increase exponentially with time because each fission produces more neutrons than the one absorbed, and the amount of energy released can become enormous. The time interval between successive generations of fissions can be a very small fraction of a second and the energy released in the chain reaction can take the form of an explosion; the result is an "atomic bomb." Under other conditions the chain reaction can be controlled, and a steady state can be attained in which just as many neutrons are produced per unit time as are used up. The rate at which fissions occur and energy is released is kept constant, and the result is a chain-reacting pile, or nuclear reactor, which can be used as a source of neutrons or of power.

These ideas can be illustrated by some simple calculations. The fission of one U^{235} nucleus liberates $200 \times 1.6 \times 10^{-6}$ erg $= 3.2 \times 10^{-4}$ erg. When this quantity is multiplied by the Avogadro number, the product gives the energy released in the fission of all the nuclei in 1 gram atom (235 gm of U^{235}), and is equal to 1.93×10^{20} ergs. The energy liberated by the complete fission of 1 kg (2.2 lb) of U^{235} would be 8.21×10^{20} ergs, or about 2×10^{10} kilocalories, which is roughly equivalent to the energy released in the explosion of 20,000 tons of TNT. Even if only a small fraction of this energy could be released explosively, the result would be a formidable bomb.

The energy release in fission can also be expressed in terms of power units with interesting results. Since 1 Mev $= 1.60 \times 10^{-6}$ erg $= 1.6 \times 10^{-13}$ watt-sec, the fission of one U^{235} nucleus frees 3.2×10^{-11} watt-sec of energy, and 3.1×10^{10} fissions/sec give one watt of power. The complete fission of 1 gm of U^{235} would release 8.2×10^{10} watt-sec of energy, or 2.3×10^{4} kilowatt-hours, or nearly 1 megawatt-day. If the energy release were spread out over the period of a day, the complete fission of one kilogram of U^{235} would produce energy in the form of heat at a rate of 1000 megawatts. If this heat could be changed to electricity at a conversion efficiency of 30%, electrical energy would be supplied at the rate of 300,000 kilowatts. This output is equivalent to that from a large power plant which consumes about 2500 tons of coal per day. The fact that 1 kg of a fissionable material such as U^{235} is equivalent to 2500 tons of coal as a source of electric power is responsible for research and development work on the use of nuclear chain reactions and for the construction of nuclear powerplants.

Although useful nuclear power is a worth-while objective, another reason why nuclear reactors have been built has to do with the military applications of the fission process. It was realized in 1940 that an atomic bomb would have to depend on fission by fast neutrons, because only in this case would the energy release be rapid enough to make a bomb effective. But experiments indicated that the cross section for the fission of U^{238} by fast neutrons is small compared with the total cross section, with the result that an explosive fast neutron chain reaction with normal uranium seemed improbable. The relatively rare isotope U^{235} appeared to have satisfactory nuclear properties for an atomic bomb, but the separation of U^{235} in amounts large enough to be useful for military purposes presented a very difficult problem. The transuranium nuclide Pu^{239}, however, offered an alternative solution. There was evidence, in 1940, which indicated that the radiative capture of neutrons by U^{238} would probably lead by two successive β-decays to the formation of a nucleus for which $Z = 94$ and $A = 239$. It was predicted from the Bohr-Wheeler theory of fission that this nuclide would be a fairly long-lived α-emitter which would probably undergo fission when bombarded by slow or fast neutrons, as does U^{235}. If a method could be devised for converting some of the U^{238} to Pu^{239}, a chemical separation of the plutonium from uranium would avoid the difficulties of the isotopic separation of U^{235} from U^{238}. It was then suggested that Pu^{239} could be made in sufficiently large amounts in a controlled chain reaction based on the fission by slow neutrons of the U^{235} in natural uranium.

It is now well known that both the isotopic separation of U^{235} and the manufacture of Pu^{239} in a chain-reacting pile have been accomplished on a large scale. Many atomic bombs have been exploded but, for various reasons, this application of nuclear physics will not be discussed further in this book. A considerable number of nuclear reactors has been built for both military and peaceful purposes. Their designs depend on the nuclear properties of materials and involve the application of the methods and ideas of nuclear physics. These applications often serve to sharpen the distinctions among different nuclear properties and among different reactions and make use of the available knowledge about heavy atomic nuclei in surprising and fascinating ways.

20–2 The chain-reacting system or nuclear reactor. The achievement of a chain reaction with uranium depends on a favorable balance among four competing processes:

1. fission of uranium nuclei, with the emission of more neutrons than are captured,
2. nonfission capture of neutrons by uranium,
3. nonfission capture of neutrons by other materials,
4. escape of neutrons without being captured.

If the loss of neutrons by the last three processes is less than or equal to the surplus produced by the first, the chain reaction occurs; otherwise it does not.

The need for a favorable neutron balance sets certain conditions on any system in which a chain reaction is sought; one condition is on the size. If the uranium is distributed in a regular way throughout the assembly, the neutron production depends on the volume of the system, while the probability of escape depends on the surface area. If the system is very small and the surface-to-volume ratio is large, most of the neutrons can escape, and this process alone may make it impossible to achieve a chain reaction. The loss of neutrons by nonfission capture is a volume effect like the production, and its relative importance does not change with the size of the system. The result is that the greater the size of the assembly, the less likely it is that the escape of neutrons and their loss by absorption will outweigh their production and prevent a chain reaction. There is a certain size, called the *critical size*, for which the production of neutrons by fission is just equal to their loss by nonfission capture and escape, and a chain reaction is possible. If the size of the system is smaller than the critical size, a chain reaction cannot be sustained. The existence of a critical size below which a chain reaction will not "go" contrasts sharply with systems based on chemical reactions in which the possibility of a reaction is independent of the size of the system.

The relative probabilities of fission, radiative capture, and scattering are expressed and discussed in terms of cross sections. Values of the cross sections of U^{235} for thermal neutrons are listed in Table 19–1; the variation with energy of the fission cross section of U^{235} in the epithermal or "intermediate" energy range is shown in Fig. 19–1(b). Cross sections of natural uranium for thermal neutrons are listed in Table 19–1. The variation of the total cross section of U^{238} in the epithermal energy range is shown in Fig. 20–1; there are many resonances with strong absorption, that is, radiative capture. Average values of the cross sections of U^{238} for neutrons with energies above the fission threshold (taken for practical purposes to be 1.4 Mev) are listed in Table 20–1. Some of the basic requirements for the achievement of chain reactions can be deduced from this information about cross sections.

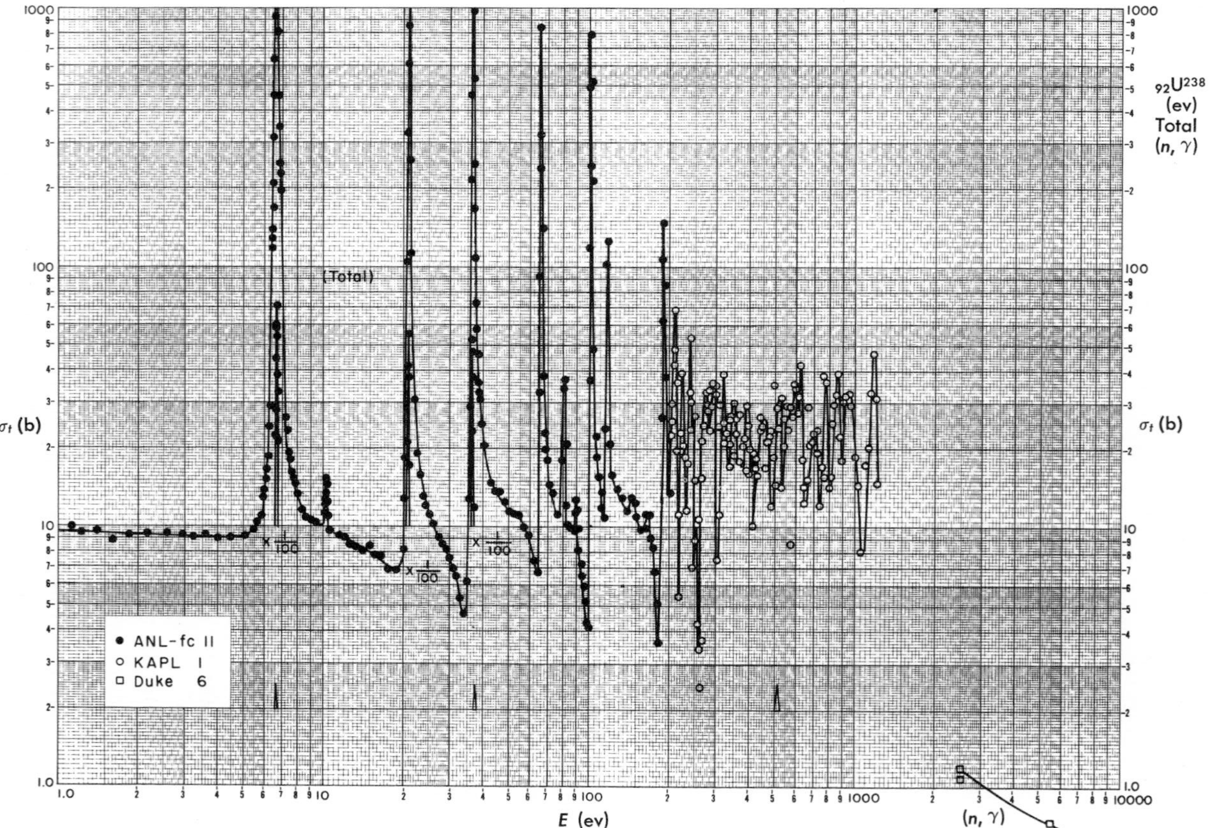

FIG. 20–1. The total cross section of U^{238} as a function of neutron energy, showing the absorption resonances (BNL-325, 2nd ed.).

TABLE 20–1

AVERAGE VALUES OF CROSS SECTIONS OF U^{238} FOR NEUTRONS WITH ENERGIES GREATER THAN THE FISSION THRESHOLD

Fission	$\bar{\sigma}_f = 0.50$b
Radiative capture	$\bar{\sigma}_r = 0.05$b
Elastic scattering	$\bar{\sigma}_e = 4.55$b
Inelastic scattering	$\bar{\sigma}_i = 2.10$b
Total	$\bar{\sigma} = 7.2$ b
Average number of neutrons per fission	$\bar{\nu} = 2.7$ b

We consider first the question of the possibility of a chain reaction in a mass of natural uranium large enough so that the loss of neutrons by leakage may be neglected. Some spontaneous fissions will occur, releasing about 2.5 fast neutrons per fission, with an average energy of about 2 Mev. Now, for practical purposes, natural uranium consists of U^{238} with an abundance of 99.28 atom percent and U^{235} with an abundance of 0.72 atom percent; the fission cross sections of the two isotopes in the energy range of the fission neutrons are not very different. Hence, if the fission neutrons cause further fissions, these will occur chiefly in U^{238}, with a negligibly small amount of fast fission in U^{235}. It is only necessary, therefore, to consider the interaction of the initial fission neutrons with U^{238}. For these neutrons, fission is more probable than radiative capture but less probable than either elastic or inelastic scattering. Neutrons which undergo radiative capture are lost. Elastic scattering by a uranium nucleus has very little effect on the energy of a fission neutron and, after such a collision, the neutron is free to make another collision almost as though nothing had happened. Inelastic scattering, however, has a serious effect in that a single process of this kind can reduce the neutron energy to a value below the fission threshold, to about 0.3 Mev on the average. Consequently, the original fission neutrons are only slightly increased in number by fast fission in U^{238}, and nearly all of the neutrons are left with energies below the fission threshold. Inelastic scattering is also a threshold reaction, and doesn't occur for neutrons with energies below about 1 Mev. The neutrons which have been degraded in energy below the thresholds for fission and inelastic scattering can undergo either radiative capture or elastic scattering. Those which undergo radiative capture are, of course, lost. Neutrons which are scattered elastically lose energy very slowly, and many collisions are needed to reduce the energy from about 10^5 ev to 1 ev. In this energy range, the probability of resonance capture in U^{238}

is much greater than the probability of fission in U^{235}, as can be seen by comparing Figs. 19–1 and 20–1 and remembering the small abundance of the U^{235}. Since the neutrons lose energy very slowly, the probability is very small that they will reach thermal energies without being captured. Although the fission cross section of natural uranium at thermal energies is greater than the radiative capture cross section, very few neutrons can reach these low energies and cause fission, with the result that a chain reaction is not possible in uranium alone.

A chain reaction can be achieved if the U^{238} is removed, that is, in pure U^{235}, or in Pu^{239}. Neutrons of any energy can cause the fission of U^{235}, so that inelastic scattering of fission neutrons does not reduce the fission probability. In fact, even neutrons which might reach resonance energies would contribute to the chain reaction because resonance fission of U^{235} is much more probable than radiative capture. Analogous arguments hold for Pu^{239}. Thus, bombs made of U^{235} or Pu^{239} have been exploded, and nuclear reactors have been operated which have U^{235} or Pu^{239} as the fuel and in which the fissions are induced by fast neutrons. In such a "fast" reactor, there is very little slowing-down of neutrons, and the presence of materials which can cause moderation is avoided as far as possible.

A chain reaction can be achieved with natural uranium and a suitable moderator in an appropriate arrangement. The fission cross section of U^{235} is so large that, in spite of the small abundance of U^{235}, fission by thermal neutrons competes favorably with radiative capture. The requirement is then that enough neutrons reach thermal energies. This result can be achieved by using a moderator; a neutron can lose enough energy in a single collision with a moderator nucleus so that it skips over many resonances, with the result that the resonance absorption of neutrons can be made small. Heavy water and graphite have been used successfully as moderators with natural uranium. When D_2O is the moderator, the uranium can be in the form of a solution of a salt such as uranyl sulfate, or very small particles of uranium oxide can be suspended uniformly in the D_2O; this type of reactor is said to be *homogeneous*. When graphite is the moderator, the uranium must be in the form of large lumps for reasons which will be seen later, and rods of uranium can be distributed in a regular way throughout the graphite, forming a lattice. This kind of assembly is referred to as *heterogeneous* because of the separation of the fuel and the moderator. Natural uranium and ordinary water cannot be made to sustain a chain reaction in either a homogeneous or a heterogeneous system because the absorption cross section of hydrogen for thermal neutrons is too large.

When fission was discovered (1939), the only material available for chain reacting systems was natural uranium. The enrichment of uranium in the isotope U^{235} in gaseous diffusion plants and the production of Pu^{239} in nuclear reactors have provided highly purified nuclear fuels and have greatly extended the range of possible chain-reacting systems. For example, a solution in water of a salt containing enriched uranium can form a small, homogeneous reactor. Enriched uranium or plutonium can also be used in partially moderated assemblies in which the fissions are caused mainly by neutrons of intermediate energy. In these "intermediate" reactors there is only enough moderator to slow neutrons down part of the way, but not all of the way, to thermal energies. The nuclide U^{233}, which can be made in a reactor by irradiating Th^{232} with neutrons, can also be used in thermal, intermediate, or fast reactors.

The rate at which fissions occur in a reactor determines the number of neutrons produced per unit time and the rate at which heat is produced —the power level. For a reactor to operate at a steady power level, the energy released in fission must be removed from the assembly. The fission energy, originally in the form of the kinetic energy of fission fragments, neutrons, β-rays, and γ-rays, is converted to heat when these particles are stopped in the reactor materials. The heat is removed by circulating a coolant through the reactor, and water, gas, or a liquid metal can be used as the coolant. The choice of the coolant depends on the purpose of the reactor and is limited by nuclear and engineering considerations. The coolant is an impurity from the standpoint of the nuclear properties of the system, and adds to the nonfission capture of neutrons; the resulting damage to the neutron economy must be balanced against the efficiency of heat removal. In a reactor designed to serve as a research tool, to supply neutrons for experiments, there is no premium on the total power output; a coolant which is relatively inefficient but absorbs few neutrons can be used, and air is good enough. In reactors for the production of plutonium, the rate of production depends on the power level and more efficient cooling is needed; water is often used because it is a better coolant than air, even though it absorbs many more neutrons. When the emphasis is on making useful power, liquid metals may be used to increase the efficiency of heat removal still further.

Nuclear reactors can be classified according to the characteristics of the chain-reacting system, and the following features are used to describe them.[(1)]

1. The neutron energies at which most of the fissions occur.
   (a) High energy, that is, most of the fissions are induced by the fast neutrons from fission.
   (b) Intermediate energy.
   (c) Low energy (thermal).

2. The fuel.
   - (a) Natural uranium, containing 0.72% of U^{235}.
   - (b) Enriched uranium, containing more than 0.72% of U^{235}.
   - (c) Pu^{239}.
   - (d) U^{233}.
3. The fuel-moderator assembly.
   - (a) Heterogeneous.
   - (b) Homogeneous.
4. The moderator.
   - (a) Graphite.
   - (b) Water.
   - (c) Heavy water (D_2O).
   - (d) Beryllium or beryllium oxide.
5. The coolant.
   - (a) Air, CO_2, or He.
   - (b) Water or other liquid.
   - (c) Liquid metal.
6. The purpose.
   - (a) Research tool.
   - (b) Production of fissile material.
   - (c) Power.

The description of nuclear reactors based on the factors listed above can be illustrated by the following examples. The reactor at the Brookhaven National Laboratory was originally a thermal, natural uranium, graphite-moderated, heterogeneous, air-cooled research reactor; to obtain a higher thermal neutron flux, the fuel has recently been changed to highly enriched uranium. The NRX reactor at Chalk River, Canada, is a thermal, normal uranium, heterogeneous, heavy water-moderated, ordinary water-cooled, research reactor. The EBR (Experimental Breeder Reactor at Arco, Idaho) is a fast, enriched uranium, heterogeneous, liquid metal-cooled, power-breeder reactor on a pilot-plant scale. (The term *breeder* will be discussed in Section 20–7.)

It is evident that a detailed treatment of nuclear reactors would have to cover a wide variety of problems, and such a treatment is beyond the scope of this book. It is possible, however, to indicate by means of some relatively simple examples how the conditions set by nuclear requirements affect the design of nuclear reactors. The examples will be limited to thermal reactors because most of the reactors built so far are of this type, because they are the best understood reactors, and because most of the available information is about them.

20–3 Thermal nuclear reactors. The neutron cycle. The neutron balance in a thermal reactor can be described in terms of a cycle that

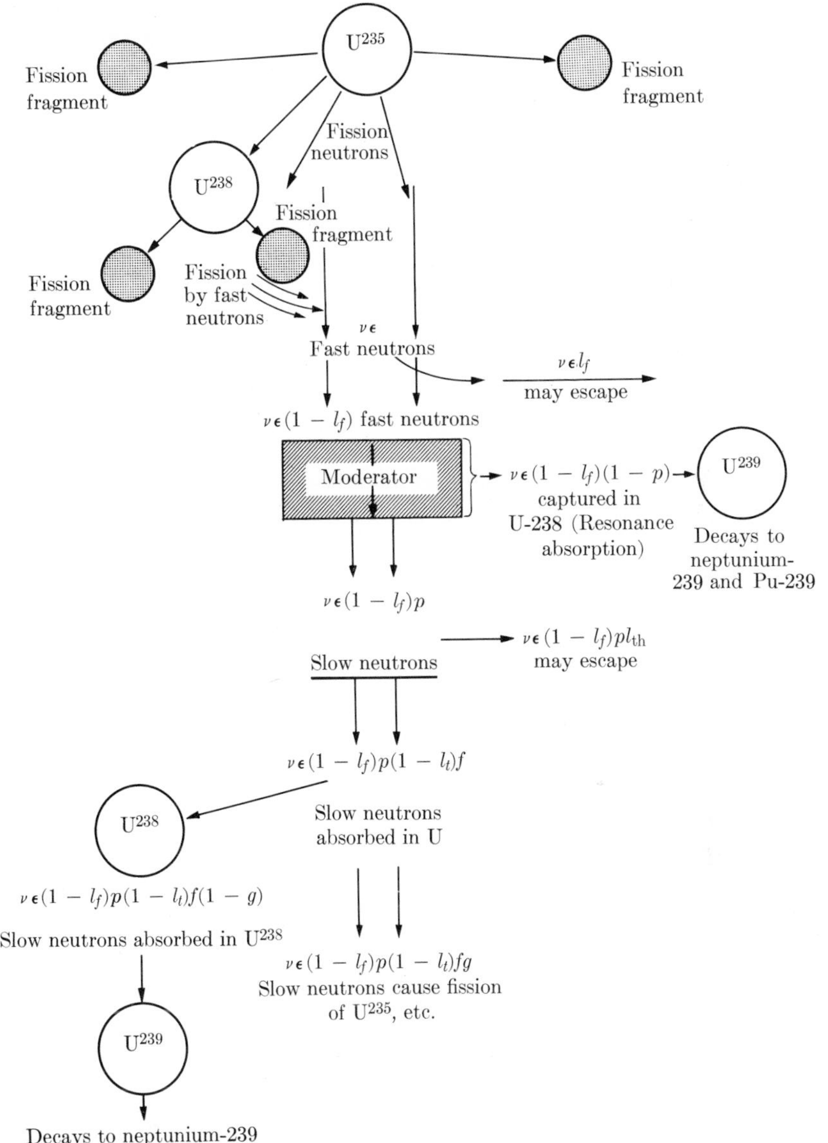

FIG. 20–2. Schematic representation of a chain reaction based on the fission of uranium nuclei by thermal neutrons.

shows what happens to the neutrons. Consider this cycle, as illustrated in Fig. 20–2, starting with the fission of a U^{235} nucleus by a thermal neutron. In this fission process, ν fast neutrons are emitted. These neutrons have an average energy above the fission threshold for U^{238}, and some of them can cause the fission of U^{238} nuclei. The chance for such additional fissions depends on whether the fission neutrons can collide with U^{238} nuclei before colliding with moderator atoms. A small fraction of the neutrons that do collide with U^{238} nuclei can cause fissions, and for a given number of neutrons from the thermal fission of U^{235}, some additional neutrons result from the fast fission of U^{238} nuclei. The total number of fast neutrons from fission is increased from ν to $\nu\epsilon$, where ϵ is a quantity which may be equal to, or greater than, unity. The quantity ϵ is called the *fast fission factor;* it is very close to unity in a homogeneous reactor, but can be as high as 1.1 in certain heterogeneous reactors. In several natural-uranium, graphite-moderated reactors, ϵ is 1.03; the fission of U^{238} nuclei by fast neutrons from the thermal neutron fission of U^{235} nuclei increases the total number of fission neutrons by 3%, and this effect can be very important.

The $\nu\epsilon$ neutrons diffuse through the pile; most of them are slowed down, but a fraction l_f escapes before being slowed down to thermal energies; $\nu\epsilon l_f$ neutrons are lost, so far as a chain reaction is concerned. The remaining $\nu\epsilon(1 - l_f)$ neutrons are slowed down by collisions with moderator atoms, but during the slowing-down process, some of them may be captured by U^{238} to form U^{239}, which decays to Np^{239} and then to Pu^{239}. The capture process is a resonance reaction, as can be seen from Fig. 20–1, which shows the total cross section of uranium as a function of energy for neutron energies from one to several thousand electron volts. The resonances have been shown to include radiative capture reactions. Of the $\nu\epsilon(1 - l_f)$ neutrons which start the slowing-down process, a fraction $\nu\epsilon(1 - l_f)p$ escapes resonance capture while the fraction $\nu\epsilon(1 - l_f)(1 - p)$ is captured and goes to form Pu^{239}. The quantity p is called the *resonance escape probability.* Although the absorbed neutrons cannot cause further fissions of U^{235}, and in this sense are lost, they are important because they contribute to the making of Pu^{239}.

The neutrons which escape resonance absorption are slowed down to thermal energies, where either of two processes occurs. Some of the neutrons diffuse without being captured and eventually escape from the system. If the fraction escaping is denoted by l_t, the number of thermal neutrons, per fission of U^{235}, that escape is equal to $\nu\epsilon(1 - l_f)pl_t$. Of the remaining $\nu\epsilon(1 - l_f)p(1 - l_t)$ thermal neutrons, a fraction f is absorbed in uranium; the fraction $(1 - f)$ is absorbed in other materials, such as the moderator or structural materials, and these neutrons are lost. The number of neutrons still available for carrying on the chain reaction is then

$\nu\epsilon(1 - l_f)p(1 - l_t)f$. The quantity f is called the *thermal utilization.* Not all of the neutrons absorbed in the uranium cause fission of U^{235} nuclei. Some of the neutrons are absorbed by U^{238} to form U^{239} and eventually Pu^{239}, while others are absorbed in U^{235} to form U^{236}. The fraction of the thermal neutrons absorbed in uranium that cause fission of U^{235} is just $\sigma_f(\text{U})/\sigma_a(\text{U})$.

The number of second generation fissions in U^{235} per fission of U^{235} by a first generation neutron is called the *reproduction factor*, or *multiplication factor*, and is denoted by k. It follows that

$$k = \nu\epsilon(1 - l_f)p(1 - l_t)f\,\frac{\sigma_f(\text{U})}{\sigma_a(\text{U})}\,. \tag{20–1}$$

The product $\nu[\sigma_f(\text{U})/\sigma_a(\text{U})]$ represents the number of (fast) fission neutrons produced per thermal neutron absorbed in uranium; it is called *eta* (η). The multiplication factor can be written as

$$k = \eta\epsilon p f(1 - l_f)(1 - l_t). \tag{20–2}$$

The quantity k, as given by Eq. (20–2), is usually called the *effective* multiplication factor and corresponds to an assembly of finite size. To have a chain-reacting system operating in a steady state, it is necessary that k be equal to unity; if k is less than unity there can be no chain reaction; if k is greater than unity the numbers of neutrons and fissions increase from cycle to cycle and the chain reaction is said to be divergent. The critical size of a chain-reacting system is the size for which k is equal to unity; for $k < 1$, the system is *subcritical,* and for $k > 1$ it is *supercritical.*

The value of k for a system which is infinitely large is especially useful in the theory and design of nuclear reactors. There can be no leakage of neutrons from such a system, and the quantities l_f and l_t are both zero. The multiplication factor is then denoted by k_∞ and is given by

$$k_\infty = \eta\epsilon p f. \tag{20–3}$$

This formula is called the *four-factor formula,* and the calculation of k_∞ is one of the main problems in reactor design theory.

In the special case of a reactor in which the fuel contains only U^{235} and no U^{238}, both the fast fission factor ϵ and the resonance escape probability p are practically unity. The formula for k_∞ reduces to

$$k_\infty = \eta f. \tag{20–4}$$

Of the four factors which make up k_∞, η depends only on the nuclear properties of the fuel; ϵ depends on the nuclear properties of the fuel as well as on its size and shape. The remaining quantities, p and f, depend on the nuclear properties of the fuel, moderator, and any other materials present in the reactor, and also on the way in which all of these materials are arranged. One of the basic design problems is that of finding the relative amounts of all the materials and the way in which they can be assembled to yield the largest value of the product pf. For a given fuel, this arrangement gives the largest value of k_∞. The above discussion of the neutron cycle, leading to the four-factor formula, is an over-simplified description of the interactions between neutrons and matter in a chain-reacting system, because many details have been omitted. But this treatment provides a convenient way of looking at the physics of a reactor and, for homogeneous thermal reactors, the four factors η, ϵ, p, f can be calculated without too much trouble. Some simple examples will be considered in the next section.

20–4 The calculation of the multiplication factor for a homogeneous thermal reactor. The multiplication factor will be calculated for three homogeneous systems consisting of natural uranium moderated with graphite, ordinary water, and heavy water, respectively. In the first of these, very small particles of uranium are assumed to be dispersed uniformly throughout a large block of graphite. When H_2O or D_2O is the moderator, the uranium is assumed to be in the form of a suspension or slurry of UO_2 in the fluid. It will be seen that in each case the independent variable, or design parameter, may be taken to be the ratio of the numbers of moderator atoms and uranium atoms.

The value of η for natural uranium and thermal neutrons is obtained from the formula

$$\eta = \nu\,[\sigma_f(\mathrm{U})/\sigma_a(\mathrm{U})], \tag{20–5}$$

and is given in Table 19–1 as 1.34 ± 0.02. For the purposes of later calculations, it is desirable to use consistent values of various constants and to calculate η to three decimal places. For example, the value of η changes when the uranium is enriched with respect to the isotope U^{235}, and it is useful to express η in a form which shows its dependence on the concentration of U^{235}. Let N_{28} and N_{25} represent the numbers of atoms of U^{238} and U^{235}, respectively, per cubic centimeter of uranium. Equation (20–5) can be written, in greater detail, as

$$\eta = \nu\,\frac{N_{25}\sigma_f(\mathrm{U}^{235})}{N_{25}\sigma_f(\mathrm{U}^{235}) + N_{25}\sigma_r(\mathrm{U}^{235}) + N_{28}\sigma_a(\mathrm{U}^{238})}, \tag{20–6}$$

where $\sigma_r(U^{235})$ stands for the cross section for radiative capture by U^{235}. When the numerator and denominator on the right side of Eq. (20–6) are divided by $N_{25}\sigma_f(U^{235})$, the result is

$$\eta = \frac{\nu}{1 + [\sigma_r(\mathrm{U}^{235})/\sigma_f(\mathrm{U}^{235})] + [N_{28}\sigma_a(\mathrm{U}^{238})/N_{25}\sigma_f(\mathrm{U}^{235})]}. \tag{20–7}$$

The value of $\sigma_r(\text{U}^{235})/\sigma_f(\text{U}^{235})$ is 0.180, and that of $\sigma_a(\text{U}^{238})/\sigma_f(\text{U}^{235})$ is 0.00470, so that Eq. (20–7) becomes

$$\eta = \frac{2.44}{1.180 + 0.00470 N_{28}/N_{25}}. \tag{20–8}$$

The ratio N_{28}/N_{25} is obtained from experimental values of the concentration of U^{235} in uranium, usually expressed in terms of weight percent. For natural uranium, this concentration is 0.00715 weight percent, or

$$\left(\frac{234.1 N_{25}}{238.1 N_{28} + 235.1 N_{25}}\right)_{\text{Nat. U}} = 0.00715,$$

and

$$N_{28}/N_{25} = 137.8.$$

Hence, for natural uranium, $\eta = 1.335$. When the ratio N_{25}/N_{28} is increased, as in enriched uranium, the ratio N_{28}/N_{25} is decreased; η is increased and more neutrons are produced per thermal neutron absorbed in uranium.

In the homogeneous systems to be considered the fuel is in the form of very small particles, of the order of microns in diameter. It is almost certain that the fission neutrons will escape from the uranium before colliding with a U^{238} nucleus. The neutrons will enter the moderator and will be slowed down before they can cause fission of U^{238}, and the value of ϵ will therefore be unity. This loss of the additional neutrons contributed by the fast effect is one of the disadvantages of a homogeneous system.

The condition that a chain reaction be possible in a homogeneous system of finite size can be written as $k_\infty > 1$, or $\eta f p > 1$, or

$$pf > \frac{1}{\eta} = 0.75. \tag{20–9}$$

This condition expresses a difficult design problem. The resonance escape probability and the thermal utilization must both be in the neighborhood of 0.85. In other words, not more than about 15% of the neutrons may be lost to resonance absorption during the slowing-down process, and not more than about 15% of the thermal neutrons may be absorbed in materials other than the nuclear fuel. It will be seen that when the fuel is natural uranium, the condition (20–9) may be very hard to meet.

The thermal utilization is defined by the equation

$$f = \frac{\text{Thermal neutron absorption in uranium}}{\text{Total thermal neutron absorption}}. \tag{20–10}$$

If we consider an assembly consisting only of fuel and moderator, and if both of these materials are $1/v$-absorbers, we may write

$$f = \frac{N_0\sigma_{a0}}{N_0\sigma_{a0} + N_1\sigma_{a1}} = \frac{1}{1 + (N_1\sigma_{a1}/N_0\sigma_{a0})}, \tag{20–11}$$

where the subscripts 0 and 1 stand for fuel and moderator, respectively, and N_0 and N_1 represent the number of atoms of fuel and moderator per cubic centimeter of reactor material, respectively. When absorption in other materials must be taken into account, additional terms must be included in the denominator of Eq. (20–11). This equation is actually a simplified form for the thermal absorption ratio, in which the values of the cross sections correspond to neutron velocities of 2200 m/sec. The thermal absorption should really be expressed in terms of the thermal neutron flux and the absorption cross section, and depends on the Maxwellian velocity distribution of the neutrons; each term in f should be an integral of the flux and cross section weighted with the velocity distribution, with the integral taken over all possible neutron velocities. It was shown, however, in Section 18–4, that if the cross section follows the $1/v$-law, each absorption term reduces to $N_i n v^0 \sigma^0_{ai}$, where a superscript zero has been written to denote the cross section for 2200 m/sec neutrons, and i stands for either 0 or 1. The superscript has been omitted in Eq. (20–11) to simplify the expression. In a homogeneous reactor, the flux is the same for the fuel and the moderator, and the ratio of the absorption in two $1/v$-absorbers reduces simply to the ratio of the macroscopic absorption cross sections. The absorption cross section of graphite follows the $1/v$-law exactly, while that of uranium deviates from $1/v$-behavior by about one percent. If we neglect this deviation, the error introduced is again small in illustrative calculations, although a correction would be made in real design calculations.

The values of the parameters needed for the calculation of the thermal utilization are listed in Table 20–2. The determination of the absorption cross section of heavy water presents a problem because, in practice, heavy water always contains some H_2O; it may, for example, consist of 99.75 atom percent of D and 0.25 atom percent of H. But the absorption cross section of H is so much greater than that of D that the effect of the H impurity must be taken into account. We, therefore, write for the macroscopic absorption cross section of the heavy water

$$\sum_a^{\text{pract}} (D_2O) = mN(D_2O)\sigma_a(D_2O) + (1 - m)N(H_2O)\sigma_a(H_2O), \tag{20–12}$$

where m is the fraction, in atom percent, of D in the heavy water, and $(1 - m)$ is the fraction, in atom percent, of H; $N(D_2O)$ and $N(H_2O)$ are the numbers of molecules in one cubic centimeter of D_2O and H_2O, re-

TABLE 20–2

QUANTITIES NEEDED FOR THE CALCULATION OF THE THERMAL UTILIZATION IN CERTAIN HOMOGENEOUS REACTORS

Material	Absorption cross section per atom for 2200 m/sec neutrons, b	Number of atoms/cm^3 of material, $\times 10^{-24}$
U	7.68	0.0478 (density: 18.9 gm/cm^3)
H	0.332	0.0668 (in H_20)
D	0.00046	0.0662 (in D_20)
C	0.0045	0.0803 (density: 1.60 gm/cm^3)
O	<0.0002 (≈ 0)	

spectively; $\sigma_a(D_2O)$ and $\sigma_a(H_2O)$ are the absorption cross sections per molecule for 2200 m/sec neutrons. On inserting values from Table 20–2 into the expression (20–12), we get

$$\begin{aligned}\sum_a^{\text{pract}} (D_2O) &= (0.9975)(0.0331)(9.2)10^{-4}\ \text{cm}^{-1} \\ &\quad + (0.0025)(0.0334)(0.664)\ \text{cm}^{-1} \\ &= (0.304 + 0.554)10^{-4}\ \text{cm}^{-1} = 0.858 \times 10^{-4}\text{cm}^{-1}.\end{aligned}$$

Thus, the small impurity of H_2O in the D_2O contributes more to the total macroscopic cross section of the heavy water than does all of the D_2O.

It is convenient to introduce an *effective* value for the absorption cross section per molecule of heavy water, defined by the relation

$$N(D_2O)\sigma_a^{\text{eff}}(D_2O) = 0.858 \times 10^{-4}\ \text{cm}^{-1},$$

so that

$$\sigma_a^{\text{eff}}(D_2O) = 0.00259\text{b}.$$

Since the absorption cross section of oxygen is practically equal to zero, we may write

$$\sigma_a^{\text{eff}}(D) = 0.00130\text{b}.$$

This cross section has been introduced because it is convenient to use an effective absorption cross section per atom of deuterium together with the number of atoms of deuterium per cubic centimeter of reactor material when we compare the results for the three reactors. Similarly, for the H_2O-moderated reactor we shall use the absorption cross section per hydrogen atom and the number of hydrogen atoms per cubic centimeter.

We may finally write for the thermal utilization for the three reactors, uranium-graphite,

$$f = \frac{1}{1 + (N_1/N_0)(0.000579)}; \tag{20–13}$$

uranium-heavy water,

$$f = \frac{1}{1 + (N_1/N_0)(0.000169)}; \tag{20–14}$$

uranium-water,

$$f = \frac{1}{1 + (N_1/N_0)(0.0432)}. \tag{20–15}$$

The resonance escape probability can be calculated by means of a formula[2] the derivation of which is beyond the scope of this book,

$$p = \exp\left[-\frac{1}{\xi\sigma_{s1}}\frac{N_0}{N_1}\left(\int \sigma_a \frac{dE}{E}\right)_{\text{eff}}\right]. \tag{20–16}$$

The quantity $(\int\sigma_a[dE/E])_{\text{eff}}$ is called the *effective resonance integral* and is a measure of the total neutron absorption in the resonance energy region. It has been determined experimentally for homogeneous mixtures of uranium and scattering materials and may be expressed by the empirical formula

$$\left(\int \sigma_a \frac{dE}{E}\right)_{\text{eff}} = 3.9\left(\frac{\sum_s}{N_0}\right)^{0.415} \text{b}. \tag{20–17}$$

In this formula, $\sum_s/N_0$ represents the total macroscopic scattering cross section of the reactor material per atom of uranium, in the resonance energy range. For the uranium-graphite reactor we have

$$\frac{\sum_s}{N_0} = \frac{N_1\sigma_s(C) + N_0\sigma_s(\text{U})}{N_0} = \frac{N_1}{N_0}\sigma_s(C) + \sigma_s(\text{U}); \tag{20–18}$$

for the uranium-D_2O reactor,

$$\frac{\sum_s}{N_0} = \frac{N(\text{D}_2\text{O})\sigma_s(\text{D}_2\text{O}) + N(\text{UO}_2)\sigma_s(\text{UO}_2)}{N_0}$$

$$= \frac{N_1}{N_0}\sigma_s(\text{D}) + \sigma_s(\text{U}) + \left(\frac{1}{2}\frac{N_1}{N_0} + 2\right)\sigma_s(\text{O}); \tag{20–19}$$

TABLE 20–3

SCATTERING CROSS SECTIONS IN THE RESONANCE ENERGY REGION

Material	Scattering cross section per atom or molecule, b	Average logarithmic energy loss per collision
U	8.3	
C	4.7	0.158
H	20.5	1.00
D	3.3	0.725
O	3.8	0.120
H_2O	44.8	0.926
D_2O	10.4	0.508

for the uranium-H_2O reactor,

$$\frac{\sum_s}{N_0} = \frac{N_1}{N_0}\,\sigma_s(\text{H}) + \sigma_s(\text{U}) + \left(\frac{1}{2}\,\frac{N_1}{N_0} + 2\right)\sigma_s(\text{O}). \tag{20–20}$$

In the last three equations, the scattering cross sections are those in the resonance region, about 1 ev to 10^5 ev; the appropriate values are listed in Table 20–3.

The results of the calculations are collected in Table 20–4, where the values of f, p, pf and k_∞ are listed for different values of N_1/N_0, the ratio of the number of moderator atoms to that of uranium atoms. For small values of N_1/N_0, f approaches unity, while p is small. This result is expected because both thermal and resonance absorption should increase as the amount of uranium in the reactor increases. We expect, therefore, that an arrangement which increases f should decrease p, and that the product pf should pass through a maximum for a given moderator. Hence, k_∞ should also pass through a maximum as N_1/N_0 is varied, and the calculations show that this is indeed the case. The maxima are quite flat, occurring in the range $N_1/N_0 = 4$ to 10 for H_2O, 150 to 500 for D_2O and 300 to 600 for graphite.

The most important result is that values of k_∞ greater than unity can be attained with homogeneous mixtures containing natural uranium and heavy water, but not with homogeneous mixtures of natural uranium and graphite, or natural uranium and ordinary water. It is this result, especially in the case of graphite, which led to the consideration of heterogeneous systems. When graphite is the moderator, the failure results from the low value of the resonance escape probability. Adequate values of f can be

TABLE 20–4

VALUES OF THE THERMAL UTILIZATION, RESONANCE ESCAPE PROBABILITY AND MULTIPLICATION FACTOR FOR THREE HOMOGENEOUS FUEL-MODERATOR SYSTEMS

$\frac{N_1}{N_2}$	Uranium-graphite				UO_2–D_2O				UO_2–H_2O			
	f	p	pf	k_∞	f	p	pf	k_∞	f	p	pf	k_∞
1	0.999	0			1.000	0.018	0.018	0.02	0.958	0.317	0.304	0.41
2	0.999	0			1.000	0.102	0.102	0.14	0.920	0.510	0.469	0.63
4	0.998	0.006			0.999	0.267	0.267	0.36	0.853	0.680	0.580	0.78
5	0.997	0.012			0.999	0.333	0.333	0.45	0.822	0.724	0.595	0.80
10	0.994	0.062	0.062	0.08	0.998	0.509	0.508	0.63	0.698	0.832	0.581	0.78
20	0.989	0.167	0.165	0.22	0.997	0.651	0.649	0.87	0.536	0.903	0.484	0.65
50	0.972	0.364	0.354	0.47	0.991	0.787	0.780	1.04	0.316	0.974	0.308	0.41
100	0.945	0.523	0.494	0.66	0.984	0.853	0.839	1.12				
150	0.920	0.586	0.539	0.72	0.976	0.882	0.861	1.15				
200	0.896	0.641	0.574	0.77	0.968	0.900	0.871	1.16				
300	0.852	0.704	0.600	0.80	0.952	0.920	0.876	1.17				
400	0.812	0.744	0.604	0.81	0.938	0.932	0.874	1.17				
500	0.775	0.772	0.598	0.80	0.924	0.940	0.869	1.16				
600	0.742	0.793	0.588	0.79	0.909	0.946	0.850	1.13				
800	0.683	0.820	0.560	0.75	0.882	0.954	0.841	1.12				
1000	0.633	0.841	0.532	0.71	0.857	0.960	0.823	1.10				

attained for values of N_1/N_0 up to about 300 or 400; but a value of p of 0.82 is not reached until $N_1/N_0 = 800$, with the result that the maximum value of pf is about 0.6, and the maximum value of k_∞ is about 0.8. Fortunately, the properties of slowing-down and resonance absorption are such that the resonance escape probability can be increased significantly by lumping the uranium. In the case of water, the difficulty is in the small value of the thermal utilization obtained for all but the smallest values of N_1/N_0; to get useful values of f there must be so much uranium that there is too much resonance absorption. The preceding arguments can be put in another way. With D_2O, the thermal utilization is favorable, say 0.85 or more, for values of N_1/N_0 up to about 1000, because of the very small absorption cross section of D_2O. In this wide range it is possible to find a fairly wide range of values over which p is also favorable, and a chain reaction can be attained. With graphite as moderator, the slowing-down is in such small steps that the probability is high that neutrons will have energies corresponding to the resonance energies of U^{238}; if U^{238} atoms are readily available to the neutrons, the resonance capture is high, and the escape probability is low. The resonance escape is small until the uranium is very dilute, but by then the thermal utilization has decreased too much.

When uranium enriched in U^{235} is available, a chain reaction can be achieved with either graphite or water as the moderator. With enriched uranium, both f and η are increased, the increase in η being more important. We consider the graphite-moderated reactor. The maximum value of k_∞ with natural uranium and graphite was about 0.8, so that an increase of about 20% in η would be needed to bring k_∞ up to unity, or a value of η between 1.6 and 1.7. Since $\eta = 2.07$ for pure U^{235}, it is clear that the necessary value of η can be attained for some concentration of U^{235} between that in natural uranium and that of pure U^{235}. The value of η can be computed from Eq. (20–7). The absorption cross section of uranium also increases with increasing U^{235} concentration so that f will increase; σ_a can be calculated from the formula

$$\sigma_a(\mathrm{U}) = \frac{N(\mathrm{U}^{235})\sigma_a(\mathrm{U}^{235}) + N(\mathrm{U}^{238})\sigma_a(\mathrm{U}^{238})}{N(\mathrm{U}^{235}) + N(\mathrm{U}^{238})} \tag{20–21}$$

$$= \frac{\sigma_a(\mathrm{U}^{235}) + (N_{28}/N_{25})\sigma_a(\mathrm{U}^{238})}{1 + (N_{28}/N_{25})}. \tag{20–22}$$

The values of the cross sections to be used are

$$\sigma_a(\mathrm{U}^{235}) = 683 \text{ b}; \qquad \sigma_a(\mathrm{U}^{238}) = 2.71 \text{ b}.$$

The calculation of k_∞ has been made for $N_1/N_0 = 400$, with p equal to 0.744, the value for natural uranium. It is assumed that the decrease in

TABLE 20–5

THE MULTIPLICATION FACTOR FOR A HOMOGENEOUS MIXTURE OF ENRICHED URANIUM AND GRAPHITE, $N_1/N_0 = 400$

$\dfrac{N_{28}}{N_{25}}$	σ_a, b	f	pf	η	k_∞
137.8	7.68	0.810	0.603	1.335	0.80
130	8.01	0.816	0.607	1.362	0.83
120	8.45	0.824	0.613	1.399	0.86
110	9.04	0.834	0.620	1.439	0.89
100	9.57	0.841	0.626	1.479	0.93
90	10.3	0.852	0.634	1.523	0.97
80	11.3	0.862	0.641	1.569	1.01
70	12.5	0.874	0.650	1.618	1.05
60	14.1	0.887	0.660	1.670	1.10
50	16.3	0.901	0.670	1.726	1.16

the concentration of U^{238} over the range of U^{235} concentration is too small to affect the resonance escape probability; a more careful calculation would take this effect into account. The results are shown in Table 20–5. It is seen that a chain reaction can be achieved when the U^{235} atom concentration is increased from 1:138 to about 1:80. Similarly, a chain reaction can be achieved in a homogeneous system with partially enriched uranium and ordinary water.[(3)]

20–5 The heterogeneous thermal reactor. It was seen in the last section that a homogeneous natural uranium-graphite assembly cannot support a chain reaction no matter how large it is, but that when the uranium is sufficiently enriched with respect to the isotope U^{235}, a chain reaction is possible. In the absence of enriched uranium, the question arose as to whether natural uranium and graphite could be arranged in some way which could lead to a chain reaction. The results listed in Table 20–3 indicate that the main difficulty is the small value of the resonance escape probability. The value of p does not become appreciable until the uranium concentration becomes very low, but it is hard to conceive of a chain reaction in a system with only a very small amount of natural uranium. It was apparent, therefore, that something had to be done to increase the value of the resonance escape probability in an assembly which has enough uranium to yield an adequate value (0.8 to 0.9) for the thermal utilization factor. The solution to this problem was suggested by Fermi and Szilard, who proposed the use of lumps of uranium of considerable size embedded in the moderator. The use of lumps also affects the thermal utilization and the fast fission factor, and a qualitative discussion of the various effects of lumping will be useful.

A consideration of the way in which resonance absorption takes place shows that the lumping of the uranium increases the resonance escape probability. Resonance absorption occurs when neutrons, during the course of the slowing-down process, reach energies which, together with the binding energy, correspond to levels of the compound nucleus [U^{239}]. When a neutron with such an energy collides with a U^{238} nucleus, there is an appreciable chance that it will be absorbed. In a homogeneous uranium-graphite assembly, each uranium particle has the same chance of capturing a certain fraction of the resonance neutrons, but when the uranium is present in relatively large lumps this is no longer the case. For, suppose that neutrons collide with a lump of uranium; those neutrons which happen to have energies within a resonance energy band may be absorbed at the surface of the lump, while neutrons with energies which do not correspond to a resonance band pass into the lump. Some of these neutrons may undergo elastic collisions with the uranium nuclei and, although the energy loss per collision is very small, a fraction of them may be scattered into resonance energy bands and absorbed. Many of the neutrons, however, can pass through the lump without being absorbed; they may either escape elastic collisions with uranium nuclei altogether, or they may be scattered, but not into a resonance band. The probability of resonance absorption inside the lump is therefore smaller than that at the surface of the lump. In other words, U^{238} nuclei inside the lump have a smaller chance of capturing neutrons during the slowing-down process than U^{238} nuclei near the surface of the lump, and there is some "self-shielding" of the uranium against resonance absorption. The neutrons which re-enter the moderator undergo further slowing down in a region devoid of uranium, and they may lose enough energy to miss several resonance bands before entering another lump or reaching thermal energies. The magnitude of this effect depends on the spacing of the uranium lumps in the moderator matrix, and its probability is appreciably greater than in a homogeneous assembly. Thus, one advantage of the heterogeneous assembly is the lower resonance absorption owing to filtering at the lump surface of neutrons whose energies lie within the resonance bands of uranium.

It follows from the above discussion that the resonance absorption may be divided into two parts, a surface absorption and a volume absorption, and the size and shape of the uranium lumps can be chosen to make the resonance escape probability relatively high. Some qualitative notions about the size of the lumps can be obtained quite easily. To cut down the volume absorption, the size of the lump should not be too large compared with the mean free path of *resonance* neutrons in the lump. Although this mean path is not known accurately, a rough estimate can be made by considering thermal and fission neutrons as limiting cases. The total thermal cross section of normal uranium is 16.0 b and uranium has 0.048×10^{24} atoms/cm^3; the total mean free path for thermal neutrons is then

$$\lambda_t = \frac{1}{N\sigma_t} \approx \frac{1}{(.048)(16.0)} \approx 1.3 \text{ cm}.$$

The average total cross section for fast neutrons is 7.2 b (from Table 20–1), and the mean free path is about 3 cm. A good lump size corresponds, therefore, to a diameter between 1.3 cm and 3 cm, or about one inch.

The flux of resonance neutrons is lower inside the lump than near the surface because a considerable part of the resonance absorption occurs near the surface of the lump. This effect favors large lumps but the choice of lump size depends also on the effect of lumps on the fast fission factor and the thermal utilization. The value of the fast fission factor increases with the size of the lump because the probability of a collision between a fast neutron and a U^{238} atom increases with the lump size. This effect also tends to favor larger sizes.

The main disadvantage of a heterogeneous assembly of uranium and a moderator is the decrease in the thermal utilization, as compared with that in a homogeneous mixture with the same fuel concentration. Suppose that monoenergetic thermal neutrons are incident on the surface of a uranium lump. Some of these neutrons are absorbed near the surface and the thermal flux is decreased; as the remaining neutrons penetrate the lump, more are absorbed. The thermal neutron flux is then reduced inside the lump and the probability of further absorption is decreased. The effect is directly analogous to that in resonance absorption, except that it is not desirable in the case of thermal absorption. In other words, the *self-shielding* of the uranium lump against the absorption of thermal neutrons results in a decrease in the thermal utilization.

The choice of a suitable lump size depends on the balance among the effects on p and ϵ which tend to increase the value of k_∞ and the decrease in f which tends to decrease the value of k_∞. The actual calculation of ϵ, p, and f is much more complicated for a heterogeneous assembly than for a homogeneous one. Appropriate methods can be found in the general references listed at the end of this chapter. Values up to $k_\infty = 1.075$ have been obtained for natural uranium, graphite-moderators,[(4)] and values up to 1.21 have been obtained[(5)] for natural uranium, heavy water-moderated reactors.

Cylindrical rods of natural uranium have been used in many reactors. At Brookhaven, the fuel elements were originally 1.10 in. in diameter (1.40 cm in radius) and 11 ft long; there were two such rods, covered with

TABLE 20–6

SLOWING-DOWN LENGTH OF FISSION NEUTRONS IN SEVERAL MODERATORS

Moderator	Slowing-down length, L_s		Slowing-down area, L_s^2, cm^2
	cm	in	
H_2O	5.3	2.1	28
D_2O	11.2	4.4	125
Be (density: 1.85 gm/cm^3)	9.8	3.9	96
C (density: 1.60 gm/cm^3)	19.1	7.5	364

thin aluminum jackets, in a channel through which air could be pumped to cool the uranium. The French heavy water-moderated reactor, designated as P2, has natural uranium rods 1.3 cm in radius, also clad in aluminum. The British Experimental Pile (BEPO), which is a heterogeneous, natural-uranium, air-cooled, graphite pile has rods 0.9 in. in diameter.

In a heterogeneous fuel-moderator assembly (lattice), the choice of the lattice spacing, or the distance between fuel elements, is also important. From the standpoint of neutron economy, the choice involves a balance between the slowing-down and diffusion properties of the assembly. If the moderator is a very weak absorber (D_2O or graphite), the distance between lumps should be large enough so that, on the average, the fast neutrons from one lump are slowed down to thermal energies before reaching another lump. There is a quantity L_s, called the *slowing-down length*, which can be either determined experimentally or calculated theoretically, and which is a measure of how far a fission neutron travels on the average while being slowed down to thermal energies. The values of the slowing-down length for some moderators are listed in Table 20–6, together with L_s^2, the *slowing-down area*. For reactors moderated with D_2O or graphite, we would expect the lattice spacing to be at least one slowing-down length. In the BNL reactor, the spacing is 8 in, in BEPO it is $7\frac{1}{2}$ in, in P2 it is 5.9 in, so that our estimate of the lattice spacing is a reasonable one. Ordinary water is a much stronger absorber than heavy water or graphite and a lattice spacing of one slowing-down length would permit too much absorption of slow neutrons. Hence, in a reactor with uranium enriched in U^{235} and moderated with H_2O, the rods would be quite close together and N_1/N_0 would have to be much smaller than for D_2O or graphite.[(6)]

In detailed design calculations, the choices of the lump size and the lattice spacing are not independent and must be considered together. The final choice of design parameters may also be strongly influenced by factors other than the nuclear properties of the fuel and moderator. Those factors may include the ease or efficiency of heat removal, structural properties, and effects of emphasis on a special purpose for the reactor.

20–6 The critical size of a thermal reactor. The value of the multiplication factor alone does not determine the critical size of a chain-reacting system. The multiplication factor k_∞ tells how the neutron concentration would increase from generation to generation if no neutrons leaked out of the system, and it must be combined in some way with a quantity which tells something about the neutron leakage. The mathematical theory of the critical size is an important part of reactor theory and it will be illustrated by means of a very simple problem. We shall consider a homogeneous reactor in the form of a slab, infinite in the y- and z-directions but of finite thickness in the x-direction. It will be assumed that the nuclear properties, including k_∞, are known, and the problem is to determine the critical thickness a at which the slab will sustain a chain reaction. The treatment is an extension of that of Section 18–5, where we considered the diffusion of thermal neutrons from an infinite plane source, and some of the basic ideas of the earlier treatment can be used again. In the new problem, the source is distributed uniformly throughout the slab; neutrons are both produced and absorbed throughout the slab and can leak out of the boundary of the slab. We shall assume at the start that all neutrons are thermal and monoenergetic, that the neutrons released in fission have a Maxwellian velocity distribution which can be represented by the most probable velocity. This assumption may be made if the reactor materials are $1/v$-absorbers, because of the arguments of Section 18–4. The treatment based on this assumption is called the *one-group model* because all the neutrons are taken to have the same energy. The use of the one-group model simplifies the mathematics, and the model can easily be improved to give satisfactory results for certain types of reactors.

We seek the thickness needed for the reactor to operate in a steady state. The condition for this equilibrium may be written as a neutron balance equation,

$$\text{Rate of neutron leakage} + \text{rate of neutron absorption} = \text{rate of neutron production.} \tag{20–23}$$

If we apply this balance condition to a slab of thickness dx between x and $x + dx$, we have

$$\text{Net rate of leakage } \textit{out} \text{ of element } dx = -\frac{\lambda_{\text{tr}} v}{3}\frac{d^2 n}{dx^2}\,dx, \tag{20–24}$$

where n is the neutron density, v is the constant neutron velocity; λ_{tr} is the transport cross section for the reactor material, and is given by

$$\lambda_{tr} = \frac{1}{N_1\sigma_{tr1} + N_0\sigma_{tr0}}. \tag{20–25}$$

The leakage rate gives the net number of neutrons leaving the slab of thickness dx per second across a unit area (1 cm^2) normal to dx.

The absorption rate is given by

$$\text{Rate of absorption of neutrons in element } dx = nv \textstyle\sum_a dx, \tag{20–26}$$

where

$$\textstyle\sum_a = N_0\sigma_{a0} + N_1\sigma_{a1} \tag{20–27}$$

is the total absorption cross section of the reactor material. The expression on the right of Eq. (20–26) gives the number of neutrons absorbed per second in a volume element one square centimeter in area and with thickness dx. The production rate is obtained if we note that each thermal neutron absorbed results in the emission of k_∞ new neutrons,

$$\text{Rate of production of neutrons in the element } dx = nv \textstyle\sum_a k_\infty\, dx. \tag{20–28}$$

The neutron balance equation is then

$$-\frac{\lambda_{tr}v}{3}\frac{d^2n}{dx^2} + nv \textstyle\sum_a = nv \textstyle\sum_a k_\infty, \tag{20–29}$$

or

$$\frac{d^2n}{dx^2} + \frac{3\sum_a}{\lambda_{tr}}(k_\infty - 1)n = 0. \tag{20–30}$$

We define a diffusion length L for the reactor by analogy with the definition used in Section 18–5 for a single substance,

$$L^2 = \frac{\lambda_{tr}}{3\sum_a}, \tag{20–31}$$

where λ_{tr} and $\sum_a$ are defined by Eqs. (20–25) and (20–27), respectively. The neutron balance equation then becomes

$$\frac{d^2n}{dx^2} + \frac{k_\infty - 1}{L^2}\, n = 0. \tag{20–32}$$

It is now possible to take into account the fact that fission neutrons are released as fast neutrons and must be slowed down to thermal energies.

If the reactor is *well moderated*, that is, if nearly all the fissions (say, 95%) are produced by thermal neutrons, L^2 can be replaced by

$$M^2 = L^2 + L_s^2, \tag{20–33}$$

where L_s is the slowing down length introduced in Section 20–5; M^2 is called the migration area. The balance equation then becomes

$$\frac{d^2n}{dx^2} + \frac{k_\infty - 1}{M^2}\, n = 0. \tag{20–34}$$

Equation (20–32) is the *modified one-group equation*. A more detailed theoretical treatment[7] shows that this model gives reasonably good results for well moderated reactors for which k_∞ is close to unity, or $k_\infty - 1 \ll 1$; natural uranium, graphite moderated reactors are an example.

The solution of Eq. (20–34) requires boundary conditions, and the accurate formulation of these conditions is a serious problem. We adopt the simple, although not quite accurate, condition that the neutron density must vanish at the outer boundary of the reactor. If the center, $x = 0$, of our coordinate system is taken at the central plane of the reactor, the condition is that

$$n(x) = 0 \quad \text{at} \quad x \pm \tfrac{1}{2}a. \tag{20–35}$$

If the reactor is uniform, the neutron density must be symmetric about $x = 0$; physical intuition also tells us that the neutron density should be positive throughout the interior of the reactor. The general solution of Eq. (20–35) is

$$n(x) = A \cos Bx + C \sin Bx, \tag{20–36}$$

where

$$B^2 = \frac{k_\infty - 1}{M^2}, \tag{20–37}$$

and A and C are arbitrary constants. Now, sin Bx is not symmetric about $x = 0$ so that C must vanish, and the solution is just

$$n(x) = A \cos Bx. \tag{20–38}$$

The condition (20–35) requires that

$$\frac{B_a}{2} = \frac{\pi}{2},$$

or

$$a = \frac{\pi}{B} = \frac{\pi M}{(k_\infty - 1)^{1/2}}\cdot \tag{20–39}$$

Equation (20–39) gives the critical thickness of the reactor in terms of the nuclear properties as expressed in k_∞ and $M = (L^2 + L_s^2)^{1/2}$. The neutron density distribution as a function of position is given by

$$n(x) = A \cos \frac{\pi x}{a}, \tag{20–40}$$

where A is an arbitrary constant whose value is determined by the rate of heat removal or the power level of the reactor. Equation (20–39) shows that if k_∞ is less than unity, there is no real value of the critical thickness; if $k_\infty = 1$, the critical thickness is infinite (a result consistent with the definition of k_∞); if $k_\infty > 1$, a chain reaction can be achieved in an assembly of finite size. The critical radius R_c of a spherical reactor is

$$R_c = \frac{\pi M}{(k_\infty - 1)^{1/2}}, \tag{20–41}$$

and the side of a critical cube is given by

$$A_c = \frac{\sqrt{3}\,\pi M}{(k_\infty - 1)^{1/2}}. \tag{20–42}$$

The sphere has a smaller surface-to-volume ratio than the cube, and a smaller critical volume. For the sphere, the critical volume is

$$V_c = \frac{4}{3}\pi R_c^3 = 130\left(\frac{M^2}{k_\infty - 1}\right)^{3/2}, \tag{20–43}$$

while for the cube,

$$V_c = A_c^3 = 161\left(\frac{M^2}{k_\infty - 1}\right)^{3/2}, \tag{20–44}$$

or about 24% greater than the critical volume of the sphere. The critical mass of uranium can be determined from the critical volume and the mass of uranium per unit volume; the latter is, of course, one of the basic design parameters.

It is usual to build a reactor so that its size is greater than the critical size, that is, so that $k = 1 + \delta$, where δ is called the *excess reactivity*. The excess reactivity is needed to allow neutrons to be used, either in the form of beams coming out of the reactor or by being absorbed inside the reactor to make new radionuclides, and for other reasons too involved to discuss here. The existence of the excess reactivity makes it necessary to have a *control system*, usually in the form of strong neutron absorbers such as boron-steel rods or cadmium strips. The value of the actual multiplication factor in the operating reactor can be kept at unity by proper adjustment of the control rods.

FIG. 20–3. Photograph of the model of the uranium-graphite reactor at the Brookhaven National Laboratory.

The leakage of neutrons from a reactor can be reduced by surrounding the reactor with a *reflector* made of a weak neutron absorber; graphite is often used for this purpose. The reflector knocks neutrons which leave the reactor back into it, and the decreased leakage reduces the critical size of the reactor, with a resulting saving in fuel. The reactor system, including the reflector, must be enclosed in a shield, usually of concrete, to reduce the intensity of the radiations (neutrons and γ-rays) leaving the reactor to values below the biological tolerance limits.

Figure 20–3 is a schematic diagram of the uranium-graphite reactor at the Brookhaven National Laboratory. The diagram shows how some of the features which have been discussed are incorporated into the design of an actual reactor.

20–7 Power and breeding. The possibility that a power industry based on nuclear fission will be developed raises the question of the availability and cost of fissionable material.[(8)] In the discussion so far, the fissionable materials considered have been U^{235}, Pu^{239}, and U^{233}, none of which is available naturally in large amounts. It would be inefficient to get power on a large scale from the fission of U^{235} because only 0.7% or less of the available uranium could be used directly. Plutonium has been made by the fission of U^{235} in a reactor containing U^{238}, but this process is involved and expensive. Similarly, U^{233}, which does not occur in nature, can be made in a reactor containing Th^{232}. Although the problem of availability and cost of nuclear fuels is a serious one, a unique and remarkable solution is possible because of one of the nuclear properties of the fission process. This property is the emission, on the average, of more than *two* neutrons per fission.

Suppose, for the sake of example, that 3 neutrons are emitted per fission. One of these is needed to keep the chain reaction going, to induce fission in another fuel atom, say an atom of U^{235}. Of the two neutrons that are left, one may be used to convert an atom of uranium to one of plutonium, leaving one neutron. If this last neutron could be used to produce another atom of plutonium from another atom of U^{238}, one atom of U^{235} would have been used and two atoms of Pu^{239} would be made, leaving a profit of one atom of fissionable material. In other words, more fissionable material would be produced than consumed. This example is, of course, an oversimplified one.[(9)] It is inevitable that some neutrons should be lost by leakage or by parasitic absorption, but so long as more than two neutrons are produced in fission there is a possibility that more nuclear fuel may be produced than is consumed. This process is called *breeding;* if it could be made to work, a reactor could make new fuel for itself and, in addition, a stockpile of fissionable material could be built up for use in new reactors.

The process of breeding fissionable material depends on two materials, one of which is fissionable and one of which is *fertile.* Uranium-238 and Pu^{239} form such a pair, and Th^{232} and U^{233} form another pair. Consider the first pair and suppose that enough Pu^{239} is available to achieve a chain reaction in a system which can be used to produce power. The thermal fission of Pu^{239} yields 3 neutrons, so that if some of these neutrons could be absorbed in U^{238} to form more Pu^{239}, a breeding cycle might be achieved. In this case, U^{238} is the fertile material because although it is not itself usefully fissionable, it can be converted into a good fissionable material, and it is conceivable that a reactor system could be built in which more plutonium is produced than is consumed. The nuclear property that determines whether the possibility of breeding exists is not the value of the number of neutrons emitted per fission, ν, but the number produced per neutron absorbed in the fuel, η, because the nonfission capture of neutrons by the fuel can be large enough to interfere seriously with breeding. The value of η for plutonium and thermal neutrons (Table 19–1) is 2.08 ± 0.02 and turns out to be only slightly greater than two. This result seems to make impractical the breeding of Pu^{239} in a thermal reactor. It is expected, on theoretical grounds, that the ratio of the fission cross section to the total absorption cross section should be closer to unity at high neutron energies than for slow neutrons, and that η should be close to ν in a fast reactor. Consequently, breeding of Pu^{239} should be possible in a fast neutron, Pu^{239}-U^{238} system. A reactor, the Experimental Breeder Reactor, or EBR, has been built to study the feasibility of power production and breeding in such a system. A large power-breeder reactor, the Enrico Fermi Atomic Power Plant has been built near Detroit, Michigan to combine production of electricity with the breeding of Pu^{239}.[10]

The value of η for U^{233} and thermal neutrons (Table 19–1) is 2.28 ± 0.02 which makes the breeding of U^{233} from Th^{232} in thermal reactor systems appear feasible. Research and development work are in progress on such systems.[11] One design involves the use of a solution of $U^{233}O_2SO_4$ in D_2O; another involves the use of a solution of U^{233} Bi in liquid bismuth, with graphite as moderator. The U^{233}-Th system could also be used in a fast reactor, but the greater value of ν for Pu^{239} favors the use of the latter material in fast breeder reactors.

If breeding could be achieved, U^{238} and Th^{232} could be converted into fissionable materials. The resulting increase in available nuclear fuels would be a step in the direction of economic nuclear power. Much information is needed concerning the cross sections for reactions of fast and intermediate energy neutrons with fissionable and fertile materials, and much work is being done in this field.

20–8 Energy production in stars. Thermonuclear reactions. One of the most intriguing problems in physical science is that of the source of the energy emitted by stars. The sun, which is not a particularly impressive star, emits electromagnetic energy at a rate of about 4×10^{23} ergs/sec or 2 ergs/gm/sec, and astronomical and geological evidence shows that the sun has been radiating energy at about its present rate for several billion years. Chemical reactions cannot possibly be the source of this energy, because even if the sun consisted of pure carbon its complete combustion would supply enough energy to maintain the radiation for only a few thousand years. Another possible source of energy is the conversion of gravitational energy into heat energy, a process that is analogous to the production of electricity from the falling of water, as in a hydroelectric plant. But it has been shown that if contraction were taking place it could supply not more than one percent of the total energy output needed, and if it were the only energy source, the sun could be not more than 20 million years old.

The inability to account for the energy emitted by the sun on the basis of ordinary energy sources led to the suggestion that *subatomic* or nuclear processes provide the necessary energy. Many exothermic nuclear reactions liberate energies of several Mev per nuclear particle and if all the nucleons in a gram of matter could be used, about 10^{19} ergs would be produced per gram of matter. At the rate of 2 ergs/gm/sec, reactions of this kind could supply energy for about 10^{11} years and are the only kind of reactions which could do so. The problem is to find a nuclear reaction or set of reactions which result in the observed rate of energy emission under the conditions of temperature and density inside the sun, and which are consistent with the available information about the chemical composition of the sun. When the cross sections for the reactions are known, either from experiment or theory, the rates of the reactions and the rate of energy production can be calculated.

Stars can be classified according to their luminosity, or brightness, and their radiation spectra. The spectral properties of a star are related to the *effective temperature* at which the star radiates energy; the effective temperature is a convenient way of describing the total radiation per unit area. The sun is an example of one class of stars, the so-called main sequence stars. The effective surface temperature of the sun is about 6000°K, and the internal temperature may be as high as 2×10^7 °K. The stars of the main sequence have effective surface temperatures from about 2000°K to about 50,000°K, and densities between about $\frac{1}{10}$ and 10 times that of the sun. In addition to the main sequence stars there are white dwarfs, red giants, variables, novae, and supernovae. The white dwarfs have extremely great densities, possibly 100,000 times as great as the sun, and are very faint, that is, have very small luminosities. The red giants have very low densities and high luminosities. The variable stars show recurring variations in luminosity and surface temperature; the variations may be periodic or irregular. Finally, there are novae and supernovae, which show sudden large increases in luminosity. There are wide variations in internal conditions among the different kinds of stars, in their temperature, density, pressure, and chemical composition. There may, therefore, be different mechanisms of energy production; different nuclear processes may occur in different types of stars, and more than one set of reactions may be responsible for the energy production in stars. The sun is the best known and most carefully studied star, and its energy production as calculated from nuclear reactions can be compared with well-known astrophysical data.

It can be stated first that the fission of heavy elements cannot supply the sun's energy because the abundance of these elements in the sun is much too small[(12)] to account for the rate of emission and for the lifetime of the sun. The light elements predominate, and hydrogen and helium together form about 90% by weight of the sun's matter, with approximately equal fractions of those two elements. It seems likely, from the chemical composition, that the nuclear processes should involve hydrogen and helium, and the nuclear properties of these elements support this idea. Assume, for example, that 4 protons can be combined to form a helium nucleus. The process would be exothermic, as can be seen from the mass balance,

$$
\begin{aligned}
\text{atomic mass of 4 hydrogen atoms} &= 4.13258 \text{ amu} \\
\text{atomic mass of 1 helium atom} &= \underline{4.00387} \\
\text{difference in mass} &= 0.02871 \text{ amu} \\
&= 26.7 \text{ Mev} \\
&= 42.7 \times 10^{-6} \text{ erg.}
\end{aligned}
$$

Thus, about 10^{-5} erg could be produced for each proton destroyed and, since one gram of the sun's matter contains about 2×10^{23} protons, the available energy supply would be about 2×10^{18} ergs/gm—of the order of magnitude needed.

The possibility that four protons could collide to form a helium nucleus is ruled out because the probability for such a reaction under solar conditions is too low. It seems more likely that the four protons are formed into a helium nucleus by a series of reactions, that is, in a *cyclic nuclear reaction*. The rates of the reactions depend on the number of nuclei present per unit volume and also upon the temperature; the higher the temperature, the faster the thermal motion of the particles, and the more frequent and energetic are the collisions. At *stellar* temperatures of 10 to 20 million degrees, the kinetic energies resulting from thermal motion are in the neighborhood of 1 kev, as compared with $\frac{1}{40}$ ev for particles at room temperature on earth, and the nuclei in the sun's interior are assumed to have a Maxwellian velocity distribution with a kT energy of about 1 kev. The reactions that occur under these conditions are called *thermonuclear reactions*.

Two sets of thermonuclear reactions have been proposed as sources of energy in the sun and other stars of the main sequence. One set, sometimes called the *proton-proton chain*, consists[13,14,15] of the reactions

$$
\begin{aligned}
&(\mathrm{H}^1 + \mathrm{H}^1 \rightarrow \mathrm{H}^2 + \beta^+ + \nu + 0.42 \text{ Mev}) \times 2, \\
&(\mathrm{H}^1 + \mathrm{H}^2 \rightarrow \mathrm{He}^3 + \gamma + 5.5 \text{ Mev}) \times 2, \\
&\mathrm{He}^3 + \mathrm{He}^3 \rightarrow \mathrm{He}^4 + 2\mathrm{H}^1 + 12.8 \text{ Mev.}
\end{aligned}
\tag{20–45}
$$

For the third reaction to occur, each of the first two reactions must occur twice. The effect of the reactions is

$$
4\mathrm{H}^1 \rightarrow \mathrm{He}^4 + 2\beta^+ + 2\gamma + 2\nu, \tag{20–46}
$$

with a total energy release of 26.7 Mev, as calculated previously; when the kinetic energy of the neutrinos is subtracted, the energy is 26.2 Mev.

The positrons emitted are annihilated by free electrons with the production of γ-rays.

Another proton-proton chain consists of the reactions

$$\begin{aligned} &H^1 + H^1 \rightarrow H^2 + \beta^+ + \nu, \\ &H^1 + H^2 \rightarrow He^3 + \gamma, \\ &He^3 + He^4 \rightarrow Be^7 + \gamma. \end{aligned} \tag{20–47a}$$

$$\begin{aligned} &Be^7 + \beta^- \rightarrow Li^7 + \nu + \gamma, \\ &Li^7 + H^1 \rightarrow He^4 + He^4. \end{aligned} \tag{20–47b}$$

$$\begin{aligned} &Be^7 + H^1 \rightarrow B^8 + \gamma, \\ &B^8 \rightarrow Be^8 + \beta^+ + \nu, \\ &Be^8 \rightarrow He^4 + He^4. \end{aligned} \tag{20–47c}$$

It is thought[(15)] that the chain (20–45) is important at lower temperatures, corresponding to those in the sun when it was first formed, and that the chain (20–47) is more important in the present state of the sun, with its higher central temperature and larger He^4 concentration.

The production of one helium nucleus from four protons is an example of a process called *fusion* in which a heavier element is built up from one or more lighter elements. When this process occurs among the light elements, energy is usually released because the mass of the product nucleus is less than the sum of the masses of the nuclei which are fused. This property of the light elements is shown by the binding-energy curve, Fig. 9–11, in which the binding energy rises rapidly with mass number in the region of small mass numbers. Under appropriate conditions, reactions such as those discussed can liberate vast amounts of energy, amounts much greater than that released in an atom-bomb explosion, and thermonuclear reactions form the basis of the so-called thermonuclear or hydrogen bomb.

Another set of reactions, the well-known *carbon-nitrogen cycle* was proposed by Bethe[(16)] to account for energy production in the sun and other stars of the main sequence. Bethe showed that reactions starting with Li, Be, B, C, N, and F have mean reaction times less than 10^9 years and had to be ruled out, while the calculated rates of energy production involving O, Ne, Mg, and other light elements are too low. But the reactions involving carbon and nitrogen were found to have the remarkable property that they could be formed into a cycle in which the carbon and nitrogen nuclei are not used up, but are regenerated. These nuclei act as catalysts in a series of reactions in which 4 protons are converted into a helium nucleus and about 26 Mev of energy are liberated. The sequence of reactions is

$$\begin{aligned}
&C^{12} + H^1 \rightarrow N^{13} + \gamma, \\
&N^{13} \rightarrow C^{13} + \beta^+ + \nu, \\
&C^{13} + H^1 \rightarrow N^{14} + \gamma, \\
&N^{14} + H^1 \rightarrow O^{15} + \gamma, \\
&O^{15} \rightarrow N^{15} + \beta^+ + \nu, \\
&N^{15} + H^1 \rightarrow C^{12} + He^4.
\end{aligned} \tag{20–48}$$

The mean lifetime of hydrogen in the cycle as a whole varies rapidly with the temperature but is in the correct range for the sun and other main sequence stars.

For some years it was thought that the C-N cycle was responsible for nearly all of the solar energy production but, because of recently obtained nuclear data, the proton chain is now considered to be more important in the sun than the carbon cycle. It is thought that the carbon cycle produces more energy in main sequence stars which are much more luminous than the sun and whose central temperatures are higher, while the proton chain is more important for main sequence stars less luminous than the sun.

The application of nuclear physics to the problem of energy production in stars depends on nuclear data and on astrophysical data and calculations. The latter have to do with such matters as structure, temperature distribution, chemical composition, and density. Knowledge of the nuclear reactions has reached the stage where the nuclear calculations are probably more reliable than the astrophysical calculations. If the energy production is calculated for different stellar models and the results compared with observations, deductions can be made about the astrophysical aspects of stars, and some of the present uncertainties about stellar models should be removed in the near future.

Nuclear reactions have been suggested for stars which differ greatly from the sun. It is probable that there are stars which were like the sun but had such a high luminosity and rate of conversion of hydrogen into helium that they have already exhausted their hydrogen supply. A star of this kind would be expected to undergo gravitational contraction until its central density and temperature become very great ($T \sim 2 \times 10^8$ °K). The following reactions could then occur:

$$\begin{aligned}
&He^4 + He^4 + 95 \text{ kev} \rightarrow Be^8, \\
&He^4 + Be^8 \rightarrow C^{12} + 7.4 \text{ Mev}.
\end{aligned} \tag{20–49}$$

When a C^{12} nucleus has been produced, it can undergo another (α, γ) reaction, giving an O^{16} nucleus and about 7 Mev, and so on. In this way, fusion reactions among nuclei heavier than helium may produce energy in some stars.

20–9 Controlled thermonuclear reactions. The fact that nuclear fusion reactions can release great amounts of energy, as in stars, has led to the search for practical means for the controlled release of such energy. The reactions of the proton-proton chain (20–45) occur extremely slowly, which is fortunate because the sun can then have a reasonably long life. There are thermonuclear reactions which occur much more rapidly and depend on an abundant material, deuterium. The commercial separation of deuterium (H^2 or D) from hydrogen permits the use of the following reactions:

$$H^2 + H^2 \rightarrow H^3 + H^1 + 4\ \text{Mev}, \qquad (20\text{–}50)$$

$$H^2 + H^2 \rightarrow He^3 + n + 3.3\ \text{Mev}, \qquad (20\text{–}51)$$

$$H^2 + H^3 \rightarrow He^4 + n + 17.6\ \text{Mev}, \qquad (20\text{–}52)$$

$$H^2 + He^3 \rightarrow He^4 + H^1 + 18.3\ \text{Mev}. \qquad (20\text{–}53)$$

The D-D reaction can go in two equally probable ways, the first of which produces tritium (H^3 or T) while the second produces He^3. If the D-D reactions take place in a chamber, and if the products remain in the chamber, deuterium can react with tritium, Eq. (20–52), to give an α-particle and a neutron, or with He^3 to give an α-particle and a proton. The net result of the reactions is the burning of six deuterium nuclei and the formation of two helium nuclei, two neutrons, and two protons. The detailed energy balance shows that of the 17.6 Mev of kinetic energy released in the reaction (20–52), 14.1 Mev are carried off by the neutron and 3.5 Mev by the α-particle; of the 18.3 Mev released in the reaction (20–53), 14.7 Mev are carried off by the proton and 3.6 by the α-particle. The total energy output per deuteron burned is approximately 7.2 Mev of which the neutrons carry off 2.76 Mev.

The energy released in the above fusion reactions is much less than that released in a fission reaction but the energy yield per unit mass of material, e.g., per gram, is slightly greater from fusion than from fission. Deuterium occurs in nature with an abundance of about one part in six thousand of hydrogen, and can be separated from the lighter isotope quite cheaply. One gallon of water contains about one-eighth of a gram of deuterium, and the cost of its separation is less than 4 cents; but its energy content if it could be burned as a fuel in a thermonuclear reactor would be the equivalent of about 300 gallons of gasoline. The total amount of deuterium in the oceans is estimated to be about 10^{17} lb, and the energy content of this fuel would be 10^{20} kilowatt-years, so that the controlled use of thermonuclear energy might provide a practically infinite source of energy with a fuel of negligible cost. There are, however, some difficult problems to be solved, and some of these will be discussed briefly.

The nuclei which interact in the fusion processes are positively charged and, therefore, repel one another. They must be made to collide with enough relative velocity to overcome the Coulomb barrier which tends to keep them apart. It can be shown that this can occur when the interacting particles have energies of about 0.1 Mev or more. Since the energy released in the reactions is between 3 Mev and 18 Mev, the net energy gain in any one fusion process is appreciable and most of the gain appears in the kinetic energies of the neutron and proton. The problem that must be solved is that of making the reactions go in such a way that useful energy can actually be obtained. Thus, the nuclear reactions can be produced in the laboratory by accelerating deuterons or tritons on to a target; but this process will not yield a net gain of energy because the energy is lost in ionizing the target material. To produce a net gain, the target must be ionized before the reactions occur. But, if a beam of deuterons is incident on an ionized target, the probability that the deuterons will undergo elastic scattering is much greater than the probability of fusion. To achieve fusion, the deuterium nuclei (or deuterium and tritium nuclei) must be confined somehow in a region where they can undergo many collisions, enough so that fusion eventually occurs. It is necessary, therefore, for the nuclei to be in a container and moving at high kinetic energies with respect to one another; in other words, the particles must be in a container at very high temperature. Some idea of the temperatures required can be obtained from Fig. 20–4 which shows the cross sections for the D-D and D-T reactions as functions of the deuteron energy. The probability of fusion is very small at low energies but rises rapidly as the energy of the interacting particles increases toward 100 kev. The relation $E = kT$ implies that 1 ev of energy corresponds to a temperature of 1.16×10^4 °K, and 10 kev corresponds roughly to 10^8 °K. Consequently, the interacting particles must be contained under conditions at which their thermal energies correspond to temperatures of the order of 10^8 °K.

At the temperatures required for fusion, the atoms are entirely stripped of their electrons. The result is a completely ionized gas, or *plasma* consisting of atomic nuclei (deuterons, tritons, protons, etc.,) and electrons in rapid random motion. No wall of ordinary material can, of course, contain such a plasma. In the sun, the proton fusion reactions are contained by a tremendous gravitational pressure. Such a high pressure is not available in experiments on controlled thermonuclear reactions, but the plasma can be contained in a magnetic field. The problem then is to

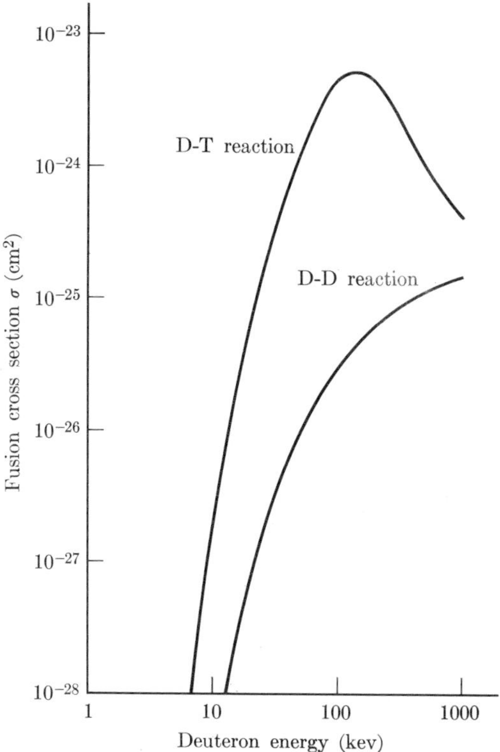

Fig. 20–4. Fusion cross section as a function of deuteron energy. (From A. S. Bishop, *Project Sherwood—The U.S. Program in Controlled Fusion*. Reading, Mass.: Addison-Wesley, 1958.)

devise an apparatus in which the plasma can be contained by means of a magnetic field at the temperature required for the fusion reactions to proceed. The plasma pressure must be kept from becoming too high, which requires that the density of the plasma be low, in the range from 10^{14} to 10^{18} particles/cm^3.

Another requirement for the controlled release of useful thermonuclear energy is that the reaction be self-sustaining; once the plasma temperature has reached the point at which fusion occurs at an appreciable rate, the energy generated in the plasma must be at least high enough to maintain that temperature. This energy includes the energy of the protons and α-particles produced by the reactions (since these particles are contained by the magnetic field), but does not include the kinetic energy of the neutrons which escape from the magnetic container. The plasma loses energy by radiation, for example, by bremsstrahlung resulting chiefly from collisions between electrons and ions. The rates of both power generation and power loss by radiation increase with temperature; but the generation rate increases faster with the result that a certain critical temperature will exist above which the reactions should be self-sustaining. According to theory, in the case of the D-D reaction, this so-called *ignition temperature* is about 4×10^8 °K; for the D-T reaction it is about 4.5×10^7 °K.

The cross sections shown in Fig. 20–4 and the ignition temperatures just cited indicate that the D-T reaction is more favorable than the D-D reaction. Although deuterium is plentiful in nature, tritium does not occur naturally. But tritium can be made, for example, by bombarding lithium with slow neutrons in a reactor,

$$\mathrm{Li}^6 + \mathrm{n} \rightarrow \mathrm{H}^3 + \mathrm{He}^4 + 4.8\ \mathrm{Mev}. \tag{20–54}$$

This reaction can serve two useful purposes. In a thermonuclear reactor burning a mixture of deuterium and tritium, the neutrons which escape from the container carry off much of the energy released. This energy may be converted to heat by slowing-down the neutrons in a *blanket* surrounding the reactor in which the plasma is contained. The slow neutrons could then be captured in lithium, producing tritium in the process. Thus the tritium would be recovered, and deuterium and lithium consumed. The blanket might consist of moderator, coolant, and lithium. The heat generated by the moderation and absorption of the neutrons could then be transferred by the coolant to external heat exchangers and equipment for generating electricity.

The problem of producing and maintaining deuterium plasmas at the temperatures required is an extremely difficult one, and the major effort in the field of thermonuclear research is the investigation of the behavior of totally ionized gases (plasmas) at ultra-high temperatures. Although the problem of confining a plasma and heating it to thermonuclear temperatures has not been solved, extensive experimental and theoretical research programs are under way. Details can be found in the general references listed at the end of the chapter.

The necessarily brief discussion of thermonuclear reactions has indicated some of the ways in which nuclear physics can be applied to astrophysics and fusion. Another interesting application, which cannot be treated here, is in the problem of the origin and abundance of the elements.[(17)]

References

GENERAL

FISSION

H. D. Smyth, *Atomic Energy for Military Purposes.* Princeton: Princeton University Press, 1945; also, *Revs. Mod. Phys.* **17,** 351 (1945).

A. M. Weinberg and E. P. Wigner, *The Physical Theory of Neutron Chain Reactors.* Chicago: University of Chicago Press, 1958.

S. E. Liverhant, *Elementary Introduction to Nuclear Reactor Physics.* New York : Wiley, 1960.

S. Glasstone and M. C. Edlund, *The Elements of Nuclear Reactor Theory.* New York: Van Nostrand, 1952.

R. L. Murray, *Nuclear Reactor Physics.* Englewood Cliffs, New Jersey: Prentice-Hall, 1957.

D. J. Littler and J. F. Raffle, *An Introduction to Nuclear Reactor Physics,* 2nd ed. London: Pergamon Press, 1957.

R. L. Murray, *Introduction to Nuclear Engineering.* Englewood Cliffs, New Jersey: Prentice-Hall, 1954.

R. Stephenson, *Introduction to Nuclear Engineering,* 2nd ed. New York: McGraw-Hill, 1958.

R. V. Meghreblian and D. K. Holmes, *Reactor Analysis.* New York: McGraw-Hill, 1960.

C. F. Bonilla, ed., *Nuclear Engineering.* New York: McGraw-Hill, 1957.

H. Etherington, ed., *Nuclear Engineering Handbook.* New York: McGraw-Hill, 1958.

S. Glasstone, *Principles of Nuclear Reactor Engineering.* New York: Van Nostrand, 1955.

J. R. Dietrich and W. H. Zinn, eds., *Solid Fuel Reactors.* Reading, Mass.: Addison-Wesley, 1958.

J. A. Lane, H. G. MacPherson, and F. Maslan, eds., *Fluid Fuel Reactors.* Reading, Mass.: Addison-Wesley, 1958.

Proceedings of the International Conference on the Peaceful Uses of Atomic Energy, Geneva, 1955. New York: United Nations, 1956.

Vol. 2. *Physics; Research Reactors.*
Vol. 3. *Power Reactors.*
Vol. 5. *Physics of Reactor Design.*

Proceedings of the Second International Conference on the Peaceful Uses of Atomic Energy, Geneva, 1958. Geneva: United Nations, 1958.

Vol. 8. *Nuclear Power Plants,* Part 1.
Vol. 9. *Nuclear Power Plants,* Part 2.
Vol. 10. *Research Reactors.*
Vol. 12. *Reactor Physics.*
Vol. 16. *Nuclear Data and Reactor Theory.*

FUSION

R. F. Post, "Controlled Fusion Research—An Application of the Physics of High Temperature Plasmas," *Revs. Mod. Phys.* **28,** 338 (1956).

R. F. POST, "High-temperature Plasma Research and Controlled Fusion," *Ann. Rev. Nuc. Sci.* **9,** 367 (1959).

A. S. BISHOP, *Project Sherwood—The U.S. Program in Controlled Fusion.* Reading, Mass.: Addison-Wesley, 1958.

A. SIMON, *An Introduction to Thermonuclear Research.* London: Pergamon Press, 1959.

D. J. ROSE and M. CLARK, JR., *Plasmas and Controlled Fusion.* Cambridge, Mass.: Technology Press of M.I.T.; New York: Wiley, 1961.

Proceedings of the Second International Conference on the Peaceful Uses of Atomic Energy, Geneva, 1958. Geneva: United Nations, 1958.

Vol. 31. *Theoretical and Experimental Aspects of Controlled Nuclear Fusion.*
Vol. 32. *Controlled Fusion Devices.*

W. P. ALLIS, ed., *Nuclear Fusion.* New York: Van Nostrand, 1960.

C. LONGMIRE, J. L. TUCK, and W. B. THOMPSON, eds., *Progress in Nuclear Energy, Series XI. Plasma Physics and Thermonuclear Research,* Vol. 1. London: Pergamon Press, 1959.

PARTICULAR

1. H. S. ISBIN, "Catalogue of Nuclear Reactors," *Proceedings of the Second International Conference on the Peaceful Uses of Atomic Energy, Geneva, 1958, op. cit.* gen. ref., **8,** 561 (1958).
2. A. M. WEINBERG and E. P. WIGNER, *op. cit.* gen. ref., p. 312.
3. "Research Reactors," *U.S.A.E.C. Report T.I.D. 5275.* Washington, D.C.: U.S. Government Printing Office, 1955; New York: McGraw-Hill, 1955, Chap. 1.
4. I. KAPLAN and J. CHERNICK, "Uranium-Graphite Lattices-The Brookhaven Reactor," Paper No. 606, *Proceedings of the International Conference on the Peaceful Uses of Atomic Energy, Geneva 1955, op. cit.* gen. ref., **5,** 295 (1956).
5. J. HOROWITZ, "Studies of the Neutronics of Two French Heavy Water Reactors," Paper No. 361, *Proceedings of the International Conference on the Peaceful Uses of Atomic Energy, Geneva 1955, op. cit.* gen. ref., **5,** 256 (1956).
6. KOUTS, PRICE, DOWNES, SHER, and WALSH, "Exponential Experiments with Slightly Enriched Uranium Rods in Ordinary Water," Paper No. 600, *Proceedings of the International Conference on the Peaceful Uses of Atomic Energy, Geneva 1955, op. cit.* gen. ref., **5,** 183 (1956).
7. R. L. MURRAY, *Nuclear Reactor Physics, op. cit.* gen. ref., Chapter 3.
8. J. A. LANE, "Economics of Nuclear Power," *Ann. Rev. Nuc. Sci.* **9,** 473 (1959).
9. F. T. MILES and I. KAPLAN, "Optimizing and Comparing Reactor Designs," *Chemical Engineering Progress Symposium Series,* Vol. 50, No. 11, p. 159 (1954).
10. J. R. DIETRICH and W. H. ZINN, eds., *op. cit.* gen. ref., Chapters 2, 3, 4.
11. LANE, MACPHERSON, and MASLAN, eds., *op. cit.* gen. ref., Parts I, III.
12. H. E. SUESS and H. C. UREY, "Abundances of the Elements," *Revs. Mod. Phys.* **28,** 53 (1956).
13. H. A. BETHE and C. L. CRITCHFIELD, "The Formation of Deuterons by Proton Combination," *Phys. Rev.* **54,** 248 (1938).

14. E. E. SALPETER, "Nuclear Reactions in the Stars. I. Proton-Proton Chain," *Phys. Rev.* **88**, 547 (1952).

15. A. G. W. CAMERON, "Nuclear Astrophysics," *Ann. Rev. Nuc. Sci.* **8**, 299 (1958).

16. H. A. BETHE, "Energy Production in Stars," *Phys. Rev.* **55**, 434 (1939).

17. BURBIDGE, BURBIDGE, FOWLER, and HOYLE, "Synthesis of the Elements in Stars," *Revs. Mod. Phys.* **29**, 547 (1957).

PROBLEMS

1. A thermal nuclear reactor containing a mixture of U^{235} and U^{238} operates at a power level of 1000 megawatts. At what rate is U^{235} consumed by fission? What is the total rate of consumption of U^{235}? What is the maximum rate at which Pu^{239} might be produced if all neutrons released in fission over those required to maintain fission were absorbed by U^{238}?

2. What is the smallest value of the ratio of the number of U^{235} atoms to moderator atoms, in a homogeneous mixture of uranium and graphite, for which a thermal neutron chain reaction will be possible in an infinitely large assembly?

3. What is the smallest value of the ratio of the number of U^{235} atoms to moderator atoms, in a homogeneous mixture of U^{235} and beryllium, for which a thermal neutron chain reaction will be possible in an infinitely large assembly?

4. What is the smallest value of the ratio of the number of U^{235} atoms to water molecules in a homogeneous mixture of U^{235} and water for which a thermal neutron chain reaction will be possible in an infinitely large system?

5. What is the smallest value of the ratio of the number of U^{235} atoms to D_2O molecules in a homogeneous mixture of U^{235} and D_2O for which a thermal neutron chain reaction will be possible in an infinitely large system?

6. The migration area in a homogeneous mixture of uranium and moderator may be written as

$$M^2 = \tau + \frac{L_M^2}{1 + (N_u\sigma_{au}/N_M\sigma_{aM})},$$

where τ, the "age" of the thermal neutrons, is the square of the slowing-down length in the moderator, and L_M is the thermal neutron diffusion length in the moderator. From the known values of the cross section, the diffusion lengths, and the slowing-down lengths, calculate the migration area corresponding to the solution to each of Problems 2, 3, 4, and 5.

7. Calculate the ratio of U^{235} atoms to moderator atoms needed to make $k_\infty = 1.10$ in each of the cases of Problems 2, 3, 4, and 5. Then find the corresponding values of the migration area, critical radius, and critical mass of a spherical reactor in each case.

8. If n_0 is the value of the neutron density at the central plane of an infinite slab reactor and $\bar{n}$ is the average value, show that

$$\bar{n} = \frac{2}{\pi} n_0.$$

9. Derive the differential equation for the neutron density in a spherical reactor on the modified one-group model. Then show that the neutron density is given by

$$n(r) = A \frac{\sin \pi r/R}{r},$$

where A is an arbitrary constant. Evaluate A in terms of n_0, the neutron density at the center of the sphere, and calculate the value of the ratio $\bar{n}/n_0$. [*Hint:* See Problem 9, Chapter 18.]

10. It can be shown that the neutron density in a reactor which is a rectangular parallelopiped is given by

$$n(x_1 y_1 z) = n_0 \cos \frac{\pi x}{a} \cos \frac{\pi y}{b} \cos \frac{\pi z}{c},$$

where a, b, c are the sides, and the center of the reactor is at (0, 0, 0). Find the ratio of the average neutron density to the maximum neutron density. Compare the result with that for a spherical reactor.

11. Calculate the Q-value of each reaction in the carbon-nitrogen cycle and find, from the results, the amount of energy, in Mev, liberated in the formation of one helium atom.

CHAPTER 21

THE ACCELERATION OF CHARGED PARTICLES

The development of machines for the acceleration of charged particles is closely related to the advance of nuclear physics and is an example of the interplay between the invention of new instruments and the progress of physical science. Inventions like the cloud chamber, the Geiger-Mueller counter, the cyclotron, and the proton synchrotron extend the range of phenomena which can be studied quantitatively, and the new information obtained leads to new developments in theory. The theoretical advances lead, in turn, to predictions about phenomena which lie outside the available range of experiment and inspire improvements in existing instruments as well as the invention of new ones.

The design and construction of accelerators depend strongly on electrical and mechanical engineering, and are not, in themselves, a part of nuclear physics. But accelerators are intimately connected with nuclear physics and are far enough from ordinary industrial engineering so that they are usually regarded as a part of nuclear physics. The subject of particle accelerators will therefore be discussed briefly, with the object of presenting some of the basic ideas underlying the design of accelerators rather than a detailed treatment of accelerator design.

21–1 The Cockcroft-Walton machine. Charged particles can be accelerated by applying a steady potential between a source of particles and a suitable electrode. A hot filament can be used as a source of electrons, while a gaseous discharge tube containing hydrogen, deuterium, or helium can supply protons, deuterons, or α-particles, respectively. When a discharge tube is used, the particles can emerge from a small hole in the tube. The ion source is placed at one end of an evacuated accelerating tube and the target or exit hole for the beam is at the other end. The problem is to make the potential across the accelerating tube as large as feasible. One of the simplest and cheapest methods of getting a potential which can accelerate charged particles to energies great enough to cause nuclear transmutations was devised by Cockcroft and Walton,[(1)] and protons accelerated by means of their machine were used in the first artificial disintegration induced by laboratory-accelerated particles, the $Li^7(p, \alpha)He^4$ reaction.[(2)]

Cockcroft and Walton used a condenser-rectifier voltage multiplier based on a principle illustrated in Fig. 21–1. Three condensers, K_1, K_2, and K_3, each of capacity C, are connected in series, and condenser K_3 is connected to a source of steady potential E. If two other condensers X_1 and X_2 are connected to condensers K_1, K_2, and K_3, first as shown by the dotted lines S_1, S_2, and S_3, and then as shown by the full lines S_1', S_2', S_3', then in the first cycle, when X_1 and X_2 are connected to K_2 and K_3, condenser X_2 is charged to voltage E. When the switches are moved to the upper position, condenser X_2 shares its charge with condenser K_2 and both are charged to $E/2$ if they have equal capacity. In the next reversal of the switches, condensers K_2 and X_1 are connected and take up potentials $E/4$, while condenser X_2 is recharged to potential E. The process can be continued and charge can be transferred gradually to all the condensers until, in the absence of loss, a potential $3E$ is developed across the condensers K_1, K_2, K_3 in series. By adding more condensers, large multiples of a given steady voltage can be obtained.

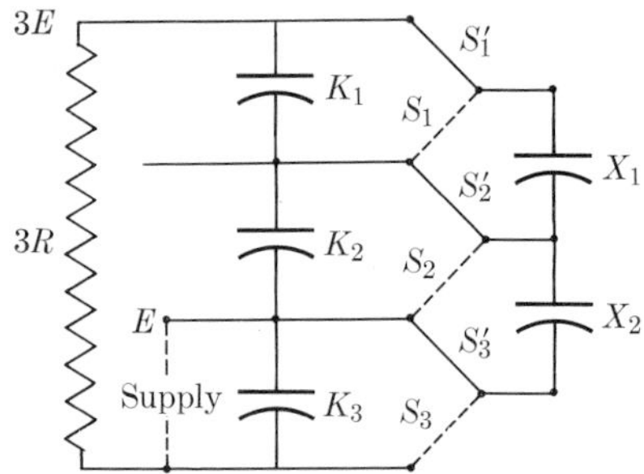

FIG. 21–1. The principle of the Cockcroft-Walton condenser-rectifier voltage multiplier.[(1)]

Although the system just described shows the idea underlying the Cockcroft-Walton machine, the actual apparatus they used was more complicated. An alternating voltage from a transformer was applied and the switching action was accomplished by the use of vacuum-tube rectifiers. By using their voltage multiplier in conjunction with an accelerating tube, Cockcroft and Walton succeeded in getting protons with energies up to 800 kev in their early apparatus. Particle energies greater than 1 Mev have been obtained with later forms of the apparatus.

The Cockcroft-Walton machine is a relatively simple one, with no moving parts; the maximum energies which can be obtained are low compared with those from other accelerators, but it provides fairly large ion currents at constant voltage, and is very useful for experimental work at moderate particle energies, especially below 1 Mev. As an example, the $H^2(d, n)H^3$ reaction, in which "heavy ice" is bombarded with deuterons accelerated in a machine of this kind, is a good source of fast neutrons because good yields can be obtained with bombarding energies as low as 200 kev.

Newer types of high voltage rectifier equipment have also been developed[3] because of the continued usefulness of relatively low-energy, but simple, machines.

21–2 The electrostatic generator or Van de Graaff machine. To increase the probability of nuclear transmutations and the yield of artificial nuclides, it is necessary that the charged particles used as the projectiles be available in adequate amounts and with enough energy to penetrate the nucleus. The quantitative study of nuclear reactions requires, in addition, that the particles be homogeneous in energy, and that they emerge from the apparatus in which they are produced in a parallel beam and with little accompanying stray radiation. Furthermore, it should be possible to measure the energy of the particles accurately and to vary it over a wide range. These requirements can be met, in the energy range up to about 10 Mev, by the combination of an electrostatic generator, or Van de Graaff machine, and an accelerator tube.

The design of the Van de Graaff generator[4,5,6] is based on the principle that if a charged conductor is brought into internal contact with a hollow second conductor, all of its charge transfers to the hollow conductor no matter how high the potential of the latter may be.[7] Except for insulation difficulties, the charge, and hence the potential, of a hollow conductor could be raised to any desired value by successively adding charges to it by internal contact. The conductor must be supported in some way, and its maximum potential is limited to that at which the rate of leakage of charge from it through its supports or through the surrounding air equals the rate at which charge is delivered to it.

A schematic diagram of a Van de Graaff generator operated to accelerate positive ions is shown in Fig. 21–2. One terminal, which consists of a number of sharp points projecting from a horizontal rod, is maintained at a positive potential of about 5000 to 20,000 volts relative to ground. This terminal is indicated in the diagram by the *spray points*. Because of the large electrostatic field in the air near the points, ions (positive and negative) are formed, and a spray of positive ions is repelled from the points. These positive ions attach themselves to the surface of a flexible belt made of paper, silk, rayon, or some other nonconducting material. The belt passes over pulleys, the lower pulley being driven by a motor, while the upper pulley is an idler. The positive ions are carried upward by the belt toward the metal shell that forms the upper terminal. Another set of points draws off the charge and transfers it to the shell. Negative ions are carried downward by the belt and migrate to the spray points. The charge-transfer process is repeated, and the potential of the shell rises

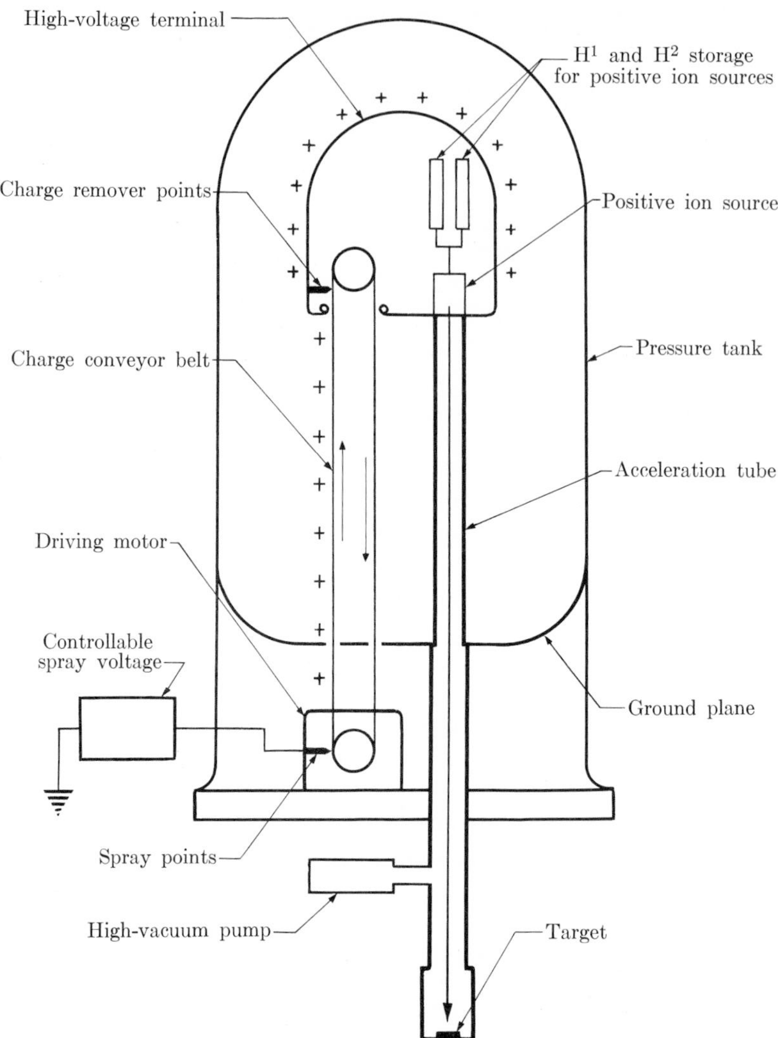

FIG. 21–2. Schematic diagram of a Van de Graaff generator operated to accelerate positive ions.

until the rate of loss of charge by leakage equals the rate at which charge is introduced. The leakage can be minimized and the maximum attainable voltage increased by enclosing the apparatus in a gas-tight steel chamber and operating at pressures up to about 15 atmospheres. The gas in the chamber may be air, nitrogen, methane, or freon.

A gas discharge tube, which is the source of the positive ions used as projectiles, is inside the shell and the ions are accelerated down the tube onto the target located at the low-voltage end of the tube. The accelerating tubes are usually made up of sections of glass, porcelain, or other insulating material joined end to end by vacuum-tight seals, and must be long enough to avoid the possibility of a spark or other discharge passing from one end to the other when the potential is applied.

In the first Van de Graaff machine the maximum voltage attained was 1.5 Mev. In later machines[3,8,9] this voltage was increased to 5 Mev, and the newest machines should reach still higher energies, possibly 20 Mev.[45] These machines give good supplies of charged particles, at voltages which can be held constant within about 0.1%. Proton beams from these machines have been used to study proton-proton scattering, to study nuclear reactions, to make new nuclides, and to produce beams of monoenergetic neutrons. The Van de Graaff generator can also be used to accelerate electrons.[10]

21-3 The cyclotron. The methods discussed so far for obtaining energetic charged particles depend on the direct use of high voltages and are subject to certain practical limitations. The experimental problems increase rapidly with increasing voltage because there are difficulties of insulation and corona discharge. Lawrence and Livingston[11] developed a method that avoids these difficulties by means of the multiple acceleration of ions to high speeds without the use of high voltage. Their first instrument was the earliest version of the *magnetic resonance accelerator* or *cyclotron*, which is the most famous of all accelerators.[12,13,14]

The cyclotron consists of two flat, semicircular metal boxes, called "dees" or "D's" because of their shape. These hollow chambers have their diametric edges parallel and slightly separated from each other, as shown in Fig. 21-3. A source of ions is located near the midpoint of the gap between the dees. The dees are connected to the terminals of a radiofrequency oscillator so that a high-frequency alternating potential of several million cycles per second is applied between the dees, which act as electrodes. In this way, the potential between the dees is made to alternate rapidly, some millions of times per second, and the electric field in the gap is directed first toward one dee and then toward the other. The space within each dee is, however, a region of zero electric field because of the electrical shielding effect of the dees. The dees are enclosed within, but insulated from, a larger metal box containing gas at low pressure, and the whole apparatus is placed between the poles of a strong electromagnet which provides a magnetic field perpendicular to the plane of the dees.

When an ion with a positive charge q is emitted from the source, it is accelerated by the electric field between the dees toward the dee which is negative at that instant. Since there is a uniform magnetic field H acting at right angles to the plane of the dees, the ion travels in a circular path

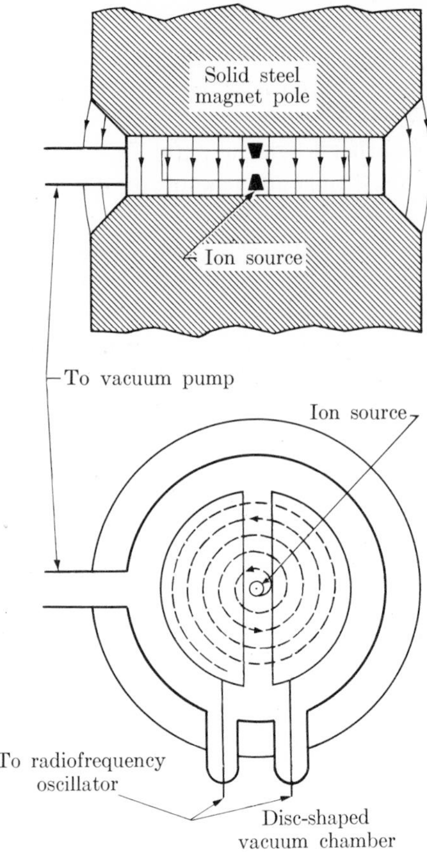

FIG. 21-3. The accelerating chamber in a cyclotron.

of radius r_1, given by

$$r_1 = \frac{mv_1}{Hq}, \tag{21-1}$$

where m is the mass of the ion, H is the strength of the magnetic field, and v_1 is the speed of the ion. While the ion is inside the dee, its speed stays constant, but after describing a semicircle through the dee, the ion reaches the gap, where it is again subject to the action of the applied potential difference. If now, in the time needed for the ion to complete the semicircle, the electric field has reversed so that its direction is toward the second dee, the ion is again accelerated as it crosses the gap between the dees and enters the second dee with an increased speed v_2. It then moves in a semicircle of larger radius r_2 within the second dee to emerge again into the gap.

The angular velocity of the ion is

$$\omega = \frac{v}{r} = H\,\frac{q}{m}\,, \tag{21–2}$$

where v is the velocity of the ion in a semicircle of radius r. It is seen that the angular velocity is independent of the speed of the ion and of the radius of the circle in which it travels, depending only on the strength of the magnetic field and the charge-to-mass ratio (q/m) of the ion. If, therefore, the electric field reverses at regular intervals, each equal to the time needed for the ion to make a half revolution, the field in the gap is always in the right direction to accelerate an ion each time the gap is crossed. In other words, if the oscillation frequency is adjusted to the properties of the given ion and to the strength of the magnetic field, the charged particles always keep in phase with the changes of electric potential between the dees. Each time the ions cross the gap they receive an additional kick, with the result that the energy of the ions is steadily increased, and the ions describe a flat spiral of increasing radius. Eventually, the ions reach the periphery of the dee, where they can be brought out of the chamber by means of a deflecting plate charged to a high negative potential. The attractive force acting on the positive ions draws them out of their spiral paths, and they can be used to bombard a properly placed target.

The energies of the ions produced by a cyclotron can be calculated from the equation for the motion of an ion in a magnetic field,

$$Hqv = \frac{mv^2}{R}\,, \tag{21–3}$$

where R is the radius at which the particles leave the machine, and H and q are expressed in electromagnetic units. After an ion has been accelerated, its energy is given by

$$qV = \tfrac{1}{2}mv^2, \tag{21–4}$$

where V is the equivalent potential difference (also in emu) through which the ion has been accelerated. When v is eliminated from Eq. (21–4) by inserting its value from Eq. (21–3), the result is

$$V = \frac{1}{2}\,H^2R^2\,\frac{q}{m}\,\cdot \tag{21–5}$$

If q is in esu and V in volts, then

$$V = \frac{1}{2\times 10^8}\,H^2R^2\,\frac{q}{mc} = H^2R^2\,\frac{q}{m}\,(16.7\times 10^{-20}), \tag{21–6}$$

since c is the velocity of light. The value of q/m for a proton is 2.87×10^{14}

esu/gm, and, since H and r are known from the design, the equivalent voltage of the particles can be calculated.

The value of the magnetic field strength cannot be chosen arbitrarily but depends on the frequency of the voltage applied to the dees, and must be chosen so as to give resonance. For resonance, the time taken for an ion, traveling with an angular velocity ω, to describe a semicircle within one of the dees must be equal to half the time period T of oscillation of the applied high frequency voltage; that is,

$$\frac{\pi}{\omega} = \frac{T}{2},$$

and the frequency n is

$$n = \frac{1}{T} = \frac{\omega}{2\pi} = \frac{H}{2\pi}\frac{q}{m}. \tag{21–7}$$

For ions with a given value of q/m, the magnetic field needed to give resonance for any given frequency of the applied voltage is given by Eq. (21–7). The higher the frequency, the stronger must be the magnetic field, and therefore from Eq. (21–6), the greater will be the equivalent voltage through which the ions will have been accelerated at any given radius. With a radiofrequency voltage of 10^7 cycles/sec, the magnetic field must be about 6500 gauss for protons and 13,000 gauss for deuterons. When these values are inserted into Eq. (21–6), and if the radius is taken to be 30 cm, the resulting values of V are 1.8 million volts for protons and 3.6 million volts for deuterons. If the radius is doubled, the equivalent voltage through which the ions are accelerated is raised by a factor of four.

The voltage applied to the dees does not appear in Eq. (21–6), so that the energy which a charged particle can acquire in a particular cyclotron is independent of this voltage. When the voltage is small, the ion makes a large number of turns before reaching the periphery; when the voltage is high, the number of turns is small. It is this property of the cyclotron that made it possible to accelerate charged particles to relatively high energies by means of small applied voltages.

Cyclotrons are usually described in terms of the diameter of the pole faces of the magnet. The first machine built by Lawrence and Livingston had a magnet with pole faces 11 in. in diameter and produced 1.2-Mev protons. A 60-in cyclotron can produce protons of about 10 Mev, deuterons of about 20 Mev, and α-particles of about 40 Mev.(15) The particle beams are strong, but the voltage is neither as constant nor as uniform as it is for the Van de Graaff generator. The cyclotron is a powerful tool for studies requiring particles of high energy but where it is not essential to know the exact value of energy. The use of particles from the cyclotron to produce beams of neutrons for experiments with a time-of-flight neutron

(a)

(b)

Fig. 21–4. The cyclotron at the Massachusetts Institute of Technology.

spectrometer is an excellent example of the way in which a cyclotron can be employed. The high-energy particles can, of course, be used directly in the study of nuclear reactions and in the production of radionuclides.

Figure 21–4(a) is a photograph of the cyclotron at the Massachusetts Institute of Technology, and Fig. 21–4(b) shows the dees removed from the gap between the poles of the electromagnet; the cover of the outer vacuum chamber has been removed in this photograph. The bar in the gap between the dees supports the ion source. Accelerated particles can be brought out of the chamber through a thin foil window in the short tube at the lower right. In this machine, R is 48 cm, H is 1.8×10^4 gauss, and 18-Mev deuterons can be obtained.

The energies to which particles can be accelerated in a cyclotron are limited by the relativistic increase of mass with velocity. The mass is given by the formula

$$m = \frac{m_0}{\sqrt{1 - (v^2/c^2)}},$$

where m_0 is the rest mass. According to Eq. (21–2), the angular velocity of the ions becomes

$$\omega = H \frac{q}{m_0} \sqrt{1 - \frac{v^2}{c^2}}, \tag{21–8}$$

and decreases as the velocity increases. The frequency of ion rotation then decreases; the ions take longer to describe their semicircular paths than the fixed period of the oscillating electric field and they arrive too late at the gap between the dees. Consequently, they lag in phase behind the voltage applied to the dees until finally they are no longer accelerated. The ion velocities at which the relativistic mass increase leads to this effect represents a limiting size and energy for the cyclotron. If the limiting energy is taken to be about one percent of the rest mass, the limits for different particles are as shown in the table below. It is clear from these values that a cyclotron operates successfully only with relatively heavy particles and cannot be used to accelerate electrons. This difficulty does not arise in the Van de Graaff generator, which can be used with electrons.

	1% of rest mass, Mev	Radius, cm, for H = 10,000 gauss
electrons	0.005	0.02
protons	10	40
deuterons	20	80
α-particles	40	80

21-4 The frequency-modulated cyclotron or synchrocyclotron. The loss of resonance caused by the relativistic increase in mass can, in principle, be balanced in two ways, as can be seen by considering the resonance condition, Eq. (21-7). That relationship can be written

$$n = \frac{H}{2\pi} \frac{q}{m_0} \sqrt{1 - \frac{v^2}{c^2}}. \tag{21-9}$$

It should be possible to compensate for the decrease in the ion frequency either by increasing the field strength H at such a rate that the product $H\sqrt{1 - (v^2/c^2)}$ remains constant, or by gradually decreasing the frequency n of the applied electric field. It was thought for some years that the experimental difficulties involved in applying these methods would make them impractical. But it was shown theoretically by Veksler(17) and McMillan(18) that these methods could be applied successfully and that this possibility depends on a property called *phase stability* possessed by certain orbits in a cyclotron.

The phase stability inherent in a variable-frequency cyclotron can be discussed by describing qualitatively the motion of the charged particles. Suppose that an ion moves in a circular path in a uniform magnetic field, and on each revolution crosses the gap between accelerating electrodes to which is applied an oscillating electric field with a frequency identical with the frequency of rotation of the ion. Under these conditions there are certain "stationary" or stable orbits in which the ion can move. Consider first ions which cross the gap at instants of time when the oscillating electric field is just passing through the value zero. The ions are then said to have zero phase, and this condition is indicated by the points 0, 2π, 4π, and so on in Fig. 21-5; these ions will neither gain nor lose energy and will continue to revolve at constant frequency and in the same orbit.

An ion which crosses the gap at an earlier instant such as t_1 has a positive phase; it will gain speed and energy, since the voltage is positive, and its rotation frequency will decrease in accordance with Eq. (21-8).

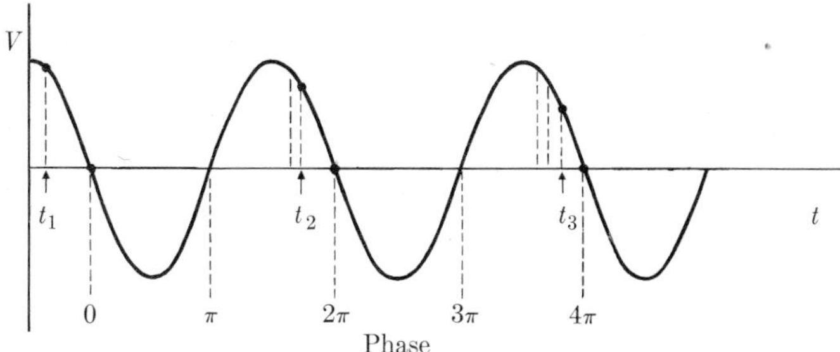

FIG. 21-5. Phase stability in a cyclic magnetic accelerator.

The ion will take a slightly longer time to return to the gap, as shown by points t_2 and t_3 in subsequent accelerations, when the voltage across the gap is smaller. Eventually the particle will cross the gap at zero phase but with the excess energy it has received, and will continue the phase shift into the decelerating part of the cycle when the voltage is negative. The particle then loses energy and speed, its frequency increases, and it is returned to the zero phase. There has been a phase oscillation about the equilibrium phase $\theta = 0$, and the ion has oscillated about a "stationary" or equilibrium orbit.

Veksler and McMillan pointed out that by increasing the magnetic field or reducing the frequency of the dee voltage, or both, the orbits of the ions could be made to expand and the energy to increase. If the variation is made sufficiently slowly, the phase stability is preserved during the acceleration, and the limits on the ordinary cyclotron can be passed. Particles can be accelerated by holding the electric field frequency fixed and varying the magnetic field; in this case the machine is called a *synchrotron.* Protons, deuterons, and α-particles can be accelerated to high energies by keeping the magnetic field constant and varying the frequency of the electric field; in this case the machine is called a *frequency-modulated (FM) cyclotron* or *synchrocyclotron.*[(14)] The name *synchrotron* was suggested by McMillan because the behavior of the machine is similar in some respects to that of a synchronous motor.

The first FM cyclotron was the 184-in machine at the University of California at Berkeley,[(18,19)] which produced 200-Mev deuterons and 400-Mev α-particles. For these particles, the frequency was modulated from 11.5 million cycles per second at the instant of injection to 9.8 million cycles per second when the ions reached the periphery of the dee. Protons with energies of about 350 Mev were obtained with an oscillator frequency modulated from 23 to 15.6 million cycles per second. The weight of the magnet is about 4000 tons and the energies cited were obtained with a dee voltage of 15 kv and a magnetic field of 15,000 gauss at the gap center. A diagram of the Berkeley machine is shown in Fig. 21–6. There is only one dee and the oscillating potential is applied between it and a ground connection. An ion path is shown, distorted to indicate only a few turns; in practice the ions make about 50,000 revolutions. The ion beam can be removed with the aid of electrostatic and magnetic deflectors.

The process of spallation[(20)] was discovered and investigated by bombarding targets of various elements with particles from this machine, and fission of bismuth and neighboring elements has been observed with 190-Mev deuterons. Mesons have also been produced as a result of the bombardment of carbon by 390-Mev α-particles or 345-Mev protons. Mesons are fundamental particles which are related to nuclear forces, and the investigation of nuclear reactions which produce mesons or are induced

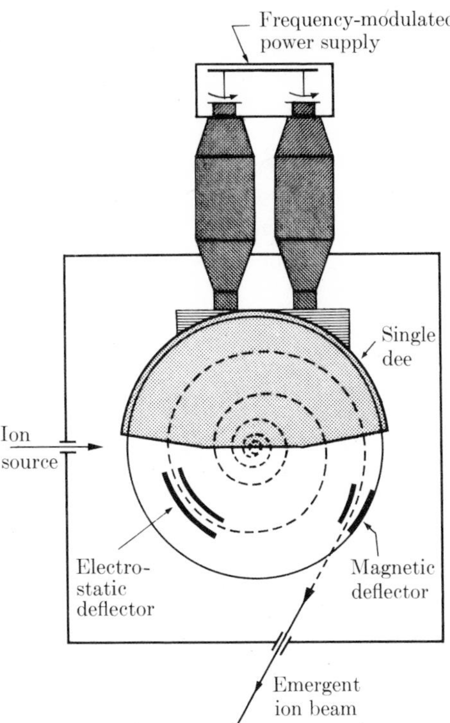

FIG. 21–6. Schematic diagram of the 184-in frequency modulated cyclotron at the University of California.

by mesons is one of the main problems of nuclear physics. Before the development of the large accelerators, of which the 184-in FM cyclotron was the first example, these reactions could be studied only in cosmic rays, where they are relatively rare, but they can now be made to occur at will and in great numbers with the aid of large accelerators.

Many synchrocyclotrons have now been built in the United States and in Europe in order to study the nature and constitution of the atomic nucleus, and many interesting and important results have already been obtained. It has been found, for example, that μ-mesons which have a rest mass of close to 200 electron masses can be used to study the structure of the nucleus.(21,22) Beams of these particles from the Columbia University 164-in synchrocyclotron have been shown to interact with nuclei in a way which is sensitive to the nuclear charge distribution. Experiments indicate that the nuclear charge is distributed in a uniform sphere of radius $1.2 \times 10^{-13} A^{1/3}$ cm, a value which seems to differ significantly from earlier values.(23) Results such as these are leading to a closer examination of the distribution of particles in the nucleus, as well as of other nuclear properties (cf. Section 17–4).

21–5 The acceleration of electrons. The betatron and the electron synchrotron. The emphasis so far in this chapter has been on the acceleration of positively charged particles because of their importance in nuclear reactions. Beams of energetic electrons are also needed for various purposes, the most important of which is the production of x-rays of very high energy. These x-rays can be used to produce nuclear reactions, such as the (γ, n), (γ, p), (γ, 2n), and (γ, np) reactions, or as highly penetrating radiations for the study of the properties of solids. The energetic electrons can also be used directly.

The voltage multiplier (Cockcroft-Walton) and the Van de Graaff electrostatic generator can both be used to accelerate electrons, but the energies are limited to a few million electron volts. The ordinary cyclotron cannot be used with electrons because of the large relativistic increase of mass at low energies; an electron with an energy of 1 Mev has a speed more than nine-tenths that of light and its mass is about 2.5 times the rest mass. It has been mentioned that this relativistic effect can be overcome by varying the magnetic field, and the design of the *electron synchrotron* is based on this idea. But before that instrument is discussed, it is convenient to consider another electron accelerator, the *betatron*, in which magnetic induction is used to accelerate the electrons.[24,14]

The action of the betatron depends on the same principle as that of the transformer, in which an alternating current applied to a primary coil induces a similar current in the secondary windings. The primary current produces an oscillating magnetic field which, in turn, induces an oscillating potential in the secondary coil. The betatron is a transformer in which a cloud of electrons, located inside an annular, doughnut-shaped vacuum chamber, takes the place of the secondary winding. The chamber is placed within the poles of an electromagnet energized by an alternating pulsed current, and the magnet produces a strong field in the central space, or hole, of the doughnut. The electrons move in a circular orbit of constant radius within the vacuum chamber, as shown in Fig. 21–7, and they gain energy by induction, because of the change with time of the magnetic flux Φ linking the orbit.

For this method to work, the induced accelerating field must have just the right strength at the stable orbit to ensure that the electrons remain in that orbit when the magnetic field increases; that is, the flux linking that orbit must have the right magnitude. The required relationship between the linking flux and the magnetic field can be derived without much trouble. The induced voltage per turn is $d\Phi/dt$, as for a transformer, and the electric field (voltage per unit length) is given by

$$E = \frac{1}{2\pi R} \frac{d\Phi}{dt}, \tag{21–10}$$

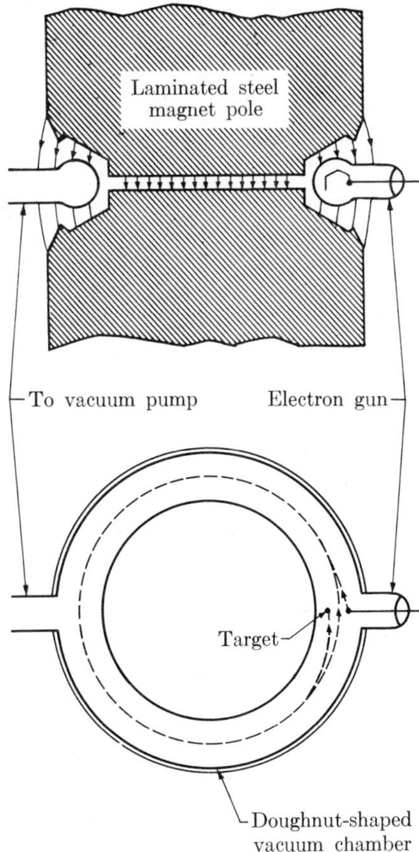

Fig. 21–7. The accelerating chamber in a betatron.

where R is the radius of the stable orbit. The force on the electron is eE, and the law of motion for the electron is

$$\frac{d}{dt}(mv) = eE = \frac{e}{2\pi R}\frac{d\Phi}{dt}. \tag{21–11}$$

To maintain motion in a circular orbit of constant radius R, the magnetic field H at the orbit must increase as the electron energy increases. The momentum of the electron in the field is given by the familiar relation

$$Hev = \frac{mv^2}{R},$$

or

$$mv = eRH,$$

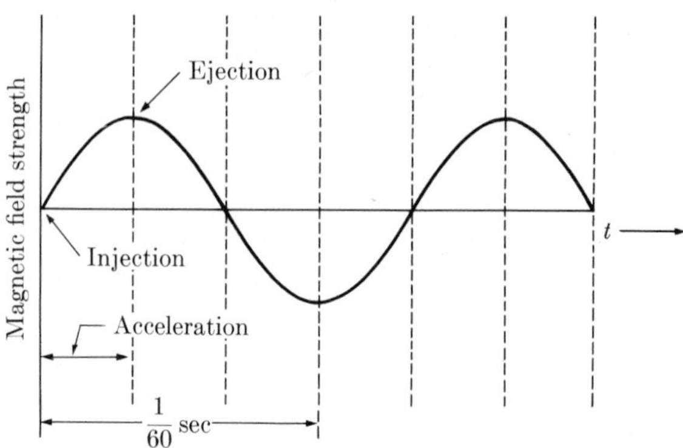

FIG. 21–8. The acceleration cycle in a betatron.

so that

$$\frac{d}{dt}(mv) = eR\,\frac{dH}{dt}\cdot \tag{21–12}$$

Acceleration by magnetic induction at constant radius is obtained when the time rate of change of momentum as given by the change in the linkage flux is the same as that given by the change in the magnetic field. In other words, the expressions on the right sides of Eqs. (21–11) and (21–12) must be equal, with the result that

$$\frac{d\Phi}{dt} = 2\pi R^2\,\frac{dH}{dt} = 2\,\frac{d}{dt}(\pi R^2 H). \tag{21–13}$$

Equation (21–13) is the *betatron condition,* which states that in any time interval the linking flux Φ must change at a rate twice that which would occur if the central magnetic field were uniform and equal to the field at the orbit. This condition holds for relativistic energies as well as in the nonrelativistic range because the law of force as written in Eqs. (21–11) and (21–12) satisfies the requirements of special relativity (Section 6–6). The betatron condition makes it necessary to have a central iron core with high flux density inside the orbit. Since the induced potential is determined by the rate of change of flux, the iron core is laminated as in a transformer, and alternating power at 60 or 180 cycles is used to produce the varying magnetic field.

Electrons are injected into the chamber from a "gun" in which thermoelectrons from a hot cathode are accelerated and focused by a potential of some thousands of volts. The electrons are injected at an instant when the magnetic field is just rising from its zero value in the first quarter cycle, as shown in Fig. 21–8. The increasing magnetic field induces a potential within the doughnut which increases the energy of the electrons. Since they are traveling in a magnetic field, the electrons move in a circular path, and the increasing magnetic field keeps them moving in a stable orbit. When the field strength passes its maximum value and starts to decrease, the direction of the induced electromagnetic force is changed and the electrons start to slow down. This effect is avoided by removing the electrons from their orbit when the magnetic field strength has its maximum value. A pulse of current is sent through an auxiliary coil and changes the magnetic field; the electrons are pulled out of their stable orbit and either strike a target, producing x-rays, or emerge from the apparatus.

The betatron at the General Electric Research Laboratories[(25)] produces 100-Mev electrons. The diameter of the pole face is 76 in and that of the stable orbit is 66 in; the magnet weighs 130 tons and the maximum magnetic field at the orbit is 4000 gauss. Electrons are injected at energies of 30–70 kev; they travel around the doughnut about 250,000 times between injection and removal, gaining 400 ev of energy at each turn. The 100-Mev electrons have a speed which is more than 0.9999 that of light and their mass is nearly 200 times the rest mass. Electrons with energies of more than 300 Mev have been obtained with the betatron at the University of Illinois.[(26,27)]

The betatron has the disadvantage that a large magnet is needed to supply the variable flux which accelerates the electrons. If a radiofrequency electric field could be used, the size of the magnet would be reduced. The radiofrequency would have to be locked to the electron frequency, and if the electrons started at low energies, the change in the radiofrequency would have to be very large, beyond practical limits. Electrons with energies of about 2 Mev, however, have speeds within a few percent of that of light. Beyond this energy the electron speed and its frequency of revolution change very little, and the radiofrequency can remain practically constant. As the electron energy increases, the electrons can be held in a stable orbit by increasing the strength of the magnetic field proportionally to the increase in the mass of the electron. Veksler and McMillan showed, independently, that the electron motion has the property of phase stability, and the theory is similar to that for positively charged particles in the FM cyclotron. Since it is not necessary to satisfy the betatron condition, a ring-shaped magnet can be used. The instrument based on these ideas is the electron synchrotron, a diagram of which is shown in Fig. 21–9.

Electrons can be injected at high enough speeds so that the orbit radius, although not constant, does not increase greatly. In practice, however, it has turned out to be more satisfactory to accelerate electrons up to 1 or 2 Mev by betatron action with the aid of *flux bars* located inside the

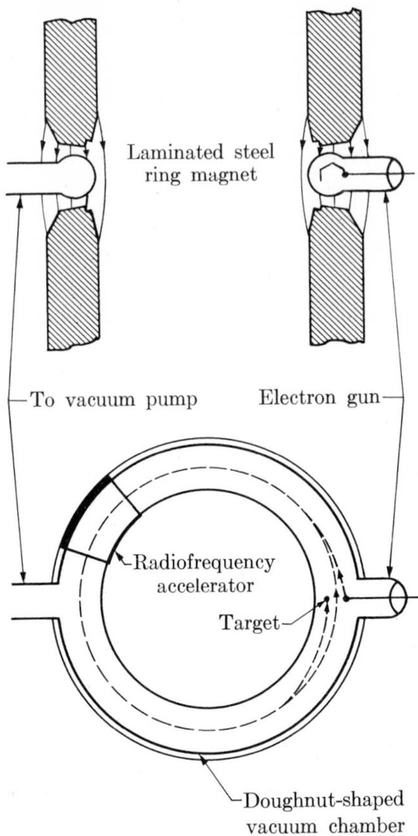

FIG. 21–9. The accelerating chamber in a synchrotron.

orbit. These bars are made of high permeability metal and do not have to be large; they *short* the magnetic field at low inductions but become saturated at high inductions, and the transition from betatron action to synchrotron action can be made smoothly.

About 20 electron synchrotrons have either been built or are under construction,[(28)] and electrons have been obtained with energies of more than 300 Mev. The machine at the Massachusetts Institute of Technology has a 50-ton magnet and an rf of 46.5 megacycles. The orbit radius is 40 in. Electrons are injected at 80 kev, accelerated to 7 Mev by betatron action, and finally reach energies of 330 Mev.

The limitation on the energies to which electrons can be accelerated in a betatron or a synchrotron is set by radiation losses. An electron loses energy by radiation when it is accelerated, and the rate increases with the fourth power of the energy. The maximum energy is reached when the energy lost per turn by radiation is equal to the maximum practical acceleration energy per turn. It is thought that the maximum electron energy that can be reached under practical conditions is about 1000 Mev or 1 Bev.

21-6 The proton synchrotron. Up to 1959, the highest energies to which charged particles were accelerated were attained in *proton synchrotrons.* In 1952, protons were accelerated to 3 Bev in the *Cosmotron* at the Brookhaven National Laboratory; in 1954, 6-Bev protons were obtained in the *Bevatron* at the Radiation Laboratory of the University of California; in 1957, proton energies of 10 Bev were reached in a Russian proton synchrotron at Dubna. Protons can be accelerated to these high energies without losing appreciable amounts of energy by radiation because the loss of energy is proportional to the fourth power of the ratio of total energy to rest energy. The rest mass of a proton is nearly 2000 times that of an electron, so that a proton would have to have an energy of 10 Bev in order to lose as much energy by radiation as a 5-Mev electron, and the loss by an electron of this energy is not important. Hence the acceleration of protons will not be limited by radiation loss until energies much higher than 10 Bev are reached.

The principles on which the proton synchrotron is designed and operated are basically the same as those of the electron synchrotron.[14,29,30] Protons revolve in an orbit of constant radius in a doughnut-shaped vacuum chamber. A ring-shaped magnet produces a magnetic field normal to the chamber. Protons do not reach a speed of 0.98c (the speed of a 2-Mev electron) until their energy is 4 Bev, so that the rotation frequency for protons in an orbit of constant radius increases by a large factor during acceleration. The frequency of the applied electric field must also increase by a large factor and this is the only real difference, apart from size, between the proton and electron synchrotrons. Protons, at low energy, are injected in periodic pulses into the synchrotron orbit. They are accelerated by an oscillating magnetic field in resonance with the motion of the ions while the magnetic field increases to its maximum. The principle of phase stability can be utilized, and if the applied frequency is correct, the ions, on crossing the accelerating gap, oscillate in phase about an equilibrium phase which provides the proper average acceleration to keep the radius of the orbit constant. The magnet is excited periodically, and the protons are accelerated during the time that the magnetic field is increasing. When they reach maximum energy, the frequency is distorted so that the orbit expands or contracts and the protons strike a target or an ejection device at the periphery.

The Brookhaven Cosmotron has an orbit radius of 30 ft with a magnetic field, at injection, of about 300 gauss and a maximum magnetic field of

FIG. 21–10. The Cosmotron at the Brookhaven National Laboratory.

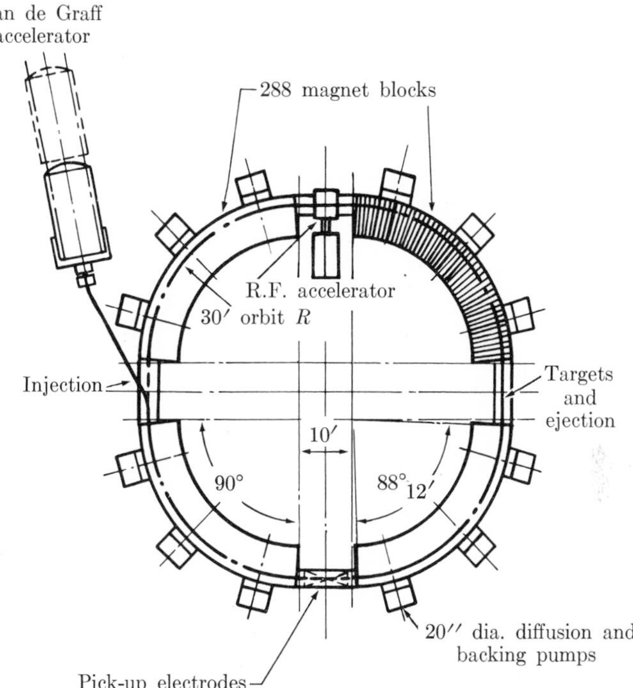

Fig. 21–11. Schematic diagram of the Cosmotron.

14,000 gauss.[31,32] The magnet, with a maximum diameter of 75 ft, is a steel ring about 8 ft by 8 ft in cross section, and weighs about 2000 tons. It is built in four quadrants separated by 10-ft gaps to allow straight sections of the vacuum chamber free from magnetic field to be used for injection, acceleration, and ejection of the ions. Pulses of protons with an energy of about 3.5 Mev accelerated in a horizontal electrostatic generator are injected into the vacuum chamber. The acceleration interval in the cosmotron is one second and about 800 ev are added per revolution; the protons make about 3 million revolutions and travel more than 100,000 miles before reaching their maximum energy. The radiofrequency changes from about 0.37 megacycle/sec to about 4 megacycles/sec, during the acceleration period. Figure 21–10 is a picture of the Cosmotron, and Fig. 21–11 is a diagram which indicates some of the main features of the machine.

The Brookhaven machine is called *Cosmotron* because it makes possible the study of some of the nuclear reactions that occur with particles with energies comparable to those of the primary cosmic ray particles. These studies are essential to an understanding of nuclear forces and nuclear structure. The Bevatron[33,34] has been used to create proton-antiproton pairs as well as neutron-antineutron pairs, and the creation and study of these particles[35] has been a major contribution to the understanding of fundamental particles.

21–7 Linear accelerators. In the linear accelerator, charged particles are accelerated by an oscillating electric field applied to a series of electrodes, with an applied frequency which is in resonance with the motion of the particles. In its earliest form, which illustrates the principles of this type of machine, the linear accelerator consisted of a set of cylindrical electrodes, of increasing length, arranged as shown in Fig. 21–12. Alternate cylinders, enclosed in a glass vacuum chamber, are connected together, the odd-numbered cylinders being joined to one terminal and the even-numbered to the second terminal of a high-frequency power supply. Ions from a discharge tube at one end move along the axis of the tubes and are accelerated on crossing the gaps between the tubes; they are not accelerated inside the tubes because the potential is constant there. Suppose that the ions are positively charged and moving from left to right, as in the diagram. If the first cylinder is positive and the second negative, the ions are accelerated and travel through the second cylinder at a speed which is constant but greater than the speed in the first cylinder. The second cylinder is just long enough so that when the ions reach the gap between the second and third cylinders, the potentials are reversed. The second cylinder is now positive and the third negative, so that the ions are again accelerated, this time in the gap between the second and third cylinders. The ions can be kept in phase with the reversals of potential by making successive cylinders longer to allow for the increasing speed of the ions, with the result that the ions gain additional energy each time they pass from one cylinder to the next.

The early linear accelerators[36] could be used only for heavy, slowly moving ions such as mercury ions, because the oscillators available did not have high enough frequencies. The intensive development of high-frequency methods in the field of radar during recent years has made possible the design of linear accelerators with which protons and even electrons can be accelerated to energies up to several hundred Mev.[37,38,39] These machines have both advantages and disadvantages as compared with circular machines. A smaller magnet can be used than is needed for

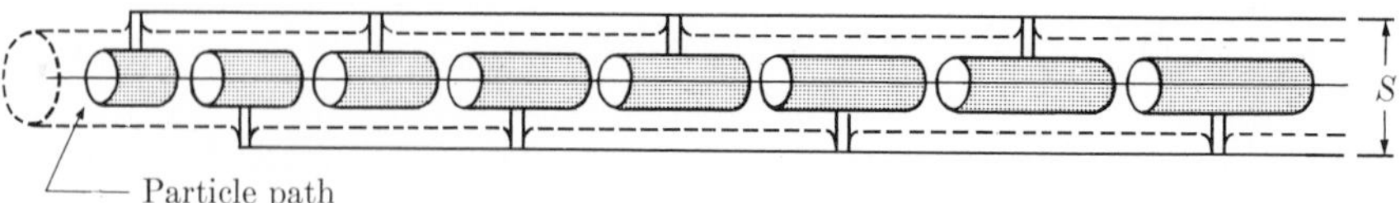

FIG. 21–12. Schematic diagram of a linear accelerator.

a circular machine producing particles of the same energy. The size and cost of the linear machine are roughly proportional to the final energy of the particles rather than to a higher power of the energy, as in circular machines, so that for very high energies the linear machines may be more economical. The particles automatically emerge from a linear accelerator in a well-collimated beam, while ejection of the particles is one of the main difficulties in the circular machines. The main disadvantage of the linear accelerator is the fact that an individual particle, instead of passing through the same alternating field again and again, and using the same power source and accelerating gap many times, as in the circular machines, must pass through a number of alternating fields and a number of power sources. This multiplicity of power sources and fields requires many pieces of high-frequency apparatus, with the resulting expense and complication. A very long and elaborate tube must be built with its many power sources nearby, and there is a difficult design problem in the adjustment of the phase of the oscillating field over a great length of accelerator, so as to ensure that the field will keep in step with the particles.

The linear accelerator can be used to obtain energies in the range above 1 Bev. In this energy range, the betatron is impractical because of the huge magnet size and the energy loss by radiation, while the synchrotron will probably be limited by radiation loss. This loss, which is caused mainly by circular motion, is not serious in the linear accelerator. The proton linear accelerator is useful at energies in the cyclotron energy range, but seems to offer few advantages in the ranges covered by the FM cyclotron and proton synchrotron.

Beams of fast electrons from a linear accelerator can be used to probe the structure of the nucleus. The elastic scattering of electrons with energies up to 125 and 150 Mev depends on the distribution of charge in the nucleus, and experiments indicate that there may be a variable density of charge in the nucleus.[40,41]

21-8 The alternating-gradient synchrotron. The increases in the energy to which charged particles have been accelerated have been accompanied by increased complexity and cost of the machines needed. Although the energies which could be reached in the proton synchrotron are much greater, in theory, than several Bev, practical limits are set by the size of the machine and by its cost. The Bevatron, which has reached 6 Bev, has a magnet weighing over 10,000 tons, an orbit radius of 80 ft, and has been estimated to cost 15 to 20 million dollars. A 30-Bev synchrotron would need a magnet of about 100,000 tons, and its cost would be exorbitant. Engineering studies indicate, for example, that direct scaling from the Cosmotron would be practical for energies up to about 20 Bev but, for higher energies, the size and cost of the ring magnet become impractical.

It is clear that new ideas are needed in order to increase the energies to which particles can be accelerated while staying within the bounds of engineering feasibility and possibly available funds. Such ideas would have to improve the efficiency of focusing the ions enough so that the size of the magnet could be greatly reduced.

The problem of improving the focusing can be illustrated in terms of some of the properties of the Brookhaven Cosmotron, in which the protons move through a pipe 36 in by 7 in. in cross section with a magnet 8 ft by 8 ft in cross section. The magnet is needed to keep the particles in the desired orbit by means of corrective forces that push the particles back when they begin to stray because of collisions with gap molecules in the tube or because of fluctuations in the accelerating voltage or frequency. Now, if the straying of the particles could be controlled so closely that the particles stay almost exactly in a circular orbit, the pipe in which they move could be very narrow and only a thin magnet would be needed around it. It has recently been found that this effect can be achieved by the use of an apparatus called the *alternating gradient synchroton* or *strong-focusing* synchrotron.[(42)] This apparatus depends on the principle that alternating focusing and defocusing ion lenses can be arranged so as to have a net focusing effect. The effect is analogous to the focusing of a beam of light by a series of alternately converging and diverging lenses. Thus, if a number of C-shaped magnets are arranged in a circle so that alternate magnets face in opposite directions, the back of one C toward the center of the circle and the back of the next toward the outside, the arrangement should keep the particles in a stable orbit. Moreover, it should be possible with proper design to make the focusing forces very strong. The first machine of this type to go into operation (1959) was that of CERN, the European Organization for Nuclear Research, near Geneva; 50-Mev protons were accelerated to 28 Bev.[(43)]

The alternating gradient synchrotron (AGS) of the Brookhaven National Laboratory recently (1960) accelerated protons to 30 Bev.[(44)] A plan drawing of this machine is shown in Fig. 21–13. The acceleration takes place in three stages. Protons are obtained from hydrogen gas by stripping the negatively charged electrons from the gas molecules, leaving bare positively charged protons in the ion source (*A* at the upper left of the figure). A high constant voltage of 750,000 volts is maintained between the ion source and ground by means of a Cockcroft-Walton generator. This voltage imparts an initial impulse to the protons by accelerating them through an evacuated ceramic tube. As they emerge, at a velocity of $0.04c$ (c is the velocity of light) they are directed into the linear accelerator *B*. This device, called the *linac,* is a long cylindrical tank about 3 ft in diameter and 110 ft long containing 124 *drift tubes* along its axis. The second stage of acceleration takes place while the protons travel through

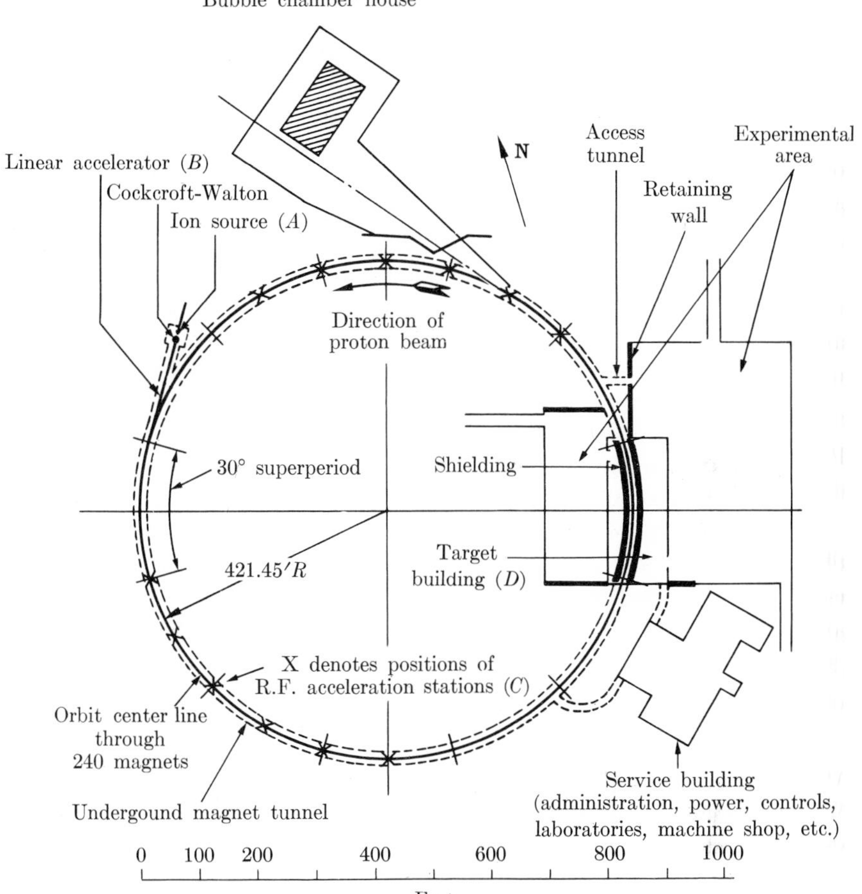

FIG. 21–13. Plan of the Brookhaven alternating gradient synchrotron (by permission of the Brookhaven National Laboratory).

these tubes. In each successive gap between adjacent drift tubes, the particles are accelerated by an electric field which is set up by feeding high frequency (2×10^8 cycles/sec) power into the tank. The strength of the electric field is such that the protons emerge from the linear accelerator at a velocity of one third that of light and an energy of 50 Mev. The proton beam is guided from the linac into the synchrotron proper through an elaborate system of debunching, deflecting, focusing, and monitoring equipment.

The circular orbit of the synchrotron, where the third stage of the acceleration is performed, is 842.9 ft in diameter (about one-half mile in circumference). The main magnet, which is divided into 240 units, each weighing some 16 tons, must bend the protons into a path of just this diameter, and apply strong focusing forces which always tend to bring the protons back toward their orbit within the vacuum chamber whenever they tend to stray away from it. Thus, as the particles increase in energy, the synchrotron's magnetic field must become stronger to hold the beam in the center of the vacuum chamber. The vacuum pipe is only 7 in wide and $2\frac{3}{4}$ in high, a small fraction of the ring's diameter. The vacuum chamber is maintained at a pressure of less than 10^{-5} mm Hg.

Besides providing a guide field, the synchrotron magnets perform another extremely important function, which embodies the alternating gradient feature of the machine. The pole pieces of each successive pair of magnets are shaped so that the magnetic field alternately increases and then decreases in the radial direction. This alternation of the magnetic field's gradient causes the circulating proton beam alternately to focus and defocus vertically and horizontally and, after many traversals of the magnets, yields a tightly focused beam. Thus, the magnets, in forming a series of alternately converging and diverging lenses, present a beam path with much greater focusing power than the corresponding effect in a conventional constant gradient machine.

As a consequence, the beam can be contained in a much smaller volume of space around the circumference of the accelerator. Hence, the magnets can be much smaller with an accompanying saving in the amount of steel and copper required. For instance, the total amount of steel in BNL's 30-Bev machine is roughly 4000 tons, compared to Russia's 10-Bev machine of conventional design whose magnet weighs some 36,000 tons.

The large magnetic gradients require that the individual magnets be very precisely built and aligned to avoid errors in the magnetic field that would cause the beam to stray from the desired path and strike the walls of the narrow vacuum chamber.

As the protons circulate around the ring, they are accelerated by electric fields produced in 12 radio-frequency acceleration stations C. At these stations, a high frequency voltage is impressed across two gaps. Protons which cross the gaps when the electric field is in the forward direction are accelerated. If the applied frequency is correct, the same protons will always be accelerated at each gap. In this way, after many traversals, very large energies can be acquired by relatively small increments. On the average each of the 12 stations accelerates the beam by about 7,500 volts each time it passes. Therefore, the protons gain about 90,000 ev of energy per revolution. To reach 30 Bev, some 325,000 revolutions around the ring are required, a distance in the neighborhood of 160,000 miles. Toward the end of this one-second acceleration period, the protons are traveling at a velocity within less than one-tenth of one per cent of the speed of light, or over 186,000 miles per second. At this velocity, the mass of the protons has been increased more than 30 times, as predicted by the theory of relativity.

After a two second recovery period, the synchrotron is ready to pulse again; 20 pulses of 30-Bev protons are produced each minute of full operation, with about 10^{10} protons per pulse.

The proton beam can be directed at appropriate target substances, and the resulting reactions can be studied by means of the emitted radiations. The primary beam can also be deflected out of the vacuum chamber and guided into the experimental area, where a number of separate experiments can be installed along the path of the beam and different portions used simultaneously. Most of the experiments will be conducted in the 25,000 square foot target building D.

Most of the elementary particles produced by the interaction of the high energy protons with the nuclei of the target atoms exist outside the atom for only a fraction of a millionth of a second before they decay into other particles or change their mass completely into energy. These brief lifetimes, however, are long enough to detect the particles by sensitive instruments (bubble chambers, counters, or photographic emulsions) and to determine their mass, electrical charge and other properties, or to observe the results when they impinge on secondary targets.

References

GENERAL

E. M. McMillan, "Particle Accelerators," *Experimental Nuclear Physics*, E. Segrè, ed., Vol. III, Part XII, pp. 639–785. New York: Wiley, 1959.

Encyclopedia of Physics (*Handbuch der Physik*), Vol. 44, *Nuclear Instrumentation I*. Berlin: Springer Verlag, 1959.

1. E. Baldinger, "Kaskaden generatoren," pp. 1–63.
2. R. G. Herb, "Van de Graaff Generators," pp. 64–104.
3. B. L. Cohen, "Cyclotrons and Synchrocyclotrons," pp. 105–169.
4. R. R. Wilson, "Electron Synchrotrons," pp. 170–192.
5. D. W. Kerst, "The Betatron," pp. 193–217.
6. G. K. Green and E. D. Courant, "The Proton Synchrotron," pp. 218–340.
7. L. Smith, "Linear Accelerators," pp. 341–389.

M. S. Livingston, *High Energy Accelerators*. New York: Interscience Publishers, 1954.

D. L. Judd, "Conceptual Advances in Accelerators," *Ann. Rev. Nuc. Sci.* **8,** 181–216, 1958.

J. J. Livingood, *Principles of Cyclic Particle Accelerators*. New York: Van Nostrand, 1961.

S. Livingston and J. P. Blewett, *Particle Accelerators*. New York: McGraw-Hill, 1961.

PARTICULAR

1. J. D. Cockcroft and E. T. S. Walton, "Experiments with High Velocity Positive Ions. I. Further Developments in the Method of Obtaining High Velocity Positive Ions," *Proc. Roy. Soc.* (London) **A136,** 619 (1932).

2. J. D. Cockcroft and E. T. S. Walton, "Experiments with High Velocity Positive Ions. II. The Disintegration of Elements by High Velocity Protons," *Proc. Roy. Soc.* (London) **A137,** 229 (1932).

3. R. L. Fortescue, "High Voltage Direct Current Generators," *Progress in Nuclear Physics*, O. R. Frisch, ed., Vol. 1, New York: Academic Press, 1950.

4. R. J. Van de Graaff, "A 1.5 Mev Electrostatic Generator," *Phys. Rev.* **38,** 1919 (1931).

5. Van de Graaff, Compton, and Van Atta, "The Electrostatic Production of High Voltage for Nuclear Investigations," *Phys. Rev.* **43,** 149 (1933).

6. Van de Graaff, Trump, and Buechner, "Electrostatic Generators for the Acceleration of Charged Particles," *Rep. Prog. Phys.* Vol. XI. London: Physical Soc., 1946, p. 1.

7. F. W. Sears, *Electricity and Magnetism*. Reading, Mass.: Addison-Wesley, 1951.

8. Parkinson, Herb, Bernet, and McKibben, "Electrostatic Generator Operating under High Air Pressure. Operation Experience and Accessory Apparatus," *Phys. Rev.* **53,** 642 (1938).

9. Wells, Haxby, Stephens, and Shoupp, "Design and Preliminary Performance Tests of the Westinghouse Electrostatic Generator," *Phys. Rev.* **58,** 162 (1940).

10. Van de Graaff, Sperduto, McIntosh, and Burrill, "Electrostatic Generator for Electrons," *Rev. Sci. Instr.* **18,** 754 (1947).

11. E. O. Lawrence and M. S. Livingston, "The Production of High Speed Light Ions without the Use of High Voltages," *Phys. Rev.* **40,** 19 (1932).

12. W. B. Mann, *The Cyclotron.* 4th ed. New York: Wiley; London: Methuen, 1953.

13. M. S. Livingston, "The Cyclotron." *J. Appl. Phys.* **15,** 2, 128 (1944).

14. J. H. Fremlin and J. S. Gooden, "Cyclic Accelerators," *Rep. Prog. Phys.* Vol. XIII. London: Physical Soc., 1950.

15. Lawrence, Alvarez, Brobeck, Cooksey, Corson, McMillan, Salisbury, and Thornton, "Initial Performance of the 60-inch Cyclotron of the William H. Crocker Radiation Laboratory, University of California," *Phys. Rev.* **56,** 124 (1939).

16. V. Veksler, "A New Method of Acceleration of Relativistic Particles," *J. Phys.* (U.S.S.R.), **9,** 153 (1945).

17. E. M. McMillan, "The Synchrotron—A Proposed High Energy Particle Accelerator," *Phys. Rev.* **68,** 143 (1945).

18. Brobeck, Lawrence, MacKenzie, McMillan, Serber, Sewell, Simpson, and Thornton, "Initial Performance of the 184-inch Cyclotron of the University of California," *Phys. Rev.* **71,** 449 (1947).

19. Henrich, Sewell, and Vale, "Operation of the 184-inch Cyclotron," *Rev. Sci. Instr.* **20,** 887 (1949).

20. D. H. Templeton, "Nuclear Reactions Induced by High Energy Particles." *Ann. Rev. Nuc. Sci.* Stanford: Annual Reviews, Inc., 1953, Vol. 2.

21. V. L. Fitch and J. Rainwater, "Studies of X-Rays from Mu-Mesonic Atoms," *Phys. Rev.* **92,** 789 (1953).

22. J. A. Wheeler, "Mu Meson as a Nuclear Probe Particle," *Phys. Rev.* **92,** 812 (1953).

23. L. N. Cooper and E. M. Henley, "Mu-Mesonic Atoms and the Electromagnetic Radius of the Nucleus," *Phys. Rev.* **92,** 801 (1953).

24. D. W. Kerst, "Acceleration of Electrons by Magnetic Induction," *Phys. Rev.* **60,** 47 (1941).

25. W. F. Westendorp and E. E. Charlton, "A 100-Million Volt Induction Electron Accelerator," *J. Appl. Phys.* **16,** 581 (1945).

26. Kerst, Adams, Koch, and Robinson, "An 80-Mev Model of a 300-Mev Betatron," *Rev. Sci. Instr.* **21,** 462 (1950).

27. Kerst, Adams, Koch, and Robinson, "Operation of a 300-Mev Betatron," *Phys. Rev.* **78,** 297 (1950).

28. Thomas, Kraushaar, and Halpern, "Synchrotrons," *Ann. Rev. Nuc. Sci.* Stanford: Annual Reviews Inc., 1952. Vol. 1.

29. Oliphant, Gooden, and Hide, "The Acceleration of Charged Particles to Very High Energies," *Proc. Phys. Soc.* (London) **59,** 666 (1947).

30. Gooden, Jensen, and Symonds, "Theory of the Proton Synchrotron," *Proc. Phys. Soc.* (London) **59,** 677 (1947).

31. LIVINGSTON, BLEWITT, GREEN, and HAWORTH, "Design Study for a 3-Bev Proton Accelerator," *Rev. Sci. Instr.* **21,** 7 (1950).

32. "The Cosmotron," *Rev. Sci. Instr.* **24,** No. 9, 723–870 (1953).

33. W. M. BROBECK, "Design Study for a 10-Bev Magnetic Accelerator," *Rev. Sci. Instr.* **19,** 545 (1948).

34. E. J. LOFGREN, "Bevatron Operational Experiences," *Proceedings of the CERN Symposium on High Energy Accelerators and Pion Physics,* Vol. 1. Geneva: CERN, 1956, p. 496.

35. E. SEGRÈ, "Antinucleons," *Ann. Rev. Nuc. Sci.* **8,** 127 (1958).

36. D. H. SLOAN and E. O. LAWRENCE, "The Production of Heavy High Speed Ions Without the Use of High Voltage," *Phys. Rev.* **38,** 2021 (1931).

37. J. C. SLATER, "The Design of Linear Accelerators," *Revs. Mod. Phys.* **20,** 473 (1948).

38. D. W. FRY and W. WALKINSHAW, "Linear Accelerators," *Rep. Prog. Phys.* Vol. XII. London: Physical Soc., 1949.

39. J. C. SLATER, "Linear Accelerators," *Ann. Rev. Nuc. Sci.* Stanford: Annual Reviews Inc., 1952, Vol. 1.

40. HOFSTADTER, FECHTER, and MCINTYRE, "High-Energy Electron Scattering and Nuclear Structure Determination," *Phys. Rev.* **92,** 978 (1953).

41. L. I. SCHIFF, "Interpretation of Electron Scattering Experiments," *Phys. Rev.* **92,** 988 (1953).

42. R. COURANT, LIVINGSTON, and SNYDER, "The Strong Focusing Synchrotron—A New High-Energy Accelerator," *Phys. Rev.* **88,** 1190 (1950).

43. *Physics Today,* **13,** No. 4, 70 (April 1960).

44. "Brookhaven National Laboratory, Information Release," Aug. 2, 1960.

45. R. J. VAN DEGRAAFF, "Tandem Electrostatic Accelerators," *Nuclear Instr. and Methods,* **8,** 195 (1960); also, J. J. Livingood, op. cit. gen. ref., Chapter 1.

Problems

1. Show that in the fixed-frequency cyclotron, the following relations hold: $H = 6.55 \times 10^{-4}\, n$, for protons; $H = 1.31 \times 10^{-3}\, n$, for deuterons, where H is the field strength in gauss and n is the frequency of the applied voltage in cycles per second. Derive a similar relationship for α-particles.

2. If the frequency of the applied voltage is 1.2×10^7 cycles/sec, find the values of the field strength for resonance when protons, deuterons, and α-particles, respectively, are to be accelerated. If the values of the radius at ejection of the particles is 50 cm, what is the energy of the particles in each case?

3. Devise reasonable sets of parameters, that is, values of the frequency, field strength, and radius, for the production of beams of the following particles. (a) 30-Mev α-particles, (b) 20-Mev deuterons, (c) 10-Mev protons.

4. Show that when a particle of charge q and rest mass M_0 is accelerated to a potential of V volts, high enough so that relativistic effects are no longer negligible, the mass ratio may be written

$$\frac{M}{M_0} = 1 + \frac{10^8}{c^2}\left(\frac{q}{M_0}\right) V = 1 + aV,$$

where a is a constant and c is the velocity of light. Find an expression for the velocity v of the particle, relative to that of light. Then show that the radius of the path of the particle in a magnetic field of strength H is given by the expression

$$H^2R^2 = 2 \times 10^8 V\left(\frac{M_0}{q}\right) + \frac{10^{16}V^2}{c^2}.$$

In a frequency-modulated cyclotron with a field of 15,000 gauss and a path radius of 80 in, what is the energy of the deuterons produced?

5. Assume that the magnetic flux linking the orbit in a betatron varies with time according to the relation

$$\Phi = \Phi_0 \sin \omega t,$$

and that the acceleration of the electrons occurs during one-fourth of the cycle, that is, during a time interval $\pi/2\omega$. Find formulas for (a) the energy per turn available when the flux changes, (b) the average value of this energy during the acceleration period, (c) the distance traveled by the electrons under the assumption that these speeds differ by only a negligible amount from that of light, (d) the number of turns, if the radius of the orbit is R, (e) the final energy of the electrons.

6. The following conditions obtain in a particular betatron: maximum magnetic field at orbit = 4000 gauss, operating frequency = 60 cycles/sec, stable orbit diameter = 66 in. Show that the average energy gained per turn is about 400 ev and the final energy is about 10^8 ev. [*Hint*: It follows from Eq. (21–13) that if the initial flux through the orbit and the initial field strength at the orbit are both zero, then $\Phi = 2\pi R^2 H$, where H is the field strength at the orbit.]

Appendixes

APPENDIX I

ALPHABETIC LIST OF THE ELEMENTS

Element	Symbol	Atomic number Z	Element	Symbol	Atomic number Z
Actinium	Ac	89	Holmium	Ho	67
Aluminum	Al	13	Hydrogen	H	1
Americium	Am	95	Indium	In	49
Antimony	Sb	51	Iodine	I	53
Argon	A	18	Iridium	Ir	77
Arsenic	As	33	Iron	Fe	26
Astatine	At	85	Krypton	Kr	36
Barium	Ba	56	Lanthanum	La	57
Berkelium	Bk	97	Lead	Pb	82
Beryllium	Be	4	Lithium	Li	3
Bismuth	Bi	83	Lutetium	Lu	71
Boron	B	5	Magnesium	Mg	12
Bromine	Br	35	Manganese	Mn	25
Cadmium	Cd	48	Mendelevium	Md	101
Calcium	Ca	20	Mercury	Hg	80
Californium	Cf	98	Molybdenum	Mo	42
Carbon	C	6	Neodymium	Nd	60
Cerium	Ce	58	Neon	Ne	10
Cesium	Cs	55	Neptunium	Np	93
Chlorine	Cl	17	Nickel	Ni	28
Chromium	Cr	24	Niobium	Nb	41
Cobalt	Co	27	Nitrogen	N	7
Copper	Cu	29	Nobelium	No	102
Curium	Cm	96	Osmium	Os	76
Dysprosium	Dy	66	Oxygen	O	8
Einsteinium	E	99	Palladium	Pd	46
Erbium	Er	68	Phosphorus	P	15
Europium	Eu	63	Platinum	Pt	78
Fermium	Fm	100	Plutonium	Pu	94
Fluorine	F	9	Polonium	Po	84
Francium	Fr	87	Potassium	K	19
Gadolinium	Gd	64	Praseodymium	Pr	59
Gallium	Ga	31	Promethium	Pm	61
Germanium	Ge	32	Protactinium	Pa	91
Gold	Au	79	Radium	Ra	88
Hafnium	Hf	72	Radon	Rn	86
Helium	He	2	Rhenium	Re	75

(Continued)

ALPHABETIC LIST OF THE ELEMENTS—*Continued*

Element	Symbol	Atomic number Z	Element	Symbol	Atomic number Z
Rhodium	Rh	45	Terbium	Tb	65
Rubidium	Rb	37	Thallium	Tl	81
Ruthenium	Ru	44	Thorium	Th	90
Samarium	Sm	62	Thulium	Tm	69
Scandium	Sc	21	Tin	Sn	50
Selenium	Se	34	Titanium	Ti	22
Silicon	Si	14	Tungsten	W	74
Silver	Ag	47	Uranium	U	92
Sodium	Na	11	Vanadium	V	23
Strontium	Sr	38	Xenon	Xe	54
Sulfur	S	16	Ytterbium	Yb	70
Tantalum	Ta	73	Yttrium	Y	39
Technetium	Tc	43	Zinc	Zn	30
Tellurium	Te	52	Zirconium	Zr	40

APPENDIX II

VALUES OF PHYSICAL CONSTANTS

(The values listed below are taken from the book, *The Fundamental Constants of Physics*, by E. R. Cohen, K. M. Crowe, and J. W. M. DuMond. New York: Interscience Publishers, Inc., 1957.)

Constant	Value
Avogadro's number	$N_0 = (6.02486 \pm 0.00016) \times 10^{23}$ per gram mole
Velocity of light	$c = (2.997930 \pm 0.000003) \times 10^{10}$ cm/sec
Standard volume of a perfect gas	$V_0 = 22420.7 \pm 0.6$ cm^3/mole
Gas constant per mole	$R_0 = (8.31696 \pm 0.00034) \times 10^7$ erg/mole°C
Boltzmann's constant	$k = R_0/N_0$ $= (1.38044 \pm 0.00007) \times 10^{-16}$ erg/°K $= (8.6167 \pm 0.0004) \times 10^{-5}$ ev/°K
Charge on the electron	$e = (4.80286 \pm 0.00009) \times 10^{-10}$ esu $= (1.60206 \pm 0.00003) \times 10^{-20}$ emu
Faraday constant (physical scale)	$F = N_0 e = (2.89366 \pm 0.00003) \times 10^{14}$ esu/(gm mole) $= (9652.19 \pm 0.11)$ emu/(gm mole)
Planck's constant	$h = (6.62517 \pm 0.00023) \times 10^{-27}$ erg-sec $h/2\pi = (1.05443 \pm 0.00004) \times 10^{-27}$ erg-sec
Rydberg constants	$R_\infty = 109737.309 \pm 0.012$ cm^{-1} $R_{\text{H}} = 109677.576 \pm 0.012$ cm^{-1} $R_{\text{D}} = 109707.419 \pm 0.012$ cm^{-1} $R_{\text{He}} = 109722.267 \pm 0.012$ cm^{-1}
Rest masses	neutron: 1.008982 ± 0.000003 amu $= (1.67470 \pm 0.00004) \times 10^{-24}$ gm proton: 1.007593 ± 0.000003 amu $= (1.67239 \pm 0.00004) \times 10^{-24}$ gm hydrogen atom: 1.008142 ± 0.000003 amu ratio: $\dfrac{\text{mass of hydrogen atom}}{\text{mass of proton}}$ $= 1.000544613 \pm 0.000000006$ electron: $(5.48763 \pm 0.00006) \times 10^{-4}$ amu $= (9.1083 \pm 0.0003) \times 10^{-28}$ gm ratio: $\dfrac{\text{proton mass}}{\text{electron mass}}$ $= 1836.12 \pm 0.02$ deuterium atom: 2.014735 ± 0.000006 amu

(Continued)

VALUES OF PHYSICAL CONSTANTS—*Continued*

Constant	Value
Mass-energy conversion factors	1 amu = 931.141 ± 0.010 Mev 1 proton mass = 938.211 ± 0.010 Mev 1 neutron mass = 939.505 ± 0.010 Mev 1 electron mass = 0.510976 ± 0.000007 Mev 1 gm = $(5.61000 \pm 0.00011) \times 10^{26}$ Mev
Energy conversion factor	1 ev = $(1.60206 \pm 0.00003) \times 10^{-12}$ erg
Energy of a 2200 m/sec neutron	$E_{2200} = 0.0252973 \pm 0.0000003$ ev $T_{2200} = 293.585 \pm 0.012$ °K $= 20.435 \pm 0.012$ °C.
First Bohr radius	$a_0 = h^2/4\pi^2 me^2$ $= (5.29172 \pm 0.00002) \times 10^{-9}$ cm
Classical electron radius	$r_0 = e^2/mc^2$ $= (2.81785 \pm 0.00004) \times 10^{-13}$ cm
Thomson cross section	$\frac{8}{3}\pi r_0^2 = (6.65205 \pm 0.00018) \times 10^{-25}$ cm^2
Compton wavelength	electron: $\lambda_{ce} = (24.2626 \pm 0.0002) \times 10^{-11}$ cm proton: $\lambda_{cp} = (13.2141 \pm 0.0002) \times 10^{-14}$ cm neutron: $\lambda_{cn} = (13.1959 \pm 0.0002) \times 10^{-14}$ cm

APPENDIX III

JOURNAL ABBREVIATIONS FOR REFERENCES

Journal	Abbreviation
American Journal of Physics	Am. J. Phys.
Canadian Journal of Physics	Can. J. Phys.
Canadian Journal of Research	Can. J. Research
Comptes rendus hebdomadaires des séances de l'académie des sciences	Compt. rend.
Journal of the American Chemical Society	J. Am. Chem. Soc.
Journal of Applied Physics	J. Appl. Phys.
Journal of Chemical Physics	J. Chem. Phys.
Journal of Inorganic and Nuclear Chemistry	J. Inorg. Nucl. Chem.
Journal of the Chemical Society (London)	J. Chem. Soc.
Journal de physique et le radium	J. phys. et radium
Journal of Physics (U.S.S.R.)	J. Phys. (U.S.S.R.)
Kongelige Danske Videnskabernes Selskab, Matematisk-fysiske Meddelelser	Kgl. Danske Videnskab, Selskab, Mat-fys. Medd.
Nature	Nature
Naturwissenschaften	Naturwiss.
Philosophical Magazine and Journal of Science	Phil. Mag.
Physical Review	Phys. Rev.
Physikalische Zeitschrift	Physik. Z.
Proceedings of the Cambridge Philosophical Society	Proc. Cambridge Phil. Soc.
Proceedings of the Physical Society (London)	Proc. Phys. Soc. (London)
Proceedings of the Royal Society (London)	Proc. Roy. Soc. (London)
Reviews of Modern Physics	Revs. Mod. Phys.
Review of Scientific Instruments	Rev. Sci. Instr.
Zeitschrift für Naturforschung	Z. Naturforsch.
Zeitschrift für Physik	Z. Physik
Zeitschrift für Physikalische Chemie	Z. Physik. Chem.

APPENDIX IV

THE RUTHERFORD SCATTERING FORMULA

The treatment to be given here of the scattering of α-particles by a positively charged atomic nucleus is more mathematical than that given in Chapter 3, but it contains some ideas which are important in nuclear physics; for example, that of the differential scattering cross section. For the purposes of this derivation we consider the infinitely heavy nucleus of charge Ze to be at the origin of a system of polar coordinates r, θ as shown in Fig. A–IV–1. The initial path of the α-particle is along the line DC, and p is the shortest distance between this line and the nucleus; the position of the α-particle is given generally by r and θ. Although the angle θ is different from what it was in Section 3–2, there should be no confusion. Other quantities are defined as in Section 3–2.

The angular momentum L about the nucleus is given by

$$L = Mr^2 \frac{d\theta}{dt}; \tag{A4–1}$$

it is a constant of the motion, and we may write

$$L = Mr^2 \frac{d\theta}{dt} = MVp. \tag{A4–2}$$

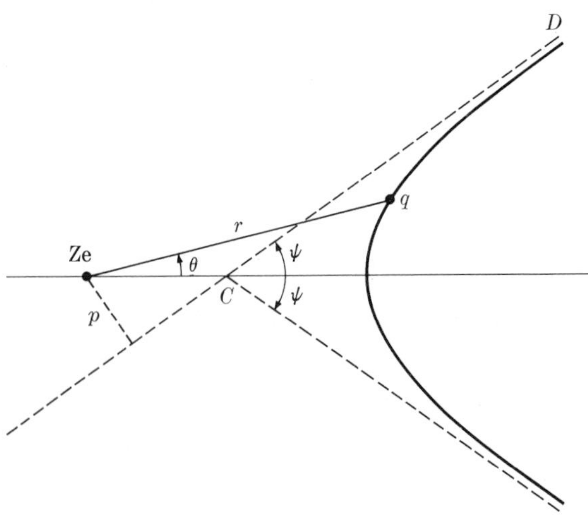

FIG. A–IV–1. Hyperbolic orbit under an inverse square law of repulsion.

The coulomb repulsive force, $F = zZe^2/r^2$, gives rise to a radial acceleration

$$a_r = \frac{d^2r}{dt^2} - r\left(\frac{d\theta}{dt}\right)^2 = \frac{F}{M} = \frac{zZe^2}{Mr^2}. \tag{A4–3}$$

Equations (A4–2) and (A4–3) are two simultaneous equations in r, θ, and t; if t is eliminated between them, we get a single differential equation in r and θ which gives the orbit of the α-particle. For convenience, we put

$$r = \frac{1}{u}, \tag{A4–4}$$

and get, from Eq. (A4–1),

$$\frac{d\theta}{dt} = \frac{Lu^2}{M}. \tag{A4–5}$$

It follows that

$$\frac{dr}{dt} = \frac{dr}{d\theta}\frac{d\theta}{dt} = -\frac{1}{u^2}\frac{du}{d\theta}\frac{Lu^2}{M} = -\frac{L}{M}\frac{du}{d\theta}, \tag{A4–6}$$

and that

$$\begin{aligned}\frac{d^2r}{dt^2} &= \frac{d\theta}{dt}\frac{d}{d\theta}\left(\frac{dr}{dt}\right)\\ &= \frac{Lu^2}{M}\left(-\frac{L}{M}\frac{d^2u}{dt^2}\right)\\ &= -\frac{L^2u^2}{M^2}\frac{d^2u}{d\theta^2}.\end{aligned} \tag{A4–7}$$

If we now substitute for d^2r/dt^2 and $d\theta/dt$ in Eq. (A4–3), the result is

$$\frac{d^2u}{d\theta^2} + u = -\frac{zZe^2M}{L^2} = -\frac{b}{2p^2}, \tag{A4–8}$$

where

$$b = \frac{zZe^2}{\frac{1}{2}MV^2}. \tag{A4–9}$$

Equation (A4–8) is the differential equation of the orbit; it has the general solution

$$u = \frac{1}{r} = A\cos(\theta - \theta_0) - \frac{b}{2p^2}, \tag{A4–10}$$

where A and θ_0 are arbitrary constants. The constant θ_0 just determines the orientation of the orbit in the (r, θ) plane. We can choose the x-axis so that $\theta_0 = 0$, and the equation of the orbit becomes

$$u = \frac{1}{r} = -\frac{b}{2p^2} + A\cos\theta. \tag{A4–11}$$

The equation

$$u = \frac{1}{r} = B + A\cos\theta \tag{A4–12}$$

is known, from analytic geometry, to represent a conic section (parabola, ellipse, or hyperbola) with eccentricity given by

$$\epsilon = \frac{A}{|B|}\cdot \tag{A4–13}$$

The orbit is a hyperbola if $\epsilon > 1$. Since B is known and is equal to $-b/2p^2$, it remains only to determine A or ϵ. On setting

$$A = \epsilon|B| = \frac{b}{2p^2}, \tag{A4–14}$$

we get

$$u = \frac{1}{r} = -\frac{b}{2p^2}(1 - \epsilon\cos\theta). \tag{A4–15}$$

We can determine ϵ (or A) by making use of the equation for the conservation of energy,

$$\frac{1}{2}MV^2 = \frac{1}{2}Mv^2 + \frac{zZe^2}{r}, \tag{A4–16}$$

where v is the speed of the α-particle when it is at a distance r from the nucleus. The last equation may be written

$$\frac{v^2}{V^2} = 1 - \frac{b}{r}\cdot \tag{A4–17}$$

The quantity v^2 is given by

$$v^2 = \left(\frac{dr}{dt}\right)^2 + \left(r\frac{d\theta}{dt}\right)^2 = \left[\left(\frac{dr}{d\theta}\right)^2 + r^2\right]\left(\frac{d\theta}{dt}\right)^2. \tag{A4–18}$$

From Eq. (A4–15), we get

$$\frac{dr}{d\theta} = \frac{\epsilon b}{2p^2}r^2\sin\theta; \tag{A4–19}$$

and, from Eq. (A4–2),

$$\frac{d\theta}{dt} = \frac{Vp}{r^2}, \tag{A4–20}$$

so that

$$\frac{v^2}{V^2} = \frac{\epsilon^2 b^2}{4p^2} \sin^2 \theta + \frac{p^2}{r^2} = 1 - \frac{b}{r}.$$

If we now insert for r the expression (A4–15), and then simplify, we get

$$\epsilon^2 = 1 + \frac{4p^2}{b^2},$$

or

$$\epsilon = \left(1 + \frac{4p^2}{b^2}\right)^{1/2}. \tag{A4–21}$$

Since b and p are positive, $\epsilon > 1$, and the orbit is a hyperbola. The constant A is then given by

$$A = \frac{1}{p}\left(1 + \frac{b^2}{4p^2}\right)^{1/2},$$

and

$$\frac{1}{r} = -\frac{b}{2p^2} + \frac{1}{p}\left(1 + \frac{b^2}{4p^2}\right)^{1/2} \cos \theta. \tag{A4–22}$$

The orbit is thus expressed in terms of the two quantities p, the impact parameter, and b, the closest distance which an α-particle of initial speed V can approach to a nucleus of charge Z. When $r \to \infty$, $\theta \to \psi$, the angle between the x-axis and the asymptote of the hyperbola, and

$$\cos \psi = \frac{b}{2p[1 + (b^2/4p^2)]^{1/2}}.$$

On solving for b, we get

$$b = 2p \cot \psi,$$

or

$$p = \frac{b}{2} \tan \psi.$$

But, the scattering angle $\phi = \pi - 2\psi$, and

$$p = \frac{b}{2} \cot \frac{\phi}{2}, \tag{A4–23}$$

which is Eq. (3–11).

If collisions occur for values of the impact parameter within a range dp at p, they will result in deflections within the angular range $d\phi$ at ϕ, with p

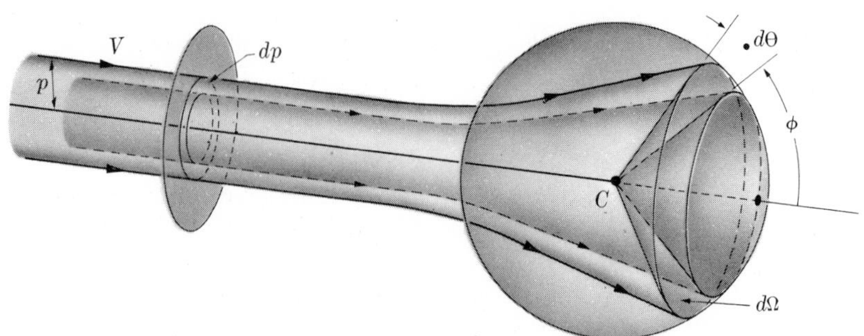

FIG. A–IV–2. Effective cross section for scattering into the solid angle $d\Omega$ (by permission from S. W. McCuskey, *Introduction to Advanced Mechanics*, Addison-Wesley Publishing Co., Inc., 1959).

and ϕ related through Eq. (A4–23). The probability of such collisions is proportional to the area of an annulus of radius p and width dp, because this is the effective target area presented by the nucleus to the incident beam for producing deflections of this particular type. We can therefore define a collision area, or cross section $d\sigma(\phi)$ for scattering into $d\phi$:

$$\begin{aligned} d\sigma(\phi) &= 2\pi p\, dp \\ &= -\frac{\pi b^2}{4}\frac{\cos\phi/2}{\sin^3\phi/2}\,d\phi. \end{aligned} \qquad \text{(A4–24)}$$

In an experiment, the scattering into a detector which subtends a fixed solid angle is usually observed. The element of solid angle is

$$\begin{aligned} d\Omega &= 2\pi \sin\phi\, d\phi \\ &= 4\pi \sin\frac{\phi}{2}\cos\frac{\phi}{2}\,d\phi, \end{aligned}$$

and

$$\frac{d\sigma(\phi)}{d\Omega} = \frac{b^2}{16\sin^4\phi/2} = \frac{Z^2e^4}{(MV^2)^2\sin^4\phi/2}. \qquad \text{(A4–25)}$$

The Rutherford scattering formula is often written in this form; $d\sigma(\phi)/d\Omega$ is called the *differential scattering cross section* and is the effective collision area for producing scattering into a unit solid angle at ϕ.

The Rutherford formula seems to show that the probability that an α-particle is scattered through an angle ϕ becomes infinite as $\phi \to 0$. The reason for this result is that the presence of negatively charged particles in the atom has been neglected. These particles have the effect of "screening" the nucleus from the α-particles so that the latter move in a modified, or "screened" Coulomb potential. The more rigorous calculation of the scattering cross section for this potential removes the infinity at $\phi = 0$ without changing any of the results of Chapter 3.

It is important to note that the Rutherford scattering formula does not depend on the geometrical details of the orbit of an α-particle even though the derivation involves some of those details. The formula contains only quantities which can be determined experimentally, such as the charge and incident speed of the α-particle, the nuclear charge of the target atoms (see Section 4–4), macroscopic properties of the scattering material and the scattering angle.

The effective cross section $2\pi p\, dp$ for scattering into $d\phi$ is illustrated in Fig. A–IV–2.

Answers to Problems

ANSWERS TO PROBLEMS

Chapter 1

1. Eight parts by weight of oxygen combine with 14.00, 7.00, 4.67, 3.50 and 2.80 parts of nitrogen, respectively. Then, 14.00 parts by weight of nitrogen combine with $8(= 8 \times 1)$, $16(= 8 \times 2)$, $24(= 8 \times 3)$, $32(= 8 \times 4)$, and $40(= 8 \times 5)$ parts of oxygen, respectively.
2. The weight of nitrogen in one molecular weight of each compound, as calculated from the data, is listed below.

$$\text{nitrous oxide } 28.010 = 2 \times 14.005, \text{ nitric oxide } 14.006 = 1 \times 14.006,$$
$$\text{nitrous anhydride } 28.006 = 2 \times 14.003, \text{ nitrogen tetroxide } 14.005 = 1 \times 14.005,$$
$$\text{nitric anhydride } 28.010 = 2 \times 14.005.$$

The smallest weight of nitrogen present in one molecular weight of these compounds is (averaged) 14.005, which is then the atomic weight of nitrogen as obtained from the data of Problems 1 and 2.

3. (a) 10.32, (b) 31.0, (c) 31.0, (d) 3, (e) 30.96, (f) 137.33. The element is phosphorus; the compound is PCl_3.
4. 195.2
5. (a) 57.1, (b) 18.617, (c) 3, (d) 55.85. The metal is iron, the compound $FeCl_3$.
6. The atomic volume curve has sharp maxima at sodium, potassium, rubidium, and cesium—the alkali metals. The atomic volume decreases in passing from lithium to boron, after which it increases through carbon, oxygen, and fluorine to sodium, when it again decreases through magnesium down to aluminum, and then increases to potassium. The elements boron, aluminum, cobalt and nickel, rhodium, etc., lie at the troughs of the curve. The melting point curve shows minima at sodium, potassium, cesium, and rubidium, and maxima at elements of groups between III and V, approximately midway between the alkali metals.
7. The new element is technicium, with an approximate atomic weight of 99. It is in group VII of the periodic table and resembles manganese in its chemical properties.

Chapter 2

1.

Element	Valence	(a) gm/amp-hr	(b) coul/mg	(c) charge, esu/gm
Cr	6	0.3234	10.132	3.34×10^{13}
Cu	2	1.186	3.036	0.910×10^{13}
In	3	1.427	5.223	0.756×10^{13}
Fe	3	0.6945	5.184	1.55×10^{13}
Hg	2	3.742	0.9621	0.288×10^{13}
Ni	2	1.095	3.288	0.986×10^{13}
Ag	1	4.0245	0.8945	0.268×10^{13}

2. $v = 3.20 \times 10^9$ cm/sec; $e/m = 1.73 \times 10^7$ emu/gm.
3. $H = 15.34$ gauss; displacement $= 19.4$ cm. 4. $R = 14.8$ cm.
5. $v = 2.60 \times 10^9$ cm/sec; $e/m = 1.74 \times 10^7$ emu/gm.
6. $H = 55$ gauss. 7. 56.8 volts/cm.
8. Equation (2–8) may be written $\Delta q = \text{Constant} \times (V_{E'} - V_E)$, where the V's are obtained by dividing 0.5222 cm by the time of rise or fall. The successive rates are listed in the second column of the following table.

Time	Rate of rise or fall, cm/sec	Rate ÷ 0.0030 cm/sec
12.45	0.0419	14.0
21.5	0.0243	8.1
34.7	0.0150	5.0
85.0	0.00614	2.0
34.7	0.0150	5.0
16.0	0.0326	10.9
34.7	0.0150	5.0
21.85	0.0239	8.0

The third column shows that the rates are all integral multiples of the greatest common divisor 0.0030 cm/sec. Hence the differences $V_{E'} - V_E$ are integral multiples of a certain constant quantity, and Δq is always an integral multiple of a certain quantity of charge.
9. 0.08323 cm/sec; 0.08571 cm/sec.
10. (a) 4.98×10^{-10} esu; (b) 4.76×10^{-10} esu.

Chapter 3

1.

ϕ	Counts/hr (rounded off)
5°	67,630
10°	4,240
15°	840
30°	54
45°	11
60°	4

2.

ϕ	Counts/hr (rounded off)
5°	28,700
10°	1,960
15°	370
30°	24
45°	5
60°	2

3.

ϕ	Counts/hr (rounded off)
5°	23,800
10°	1,490
15°	295
30°	19
45°	4
60°	1

4. $Z(\text{Cu}) = 29$.

5.

Element	Z	Distance of closest approach, cm	
		5.30 Mev	7.0 Mev
Gold	79	4.3×10^{-12}	3.2×10^{-12}
Silver	47	2.6	1.9
Copper	29	1.6	1.2
Lead	82	4.5	3.4
Uranium	92	5.0	3.8

6. 8.56×10^{-6} 7. 5.7×10^{-12} cm 8. 1.3×10^{-4} cm

Chapter 4

1. (a) $n = 4.3 \simeq 4$, (b) $n = 4.5$, or between 4 and 5, $\mu = 0.34\ \text{cm}^{-1}$.
2. $\lambda = 0.842$ A, $\theta_2 = 17° 22'$; $\theta_3 = 26° 36'$.
3. Rock salt: 10° 13′, calcite: 9° 29′, quartz: 6° 45′, mica: 2° 53′.

4.

Element	θ_1	Element	θ_1
Cr	7° 52′	Co	6° 8′
Mn	7° 13′	Ni	5° 40′
Fe	6° 39′	Cu	5° 17′
		Zn	4° 56′

5.

Element	Wavelength, A	Wave number	Frequency
Al	8.364	1.196×10^7	3.586×10^{17}/sec
Si	7.142	1.400	4.197
Cl	4.750	2.105	6.311
K	3.759	2.660	7.974
Ca	3.368	2.969	8.901
Ti	2.758	3.626	1.087×10^{18}
V	2.519	3.970	1.190
Cr	2.301	4.346	1.303
Mn	2.111	4.737	1.420
Fe	1.946	5.184	1.554
Co	1.798	5.562	1.667
Ni	1.662	6.017	1.804
Cu	1.549	6.456	1.935
Zn	1.445	6.920	2.074
Y	0.838	1.193×10^8	3.577
Zr	0.794	1.259	3.774
Nb	0.750	1.333	3.996
Mo	0.721	1.387	4.158
Ru	0.638	1.567	4.698
Pd	0.584	1.712	5.132
Ag	0.560	1.786	5.354

6. $\lambda = 0.682$ A, $Z = 43$ 7. $d = 2.816$ cm; $N_0 = 6.048 \times 10^{23}$
8. $Z = 90$

Chapter 5

1.

T, °C	W, erg/cm² sec	W, gm-cal/cm² sec
0	3.15×10^5	7.53×10^{-3}
100	1.10×10^6	2.63×10^{-2}
300	6.11×10^6	0.146
500	2.02×10^7	0.483
1000	1.49×10^8	3.56
1500	5.60×10^8	13.4
2000	1.51×10^9	36.1
3000	6.51×10^9	156

2.

T, °C	Energy loss, gm-cal/cm² sec
0	0
100	0.0188
300	0.138
500	0.475
1000	3.552
1500	13.39
2000	36.1
3000	156

3. $h = 6.625 \times 10^{-27}$ erg sec.

4. $u = \dfrac{8}{15}\dfrac{\pi^5 k^4}{c^3 h^3} T^4 = aT^4$, $h = 6.625 \times 10^{-27}$ erg sec.

5.

λ	ν, sec^{-1}	Energy	
		ergs	ev
1 km	2.99793×10^5	1.9862×10^{-21}	1.25×10^{-9}
1 m	2.99793×10^8	1.9862×10^{-18}	1.25×10^{-6}
1 cm	2.99793×10^{10}	1.9862×10^{-16}	1.25×10^{-4}
1 mm	2.99793×10^{11}	1.9862×10^{-15}	1.25×10^{-3}
10^{-4} cm	2.99793×10^{14}	1.9862×10^{-12}	1.25
5000 A	5.99586×10^{14}	3.9724×10^{-12}	2.50
1000 A	2.99793×10^{15}	1.9862×10^{-11}	12.5
1 A	2.99793×10^{18}	1.9862×10^{-8}	1.25×10^4
10^{-3} A	2.99793×10^{21}	1.9862×10^{-5}	1.25×10^7
10^{-13} cm	2.99793×10^{23}	1.9862×10^{-3}	1.25×10^9
10^{-15} cm	2.99793×10^{25}	1.9862×10^{-1}	1.25×10^{11}

6.

	(a) λ_{thr}, A	(b) V_{max}, cm/sec	(c) V, ev
Ag	2760	3.88×10^7	1.7
Ba	4970	5.71×10^7	3.7
Li	5400	5.86×10^7	3.9
Pt	3030	4.3×10^7	2.1
Ni	2485	3.26×10^7	1.2

7. $h = 6.602 \times 10^{-27}$ erg sec.
8. $v_{max} = 5.93 \times 10^7\sqrt{V}$ (volts).
9. (a) 3.0 ev, (b) 4167 A, (c) 9.5 ev

10.

Voltage	λ_m, A	ν_m, sec^{-1}	$h\nu_m$, ev
50,000	0.248	1.21×10^{19}	5×10^4
10^5	0.124	2.42×10^{19}	10^5
5×10^5	0.0248	1.21×10^{20}	5×10^5
10^6	0.0124	2.42×10^{20}	10^6

Chapter 6

1. $\Delta t' = \dfrac{\Delta t}{\sqrt{1 - (v^2/c^2)}}$; $\lambda = \dfrac{\lambda'}{\sqrt{1 - (v^2/c^2)}}$.

2. $$u'^2 = \frac{1}{[1 - (vu_x/c^2)]^2}\left[(u_x - v)^2 + \left(1 - \frac{v^2}{c^2}\right)(u_y^2 + u_z^2)\right].$$

If $u = c$, then $u_y^2 + u_z^2 = c^2 - u_x^2$ and the expression in brackets becomes

$$c^2\left(1 - \frac{2vu_x}{c^2} + \frac{v^2u_x^2}{c^4}\right) = c^2\left(1 - \frac{vu_x}{c^2}\right)^2,$$

so that $u'^2 = c^2$, and $u' = c$.

3. $1.8c$; $0.9945c$.

4.

v/c	m/m_0	v/c	m/m_0	v/c	m/m_0
0.1	1.005	0.7	1.400	0.99	7.089
0.2	1.021	0.9	2.294	0.995	10.013
0.5	1.155	0.95	3.203	0.999	22.366

5.

v/c	Electron energy		Proton energy
	ergs	Mev	
0.1	4.10×10^{-9}	2.56×10^{-3}	4.70 Mev
0.2	16.75	1.05×10^{-2}	19.3 Mev
0.5	125.7	7.85×10^{-2}	1.44×10^2 Mev
0.7	325.2	0.203	3.73×10^2 Mev
0.9	1,052	0.657	1.21 Bev
0.95	1,790	1.12	2.06 Bev
0.99	4,950	3.09	5.67 Bev
0.995	7,320	4.57	8.39 Bev
0.999	17,360	10.84	19.90 Bev

6. 0.115; 0.365; 0.816.

7.

	20 Mev	400 Mev	2.5 Bev
M/M_0	1.0002	1.43	3.66
v/c	0.020	0.713	0.962
E	958 Mev	1338 Mev	3438 Mev

8. It follows from Eq. (6–26) that

$$m_1 = \frac{m_0}{\sqrt{1-(u_1^2/c^2)}} = \frac{m_0}{\sqrt{1-(U^2/c^2)}},$$

$$m_2 = \frac{m_0}{\sqrt{1-(u_2^2/c^2)}} = \frac{m_0}{\sqrt{1-(1/c^2)v^2+U^2[1-(v^2/c^2)]}}$$

$$= \frac{m_0}{\sqrt{1-(v^2/c^2)}\sqrt{1-(U^2/c^2)}} = \frac{m_1}{\sqrt{1-(v^2/c^2)}}.$$

9. (a) Square the relation

$$m = \frac{m_0}{1-(v^2/c^2)};$$

the result is

$$m^2\frac{c^2-v^2}{c^2} = m_0^2, \qquad \text{and} \qquad (mc)^2 - (m_0c)^2 = (mr)^2 = p^2,$$

or

$$(pc)^2 = (mc^2)^2 - (m_0c^2)^2.$$

(b) Combining the last result with the relation $T = mc^2 - m_0c^2$ gives the required relation. Solution of (b) for T gives (d).

(c) Use of the result of (a) for $(pc)^2$ gives the required formula.

10. Energy and momentum conservation give

$$Mc^2 = E_1 + E_2; \qquad p_1 + p_2 = 0, \qquad \text{or} \qquad p_1^2 = p_2^2.$$

Hence $E_1^2 - E_2^2 = M_1^2c^4 - M_2^2c^4$, and solution of the two equations for E_1 and E_2 gives the required result.

11. The mass is $M = (1/c^2)\sqrt{E^2 - p^2c^2}$, where E and p are the energy and momentum of the composite particle, equal respectively to the sum of the energy and momentum of the two colliding particles. The velocity is $V = pc^2/E$, and the result follows.

12. (a) The maximum energy transfer to the electron occurs when $\phi = \pi$, and the result (a) follows directly.

(b) For $\phi \neq 0$, e.g., for $\phi = \pi$, and $\alpha \gg 1$,

$$T_{\text{max}} = h\nu_0\left(1 - \frac{1}{2\alpha} + \cdots\right),$$

and

$$(h\nu')_{\text{min}} = h\nu_0 - T_{\text{max}} \approx h\nu_0 \cdot \frac{1}{2\alpha} = \frac{1}{2}\,m_0c^2.$$

(c) Solve Eq. (6–30) for

$$\sin\theta = \frac{h\nu}{c}\,\frac{\sqrt{1-\beta^2}}{m_0\beta c}\sin\phi;$$

divide Eq. (6–29) by $\sin\theta$ and collect.

13. $\lambda_0 = 2h/m_0c$; $h\nu_0 = \frac{1}{2}m_0c^2 = 0.255$ Mev; $v = \frac{3}{5}c = 1.8 \times 10^{10}$ cm/sec.

14. (a) $\lambda = 0.1000$ A.

(b) $h\nu_0 = 137.9$ kev; $h\nu = 124.1$ kev.

(c) 7.37×10^{-18} gm cm/sec; 6.63×10^{-18} gm cm/sec.

(d) $T = 13.8$ kev; $p = 6.39 \times 10^{-18}$ gm cm/sec; $v = 6.84 \times 10^9$ cm/sec.

(e) $\theta = 57°$.

Chapter 7

1. $n = 20, 25, 30$.
2. $R_D = 109707.42\ \text{cm}^{-1}$. $\Delta\lambda = 1.79$ A for H_α; 1.32 A for H_β; 1.18 A for H_γ.
3. In Eqs. (7–9) and (7–10) replace v by $r\omega$, where ω is the angular velocity; this gives

$$mr^2\omega = n\frac{h}{2\pi}, \tag{7–9$'$}$$

and

$$mr\omega^2 = \frac{Ze^2}{r^2}. \tag{7–10$'$}$$

The frequency of orbital revolution is then found by solving these equations for ω; it is

$$\nu_{\text{orb}} = \frac{\omega}{2\pi} = \frac{4\pi^2me^4Z^2}{h^3}\,\frac{1}{n^3}.$$

For comparison, Eq. (7–16) may be written

$$\nu = \frac{4\pi^2 me^4 Z^2}{h^3} \frac{(n_1 + n_2)}{2n_1^2 n_2^2} (n_1 - n_2).$$

Since $n_1 > n_2$, it follows that

$$\frac{1}{n_1^3} < \frac{n_1 + n_2}{2n_1^2 n_2^2} < \frac{1}{n_2^3}.$$

Hence, if $n_1 - n_2 = 1$, the frequency ν of the emitted radiation is intermediate between the frequencies of orbital revolution in the initial and final states. For very large n, the orbital frequencies in successive orbits are practically equal and the emitted and orbital frequencies practically coincide.

4. $n = 1$: energy $= -109{,}678\ \text{cm}^{-1}$; -217.8×10^{-13} erg; -13.60 ev; -313.4 Kcal/gm atom.
   $n = 5$: energy $= -4.387\ \text{cm}^{-1}$; -8.71×10^{-13} erg; -0.544 ev; -12.52 Kcal/gm atom.
5. $M_H/m = 1836.0$; $e/m = (e/M_H)(M_H/m) = 5.27 \times 10^{17}$ esu/gm.
6.

Excitation energy, ev	Wavelength, A	Quantum numbers, n	
		Initial state	Final state
10.19	1216.6	2	1
12.08	1026.3	3	1
12.75	972.4	4	1
13.05	950.0	5	1
13.12	944.9	6	1
13.31	931.4	7	1

The lines belong to the Lyman series of hydrogen.

7. For the derivations, which are somewhat involved, see Born, *Atomic Physics*, Appendix XIV.
8. The tables asked for can be worked out directly from Table 7–4. The occurrence of the rare earths can be explained in terms of the filling of the 4f subshell. At xenon ($Z = 54$), the 4s, 4p, 4d, 5s, and 5p subshells are filled, which accounts for the chemical inertness of that element. The filling in of the 5d subshell is started with lanthanum, which has one 5d electron. At this point, the 4f subshell with room for 14 electrons is still empty. The successive addition of electrons up to the number of 14 would fill this shell; the resulting elements should resemble each other in chemical properties because these properties are determined mainly by the outermost electrons, i.e., those corresponding to $n = 5$. The elements from lanthanum ($Z = 57$) to lutecium ($Z = 71$) correspond to the filling of the 4f subshell, as shown in Table 7–4, and are the 15 rare earths.

9.

Element	Wavelength of K_α line, Eq. (7–39)	Element	Wavelength of K_α line, Eq. (7–39)
Al	8.438 A	Ni	1.667 A
K	3.750	Cu	1.550
Fe	1.944	Mo	0.723
Co	1.797	Ag	0.574

10. $\lambda = \dfrac{h}{mv} = \dfrac{h}{mc(v/c)} = \dfrac{h\sqrt{1 - (v/^2c^2)}}{m_0c(v/c)}$.

Compton wavelength of the electron $= h/m_0c = 2.4263 \times 10^{-10}$ cm.

Electron speed, v/c	Compton wavelength	Electron speed, v/c	Compton wavelength
0.1	2.414×10^{-9}	0.95	7.973×10^{-11}
0.2	1.188×10^{-9}	0.99	3.457×10^{-11}
0.5	4.202×10^{-10}	0.995	2.436×10^{-11}
0.7	2.475×10^{-10}	0.999	1.086×10^{-11}
0.9	1.175×10^{-10}		

The Compton wavelength of the proton is 1.3214×10^{-13} cm.

11. $\lambda = h/mv = h/m_0v$ when $v \ll c$. From the result of Problem 8, Chapter 5, $v = 5.93 \times 10^7\sqrt{V}$, where V is the voltage. Substitution of numerical values of h and m_0 gives the required formula.

Voltage, volts	λ, cm
1	12.268×10^{-8}
10	3.880×10^{-8}
100	1.2268×10^{-8}
1000	0.3880×10^{-8}

12. The formula $v = 5.93 \times 10^7\sqrt{V}$ is obtained from the relation $eV = \frac{1}{2}mv^2$. Hence, $v \sim m^{-1/2}$, and $\lambda \sim m^{-1}v^{-1}$ or $m^{-1/2}$. The relativistic correction, therefore, introduces the factor $(m/m_0)^{-1/2}$ into the formula for the wavelength.

Voltage	λ, cm
10^4	1.215×10^{-9}
10^5	3.457×10^{-10}
10^6	7.134×10^{-11}
10^7	8.554×10^{-12}

13. (a) 0.82 A, (b) 8.3×10^{-24} A.

CHAPTER 9

1. (a) 1.099×10^7 cm/sec, (b) 2.652×10^{-23} gm; 15.98 amu, (c) 16.
2. When the velocity v is eliminated between Eqs. (9–4) and (9–5), the equation

$$MV = \tfrac{1}{2}qH^2R^2$$

is obtained. In this equation, M is in grams, R in cm, and V, q, and H in electromagnetic units. For a singly-charged ion $q = 1.60207 \times 10^{-20}$ emu; 1 volt $= 10^8$ emu, and 1 amu $= 1.65982 \times 10^{-24}$ gm. When these conversion factors are inserted, the result is

$$M(\text{amu})V(\text{volts}) = 4.826 \times 10^{-5}H^2(\text{gauss}^2)R^2(\text{cm}^2).$$

For the data of Problem 1, the formula gives $M = 15.99$ amu.
3. At $R = 17.6$ cm and $R = 18.9$ cm, respectively.
4. Maxima are observed at $Z = 8$, 20, 28, 50, and 82.
5. Maxima are observed at $A–Z = 20$, 28, 50, and 82.
6. C^{12}, 16.88 Mev; N^{14}, 7.54 Mev; F^{19}, 7.95 Mev; Ne^{20}, 12.87 Mev; Si^{28}, 11.59 Mev.
7. C^{13}, 4.96 Mev; N^{13}, 1.95 Mev; N^{14}, 10.27 Mev; O^{16}, 7.15 Mev.
8. $H^1 = 1.008142$ amu; $H^2 = 2.014736$ amu; $C^{12} = 12.003821$ amu.
9. $H^1 = 1.007826$ amu; $H^2 = 2.014104$ amu; $O^{16} = 15.994904$ amu.
10. $C^{12} = 12.003816$ amu; $S^{32} = 31.982238$ amu.
12. 96495.0 coul/gm mole.

CHAPTER 10

1. $T = 137$ min; $\lambda = 0.005058$ min^{-1}. The counting rate at $t = 0$ would have been 11,100 min^{-1}.
2. $T_1 = 137$ min, $A_{01} = 11{,}100$ min^{-1}; $T_2 = 45.7$ min, $A_{02} = 8000$ min^{-1}.
3.

Time, days	Fraction of sample disintegrated, %
1	16.8
2	30.4
3	42.0
4	51.6
5	59.6
10	83.7

A one-microgram sample of radon contains initially

$$\frac{10^{-6}(6.025)10^{23}}{222} = 2.71 \times 10^{15} \text{ atoms.}$$

Then, 4.55×10^{14} atoms disintegrate in the first day; 2.17×10^{14} in the fifth day, and 0.87×10^{14} in the tenth day.

4.

Fraction changed to UX_2, %	Time, days	Fraction changed to UX_2, %	Time, days
90	80.1	99	160.2
95	104.2	63.2	24.1

5. (a) Ra: 1 curie weighs 1 gm; 1 rd weighs 2.70×10^{-5} gm.
   (b) Rn: 1 curie weighs 6.45×10^{-6} gm; 1 rd weighs 1.74×10^{-10} gm.
   (c) RaA: 10^{-6} cu weighs 3.54×10^{-15} gm; 1 rd weighs 9.56×10^{-14} gm.
   (d) RaC′: 10^{-6} cu weighs 3.1×10^{-21} gm; 1 rd weighs 8.4×10^{-20} gm.
6. $T_{1/2} = 8.0 \times 10^4$ years; 0.049 rd.

7.

Time, days	No. of disintegrations per second	Activity, millicuries	Activity, rd
0	1.664×10^8	4.50	166.4
10	1.58×10^8	4.27	158
30	1.43×10^8	3.86	143
50	1.29×10^8	3.49	129
70	1.17×10^8	3.16	117
100	1.01×10^8	2.73	101
300	3.69×10^7	1.00	37

8. 10.8 hr.
9. Neglect Tl^{206} because of its short half-life. Then RaE, RaF, and RaG form a 3-member chain which can be treated by Eqs. (10–16), (10–22), and (10–23). The greatest amount of RaF will be formed after 24.85 days, and will be 5.06×10^{11} atoms. At that time the α-activity will be 2.93×10^4 disintegrations/sec, as will be the β-activity.
10. (a) $\lambda_\beta = 1.22 \times 10^{-4}\ \text{sec}^{-1}$, $\lambda_\alpha = 0.69 \times 10^{-4}\ \text{sec}^{-1}$
    (b) $T_{1/2}(\beta) = 94.8$ min, $T_{1/2}(\alpha) = 167.1$ min
    (c) α, $1.96 \times 10^{10}\ \text{sec}^{-1}$; β, $3.46 \times 10^{10}\ \text{sec}^{-1}$
    (d) α, $2.49 \times 10^9\ \text{sec}^{-1}$; β, $4.39 \times 10^9\ \text{sec}^{-1}$
11. 1.3×10^{11}y
12. Secular equilibrium.
13. During a 10-min period, the activities of Th^{234} and Pa^{234} seem to be in secular equilibrium; but over a 10-day period, the Pa^{234} activity would decrease with a half-life of 24.1 days, in transient equilibrium with the Th^{234}.
14. The number of atoms N_x of a given member of a radioactive series in secular equilibrium is related to the number of atoms N_1 of the parent substance by the formula

$$N_x = N_1 \frac{T_x}{T_1} = N_1 \frac{\lambda_1}{\lambda_x}.$$

If the parent substance is U^{238} with a half-life of 4.50×10^9 years, the number of atoms of each daughter substance per 10^9 atoms of U^{238} is given in the table which follows:

Substance	Atoms/billion of U^{238}	Substance	Atoms/billion of U^{238}
Th^{234}	1.47×10^{-2}	Em^{222}	2.32×10^{-3}
U^{234}	5.54×10^{4}	Pb^{210}	4.88
Th^{230}	1.77×10^{4}	Bi^{210}	3.05×10^{-3}
Ra^{226}	3.59×10^{2}	Po^{210}	8.41×10^{-2}

The results for the series starting with U^{235} are

Substance	Atoms/billion of U^{235}
Th^{231}	3.95×10^{-3}
Pa^{231}	4.83×10^{4}
Ac^{227}	30.9
Th^{227}	7.17×10^{-2}
Ra^{223}	4.32×10^{-2}

15. The disintegration constant of U^{238} is 1.54×10^{-10} year^{-1}. Hence, one gram of uranium produces, in one year, an amount of Pb^{206} equal to

$$1.54 \times 10^{-10} \times \frac{206}{238} = 1.33 \times 10^{-10} \text{ gm.}$$

Hence, the age in years can be found by dividing the amount of Pb^{206} corresponding to one gram of uranium by 1.33×10^{-10}, or by multiplying it by 7.5×10^9. The formula

$$\text{Age} = (Pb^{206}/U) \times 7.5 \times 10^9 \text{ years, follows directly.}$$

(a) 10^8 years; corrected 9.9×10^7 years.
(b) 7.5×10^8 years; corrected 7.1×10^8 years.

16. One gram of uranium produces in one year

$$1.54 \times 10^{-10} \times \frac{32}{238} = 2.07 \times 10^{-11} \text{ gm of helium} = 1.16 \times 10^{-7} \text{ cm}^3 \text{ of}$$

helium at N.T.P., since the density of helium at N.T.P. is 0.1785 gm/liter. Then the age is equal to

$$(He/U) \div 1.16 \times 10^{-7} \text{ year} = (He/U) \times 8.6 \times 10^6 \text{ years.}$$

This formula may also be written

$$\text{Age} = (He/Ra)(Ra/U) \times 8.6 \times 10^6 \text{ years,}$$

where (Ra/U) is the number of grams of radium per gram of uranium and is equal to $1620/4.50 \times 10^9$. Then

$$\text{Age} = (\text{He/Ra})(1620/4.50 \times 10^9)8.6 \times 10^6 = 3.1(\text{He/Ra}).$$

The age of the mineral is approximately 1.5×10^9 years.

CHAPTER 11

1. and 2.

Reaction	Q-Value, Mev	Type	Reaction	Q-Value, Mev	Type
$H^1(n, \gamma)H^2$	2.230	exoergic	$Be^9(d, p)Be^{10}$	4.585	exoergic
$H^2(n, \gamma)H^3$	6.251	exoergic	$B^{11}(d, \alpha)Be^9$	8.016	exoergic
$Li^7(p, n)Be^7$	−1.645	endoergic	$C^{12}(d, n)N^{13}$	−0.280	endoergic
$Li^7(p, \alpha)He^4$	17.339	exoergic	$N^{14}(\alpha, p)O^{17}$	−1.198	endoergic
$Be^9(p, d)Be^8$	0.559	exoergic	$N^{15}(p, \alpha)C^{12}$	4.961	exoergic
$Be^9(p, \alpha)Li^6$	2.133	exoergic	$O^{16}(d, n)F^{17}$	−1.631	endoergic

3.

Reaction	Mass of product nucleus
(a) $Al^{27}(n, \gamma)Al^{28}$	27.99076
(b) $Al^{27}(d, p)Al^{28}$	27.99076
(c) $Al^{27}(p, \alpha)Mg^{24}$	23.99263
(d) $Al^{27}(d, \alpha)Mg^{25}$	24.99374

4. $Q = 8.04$ Mev. 5. $E_p = 2.368$ Mev.
6. $Be^9(p, n)B^9$: $Q = -1.853$ Mev; $C^{13}(p, n)N^{13}$: $Q = -3.003$ Mev; $O^{18}(p, n)F^{18}$: $Q = -2.453$ Mev.

7.

Reaction	Threshold energy, Mev
$B^{11}(p, n)C^{11}$	3.015
$O^{18}(p, n)F^{18}$	2.590
$Na^{23}(p, n)Mg^{23}$	5.091

8.

Reaction	Energy change	Reaction	Energy change
$C^{12}(n, \gamma)C^{13}$	exoergic	$C^{12}(d, p)C^{13}$	exoergic
$C^{12}(n, p)B^{12}$	endoergic	$C^{12}(\alpha, n)O^{15}$	endoergic
$C^{12}(n, \alpha)Be^9$	endoergic	$C^{12}(\alpha, p)N^{15}$	endoergic
$C^{12}(p, n)N^{12}$	endoergic	$C^{12}(\gamma, n)C^{11}$	endoergic
$C^{12}(d, n)N^{13}$	endoergic	$C^{12}(\gamma, p)B^{11}$	endoergic

9.

Reaction	Energy change	Reaction	Energy change
$B^{11}(p, \gamma)C^{12}$	exoergic	$Be^9(d, n)C^{12}$	exoergic
$N^{15}(p, \alpha)C^{12}$	exoergic	$B^{10}(\alpha, d)C^{12}$	exoergic
$B^{11}(d, n)C^{12}$	exoergic	$C^{13}(\gamma, n)C^{12}$	endoergic
$N^{14}(d, \alpha)C^{12}$	exoergic	$O^{16}(\gamma, \alpha)C^{12}$	endoergic
$C^{13}(d, t)C^{12}$	exoergic	$N^{14}(n, t)C^{12}$	endoergic

10. Equation (11–9) can be written in the form

$$E_y = \frac{Q + E_x[1 - (m_x/M_Y)]}{1 + (m_y/M_Y)}.$$

The value of Q can be calculated from the masses of the target and product nuclei; $E_x = 1.8$ Mev. The mass numbers of m_x, m_y, and M_Y are known. Hence, the value of E_y can be calculated. The results are

Reaction	Q, Mev	E_y, Mev
$Si^{28}(d, p)Si^{29}$	6.246	7.66
$Si^{29}(d, p)Si^{30}$	8.386	9.75
$Si^{30}(d, p)Si^{31}$	4.367	5.86
$Si^{29}(d, \alpha)Al^{27}$	5.994	6.66
$Si^{30}(d, \alpha)Al^{28}$	3.120	4.19

11. When the reactions are added and when quantities which appear on both sides of the resulting equation are canceled, the result is

$$O^{16} + 4d \rightarrow 6\alpha + 33.224 \text{ Mev}$$

or

$$\alpha = \frac{16.000000 + 4(2.014723) - (33.224/931.162)}{6} = 4.003869 \text{ amu}.$$

12.

	n	p	d	α
(a)	14.96 Mev	11.95	20.27	17.15
(b)	14.19	11.18	18.84	14.65

13. (a) C^{12}, 18.72 Mev; C^{13}, 4.96; C^{14}, 8.17
(b) N^{14}, 10.55; N^{15}, 10.84; N^{16}, 2.65
(c) O^{16}, 15.60; O^{17}, 4.15; O^{18}, 8.07
(d) Ne^{20}, 16.91; Ne^{21}, 6.75; Ne^{22}, 10.36

14. (a) C^{11}, 8.69 Mev; C^{12}, 16.88; C^{13}, 17.54
(b) N^{13}, 1.95; N^{14}, 7.54; N^{15}, 10.30
(c) O^{14}, 4.61; O^{15}, 7.35; O^{16}, 14.82; O^{17}, 13.60
(d) Ne^{19}, 6.36; Ne^{20}, 12.87; Ne^{21}, 13.03

16. 1.88 Mev 17. $Al^{28} - Si^{28} = 4.65$ Mev $= 0.00499$ amu.

18. From 5.7 Mev ($\theta = \pi$) to 11.2 Mev ($\theta = 0$).

Chapter 12

1.

Nuclide	Type of emission	Nuclide	Type of emission
Ga^{73}	β^-	V^{48}	β^+
Nb^{96}	β^-	Sc^{47}	β^-
Cs^{127}	β^+	Ag^{110}	β^-
Ir^{197}	β^-	Xe^{137}	β^-
Au^{198}	β^-	Xe^{123}	β^+
Br^{78}	β^+	Zn^{63}	β^+

2.

Nuclide	Type of emission	Nuclide	Type of emission
Na^{23}	stable	Na^{22}	β^+
P^{32}	β^-	F^{18}	β^+
Si^{31}	β^-	Be^{10}	β^-
Mg^{27}	β^-	He^6	β^-

3.

Decay process	Q-Value, Mev	Decay process	Q-Value, Mev
$He^6(\beta^-)Li^6$	3.215	$F^{20}(\beta^-)Ne^{20}$	7.052
$C^{14}(\beta^-)N^{14}$	0.155	$Na^{22}(\beta^+)Ne^{22}$	2.841
$N^{13}(\beta^+)C^{13}$	2.222	$Na^{24}(\beta^-)Mg^{24}$	5.531
$F^{18}(\beta^+)O^{18}$	1.671	$Al^{28}(\beta^-)Si^{28}$	4.650

4. To a first approximation, the condition for α-decay is

$$M(A) > M(B) + M(\alpha),$$

i.e., the mass of the initial nucleus must be greater than or equal to the sum of the masses of the final nucleus and an alpha-particle. Actually, the final nucleus always has some recoil kinetic energy $E(B)$, so that

$$M(A) > M(B) + E(B) + M(\alpha).$$

Since the α-particle must also have some kinetic energy $E(\alpha)$,

$$M(A) \geq M(B) + E(B) + M(\alpha) + E(\alpha).$$

5. At secular equilibrium there would be 6.65×10^{12} atoms of Mn^{56}. In general, the number N of atoms of Mn^{56} is given by the equation

$$\frac{dN}{dt} = a - \lambda N,$$

where a is the constant rate of production. This equation has the solution

$$N(t) = \frac{a}{\lambda}(1 - e^{-\lambda t}).$$

At (secular) equilibrium $N = a/\lambda$. At $t = 20$ hours, 99.5% of secular equilibrium is reached.

6. The number of atoms of Au^{198} at any time t is given by

$$N(t) = \frac{a}{\lambda}(1 - e^{-\lambda t}),$$

where $a = 10^{10}$ sec^{-1}, and $\lambda = 0.2567$ days^{-1} $= 0.2972 \times 10^{-5}$ sec^{-1}. After 100 hr there will be 2.47×10^{15} atoms of Au^{198}; after 10 days there will be 3.47×10^{15} atoms of Au^{198}. At secular equilibrium there would be 3.76×10^{15} atoms. The number of Hg^{198} atoms is equal to the total number of atoms formed minus the number of Au^{198} atoms, i.e., to the number of Au^{198} atoms that have decayed. This number is

$$at - \frac{a}{\lambda}(1 - e^{-\lambda t}),$$

under the assumption that the Hg^{198} atoms are not affected by the neutrons. (Actually, this is not the case, and the problem is more complicated.) At 100 hr there will be 1.13×10^{15} Hg^{198} atoms, and at 10 days there will be 5.17×10^{15} atoms of Hg^{198}.

7. The irradiation would have to be continued for 11.7 days to reach 95% of secular equilibrium.
8. Fe^{59}. 9. Ge^{69}. 10. β^-: F^{20}, F^{21}; β^+: Na^{20}, Na^{21}; EC: none
11. β^-: Cl^{36}, S^{35}; β^+: none; EC: Cl^{36} 12. K^{40}: β^-, EC; Sc^{40}: β^+
13. $\lambda(\beta^-) = 0.585 \times 10^{-5}$ sec^{-1}, $\lambda(\beta^+) = 0.285 \times 10^{-5}$ sec^{-1}, $\lambda(\text{EC}) = 0.630 \times 10^{-5}$ sec^{-1}, $T_{1/2}(\beta^-) = 32.8$ hr, $T_{1/2}(\beta^+) = 67.4$ hr, $T_{1/2}(\text{EC}) = 30.5$ hr
14. 1.25×10^9y
15. $He^5 \rightarrow He^4 + n$, $Q = 0.95$ Mev
$Li^5 \rightarrow He^4 + p$, $Q = 1.68$ Mev
$Be^8 \rightarrow 2He^4$, $Q = 0.093$ Mev

Chapter 13

1.

Speed, cm/sec	M/M_0	Energy, Mev (relativistic)	Energy, Mev (classical)	Hr, gauss-cm, $\times 10^{-5}$
10^8	1.00000	0.0274	0.0274	0.2073
5×10^8	1.00015	0.519	0.519	1.037
7.5×10^8	1.00034	1.117	1.117	1.555
10^9	1.00058	2.076	2.074	2.074
1.25×10^9	1.00090	3.245	3.241	2.594
1.50×10^9	1.00125	4.676	4.667	3.113
1.75×10^9	1.00170	6.293	6.277	3.634
2.0×10^9	1.00220	8.324	8.296	4.155
2.5×10^9	1.00351	13.09	12.96	5.201
3.0×10^9	1.00503	18.81	18.67	6.252
5.0×10^9	1.01420	52.94	51.85	10.51
10^{10}	1.06078	227	207	22.00
2.0×10^{10}	1.3425	1277	830	55.7

2. Since the reader has had practice in relativistic calculations in Problem 1, the present problem may be treated classically. The error introduced is small, as can be seen from the results of Problem 1. Equation (13–4) can be rewritten to read $v(\text{cm/sec}) = [E(\text{Mev})/2.074]^{1/2} \times 10^9$.

Energy, Mev	v, cm/sec	Energy, Mev	v, cm/sec
1	6.94×10^8	6	1.70×10^9
2	9.82×10^8	8	1.96×10^9
4	1.39×10^9	10	2.20×10^9
5	1.55×10^9		

3. The classical formulas may again be used, and the following results are obtained; they may be compared with values listed in Table 13–1.

Alpha emitter	Energy Mev	Speed, cm/sec, $\times 10^{-9}$	Hr, gauss-cm, $\times 10^{-5}$
Po^{218}	5.9981	1.701	3.526
Po^{216}	6.7744	1.807	3.746
Po^{215}	7.365	1.884	3.906
Po^{214}	7.680	1.924	3.988
	8.277	1.998	4.142
	9.066	2.091	4.335
	10.505	2.251	4.666
Po^{212}	8.7759	2.057	4.264
	9.488	2.139	4.434
	10.538	2.254	4.673
Po^{211}	7.434	1.893	3.924
Po^{210}	5.2984	1.598	3.313

4. When v is eliminated between Eqs. (13–3) and (13–4), the result is

$$E_\alpha(\text{Mev}) = 4.824 \times 10^{-11}\,(Hr)^2,$$

with Hr in gauss-cm. Then Q is given by

$$Q = E_y(1 + 4/25) - 2.10(1 - 2/25),$$

to a sufficiently good approximation.

Hr, gauss-cm, $\times 10^{-5}$	E_α, Mev	Q, Mev	Hr, gauss-cm, $\times 10^{-5}$	E_α, Mev	Q, Mev
3.93	7.49	6.69	3.29	5.22	4.13
3.79	6.93	6.11	3.25	5.09	3.98
3.69	6.57	5.72	3.23	5.04	3.85
3.54	6.045	5.08	3.05	4.49	3.27
3.45	5.74	4.73	2.90	4.06	2.78

5.

Energy, Mev	Range, cm
6.0	4.66
6.28	5.01
8.0	7.35
8.30	7.80
10.0	9.22

6.

Polonium isotope	Mean range of α-particle in air, cm	(a) Range in aluminum, $\times 10^3$ cm	(c) Equivalent thickness, mg/cm^2
Po^{218}	4.657	2.60	7.02
Po^{216}	5.638	3.15	8.51
Po^{215}	6.457	3.60	9.72
Po^{214}	6.907	3.85	10.40
	7.793	4.35	11.75
	9.04	5.04	13.61
	11.51	6.42	17.33
Po^{212}	8.570	4.78	12.91
	9.724	5.43	14.66
	11.580	6.46	17.44
Po^{211}	6.555	3.66	9.88
Po^{210}	3.842	2.14	5.78

7, 8, 9. The decay schemes are given in Strominger, Hollander, and Seaborg, "Table of Isotopes," *Revs. Mod. Phys.* **30**, 585 (1958): Rn p. 802, U^{233} p. 816, Am^{241} p. 826.

10.

Nucleus	Height of Coulomb barrier, Mev
Ne^{20}	3.5
Sn^{112}	10.0
Th^{232}	14.1

12. 0.6 sec. 13. 5.401 Mev; 210.0483.

Chapter 14

1. For an electron, the relationship

$$eHr = mv = p$$

holds, where p is the momentum. Then

$$Hr = p/e = (1/ec)\,(T^2 + 2m_0c^2T)^{1/2}.$$

The desired result is obtained when the values of the constants are inserted.

2. From the result of Problem 11, Chapter 6, it follows, since

$$p = eHr,$$

that

$$T = [(m_0c^2)^2 + (ec)^2(Hr)^2]^{1/2} - m_0c^2.$$

The desired result is obtained when the values of the constants are inserted.

3. Range = 7 mg/cm^2; T = 65 kev; th$_{1/2}$ = 0.6 mg/cm^2.
4. The range is about 800 mg/cm^2 by inspection. (a) ~1.75 Mev. (b) ~1.77 Mev. The measured value of the maximum energy is 1.71 Mev.
5. In each case, the extrapolated endpoint of the plot should correspond to an energy close to 1.71 Mev; the Hr-plot should extrapolate to a value close to 7300 gauss-cm.
6. N^{14}: 14.007517 amu, O^{17}: 17.004524, P^{29}: 28.990988, S^{35}: 34.98014
7. 1.483 Mev
8. (a) 0.782 Mev, (b) 0.782, (c) 0.780, (d) 0.771
9. (a) Zn^{64}, Ni^{64} are stable; Ga^{64}: β^+, Cu^{65}: β^-, β^+, EC
   (b) Cu^{64}: endpoint energy and Q-value are both 0.58 Mev for β^--emission; for β^+-emission, endpoint energy is 0.67 Mev, Q-value is 1.69 Mev; $Ga^{64}(\beta^+)$: endpoint energy is 6.22 Mev, Q-value is 7.24 Mev.
10. The average energy is given by the ratio

$$\overline{T} = \frac{\int_0^{T_0} TP(T)\,dT}{\int_0^{T_0} P(T)\,dT}.$$

The integrations can be performed analytically, and the result is $\frac{1}{3}$, under the given assumptions.

11. (a) 2.44×10^{11} sec^{-1}, (b) 5.69 kev, (c) 0.32.
12. 1 watt.
13. 6000 kilowatts.
14. Al^{25}: 3.2 Mev; Cl^{33}: 4.1 Mev.
15. The scheme is given on page 730 of the "Table of Isotopes," Strominger, Hollander, and Seaborg, *Revs. Mod. Phys.* **30**, 585 (1958).
16. See page 644 of the "Table of Isotopes."
17. See page 671 of the "Table of Isotopes."

Chapter 15

1. The values of μ obtained from the graphs should be close to those given in Table 15–1, namely, 0.182 cm^{-1}, 0.157 cm^{-1}, and 0.096 cm^{-1} for γ-ray energies of 0.835 Mev, 1.14 Mev, and 2.76 Mev, respectively.
2. The values of μ obtained from the graphs should be close to those given in Table 15–1, namely, 0.096 cm^{-1}, 0.267 cm^{-1}, and 0.478 cm^{-1} for aluminum, tin, and lead, respectively.

3. Photoelectric cross sections.

Energy		$_a\tau$, 10^{-24} cm^2/atom	
n	Mev	Aluminum, $Z = 13$	Lead, $Z = 82$
0.1	5.108	0.00013	0.675
0.125	4.086	0.00017	0.869
0.25	2.043	0.00042	2.08
0.5	1.022	0.00129	6.31
1.0	0.511	0.00691	27.7
1.5	0.341	0.0214	75.7
2.0	0.255	0.0554	161
3.0	0.170	0.196	465
4.0	0.128	0.492	985
5.0	0.102	0.991	1782

4.

Energy		Compton cross sections, 10^{-24} cm^2		
α	Mev	σ	σ_s	σ_a
0.025	0.0128	8.203	8.203	0.00
0.20	0.102	6.370	5.482	0.888
0.50	0.255	4.868	3.663	1.205
1.0	0.511	3.725	2.443	1.282
2.0	1.022	2.717	1.513	1.204
4.0	2.044	1.880	0.876	1.004
6.0	3.066	1.476	0.620	0.856
10.0	5.108	1.062	0.393	0.669
20.0	10.22	0.653	0.195	0.458
50.0	25.55	0.325	0.086	0.239

5.

Energy, Mev	$_a\kappa$, 10^{-24} cm^2	
	Aluminum	Lead
1.0	0	0
5.0	0.19	7.4
10.0	0.35	14.0
15.0	0.45	17.9
20.0	0.53	21.1
25.0	0.59	23.4

6. The values of μ for copper should come out close to the values given in Table 15–1, namely, 0.578 cm^{-1}, 0.486 cm^{-1}, and 0.316 cm^{-1} for energies of 0.835 Mev, 1.14 Mev, and 2.76 Mev, respectively.

7. 3.35 Mev, 1.53 Mev, and 0.64 Mev, respectively.
8. 0.212 Mev, 0.248 Mev, 0.377 Mev, and 0.490 Mev, respectively.
9. 0.237 Mev, 0.294 Mev, and 0.347 Mev, respectively.
10. The energies are 86.5 kev and 196.5 kev, as can be seen from the following table.

Electron energy, kev	Shell and binding energy, kev	Energy of γ-ray, kev
32.9	K (53.4)	86.3
78.0	L (8.6)	86.6
84.7	M (1.9)	86.6
86.1	N (0.4)	86.5
143.0	K (53.4)	196.4
187.0	L (8.6)	196.6
194.7	M (1.9)	196.6

11. $E_r = (h\nu)^2/2Mc^2$. With $h\nu$ in kev, and M in amu

$$E_r(\text{ev}) = \frac{5.38 \times 10^{-4}(h\nu)^2}{M}.$$

12. 21.3 min: Mn^{52m}; 5.8 d: Mn^{52}.
13, 14. See "Table of Isotopes," *op. cit.*, p. 626.
15. See Evans, *The Atomic Nucleus*, p. 233.

Chapter 16

1.

Q, Mev	Energy level, Mev	Q, Mev	Energy level, Mev
6.69	Ground state	3.98	2.71
6.11	0.58	3.85	2.84
5.72	0.97	3.27	3.42
5.08	1.61	2.78	3.91
4.73	1.96		

2. (a)

Q, Mev	E_y	Q, Mev	E_y
−6.92	6.97	−8.57	5.45
−7.87	6.09	−10.74	3.44

These results were obtained by using Eq. (11–9), with mass numbers rather than with the precise values of the masses. (b) The Q-value −6.92 Mev corresponds to the ground state. The value calculated from the masses (Table 11–1) is −6.88 Mev. (c) The different proton groups correspond to the ground state and to excited levels in B^{12} at 0.95, 1.65, and 3.82 Mev, respectively. (d) The threshold energies are 9.99 Mev, 11.36 Mev, 12.38 Mev, and 15.51 Mev, for the ground state and the three excited states, respectively.

3. 9.94 Mev. 4. (a) 1.64 Mev, (b) 2.95 Mev.

5.

Hr, kg-cm	*Q*, Mev	B^{11} Level, Mev	*Hr*, kg-cm	*Q*, Mev	B^{11} Level, Mev
448	9.235	Ground	245	1.937	7.298
399	7.097	2.138	190	0.667	8.568
338	4.776	4.459	171	0.309	8.926
322	4.201	5.034	156	0.045	9.190
266	2.477	6.758	151	—0.041	9.276
264	2.427	6.808			

6. (a) 0.003499 amu. (b) 2.24 Mev.

(c)

Resonance energy, Mev	Excitation energy, Mev	Resonance energy, Mev	Excitation energy, Mev
4.29	16.95	4.71	17.35
4.46	17.11	4.78	17.41
4.49	17.14	4.99	17.62
4.57	17.21	5.07	17.69
4.62	17.26	5.20	17.81

7. (a) 1.261×10^{16}, (b) 190.
8. (a) 3.5×10^{13} atoms of Au^{198}, (b) 2.8 millicuries.
9. (a) 1.75×10^{14} atoms of Ta^{182}, (b) 22 b.
10. (a) 4.94 Mev, (b) 4.95 Mev, (c) 4.95 Mev.

11.

Product nucleus	Binding energy of last neutron, Mev
Pb^{207}	6.71
Pb^{208}	7.37
Pb^{209}	3.87

12.

Energy of γ-ray, Mev	Reaction
6.73 ± 0.008	$Pb^{206}(n, \gamma)Pb^{207}$
7.380 ± 0.008	$Pb^{207}(n, \gamma)Pb^{208}$

13. The binding energies of the last two neutrons in Fe^{56} are

(a) 20.24 Mev, from the mass difference,

(b) 20.54 Mev from the nuclear reactions.

14. (b) The reaction $C^{14}(p, n)N^{14}$ has a Q-value of -0.628 Mev and has, therefore, a threshold energy of 670 kev.

Chapter 17

1.

Nucleus	Total binding energy, Mev	Coulomb energy, Mev	Ratio: $\dfrac{\text{Coulomb energy}}{\text{Total binding energy}}$
Be^9	58.0	3.3	0.056
Al^{27}	225	30.0	0.13
Cu^{63}	552	152	0.28
Mo^{98}	846	221	0.26
Xe^{124}	1045	362	0.35
W^{182}	1461	549	0.38
U^{238}	1803	778	0.43

2.

Nucleus	Total binding energy, Mev, from semiempirical formula
Be^9	47
Al^{27}	224
Cu^{63}	595
Mo^{98}	841
Xe^{124}	1039
W^{182}	1457
U^{238}	1819

3.

Nucleus	Atomic mass, amu (experimental)	Atomic mass from semiempirical formula
Be^9	9.0149	9.0274
Al^{27}	26.9901	26.9915
Cu^{63}	62.9486	62.9049
Mo^{98}	97.9361	97.9473
Xe^{124}	123.9458	123.9588
W^{182}	182.0033	182.0226
U^{238}	238.1234	238.1321

4.

Nuclide	Total binding energy, Mev	Coulomb energy, Mev	Ratio: $\dfrac{\text{Coulomb energy}}{\text{Total binding energy}}$
Ca^{40}	342	66	0.19
Sn^{120}	1020	294	0.29
Pb^{208}	1636	664	0.41

Nuclide	Binding energy from semiempirical formula	Atomic mass	Atomic mass from semiempirical formula
Ca^{40}	345	39.9753	39.9877
Sn^{120}	1012	119.9401	119.9945
Pb^{208}	1630	208.0422	208.0688

5. $Z_0 = 34.5A/(A^{2/3} + 66.7)$

A	Z_0 (closest integral value)
27	12
64	27
125	47
216	73

6. This problem involves just a comparison of curves.

7. (a) Mass (Ra^{226}) = 226.10543, Mass (Em^{222}) = 222.09803, Mass He^4 = 4.00387. Hence, the energy available for the α-decay of radium is 0.00343 amu or 3.2 Mev, as calculated from the semiempirical formula. The experimental value is 4.88 Mev.

(b) Mass (Po^{214}) = 214.08049, Mass (Pb^{210}) = 210.07368, Mass (He^4) = 4.00387. Hence, the energy available is 0.00304 amu or 2.8 Mev, as compared with the experimental value of 7.83 Mev.

(c) Mass (Hf^{170}) = 170.00774, Mass (W^{170}) = 170.01262. The mass difference is negative so that, according to the semiempirical mass formula, the β-decay should not take place.

(d) Mass (Au^{198}) = 198.05121, Mass (Hg^{198}) = 198.04944. Hence, 0.01777 amu or 1.65 Mev should be available for the decay process, as compared with an experimental value of 1.37 Mev.

The above results show that the semiempirical mass formula in the form used does not predict the decay energies quantitatively. But it gives semi-quantitative results which are often useful.

8. Mass (Fe^{54}) = 53.96011, Mass (n) = 1.00898, Mass (Fe^{55}) = 54.96045. Energy available is 8.97 Mev, as compared with the experimental value of 9.28 Mev.

9.

Nucleus	Mass from semi-empirical formula	Binding energy of last neutron, Mev
Pb^{206}	206.06428	—
Pb^{207}	207.06712	5.72
Pb^{208}	208.06871	6.88
Pb^{209}	209.07182	5.47

The binding energies of the last neutrons in Pb^{207}, Pb^{208}, and Pb^{209}, as obtained from (d, p) reactions have been found to be 6.70, 7.37, and 3.87 Mev, respectively.

10. $T = \dfrac{1}{2M}\left(\dfrac{\hbar}{\frac{1}{2}b}\right)^2 \approx$ 10–20 Mev

11. (a) $\Delta S_n \approx 2\delta$, (b) $\Delta S_n = 1$ Mev

Chapter 18

1. Neutron mass = 1.008984 ± 0.000003 amu.

2.

Energy, ev	T°K	v, cm/sec	λ, cm
0.001	11.6	4.37×10^4	9.04×10^{-8}
0.025	290	2.19×10^5	1.81×10^{-8}
1.0	1.16×10^4	1.38×10^6	2.86×10^{-9}
100	1.16×10^6	1.38×10^7	2.86×10^{-10}
10^4	1.16×10^8	1.38×10^8	2.86×10^{-11}
10^6	1.16×10^{10}	1.38×10^9	2.86×10^{-12}
10^8	1.16×10^{12}	1.28×10^{10}	2.79×10^{-13}
10^{10}	1.16×10^{14}	2.99×10^{10}	1.14×10^{-14}

3. (a) $Q = 5.65$ Mev. (b) 0^0: 10.8 Mev; 90^0: 8.5 Mev; 180°: 6.7 Mev.

4. $H^2(d, n)He^3$: $E_{n0} = 2.45$ Mev.
$H^3(d, n)He^4$: $E_{n0} = 14.1$ Mev.
$C^{12}(d, n)N^{13}$: $Q = -0.280$ Mev, the reaction has a threshold greater than 0.28 Mev and no neutrons would be ejected by deuterons of "zero" energy. $N^{14}(d, n)O^{15}$: $E_{n0} = 4.8$ Mev. $Li^7(d, n)Be^8$: $E_{n0} = 13.4$ Mev. $Be^9(d, n)B^{10}$: $E_{n0} = 4.0$ Mev.

5.

Reaction	$E_{n,\text{ min}}$	Reaction	$E_{n,\text{ min}}$
$H^3(p, n)He^3$	64 kev	$C^{12}(p, n)N^{12}$	118 kev
$Li^7(p, n)Be^7$	29 kev	$Na^{23}(p, n)Mg^{23}$	8.3 kev
$Be^9(p, n)B^9$	21 kev	$V^{51}(p, n)Cr^{51}$	0.58 kev

6. (a) $nv = 10^5$ neutrons/cm²sec. (b) $nv = 1.17 \times 10^5$ neutrons/cm²sec. (c) 0°:104.4, 25°: 100, 50°: 96.0, 100°: 89.4, 200°: 79.4.

7.

Element	ξ	Average number of collisions	Slowing-down power	Moderating ratio
F	0.102	178	2.74×10^{-5}	56.7
Mg	0.0807	226	0.021	8.2
Bi	0.0096	1890	0.0024	0.3

8. (a) 3.2 curies, (b) about 14,000 counts/min.
9. (a) The derivation is directly analogous to that given in Section 18–5 except that in spherical coordinates the operator d^2/dx^2 is replaced by $(d^2/dr^2) + (2/r)(d/dr)$.

(b) The general solution of the differential equation in (a) is $n = (Ae^{-r/L}/r) + (Be^{+r/L}/r)$. One boundary condition is that the total neutron current through a small sphere of radius a surrounding the source has a limit Q as the radius of the sphere approaches zero. The second condition is that n must be finite everywhere, so that $B = 0$.

10.

Energy, ev	Bragg angle for first order reflection	Energy, ev	Bragg angle for first order reflection
1	11° 15′	10	3° 32′
3	6° 28′	30	2° 2′
5	5° 0′	50	1° 35′

11.

Energy, ev	ΔE, ev	Energy, ev	ΔE, ev
1	0.023	10	0.74
3	0.121	30	3.8
5	0.26	50	8.2

12.

Δt, μ sec/min	E				
	1 ev	10 ev	100 ev	1 kev	10 kev
0.5	0.014 ev	0.44 ev	14 ev	0.44 kev	14 kev
0.1	0.0028 ev	0.09 ev	2.8 ev	88 ev	2.8 kev
0.05	0.0014 ev	0.04 ev	1.4 ev	44 ev	1.4 kev

13. (a) $\sigma_t = 26.3 \times 10^{-24}$ cm², (b) $\sigma_{\text{act}} = 21 \times 10^{-24}$ cm².
14. $I/I_0 = e^{-Nx\sigma_t} = e^{-Nx(\sigma_s+\sigma_a)} = e^{-Nx\sigma_s} \times e^{-Nx\sigma_a} = 0.575 \times 0.685 = 0.394$.
15. σ_a (0.025 ev) = 147 b, in good agreement with the measured value 150 ± 7b.
16. 29 collisions.

Chapter 19

1. (a) ${}_{30}Zn^{72} \rightarrow {}_{31}Ga^{72} \rightarrow {}_{32}Ge^{72}$ (stable).
   (b) ${}_{36}Kr^{90} \rightarrow {}_{37}Rb^{90} \rightarrow {}_{38}Sr^{90} \rightarrow {}_{39}Y^{90} \rightarrow {}_{40}Zr^{90}$ (stable).
   (c) ${}_{36}Kr^{92} \rightarrow {}_{37}Rb^{92} \rightarrow {}_{38}Sr^{92} \rightarrow {}_{39}Y^{92} \rightarrow {}_{40}Zr^{92}$ (stable).
   (d) ${}_{36}Kr^{97} \rightarrow {}_{37}Rb^{97} \rightarrow {}_{38}Sr^{97} \rightarrow {}_{39}Y^{97} \rightarrow {}_{40}Zr^{97}$ $\nearrow {}_{41}Nb^{97m} \downarrow$, $\searrow {}_{41}Nb^{97} \nearrow {}_{42}Mo^{97}$ (stable).
   (e) ${}_{42}Mo^{105} \rightarrow {}_{43}Tc^{105} \rightarrow {}_{44}Ru^{105} \rightarrow {}_{45}Rh^{105} \rightarrow {}_{45}Pd^{105}$ (stable).
   (f) ${}_{50}Sn^{126} \rightarrow {}_{51}Sb^{126} \rightarrow {}_{52}Te^{126}$ (stable).
   (g) ${}_{54}Xe^{144} \rightarrow {}_{55}Cs^{144} \rightarrow {}_{56}Ba^{144} \rightarrow {}_{57}La^{144} \rightarrow {}_{58}Ce^{144} \rightarrow {}_{59}Pr^{144} \rightarrow {}_{60}Nd^{144}$ (stable).
   (h) ${}_{61}Pm^{156} \rightarrow {}_{62}Sm^{156} \rightarrow {}_{63}Eu^{156} \rightarrow {}_{64}Gd^{156}$ (stable).
2. Sr^{91}: 0.025 gm; La^{139}: 0.031 gm.
3. (a) 205 Mev; (b) 211 Mev.
4. 179 Mev.
5. Plot the function

$$A(t) = \left(\frac{1000}{0.64}\right)(0.085e^{-0.693t/0.43} + 0.241e^{-0.693t/1.52} + 0.213e^{-0.693t/4.51} + 0.166e^{-0.693t/22.0} + 0.025e^{-0.693t/55.6})$$
$$= 133.8e^{-1.61t} + 376.6e^{-0.456t} + 332.9e^{-0.242t} + 259.5e^{-0.0315t} + 39.0e^{-0.0125t},$$

   where t is in seconds. The delayed neutron activity is here normalized to 1000 units at $t = 0$.
6. (a) 45. (b) 34, 35, 36.5, 35.5, 37, 37, 38. (c) The nuclides with the greater values of Z^2/A, namely, U^{232}, Pu^{238}, Am^{241}, and Cu^{242}, would be expected to have greater spontaneous fission rates than the other nuclides.
7.

Nuclide	Binding energy of added neutron, Mev	Fissionability	Thermal fission cross section, b
Th^{227}	7.01	yes	1500
U^{229}	7.58	yes	?
Np^{236}	7.05	yes	2800
Np^{237}	5.05	no or ?	$\approx 10^{-2}$
Np^{238}	6.35	yes	1600
Pu^{241}	6.88	yes	1080
Pu^{242}	5.32	no or ?	< 0.2
Am^{241}	5.21	no or ?	3
Am^{242}	6.45	yes	6400
Cm^{240}	6.70	yes	20,000

Since the values of the activation energy are not given in the text, it is assumed that an excitation energy of 6 Mev or greater indicates fissionability, while values of about 5 Mev indicate doubtful fissionability. The results are in good agreement with the known values of the thermal fission cross section.

10. 170 Mev

Chapter 20

1. U^{235} is consumed by fission at the rate of about 1 kg/day; the total rate of consumption is about 1.18 kg/day. The maximum conceivable rate of production of Pu^{239} would be about 1.1 kg/day.
2. When pure U^{235} is the fuel, the fast fission effect and the resonance escape probability are both unity, and

$$k_\infty = \eta f = \eta \frac{N_U \sigma_{au}}{N_U \sigma_{au} + N_M \sigma_{aM}} = \eta \frac{x}{1+x},$$

where

$$x = \frac{N_U}{N_M} \frac{\sigma_{au}}{\sigma_{aM}}.$$

The condition for criticality in an infinitely large assembly is $k_\infty = 1$, or $x = 1/(\eta - 1)$. For U^{235}, $\eta = 2.07$ and $x = 0.93$ is the condition for criticality. For graphite and U^{235}, then,

$$\frac{N_U}{N_M} = (0.93) \frac{0.0045}{687} = 5.92 \times 10^{-6}.$$

3. The thermal neutron absorption cross section of beryllium is 0.01 b. Then

$$\frac{N_U}{N_M} = (0.93) \frac{0.01}{687} = 1.35 \times 10^{-5}.$$

4. The thermal neutron absorption cross section of water is 0.66 b/molecule. Then

$$\frac{N_U}{N_{H_2O}} = (0.93) \frac{(0.66)}{687} = 8.9 \times 10^{-4}.$$

5. The thermal neutron absorption cross section of D_2O is 0.0093 b. Hence,

$$\frac{N_U}{N_{D_2O}} = \frac{(0.93)(9.3)10^{-4}}{687} = 1.2 \times 10^{-6}.$$

6.

Moderator	Migration area, cm^2	Migration length, cm
C	1708	41.3
Be	326	18.1
H_2O	36.8	6.1
D_2O	15,510	124.5

7.

Moderator	N_U/N_M	M^2, cm^2	M, cm	R_c, cm	Critical mass, kg
C	7.0×10^{-6}	1579	39.7	1247	56.5
Be	1.60×10^{-5}	304	17.4	547	16.8
H_2O	1.05×10^{-3}	36.4	6.0	188	12.3
D_2O	1.48×10^{-6}	14050	118.5	3723	13.1

Note: The values of the parameters used in this problem are a poor choice from the standpoint of efficient reactor design. They were used only to make possible very simple calculations.

9. (*a*) $\dfrac{d^2n}{dr^2} + \dfrac{2}{r}\dfrac{dn}{dr} + \dfrac{k_\infty - 1}{M^2} = 0,$ 10. $\dfrac{\bar{n}}{n_0} = \left(\dfrac{2}{\pi}\right)^3; \quad \dfrac{(\bar{n}/n_0)_{RP}}{(\bar{n}/n_0)_s} = \dfrac{8}{3\pi}.$

(c) $A = \dfrac{n_0}{\pi/R}; \quad \dfrac{\bar{n}}{n_0} = \dfrac{3}{\pi^2}.$

11.

Reaction	Q-value, Mev	Reaction	Q-value, Mev
$C^{12}(p, \gamma)N^{13}$	1.944	$N^{14}(p, \gamma)O^{15}$	7.346
$N^{13}(\beta^+)C^{13}$	2.221	$O^{15}(\beta^+)N^{15}$	2.705
$C^{13}(p, \gamma)N^{14}$	7.542	$N^{15}(p, \gamma)C^{12}$	4.961

Total: $4p \rightarrow 2\beta^+ + \alpha$, $Q = 26.72$ Mev.

Chapter 21

1. From Eq. (21–7), it follows that $H = 2\pi(q/m)^{-1} n$.
The values of (q/m) are proton: 9.581×10^3 emu/gm; deuteron: 4.794×10^3 amu/gm; α-particle: 4.823×10^3 emu/gm.
It follows that, from these values, $H = 6.55 \times 10^{-4}\, n$, for protons, $H = 1.31 \times 10^{-4}\, n$, for deuterons, $H = 1.30 \times 10^{-4}\, n$, for α-particles.

2.

Particle	H, gauss	Energy, Mev
proton	7.86×10^3	7.4 Mev
deuteron	1.572×10^4	14.8 Mev
α-particle	1.56×10^4	14.7 Mev

3. Different sets of parameters can be devised with the aid of Eq. (21–6) and the relationships of Problem 1. When the value of the energy is inserted into Eq. (21–6), a relationship is obtained between H and R. If R is chosen, H is determined from this relationship, and n is then given by the appropriate relation of Problem 1. If the frequency is chosen, H is determined, and R is then obtained from Eq. (21–6).

4. The kinetic energy of a relativistic particle is given by

$$T = (M - M_0)c^2 = M_0\left(\frac{M}{M_0} - 1\right)c^2.$$

But $T = 10^8\, qV$, if q is in emu and V in volts. When the two expressions for T are combined,

$$M_0\left(\frac{M}{M_0} - 1\right)c^2 = 10^8 qV,$$

or

$$\frac{M}{M_0} = 1 + \frac{10^8}{c^2}\frac{q}{M_0}V = 1 + aV.$$

It is also known that

$$M = \frac{M_0}{\sqrt{1 - (v^2/c^2)}},$$

so that

$$\frac{v}{c} = \left[1 - \left(\frac{M_0}{M}\right)^2\right]^{1/2} = \frac{(2aV)^{1/2}[1 + (aV/2)]^{1/2}}{1 + aV}.$$

Then, since $MV = qHR$, it follows that

$$HR = \frac{M_0 c}{q}(2aV)^{1/2}\left(1 + \frac{aV}{2}\right)^{1/2}$$

and

$$H^2R^2 = 2\left(\frac{M_0}{q}\right)^2 c^2 aV\left(1 + \frac{aV}{2}\right)$$
$$= 2 \times 10^8 V\left(\frac{M_0}{q}\right) + 10^{16}\frac{V^2}{c^2}.$$

For deuterons

$$H^2R^2 = 4.172 \times 10^4 V + 1.1127 \times 10^{-5} V^2.$$

For $H = 15{,}000$ gauss and $R = 80$ in., the deuterons have an energy of 210 Mev.

5. (a) $eV = e\dfrac{d\Phi}{dt} = e\omega\Phi_0 \cos \omega t$. (b) $\dfrac{2}{\pi}e\omega\Phi_0$. (c) $\dfrac{c\pi}{2\omega}$. (d) $\dfrac{c}{4\omega R}$.

(e) $\dfrac{ec\Phi_0}{2\pi R} = ecHR$.

6. Average energy per turn $\approx$ 400 ev; final energy $\approx 10^8$ ev.

INDEX

INDEX

ABCDE6987654